Waste-Water Engineering

Waste-Water Engineering

R. PARKER MS, PhD

N. MORRIS MSc, PhD

F. N. FAIR M Tech

S. C. BHATIA BE (Chemical), MBA

CBS

CBS Publishers & Distributors Pvt. Ltd.

New Delhi • Bengaluru • Chennai • Kochi • Kolkata • Mumbai
Hyderabad • Nagpur • Patna • Pune • Vijayawada

ISBN: 978-81-239-1643-9

First Edition: 2008
Reprint: 2010, 2014, 2018

Published by **Satish Kumar Jain** and produced by **Varun Jain** for

CBS Publishers & Distributors Pvt. Ltd.,
4819/XI Prahlad Street, 24 Ansari Road, Daryaganj, New Delhi - 110002
delhi@cbspd.com, cbspubs@airtelmail.in • www.cbspd.com
Ph.: 23289259, 23266861, 23266867 • Fax: 011-23243014

Corporate Office: 204 FIE, Industrial Area, Patparganj, Delhi - 110 092
Ph: 49344934 • Fax: 011-49344935
E-mail: publishing@cbspd.com • publicity@cbspd.com

Branches:
• *Bengaluru:* 2975, 17th Cross, K.R. Road, Bansankari 2nd Stage,
  Bengaluru - 70 • Ph: +91-80-26771678/79 • Fax: +91-80-26771680
  E-mail: cbsbng@gmail.com, bangalore@cbspd.com
• *Chennai:* No. 7, Subbaraya Street, Shenoy Nagar, Chennai - 600030
  Ph: +91-44-26681266, 26680620 • Fax: +91-44-42032115
  E-mail: chennai@cbspd.com
• *Kochi:* Ashana House, 39/1904, A.M. Thomas Road, Valanjambalam,
  Ernakulum, Kochi • Ph: +91-484-4059061-65
  Fax: +91-484-4059065 • E-mail: cochin@cbspd.com
• *Kolkata:* 6-B, Ground Floor, Rameshwar Shaw Road, Kolkata - 700014
  Ph: +91-33-22891126/7/8 • E-mail: kolkata@cbspd.com
• *Mumbai:* 83-C, Dr. E. Moses Road, Worli, Mumbai - 400018
  Ph: +91-9833017933, 022-24902340/41 • E-mail: mumbai@cbspd.com

Representatives:

• Hyderabad: 0-9885175004   • Nagpur: 0-9021734563
• Patna: 0-9334159340   • Pune: 0-9623451994
• Jharkhand: 0-9811541605   • Uttarakhand: 0-9716462459

Printed at:
India Binding House, Noida, UP (India)

Preface

Next to air, the other important requirement for human life to exist is water. It is the Nature's free gift to the human race. It is available in various forms such as rivers, lakes, streams, etc. The importance of water in human life is so much that the development of any city of the world has practically taken place near some source of water supply. It may also further be noted that the water is available in solid, liquid and gas forms. The occurrence of water in all these three forms is basically important for human beings for comfort, luxury and various other necessities of life.

Chapter 1 is devoted to 'Introduction' which discusses importance, properties and uses of water. Chapter 2 deals with quality of water along with population growth and demand aspects. Chapter 3 focuses on sources of water and their classification. Chapter 4 concentrates on 'Conveyance' of which indicates drawing off the water from the sources of water commonly known as the intakes. A number of laboratory tests are needed daily, quarterly, semi-annually and annually and other specified intervals to monitor the water quality before, during and after treatment. Keeping this in mind chapter 5 concentrates on 'Quality of water'. Chapter 6 is devoted to 'An overview of waste-water treatment'. A variety of water treatment processes involve the transfer of material from one phase to another to bring about treatment. Keeping this in mind chapter 7 focuses on 'Introduction to separation processes and mass transfer'. Chapter 8 deals with 'Aeration and gas transfer'. Aeration occupies a significant place in waste-water quality management and is an important factor in the purification of polluted water; where as gas transfer is a physical phenomenon in which gas molecules are exchanged between a liquid and a gas at a gas-liquid interface. Chapter 9 focuses on 'Screening, sedimentation, clarification, flotation and coagulation which are the major physical processes involved in purification of water.

Chapter 10 is devoted to 'Filtration' which is a fundamental unit operation that, separates suspended particle matter from water. This chapter discusses in detail filtration conditions, waste-water treatment applications and equipment selection methodology. Chapter 11 concentrates on 'Flocculation, adsorption, desalination and ion exchange' which are both physical and chemical operations, because they affect both the physical and chemical composition of the water subjected to them. Chapter 12 focuses on 'Membrane separation technologies'. Various types of membrane processes such as—electrodialysis, microfiltration, ultrafiltration, nanofiltration and reverse osmosis are discussed in detail. Chapter 13 is devoted to 'Disinfection', the chapter discusses the role disinfection plays in reducing microbial contaminants, the kinetics of disinfection process and some specific details about the design of disinfection facilities. Water distribution is the delivery of water from the source to the treatment plant and from the treatment plant to consumers through the distribution system. Keeping this in mind chapter 14 concentrates on 'Water transmission and distribution systems'. Chapter 15 focuses on 'Advance waste-water treatment' which refers to the methods and processes that remove more contaminants from waste-water than are taken out by conventional biological treatment. Chapter 16 deals with 'Biological waste treatment' which removes organic materials either by oxidation to carbon dioxide, water and other derivatives or by conversion of the organic materials into a settleable form which can be removed by gravity sedimentation. In most water treatment process the objective is to remove certain materials from the water, to purify it; these materials are referred to as residuals and consist of the liquid-solid and gaseous-phase by-products removed during the water treatment process along with any transport water treatment process. Keeping this in mind chapter 17 deals with 'Residual management'. Chapter 18 is devoted to 'Sewage' which is waste-water generated from domestic activities including kitchen, bathroom, toilet and floor washing. Chapter 19 focuses on 'Engineered systems for resource and energy, recovery'. Chapter 20 concentrates on 'Microbiology'. Microbes are of great interest to water technologies and include bacteria, viruses etc., many of which are disease–producing agents. For water treatment, it is important to understand the pressure and flow behaviour of water. Considering this Chapter 21 deals with 'Hydraulics' which is the science of fluids, such as water. Chapter 22 is devoted to 'Applications of computer in waste-water technology'. This chapter discusses the mathematical models used in water technology along with the role of computer technology and followed by a description of the various natural bodies to be simulated.

It may not be wrong to hold that the present book on '*Waste-water Engineering*' is a complete treatise on water processing techniques including recycling and reuse of waste-water. This reference/text book is essential reading for all students and teachers of graduate and postgraduate levels in engineering, environment, life sciences, water technologists, research students and industrialists.

Glossary, appendices and index have been provided at the end for quick reference. Diagrams, figures and tables supplement the text. All the topics have been covered into a cogent and lucid style to help the reader grasp the information quickly and easily.

R. PARKER
N. MORRIS
F. N. FAIR
S. C. BHATIA

Contents

Contents in Detail

7. Introduction to Separation Processes and Mass Transfer   101-142

8. Aeration and Gas Transfer   143-170

9. Screening, Sedimentation, Clarification, Flotation and Coagulation   171-218

Waste-Water Engineering

Chapter 1

Source of Water

INTRODUCTION

Next to air, the other important requirement for human life to exist is water. It is the nature's free gift to the human race. It is available in various forms such as rivers, lakes, streams, etc. The importance of water in human life is so much that the development of any city of the world has practically taken place near some source of water supply. It may also further be noted that the water is available in solid, liquid and gas forms. The occurrence of water in all these three forms is basically important for human beings for comfort, luxury and various other necessities of life.

The reliance on monsoon for the supply of water is also very important. The Indian philosophy has treated this subject from religious point of view and has led to the popular belief that there exists something like the God of Water. Many other religions of the world also support this belief. The failure of monsoon leads to many disasters such as a famine, an epidemic, etc.

The use of water by man, plants and animals is universal. As a matter of fact, every living soul requires water for its survival. It is essential for life, health and sanitation. It is the principal raw material for food production and for many other uses outside the home and on the farm. The man can live without food for about two months. But he can hardly survive for three or four days without water. In a similar way, if there is a shortage of water, there will be a decline in farm production, just like a shortage of steel will lead to the decrease in the production of automobiles.

In addition to the direct consumption of water at homes and farms, there are many indirect ways in which water affects our daily life. The water plays an important role in the manufacture of essential commodities, generation of electric power, transportation, recreation, industrial activities, etc. Thus the water can be considered as the most important raw material of civilisation because of the fact that without water, the man cannot live and industry cannot operate. With our growing population and industrial developments, the demand of water is also increasing day by day and hence every country has to take preventive measures to avoid careless pollution and contamination of the available water resources.

The water resources are certainly inexhaustible gift of nature. But to ensure their services for all the time to come, it becomes necessary to maintain, conserve and use these resources very carefully. It is an established fact that proper maintenance, conservation and use of the water resources will definitely avoid the chances of water famine for future generations for an indefinite period. It is for this reason that remedial measures will have to be found out in future to increase available water resources and to improve the quality of water. The requirement of water is also essential for the growth of crops.

1

NEED TO PROTECT WATER SUPPLIES

The water when exposed to the atmosphere contains many impurities which are harmful to any living organism. If untreated water is consumed by living organisms, it is likely to cause serious harm to their health. Hence, in order to make water potable and free from various impurities, the purification methods are found out.

The soul of purification process of present day water supply schemes is the filtration. It is preceded by pre-filtration purification methods and followed by post-filtration purification methods. The former methods make the water fit for filtration and the latter methods treat the impurities which have not been removed with the help of the process of filtration. The line of treatment to be recommended for a particular quantity of water will naturally depend upon its quality.

SOURCES OF WATER

The chief sources of water supply for industrial purposes are—ground water, surface water, sea water and rain water. Ground water may come from springs, shallow wells and from deep wells. Surface water is flowing water (rivers, streams etc.) and still water (lakes, ponds etc.). Sea water is not much used in industries.

Quality of Natural Water

Water obtained from different sources is associated with a large number of impurities. For example, water gets impurities of various kinds from ground or soil with which it comes into contact. Water also gets contaminated with sewage and industrial wastes or effluents when these are allowed to flow into running water or through percolation through the ground. The substances contained in natural or raw waters can be divided into the following three groups:

1. Coarsely dispersed or suspended substances.
2. Colloids and molecular substances.
3. Ion dispersed substances.

Suspended substances are particles of sand and clay of different size, remnants of plants and other substances entrained from the surface by rain water or thawed snow and carried into open basins-rivers, lakes and ponds. The greatest concentration of such substances in surface waters is usually observed during floods. In the analysis of raw waters, we usually determine the amount of impurities contained in a unit volume of water (ml/l) without indicating their chemical composition. Substances of both organic as well as inorganic origin are present in water in colloidal state. The humic substances (organic origin) found in great amounts in swamp waters impart a yellow or brown tinge. Contamination of raw water by organic substances is mainly due to the following important reasons.

1. Dying of and decaying of organisms dwelling in water.
2. Unpurified waste waters discharged by industries into water reservoirs.

It should be noted that all the organic substances present in water are not in colloidal state. Some of them may be present in the form of true solution.

Of the inorganic substances present in water—iron, silicon and aluminium compounds are often contained in colloidal state. These colloidal impurities hamper proper performance of boiler units by increasing the tendency of boiler water foaming. These impurities also disturb the performance of anion exchangers that are irreversibly sorbing the anions of organic substances.

The group of molecular-dispersed substances include salts and gases that are dissolved in water. Cations such as Na^+, Ca^{2+}, Mg^{2+} and anions such as Cl^-, SO_4^{2-}, HCO_3^- etc., are usually encountered in natural waters, because salts are dissociated to a large extent in aqueous solution. In other words, it may be assumed that natural waters usually contain compounds such as:

$$Ca(HCO_3)_2, Mg(HCO_3)_2, CaCl_2, MgCl_2, CaSO_4, MgSO_4, Na_2SO_4 \text{ and } NaCl$$

Tests often reveal that in addition to Na^+, Ca^{2+}, Cl^-, SO_4^{2-} and HCO_3^- (which are most widespread) natural waters also contain nitrite acid ions, NO_2^- (nitrites), nitrous acid ions, NO_3^- (nitrates) and ammonium cations (NH_4^+) in small quantities. The presence of these nitrogen compounds in water indicates that water basin is contaminated by industrial waste waters or by decomposition products of organic substances.

Natural Waters also contain ferrous compounds. In underground waters, iron usually occurs as ferrous bicarbonate, $Fe(HCO_3)_2$, which is stable only in the presence of large amount of CO_2. The removal of CO_2 from water causes the decomposition of $Fe(HCO_3)_2$, followed by the formation of ferrous hydroxide, $Fe(OH)_2$. This reaction is utilised in deferrisation of ground waters. The water containing iron is sprayed (aerated) or air is blown through it to remove CO_2. The $Fe(OH)_2$ thus formed is oxidised by air or oxygen into $Fe(OH)_2$ which is removed from the water after its precipitation. It should be noted that ferrous compounds form a precipitate capable of scaling on boiler heating surfaces.

All natural waters also contain gases, of which oxygen and carbon dioxide cause corrosion of metals. Thus treatment of boiler feed water must include degasification. Dissolved gases such as O_2, CO_2, H_2S etc., may be present in the water as impurties. For the removal of these gases it is better to classify them into two classes.

1. Gases which are removed when the solution is heated, because solubility of gases in water decreases when the temperature is increased.
2. Gases which can only be removed by chemical treatment. For example, oxygen from water may be removed by passing water over iron, as a result, iron is oxidised to ferric state and is precipitated.

$$4Fe + 3O_2 \rightarrow 2Fe_2O_3$$

Acidic gases can be removed by treating water with calcium hydroxide or calcium carbonate resulting in a precipitate. Surface water generally contains turbidity, calcium and magnesium salts, silica, bacteria, various micro-organisms, organic matter from sewage and industrial wastes.

Underground water (water from deep borings and wells) generally contains no turbidity but contains more mineral salts, free CO_2, calcium and magnesium salts, iron and manganese salts etc. The temperature of underground water is lower than that of surface water and remains almost constant.

Physical Properties of Water

The physical constants of water of vital importance to food processing and preservation, include the following:

1. Melting point, boiling point, surface tension, dielectric constant, heat capacity and heat of fusion, vapourisation and sublimation of which have unusually high values.
2. Normal viscosity and a moderately low density that has an unusually maximum value at $3.98°C$ and unusual characteristic of expanding on solidification.
3. A large thermal conductivity, (1.43 cal/second cm^2 °C/cm at $20°C$) compared to other liquids and a large thermal conductivity of ice compared to other nonmetallic solids. The thermal conductivity of ice at $0°C$ is about four times that of water at the same temperature (5.35 cal/second cm^2°C/cm)—thus ice conducts heat at a much faster rate than immobilised water in tissues.

4. Thermal diffusivity (the rate at which a substance udergoes changes in temperature) of ice ($\sim$0.011 cm^2/sec) is nine times that of water ($\sim$0.0014 cm^2/sec). This implies that tissues freeze more rapidly than they thaw due to differences in thermal conductivity and diffusivity values of water and ice.

WATER FOR INDUSTRY

The quality and quantity of available water are important in the location of a chemical plant. For this purpose surface water as well as ground water may be used, but the supply must be adequate and continuous throughout the year. The supply, should however, not disturb the municipal water supply of the area. More than 50 per cent of water supply used in chemical plants is utilised for cooling purposes, that is, to carry away heat by warming or by evaporation. A considerable amount of water is also used for solutions and dilution purposes in the chemical plants and for this purpose, pure water is extremely necessary. The location of a chemical plant in a particular area is decided by taking into consideration the following important factors.

1. Raw materials.
2. Power supply.
3. Water.
4. Transport facilities.

Each industry has its own water requirements and sometimes adequate supply of water may be very suitable for one industry but the same may be dangerous for other. It is, therefore, extremely important to take into account the uses of water for the work to be carried out, its suitability based on the results of chemical analysis and bacteriological examinations. For example, boiler feed water should be as soft as possible and should contain least amount of nitrate and organic matter in order to prevent encrustations, scales or corrosion of the boiler plates. Water used for alcoholic distilleries should be as pure as possible and should contain few micro-organisms alongwith traces of NaCl and MgCl$_2$. Water used for paper mills should not contain iron and excess of lime and magnesia, because they decompose the resin soaps. In sugar industries, the crystallisation becomes more difficult if water contains sulphates and alkaline carbonates and also the nitrates. Moreover, molasses is obtained in much greater amount and sugar formed becomes deliquescent on exposure to air and moisture. If water is rich in micro-organisms, they may decompose the sugar partially. Water used for alcoholic breweries must contain lime and magnesia in much lesser amounts. In dye industry, water used should be free from iron and should possess little hardness only. Water used for cooking purposes should contain little hardness otherwise the vegetables do not cook easily. Water used for laundries should be as soft as possible.

USES OF WATER

Water is mostly used for industrial and municipal purposes. In order to ensure the right quality and quantity of water for these purposes it is extremely important to monitor water supply thoroughly taking all the aspects into consideration. The Various factors which are to be considered for the supply of water for any purpose are:

1. The quantity of the water available.
2. Seasonal variation in quantity as well as quality.
3. Analysis of water taking into consideration—its chemical, physical, microscopical and bacteriological characteristics.
4. Influence of industrial wastes, sewage etc., on the quality of water.

5. Cost involved in getting continuous supply of required quality and quantity of water.

Water in Human Body

In human body water is of utmost physiological importance and has specific functions to perform which are given below:

1. It acts as a solvent for the secretory and excretory products.
2. It acts as a carrier of nutritive elements to tissues and removes waste materials from them.
3. Water is a best solvent for electrolytes. It helps to regulate electrolyte balance of the body and maintains a healthy equilibrium of osmotic pressure exerted by solutes dissolved in water. A state of good health is possible as long as osmotic pressure exerted by the solutes remain constant.
4. It acts as a regulator of body temperature.

Water is more important than food. Deprivation of water brings about death much more quickly than that of food. The total body water constitutes 60 per cent to 70 per cent of adult body weight.

Water as a Solvent

Water is an excellent solvent for ionic solutes. There are two important properties of water which are ,responsible for the fact that water is an excellent solvent. These are: (i) water is a polar molecule and has a dipole moment; and (ii) water has a high dielectric constant. Because water is a polar molecule, there is a force of attraction between any ion and that end of H_2O molecule which is of opposite sign. This is called ion-dipole force of attraction. The force of attraction between ions and the dipolar H_2O molecules causes the formation of hydrated ions that are lower in energy than the separated ions and water molecules. In other words, H_2O molecules tend to orient about any ion and to be loosely associated with it and ions are said to be hydrated or solvated. Water molecules and ions in solution are in constant motion and the number of water molecules close to a given ion changes with time. There is an average number, called the hydration number, of water molecules which is closely associated with a given ion. The hydration number is large in the case of small, highly charged ions.

Chapter 2

Quantity of Water

INTRODUCTION

It is very difficult to ascertain the quantity of water required for a particular town. It involves the assumptions of many variable factors and foresight of the designer plays an important role in arriving at this quantity. However, the problem of estimating the quantity of water may be tackled by studying in detail the following two factors:

1. Rate of demand: The requirements of water for various uses are properly analysed and ultimately, the rate of consumption per head is worked out.
2. Population: The persons to be served by the scheme are calculated and estimate of future population is worked out with the help of suitable method.

POPULATION GROWTH

Populations increase by births, decrease by deaths and change with migration. Communities also grow by annexation. Urbanisation and industrialisation bring about social and economic changes as well as growth. Educational and employment opportunities and medical care are among the desirable changes. Among unwanted changes are the creation of slums and the pollution of air, water and soil. Least predictable of the effects on growth are changes in commercial and industrial activity.

Were it not for industrial vagaries of the Providence type, human population kinetics would trace an S-shaped growth curve in much the same way as spatially constrained microbic populations. As shown in Fig. 2.1, the trend of seed population is progressively faster at the beginning and progressively slower towards the end as a saturation value or upper limit is approached. What the future holds for a given community, therefore, is seen to depend on where on the growth curve the community happens to be at a given time.

The growth of cities and towns and characteristic portions of their growth curves can be approximated by relatively simple equations that derive historically from chemical kinetics. The equation of a first-order chemical reaction, possibly catalysed by its own reaction products, is a recurring example. It identifies also the kinetics of biological growth and other biological reactions including population growth, kinetics or dynamics. This widely useful equation may be written

$$dy/dt = ky(L - y) \qquad\qquad ... (2.1)$$

where y is the population at time t, L is the saturation or maximum population and k is a growth or rate constant with the dimension $[t^{-1}]$. It is pictured in Fig. 2.1 together with its integral, Eq. 2.6.

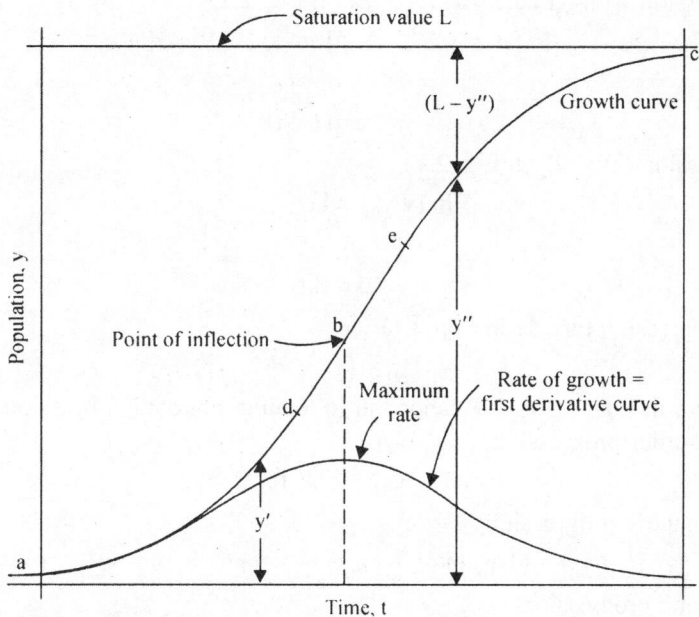

Fig. 2.1. Population growth idealised. Note geometric increase from a to d; straight-line increase from d to e (approximately); and first-order increase from e to c.

Three related equations apply closely to characteristic portions of this growth curve: (i) a first-order progression for the terminal arc 'ec' of Fig. 2.1; (ii) a logarithmic or geometric progression for the initial arc 'ad'; and (iii) an arithmetic progression for the transitional intercept, 'de' or:
For arc ec

$$dy/dt = k(L - y) \qquad\qquad ... (2.2)$$

For arc ad

$$dy/dt = ky \qquad\qquad ... (2.3)$$

For arc de

$$dy/dt = k \qquad\qquad ... (2.4)$$

If it is assumed that the initial value of k, namely k_0, decreases in magnitude with time or population growth rather than remaining constant, k can be assigned the following value:

$$k = k_0/(1 + nk_0t) \qquad\qquad ... (2.5)$$

in which n, as a coefficient of retardance, adds a useful concept to Eqs. 2.2 to 2.4.
On integrating Eqs. 2.1 to 2.4 between the limits $y = y_0$ at $t = 0$ and $y = y$ at $t = t$ for unchanging k values, they become:
For autocatalytic first-order progression (arc 'ac' in Fig. 2.1),

$$\ln[(L - y)/y] - \ln[(L - y_0)/y_0] = -kLt$$

or

$$y = L/\{1 + [(L - y_0)/y_0] \exp(-kLt)\} \qquad \ldots (2.6)$$

For first-order progression without catalysis (arc 'ec' in Fig. 2.1),

$$\ln[(L - y)/(L - Y_0)] = -kt$$

or

$$y = L - (L - y_0) \exp(-kt) \qquad \ldots (2.7)$$

For geometric progression (arc 'ad' in Fig. 2.1),

$$\ln(y/y_0) = kt$$

or

$$y = y_0 \exp(kt) \qquad \ldots (2.8)$$

For arithmetic progression (arc de in Fig. 2.1),

$$y - y_0 = kt \qquad \ldots (2.9)$$

Substituting Eq. 2.5 in Eqs. 2.2 to 2.4 yields the following retardant expressions:
For retardant first-order progression,

$$y = L - (L - y_0)(1 + nk_0t)^{-1/n} \qquad \ldots (2.10)$$

For retardant, geometric progression,

$$\ln(y/y_0) = (1/n)\ln(1 + nk_0t) \qquad \text{or} \qquad v = y_0(1 + nk_0t)^{-1/n} \qquad \ldots (2.11)$$

For retardant, arithmetic progression,

$$y - y_0 = (1/n)\ln(1 + nk_0t) \qquad \ldots (2.12)$$

These and similar equations are useful in water and waste-water practice, especially in water and waste-water treatment kinetics.

SHORT-TERM POPULATION ESTIMATES

Estimates of mid-year populations for current years and the recent past are normally derived by arithmetic from census data. They are needed perhaps most often for: (i) computing per capita water consumption and waste-water release; and (ii) for calculating the annual birth and general death rates per 1000 inhabitants or specific disease and death rates per 1,00,000 inhabitants. Understandably, morbidity and mortality rates from waterborne and otherwise water-related diseases are of deep concern to sanitary engineers.

For years between censuses or after the last census, estimates are usually interpolated or extrapolated as arithmetic or geometric progressions. If t_i and t_j are the dates of two sequent censuses and t_m is the mid-year date of the year for which a population estimate is wanted, the rate of arithmetic growth is given by Eq. 2.9 as $k_{arithmetic} = (y_j - y_i)/(t_j - t_i)$ and the mid-year populations, y_m, of inter-censal and post-censal years are respectively:

(Inter-censal)

$$v_m = y_i + (t_m - t_i)(v_j - y_i)/(t_j - t_i) \qquad \ldots (2.13)$$

(Post-censal)

$$y_m = y_i + (t_m - t_j)(v_j - y_i)/(t_j - t_i) \qquad \ldots (2.14)$$

In similar fashion, Eq. 2.8 states that $k_{geometric} = (\log y_j - \log y_i)/(t_j - t_i)$ and the logarithms of the mid-year population, $\log y_m$, for inter-censal and post-censal years are respectively:
(Inter-censal)

$$\log y_m = \log y_i + (t_m - t_i)(\log y_i - \log y_i)/(t_j - t_i) \qquad \ldots (2.15)$$

(Post-censal)

$$\log y_m = \log y_j + (t_m - t_j)(\log y_j - \log y_i)/(t_j - t_i) \qquad \ldots (2.16)$$

Geometric estimates, therefore, use the logarithms of the population parameters in the same way as the population parameters themselves are employed in arithmetic estimates; moreover, arithmetic increase corresponds to capital growth by simple interest and geometric increase to capital growth by compound interest. Graphically, arithmetic progression is characterised by a straight-line plot against arithmetic scales for both population and time on double-arithmetic co-ordinate paper and thus, geometric as well as first-order progression by a straight-line plot against a geometric (logarithmic) population scale and an arithmetic time scale on semi-logarithmic paper. The suitable equation and method of plotting is best determined by inspection from a basic arithmetic plot of available historic population information.

The Bureau of the Census estimates the current population of the whole nation by adding to the last census population the intervening differences (i) between births and deaths, that is, the natural increases; and (ii) between immigration and emigration. For states and other large population groups, post-censal estimates can be based on the apportionment method, which postulates that local increases will equal the national increase times the ratio of the local to the national inter-censal population increase.

Inter-censal losses in population are normally disregarded in post-censal estimates; the last census figures are used instead. Supporting data for short-term estimates can be derived from sources that reflect population growth in ways different from, yet related to, population enumeration. Examples are records of school enrollments; house connections for water, electricity, gas and telephones; commercial transactions; building permits; and health and welfare services. These are translated into population values by ratios derived for the recent past. The following ratios are not uncommon: population: school enrollment = 5 : 1; population: number of water, gas or electricity services = 3 : 1; and population: number of telephone services = 4 : 1.

At best, since forecasts of population involve great uncertainties, the probability that the estimated values turn out to be correct can be quite low. Nevertheless, the engineer must select values in order to proceed with planning and design of works. To use uncertainty as a reason for low estimates and short design periods can lead to capacities that are even less adequate than they otherwise frequently turn out to be. Because of the uncertainties involved, population are sometimes projected at three rates—high, medium and low. The economic and other consequences of designing for one rate and having the population grow at another can then be examined.

Population Distribution and Areal Density

Capacities of water collection purification and transmission works and of waste-water outfall and treatment works are a matter of areal as well as population size. Within communities individual service areas, their population and their occupancy are the determinants. A classification of areas by use and of expected population densities in persons per acre is shown in Table 2.1.

Values of this kind are founded on analyses of present and planned future sub-divisions of typical blocks. Helpful, in this connection, are census tract data; land-office, property, zoning, fire-insurance and aerial map; and other information collected by planning agencies.

Table 2.1. Common population densities.

Description	Persons per acre
Residential areas	
Single-family dwellings, large lots	5–15
Single-family dwellings, small lots	15–35
Multiple-family dwellings, small lots	35–100
Apartment or tenement houses	100–1000 or more
Mercantile and commercial areas	15–30
Industrial areas	5–15
Total, exclusive of parks, playgrounds and cemeteries	10–15

The design period is the useful of the water-supply scheme. A water-supply scheme is generally designed to meet the requirements over a period of 30 years after its completion.

Design Period

New water and waste-water works are normally made large enough to meet the needs and wants of growing communities for an economically justifiable number of years in the future. Choice of a relevant design period is generally based on: (i) the useful life of component structures and equipment, taking into account obsolescence as well as wear and tear; (ii) the ease or difficulty of enlarging contemplated works, including consideration of their location; (iii) the anticipated rate of population growth and water use by the community and its industries; (iv) the going rate of interest on bonded indebtedness; and (v) the performance of contemplated works during their early years when they are expected to be under minimum load. Design periods often employed in practice are shown in Table 2.2.

Table 2.2. Design periods for water and waste-water structures.

Type of structure	Special characteristics	Design period, years
Water supply		
Large dams and conduits	Hard and costly to enlarge	25–50
Wells, distribution systems	Easy to extend	
and filter plants	When growth and interest rates are low[a]	20–25
	When growth and interest rates are high[a]	10–15
Pipes more than 12 inch in diameter	Replacement of smaller pipes is more costly in long run	20–25
Laterals and secondary mains less than 12 inch in diameter	Requirements may change fast in limited areas	Full development
Sewerage		
Laterals and submains less than 15 inch in diameter	Requirements may change fast in limited areas	Full development
Main sewers, outfalls and intercepters	Hard and costly to enlarge	40–50
Treatment works	When growth and interest rates are low[a]	20–25
	When growth and interest rates are high[a]	10–15

[a] The dividing line is in the vicinity of 3% per annum.

In this chapter, the above two topics, namely, the rate of demand of water and the population to be served will now be discussed in detail.

RATE OF DEMAND

Water Consumption

Although the draft of water from distribution systems is commonly referred to as water consumption, little of it is, strictly speaking, consumed; most of it is discharged as spent or waste-water. Use of water is a more exact term. True consumptive use refers to the volume of water evaporated or transpired in the course of use—principally in sprinkling lawns and gardens, in raising and condensing steam and in bottling, canning and other industrial operations.

Service pipes introduce water into dwellings, mercantile and commercial properties, industrial complexes and public building. Table 2.5 shows approximate per capita daily uses in the United States. Wide variations in these figures must be expected because of differences in: (i) climate; (ii) standards of living; (iii) extent of sewerage; (iv) type of mercantile, commercial and industrial activity; (v) water pricing; (vi) resort to private supplies, (vii) water quality for domestic and industrial purposes; (viii) distribution-system pressure; (ix) completeness of meterage; and (x) systems management.

Extremes of heat and cold increase water consumption: hot and arid climates by frequent bathing, air conditioning and heavy sprinkling and cold climates by bleeding water through faucets to keep service pipes and internal water piping from freezing during cold spells. In metered and sewered residential areas, the observed average, daily use of water for lawns and gardens, $Q_{sprinkling}$, in gpd during the growing season is about 60 per cent of the estimated average potential evapotranspiration E, reduced by the average daily precipitation, P, effective in satisfying evapotranspiration during the period, or

$$Q_{sprinkling} = 1.63 \times 10^4 \, A(E - P) \qquad \ldots (2.17)$$

Here $1.63 \times 10^4 = 0.6 \times 2.72 \times 10^4$, the number of gallons in an acre-inch; A is the average lawn and garden acreage per dwelling unit and E and P are expressed in inches. The average lawn and garden area is given by the observational relationship

$$A = 0.803 \, D^{-1.26} \qquad \ldots (2.18)$$

where D is the gross housing density in dwelling units per acre.

High standards of cleanliness, large numbers of water-connected appliances, oversized plumbing fixtures and frequent lawn and garden sprinkling, all associated with wealth, result in heavy drafts. For sewered properties, the average domestic use of water $Q_{domestic}$ in gpd for each dwelling unit is related to the average market value M of the units in thousands of dollars by the following observational equation:

$$Q_{domestic} = 157 + 3.46 \, M \qquad \ldots (2.19)$$

General Urban Water Demands

Some commercial enterprises—hotels and restaurants, for instance—draw much water; so do industries such as breweries, canneries, laundries, paper mills and steel mills. Industries, in particular, draw larger volumes of water when it is cheap than when it is dear. Industrial draft varies roughly inversely as the manufacturing rate and is likely to drop by about half the percentage increase in cost when rates are raised. Hospitals, too, have high demands.

Although the rate of draft in fire fighting is high the time and annual volume of water consumed in extinguishing fires are small and seldom identified separately for this reason.

Water of poor quality may drive consumers to resort to uncontrolled; sometimes dangerous, sources, but the public supply remains the preferred source when the product water is clean, palatable and of unquestioned safety; soft for washing and cool for drinking and generally useful to industry. The availability of groundwater and nearby surface sources may persuade large industries and commercial enterprises to develop their own process and cooling water.

Hydraulically, leaks from mains and plumbing systems and flows from faucets and other regulated openings behave like orifices. Their rate of flow varies as the square root of the pressure head and high distribution pressures raise the rate of discharge and with it the waste of water from fixtures and leaks. Ordinarily, systems pressures are not raised above 60 psig (lb per sq in. gauge) in American practice; even though it is impossible to employ direct hydrant streams in fire fighting when hydrant pressures are below 5 psig.

Meterage encourages thrift and normalises the demand. The cost of metering and the running expense of reading and repairing meters are substantial. They may be justified in part by accompanying reductions in waste and possible postponement of otherwise needed extensions. Under study and on trial here and there is the encouragement of off-peak-hour draft of water by large users. To this purpose, rates charged for water drawn during off-peak hours are lowered preferentially. The objective is to reap the economic benefits of a relatively steady flow of water within the system and the resulting proportionately reduced capacity requirements of systems components (Fig. 2.2). The water drawn during off-peak hours is generally stored by the user at ground level even when this entails re-pumping.

Distribution networks are seldom perfectly tight. Mains, valves, hydrants and services of well-managed systems are therefore regularly checked for leaks. Superficial signs of controllable leakage are: (i) high night flows in mains; (ii) water running in street gutters; (iii) moist pavements; (iv) persistent seepage; (v) excessive flows in sewers; (vi) abnormal pressure drops; and (vii) unusually green vegetation (in dry climates). Leakage is detected by (i) driving sounding rods into the ground to test for moist earth; (ii) applying listening devices that amplify the sound of running water; and (iii) inspecting premises for leaky plumbing and fixtures. Leakage detection of well-managed waterworks may be complemented by periodic and intensive but, preferably, routine and extensive water-waste surveys. Generally involved is the isolation of comparatively small sections of the distribution system by closing valves on most or all feeder mains and measuring the water entering the section at night through one or more open valves or added piping on fire hoses. Common means of measurement are pitot tubes, bypass meters around controlling valves or meters on one or more hose lines between hydrants that straddle closed valves.

A small quantity of water is required by a man under normal conditions for his personal use. But his demand of water for other purposes will naturally depend upon the standard of living and degree of culture. In order to arrive at a reasonable value of rate of demand for any particular town, the demand of water for various purposes is divided under the following five categories:

1. Domestic purposes.
2. Civic or public purposes.
3. Industrial purposes.
4. Business or trade purposes.
5. Loss and waste.

We will briefly analyse each category and will discuss how the quantity of water under each category is worked out for the purpose of estimating the rate of demand of water.

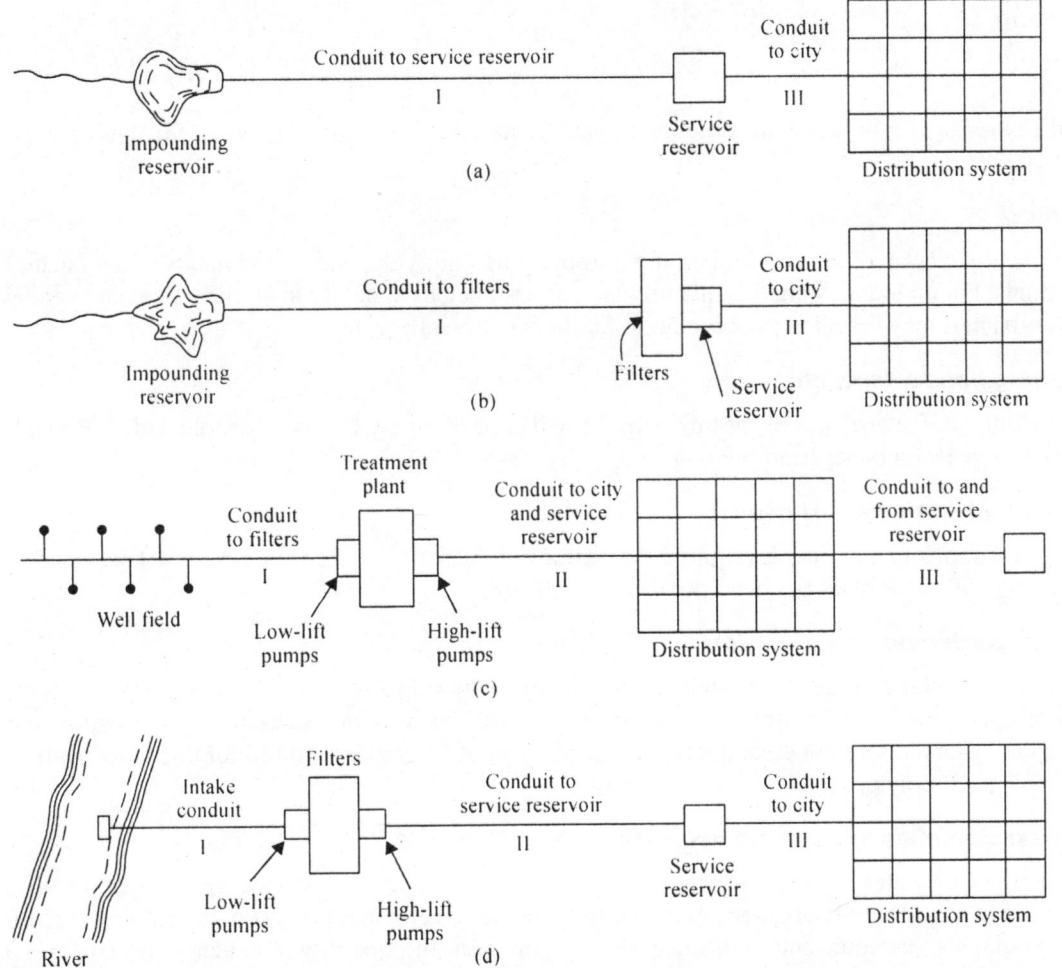

Fig. 2.2. Required capacities of four typical waterworks systems. The service reservoir is assumed to compensate for fluctuations in draft and fire drafts and to hold an emergency reserve.

Domestic Purposes

The quantity of water required for domestic purposes can be sub-divided as follows:

Drinking

A human body contains about 70 per cent of water. The consumption of water by a man is required for various physiological processes such as blood formation, food assimilation, etc. The quantity of water which a man would require for drinking depends on various factors. But on the average and under normal conditions, it is about 2 liters per day. This amount, as will be seen, is very small as compared to various other uses of water. But it is most essential to supply water for drinking purposes with a high degree of purity. If water for drinking contains undesirable elements, it may lead to epidemic. In fact, the drinking water should be protected, potable and palatable.

Cooking

Some quantity of water will also be required for cooking. The quantity of water required for this purpose will depend upon the stage of advancement of the family in particular and society in general. However, for the purpose of estimation, the amount of water required for cooking may be assumed as about 5 liters per head per day.

Bathing

The quantity of water required for bathing purpose will mainly depend on the habits of people and type of climate. For an Indian bath, this quantity may be assumed as about 30 to 40 liters per head per day and for tub-bath, it may be taken as about 50 to 80 liters per head per day.

Washing hands, face, etc.

The quantity of water required for this purpose will depend on the habits of people and may roughly be taken as 5 to 10 liters per head per day.

Household sanitary purposes

Under this division, the water is required for washing clothes, floors, utensils, etc., and it may be assumed to be about 50 to 60 liters per head per day.

Private gardening and irrigation

In case of developed cities there will be practically no demand of water for this purpose. In case of undeveloped cities, the private wells are generally used to provide water for private gardening and irrigation. It is therefore not essential to include the quantity of water required for this purpose in case of public water supply project.

Domestic animals and private vehicles

The amount of water required for the use of domestic animals and private vehicles is not of much concern to a water supply engineer. With the growth and development of town, the cattle disappear and commercial stables come into existence. For information, the quantity of water required for various types of domestic animals is mentioned in Table 2.2.

The quantity of water required for washing cars and private vehicles is very low especially in case of poor countries. The number of cars in relation to population decides the quantity of water for this purpose per head per day. For information, let us assume that a car requires 150 to 200 liters of water for washing and that a car is washed everyday. Then the approximate demand of water for washing a car per head per day would be 50 liters in USA, 10 liters in England and only 0.30 liter in India.

Table 2.2. Water required for domestic animals.

Name of domestic animal	Quantity of water required in liters per number per day
Cow or buffalo	40
Dog	10
Horse	50
Mule or pony	30
Sheep	5

Thus, the quantity of water required for domestic purposes can be worked out with the help of tentative figures mentioned above. The requirements of water for domestic purposes roughly forms about 40 per cent to 50 per cent of the total water requirements per capita per day.

Civic or Public Purposes

The quantity of water required for civic or public purposes can be sub-divided as follows:

Road washing

The roads with heavy amount of dust are to be sprinkled with water to avoid inconvenience to the users. Even in case of dust-proof roads, the periodical washing is necessary. On an average, the quantity of water required for this purpose may be taken as about 5 liters per head per day.

Sanitation purposes

In this division, the water is required for cleaning public sanitary blocks, large markets, etc., and for carrying liquid wastes from houses. The quantity of water required for this purpose will depend on the growth of civilisation and may be assumed to be about 2 to 3 liters per head per day.

Ornamental purposes

In order to adorn the town with decorative features, the fountains or lakes or ponds are sometimes provided. These objects require huge quantity of water for their performance. As far as Indian towns are concerned, the quantity of water required for this purpose may be treated as quite negligible since in most of the towns, the quantity of water available is not enough even to meet with the most urgent needs of the society.

Fire demand

Usually a fire occurs in factories and stores. The quantity of water required for fire fighting purposes should be easily available and always kept stored in the storage reservoir. The fire hydrants are located in the mains at distances of not more than 150 meters or so. When a fire occurs, the pumps installed on trucks are immediately rushed to the site of fire occurrence and these pumps, when connected to the fire hydrants, are capable of throwing water with high pressure. The fire is thus brought under control. The requirement of water for fire demand can be worked out in a logical way as follows:

1. Minimum number of streams required.
2. Discharge of each stream.
3. Duration of a fire.
4. Number of simultaneous fires.

Industrial Purposes

The quantity of water required for industrial or commercial purposes can be sub-divided as follows:

Factories

The quantity of water required for the processes involved in factories will naturally depend on the nature of products, size of factory, etc. and it has no relation with the density of population. It is quite likely that the demand of water for factories may equal or even exceed the demand of water for domestic purposes. The possibility of recycling of water in the plant will also have appreciable effect on the demand of water for a particular product. Table 2.3 shows the typical water demand per kg product in liters for some of the factories.

Table 2.3. Water demand for factories.

Product	Water demand per kg product in liters
Aluminium smelting	1350
Butter	11
Coal mining	4
Cotton bleaching	0.25–0.3
Cheese	20
Glass	70
Oil refining	10
Paper	160–175
Rayon	1–2
Soap	4.5–5.5
Steel	45–60
Sugar	9
Synthetic fibres	200
Wool scouring	12–14

Rising water use can be arrested by conserving plant supplies and introducing efficient processes and operations. Most important, perhaps, are the economies of multiple re-use through countercurrent rinsing of products recirculation of cooling and condensing water and re-use of otherwise spent water for secondary purposes after their partial purification or re-purification.

Industry often develops its own supply. Chemical plants, petroleum refineries and steel mills, for example, draw on public or private utilities for less than 10 per cent of their needs. Food processors, by contrast, purchase about half their water from public supplies, largely because the bacterial quality of drinking water makes it *de facto* acceptable.

About 90 per cent of the industrial draft is taken from surface sources. Groundwater may be called into use in the summer because their temperature is then seasonally low. They may be prized, too, for their clarity and their freedom from colour, odour and taste. Figure 2.2 shows the required capacities of four typical waterworks systems.

Available sources may be drawn on selectively: municipal water for drinking, sanitary purposes and delicate processes, for example and river water for rugged processes and cooling and for emergency uses such as fire protection. Treatment costs as well as economic benefits are the determinants.

Power stations

A huge quantity of water will require for working of power stations. But generally the power stations are situated away from the cities and they do not represent a serious problem to public water supply.

Railways and airports

In most of the cases, the railways and airports make their own arrangements regarding their water requirements and hence, the quantity of water to be consumed by railways is not ordinarily included in any public water supply scheme. For the purpose of estimate, the railways provide 25 to 70 liters of water per head per day depending upon the nature of station and facilities like bathing, etc. The airport authorities usually make the provision of water at about 70 liters of water per head per day.

It is thus not possible to connect the requirement of water for industrial purposes to the population of the city. It is therefore advisable to study each case independently in this regard and decide the quantity of water required for industrial purposes accordingly. For a city with moderate factories, it is estimated that about 20 to 25 per cent of per capita consumption will be required for industrial purposes.

Business or Trade Purposes

Some trades such as dairies, hotels, laundries, motor garages, restaurants, stables, schools, hospitals, cinema halls, theatres, etc., require a large quantity of water. Such trades are to be maintained in hygienic conditions and sanitation of such places should be strictly insisted.

The number of such business centers will depend upon the population and for a moderate city, an average value of about 15 to 25 liters of water per head per day may be taken as water requirement for this purpose.

Loss and Waste

The quantity of water required under this category is sometimes termed as the unaccounted requirement. It includes careless use of water, leakage in mains, valves, other fittings, etc., unauthorised water connections and waste due to other miscellaneous reasons. The quantity of water lost due to all these reasons is uncertain and cannot be effectively predicted. However, for the purpose of calculating the average rate of demand, it may be estimated to be about 30 to 40 per cent of per capita consumption.

If the distribution of water is entirely through meters only and if the distribution system is well-maintained, it is possible to bring down the percentage of unaccounted water to the extent of about 10 to 15.

Variations in Water Demand

Water consumption changes with the seasons, the days of the week and the hours of the day. Fluctuations are greater: (i) in small than in large communities; and (ii) during short rather than during long periods of time. Variations are usually expressed as ratios to the average demand.

Factors affecting rate of demand

There are various factors which influence the rate of demand of water. These factors are to be analysed carefully and properly before arriving at the rate of demand for a particular locality. Following are the factors affecting the rate of demand of water:

1. Climatic conditions.
2. Cost of water.
3. Distribution pressure.
4. Habits of population.
5. Industries.
6. Policy of metering.
7. Quality of water.
8. Sewerage.
9. Size of city.
10. System of supply.

We will now briefly discuss each of the above factor affecting the rate of demand of water.

Climatic conditions

The requirement of water in summer is more than that in winter. So also is the case with hotter and cooler places. In extreme cold, the people may keep water taps open to avoid freezing of pipes. This may result in increased rate of consumption.

Cost of water

The rate at which water is supplied to the consumers may also affect the rate of demand. The higher the cost, the lower will be the rate of demand and vice versa.

Distribution pressure

The consumption of water increases with the increase in the distribution pressure. This is due to increase in loss and waste of water at high pressure. For instance, an increase of pressure from 2 to 3 kg/cm^2 may lead to an increase in consumption to the extent of about 25 to 30 per cent. The designer therefore should only provide for distribution pressure which is necessary for rendering satisfactory service.

Habits of population

For high-value premises, the consumption rate of water will be more due to better standard of living of persons. For middle-class premises, the consumption rate will be average while in case of slum areas, it will be much lower. A single water tap may be serving several families in low-value areas.

Industries

The presence or absence of industries in a city may also affect its rate of demand. As there is no direct relation between the water requirement for industries and population, it is necessary to calculate carefully present and future requirements of industries.

Policy of metering

The quantity of water supplied to a building is recorded by a water meter and the consumer is then charged accordingly. The installation of meters reduces the rate of consumption. But the fact of adopting policy of metering is a disputable one as seen from the following arguments which are advanced for and against it.

Arguments for policy of metering:
1. It becomes very easy to locate the points of leakage when meters are installed.
2. The consumer is charged in proportion to the quantity of water which he uses.
3. The reduction in consumption of water results in decrease in loads on purification plants, pumps, sewers, etc.
4. The wastage of water is decreased.
5. The careful consumer pays less and the careless consumer pays more.

Arguments against policy of metering:
1. There is loss of pressure due to installation of meters and it adds to the pumping cost.
2. The use of water for gardens, fountains, etc., is greatly diminished. This decreases the beauty of the locality.
3. The limited use of water may lead to unhygienic conditions and may cause epidemic.
4. The policy of metering is expensive in the sense that the cost is to be incurred to buy, to install, to read and to maintain the meters.

5. It is suggested that the amount spent after introducing the policy of metering may well be spent in improvement of water supply scheme itself.

In conclusion, it may be stated that engineers dealing with water supply schemes, recommend installation of water meters, when the following two conditions are existing:

1. The quantity of water that is available from the source is limited.
2. The total cost of water supply scheme is an important consideration.

Quality of water

The improvement in quality of water may result in the increase of rate of consumption. The public using the improved water will consider it safe and may make various uses of the available water. On the other hand, if water has unpleasant taste or adour, the rate of consumption will come down.

Sewerage

The existence of sewerage system in a locality will lead to an increase in use of water for civic or public purposes. The people will also use more quantity of water for flushing sanitary units such as urinals and water closets.

Size of city

Generally, the smaller the city, the lower is the rate of demand. But the presence of a water-consuming industry in a small town may result in a higher rate of demand, even if the town is small. Table 2.4 gives the rates of demand for the Indian towns of various population.

Table 2.4. Rates of demand with respect to size of town.

Population	Rate of demand in liters per capita per day
Upto 20000	110
20000–50000	110–150
50000–2,00,000	150–180
2,00,000–5,00,000	180–210
5,00,000–10,00,000	210–240
Above 10,00,000	240–270

System of supply

The supply of water may be continuous or intermittent. In the former case, the water is supplied for 24 hours and in the latter case, it is supplied for certain duration of day only. It is claimed that the intermittent supply system will reduce the rate of demand. But sometimes, the results are proved to be disappointing, mainly for the following two reasons:

1. During non-supply period, the water taps are kept open and hence, when the supply starts, the water flowing through open taps is unattended and this results in waste of water.
2. There is a tendency of many people t o throw away water stored previously during non-supply hours and to collect fresh water. This also results in waste of water.

Measurement of water

During the process of water supply scheme, it becomes necessary to measure the quantity of water for the following reasons:
1. To ascertain the quantity of water to be supplied free of charge.
2. To determine the costs of treated water at different stages of operations.
3. To give an idea of operation efficiencies of various units of the water supply scheme.
4. To maintain records for administrative purposes.
5. To measure the quantity of water to be sold either on retail basis to the individuals or on wholesale basis to another supply.
6. To provide a control on operations of various units such as pumps, elevated reservoirs, chemical feeding devices, etc.

For measuring water, the meters are employed. The meters can broadly be divided into the following two categories:

Displacement type

Such meters contain a vessel of known volume and the number of times it is filled and emptied is automatically recorded. This type of meter is useful for small installations to measure relatively small quantity of flow as in case of hotels, residential buildings, etc.

Velocity type

Such meters are turbine or venturi type and they contain a device by which a vane or propeller turns in direct ratio to the quantity of flow passing through the propeller. This type of meter is useful for big installations such as on pumps, water main lines, etc.

In any case, the following points should be considered while making selection for any water meter:
1. Accuracy of measurement.
2. Availability of spare parts.
3. Capacity with minimum head loss.
4. Cost.
5. Durability.
6. Ease of repair.
7. Noise during working.
8. Quality of workmanship.
9. Registration with varying discharges.
10. Self-cleansing property; etc.

Variations in rate of demand

The average daily rate of demand per head is the ratio of total quantity of water supplied during the year to the number of persons served multiplied by the days of year. This average daily rate of demand per head is likely to deviate, if period of observation is shortened. Thus, if average daily rate of demand is say 100, then:
1. Seasonal maximum demand may be 130.
2. Monthly maximum demand may be 140.
3. Daily maximum demand may be 180 or so.

These variations are due to many factors such as habits of people, climatic conditions, types of industries, etc. The above figures for deviation from the average for seasonal, monthly and daily demand, are taken for illustration only. Every city possesses peculiarities of its own and hence, before arriving at a particular decision, the detailed studies become essential. Every case is therefore studied separately and the variation from the average rate of demand is worked out accordingly.

In practice, the maximum daily rate of consumption is very important. This maximum daily consumption is to be consumed in 24 hours. But demand during 24 hours will not be uniform and it will vary according to hour of day. The peak demands occur in the morning and evening. The slack periods occur early in the morning and late at night.

The demand of water from hour to hour is thus variable and the maximum hourly demand will be much higher than the average daily demand. It may be assumed as about 150 per cent of the average daily demand. In order to meet with the maximum hourly demand, the pumps are either to run at variable speeds or to run at an average speed. In the former case, the speeds of pumps are changed as per hourly requirements. This method of working the pumps results in great inconvenience. In the latter case, the pumps are run at an average speed in such a way that the surplus water stored at slack demand period is used at peak demand period.

Effects of variations on design

The water supply units are designed in accordance with the fluctuations or variations in rate of demand. The effects of these variations of various units on water supply scheme are as follows:

1. The pumps and filters are generally designed for 1.50 times the average rate of daily demand.
2. If pumps are working in shifts, the above rate is still to be multiplied by the ratio of 24 hours to pumping hours. Thus, if the pumps are working for 12 hours in a day, the multiplying factor for the design of pumps will be $1.50 \times \dfrac{24}{12} = 3.00$
3. The distribution mains are to be designed for the maximum hourly demand on maximum day. The multiplying factor is taken as about $1.80 \times 1.50 \, \Omega \, 2.50$. This provision is quite sufficient for towns with a population varying from 50,000 to 2,00,000. For towns with a population upto 50000 and those with a population exceeding 2 lakhs, the multiplying factors for the distribution mains are respectively 3.00 and 2.00.
4. The other units of water supply scheme such as sedimentation tanks and overhead tanks are designed for the average daily rate of demand only.

Water requirements for buildings other than residences

The buildings like cinema, school, hotel, hostel, etc., will require water as per their requirements. Table 2.5 shows the water requirements for such buildings. It may be noted that the requirements mentioned in Table 2.5 are for average type of constructions and further they do not include water required for civic or public purposes.

Table 2.5. Water requirements for buildings other than residences.

Type of buildings	Water requirement in liters per day
Cinema and concert halls	15 per seat
Factories	50 per worker
Hospitals with less than 100 beds	340 per bed

(Contd...)

Type of buildings	Water requirement in liters per day
Hospitals with more than 100 beds	450 per bed
Hostels	135 per hed
Hotels	180 per bed
Medical quarters	135 per head
Offices	45 per head
Restaurants	70 per seat
Schools	45 per student

Estimating population

The term population is used to indicate the total number of human beings residing in a certain area at any particular time. The present population is obtained by referring to the statistics of census records prepared by the local body. The water supply project is not designed only for present population. But it is made to accommodate the future population at the end of three or four decades. The growth of population may be sharp, slow or even stationary depending upon the factors contributing to the future development of the locality such as coming up of new industries, trade expansion, etc.

The future period for which various service units of water supply or sanitary engineering are designed is known as the period of design. It should neither be too short to make the plant obsolete and uneconomical in near future nor should it be too long to throw unnecessary financial burden on the present generation. The period of design varies from 20 to 40 years or even upto 50 years. But for normal projects, the period of design is taken as 20 to 30 years. Table 2.6 gives the suggested design periods for some of the important components of water supply project.

Table 2.6. Periods of design for important components of water supply project.

Component	Period of design in years
Clear water service reservoirs	15
Conveying pipes for raw water and clear water	30
Distribution system	30
Electric motors and pumps	15
Infiltration works	30
Storage reservoirs	50
Water treatment units	15

The entire population of the whole city is usually distributed unevenly in various parts of the city depending upon land-use pattern, available facilities, etc. The term population density is used to indicate the number of persons per unit area and the distribution of population is well studied by finding out the population densities of various parts of the city. Such a study will be of considerable assistance in the design of various units of the public utility services.

It may be noted that the population forecasts are useful and necessary not only to the public health engineers, but they are also required by various organisations for the following purposes:

1. To assist government agencies for the preparation of economic, employment and social programmes.

2. To collect information for location of an industry, its future expansion, availability of labour, marketing and distribution of the product, etc.

3. To provide data to the transportation industry.

4. To work out requirements for other public utilities such as telephones, electric power, etc.

Following are various methods of population forecasts or population projections and the selection of method will naturally depend on the available data:

1. Arithmetical increase method.
2. Geometrical increase method.
3. Incremental increase method.
4. Graphical method.
5. Comparative method.
6. Zoning method.
7. Ratio and correlation method.
8. Growth composition analysis method.

A brief description of each of the above method of forecasting population will now follow.

Arithmetical increase method

In this method, the average increase of population for the last three or four decades is worked out and then for each successive future decade, this average is added. This method gives low results and it is to be adopted for large cities which have practically reached their maximum development.

Geometrical increase method

In this method, it is assumed that the percentage increase in population from decade to decade remains constant. From the available census records, this percentage is fixed and then population of each future successive decade is worked out. The fixation of percentage in case of developing cities should be done carefully. Otherwise this method is likely to give very high results. This method gives better results for old cities which are not undergoing further development.

As the increase in population is compounded over the existing population every decade, this method is sometime also referred to as the uniform increase method.

The assumed average growth rate can be computed in the following two ways:

Arithmetic average

The average of the percentage growth rates of the several known decades of the past are worked out and then, by taking the arithmetic mean, the constant increase per decade is obtained.

Geometric average

In this method, the average growth rate is obtained by the geometric average.

Incremental increase method

This method combines the above two methods. The population of each successive future decade is first worked out by the arithmetical increase method and to these values, the incremental average per decade is added. It thus combines the advantages of both the above methods and hence it gives satisfactory results.

Graphical method

In this method, a curve of population against time is drawn for the city under consideration. The known census records are put up on the graph to get the shape of the curve. The curve is then carefully extended from present to future decades and the population after each successive future decade is read from the curve. The extension of curve in future decades should be based on personal judgement of the designer and it should be assisted by probable future conditions and past history of the city.

It may be noted that graphical method indicates the graphic representation of previous mathematical methods. The nature of extension will determine the mathematical method.

The above four methods are easy and simple. They are based on the main assumption that factors and conditions which were responsible for population growth in the past will continue to exist in future with the same intensity.

In view of the recent changes in birth rates, life expectancy, mobility of labour and various other factors affecting population growth, it is clear that the above assumption may prove to be wrong and hence the results obtained by these methods in such cases may become less reliable. However, these methods are useful for areas having slow and steady population growth. They are also useful for providing a check on the results obtained by other methods of population forecast.

Comparative method

In this method, it is assumed that the city under consideration will develop as similar cities have developed in the past. It is thus assumed that the future population growth of the city under consideration will parallel the past growth of similar cities. In practice, however, it is rather difficult to find identical cities with respect to population growth. For such comparisons, the complicated topics of such cities are to be carefully examined and historical periods involved in such comparisons are also to be properly analysed.

The comparative method has some logical background for its accuracy and hence if statistics of development of similar cities are available, the results obtained by this method may prove to be reliable and satisfactory.

Zoning method

This is rather the modern and the most useful method of population forecast. In this method, the master plan of the city is prepared and it is divided into several zones such as residential zone, industrial zone, commercial zone, etc. The city is allowed to develop in a definite way only. The provisions in the master plan control the character of various zones. Thus the future population of the city when fully developed can easily be worked out. It is also possible to assume different density of population and different rates of consumption for different zones of the city.

Suppose, in a residential zone of the city, the maximum number of tenements to be allowed is 1000 per hectare. Assuming 4 persons per tenement, the population of this zone, when fully developed, will be about 4000 per hectare.

Ratio and correlation method

It is evident that the population growth of a small area is related to some extent to the population growth of a wide area. Thus the rate of population growth of a town is related to some extent to the rate of population growth of state or nation. Hence it is possible to estimate the population of town under consideration by considering the rate of population growth of state or nation.

The various techniques may be adopted for this purpose. The most simple procedure is to adopt a constant ratio, which is equivalent to the population of town at last census to the population of larger region at last census. However, depending upon the merit of the case, the changing ratios derived from correlation studies may also be adopted.

Following are the advantages of this method:
1. In addition to local factors affecting population growth, the national or regional factors affecting population growth are also included in this method.
2. The population forecasts for state or nation are usually prepared by careful considerations.
3. The population forecasts of large areas are generally more reliable.

It is clear that the ratio and correlation method would be useful for areas whose population growth in past is fairly consistent with that of state or nation.

Growth composition analysis method

There are only three ways by which change in population occurs:
1. Through births.
2. Through deaths.
3. Through migration.

If the above three factors for area under consideration are properly analysed, the estimated population of some future period can be obtained by the following equation:

Estimated population = Present population + natural increase or decrease + migration.

The difference between births and deaths is known as the natural increase and it will be positive, if births exceed deaths and it will be negative, if deaths exceed births. The migration also affects natural increase and hence, the migration trends for future should be determined carefully by considering past migration trends, reasons for migration, etc.

Factors affecting estimated population

The eight methods, discussed above, are more or less the standard methods of estimating population after each successive future decade. Generally, two or more methods are adopted in each case and a mean value is adopted. However, the difference between the actual and estimated population is likely to be more marked with smaller cities than with larger cities because of the fact that the former are more susceptible to the various conditions and factors affecting the rate of growth.

Also the accuracy of population estimates decreases as the time period of the forecast increases or as the population of the area decreases. But the following factors affect considerably the values of the estimated population:
1. Accident in the nature of big fires, epidemic, floods, earthquakes, war, etc.
2. Changes in education, politics, recreation, etc.
3. Economic changes, development of new industries, etc.
4. Increase in transport and conveyance facilities.
5. Unforeseen circumstances such as discovery of oil, mine, etc.
6. Sudden increase in religious importance of the city.
7. Starting of a project of national importance in or around the city.
8. Political changes in the adjoining country and nearness to the national borders.

Chapter 3

Sources of Water

INTRODUCTION

In any water-supply scheme for a town or city, a source of water has to be found first. The source may be a river, stream, lake, well, etc. But it is very important that the source should be able to provide adequate supply of water to meet the demand of the town or city.

The chief source of all water supply schemes at present is rainfall. As time may pass, it may become necessary to find out substitutes for rainfall as sources of water supply schemes. The scientists have already started experiments in this line and attempts are being made to find out feasibilities of converting ocean water and sewage effluent into potable water.

Another development of present day is that of creating artificial rainfall in a particular locality. The tests are carried out by spraying silver iodide or compressed carbon dioxide in vapourised form on water bearing clouds. The temperature required for precipitation of natural nuclei in the form of ice crystals or dust particles is about $-15°C$. Thus the water vapour will condense and freeze around the solid nuclei present in the clouds and start falling in the shape of rain or snow.

PROPERTIES OF WATER

Pure water is a colourless, odourless and tasteless liquid. The depth and light give it a blue or bluish-green tint. Tastes and odours in water are due to dissolved gases, such as sulphur dioxide and chlorine and minerals. Water, a unique substance, exists in nature simultaneously as a solid (ice), liquid (water) and a gas (vapour). Its density is 1 g/ml or cubic centimeter. It freezes at $0°C$ and boils at $100°C$. When frozen, water expands by one-ninth of its original volume.

Water is in continuous circulation through the water or hydrologic cycle, which is formed of three phases: atmospheric water, surface water and groundwater.

CLASSIFICATION OF SOURCES

The source of water-supply may be classified as:
1. Surface water.
2. Groundwater or sub-surface water.

The quantity of water remaining on the surface after all the losses is called run-off and forms the source for all surface water. The portion of water that percolates into the soil and flows or collects underground is the source of groundwater.

Surface water source may be further classified as:

1. Lakes.
2. Impounding reservoirs.
3. Rivers/streams, irrigation canals.

Surface Run-off

The rainfall on an area is expressed as so many millimeters over the entire area for a certain fixed interval of time i.e., day, month, season or year. Thus the quantity of water obtained from rainfall during a certain interval of time can be easily worked out by the multiplication of the area and depth of rainfall.

But all the water coming down from the rainfall is not available for further use. Some quantity of it is lost either in evaporation or percolation or transpiration. The evaporation is the loss of water from land and water surfaces back to the atmosphere due to action of heat of the sun. The percolation indicates the loss of water penetrated into the soil and it may join some underground source of water. The transpiration is the loss of water caused by the leaves of the growing vegetation. The net quantity of rain water which remains on surface after all these losses is termed as the surface run-off. This surface run-off is seen in the form of various streams which ultimately join and form a river.

The surface run-off is harmful because of the following reasons:

1. Economic use: If surface run-off is to be used economically, it requires costly reservoirs or land improvement schemes.
2. Erosion: It takes away top soil and the soil erosion due to surface run-off causes serious economic losses.
3. Loss of water: It takes away the water which might have been used for agriculture. It thus leads to the loss of water for agriculture.
4. Occurrence of floods: It leads to floods. The rivers during floods overflow their banks and inundate the surrounding land area.

The upstream area contributing to the water of a river is termed as its catchment area. The term run-off coefficient is used to indicate the ratio of surface run-off from an area to the total rainfall on that area in a fixed interval of time. Thus it indicates the percentage of rainfall water which is available on surface for consumption. The run-off coefficient depends on the following factors:

1. Area of catchment: The smaller the catchment area, the smaller will be the coefficient of run-off and vice versa.
2. Characteristics of catchment: It is very essential to study in detail the characteristics of catchment area as they considerably affect the value of run-off coefficient. The matters to be studied are size, slope, vegetation, porosity, climate, shape, etc.
3. Condition of ground at the time of rainfall: If ground is dry at the time of rainfall, it will absorb more water and coefficient of run-off will be small. For ground, wet at the time of rainfall, the reverse will be the case.
4. Intensity of rainfall: If it rains heavily in short duration of time, the soil does not get opportunity to absorb all water. It thus increases surface flow and consequently, the coefficient of run-off is also increased.
5. Interval between successive showers: The smaller the interval between successive rainfall showers, the greater will be the coefficient of run-off and vice versa.
6. Season of rainfall: The rainfall during hot season gives less surface flow than that during cold season.
7. Yearly rainfall: The greater the annual rainfall, the greater is the run-off coefficient and vice versa.

Precipitation

The term precipitation is used to indicate the water which returns to the surface of earth in various form like rain, snow, etc. The major part of precipitation occurs in the form of rain and only a small portion of it occurs in the form of snow, etc.

RAINFALL

Following three terms will be discussed in connection with the rainfall of a locality:
1. Average annual rainfall.
2. Index of wetness.
3. Minimum annual rainfall.

Average Annual Rainfall

The annual rainfall at a given rain gauge station is recorded for a number of years and the mean of annual rainfall from the records of 35 years or so is worked out. This is known as the average annual rainfall at the given rain gauge station. It is to be remembered that whenever the rainfall of a particular locality is mentioned, it indicates the average annual rainfall of that place.

Index of Wetness

The ratio of the actual rainfall in a given particular year at a given place to its average annual rainfall is known as the index of wetness. Thus,

$$\text{Index of wetness} = \frac{\text{Actual rainfall in a particular year}}{\text{Average annual rainfall}}$$

The index of wetness thus gives an idea about the wetness of the year and if it less than 100 per cent, it indicates the deficiency of rain. For instance, if the index of wetness is 60 per cent, it means that there is a rain deficiency of 40 per cent.

If the deficiency is about 30 per cent to 45 per cent, it is known as large deficiency; if it is about 45 per cent to 60 per cent, it is known as serious deficiency; and if it exceeds 60 per cent, it is referred to as disastrous deficiency.

Minimum Annual Rainfall

The term bad year or dry year or sub-normal year is used to mean the year in which the rainfall is less than the average annual rainfall. The study of rainfall records of the locality is made say for 35 years or so and the minimum of all the bad years is obtained. This is known as the minimum annual rainfall of the locality and in rare cases only, there are three successive bad years indicating minimum annual rainfall. The provision of water in storage reservoirs is therefore usually made for two or three successive bad or dry years.

TYPES OF SOURCES

The sources from which water is available for water supply schemes can conveniently be classified into the following two categories according to their proximity to the ground surface:
1. Surface sources.
2. Underground sources.

We will now discuss various forms of surface sources and underground sources. But it will be necessary to consider the following important factors while making choice of source of water supply for a particular town or city:

1. Cost: The selection of source should be such that the overall cost of the water supply project is brought down to the minimum.
2. Elevation: The source of water supply should be at a higher level so that it becomes possible to supply water by the gravity flow only. If water source is at a lower level, it will involve huge expenditure on the operational and maintenance costs of pumping.
3. Location: The source whether surface or underground should be situated as near to the town or city as possible because such a location will require less lengths of pipes and few associated appurtenances.
4. Quality of water: The source should contain water which is free from pollution or other undesirable impurities and capable of being easily and cheaply treated.
5. Quantity of water: The source should be able to supply enough quantity of water to meet the demands of town or city for various purposes like domestic, industrial, fire fighting, etc. In some cases, part of the available source may be used to meet with the present demand and additional units may be brought into use as demand increases with passage of time.

Surface Sources

In this type of source, the surface run-off is available for water supply schemes. The usual forms of surface sources are as follows:

1. Lakes and streams.
2. Ponds.
3. Rivers
4. Storage reservoirs.

Each of the above form of surface sources will now be briefly discussed.

Lakes and streams

A natural lake represents a large body of water within land with impervious bed. Hence it may be used as a source of water supply scheme for nearby localities. The quantity of run-off that goes to the lake should be accurately determined and it should be seen that it is at least equal to the expected demand of locality. Similar is the case with streams which are formed by the surface run-off.

It is found that the flow of water in streams is quite ample in rainy season. But it becomes less and less in hot season and sometimes the stream may even become absolutely dry.

The catchment area of lakes and streams is very small and hence the quantity of water available from them is also very low. Hence the lakes and streams are not considered as principal sources of water supply schemes for large cities. But they can be adopted as sources of water supply schemes for hilly areas and small towns. The water which is available from lakes and streams is generally free from undesirable impurities and can therefore be safely used for drinking purposes.

Ponds

A pond is a man-made body of standing water smaller than a lake. Thus the ponds are formed due to excessive digging of ground for the construction of roads, houses, etc. and they are filled up with water in rainy season. The quantity of water in pond is very small and it contains many impurities.

A pond cannot be adopted as a source of water supply and its water can only be used for washing of clothes or for animals only.

Rivers

It is observed that rivers are studied more thoroughly than other sources of water. Since the dawn of civilisation, the ancient man settled on the banks of river, drank river water, ate fish caught from river water and sailed down rivers to find out unknown lands. Even the occurrence of floods did not disappoint the man and he tried to study the regularities of floods and make use of flood water for irrigation of his fields. As a matter of fact, many ancient civilisations such as India, Egypt, etc. were inseparably bound up with rivers.

The large rivers constitute the principal source of water supply schemes for many cities. Some rivers are perennial while others are non-perennial. The former rivers are snowfed and hence the water flows in such rivers for all the seasons. The latter type of rivers dries in summer either wholly or partly and in monsoon, the heavy flood visits them.

For such types of rivers, it is desirable to store the excess water of flood in monsoons by constructing dams across such rivers. This stored water may then be used in summer.

The principal uses of a river can be summarised as follows:

1. It can be developed as the chief source of water supply for a town or a city.
2. It can be used for navigation.
3. It can be used to supply water for irrigation purposes.
4. It can serve as an agent of purification of wastes.
5. It can serve as a center of recreational activities such as bathing, boating, fishing, fountains, etc.

In order to ascertain the quantity of water available from the river, the discharges at various periods of the year are taken and recorded. The observations over a number of years serve as a good guide for estimating the quantity of water available from the river in any particular period of the year.

Generally the quantity of water available from non-perennial rivers is variable throughout the year and it is likely to fall down in hot season when demand of water is maximum. It becomes therefore essential to augment such source of water supply by some other sources so as to make the water supply scheme successful.

The quality of surface water obtained from rivers is not reliable. It contains silt and suspended impurities. When completely or partly treated sewage is being discharged into the river at some upstream point, the river water is to be suspected for high contamination. The river water requires to be properly analysed as regards to the contents of disease bacteria, harmful impurities, etc. The presence of all such undesirable elements in river water requires an exhaustive treatment of water before it can be made fit for drinking purposes.

It should, however, be noted that the quality of river water is subject to the widest variations because it depends on various uncertain factors such as character of the catchment area, discharges of sewage and industrial wastes, climatic conditions, season of the year, etc. The character of the water differs not only with each individual river, but also at many points along the course of the same river. It is usually found that the quality of river water at its head is good, but it goes on deteriorating as the river proceeds along its course. The main reasons why the river pollution is undesirable are as follows:

1. Contamination of water supplies resulting in additional load on treatment plants.
2. Creation of nuisances in the form of appearance and odour.
3. Detrimental effect on fish life.

4. Hindrance to the navigation by banks of deposited solids.
5. Restriction of recreational use; etc.

The chief points to be considered in investigating a river supply of water are as follows:

1. Adequacy of storage of purified water so as not to disturb the distribution system during periods of flood when the river water is turbid.
2. Efficiency of the subsequent stages of purification system adopted.
3. General nature of river, the rate of flow and the distance between the sources of pollution and the intake of the water.
4. Relative proportions of the polluting matter and the flow of river when at its minimum.

Storage reservoirs

An artificial lake formed by the construction of dam across a valley is termed as a storage reservoir. Whatever may be the size or use of a reservoir, the main object or function of a reservoir is to store water and thus it stabilises the flow of water. The most important physical characteristic of a reservoir is therefore its storage capacity. The topographic survey of the dam site is carried out and a contour map is prepared. The capacity of reservoir is then worked out with the help of the contour map.

The discharge in a river or stream decreases during summer. To obtain a continuous supply of water, a dam is constructed across the river and the surplus discharge in the river during rains is impounded in the reservoir.

A reservoir is subjected to almost the same conditions as a lake. Algae (Minute plankton) is likely to grow in the top layers, while the bottom may be turbid and contain minerals such as iron, manganese and gases such as carbon dioxide, hydrogen sulphide: The removal of the top-soil before impounding water will reduce the organic matter.

A storage reservoir essentially consists of the following three parts:

1. A dam to hold water.
2. A spillway to allow the excess water to flow.
3. A gate chamber containing necessary valves for regulating the flow of water.

At present, this is rather the chief source of water supply schemes for very big cities. The multi-purpose reservoirs also make provisions for other uses in addition to water supply such as irrigation and power generation. The subject of reservoir design is a topic by itself. Its salient features in brief are discussed below.

Salient features of reservoir

Selection of site: Following are the factors which are to be taken into consideration while selecting the site for a storage reservoir:

1. Area of land to be submerged by the construction of reservoir.
2. Availability of construction materials and possibilities of using local materials for the construction of dam.
3. Availability of good foundation bed for dam.
4. Availability of skilled labour for the construction of dam.
5. Chances of biological troubles.
6. Characteristics of catchment area.
7. Density of population over the catchment area.
8. Distance between the proposed site and the point of distribution.
9. Elevation of reservoir level.

10. Facilities of transport for men and materials.
11. Geological conditions of basin of storage area.
12. Nature of land to be acquired.
13. Possibilities of earthquake occurrences due to the storage of water.
14. Quality of water available.
15. Quality of water likely to come to the reservoir site.
16. Watertightness of the reservoir area, etc.

Storage capacity of the reservoir

Following are the two methods which are used to compute the storage capacity of reservoir:
1. Analytical method: In this method, an analysis of demand and supply of water per month of the year is made. Following procedure is adopted:
   (a) The average monthly rainfall for every month of year is determined.
   (b) The average coefficient of run-off for different months of year is worked out by suitable method.
   (c) The multiplication of 1 and 2 indicates the total surface flow in the stream for different months of the year.
   (d) From the available surface run-off, the quantity representing various losses such as evaporation loss, penetration loss, etc. is subtracted. This gives the net supply of water from the stream for different months of the year.
   (d) Now the demand of water for every month of year is worked out.
   (f) The surplus or deficiency of water for each month is obtained by manipulation of above results. When supply is more, it indicates surplus and when supply is less, it indicates deficiency.
   (g) The total deficiency during successive months gives the storage capacity of reservoir.
   (h) If provision is to be made for two or three successive dry years, the capacity obtained in (g) above is increased accordingly.
2. Graphical or mass curve method: In this method, the required storage capacity of the reservoir is worked out graphically.

Rivers, Streams, Irrigation Canals

The quality of water from a river or stream depends on the:
1. Character and area of the catchment.
2. Topography.
3. Extent and nature of development of catchment by human beings.
4. Seasonal and weather conditions.

When the catchment is thinly populated, the stream water would carry suspended impurities, mineral salts and organic debris due to erosion of the catchment. In catchments where the population is heavy, the water will be polluted by sewage and industrial waste. These produce colour, turbidity, taste, odour and hardness and also contains bacteria and micro-organisms. Industrial waste may contain poisonous chemicals which are difficult to be removed.

GROUNDWATER

The sources of water which supply water from below the earth's surface are called sub-surface sources or groundwater sources.

Definition of Terms with Respect to Groundwater

Aquifier

It is observed that the surface of earth consists of alternate layers of pervious and impervious soils. A portion of the rainfall on the ground percolates into the soil by gravity until it reaches an impervious stratum. It then moves in a lateral direction towards some outlet. The pervious layer in which the water moves laterally is known as aquifier. If the aquifier is composed of sand and gravel, water can be easily drawn. Hard clayey soils holding water do not, however, give up their water easily and the layer is called aquiclude.

Water table

The free surface of water in the aquifier is called water table. When the aquifier is overlaid by an impervious layer such as clayey soil, then the water table is the free surface of the water in the topmost layer of soil. The water table varies considerably in summer it is at a low level and during rains it rises. Figure 3.1 show the profile of groundwater table.

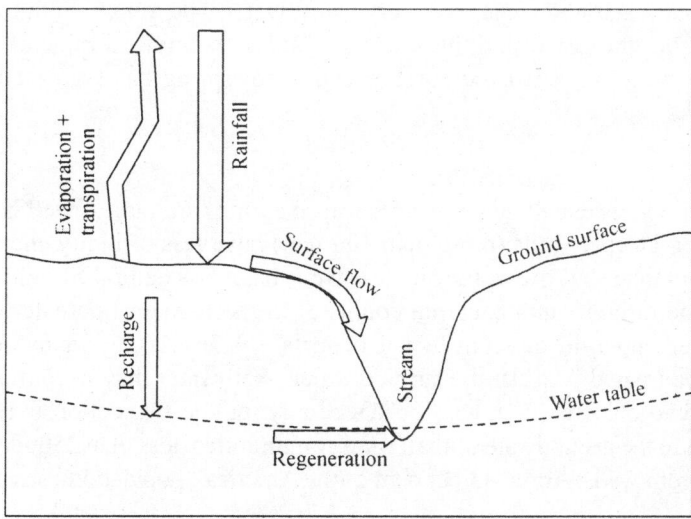

Fig. 3.1. Profile of a groundwater table.

General Characteristics of Groundwater

Generally groundwater is clear and colourless. When water seeps down into ground, it dissolves inorganic salts. This water is, therefore, harmful than surface water of the area in which it occurs. Groundwater is also generally free from bacteria and other living organisms as they get filtered out while percolating through the sub-soil.

Groundwater Supply

Underground water is supposed to be the purest form of natural water. Sometimes, it is so pure that it does not need any further treatment for drinking purposes. It is the least contaminated and has very low turbidity due to natural filtration of the rain water. Generally, it is cool, clear and odourless. It can be contaminated by underground streams in areas with limestone deposits, septic tanks discharge and underground deep well leaks. Therefore, it may need disinfection. It needs only mineral removal treatment when compared to surface water supply. It contains more dissolved minerals such as calcium, magnesium, iron, manganese and sulphur compounds than the surface supply. There are two sources of groundwater: springs and wells.

Springs

Whenever an aquifer or an underground channel reaches the ground surface such as a valley or a side of a cliff, water starts flowing naturally. This natural flow is known as a spring. A spring may form a lake (at the bottom of a valley), a creek or even a river. The quantity and velocity of a spring flow depend on the aquifer size and the position of the spring relative to the highest level of the water table. The larger the aquifer, the more is the flow; the lower the spring site (than the water table), the higher is the velocity, and vice versa. Regions with limestone deposits have large springs as the water flows in underground channels, formed by the erosion of limestone.

Some springs have a special quality of water; some are believed to have medicinal value due to certain minerals in their water and some have very warm water. The quality of the spring water depends on the nature of the soil through which the water flows. For example, a mineral spring has dissolved minerals, a sulphur spring has dissolved sulphur and a hot spring has hot water as the water flows through volcanic rocks.

Wells

Public groundwater supply is usually well water because springs are rare. A well is a device to draw the water from the aquifer. Deeper wells (more than 100 feet) have less turbidity, more dissolved minerals and less bacterial count than shallow wells. Shallow wells have less natural filtration of water due to less depth of the soil. If the turbidity and bacterial counts of the well water fluctuate with the rainfall, then this water is considered under the direct influence of surface water. According to the SDWA, water from such wells needs to be treated just like the surface water. Well water may need only some disinfection; commonly, free residual chlorine does the job. Generally, public perception is that the treated water should be comparable to the groundwater. Small rural communities (less than 25000 population) generally use the groundwater from wells. About 35 per cent of the American population uses groundwater supply.

Well protection

Protection of well water is important. First, select the proper site by drilling test holes and keeping in mind various sources of well contamination. For proper sanitary protection, specifications are devised by the engineering division of the state health department. Here are some guidelines:

Distance from human wastes

A well should be properly located in regard to sewer lines and septic tank drain fields. The distance should be 50 feet. Sewage treatment plant, animal feed lots, sanitary landfill and habitation should all be 500 feet away from a well. These are some generally recommended distances. Due to geological variations, each state has its own guideline for the well location.

Protection from geological leakage

Only water table wells need good protection. This can be accomplished, by having a two-inch-thick cement layer around the well pipe.

Developing a well

A well is developed by pumping out the mud and washing the well, until the discharge water is clear, as shown in Fig. 3.2. The height to which water rises when not pumping is called the static level. It is the distance between the water level and the ground surface. During pumping, the water level drops to a point called the pumping level. The difference between the static level and the pumping level is called the drawdown. The capacity of the well per foot drawdown is called the specific capacity. If a well capacity is 200 gallons per minute (gpm) and its drawdown is 10 feet, specific capacity of the well is 200 gpm/10 ft or 20 gpm/ft. The radius of the circular area of aquifer, which is dewatered around the well while pumping, is known as radius of influence. This dewatered area becomes a depression like an inverted cone called the cone of depression. If the water level does not reach the original static level when pumping is stopped, the distance it falls short is called the residual drawdown.

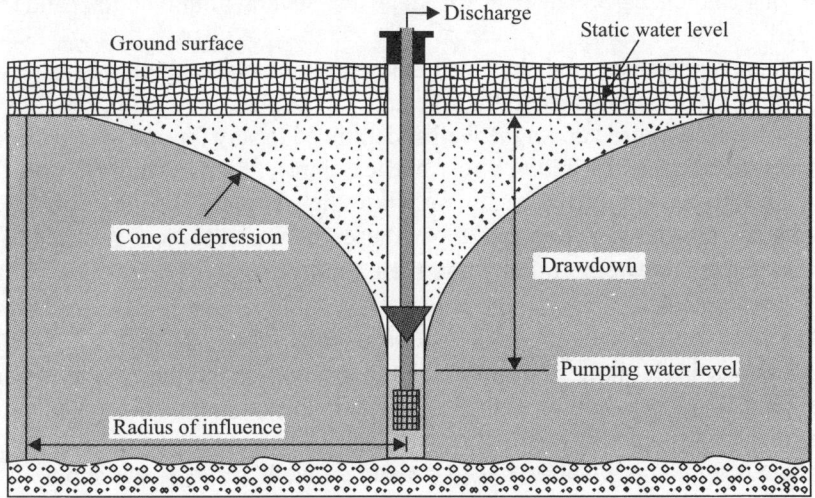

Fig. 3.2. A well while pumping.

Factors affecting the well yield

Well yield or capacity is the rate of water production by the well expressed as gallons per minute.

Drawdown

The yield is proportional to the drawdown. The more the drawdown, the greater the yield.

Pump diameter

Doubling the diameter size of the well pipe will increase the yield by about 15 per cent.

Aquifer

The yield is directly proportional to the depth of the well in the aquifer. If a well penetrates 20 feet into the aquifer, the yield will be doubled if the penetration depth is increased to 40 feet with the same drawdown.

Nature of the sand particles

The coarser the particles, the easier the movement of the water and the higher is the yield.

Distance from another well

The more the distance, the better it is. There should not be any overlapping of the areas of the influence of different wells.

Stimulation of a less productive well

Common causes of lower yield of a well are overpumping, clogging of the screen or aquifer depletion. Here are some commonly used corrective measures:

1. Surging: This is alternately pumping and backwashing to force the clogging material out. It is also done by forcing compressed air against the screen or by using a surge block up and down in the well casing.
2. Pumping at a higher rate: The higher rate of pumping will clear the sand from the screen.
3. Chlorination: Chlorine treatment of the wells has been successful for removing bacterial slimes and dissolving calcium carbonate ($CaCO_3$) deposits. Several hours' contact and repeat treatment are required.
4. Sodium hexametaphosphate and chlorine treatment: For wells with high hardness and high iron, a mixture of 15 to 30 pounds of sodium hexametaphosphate and 1 to 2 pounds of bleach/100 gallons of solution is poured down the well and allowed to stay for one to two days. This solution kills the iron bacteria and dissolves rust and calcium carbonate build-up on the screen and improves the well capacity. After treatment, wash the well until the water quality is acceptable.

Water is the most abundant, most used, and most abused natural resource. We need to be water wise to enjoy it ourselves and to allow our children and their children to enjoy it.

WELLS

A well is defined as an artificial hole or pit made in the ground for the purpose of tapping water. The holes made for tapping oil are also known as the wells. But in the general sense, a well indicates a source of water. In India, the chief source of water supply for most of its population is wells and it is estimated that 75 to 85 per cent of Indian population has to depend on wells for its water supply.

The three factors which form the bases of theory of wells are as follows:

1. Geological conditions of the earth's surface.
2. Porosity of various layers.
3. Quantity of water which is absorbed and stored in different layers.

The geological conditions of the earth's surface indicate the slope of water bearing strata. If the slope of water bearing layers is towards the well, there will be some quantity of water in the well even during the severe hot season. On the other hand, if the slope of water bearing layers is away from the well, such well will soon get dry and it will only give some quantity of water only in monsoon.

The porosity of aquifers will also play a great role in determine the quantity of water in the well. If the porosity of aquifers is more, the well will easily collect more quantity of water in less time. The capacity of aquifers to absorb and store water will determine the supply rate of water to the well. If the aquifers are capable of storing more water, the well will get more quantity of water and practically at a constant rate.

Classification of Wells

Following is the general classification of different types of wells:

1. Shallow wells.
2. Deep wells.

3. Tube wells.
4. Artesian wells.

We will now discuss each of the above type of well at length.

Shallow wells

Construction of shallow wells

The shallow wells are constructed in the uppermost layer of the earth's surface. They obtain their quota of water supply from the groundwater table as shown in Fig. 3.3. The diameter of shallow wells varies from 2 to 6 meters. They may be lined or unlined from inside. The lining is also called the steining and its thickness varies from 30 cm to 50 cm.

Figure 3.3 shows a shallow well with steining. The unlined wells are generally constructed upto a maximum depth of about 7 meters or so. But for greater depths, the soil cannot stand vertically and hence the steining becomes essential for such wells. These wells are also sometimes referred to as the draw wells or gravity wells or open wells or dug wells or percolation wells.

Quantity of water from shallow wells

The quantity of water available from shallow wells is generally limited as their source of supply is the uppermost layer of earth only. They sometimes even dry up in summer. In order to ensure the supply of water from shallow wells, even in dry years, they are taken much below the surface of groundwater table. The depth below water table is kept as about 6 to 8 meters so that even if the water table falls by 3 to 5 meters, some water will be available from the shallow wells. In any case, the discharge of shallow wells does not exceed 5 liters per second and hence, they are not suitable for public water supply schemes.

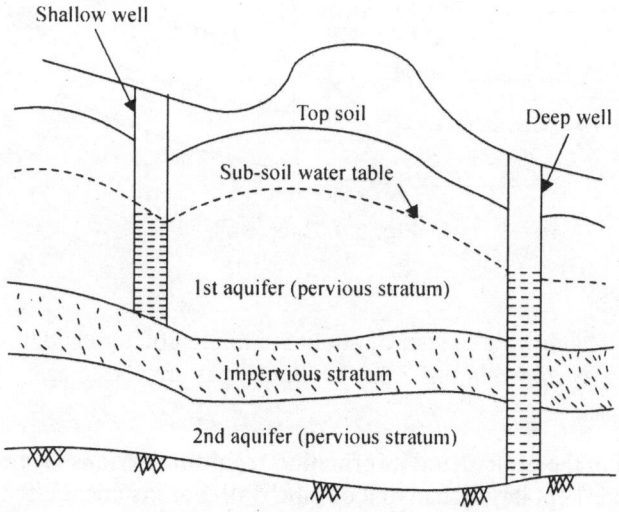

Fig. 3.3. Shallow and deep wells.

Quality of shallow well water

The quality of water obtained from shallow wells is better than river water. But it is not reliable and requires purification. The main source of contamination is the effluent from nearby septic tanks, soak wells, etc. It is therefore desirable to construct shallow wells away from such possible sources of contamination. It may also be noted that the shallow well water is notoriously liable to intermittent pollution and hence the samples of water should always be collected after heavy rainfall, when marked deterioration in purity may be revealed.

Use of shallow wells

Looking to the uncertain supply of water and bad quality of water, the shallow wells are used as source of water supply for small villages, undeveloped municipal towns, isolated buildings, camps, etc.

Deep wells

The deep wells obtain their quota of water from an aquifer below an impervious layer as shown in Fig. 3.4. The theory of deep well is based on the travel of water from the outcrop to the site of deep well. The outcrop is the place where aquifer is exposed to the atmosphere as shown in Fig. 3.4. The entry of rain water takes place at outcrop and it reaches the site of deep well. During travel, the water gets thoroughly purified. But it dissolves certain salts and may therefore become hard. In such cases, some treatment would be necessary to remove the hardness of water.

The depth of deep well should be decided in such a way that the location of outcrop is not very near to the site of well. The water of deep wells is contained in lower embedded aquifers and hence it is always available at a pressure greater than the atmospheric pressure. The deep wells are therefore referred to as the pressure wells.

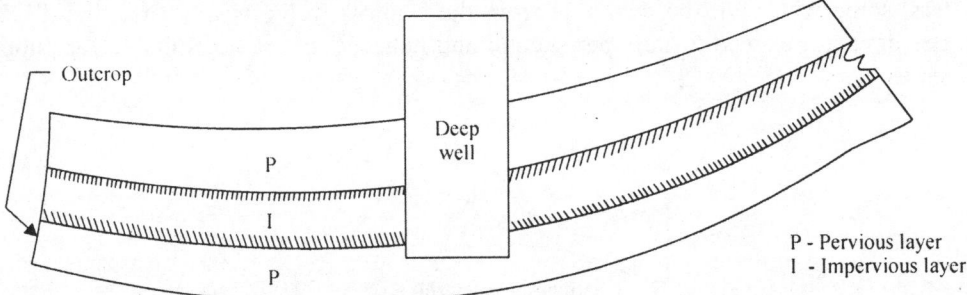

Fig. 3.4. Deep well.

Tube wells

A tube well is a deep well having a diameter of about 50 mm to 200 mm and it obtains its quota of water from a number of aquifers as shown in Fig. 3.5. The blind pipes are placed against the impervious layers.

Construction of tube wells

1. A bore is drilled in the ground and information regarding various layers of soil is obtained. The diameter of bore is kept larger than that of tube well. For instance, if the diameter of tube well is 150 mm, the diameter of bore may be kept as 300 mm.
2. The depth of tube well is decided with respect to the quantity of water required. The discharge of various aquifers composing tube well depends on the material of which they are formed. The usual depth of a tube well is about 30 to 50 meters. But in some dry areas, it may even go upto 300 meters or so. The aquifers composed of coarse sand or gravel are very good suppliers of

water while aquifers composed of limestone or marble will give good quantity of water only if cracks or fissures are present in them.

3. The pipe for tube well is then inserted in the bore hole. It consists of strainers and blind sections. A strainer is a perforated pipe which is provided with an arrangement such that only water will be admitted to the inside of pipe. The various types of patented strainers are available in the market. Figure 3.6 shows the Ashford strainer. It consists of a perforated pipe with wire mesh. In order to maintain some distance between the pipe and wire mesh, a thick wire is spirally wound around the perforated pipe. The strainers are usually manufactured from pipes of diameter 75 mm and onwards and in 2.50 m lengths. The total length of the strainer is obtained by suitably joining the individual pieces.

4. The pumping is then started. It should be done gradually to avoid sticking of fine sand particles on the external surface of the strainers. The strainers can be cleaned through perforations by allowing water from a high tank under pressure or by blowing air under pressure or by reversing the direction of flow. The process of removing the finer particles surrounding the strainers is known as the well development and it grants the following advantages:

(a) It increases the specific yield of well.

(b) It prevents the entry of fine sand particles in well pipe along with water.

(c) The economic life of well is increased.

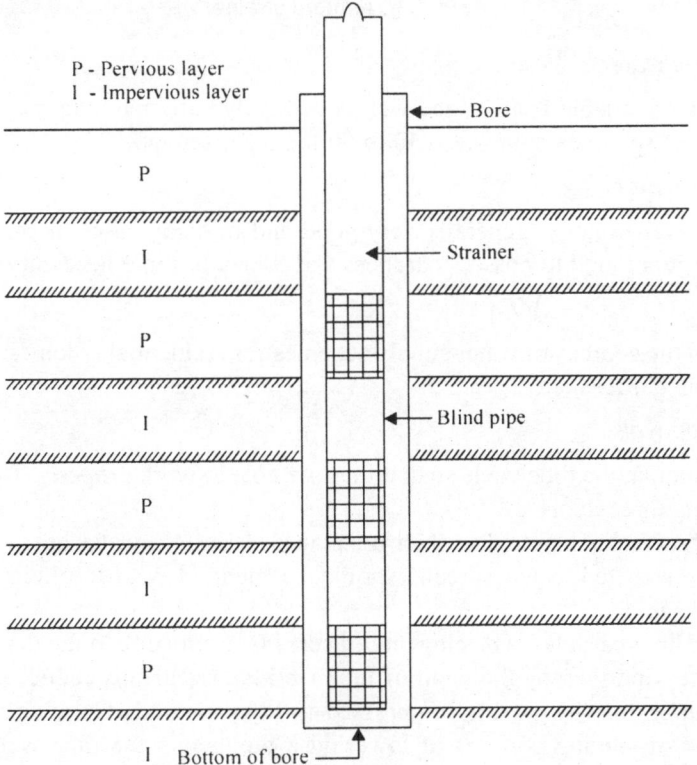

Fig. 3.5. Tube well.

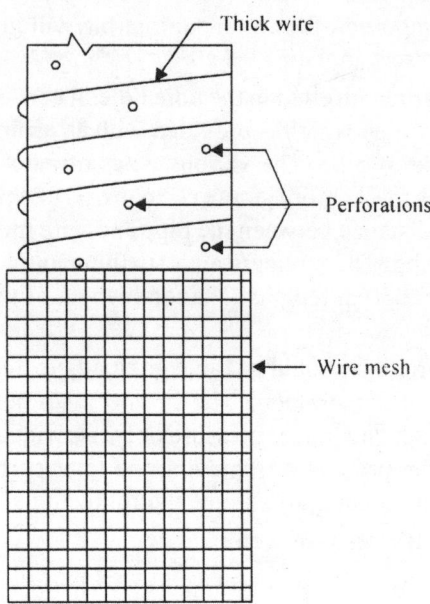

Fig. 3.6. Ashford strainer

Quantity of tube well water

The quantity of water available from a tube well is generally sufficient and more or less reliable. The discharge from a tube well does not exceed 40 to 50 liters per second.

Quality of tube well water

The quality of tube well water is generally very good and in many cases, it can be used without any treatment. However it is found to possess hardness and it may become necessary to remove it.

Use of tube well

The tube wells form the sources of water supply schemes for residential colonies, small towns, isolated portions of cities, big gardens, etc.

Maintenance of tube well

It is necessary to maintain the tube wells so that they are able to work properly. The cleaning operations include the following three items:

1. Cleaning of screens: The screens which are placed against the water-bearing strata are sometimes corrugated or rusted. Such screens should be cleaned with the help of sulphuric acid or hydrochloric acid.

2. Removal of lime particles: The clogging of screens occurs due to the deposition of fine clay or sand particles and thereby the yield of tube well is also greatly reduced. The removal of fine particles can be done by passing compressed air or by method of surging or by dropping dry ice. The method of surging consists of lowering a plunger in the tube well and giving rushing motion to the series of water waves inside the strainer pipe. In another method, the dry ice which is solid carbon dioxide is dropped inside the tube well and it is closed from the top by

tightening cap. The dry ice vapourises in a short time and the pressure developed inside the tube well proves to be sufficient to remove fine particles from the screens.

3. Replacement of parts: The parts of tube well which are totally rusted or corrugated should be replaced with new parts.

Failure of tube well

Following are the two main reasons for causing the failure of a tube well:

Corrosion

The groundwater contains acids, chlorides and sulphates. The tube well materials are therefore corroded in due course of time due to continuous effect of such groundwater. The damaged strainer screens permit sand particles to come out along with water. To avoid corrosion, the following precautions should be taken:

1. The diameter of tube well should be kept more.
2. The pipes should be galvanised or provided with coating of other corrosion resisting material.
3. The pumping rate should be reduced so as to decrease the quantity of coming out sand particles.
4. The stainless steel pipes, though very costly, are desirable for avoiding failure of tube well due to corrosion.
5. The thickness of pipes should be more.

Incrustation

The deposition of alkali salts on the inside walls of the tube well is known as the incrustation and it reduces diameter of the pipe as well as the effective area of the screens. The measures to reduce the incrustation are as follows:

1. If acid resistant material screens are used, the incrustating deposits can be removed at a later stage by acids.
2. The area of the screens should be kept large to accommodate future incrustation.
3. The incrustating deposits should be periodically removed during maintenance of tube well.
4. The incrustation is reduced by pumping at low rates or by using bigger capacity tube well.

Typical tube well station

A typical tube well station is shown in Fig. 3.7. The pump room is constructed above ground level and pump is installed below the pump room. The suction pipe of pump is connected to the tube well pipe through a reflux valve. This valve opens in one direction only and hence it allows water from the tube well pipe. But when pump fails or when it is removed for repairs, it does not allow the pumped water to go back into the tube well pipe. The pumped water is taken through the delivery pipe. A stop valve is provided to control the discharge of water through the delivery pipe. A bottom plug is constructed at the foundation level of the station to prevent the entry of water from groundwater table. A ladder is provided to connect the floor of pump room with the bottom of the station.

Advantages of tube well

Following are the advantages of a tube well:

1. The tube well water is not liable to be contaminated by any outside source.
2. The water can be brought on surface by the installation of suitable pumps.
3. The entry of water from top porous layers can easily be prevented by inserting blind pipes of galvanised iron or steel or wrought iron.

4. The overall construction of a tube well is easier and cheaper than that of an ordinary well.
5. The installation of tube wells helps in lowering the groundwater table and it is thus possible to reclaim the water-logged areas.

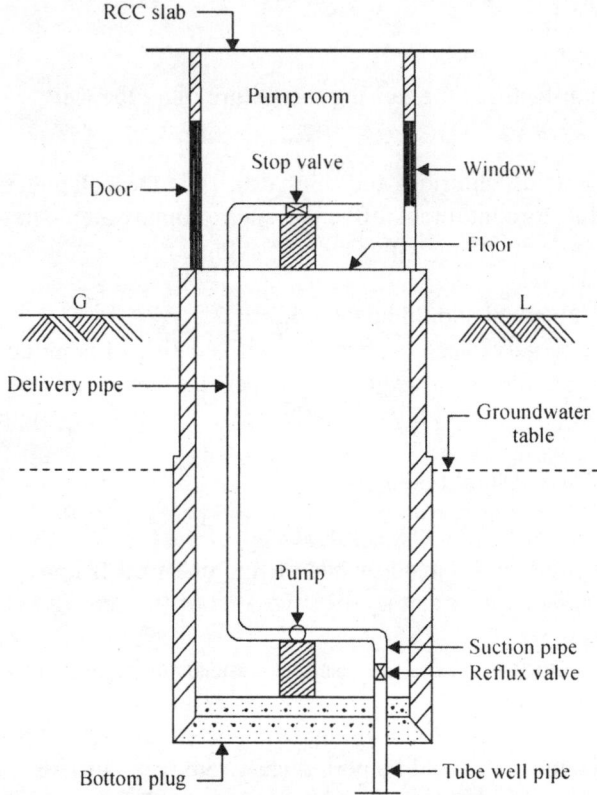

Fig. 3.7. Typical tube well station.

Artesian wells

This type of well derives its name from the fact that the first such well was sunk in the province of Artois in France. One of the earliest artesian well was sunk in England at London. It was meant for supplying water to fountain in Trafalgar Square. Its depth is about 118 meters. In India, very few artesian wells exist.

The theory of working of an artesian well is based on the principle of hydraulics, namely, that water tends to remain at the same level. As seen in Fig. 3.8, the artesian condition develops when an aquifer is enclosed between two impervious layers. The hydraulic gradient line is above the ground level at the site of artesian well and hence, when a hole is made in the ground, the water comes out with force under pressure. The pumping is usually not required in the beginning for such wells. But later on, when pressure falls down, the pumping may become necessary.

Thus the artesian wells can be grouped into the following two categories:
1. Fully artesian well or flowing well.
2. Semi-artesian well or sub-artesian well.

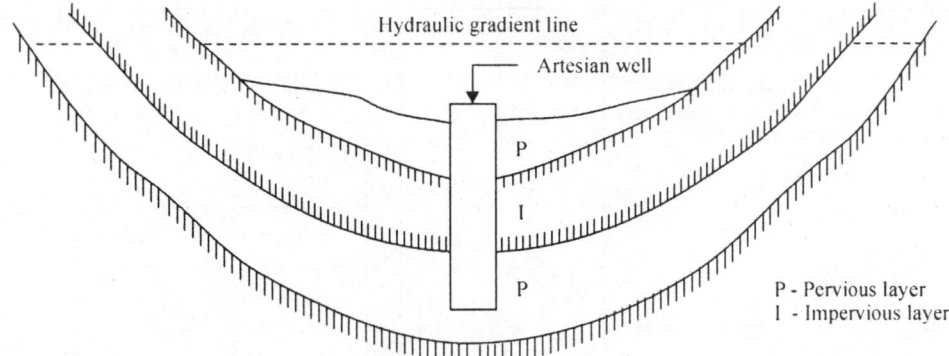

Fig. 3.8. Artesian well.

In case of fully artesian well, the water comes up the bore under pressure and it is available above ground. In case of semi-artesian well, the water is available in the bore below ground level and it has to be lifted up by pump or some such device.

Quantity of water from artesian wells

The quantity of water available from an artesian well is plentiful and it can be used with advantage when artesian conditions exist.

Quality of artesian well water

The quality of water available from artesian well is found to be very pure and it usually does not require any treatment. The pumped water is collected in storage tanks and then it is thrown into the distribution system of water supply.

Use of artesian well

It is rare to find artesian conditions and hence the artesian wells are not of much importance as source of public water supply.

Types of well construction

The three common methods of well construction are digging, driving and boring or drilling. Accordingly, the wells may be classified as follows:

1. Dug wells.
2. Driven wells.
3. Bored or drilled wells.

Each of the above method of well construction will now be briefly described.

Dug wells

In this method of construction, a hole is made in the ground by manual labour till a flow of water is reached. It may be necessary in some cases to adopt drilling and blasting for the construction of these wells. To prevent the entry of water directly from the surface, a circular wall of small height is sometimes constructed at the top. This method is adopted for shallow wells (Fig. 3.9).

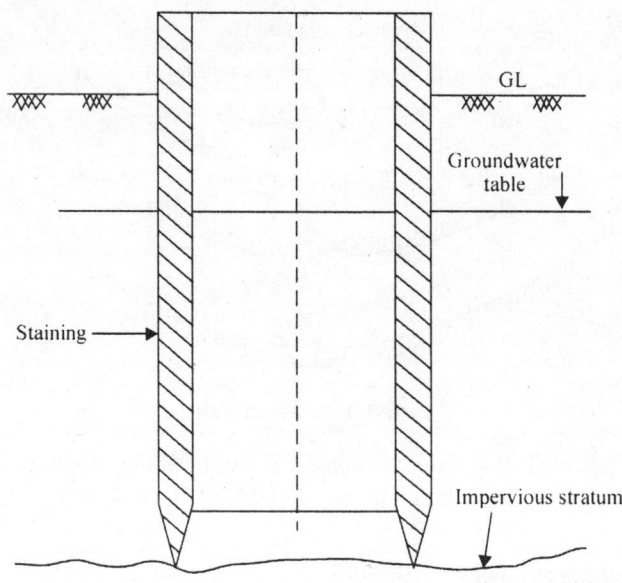

Fig. 3.9. A dug well.

Driven wells

In this method of construction, a specially designed well point is driven into the ground to tap water from an aquifer below an impervious layer. This method is useful for the construction of deep wells in unconsolidated soils. The diameter of driven wells varies from 25 mm to 80 mm and the driving is carried out by mauls or heavy wooden hammers (Fig. 3.10).

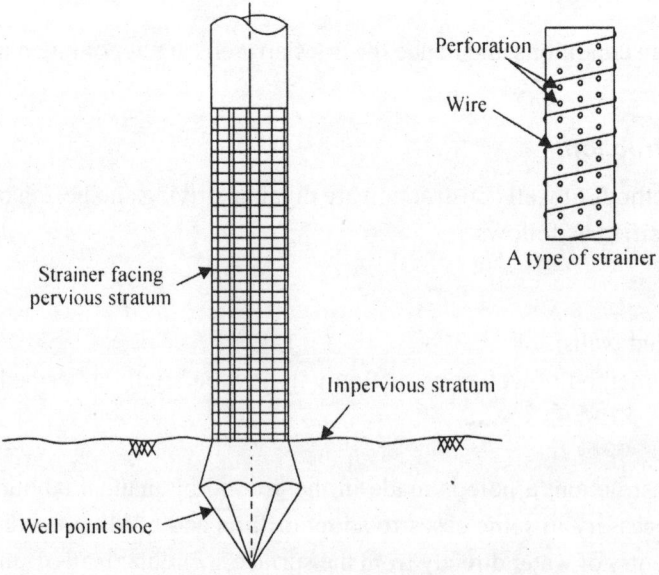

Fig. 3.10. A driven well showing strainer.

Bored or drilled wells

In this method of construction, the wells are bored or drilled into the ground with the help of special boring or drilling equipment. This method is adopted for the construction of tube wells (Fig. 3.11).

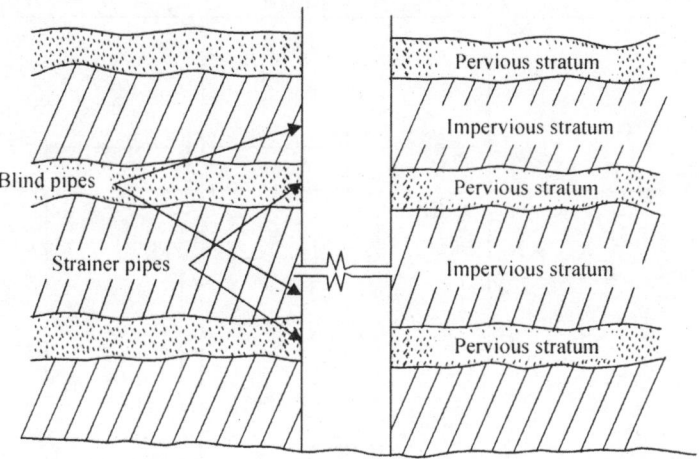

Fig. 3.11. A bored well.

Yield of a well

It is necessary to estimate the quantity of water available from a well. The term yield of a well is used to indicate the rate of withdrawal or pumping of water from wells without causing failure or drying of wells. It is thus a rate at which a well delivers water (Fig. 3.12).

Following are the six factors on which yield of a well depends:
1. Dimensions of well.
2. Location of nearby wells.
3. Porosity of aquifer.
4. Quantity of water present in aquifer.
5. Rate of pumping water.
6. Slope of water table.

When water is drawn or pumped from a well, the water around the well enters it under the action of head (H – h) as shown in Fig. 3.12. This head is known as the depression head or depletion head or infiltration head or percolation head or drawdown. Due to this drawdown, the water table near the well assumes the shape of an inverted cone. It is known as the cone of depression and the surface of it is known as the drawdown curve as shown in Fig. 3.12. The base of cone is known as the circle of influence.

The fundamental principles involved in the yield of a well are as follows:
1. It decreases if there is interference from the neighbouring well.
2. It decreases with the increase in radius of influence.
3. It increases very rapidly with the coarseness of particles of the water-bearing layer of sub-soil.
4. It is approximately directly proportional to the depth of penetration of well in a water-bearing layer of sub-soil.

5. It is approximately proportional to the depression head or drawdown.
6. It is not increased appreciably by the increase of the diameter of well.

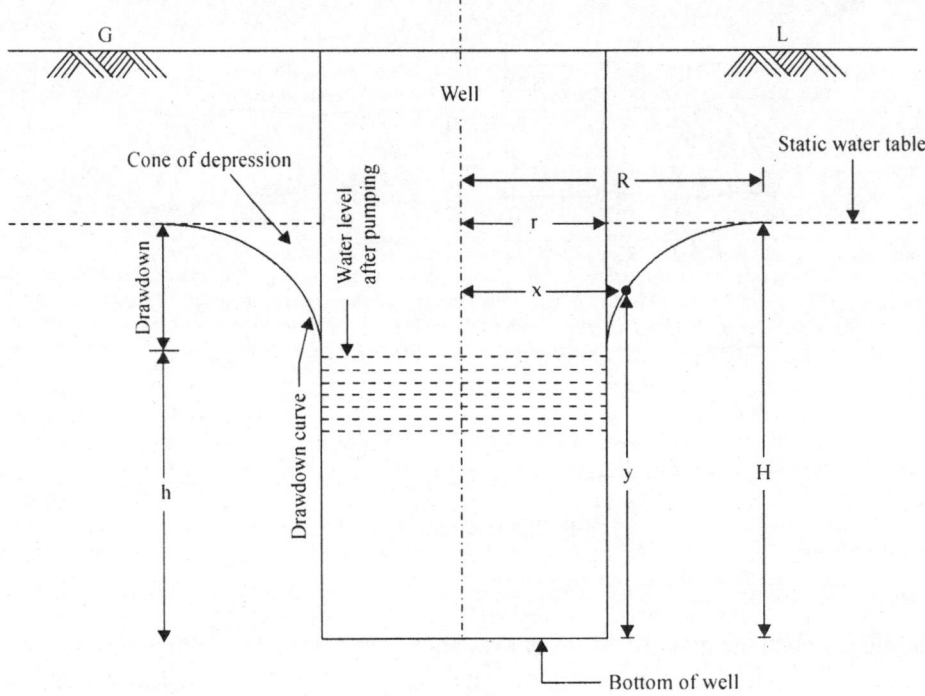

Fig. 3.12. Yield of a well.

SPECIFIC CAPACITY OF A WELL

From above equation, it is clear that the yield of a well varies directly as drawdown. This has been found experimentally true and hence it is possible to connect the yield of a well with the drawdown. The term specific capacity of a well is used to indicate the yield of a well per meter of drawdown. It helps in finding a common base for comparison of yields of different wells.

Tests for Yield of a Well

Following are the two tests by which the yield of a well can be determined:
1. Pumping test or constant level test.
2. Recuperation test.

Pumping test or constant level test

In this test, the water level of the existing open well is depressed to such an extent which represents the safe working head for sub-soil at the site of well. The rate of pumping is so adjusted that the water level in well remains constant which means that whatever water percolates into the well is taken out by the regulated pumping. Thus the rate of pumping indicates the yield of a well.

The test may also be carried out for working out the yield of a proposed well. For this purpose, a bore is driven into the ground and by regulated pumping, the yield of well per hour per base area of the bore is calculated and by knowing the diameter of proposed open well, its probable yield can then be calculated.

It is, however, very difficult to maintain the constant level in the well and hence this test is generally not adopted to ascertain the quantity of water available from a well.

Recuperation test

In this test, the water is first pumped to a certain depressed head which should be less than the safe working depressed head. The pumping is then stopped and time required by water to come to the original level is recorded. From this data, it is possible to work out the recuperative power of well at different drawdowns.

Spacing of Wells

When wells are spaced closely, their circles of influence formed during their pumping overlap each other and consequently, the yield of each of them is reduced. The radius of circle of influence is closely related to the rate of pumping. The lower the rate of pumping, the smaller will be the radius of circle of influence and vice versa. For satisfactory working of wells, they should be placed apart at least by a distance equal to twice the radius of circle of influence.

Sanitary Protection of Wells

A well should be properly protected from the possible contamination from undesirable water entering into it. Following are the precautions to be taken for the sanitary protection of wells:

1. Connection of pump: The connection between casing and pumping unit should be watertight.
2. Covered top: The top of well should be properly covered to prevent the entry of groundwater from top. The ground must slope away from the well.
3. Depth of casing: The casing should extend by about 3 meters below the groundwater table.
4. Distance from source of contamination: The minimum horizontal distance of the possible source of contamination from the well should be 15 meters and preferably, it should be 90 meters.
5. Drilled wells: In case of drilled wells, the hollow space between the well hole and casing should be filled up by cement grout to a depth of at least 3 meters.
6. Presence of trees: No tree should be allowed to be grown over or near the well. There are chances of well water to be contaminated by the fallen leaves, etc.
7. Priming of pumps: The priming of pumps should not be carried out by unsafe water.
8. Pump house: The pump house should be adequately drained and it should be protected against flooding.
9. Rate of pumping: The pumping rate from the well should be normal and not excessive.
10. Vents: The wells should be provided with enough vents so as to prevent the suction of contaminated water into the wells.
11. Washing of clothes: It is desirable to disallow or prevent the washing of clothes, utensils, etc., at or near the wells.
12. Well pits: The pumping machinery should not be installed below ground in pits.

Table 3.1 shows the comparative study of the surface and underground sources of water supply with respect to certain features.

Table 3.1. Comparison of surface and underground sources of water.

Item	Surface sources	Underground sources
Forms in which available	Lakes, streams, ponds, rivers and storage reservoirs	Infiltration galleries, infiltration wells, springs and wells
Quality of water	They are sometimes highly polluted and unsafe to consume. They contain inorganic impurities, organic impurities and industrial wastes	They are generally free from impurities because of natural filtration but may contain large amounts of dissolved salts, minerals and gases
Quantity of water	Huge quantity of water is available during monsoon, but is considerably reduced during summer	The quantity of water available is generally limited
Treatment	They are to be suitably tested and a line of treatment is to be decided before they are adopted for public use	They can be supplied to the public with no or minor treatment only
Use	They are useful for big towns and cities. They can be adopted for irrigation facilities also	They are useful for small towns and villages only

Chapter 4

Conveyance of Water

INTRODUCTION

The water required for a water-supply scheme has to be drawn from a surface source such as a river, lake, impounding reservoir, canal or from an underground source such as a well or spring. When the source is a surface source, a structure called intake is needed to collect and draw the water. The water is conveyed to the treatment plant by gravity if the level permit or pumping is resorted to both at the source and after treatment. The water from the treatment plant is conveyed to storage reservoir from where it is supplied to the town through the distribution system consisting of mains, valves, service pipes, etc. Sometimes the water is pumped directly into the mains of the distribution system to build up pressure.

If the source is underground, then no intake is necessary as the well or spring acts as intake. As treatment plants are generally not necessary for an underground source, the water is conveyed to the pumping station from where it is pumped to the reservoir or directly into pipes.

The term conveyance of water is used to indicate the following two arrangements: (i) drawing off the water from the sources of water, commonly known as the intakes; and (ii) leading the water from intakes to the purification plants and then leading the treated water to the consumers through distribution pipes.

An intake is a structure which is constructed across the surface of water so as to permit the withdrawal of water from the source. The structure may be of stone masonry, brick masonry, RCC or concrete blocks. It is to be constructed watertight and it should be designed for all forces likely to come upon it including the pressures due to wave action, wind, floating debris, etc.

LOCATION OF INTAKE

The location of an intake should be carefully made. Following are the important considerations which govern the selection of site of an intake.

Controlling Devices

The controlling devices of an intake should be located at a place which is accessible even during floods.

Cost

The intake should be constructed of locally available materials and labour so as to make its cost reasonable.

Navigation Channels

The intakes should never be located near the navigation channels. There are chances of water of intake being polluted due to discharge of refuse and wastes from ships and boats.

Permanancy of Supply

The intake should be located so as to ensure the supply of water even under the worst conditions and thus its location should be such that permanancy of water can be anticipated and it can be relied upon to meet with the future requirements.

Quality of Water

It is desirable to install a submerged intake. The water from such an intake is drawn off from a level which is lower than the surface level of water. If surface level of water is varying, the openings at different heights should be provided to draw off water. The location of an intake should be such that polluted water does not get entry into it. It is desirable to locate the intake on the upstream side of the town because such location will prevent the contamination of water by sewage disposal of the town.

Situation

The situation of an intake should be so selected that it is least affected by floods, scouring, silting and storms. The site of intake should be well connected by good approach roads and it should be free from the attack of heavy water currents.

The location of the intake on the curve should be avoided and natural causes such as wind currents, seasonal variations in quantity and quality of water, climatic conditions, etc., should be studied to grant the maximum stability and safety to the intake works. The location of intake at the downstream or in the vicinity of the point of waste-water should also be avoided.

DESIGN OF INTAKES

If an intake has to work satisfactorily, it has to be properly designed. Following are the factors which are to be considered in the design of intakes.

Factor of Safety

The intake should be designed with sufficient factor of safety so that it can effectively resist the external forces caused by heavy waves and currents, ice pressures, impact of floating objects, etc.

Foundations

The depth of foundations for an intake should be sufficient so that no damage is done by the current of water. If this factor is not considered, the undermining of foundation may take place and the structure may overturn.

Protection of Sides

If the intake is situated in navigational channels, its sides should be protected by a cluster of piles all around from the blows of moving ships.

Screens or Strainers

The screens or strainers should be provided at the entry level of an intake. The screens avoid the entry of floating matter and fish. The screens may be of coarse type or of fine type. The coarse screens usually consist of metallic rods placed about 25 mm to 50 mm apart and they remove large objects. The fine screens usually consist of wire net having openings of size 6 mm × 6 mm or less and they remove small objects.

Self-Weight

The intake should be of adequate self-weight so that the chances of its floating or washing by the upthrust of water are minimised. It is necessary to make intake with massive masonry work and broken stones should be filled in the bottom to grant additional safety.

Size and Number of Inlets

The size of inlets to an intake should be sufficient so that the required quantity of water is allowed to enter. The number of inlets should also be more so that the difficulty of drawing water does not arise even if some of them are blocked due to any reason.

TYPES OF INTAKES

There are mainly four types of intakes:
1. Canal intake.
2. Reservoir.
3. River intake.
4. Lake intake.

Canal Intake

A canal intake consists of a well constructed with concrete (Fig. 4.1). A pipe with a bell-mouth is placed in the well with its mouth opening upwards. (Since the level of water in the canal is nearly constant, pipes are not provided at different heights). A fine screen is provided on top of the bell-mouth to prevent small fish and floating matter from entering the pipe. A coarse screen is provided in the chamber.

The outlet pipe is regulated by means of a valve which is operated from top. To protect the bed and banks from erosion due to the increased velocity of water at the intake, bed-pitching and revetment are provided for some distance upstream and downstream of intake.

Reservoir Intake

Figure 4.2 shows the details of a reservoir intake. It consists of an intake well which is placed near the dam. It is connected to the top of dam by a foot bridge.

The intake pipes are located at different levels with a common vertical pipe. The valves of intake pipes are operated from the top and they are installed in a valve room. Each intake pipe is provided with bell-mouth entry with perforations of fine screen on its surface. The outlet pipe is taken out through the body of dam. The outlet pipe should be suitably supported. The location of intake pipes at different levels ensures supply of water from a level lower than the surface level of water.

When the valve of an intake pipe is opened, the water is drawn off from the reservoir to the outlet pipe through the common vertical pipe. To reach upto the bottom of intake from the floor of valve room, the steps should be provided in zigzag manner.

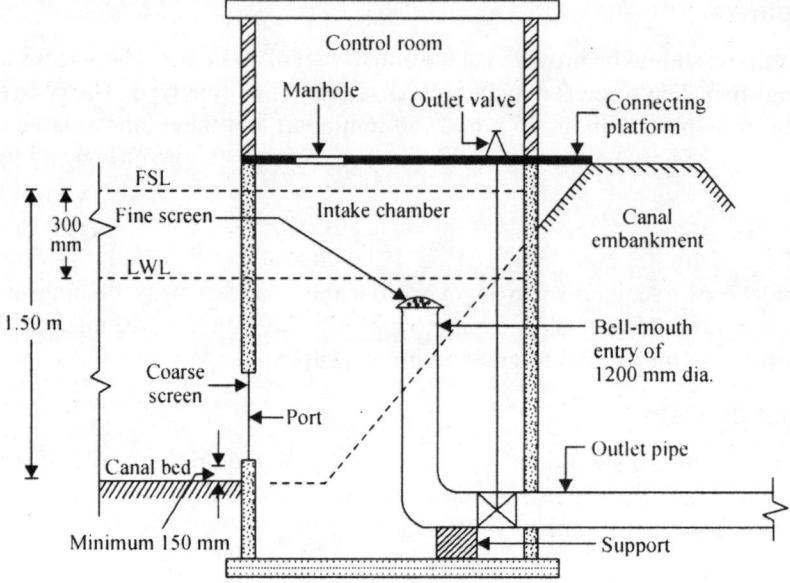

Fig. 4.1. Canal intake.

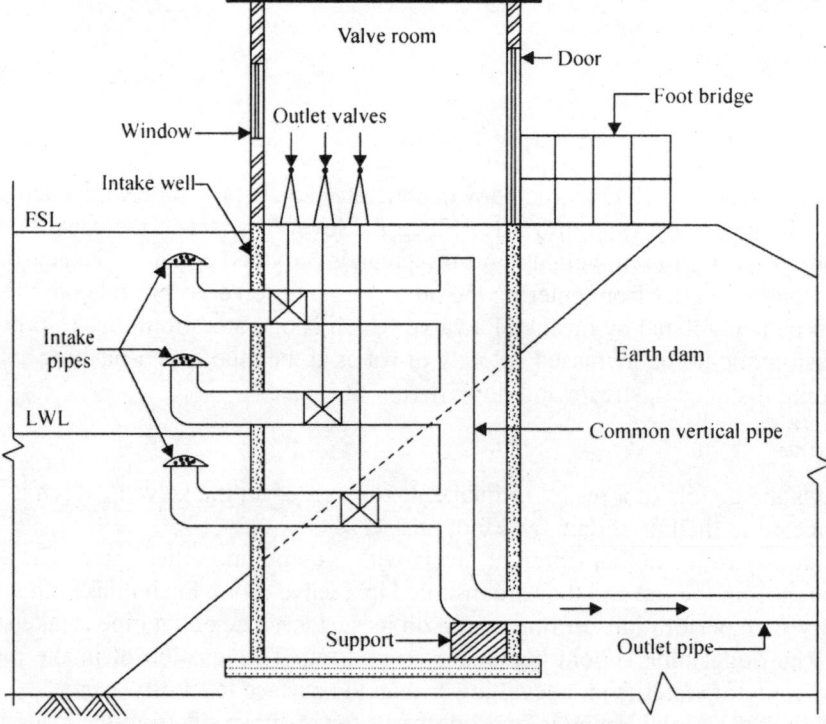

Fig. 4.2. Reservoir intake.

River Intake

Figure 4.3 shows the details of a typical river intake. An approach channel is constructed to lead the water from the upstream side of the river to the jack well. The penstocks with screens are provided at different levels. The suction pipe is provided with strainer at its lower end.

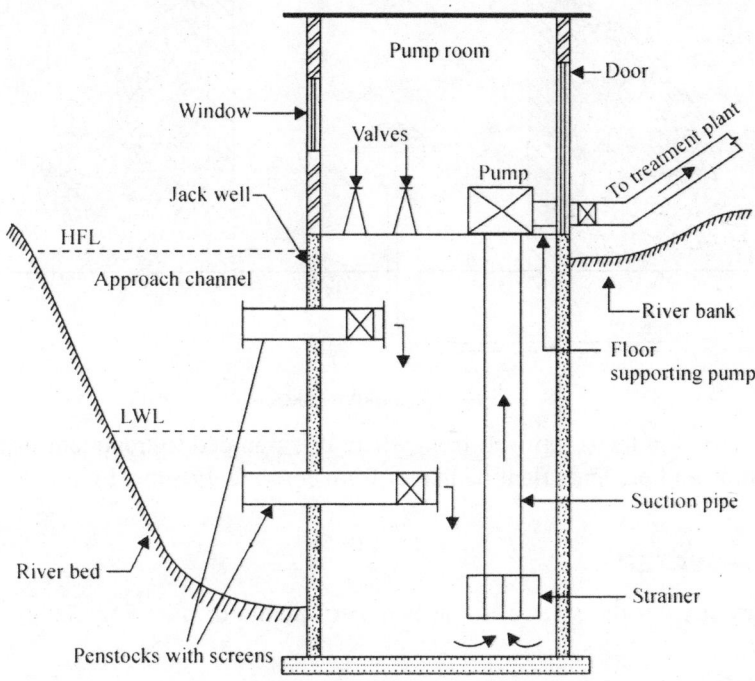

Fig. 4.3. River intake.

The water from jack well is pumped and sent to the treatment plant. To prevent the back flow of water due to gravity, a valve should also be provided on the rising main leading to the treatment plant. To reach upto the bottom of intake from the floor of pump room, the ladder or steps in zigzag manner should be provided.

If there is great variation in water levels of summer and winter, a small pick-up weir may be constructed across the river bed.

A cross approach channel is excavated upto the jack well, if the river has a wide basin. Sometimes only pipes are laid horizontal across the river and they are provided with strainers on their ends. The exact details of a river intake will depend on the nature of river itself.

Lake Intake

Lake intakes are similar to reservoir intakes. However, as the depth of water is maximum in the center of the basin, submersible intakes are provided. These are sometimes provided on rivers also.

This consists of one or more bell-mouthed pipes placed on the bed of the lake (Fig. 4.4). The bell-mouth is covered by means of a screen. The whole arrangement is covered by square or octagonal crib or cage made of wood and protected by rip-rap (broken stones).

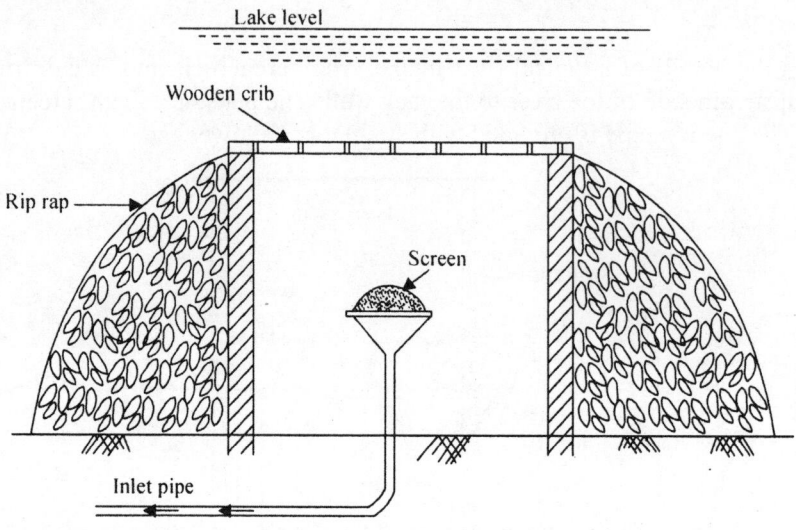

Fig. 4.4. Lake intake.

The water is collected in the sump well from where it is pumped to treatment plant. This type does not obstruct navigation and prevents floating matter from entering the pipe. In dry weather, all the water can be tapped for use.

CONVEYANCE OF WATER

The water is conveyed from the source to the treatment plants or pumping station by anyone of the following methods:

1. Open channels.
2. Aqueducts.
3. Pipes.

Open Channels

Open channels are excavated in earth and are generally trapezoidal (Fig. 4.5). Where the water is valuable and seepage is to be kept to a minimum, the channel is lined with concrete or masonry.

Open channels have the following advantages:

1. They are cheap.
2. Local materials can be used for construction.

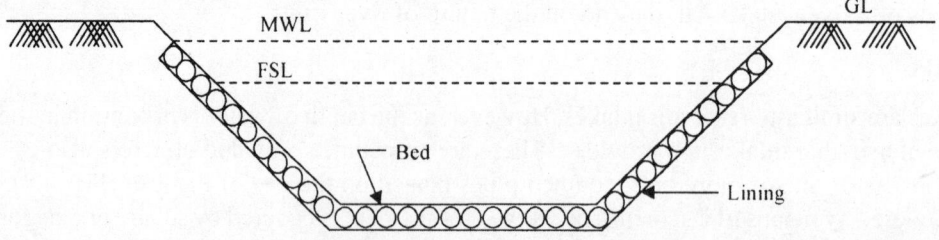

Fig. 4.5. Open channel.

However, open channels are not generally used in waterworks practice as they have the following disadvantages:

1. They can be used to carry water only under gravity and hence the hydraulic gradient has to followed.
2. There will be loss of water due to seepage and evaporation.
3. Contamination of water in thickly populated areas.
4. Tree roots, burrowing by animals cause damage to the channels.

Aqueducts

An aqueduct is a closed conduit constructed in masonry or concrete (Fig. 4.6). The advantages of an aqueduct over a pipeline are:

1. Local materials which are cheap, could be used.
2. It has longer life than metal conduits as the latter are subjected to corrosion.
3. Loss of water is less with age.
4. The head loss is less in gravity-flow tunnels.

The disadvantages are:

1. The full size of conduit should be provided initially while in a pipeline a small-sized pipe could be provided and later a parallel pipe added.
2. The balancing of cuts and fills in excavation interferes with the natural drainage.

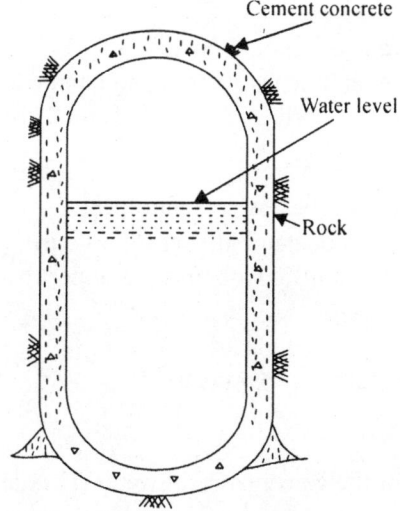

Fig. 4.6. Aqueduct.

The shape of the aqueduct is generally that of a horse-shoe as it resists the earth pressure and has good hydraulic properties. The bottom is shaped like a dish. If it carries water under pressure, it is made circular in shape. Aqueducts are usually lined, but in stable rock lining is not necessary.

Pipeline

The pipe material is selected while keeping in view the forces to be resisted by it.

The usual stresses to which pipes are subjected are as follows:
1. Stress due to change of direction.
2. Stress due to internal water pressure.
3. Stress due to soil above the pipes.
4. Stress due to water hammer.
5. Stress due to yielding of soil below pipes.
6. Temperature stresses.

Following are the various types of pipes used:

Asbestos cement pipes

These pipes are made from a mixture of asbestos fibers and cement. These pipes are used to convey water under very low pressure and their use in conveying and supplying water is very much restricted.

Advantages

Following are the advantages of asbestos cement pipes:
1. The inside surface of pipes is very smooth.
2. The joining of pipes is very good and flexible. The joints are easily formed. In fact, it consists of a collar with two rubber rings inside it. The collar is inserted upto one ring at one end of pipe where specially prepared grease is applied. The other end of next pipe is inserted in a similar way from other side of collar upto the other rubber ring. The rubber rings make the joint watertight.
3. The pipes are anti-corrosive and cheap in cost.
4. The pipes are light in weight and hence it is easy to handle and transport them.
5. The pipes are very suitable for distribution pipes of small size.

Disadvantages

Following are the disadvantages of the asbestos cement pipes:
1. The pipes are brittle. They cannot stand impact forces during handling and they cannot be also expected to resist vibrations of traffic when placed under road.
2. The pipes are not durable.
3. The pipes cannot be laid in exposed places.
4. The pipes can be used only for very low pressures.

Cast-iron pipes

These pipes are extensively used for the conveyance of water. They are available in sizes upto diameter about 1200 mm or more. The pipes are joined either by bell and spigot joint or by expansion joint or by flanged joint.

Advantages

Following are the advantages of the cast-iron pipes:
1. The cost is moderate.
2. The pipes are easy to join.
3. The pipes are not subject to corrosion.
4. The pipes are strong and durable.

5. The service connections can be easily made.
6. The usual life of cast-iron pipes under normal conditions is about 100 years or so.

Disadvantages

Following are the disadvantages of the cast-iron pipes:
1. The breakages of these pipes are large.
2. The carrying capacity of these pipes decreases with the increase in life of pipes. The decrease in capacity may be as high as 30 to 40 per cent.
3. The pipes are not used for pressures greater than 0.7 N/mm^2.
4. The pipes become heavier and uneconomical, specially when their size increases beyond 1200 mm diameter.

Cement concrete pipes

The cement concrete pipes may be plain, reinforced or pre-stressed with diameters varying from 500 mm to 2500 mm or more. The plain cement concrete pipes are used for low heads upto about 15 m. The reinforced cement concrete pipes are adopted for heads upto about 75 m and for larger heads, pre-stressed cement concrete pipes are used. The reinforcement in RCC pipes consists of rings or hoops and longitudinal steel bars. The reinforcement is placed in the mould and the cement concrete is then poured into it. The mould is rotated with great speed around its longitudinal axis. These are known as the hume pipes or spun concrete pipes and they are widely adopted in the conveyance of water.

Advantages

Following are the advantages of the cement concrete pipes:
1. The inside surface of pipes can be made smooth.
2. The maintenance cost is low.
3. The pipes are durable with useful life of about 75 years.
4. The pipes can be cast at site of work and thus there is reduction in transport charges.
5. These pipes are heavy in weight and hence they will not be affected by the force of buoyancy when placed under water even when they are empty.
6. These pipes do not require expansion joint because they possess the least coefficient of thermal expansion as compared to other types of pipes.
7. These pipes under normal conditions are not affected by the atmospheric actions or by ordinary soil.
8. These pipes will not collapse or fail under normal traffic loads when placed below roads.
9. When these pipes are used, there is no danger of rusting and incrustation.

Disadvantages

Following are the disadvantages of the cement concrete pipes:
1. If no reinforcement is provided, the pipes possess no tensile strength and they cannot withstand high pressures.
2. The pipes are heavy and difficult to transport.
3. The pipes are likely to crack during transport and handling operations.
4. The repairs of these pipes are difficult. This is due to the fact that the pipe line becomes rigid as joints are made in cement mortar. Hence when one joint is to be repaired, considerable length on either side is to be disturbed.

5. These pipes are affected by acids, alkalies and salty water.
6. These pipes are likely to cause leakage due to porosity.

Copper pipes

The copper pipes do not sag or bend due to hot water. Hence their use is restricted for conveyance of hot water in buildings and steam boilers. They are not liable to corrosion and can be bent easily. But as they are costly, they are not used for distribution of water.

Galvanised iron pipes

These pipes are widely used for service connections and their diameters vary from 6 mm to 75 mm.

Advantages

Following are the advantages of the galvanised iron pipes:
1. The pipes are cheap, light in weight and easy to handle and transport.
2. The pipes are easy to join.

Disadvantages

Following are the disadvantages of the galvanised iron pipes:
1. These pipes are liable to incrustation and can be easily affected by acidic or alkaline water.
2. The useful life of pipes is short about 7 to 10 years or so.

Lead pipes

These pipes are usually not adopted for the conveyance of water. If proper care is not taken, the lead pipes may cause lead poisoning. They can be easily bent and hence, when these pipes are used, less number of specials will be required. The acidic water react on lead pipes and therefore they cannot be used to convey acidic water. The lead pipes are usually adopted for apparatus required for alum and chlorine dosages. They cannot be used to carry hot water as they sag or bend due to heat.

Plastic pipes

The plastic is a new material of the modern age. Its various uses in many fields have made it very popular. The use of plastic in the conveyance of water has increased due to various types of plastic pipes available in the market.

The low density polyethylene pipes (LDPE) are flexible and can be used for point to point conveyance of water in long runs. The high density polyethylene (HDPE) pipes are tough and can be used for conveyance of water in long runs from point to point with large diameter.

The polyethylene pipes are black in colour and are resistant to most of the chemicals except nitric acid and very strong acids, fats and oils and certain other solvents. The rigid PVC pipes are three times as rigid as polyethylene and are widely used in the water mains.

Advantages

Following are the advantages of the plastic pipes:
1. There is freedom from damage due to thawing and freezing of water in closed pipes.
2. The pipes are cheap.
3. The pipes are durable and they possess enough strength to resist impact, sunlight and atmospheric actions.

4. The pipes are flexible and possess low hydraulic resistance.
5. The pipes are free from corrosion.
6. The pipes are good electric insulators.
7. The pipes are light in weight and it is easy to bend, join and install them.
8. The pipes upto certain sizes are available in coils and therefore it becomes easy to transport them.

Disadvantages

Following are the disadvantages of the plastic pipes:
1. The coefficient of expansion for plastic is high.
2. It is difficult to obtain the plastic pipes of uniform composition.
3. The pipes are less resistant to heat.
4. Some types of plastics may impart taste to the water.

Steel pipes

The mild steel is used for the manufacture of steel pipes. The joints of steel pipes are either riveted or welded and hence they vary both in length and diameter.

The steel pipes are generally used for pipes having diameter greater than 1200 mm. The inside and outside surfaces of steel pipes are generally galvanised.

Advantages

Following are the advantages of the steel pipes:
1. The pipes are available in long lengths and hence the number of joints becomes less.
2. The pipes are cheap in first cost.
3. The pipes are durable and strong enough to resist high internal water pressure.
4. The pipes are flexible to some extent and they can therefore be laid easily on curves.
5. The pipes are light in weight and therefore it becomes easy to transport them.

Disadvantages ,

Following are the disadvantages of the steel pipes:
1. The maintenance cost is high.
2. The pipes are likely to be rusted by slightly acidic or alkaline water. The outside earth also helps in this process in some cases. The rivets used in the joints form nucleus for the process of rusting.
3. The pipes require more time for repairs during breakdown and hence they are not suitable for distribution pipes.
4. The steel pipes are likely to deform in shape under the combined actions of external and internal loads.

Wrought-iron pipes

These pipes are light in weight and they can be easily cut, threaded and worked. But they are found to be costly and less durable as compared to the cast-iron pipes and hence they are not generally used in the conveyance of water.

CORROSION OF PIPE

The term pipe corrosion is used to indicate the loss of pipe material due to the action of water. The metallic structure of pipe is attacked and dissolved by water. The pipe corrosion may be internal or external and it leads to the disintegration of metal. The former is due to the action of water flowing in the pipe. The latter is due to the action of waterlogged soil above the pipe surface.

The metals chiefly concerned with corrosion are iron and steel of which mains and distribution pipes are usually composed. The results of pipe corrosion are apparent and troublesome to both, the water authority and the consumer.

Factors Contributing to Corrosion

The various factors contributing to the pipe corrosion are as follows:

Acidity

This is rather the most important factor in corrosion and the water having low pH value due to the presence of carbonic acid or other acids is invariably corrosive.

Alkalinity

The water possessing sufficient calcium bicarbonate alkalinity is anti-corrosive in nature.

Biological action

The growths of iron-bacteria and sulphur bacteria may develop aerobic and anaerobic corrosion respectively.

Chlorination

The presence of free chlorine or chloramines makes the water corrosive and it may be responsible for making the water corrosive in nature which if unchlorinated, would have caused no trouble of corrosion.

Electrical currents

It is quite likely that the corrosion can also be developed by the union of dissimilar metals or by the earthing of electrical systems to the water pipes.

Mineral and organic constituents

The presence of high total solids in water accelerates the process of corrosion. The calcium and magnesium chlorides are particularly active in hot-water systems. The nitrates play a secondary role in corrosion processes and high ammonia contents are objectionable in boiler feed-water. The organic matter plays a great role in anti-corrosion treatment processes.

Oxygen

The presence of oxygen is found in both the corrosive and non-corrosive water and under ordinary conditions, it is not the sole or primary cause of pipe corrosion. The aeration in fact is employed in some cases for the prevention of corrosion.

The corrosion may be also local or uniform in character. In case of local corrosion, only separate areas of pipe surface are attacked. In case of uniform corrosion, the pipe metal is attacked evenly over its entire surface.

Effects of Pipe Corrosion

Following are the effects of pipe corrosion:
1. The pipe corrosion may lead to the tuberculation which is the phenomena of the formation of small projections on the inside surface of pipes. The small projections or turbercules are cone-

shaped and they decrease the cross-sectional area of pipe. The carrying capacity of pipe is thus affected.

2. The pipe corrosion leads to the disintegration of pipe line and it demands heavy repairs.
3. The pipe corrosion imparts colour, taste and odour to the flowing water.
4. The pipe connections are also seriously affected by the action of pipe corrosion.
5. The pipe corrosion may make the water dangerous for drinking and other purposes.

Theories of Pipe Corrosion

In order to explain the phenomena of pipe corrosion, the various theories are advanced. Following are the five theories of pipe corrosion.

Action of water motion

According to this theory, the pipe corrosion takes place due to the action of physical movement of water. If the flow is turbulent and not laminar, the cross currents developed during such a flow impart great pressure on pipe material and such repeated attacks on pipe material leads to the pipe corrosion.

Bimetallic action

According to this theory, the pipe corrosion is due to the bimetallic electrolytic action. When two dissimilar metals are placed in water and kept in contact with each other, the hydrogen ions from water are deposited on the metal having a lower solution potential and positively charged metallic ions are liberated from the metal having a higher solution potential.

The term solution potential is used to indicate the ability of metal to emit ions to go in solution with water. The sources of bimetallic or galvanic action are the impurities in water, copper and lead fixtures and electric current.

Biological action

According to this theory, the bacteria are considered to be responsible for the pipe corrosion. There are mainly two types of bacteria which cause pipe corrosion:

1. Iron consuming bacteria.
2. Sulphate reducing bacteria.

The iron consuming bacteria are aerobic i.e., they operate in the presence of oxygen. They consume iron ions to support their activities. The metallic iron is thus continuously eaten away by bacteria. This type of bacteria will be predominant in water having a higher iron content.

The sulphate reducing bacteria are anaerobic they i.e., operate in the absence of oxygen. They react with sulphur and produce hydrogen sulphide which attacks on pipe material to cause the pipe corrosion.

Chemical reaction

According to this theory, the chemical characteristics are considered to be responsible for pipe corrosion. The important factors are acidity or alkalinity of water and presence or absence of carbon dioxide. The chemical reaction takes place between pipe material and water. It then results into pipe corrosion.

Electrolysis

According to this theory, the flow of electric current is treated as responsible for pipe corrosion. The passage of electric current may be due to attachment of wires of electric light, telephone, etc., to the pipe material. The bimetallic action takes place due to electric current and pipe corrosion occurs.

Prevention of Pipe Corrosion

In practice, it is not possible to completely eliminate the pipe corrosion. But the following are the measures which are commonly adopted for minimising pipe corrosion:

Cathodic protection

It is found that if entire pipe line acts as cathode, the pipe corrosion may be minimised. This is achieved by connecting the pipe line either to the negative pole of a DC generator or to the anodic metals like magnesium. The emerging currents from anodic areas are neutralised and the corrosion is prevented. The cathodic treatment is most effective. But it is too expensive and involves many practical difficulties.

Proper pipe material

The pipe material, if metallic, should be able to resist the dissolving effect of water. The alloys of iron or steel with chromium, copper or nickel are found to be more resistant to the corrosion.

Protective linings

The pipe surfaces should be coated with anti-corrosive linings. The usual coatings employed are those of asphalt, bitumen, cement mortar, paints, resins, tar, zinc, etc. The degree of prevention achieved will depend on the individual properties of the coating material.

Treatment of water

The water should be given proper treatment to prevent pipe corrosion. The usual treatments employed are adjustment of pH value, control of calcium carbonate, removal of dissolved oxygen and carbon dioxide, addition of sodium silicate, etc.

Chapter 5

Quality of Water

INTRODUCTION

A laboratory is a place for precise work to determine appropriate treatment of raw water and the quality of the finished water. It must be kept properly organised, well maintained and scrupulously clean. All instruments must be kept clean and routinely calibrated with proper records. A number of laboratory tests are needed daily, quarterly, semi-annually, annually and at other specified intervals to monitor the water quality before, during and after the treatment. A test is not better than the sample and the sample is not better than the manner in which it is collected.

SAMPLING

Valid testing starts with an adequate and representative sampling. A sample is either a grab or a composite. A grab sample, as the name indicates, is a specific volume collected at one site at one time. These samples indicate the quality of water at that time and at that site. Grab samples are taken for bacteriological and disinfection residual tests. A composite sample is a mixture of a number of portions taken at the specific intervals (e.g., samples for mineral analysis because minerals in water are more stable). This reduces the number of tests. Each portion can be proportionate to the flow or volume.

For each test operators should follow the prescribed sampling size, collecting and preserving procedure given in the standard methods for the examination of water and waste-water (standard methods). Testing must be done as soon as possible and not later than the specified holding time.

QUALITY ASSURANCE OR QUALITY CONTROL

There are two types of quality control measures internal and external quality control measures. Internal quality control is like running a standard with the sample, testing duplicate samples and having standard curves. External quality control has laboratory certification by the state for all parameters performed by the laboratory. It requires successfully analysing the performance samples twice a year. Furthermore, there is an onsite visit by the State Certification Personnel to check the equipment, procedures, laboratory personnel and records.

TESTS

Various regularly performed common tests by the operating staff are for tastes and odours, turbidity, jar test, pH, alkalinity, hardness, disinfection residual, coliform bacteria and the heterotrophic plate count.

All other tests are run either by highly trained chemists and microbiologists of the laboratory or by certified contract laboratories.

Tastes and Odours

Testing for taste and odour is important because of aesthetic value. The majority of water quality complaints are of this type. Most of the organic and some inorganic chemicals cause tastes and odours. These chemicals come from the decaying organic matter, run-offs, industrial wastes and municipal sewage discharges. Geosmin and methyl-isobarneol (MIB) are the serious odour-causing chemicals; they are produced by bacteria—particularly actinomycetes—while decomposing dead organic matter at the bottom of the water bodies. Even a very low concentration of these chemicals can cause earthy-musty odours. These odours are common in spring and fall due to the turn over of the lakes and reservoirs. In the groundwater, the tastes and odours can be due to iron, manganese and hydrogen sulphide (H_2S).

The general classes of odours are:
1. Aromatic (spicy).
2. Balsamic (flowery).
3. Chemical.
4. Disagreeable.
5. Earthy.
6. Musty.
7. Grassy.
8. Vegetable.

These are called the reference odour in the water samples.

Principle

The sample is diluted with odour-free water until a dilution is found with the least detectable odour; this dilution factor is the threshold odour number. The sense of smell varies widely—even in the same individual and from person to person. Therefore, it is recommended that a panel of individuals with keen sense of smell be used for an accurate determination of the odours in the water. Water samples are heated in a water bath at 60°C to vapourise odour-causing chemicals. These odours are sniffed after sniffing the odour-free water.

Procedure

Add 200 ml, 100 ml, 50 ml, 12 ml and 2.8 ml portions of the representative sample sequentially to (thoroughly cleaned and rinsed with odour-free water) correspondingly marked 500 ml Erlenmeyer flasks and dilute each portion to 200 ml with odour-free water. Apply a glass stopper to each of them and heat in a water bath for 30 minutes at 60°C. Agitate the flask with 0 ml sample (control or odour-free), remove the stopper and sniff the vapours. There should not be any odours in the control sample. Repeat this with the flask containing the undiluted water. If it has an odour, this is the reference odour. Repeat the sniffing procedure by first sniffing odour—free water and then the flask with the lowest volume (2.8 ml) of the sample. Continue progressively to the highest sample volume until the dilution with the detectable odour is determined. Dilution factor of the dilution with the detectable odour is the threshold odour number. For example, we are testing raw river water with musty-earthy odour and 50 ml sample is the dilution with some musty odour. Threshold odour number of this sample is 200 ml/50 ml or 4 and the reference odour is musty-earthy. This means that people can easily detect these odours in the water.

Turbidity

Turbidity is the murkiness in the water caused by colloidal (1 to 100 nanometer (nm) particles) and other suspended particles, such as clay, sand, silt, organic matter of plant and animal origin, planktons and other microscopic organisms. Turbidity particles can be waterborne pathogens or particles harbouring them. The lower the turbidity, the less is the amount of the particulate matter. It means there is less probability of the presence of waterborne pathogens and the water is safer. Therefore, turbidity is one of the primary standards for the drinking water. The finished water turbidity is tested at least every four hours. Turbidity of the finished water should be equal to or less than 0.3 nephalometric turbidity unit (NTU) in 95 per cent of the samples/month.

Principle

Turbidity is measured as the amount of scattered light by the suspended particles in the sample. The turbidity unit, NTU, is based on the amount of light scattered by particles of formazine, a polymer, used as a reference standard due to the reproducibility of the results. One mg/l of formazine equals 1 NTU.

Procedure

Collect a representative grab sample below the surface of water, mix it and pour into a clean and scratch—free measuring cell. Pour the sample slowly to avoid air bubble formation. To avoid any smudges from the hands, hold the cell from the top part. Scratches, smudges and air bubbles give false higher readings.

Calibrate the turbidity meter and determine the turbidity. Follow the procedure according to the manufacturer's instructions.

Jar Testing

Jar testing is a useful tool to determine the practical optimum dose of a chemical under the simulated plant conditions. It uses a range of increasing dose of a particular chemical in a series of six jars with a stirring and illumination mechanism. Most of the problems in the source water (particularly in the surface water) quality are due to seasonal variations or other unusual circumstances, such as drought, heavy rains, unexpected discharge of raw sewage or run-offs from farm land. These problems can be solved by this test, which is important for coagulation, softening, sedimentation, removal of synthetic organics (such as atrazine) and for tastes and odour control. It makes the water treatment more effective, easy and economical.

Prepare the required dosing solutions and have the jars clean and washed.

Dosing solutions

Weigh the chemical as accurately as required to make a dosing solution: 1 g/l of the chemical equals 1 mg in 1 ml of the dosing solution; 5 g/l, 5 mg in 1 ml and so on. When we apply 1 ml of the dosing solution to a liter of sample, it gives the corresponding number of mg/l of the chemical (e.g., 1 ml of the 5 g/l of dosing solution gives 5 mg/l dose when applied to 1 liter of water).

For rough estimates, 1 ml equals 20 drops; thus, 2 drops mean 0.1 ml of the dosing solution. Sometimes drops are used for lower doses.

General procedure

Simulate the treatment plant conditions for each jar test and pour 1 liter of representative water sample into each of six jars and apply a series of six doses of the required chemical, starting with a control

(0 dose) and ending with the highest dose. Mix the chemical and the sample by using the mixing paddles. Treat it under the plant conditions and check the results.

pH

pH, hydronium ion index, is the measurement of acidity (H^+). Acidity in water is usually due to carbon dioxide (CO_2) from rain water, mineral acids, chlorine and heavy metal salts, such as alum. pH is an important parameter in the water utility. It is used to determine the condition of water for proper coagulation, softening and stabilisation.

Principle and procedure

The potentiometric or pH meter method is a convenient and accurate method for this test. The meter measures the voltage difference developed between the electrodes due to hydronium ions in the solution. This differential is read as the pH of the sample. Testing procedure varies somewhat with different brands and models of the pH meter. Follow the instructions provided by the manufacturer. Proper calibration by using an appropriate buffer solution (solution with the stable pH), such as 6 or 9, is important. Rinse the electrode with the deionised (DI) water and wipe with a soft paper to remove the DI water. Immerse the electrode in the sample and swirl the sample. Record the pH after the reading is stabilised. The electrode should be rinsed with DI water after every use. Leave the electrode in the DI water when the meter is not in use.

Alkalinity

Alkalinity of water is its capacity to neutralise acidity. Carbonates, bicarbonates and hydroxides are the most common forms of alkalinity in natural water. These chemicals are mostly compounds of calcium and magnesium coming from mineral deposits such as limestone and dolomite. Industrial discharges can also cause alkalinity. Bicarbonate alkalinity is present between pH 4.3 and 8.3. Carbonate and bicarbonate alkalinity is present between pH 8.3 and 9.4 and carbonates and hydroxides are present between pH 9.4 and 14. Alkalinity does not exist below pH 4.3. Alkalinity test is important to determine proper coagulation and the stability of water.

Principle and procedure

The commonly used method is titration. Alkalinity of a sample is determined by titrating the sample with 0.02 normal (N) sulphuric acid (H_2SO_4). Results are expressed in mg/l as calcium carbonate. This normality simplifies the calculations of results by giving 1 mg/l of alkalinity for 1 ml of titrant per liter of sample.

Alkalinity is expressed as P (phenolphthalein an indicator) when hydroxides and carbonates are present; they produce a pink colour with the indicator phenolphthalein. Alkalinity is called M (methyl orange) when carbonates and bicarbonates are present. One half of carbonate alkalinity is titrated as P and other half as M. The sum of P and M is known as T or total alkalinity. For an alkalinity test, water sample should be free of turbidity and colour because they obscure the indicator colour. The chlorine residual above 1.8 mg/l interferes by bleaching the colour of the indicator.

Carefully measure the appropriate volume of the sample (normally 50 ml) and neutralise the residual chlorine, if required, by adding two drops of 10 per cent sodium thiosulphate. Add two drops of phenolphthalein indicator and swirl. If the sample turns pink, hydroxide, carbonate or both of them are present. Titrate with sulphuric acid until the pink colour disappears. Mark the buret reading (ml used) as 'P'. If there is no colour with phenolphthalein, then P alkalinity is 0, which means hydroxides and carbonates are absent. After P alkalinity titration, add 3 to 6 drops of the indicator methyl orange or bromcreosol green methyl red to the same sample. If M alkalinity is present, the sample will turn yellow with methyl orange and purple with bromcreosol green methyl red. These changes in colour indicate the presence of carbonates, bicarbonates or both. Titrate with sulphuric acid to orange when methyl orange is the indicator and to salmon colour for the indicator bromcreosol green methyl red. Record the reading from the start of the test to the end as T.

The difference between T and P is the M alkalinity. Calculate the alkalinity as mg/l of calcium carbonate by multiplying the total number of ml of the titrant (T) with factor 20, because 50 ml sample is 1/20 of 1 liter or 1000 ml. Similarly, for 25 ml sample the factor is 40 and for 100 ml it is 10. Table 5.1 gives the types and calculations of alkalinity for different titration results.

Table 5.1. Types and calculations of alkalinity.

Titration results	Hydroxide	Carbonate	Bicarbonate
P = 0	0	0	T
P = less than ½ T	0	2P	T–2P
P = ½ T	0	T	0
P = more than ½ T	2P–T	2T–2P	0
P = T	T	0	0

Hardness

Hardness of water is the total concentration of calcium and magnesium ions expressed as calcium carbonate. Hardness is caused by the soluble bicarbonates, sulphates, nitrates and chlorides of calcium and magnesium. Hardness, due to bicarbonates, is known as carbonate hardness. When hardness is due to other compounds, it is known as non-carbonate hardness. The carbonate and non-carbonate hardness are also known as temporary and permanent hardness, respectively. Other bivalent and trivalent (Fe^{+2} and Al^{+3}) metal ions, such as iron, aluminium, manganese and zinc will also cause hardness, but their concentrations are mostly insignificant.

Principle and procedure

The most commonly used method is titration. Calcium and magnesium ions are titrated with disodium ethylenediamine tetra acetate (EDTA). Hardness-causing ions form a chelated compound with EDTA and the colour of the indicator eriochrome black T is changed from wine red to blue at the end point.

Measure 50 ml of sample, buffer it, add indicator and swirl to mix. If the sample turns wine red, then hardness is present. Titrate the sample with 0.02 N EDTA until the colour changes to blue. To avoid over-titrating, pour EDTA drop by drop as soon as the colour starts changing to purple. After the purple colour, it takes only a few more drops of the titrant to reach the end point. After titration the colour may change back to purple due to the reversal of the reaction. For this reason, the first reading is considered a rough reading. The second titration is done carefully to have an accurate reading. For a 50 ml sample, multiply the number of milliters of titrant used with 20 to determine mg/l of hardness as calcium carbonate.

Principle and procedure for calcium hardness

A calcium test is required to determine the stability of the water. Calcium is present in the natural water at levels ranging from 0 to several hundreds mg/l. Another source of calcium in water is the use of lime [CaO or $Ca(OH)_2$] for softening purposes.

This method is a modification of the total hardness EDTA titration method. The pH of the sample is raised to 12–14 to prevent the magnesium interference. Eriochrome blue-black R is used as an indicator that turns red if calcium is present and changes to sharp blue at the end point. The alkalinity of the sample above 300 mg/l obscures the end point. This problem is resolved by diluting the sample.

Relationship between alkalinity, carbonate and non-carbonate hardness

Normally, the alkalinity in water is due to carbonates and bicarbonates of calcium and magnesium. When testing two different parts of the same compounds in the hardness and alkalinity tests, both results are expressed as calcium carbonate. Therefore, the total alkalinity of a sample is equal to its carbonate hardness. Thus, non-carbonate hardness is the difference between total hardness and alkalinity.

$$\text{Calcium carbonate} = \underset{\text{Hardness}}{Ca^{+2}} \qquad \underset{\text{Alkalinity}}{CO_3^{-2}}$$

Disinfection Residual

Chlorine is one of the most effective disinfectants and is quite commonly used for water disinfection. Chlorine, combined with ammonia, forms chloramines, which are called combined residual chlorine. Total residual chlorine is the sum of the free residual chlorine and combined residual chlorine.

Principle

The most commonly used method is an amperometric method, which is a modification of the polarographic (automatic measuring and recording) principle. It uses 0.00564 N phenylarsine oxide (PAO) as a titrant. By using this normality, 1 ml of the titrant equals 1 mg/l of chlorine for the sample volume 200 ml. PAO is quite stable and available from venders in a form of accurate strength and ready for use. A special amperometric cell is used to detect the end point. Free residual chlorine is titrated at pH 7 and combined residual at pH 4.

Procedure

Rinse the titration cell of the analyser by filling the cup with a sample and turning on the mixer. Discard the sample and refill the cup to the 200 ml mark with a grab sample and turn the mixer on. Add 1 ml of phosphate buffer to adjust the pH to 7 and adjust the needle to the middle of the scale. Add small amounts of PAO until deflection of the needle to the left stops. This is the end point for free residual chlorine; the milliliters of the titrant used are equal to free residual chlorine as mg/l. Add to the same sample the potassium iodide (KI) solution and 1 ml of the acetate buffer to adjust the pH to 4. If the combined residual chlorine is present, the needle will move to the right. Adjust the needle to the middle of the scale and titrate until the deflection to the left stops. This second reading is the combined residual chlorine. Assume the first end point comes after 2 ml and the second comes after 3 ml. Then 2 mg/l is free; 1 mg/l is combined and 3 mg/l is total residual chlorine. After testing, discard the sample and rinse the cell with tap water and turn off the mixer.

Colormetric method by using diethyl phenylenediamine (DPD) is a common field method. The colormetric method is based on the principle that the darker the colour of the indicator, the higher is the concentration and vice versa. Being less accurate, it is not used for compliance purposes. It is quite simple and an easy test. Free-residual chlorine reacts with DPD to produce a pinkish red colour. The colour intensity is matched with the standard colour on the colour wheel to determine the amount of the residual chlorine. For total residual chlorine, use potassium iodide along with DPD to liberate iodine from the combined residual chlorine. Iodine reacts with DPD to produce the red colour. Use of a colour comparator test kit by the Hach Company is quite common. The company provides powder pillows of various chemicals used in the test and simple instructions for testing.

Coliform Bacteria Tests

Bacteriological quality of water is important to determine the degree of disinfection and possible presence of waterborne pathogens. Bacteria, being small, are present almost everywhere, such as in air, water and on laboratory equipment. Therefore, all equipment and handling is done in a sterile environment to ensure the accuracy of data.

Preparation

All glassware to be used in the bacteriological tests should be thoroughly cleaned in a glass-washing machine and sterilised. If a sample has chlorine, then use two drops of 10 per cent sodium thiosulphate solution per sample bottle before sterilising. For thorough sterilisation in an autoclave, bottles should be loosely capped to allow the steam to penetrate. Caps are covered with aluminium foil to prevent the contamination by hands while handling the sample. Sterilise the capped bottles in an autoclave for 15 minutes at 15 psi pressure, which equals 121°C. Cool the sterilised bottles to room temperature and tighten the caps. If caps are tightened when bottles are still hot, condensation of the steam inside the bottle will create a vacuum, which will make the cap hard to open. Moreover, when opened, the bottle will suck air and will be contaminated before collecting a sample.

Growth media

Media are the food for the bacteria to culture them in the laboratory. Different bacteria have different food requirements; therefore, each medium will allow certain types of bacteria to grow. Media are either liquid, known as broths or semi-solid (gelatinous), which are called agars. Agar, an extract of seaweeds, is used as a solidifying agent in the media to facilitate the growth of individual bacteria into colonies by keeping the media solid at incubation temperature below 45°C. The number of colonies corresponds to the number of individual bacteria in the sample.

A medium is called selective if it contains certain chemicals that allow only certain types of bacteria to grow. For example, brilliant green bile broth (BGB) allows only fecal coliforms to grow from the coliform group. A medium is known as differential if different types of bacterial colonies are differentiated by different colours, shapes and sizes. For example, M endo broth gives coliform colonies the golden green metallic sheen.

Preferably, media should be freshly prepared daily. Prepare and sterilise media according to the instructions from the supplier. Dilution water, for preparation of media, is buffered to pH 7.2 with phosphate buffer solution.

Label all the sample bottles and petri dishes properly. Petri dishes should be labelled on the underside, as they are incubated upside down.

Sampling

Proper sampling is important. Avoid contamination of a sample from hands, air and sneezing. To collect a representative grab sample from a tap, open the faucet fully and allow the water to flow for 2 to 3 minutes to flush out the stagnant water. If the sample is collected from a river, stream or lake, hold the bottle upside down below the surface to avoid any floating material from entering the bottle, turn the bottle upward facing the current and fill it. (Hold the cap upside down in the other hand.) If there is no current, push the bottle forward to fill it. If the sample comes from a well, run the well water for 5 minutes before collecting the sample. Collect at least 100 ml of sample by filling only two-thirds or three quarters of the sample bottle or container to leave enough space for the proper mixing of the sample before testing. Take the sample immediately to the laboratory or ship it to its testing destination. During the holding time, which should not be more than eight hours for coliform bacteria, keep the samples at 5 to 10°C to avoid any change in the bacterial density.

Coliform test

To ensure the absence of waterborne pathogens, the water is tested for coliform bacteria. Coliform bacteria are present in human wastes and in soil contaminated with human wastes. These bacteria in human wastes are known as fecal coliform bacteria. Those in the soil are called non-fecal coliforms. Fecal coliforms are represented by *Eschrechia coli* (*E. coli*) and non-fecal coliforms are represented by *Enterobacter aerogenes (E. aerogenes)*. Both fecal coliforms and non-fecal coliforms are called the total coliform group. This group is used as an indicator of the presence of human wastes in water and the possible presence of waterborne pathogens.

For compliance, use a state-approved sampling site plan of the distribution system. The number of samples/month is based on the population being served. Testing methodology for coliform bacteria has four different techniques:

1. Membrane filter technique.
2. Multiple tube fermentation (MTF)/most probable number (MPN) technique.
3. Presence-absence technique.
4. Defined substrate technology technique (e.g., Colilert).

Membrane filter technique is the filtration of the sample through a 0.45 micrometer pore-size filter, which retains the coliform bacteria that are then provided with a selective and differential media pad in a tight-fitting petri dish (see Fig. 5.1). M. endo broth is used for total coliforms and M FC broth is used for the fecal coliforms. A pad is soaked with about 2 ml of media and the filter is placed on it. The dishes are incubated upside down (to prevent drying of the filter) at 35.5 ± 0.5°C for total coliforms and 44.5 ± 0.2°C for fecal coliforms for 24 ± 2 hours. Each coliform bacterium multiplies every half hour and forms a colony. Total coliform bacterial colonies have a distinct golden green metallic sheen and fecal coliform colonies are blue. This technique gives both qualitative and quantitative results. Therefore, it is a commonly used technique in most water treatment laboratories. Results are reported as colonies per 100 ml of sample.

Fig. 5.1. Placing a membrane filter on a pad soaked with medium in a petri dish.

For accuracy, use the results of the filters with a colony count of 20 to 80 for coliforms and 20 to 60 for fecal coliforms (fecal coliform colonies are bigger). If all filters have a colony count less than 20 each, then total all such counts and compute the results from total volume of the sample. If all filters have a colony count higher than the upper limit (80 for coliforms and 60 for fecal coliforms) or more than 200 total colonies (all kinds), then report as too numerous to count (TNTC). If colonies are fused and not well defined, then report as confluent and repeat the test.

The multiple tube fermentation technique is based on the principle of selecting coliform bacteria by using a selective medium and culturing them to produce gas. It is a qualitative test.

A set of five or multiple of five culture tubes (test tubes with an inverted small tube to trap the gas) containing the required selective media is inoculated with the water sample. Sample and media are mixed for the proper dispersal of microbes. These tubes are then incubated at the required temperature for 24 to 48 hours. Coliform bacteria produce gas by fermenting lactose. Presence of gas in the insert tube indicates a positive test (see Fig. 5.2).

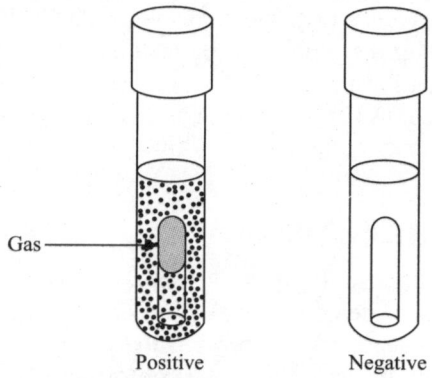

Gas

Positive   Negative

Fig. 5.2. Culture tubes.

This test is divided into three parts: presumptive, confirmed and completed tests. Laurel tryptose broth, brilliant green bile broth (BGB) and lactose broth are the respective media for these parts. The presence of gas in the laurel tryptose broth shows a positive presumptive test indicating the possible presence of coliform bacteria; the presence of gas in the BGB tubes means a positive confirmed test that confirms the presence of coliform bacteria; and a completed test reconfirms their presence by showing the presence of gas, gram negative (red colour) and non-sporulating bacilli. The last three characteristics of coliform bacteria are determined by preparing a slide of the culture, staining it with gram stain and observing the culture under a microscope. It takes 2 to 5 days to have the final results.

Use EC broth and incubation at 44°C for the fecal coliforms completed test. The density per 100 ml sample is estimated from the MPN table, which is based on the number of the positive tubes. This technique, being lengthy and laborious, is less popular than the membrane filter technique. It is normally used as a back-up test with the membrane filter technique.

Presence–absence technique is a qualitative test based on the incubation of 100 ml of water sample and laurel tryptose broth to indicate the presence of coliforms by fermenting the media. The presence of yellow colour after 24 to 48 hours incubation at 35 ± 0.5°C indicates a positive presumptive test for coliform bacteria, which is confirmed by using BGB broth. It is a modification of the MTF technique to encourage even a single coliform bacterium to grow and give a positive result.

Defined substrate technique is based on the principle that each organism has some specific enzyme. An enzyme is a chemical that accelerates the rate of a specific reaction of a chemical called substrate, without undergoing any change in itself. The presence of the organism is tested by using the specific substrate for that particular enzyme. If a reaction takes place, the organism is present and the test is positive and vice versa. An indicator chemical is used to show the reaction. This test is a biochemical test for specific organisms. Enzyme beta-galactosidase is specific to the total coliforms and enzyme beta-glucuronidase is found only in *E. coli*. In the Colilert test the medium contains substrate bonded with indicators ortho nitrophenyl-beta-d-galactopyranoside (ONPG) for total coliforms and 4-methyl-umbelliferyl-beta-d-glucuronide (MUG) for *E. coli*. This medium is known as ONPG-MUG or minimal media, MMO-MUG. When the sample is incubated with these chemicals in the media, coliform bacteria produce a yellow colour that fluoresces in ultraviolet light if *E. coli* is present.

Procedure

Add 100 ml of sample to the container with Colilert media and incubate at $35 \pm 0.5°C$ for 24 hours. Check for yellow colour and fluorescence with 366 nm wavelength UV light. The density of the coliform bacteria can be determined by using the multiple tube technique with ONPG-MUG and applying the MPN table. This test gives total coliform and *E. coli* results within 24 hours, which is a useful tool for the water utility under crisis conditions (like possible cross connection) for a quick response. This fast, sensitive and specific technique is becoming popular with water utilities. The only drawback is that it is still an expensive technique.

Heterotrophic or Standard Plate Count (HPC)

This test gives the total count of almost all types of bacteria in the water sample that can grow on a general medium called the standard plate count agar or nutrient agar. A count less than 500 colonies/ml of the sample means that the water is properly disinfected and vice versa. Furthermore, a count of higher than 500 colonies/ml interferes with the growth of total coliform bacteria. It is a supplementary test for process control. This test uses only 0.1 to 1 ml of a sample. Use a straight sample for treated water and a dilution for raw water.

Principle

Standard plate count agar allows most of the bacteria (heterotrophic bacteria) in the water to grow and form colonies of different shapes and sizes, in 48 hours at 35°C, incubation temperature. Colony count/ml of sample determines the quality of water. The higher the count, the more polluted the water and vice versa. This test is used to determine the quality of source water, level of disinfection in the process control and adequate disinfection of water in the distribution system. If the number is more than 500/ml of sample from the distribution system, the system is deficient of disinfectant.

Procedure

Mix a sample or dilution thoroughly in a capped bottle and transfer 0.1 and 1 ml to two sterilised 100 mm × 15 mm petri dishes, respectively. Add to each dish 10 to 15 ml of the tryptone glucose extract agar (standard plate count agar) that has been liquefied, sterilised and cooled to $44 \pm 1°C$ and rotate the dish gently to mix the sample and medium evenly. Allow the mixture to solidify; invert the plates and incubate them at $35 \pm 0.5°C$ for 48 ± 3 hours. Use a Quebec Colony Counter and count all bacterial colonies (even pin/point size). Report results as number of bacteria per milliliter of sample.

Valid and accurate data from the quality control laboratory ensure the adequate water quality and compliance with the SDWA.

Chapter 6

An Overview of Waste-Water Treatment

INTRODUCTION

We may organise water treatment technologies into three general areas: physical methods, chemical methods and energy intensive methods. Physical methods of waste-water treatment represent a body of technologies that we refer largely to as solid-liquid separations techniques, of which filtration plays a dominant role. Filtration technology can be broken into two general categories—conventional and non-conventional. This technology is an integral component of drinking water and waste-water treatment applications. It is, however, but one unit process within a modern water treatment plant scheme, whereby there are a multitude of equipment and technology options to select from depending upon the ultimate goals of treatment. To understand the role of filtration, it is important to make distinctions not only with the other technologies employed in the cleaning and purification of industrial and municipal water, but also with the objectives of different unit processes.

Chemical methods of treatment rely upon the chemical interactions of the contaminants we wish to remove from water and the application of chemicals that either aid in the separation of contaminants from water or assist in the destruction or neutralisation of harmful effects associated with contaminants. Chemical treatment methods are applied both as stand-alone technologies and as an integral part of the treatment process with physical methods.

Among the energy intensive technologies, thermal methods have a dual role in water treatment applications. They can be applied as a means of sterilisation, thus providing high quality drinking water and/or these technologies can be applied to the processing of the solid wastes or sludge, generated from water treatment applications. In the latter cases, thermal methods can be applied in essentially the same manner as they are applied to conditioning water, namely to sterilise sludge contaminated with organic contaminants and/or these technologies can be applied to volume reduction. Volume reduction is a key step in water treatment operations, because ultimately there is a trade-off between polluted water and hazardous solid waste. Energy intensive technologies include electrochemical techniques, which by and large are applied to drinking water applications. They represent both sterilisation and conditioning of water to achieve a palatable quality.

All three of these technology groups can be combined in water treatment or they may be used in select combinations depending upon the objectives of water treatment. Among each of the general technology classes, there is a range of both hardware and individual technologies that one may select from. The selection of not only the proper unit process and hardware from within each technology group, but the optimum combinations of hardware and unit processes from the four groups depends upon such factors as:

1. How clean the final water effluent from our plant must be.
2. The quantities and nature of the influent water we need to treat.
3. The physical and chemical properties of the pollutants we need to remove or render neutral in the effluent water.
4. The physical, chemical and thermodynamic properties of the solid wastes generated from treating water.
5. The cost of treating water, including the cost of treating, processing and finding a home for the solid wastes.

To understand this better, let us step back and start from a very fundamental viewpoint. All processes are comprised of a number of unit processes, which are in turn made up of unit operations. Unit processes are distinct stages of a manufacturing operation. They each focus on one stage in a series of stages, successfully bringing a product to its final form. In this regard, a waste-water treatment plant, whether industrial, a municipal waste-water treatment facility or a drinking water purification plant, is no different than, say, a synthetic rubber manufacturing plant or an oil refinery. In the case of a rubber producing plant, various unit processes are applied to making intermediate forms of the product, which ultimately is in a final form of a rubber bale, that is sold to the consumer. The individual unit processes in this case are comprised of: (i) a catalyst reparation stage—a pre-preparation stage for monomers and catalyst additives; (ii) polymerisation—where an intermediate stage of the product is synthesised in the form of a latex or polymer suspended as a dilute solution in a hydrocarbon diluent; and (iii) followed by finishing— where the rubber is dried, residual diluent is removed and recovered and the rubber is dried and compressed into a bale and packaged for sale. Each of these unit process operations are in turn comprised of individual unit operations, whereby a particular technology or group of technologies are applied, which, in turn, define a piece of equipment that is used along the production line. Drinking water and waste-water treatment plants are essentially no different. There are individual unit processes that comprise each of these types of plants that are applied in a succession of operations, with each stage aimed at improving the quality of the water as established by a set of product-performance criteria. The criteria focuses on the quality of the final water, which in the case of drinking water is established based upon legal criteria (e.g., the Safe Drinking Water Act, SDWA) and if non-potable or process plant water, may be operational criteria (e.g., non-brackish water to prevent scaling of heat exchange equipment).

The number and complexity of unit processes and in turn unit operations comprising a water purification or waste-water treatment facility are functions of the legal and operational requirements of the treated water, the nature and degree of contamination of the incoming water (raw water to the plant) and the quantities of water to be processed. This means then, that water treatment facilities from a design and operational standpoints vary, but they do rely on overlapping and even identical unit processes.

If we start with the first technology group, then filtration should be thought of as both a unit process and a unit operation within a water treatment facility. As a separate unit process, its objective is quite clear: namely, to remove suspended solids. When we combine this technology with chemical methods and apply sedimentation and clarification (other physical separation methods), we can extend the technology to removing dissolved particulate matter as well. The particulate matter may be biological, microbial or chemical in nature. As such, the operation stands alone within its own block within the overall manufacturing train of the plant. Examples of this would be the roughening and polishing stages of water treatment. In turn, we may select or specify specific pieces of filtration equipment for these unit processes.

The above gives us somewhat of an idea of the potential complexity of choosing the optimum group of technologies and hardware needed in treating water. To develop a cost-effective design, we need to understand not only what each of the unit processes are, but obtain a working knowledge of the operating basis and ranges for the individual hardware. That, indeed, is the objective of this chapter; namely, to take a close look at the equipment options available to us in each technology group, but not individually. Rather, to achieve an integrated and well thought out design, we need to understand how unit processes and unit operations compliment each other in the overall design.

This chapter is for orientation purposes. Its objectives are to provide an overview of water treatment and purification roles and technologies and to introduce terminology that will assist you in understanding the relation of the various technologies to the overall schemes employed in waste treatment applications. Recommended resources that you can refer to for more in-depth information are included at the end of each chapter. These will assist in reinforcing some of the principles and concepts presented in each chapter, if the book is used as a primary or supplement textbook. We should recognise that the technology options for water treatment are great and quite often the challenge lies with the selection of the most cost-effective combinations of unit processes and operations. In this regard, cost-factors are examined where appropriate in our discussions within later chapters.

WATER PURIFICATION

When we refer to water purification, it makes little sense to discuss the subject without first identifying the contaminants that we wish to remove from water. Also, the source of the water is of importance. Our discussion at this point focuses on drinking water. Groundwater sources are of a particular concern, because there are many communities throughout the world that rely on this form. The following are some of the major contaminants that are of concern in water purification applications, as applied to drinking water sources, derived from groundwater (Fig. 6.1).

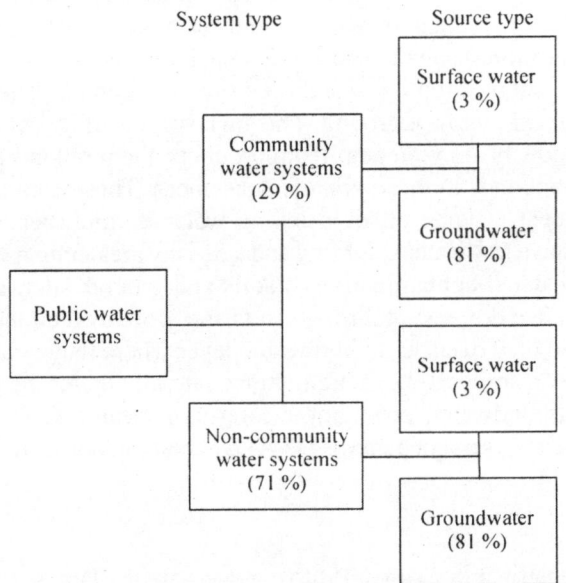

Fig. 6.1. Major contaminants in water purification applications, as applied to drinking water sources, derived from groundwater.

Heavy Metals

Heavy metals represent problems in terms of groundwater pollution. The best way to identify their presence is by a laboratory test of the water or by contacting county health departments. There are concerns of chronic exposure to low levels of heavy metals in drinking water.

Turbidity

Turbidity refers to suspended solids, i.e. muddy water, is very turbid. Turbidity is undesirable for three reasons:
1. Aesthetic considerations.
2. Solids may contain heavy metals, pathogens or other contaminants.
3. Turbidity decreases the effectiveness of water treatment techniques by shielding pathogens from chemical or thermal damage or in the case of UV (ultra violet) treatment, absorbing the UV light itself.

Organic Compounds

Water can be contaminated by a number of organic compounds, such as chloroform, gasoline, pesticides and herbicides from a variety of industrial and agricultural operations or applications. These contaminants must be identified in a laboratory test. It is unlikely groundwater will suddenly become contaminated, unless a quantity of chemicals is allowed to enter a well or penetrating the aquifer. One exception is when the aquifer is located in limestone. Not only will water flow faster through limestone, but the rock is prone to forming vertical channels or sinkholes that will rapidly allow contamination from surface water. Surface water may show great variations in chemical contamination levels due to differences in rainfall, seasonal crop cultivation and industrial effluent levels. Also, some hydrocarbons (the chlorinated hydrocarbons in particular) form a type of contaminant that is especially troublesome. These are a group of chemicals known as dense non-aqueous phase liquids (DNAPLs). These include chemicals used in dry cleaning, wood preservation, asphalt operations, machining and in the production and repair of automobiles, aviation equipment, munitions and electrical equipment. These substances are heavier than water and they sink quickly into the ground. This makes spills of DNAPLs more difficult to handle than spills of petroleum products. As with petroleum products, the problems are caused by groundwater dissolving some of the compounds in these volatile substances. These compounds can then move with the groundwater flow. Except in large cities, drinking water is rarely tested for these contaminants. Disposal of chemicals that have low water solubility and a density greater than water result in the formation of distinct areas of pure residual contamination in soils and groundwater. Because of their relatively high density, they tend to move downward through soils and groundwater, leaving small amounts along the migratory pathway, until they reach an impermeable layer where they collect in discrete pools. Once the DNAPLs have reached an aquitard they tend to move laterally under the influence of gravity and to slowly dissolve into the groundwater, providing a long-term source for low level contamination of groundwater. Because of their movement patterns DNAPL contamination is difficult to detect, characterise and remediate.

Pathogens

These include protozoa, bacteria and viruses. Protozoa cysts are the largest pathogens in drinking water and are responsible for many of the waterborne disease cases in all over the world. Protozoa cysts range is size from 2 to 15 µm (a micron is one millionth of a meter), but can squeeze through smaller openings.

In order to insure cyst filtration, filters with a absolute pore size of 1 μm or less should be used. The two most common protozoa pathogens are *Giardia lamblia* (Giardia) and *Cryptosporidium* (Crypto). Both organisms have caused numerous deaths in recent years, the deaths occurring in the young and elderly and the sick and immune compromised.

Bacteria are smaller than protozoa and are responsible for many diseases, such as typhoid fever, cholera, diarrhea and dysentery. Pathogenic bacteria range in size from 0.2 to 0.6 μm and a 0.2 μm filter is necessary to prevent transmission. Contamination of water supplies by bacteria is blamed for the cholera epidemics, which devastate undeveloped countries from time to time. Even in the US, *E. coli* is frequently found to contaminated water supplies. Fortunately, *E. coli* is relatively harmless as pathogens go and the problem isn't so much with *E. coli* found, but the fear that other bacteria may have contaminated the water as well. Never the less, dehydration from diarrhea caused by *E. coli* has resulted in fatalities.

One of hundreds of strains of the bacterium *Escherichia coli*, *E. coli* O157:H7 is an emerging cause of food borne and waterborne illness. Although most strains of *E. coli* are harmless and live in the intestines of healthy human and animals, this strain produces a powerful toxin and can cause severe illness. *E. coli* O157: H7 was first recognised as a cause of illness during an outbreak in 1982 traced to contaminated hamburgers. Since then, most infections are believed to have come from eating undercooked ground beef. However, some have been waterborne. The presence of *E. coli* in water is a strong indication of recent sewage or animal waste contamination. Sewage may contain many types of disease-causing organisms. Since *E. coli* comes from human and animal wastes, it most often enters drinking water sources via rainfalls, snow melts or other types of precipitation, *E. coli* may be washed into creeks, rivers, streams, lakes or groundwater. When these water are used as sources of drinking water and the water is not treated or inadequately treated, *E. coli* may end up in drinking water. *E. coli* O157:H7 is one of hundreds of strains of the bacterium *E. coli*. Although most strains are harmless and live in the intestines of healthy human and animals, this strain produces a powerful toxin and can cause severe illness. Infection often causes severe bloody diarrhea and abdominal cramps; sometimes the infection causes non-bloody diarrhea. Frequently, no fever is present. It should be noted that these symptoms are common to a variety of diseases and may be caused by sources other than contaminated drinking water. In some people, particularly children under 5 years of age and the elderly, the infection can also cause a complication, called hemolytic uremic syndrome, in which the red blood cells are destroyed and the kidneys fail. About 2–7 per cent of infections lead to this complication. In the US hemolytic uremic syndrome is the principal cause of acute kidney failure in children and most cases of hemolytic uremic syndrome are caused by *E. coli* O157:H7. Hemolytic uremic syndrome is a life-threatening condition usually treated in an intensive care unit. Blood transfusions and kidney dialysis are often required. With intensive care, the death rate for hemolytic uremic syndrome is 3–5 per cent. Symptoms usually appear within 2 to 4 days, but can take up to 8 days. Most people recover without antibiotics or other specific treatment in 5–10 days. There is no evidence that antibiotics improve the course of disease and it is thought that treatment with some antibiotics may precipitate kidney complications. Antidiarrheal agents, such as loperamide (Imodium), should also be avoided. The most common methods of treating water contaminated with *E. coli* is by using chlorine, ultraviolet light or ozone, all of which act to kill or inactivate *E. coli*. Systems, using surface water sources, are required to disinfect to ensure that all bacterial contamination is inactivated, such as *E. coli*. Systems using ground water sources are not required to disinfect, although many of them do. According to EPA regulations, a system that operates at least 60 days per year and serves 25 people or more or has 15 or more service connections, is regulated as a public water system under the Safe Drinking Water Act (SDWA). If a system is not a public water system as defined by EPA's regulations, it is not regulated under the SDWA, although it may be regulated by state or local authorities. Under the SDWA, EPA requires public water systems to monitor for coliform bacteria. Systems analyse first for total coliform, because this test is faster to produce results. Any time that a sample is positive for total coliform, the same sample must be analysed for either fecal coliform or *E. coli.* Both are indicators of contamination with animal waste or human sewage. The largest public water systems (serving millions of people) must take at least 480 samples per month. Smaller systems must take at least five samples a month, unless the state has conducted a sanitary survey—a survey in which a state inspector examines system components and ensures they will protect public health—at the system within the last five years.

Viruses are the 2nd most problematic pathogen, behind protozoa. As with protozoa, most waterborne viral diseases don't present a lethal hazard to a healthy adult. Waterborne pathogenic viruses range in size from 0.020–0.030 μm and are too small to be filtered out by a mechanical filter. All waterborne enteric viruses affecting human occur solely in human, thus animal waste doesn't present much of a viral threat. At the present viruses don't present a major hazard to people drinking surface water in the US, but this could change in a survival situation as the level of human sanitation is reduced. Viruses do tend to show up even in remote areas, so a case can be made for eliminating them now.

DRINKING WATER STANDARDS

When the objective of water treatment is to provide drinking water, then we need to select technologies that are not only the best available, but those that will meet local and national quality standards. The primary goals of a water treatment plant for over a century have remained practically the same: namely to produce water that is biologically and chemically safe, is appealing to the consumer and is non-corrosive and non-scaling. Today, plant design has become very complex from discovery of seemingly innumerable chemical substances, the multiplying of regulations and trying to satisfy more discriminating palates. In addition to the basics, designers must now keep in mind all manner of legal mandates, as well as public concerns and environmental considerations, to provide an initial prospective of water works engineering planning, design and operation.

Today resource limitations have caused the United States Environmental Protection Agency (USEPA) to re-assess schedules for new rules. Small systems are the most frequent violators of federal regulations. Microbiological violations account for the vast majority of cases, with failure to monitor and report. Among others, violations exceeding SDWA maximum contaminant levels (MCLs) are quite common. Bringing small water systems into compliance requires applicable technologies, operator ability, financial resources and institutional arrangements.

For turbidity colour and microbiological control in surface water treatment filtration. Common variations of filtration are conventional, direct, slow sand, diatomaceous earth and membranes.

For inactivation of micro-organisms: disinfection. Typical disinfectants are chlorine, chlorine dioxide, chloramines and ozone.

For organic contaminant removal from surface water: packed-tower aeration, granular activated carbon (GAC), powdered activated carbon (PAC), diffused aeration, advanced oxidation processes and reverse osmosis (RO).

For inorganic contaminants removal: membranes, ion exchange, activated alumina and GAC.

For Corrosion Control

Typically, pH adjustment or corrosion inhibitors. The implications of the 1986 amendments to the SDWA and new regulations have resulted in rapid development and introduction of new technologies and equipment for water treatment and monitoring over the last two decades. Biological processes in particular have proven effective in removing biodegradable organic carbon that may sustain the re-growth of potentially harmful micro-organisms in the distribution system, effective taste and odour control and reduction in chlorine demand and DBP formation potential. Both biologically-active sand or carbon filters provide cost effective treatment of micro-contaminants than do physico-chemical processes in many cases. Pertinent to the subject matter cover in this volume, membrane technology has been applied in drinking water treatment, partly because of affordable membranes and demand to removal of many contaminants. Microfiltration, ultrafiltration, nanofiltration and others have become common names in the water industry. Membrane technology is experimented with for the removal of microbes, such as *Giardia* and *Cryptosporidium* and for selective removal of nitrate. In other instances, membrane technology is applied for removal of DBP precursors, VOCs and others.

Other treatment technologies that have potential for full-scale adoption are photochemical oxidation using ozone and UV radiation or hydrogen peroxide for destruction of refractory organic compounds. This process combines contact flocculation, filtration and powdered activated carbon adsorption to meet a wide range of requirements for surface water and groundwater purification.

Utilities are seeking not only to improve treatment, but also to monitor their supplies for microbiological contaminants more effectively. Electro-optical sensors are used to allow early detection of algal blooms in a reservoir and allow for diagnosis of problems and guidance in operational changes. Gene probe technology was first developed in response to the need for improved identification of microbes in the field of clinical microbiology. Attempts are now being made by radiolabelled and non-radioactive gene-probe assays with traditional detection methods for enteric viruses and protozoan parasites, such as *Giardia* and *Cryptosporidium*. This technique has the potential for monitoring water supplies for increasingly complex groups of microbes.

In spite of the multitudinous regulations and standards that an existing public water system must comply with, the principles of conventional water treatment process have not changed significantly over half a century. Whether a filter contains sand, anthracite or both, slow or rapid rate, constant or declining rate, filtration is still filtration, sedimentation is still sedimentation and disinfection is still disinfection. What has changed, however, are many tools that we now have in our engineering arsenal. For example, a supervisory control and data acquisition (SCADA) system can provide operators and managers with accurate process control variables and operation and maintenance records. In addition to being able to look at the various options on the computer screen, engineers can conduct pilot plant studies of the multiple variables inherent in water treatment plant design. Likewise, operators and managers can utilise an ongoing pilot plant facility to optimise chemical feed and develop important information needed for future expansion and upgrading.

Technology and ultimately equipment selection depends on the standards set by the regulations. Drinking water standards are regulations that EPA sets to control the level of contaminants in the nation's drinking water. These standards are part of the Safe Drinking Water Act's 'multiple barrier' approach to drinking water protection, which includes assessing and protecting drinking water sources; protecting wells and collection systems; making sure water is treated by qualified operators; ensuring the integrity of distribution systems; and making information available to the public on the quality of their drinking water. With the involvement of EPA, states, tribes, drinking water utilities, communities and citizens, these multiple barriers ensure that tap water in the US and territories is safe to drink. In most cases, EPA delegates responsibility for implementing drinking water standards to states and tribes. There are two categories of drinking water standards:

1. A National Primary Drinking Water Regulation (NPDWR or primary standard) is a legally-enforceable standard that applies to public water systems. Primary standards protect drinking water quality by limiting the levels of specific contaminants that can adversely affect public health and are known or anticipated to occur in water. They take the form of maximum contaminant levels (MCL) or treatment techniques (TT).

2. A National Secondary Drinking Water Regulation (NSDWR or secondary standard) is a non-enforceable guideline regarding contaminants that may cause cosmetic effects (such as skin or tooth discolouration) or aesthetic effects (such as taste, odour or colour) in drinking water. EPA recommends secondary standards to water systems but does not require systems to comply. However, states may choose to adopt them as enforceable standards. This information focuses on national primary standards.

Non-carcinogens (excluding microbial contaminants)

For chemicals that can cause adverse non-cancer health effects, the MCLG is based on the reference dose. A reference dose (RFD) is an estimate of the amount of a chemical that a person can be exposed to on a daily basis that is not anticipated to cause adverse health effects over a person's lifetime. In RFD calculations, sensitive sub-groups are included and uncertainty may span an order of magnitude. The RFD is multiplied by typical adult body weight (70 kg) and divided by daily water consumption (2 liters) to provide a drinking water equivalent level (DWEL). Note that the DWEL is multiplied by a percentage of the total daily exposure contributed by drinking water to determine the MCLG. This empirical factor is usually 20 per cent, but can be a higher value.

Chemical contaminants (carcinogens)

If there is evidence that a chemical may cause cancer and there is no dose below which the chemical is considered safe, the MCLG is set at zero. If a chemical is carcinogenic and a safe dose can be determined, the MCLG is set at a level above zero that is safe.

Microbial contaminants

For microbial contaminants that may present public health risk, the MCLG is set at zero because ingesting one protozoa, virus or bacterium may cause adverse health effects. EPA is conducting studies to determine whether there is a safe level above zero for some microbial contaminants. So far, however, this has not been established.

Once the MCLG is determined, EPA sets an enforceable standard. In most cases, the standard is a maximum contaminant level (MCL), the maximum permissible level of a contaminant in water which is delivered to any user of a public water system. The MCL is set as close to the MCLG as feasible, which the Safe Drinking Water Act defines as the level that may be achieved with the use of the best available technology, treatment techniques and other means which EPA finds are available (after examination for efficiency under field conditions and not solely under laboratory conditions) are available, taking cost into consideration. When there is no reliable method that is economically and technically feasible to measure a contaminant at particularly low concentrations, a treatment technique (TT) is set rather than an MCL. A treatment technique (TT) is an enforceable procedure or level of technological performance which public water systems must follow to ensure control of a contaminant. Examples of Treatment Technique rules are the Surface Water Treatment Rule (disinfection and filtration) and the Lead and Copper Rule (optimised corrosion control). After determining a MCL or TT based on affordable technology for large systems, EPA must complete an economic analysis to determine whether the benefits of that standard justify the costs. If not, EPA may adjust the MCL for a particular class or group of systems to a level that 'maximises health risk reduction benefits at a cost that is justified by the benefits.'

NATIONAL SECONDARY DRINKING WATER REGULATIONS

National Secondary Drinking Water Regulations (NSDWRs or secondary standards) are non-enforceable guidelines regulating contaminants that may cause cosmetic effects (such as skin or tooth discolouration) or aesthetic effects (such as taste, odour or colour) in drinking water. EPA recommends secondary standards to water systems but does not require systems to comply. However, states may choose to adopt them as enforceable standards. Table 6.1 summarises the secondary standards.

Table 6.1. Summary of National Secondary Drinking Water Regulations.

Contaminant	Secondary standard
Aluminium	0.05 to 0.2 mg/l
Chloride	250 mg/l
Colour	15 (colour units)
Copper	1.0 mg/l
Corrosivity	Non-corrosive
Fluoride	2.0 mg/l
Foaming agents	0.5 mg/l
Iron	0.3 mg/l
Manganese	0.05 mg/l
Odour	3 threshold odour number
pH	6.5–8.5
Silver	0.10 mg/l
Sulphate	250 mg/l
Total dissolved solids	500 mg/l
Zinc	5 mg/l

PHYSICAL TREATMENT METHODS

The following technologies are among the most commonly used physical methods of purifying water:

Heat Treatment

Boiling is one way to purify water of all pathogens. Most experts feel that if the water reaches a rolling boil it is safe. A few still hold out for maintaining the boiling for some length of time, commonly 5 or 10 minutes, plus an extra minute for every 1000 feet of elevation. One reason for the long period of boiling is to inactivate bacterial spores (which can survive boiling), but these spore are unlikely to be waterborne pathogens. Water can also be treated at below boiling temperatures, if contact time is increased. Commercial units are available for residential use, which treat 500 gals of water per day. The process is similar to milk pasteurisation and holds the water at 161°F for 15 seconds. Heat exchangers recover most of the energy used to warm the water. Solar pasteurisers have also been built that can heat three gallons of water to 65°C and hold the temperature for an hour. A higher temperature could be reached, if the device was rotated east to west during the day to follow the sunlight. Regardless of the method, heat treatment does not leave any form of residual to keep the water free of pathogens in storage.

Reverse Osmosis

Reverse osmosis forces water, under pressure, through a membrane that is impermeable to most contaminants. The membrane is somewhat better at rejecting salts than it is at rejecting non-ionised weak acids and bases and smaller organic molecules (molecular weight below 200). In the latter category are undissociated weak organic acids, amines, phenols, chlorinated hydrocarbons, some pesticides and low molecular weight alcohols. Larger organic molecules and all pathogens are rejected. Of course, it is possible to have a imperfection in the membrane that could allow molecules or whole pathogens to pass through. Using reverse osmosis to desalinate seawater requires considerable pressure (1000 psi) to operate. Reverse osmosis filters are available that will use normal municipal or private water pressure to remove contaminates from water. The water produced by reverse osmosis, like distilled water will be close to pure H_2O. Therefore mineral intake may need to be increased to compensate for the normal mineral content of water in much of the world.

Distillation

Distillation is the evaporation and condensation of water to purify water. Distillation has two disadvantages: (i) a large energy input is required; and (ii) if simple distillation is used, chemical contaminants with boiling points below water will be condensed along with the water. Distillation is most commonly used to remove dissolved minerals and salts from water. The simplest form of a distillation for use in the home is a solar still. A solar still uses solar radiation to evaporate water below the boiling point and the cooler ambient air to condense the vapour. The water can be extracted from the soil, vegetation piled in the still or contaminated water (such as radiator fluid or salt water) can be added to the still. While per still output is low, they are an important technique if water is in short supply. Other forms of distillation require a concentrated heat source to boil water which is then condensed. Simple stills use a coiling coil to return this heat to the environment. Efficient distillations plants use a vapour compression cycle where the water is boiled off at atmospheric pressure, the steam is compressed and the condenser condenses the steam above the boiling point of the water in the boiler, returning the heat of fusion to the boiling water. The hot condensed water is run through a second heat exchanger, which heats up the water feeding into the boiler. These plants normally use an internal combustion engine to run the compressor. Waste heat from the engine, including the exhaust, is used to start the process and make up any heat loss.

Microfilters

Microfilters are small-scale filters designed to remove cysts, suspended solids, protozoa and in some cases, bacteria from water. Most filters use a ceramic or fiber element that can be cleaned to restore performance as the units are used. Most units and almost all made for camping use a hand pump to force the water through the filter. Others use gravity, either by placing the water to be filtered above the filter (e.g., the Katadyn drip filter) or by placing the filter in the water and running a siphon hose to a collection vessel located below the filter (e.g., Katadyn siphon filter). Microfilters are the only method, other than boiling, to remove Cryptosporidia. Microfilters do not remove viruses. Microfilters share a problem with charcoal filter in having bacteria grow on the filter medium. Some handle this by impregnating the filter element with silver, such as the Katadyn, others advise against storage of a filter element after it has been used. Many microfilters may include silt pro-filters, activated charcoal stages or an iodine resin. Most filters come with a stainless steel prefilter, but other purchased or improvised filters can be added to reduce the loading on the main filter element. Allowing time for solids to settle and/or pre-filtering will also extend filter life. Iodine matrix filters will kill viruses that will pass through the filter and if a charcoal stage is used it will remove much of the iodine from the water. Charcoal filters will also remove other dissolved natural or manmade contaminates. Both the iodine and the charcoal stages do not indicate when they reach their useful life, which is much shorter than the filter element.

Slow Sand Filter

Slow sand filters pass water slowly through a bed of sand. Pathogens and turbidity are removed by natural die-off, biological action and filtering. Typically the filter will consist of a layer of sand, then a gravel layer in which the drain pipe is embedded. The gravel doesn't touch the walls of the filter, so that water can't run quickly down the wall of the filter and into the gravel. Building the walls with a rough surface also helps. A typical loading rate for the filter is 0.2 meters/hour day (the same as 0.2 m^3/m^2 of surface area). The filter can be cleaned several times before the sand has to be replaced. Slow sand filters should only be used for continuous water treatment. If a continuous supply of raw water can't be insured (say, using a holding tank), then another method should be chosen. It is also important for the water to have as low turbidity (suspended solids) as possible. Turbidity can be reduced by changing the method of collection (for example, building an infiltration gallery, rather than taking water directly from a creek), allowing time for the material to settle out (using a raw water tank), pre-filtering or flocculation (adding a chemical, such as alum to cause the suspended material to floc together.) The SSF filter itself is a large box. The walls should be as rough as possible to reduce the tendency for water to run down the walls of the filter, by-passing the sand. The bottom layer of the filter is a gravel bed, in which a slotted pipe is placed to drain off the filtered water. The slots or the gravel should be no closer than 20 cm to the walls, again, to prevent the water from by-passing the sand. The sand for a SSP needs to be clean and uniform and of the correct size. The sand can be cleaned in clean running water, even if it is in a creek. The ideal specs on sand are effective size (sieve size through which 10 per cent of the sand passes) between 0.15 and 0.35 mm, uniformity coefficient (ratio of sieve sizes through which 60 per cent pass and through which 10 per cent pass) of less than 3; maximum size of 3 mm and minimum size of 0.1 mm. The sand is added to a SSP to a minimum depth of 0.6 meters. Additional thickness will allow more cleanings before the sand must be replaced. 0.3 to 0.5 meters of extra sand will allow the filters to work for 3–4 years. An improved design uses a geotextile layer on top of the sand to reduce the frequency of cleaning. The outlet of a SSP must be above the sand level and below the water level. The water must be maintained at a constant level to insure an even flow rate throughout the filter. The flow rate can be increased by lowering the outlet pipe or increasing the water level. While the SSF will begin to work at once, optimum treatment for pathogens will take a week or more. During this time the water should be chlorinated, if at all possible (iodine can be substituted). After the filter has stabilised, the water should be safe to drink, but chlorinating of the output is still a good idea, particularly to prevent re-contamination. As the flow rate slows down the filter will have to be cleaned by draining and removing the top few inches of sand. If a geotextile filter is used, only the top ½" may have to be removed. As the filter is re-filled, it will take a few days for the biological processes to re-establish themselves.

Activated Charcoal Filter

Activated charcoal filters water through adsorption; chemicals and some heavy metals are attracted to the surface of the charcoal and are attached to it. Charcoal filters will filter some pathogens, though they will quickly use up the filter adsorptive ability and can even contribute to contamination, as the charcoal provides an excellent breeding ground for bacteria and algae. Some charcoal filters are available impregnated with silver to prevent this, though current research concludes that the bacteria growing on the filter are harmless, even if the water wasn't disinfected before contacting the filter. Activated charcoal can be used in conjunction with chemical treatment. The chemical (iodine or chlorine) will kill the pathogens, while the carbon filter will remove the treatment chemicals. In this case, as the filter reaches its capacity, a distinctive chlorine or iodine taste will be noted. The more activated charcoal in a filter, the longer it will last. The bed of carbon must be deep enough for adequate contact with the water. Production designs use granulated activated charcoal (effective size or 0.6 to 0.9 mm for maximum flow rate). Home or field models can also use a compressed carbon block or powered activated charcoal (effective size 0.01) to increase contact area. Powered charcoal can also be mixed with water and filtered out later. As far as life of the filter is concerned, carbon block filters will last the longest for a given size, simply due to their greater mass of carbon. A source of pressure is usually needed with carbon block filters to achieve a reasonable flow rate.

CHEMICAL TREATMENT

Chlorine

Chlorine is familiar to most people as it is used to treat virtually all municipal water systems. Chlorine has a number of problems when used for field treatment of water. When chlorine reacts with organic material, it attaches itself to nitrogen containing compounds (ammonium ions and amino acids), leaving less free chlorine to continue disinfection. Carcinogenic trihalomethanes are also produced, though this is only a problem with long-term exposure. Trihalomethanes can also be filtered out with a charcoal filter, though it is more efficient to use the same filter to remove organics before the water is chlorinated. Unless free chlorine is measured, disinfection can not be guaranteed with moderate doses of chlorine. One solution is superchlorination, the addition of far more chlorine than is needed. This must again be filtered through activated charcoal to remove the large amounts of chlorine or hydrogen peroxide can be added to drive the chlorine off. Either way there is no residual chlorine left to prevent recontamination. This isn't a problem, if the water is to be used at once.

Chlorine is sensitive to both the pH and temperature of the treated water. Temperature slows the reaction for any chemical treatment, but chlorine treatment is particularly susceptible to variations in the pH as at lower pHs, hypochlorous acid is formed, while at higher pHs, it will tend to dissociate into hydrogen and chlorite ions, which are less effective as a disinfectant. As a result, chlorine effectiveness drops off when the pH is greater than 8. Ordinary household bleach (such as Clorox) in the US contains 5.25 per cent sodium hypochlorite ($NaOCl$) and can be used to purify water if it contains no other active ingredients, scents or colourings. Some small treatment plants in Africa produce their own sodium hypochlorite on-site from the electrolysis of brine. Another system, designed for China, where the suitable raw materials were mined or manufactured locally, used a reaction between salt, manganese dioxide and sulphuric acid to produce chlorine gas. The gas was then allowed to react with slaked lime to produce a bleaching powder that could then be used to treat water. A heat source is required to speed the reaction up. Bleaching powder (or chlorinated lime) is sometimes used at the industrial scale. Bleaching powder is 33–37 per cent chlorine when produced, but losses its chlorine rapidly, particularly when exposed to air, light or moisture.

Calcium hypochlorite, also known as High test hypochlorite (HTH) is supplied in crystal form; it is nearly 70 per cent available chlorine. One product, the sanitiser (formally the Sierra water purifier) uses these crystals to superchlorinate the water to insure pathogens were killed off, then hydrogen peroxide is added to drive off the residual chlorine. This is the most effective method of field chlorine treatment.

IODINE

Iodine was found to be in many ways superior to chlorine for use in treating small batches of water. Iodine is less sensitive to the pH and organic content of water and is effective in lower doses. Some individuals are allergic to iodine and there is some question about long term use of iodine. Cloudy water needs twice as much iodine or twice as much contact time. In cold water (below 41°F or 5°C) the dose or time must also be doubled. In any case doubling the treatment time will allow the use of half as much iodine. These doses are calculated to remove all pathogens (other than cryptosporida) from the water. Of these, giardia cysts are the hardest to kill and are what requires the high level of iodine. If the cysts are filtered out with a microfilter (any model will do since the cysts are 6 μm), only 0.5 ppm is needed to treat the resulting water.

Water treated with iodine can have any objectionable taste removed by treating the water with vitamin C (ascorbic acid), but it must be added after the water has stood for the correct treatment time. Flavoured beverages containing vitamin C will accomplish the same thing. Sodium thiosulphate can also be used to combine with free iodine and either of these chemicals will also help remove the taste of chlorine as well. Usually elemental iodine cannot be tasted below 1 ppm and below 2 ppm the taste is not objectionable. Iodine ions have an even higher taste threshold of 5 ppm. Note that removing the iodine taste does not reduce the dose of iodine ingested by the body.

SILVER

Silver has only be proven to be effective against bacteria and protozoan cysts, though it is quite likely also effective against viruses. Silver can be used in the form of a silver salt, commonly silver nitrate, a colloidal suspension or a bed of metallic silver. Electrolysis can also be used to add metallic silver to a solution. Some evidence has suggested that silver deposited on carbon block filters can kill pathogens without adding as much silver to the water (Fig. 6.2).

POTASSIUM PERMANGANATE

Potassium permanganate is no longer commonly used in the developed world to kill pathogens. It is, much weaker than the other alternatives cited, more expensive and leaves a objectionable pink or brown colour. Still, some underdeveloped countries rely on it, especially in home-use applications. If it must be used, 1 gram per liter would probably be sufficient against bacteria and viruses (no data is available on it effectiveness against protozoan cysts). Hydrogen peroxide can be used to purify water if nothing else is available. Studies have shown of 99 per cent inactivation of poliovirus in 6 hours with 0.3 per cent hydrogen peroxide and a 99 per cent inactivation of rhinovirus with a 1.5 per cent solution in 24 minutes. Hydrogen peroxide is more effective against bacteria, though Fe^{+2} or Cu^{+2} needs to be present as a catalyst to get a reasonable concentration-time product.

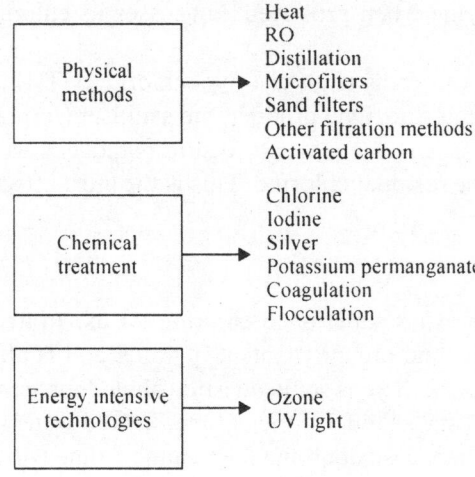

Fig. 6.2. Water treatment technologies for silver.

COAGULATION/FLOCCULATION AGENTS

While flocculation doesn't kill pathogens, it will reduce their levels along with removing particles that could shield the pathogens from chemical or thermal destruction and organic matter that could tie up chlorine added for purification. 60–98 per cent of coliform bacteria, 65–99 per cent of viruses and 60–90 per cent of giardia will be removed from the water, along with organic matter and heavy metals.

Some of the advantages of coagulation/flocculation can be obtained by allowing the particles to settle out of the water with time (sedimentation), but it will take a while for them to do so. Adding coagulation chemicals, such as alum, will increase the rate at which the suspended particles settle out by combining many smaller particles into larger floc, which will settle out faster. The usual dose for alum is 10–30 mg/l of water. This dose must be rapidly mixed with the water, then the water must be agitated for 5 minutes to encourage the particles to form flocs. After this at least 30 minutes of settling time is need for the flocs to fall to the bottom and then the clear water above the flocs may be poured off.

Most of the flocculation agent is removed with the floc, nevertheless, some question the safety of using alum due to the toxicity of the aluminium in it. There is little to no scientific evidence to back this up. Virtually all municipal plants dose the water with alum. In bulk water treatment, the alum, dose can be varied until the idea dose is found. The needed dose varies with the pH of the water and the size of the particles. Increase turbidity makes the flocs easier to produce not harder, due to the increased number of collisions between particles.

ENERGY INTENSIVE TREATMENT TECHNOLOGIES

Ozone

Ozone is used extensively in Europe to purify water. Ozone, a molecule composed of 3 atoms of oxygen rather than two, is formed by exposing air or oxygen to a high voltage electric arc. Ozone is much more effective as a disinfectant than chlorine, but no residual levels of disinfectant exist after ozone turns back into O_2. (One source quotes a half life of only 120 minutes in distilled water at 20°C.) Ozone is expected to see increased use in the US as a way to avoid the production and formation of trihalomethanes and while ozone does break down organic molecules, sometimes this can be a disadvantage as ozone treatment can produce higher levels of smaller molecules that provide an energy source for micro-organisms. If no residual disinfectant is present (as would happen if ozone were used as the only treatment method), these micro-organisms will cause the water quality to deteriorate in storage. Ozone also changes the surface charges of dissolved organics and colloidally suspended particles. This causes microflocculation of the dissolved organics and coagulation of the colloidal particles.

Ultraviolet (UV) Light

Ultraviolet light has been known to kill pathogens for a long time. A low pressure mercury bulb emits between 30 to 90 per cent of its energy at a wave length of 253.7 nm, right in the middle of the UV band. If water is exposed to enough light, pathogens will be killed. The problem is that some pathogens are hundreds of times less sensitive to UV light than others. The least sensitive pathogens to UV are protozoan cysts. Several studies show that *Giardia* will not be destroyed by many commercial UV treatment units. Fortunately, these are the easiest pathogens to filter out with a mechanical filter. The efficiency of UV treatment is very dependent on the turbidity of the water. The more opaque the water is, the less light that will be transmitted through it.

The treatment units must be run at the designed flow rate to insure sufficient exposure, as well as insure turbulent flow rather than plug flow. Another problem with UV treatment is that the damage done to the pathogens with UV light can be reversed if the water is exposed to visible light (specifically 330–500 nm) through a process known as photoreactivation. UV treatment, like ozone or mechanical filtering, leaves no residual component in the water to insure its continued disinfection. Any purchased UV filter should be checked to insure it at least complies with the 1966 HEW standard of 16 mW s/cm^2 with a maximum water depth of 7.5 cm. ANSI/NSF require 38 mW-s/cm^2 for .primary water treatment systems. This level was chosen to give better than 3 log (99.9 per cent) inactivation of *Bacillus subtillis*. This level is of little use against *Giardia* and of no use against *Crypto*.

The US EPA explored UV light for small scale water treatment plants and found it compared unfavourably with chlorine due to: (i) higher costs; (ii) lower reliability; and (iii) lack of a residual disinfectant.

WATER TREATMENT IN GENERAL

Water must have eye appeal and taste appeal before we will drink it with much relish. Instinctively we draw back from the idea of drinking dirty, smelly water. Actually far more important to our well-being is whether or not a water is safe to drink. If it holds disease bacteria, regardless of its clarity and sparkle, we should avoid it. Let's consider these two highly important aspects of water: potability and palatability.

Regardless of any other factors, water piped into the home must be potable. To be potable, it should be completely free of disease organisms. Water is the breeding ground for an almost unbelievably large variety of organisms. Water does not produce these organisms. It merely is an ideal medium in which they can grow. These organisms gain entry into water through a variety of sources. They enter from natural causes, surface drainage and sewage. Many of the organisms in water are harmless. In fact, they are extremely beneficial to man. Others have a wide nuisance value and still others are the source of disease. In general, we are primarily concerned here with organisms which are potential disease-producers. These are of five types: bacteria, protozoa, worms, viruses, fungi. The presence of certain organisms of these various types can lead to such infectious diseases as typhoid fever, dysentery, cholera, jaundice, hepatitis, undulant fever and tularaemia.

There are other diseases as well, which spread through drinking unsafe water. Tremendous strides have been made in the control of these diseases within recent years. Much of the credit must go to sanitary engineers for their careful, consistent control of public water supplies. Biologically, there are two major classifications for our purposes. We can classify water organisms either as members of the plant or animal kingdoms. The following ways are the natural ways, in which water is purified: bacteria and algae consume organic waste; Micro-organisms devour bacteria and algae; oxidation renders organic matter harmless; ultraviolet rays of sun have germicidal effects.

Under the broad heading of plant forms, we can classify the following:

Algae

These organisms are found throughout the world. They constitute the chief group of aquatic plants both in sea and fresh water. Algae range in size from microscopic organisms to giant seaweeds several hundred feet in length. They contain chlorophyll and other pigments which give them a variety of colours. They manufacture their food by photosynthesis. Algae thrive well in stagnant surface water especially during the warm weather. Algae gives water fishy, grassy and other even more objectionable odours. While algae-laden water are repulsive to man, animals will drink them and the presence of blue-green algae has been known to cause the death of cattle drinking this water.

Diatoms

Diatoms belong to the algae family. Some exist as single cells; others are found as groups or colonies. More than 15000 forms of diatoms are known to exist. Diatoms have silica-impregnated cell walls. At time they release essential oils which give water a fishy taste.

Fungi

Fungi are another large group of plant forms. Like the algae, fungi have many varieties included among these are moulds and bacteria. Fungi are not able to manufacture their own food. They exist by feeding on living things or on dead organic matter. Depending on their individual characteristics, they are usually colourless but may vary in this respect.

Moulds

One important category of fungi is moulds. This group of fungi feeds entirely on organic matter. They decompose carbohydrates such as sugars, starches and fats as well as proteins and other substances. They thrive ideally in water that has a temperature range of approximately 80 degrees to 100°F. The presence of moulds is generally a strong indicator of heavy pollution of water.

Bacteria

Bacteria are another important class of fungi. Again numerous smaller groupings are possible. Among the higher organisms in this group are the iron, manganese and sulphur bacteria. These higher bacteria gain their energy from the oxidation of simple organic substances. Lower forms of bacteria can be grouped as those that are helpful and those that are harmful to man. Those harmful to man are mainly the disease-producing organisms. Helpful organisms hasten the process of decomposing organic matter and by feeding on waste material; they aid in the purifying of water. All bacteria are sensitive to the temperature and pH of water. Some bacteria can tolerate acid water. But for the most part, they thrive best in water that have a pH between 6.5 to 7.5, that is essentially neutral water. As to temperature, most pathogenic or disease bacteria thrive best in water of body temperature. Beyond this no hard and fast statements can be made. Some bacteria are more resistant to heat than are others. Some are more sensitive to cold. At low temperatures, for example, some bacteria may become dormant for long periods of time but will still continue to exist. Interestingly enough, the waste products of their own growth can hamper bacteria and may even prove toxic to them.

Animal forms like plant life thrive in water providing conditions are right. Among the higher forms of animal life found in water are fish, amphibians (turtles and frogs), mollusks (snails and shellfish) and anthropoids (lobsters, crabs, water insects, water mites and others). Concern here is with those lower forms of animal life in water. Again, some are helpful to man as scavengers; others are injurious as possible sources of infection. Principal ones of concern are as follows:

Worms

There are three types of worms found in water. For the most part, they dwell in the bed of the material at the bottom of lakes and streams. There they do important work as scavengers. The rotifiers are the only organisms in this category at or near the surface. They live primarily in stagnant fresh water. The eggs and larvae of various intestinal worms found in man and warm-blooded animals pollute the water at times. They do not generally cause widespread infection for several reasons. They are relatively few in number and are so large they can be filtered out of water with comparative ease.

Protozoa

Another basic classification in the animal kingdom is that group of microscopic animals known as protozoa. These one-celled organisms live mainly in water either at or near the surface or at great depths in the oceans. Many live as parasites in the bodies of man and animals. Sometimes, drinking water becomes infested with certain protozoa which are not disease-producing. When present, they give the water a fishy taste and odour. Some protozoa are aerobic, that is, they exist only where free oxygen is available. Some exist where no free oxygen is available. Others can either be aerobic or anaerobic.

Nematodes

Nematodes belong to the worm family. They have long, cylindrical bodies which have no internal segments. Interestingly enough, those nematodes which are found in the bodies of men and warm-blooded animals are large enough to be visible to the naked eye; those living in fresh water or the soil are microscopic. Nematodes can be a problem in drinking water because they impart objectionable tastes and odours to water. They are also under suspicion of being carriers of the type of disease-bearing bacteria found in the intestines of warm-blooded animals though studies show that possibility is somewhat remote. Nematodes are apt to be found in municipal water derived from surface supplies.

Viruses

As yet not too well understood is that group of parasitic forms known as viruses. Too small to be seen under a microscope, viruses are capable of causing disease in both plants and animals. Viruses can pass through porcelain filters that are capable of screening out bacteria: At least one virus that produces infectious hepatitis is water born. Drinking water contaminated with this virus is hazardous.

As the reader can see from even this brief summary, there is a tremendous variety of living organisms in water. To understand and classify the countless varieties requires an immense amount of knowledge and time. These organisms, whether plant or animal forms, are pathogenic or disease-producing; they make water unsafe to drink. For obvious reasons, even where there is just a possibility that water contains pathogenic organisms, it must be considered contaminated.

While there is a large and varied number of pathogens, no single contaminated water supply is apt to contain more than a few of these countless varieties. On one hand this is fortunate but, at the same time it makes detection of pathogens extremely difficult in terms of a routine water analysis. Since both speed and accuracy are essential, laboratory scientists need a sure way to expedite detection of pathogens. They have a dependable answer in a group of readily identified organisms that indicate possible contamination. These indicator organisms are the coliform bacteria.

Study has proved that these coliform bacteria indicate the presence of human or animal wastes in water. Coliform bacteria naturally exist in the intestines of human and certain animals. Thus, the presence of these bacteria in water is accepted proof that the water has been contaminated by human or animal wastes. Although such water may contain no pathogens, an infected person, animal or a carrier of disease, could add pathogens at any moment. Thus, immediate corrective action must be taken. The presence of coliform bacteria shows water is contaminated by human wastes and is potentially contaminated with pathogens. In short, these bacteria become a measure of guilt by association. The other side is this: the mere absence of coliform bacteria does not assure there are no pathogens. However, this is considered unlikely. Just how can water be tested for the presence of coliform bacteria? These organisms cause the fermentation of lactose (the crystalline sugar compound in milk). When water containing coliform bacteria is placed in a lactose culture, it will cause fermentation resulting in the form of gas. This confirms the suspicions. The maximum acceptable concentration (MAC) for coliforms in drinking water is zero organisms detectable per 100 ml. Recognising the danger, what can be done to provide adequate protection against contamination? When a water supply becomes contaminated, correct the problem at once. This means going beyond treatment alone-important as this may be. It is a basic rule of water sanitation to get to the source of the problem and eliminate it. If a well, for example, becomes badly contaminated, it is necessary to trace the contamination to its source and if possible, remedy the situation. It may even be necessary to seek out a new source of supply.

WATER DISINFECTION ONE MORE TIME

Treatment of a water supply is a safety factor, not a corrective measure. There are a number of ways of purifying water. In evaluating the methods of treatment available, the following points regarding water disinfectants should be considered:

1. A disinfectant should be able to destroy all types of pathogens and in whatever number present in water.
2. A disinfectant should destroy the pathogens within the time available for disinfection.
3. A disinfectant should function properly regardless of any fluctuations in the composition or condition of the water.
4. A disinfectant should not cause the water to become toxic or unpalatable.
5. A disinfectant should function within the temperature range of the water.
6. A disinfectant should be safe and easy to handle.
7. A disinfectant should be such that it is easy to determine its concentration in the water.
8. A disinfectant should provide residual protection against re-contamination.

Techniques such as filtration may remove infectious organisms from water. They are, however, no substitute for disinfection.

The following are the general methods used for disinfecting water:

Boiling

This involves bringing the water to its boiling point in a container over heat. The water must be maintained at this temperature 15 to 20 minutes. This will disinfect the water. Boiling water is an effective method of treatment because no important waterborne diseases are caused by heat-resisting organisms.

Ultraviolet Light

The use of ultraviolet light is an attempt to imitate nature. As you recall, sunlight destroys some bacteria in the natural purification of water. Exposing water to ultraviolet light destroys pathogens. To assure thorough treatment, the water must be free of turbidity and colour. Otherwise, some bacteria will be protected from the germ-killing ultraviolet rays. Since ultraviolet light adds nothing to the water, there is little possibility of its creating taste or odour problems. On the other hand, ultraviolet light treatment has no residual effect. Further, it must be closely checked to assure that sufficient ultraviolet energy is reaching the point of application at all times.

Use of Chemical Disinfectants

The most common method of treating water for contamination is to use one of various chemical agents available. Among these are chlorine, bromine, iodine, potassium permanganate, copper and silver ions, alkalis, acids and ozone. Bromine is an oxidising agent that has been used quite successfully in the disinfecting of swimming pool water. It is rated as a good germicidal agent. Bromine is easy to feed into water and is not hazardous to store. It apparently does not cause eye irritation among swimmers nor are its odours troublesome.

One of the most widely used disinfecting agents to ensure safe drinking water is chlorine. Chlorine in cylinders is used extensively by municipalities in purification work. However, in this form chlorine gas (Cl_2) is far too dangerous for any home purpose. For use in the home, chlorine is readily available as sodium hypochlorite (household bleach) which can be used both for laundering or disinfecting purposes. This product contains a 5.25 per cent solution of sodium hypochlorite which is equivalent to 5 per cent available chlorine. Chlorine is also available as calcium hypochlorite which is sold in the form of dry granules. In this form, it is usually 70 per cent available chlorine. When calcium hypochlorite is used, this chlorinated lime should be mixed thoroughly and allowed to settle, pumping only the clear solution. For a variety of reasons not the least of which is convenience, chlorine in the liquid form (sodium hypochlorite) is more popular for household use. Chlorine is normally fed into water with the aid of a chemical feed pump. The first chlorine fed into the water is likely to be consumed in the oxidation of any iron, manganese or hydrogen sulphide that may be present. Some of the chlorine is also neutralised by organic matter normally present in any supply, including bacteria, if present. When the 'chlorine demand' due to these materials has been satisfied, what's left over the chlorine that has not been consumed remains as 'chlorine residual'. The rate of feed is normally adjusted with a chemical feed pump to provide a chlorine residual of 0.5–1.0 ppm after 20 minutes of contact time. This is enough to kill coliform bacteria but may or may not kill any viruses or cysts which may be present. Such a chlorine residual not only serves to overcome intermittent trace contamination from coliform bacteria but, also provides for minor variations in the chlorine demand of the water. The pathogens causing such diseases as typhoid fever, cholera and dysentery succumb most easily to chlorine treatment. The cyst-like protozoa causing dysentery are most resistant to chlorine. As yet, little is known about viruses, but some authorities place them at neither extreme in resistance to chlorination.

There are three basic terms used in the chlorination process: chlorine demand, chlorine dosage and chlorine residual. Chlorine demand is the amount of chlorine which will reduced or consumed in the process of oxidising impurities in the water. Chlorine dosage is the amount of chlorine fed into the water. Chlorine residual is the amount of chlorine still remaining in water after oxidation takes place. For example, if a water has 2.0 ppm chlorine demand and is fed into the water in a chlorine dosage of 5.0 ppm, the chlorine residual would be 3.0 ppm.

For emergency purposes, iodine may be used for treatment of drinking water. Much work at present is being done to test the effect of iodine in destroying viruses which are now considered among the pathogens most resistant to treatment. Tests show that 20 minutes exposure to 8.0 ppm of iodine is adequate to render a potable water. As usual, the residual required varies inversely with contact time. Lower residuals require longer contact time while higher residuals require shorter contact time. While such test results are encouraging, not enough is yet known about the physiological effects of iodine-treated water on the human system. For this reason, its use must be considered only on an emergency basis.

Silver in various forms has been used to destroy pathogens. It can be added to the water as a liquid or through electrolytic decomposition of metallic silver. It has also been fed into water through an absorption process from silver-coated filters. Various household systems have been designed to yield water with a predetermined silver concentration. However, fluctuations in the flow rate often result in wide variations in the amount of silver in the water. In minute concentrations, silver can be highly destructive in wiping out disease-bearing bacteria. While long contact time is essential, silver possesses residual effect that can last for days. Silver does not produce offensive tastes or odours when used in water treatment. Further organic matter does not interfere with its power to kill bacteria as in the case with free chlorine. Its high cost and the need for long periods of exposure have hindered its widespread acceptance. Copper ions are used quite frequently to destroy algae in surface water but these ions are relatively ineffective in killing bacteria.

Disease-bearing organisms are strongly affected by the pH of a water. They will not survive when water is either highly acidic or highly alkaline. Thus, treatment which sharply reduces or increases pH in relation to the normal range of 6.5 to 7.5 can be an effective means of destroying organisms.

There are numerous other agents which have proved to be successful in destroying pathogens. Many of these must still be subjected to prolonged testing with regard to their physiological effect on man. Among these are certain surfactants and chlorine dioxide. There are several types of surfactants which aid in destroying pathogens. The cationic detergents readily kill pathogens. Anionic detergents are only weakly effective in destroying pathogens. Surfactants have not been seriously considered for treating drinking water because of their objectionable flavour and possible toxic effects. Chlorine dioxide has unusually good germ killing power. Up to the present time, no valid tests for its use have been developed because of the lack of means for determining low residual concentrations of this agent. It is such a strong oxidising agent, a larger residual of chlorine dioxide would probably be needed than is the case with chlorine. At present, chlorination in one form or another is regarded as the most effective disinfectant available for all general purposes. It has full acceptance of health authorities. Still there are certain factors which affect its ability to disinfect water. These should always be kept in mind. They are:

1. 'Free' chlorine residuals are more effective than 'combined' or 'chloramine' residuals. Disinfection regardless of the type of chlorine becomes more effective with increased residuals. Chloramine is the compound formed by feeding both chlorine and ammonia to the water. This, treatment has been used for controlling bacteria growth in long pipelines and in other appliances where its slower oxidising action is of particular benefit.

2. A pH of 6.0 to 7.0 makes water a far more effective medium for chlorine as a disinfecting agent than do higher pH values of around 9.0 to 10.0.
3. The effectiveness of chlorine residuals increase with higher temperatures within the normal water temperature range.
4. The effectiveness of disinfection increases with the amount of contact time available.
5. All types of organisms do not react in the same way under various conditions to chlorination.
6. An increase in the chlorine demand of a water increases the amount of chlorine necessary to provide a satisfactory chlorine residual.

In order to ensure the destruction of pathogens, the process of chlorination must achieve certain control of at least one factor and preferably two, to compensate for fluctuations that occur. For this reason, some authorities on the subject stress the fact that the type and concentration of the chlorine residual must be controlled to ensure adequate disinfection. Only this way, they claim, can chlorination adequately take into account variations in temperature, pH, chlorine demand and types of organisms in the water. While possible to increase minimum contact times, it is difficult to do so. Five to ten minutes is normally all the time available with the type of pressure systems normally used for small water supplies. Many experts feel that satisfactory chlorine residual alone can provide adequate control for disinfection. In their opinion, superchlorination-dechlorination does the best job. Briefly, what is this technique and how does it operate?

The success of superchlorination-dechlorination system depends on putting enough chlorine in the water to provide a residual of 3.0 to 5.0 ppm. This is considerably greater than chlorine residual of 0.1 to 0.5 ppm usually found in municipal water supplies when drawn from the tap. A superchlorination-dechlorination systems consists of two basic units. A chlorinator feeds chlorine into raw water. This chlorine feed is stepped up to provide the needed residual. A dechlorinator unit then removes the excess chlorine from the water before it reaches the household taps. The chlorinator should be installed so that it feeds the chlorine into the water before it reaches the pressure tank. A general purpose chemical feed pump will do the job. The size and the placement of the dechlorinator unit depends on the type of treatment necessary. This will usually be an activated carbon filter. If pathogen kill is all that is required, a small dechlorinator can be installed at the kitchen sink. This unit then serves to remove chlorine from water used for drinking and cooking. The advantage in dechlorinating only a part of the water is obvious. A smaller filter unit does the job and since only a small portion of the total water is filtered under such conditions, the unit lasts longer before either servicing or replacement is necessary. Essentially dechlorination is not needed to ensure a safe drinking water. Once the water is chlorinated, the health hazard is gone. The chlorine residual is removed merely to make the water palatable. If the problem is compounded due to the presence or iron and/or manganese, all the water should be filtered. Under such conditions, a larger central filter is necessary and should be placed on the main line after the pressure tank. The prime advantage of the superchlorination-dechlorination process is that it saturates water with enough chlorine to kill bacteria. Simple chlorination sometimes fails of its objective because homeowners may set the chlorine feed rate too low in order to avoid giving their water a chlorine taste.

Sodium Dichloroisocyanurate

Sodium dichloroisocyanurate can sterilise drinking water, swimming pool, tableware and air or be used for fighting against infectious diseases as routine disinfection, preventive tableware and environmental sterilisation in different places or act as disinfectant in raising silkworm, livestock, poultry and fish. It can also be used to prevent wool from shrinkage, bleach the textile and clean the industrial circulating water. The product has high efficiency and constant performance with no harm to human beings. It enjoys goods reputation both at home and abroad. Table 6.2 summarises some of this chemical's properties.

Table 6.2. Properties of sodium dichloroisocyanurate.

Formula	$C_3O_3\ N_3Cl_3Na$
Physico-chemical properties	White crystalline powder, granular or tablets
Specifications	Powder or granular
Available chlorine	56 to 60%
pH value	5.5–7.0
Packing	25 or 50 kg plastic drums

Trichloroisocyanuric Acid

With strong bleaching and disinfection effects, trichloroisocyanuric acid is widely used as high effective disinfectant for civil sanitation, animal husbandry and plant protection as bleaching agent of cotton, gunny, chemical fabrics or as shrink-proof agent for woolens, battery materials, organic synthesis industry and dry-bleaching agent of clothes. Table 6.3 provides some general properties.

Table 6.3. Properties of trichloroisocyanuric acid.

Formula	$C_3O_3\ N_3Cl_3$			
Physico-chemical properties	White crystal powder, granular or tablets, with stimulant smell of hypochloric acid, slightly soluble in water, easily soluble in acetone			
Specifications	Powder	Granular	Tablet (200g)	Tablet (20g)
Available chlorine	90% minimum			
Moisture	0.5% maximum			
pH value (1% WS)	2.7–2.9			
Packing	25 or 50 kg plastic drums			

Isocyanuric Acid

Cyanuric acid is widely used for the stabilisation of available chlorine swimming pool water treatment. It is also the starting compound for the synthesis of many organic derivatives. Table 6.4 provides some general properties.

Table 6.4. Properties of isocyanuric acid.

Formula	$C_3H_3N_3O_3$	
Physico-chemical properties	White crystalline solid powder or granular, non-toxic and odourless	
Specifications	Powder	Granular
Cyanuric acid	98.5% minimum	98% minimum
Moisture	0.4% maximum	0.5% maximum

(Contd...)

Particle	0.3 mm maximum 90% through	0.6–2 mm 90% through
pH value (1% water solution)	4.0–4.6	4.0–4.6
Melting point (centigrade)	330 minimum	330 minimum
Fe	25 ppm maximum	25 ppm maximum
Packing	Woven bags	Fiber drums

Discussions thus far have focused on pathogens and methods of destroying them in the process of making water potable-safe to drink. This is highly important but it's not the whole story; for water must be palatable as well as potable. The obvious question to ask is—What makes a water palatable?

TASTES

To be palatable, a water should be free of detectable tastes and odours. Immediately, we come to a stumbling block. What constitutes a detectable taste or odour? Undoubtedly when you have travelled around the country, you have tasted water which must have had unpleasant tastes or odours. Natives in the area may be surprised to note your reaction for after drinking the water for many years, they find nothing peculiar to either the taste or odour of the water. Then, there are those water which have tastes and odours so obnoxious (hydrogen sulphide water, for example), even the long time inhabitant cannot stomach them. Turbidity, sediment and colour play important roles in determining whether a water is a delight to drink. Various odours and tastes may be present in water. They can be traced to many conditions. Unfortunately, the causes of bad taste and odour problems in water are so many, it is impossible to suggest a single treatment that would be universally effective in controlling these problems. Tastes are generally classified in four groups—sour, salt, sweet and bitter.

Classification of Tastes

Odours possess many classifications. There are 20 of them commonly used, all possessing rather picturesque names. In fact the names, in many cases, are far more pleasant than the odours themselves. To name a few of them—nasturtium, cucumber, geranium, fishy, pigpen, earthy, grassy and musty. Authorities further classify these odours in terms of their intensity from very faint, faint distinct and decided to very strong. Now your taste buds and olfactory organs are not necessarily of the same acuteness as your neighbours. So there may be some disagreement on the subject. Generally you or your neighbour should not be made aware of any tastes or odours in water if there is to be pleasure in drinking it. If you are conscious of a distinct odour, without specifically seeking for such, the water is in need of treatment. In many cases, it is difficult to detect what constitutes a taste or an odour. The reason is obvious. Both the taste buds and olfactory organs work so effectively as a team, it is hard to realise where one leaves off and the other begins. To illustrate: hydrogen sulphide gives water an 'awful' taste yet actually it is this gas's unpleasant odour that we detect rather than an unpleasant taste. Unfortunately, there is little in the way of standard measuring equipment for rating tastes and odours. Tastes and odours in water can be traced to at least six factors. They are:
1. Decaying organic matter.
2. Living organism.
3. Iron, manganese and the metallic product or corrosion.
4. Industrial waste pollution from substances such as phenol.
5. Chlorination.
6. High mineral concentrations.

In general, odours can be traced to living organisms, organic matter and gases in water. Likewise, tastes can be traced generally to the high total minerals in water. There are some tastes due to various algae and industrial wastes. Some tastes and odours, especially those due to organic substances, can be removed from water simply by passing it through an activated carbon filter. Other tastes and odours may respond to oxidising agents such as chlorine and potassium permanganate. Where these problems are due to industrial wastes and certain other substances, some of the above types of treatment may completely fail. In some cases, for example, chlorination may actually intensify a taste or odour problem. Potassium permanganate has been found to be extremely effective in removing many musty, fishy, grassy and moldy odours. Two factors make this compound valuable—it is a strong oxidising agent and it does not form obnoxious compounds with organic matter. However, a filter must be used to remove manganese dioxide formed when permanganate is reduced.

Turbidity and suspended matter are not synonymous terms although most of us use the terms more or less interchangeably. Correctly speaking, suspended matter is that material which can be removed from water through filtration or the coagulation process. Turbidity is a measure of the amount of light absorbed by water because of the suspended matter in the water. There is also some danger of confusion regarding turbidity and colour. Turbidity is the lack of clarity or brilliance in a water. Water may have a great deal of colour-it may even be dark brown and still be clear without suspended matter. The current method of choice for turbidity measurement in Canada is the nephelometric method; the unit of turbidity measured using this method is the nephelometric unit (NTU). Turbidity in excess of 5 NTU becomes apparent and may be objected to by a majority of consumers. Therefore an Aesthetic Objective (AO) of < = 5 NTU has been set for water at the point of consumption. The suspended particles clouding the water may be due to such inorganic substances as clay, rock flour, silt, calcium carbonate, silica, iron, manganese, sulphur or industrial wastes. Again the clouding may be due to a single foreign substance in water, chances are it is probably due to a mixture of several or many substances. These particles may range in size from fine colloidal materials to course grains of sand that remain in suspension only as long as the water is agitated. Those particles which quickly sink to the bottom are usually called, 'sediment'. There are no hard and fast rules for classifying such impurities.

If you take water from a swiftly flowing river or stream, you generally find that it contains a considerable amount of sediment. In contrast, you find that water taken from a lake or pond is usually much clearer. In these more quiet, non-flowing water, there is greater opportunity for settling action. Thus all but very fine particles sink to the bottom. Least apt to contain sediment are wells and springs. Sediment is generally strained from these water as they percolate through sand, gravel and rock formations. Turbidity varies tremendously even within these various groupings. Some rivers and streams have water that appears crystal clear with just trace amounts of turbidity in them especially at points near their sources. These same moving water may contain upwards of 30000 ppm of turbidity at other points in their course to the oceans. In fact, turbidity in amounts well over 60000 ppm have been registered. Again there are significant fluctuations in the amount of turbidity in a river at different times in a year. Heavy rainfalls, strong winds and convection currents can greatly increase the turbid state of both lakes and rivers. Warm weather and increases in the temperature can also add to the problem. For with warm weather, micro-organisms and aquatic plants renew their activity in the water. As they grow and later decay, these plant and animal forms substantially add to the turbid state of a water. Also, they frequently cause a heightening of taste, odour and colour problems.

Mechanical filtration will remove all forms of turbidity. Of course, the smaller the turbid particles, the finer the filter openings must be in order to strain them out. Under some circumstances, the openings may have to be so small that they cause an excessive pressure drop as the water creeps through the filter and the unit may be impractical. In many cases, filters containing specially graded and sized gravel and sand are effective in screening out turbid particles. With such units, a periodic backwashing to remove the filtered material is all the maintenance necessary. As discussed in later chapters, the use of filter aids is necessary in treating many water sources. A filter aid is a chemical that is added onto the top of the filter bed immediately after backwashing. The filter aid traps fine dirt particles producing a more a sparkling clear water and keeps dirt from penetrating the filter bed, insuring better bed cleansing during backwashing. In some cases, cartridge filters are effective.

Municipal and industrial systems frequently make use of the coagulation process to aid in the removal of turbidity. In this economical process, a coagulating agent such as aluminium sulphate is fed into the water. After rapid mixing, the coagulating agent forms a 'floc' generally in the form of a gelatinous precipitate. This floc gives the appearance of a soft, gentle snowfall. A settling period is then needed to allow the floc to fall gently through the water. As the floc forms and settles, it tends to collect or entrap the turbid particles and form them into larger particles which sink to the bottom.

On large installations, huge settling basins provide the necessary time and space for the process. After the settling period, the water flows through a filter to remove the last traces of the coagulant and any remaining turbid particles. An additional water quality parameter of importance is colour. Ordinarily we think of water as being blue in colour. When artists paint bodies of water, they generally colour them blue or blue-green. While water does reflect blue-green light, noticeable in great depths, it should appear colourless as used in the home. Ideally, water from the tap is not blue or blue-green. If such is the case, there are certain foreign substances in the water. Among these substances: Infinitely small microscopic particles add colour to water. Colloidal suspensions and non-colloidal organic acids as well as neutral salts also affect the colour of water.

The colour in water is primarily of vegetable origin and is extracted from leaves and aquatic plants. Naturally, water draining from swamps has the most intense colouring. The bleaching action of sunlight plus the ageing of water gradually dissipates this colour, however.

All surface water possess some degree of colour. Like some shallow wells, springs and an occasional deep well can contain noticeable colouring.

In general, water from deep wells is practically colourless. An arbitrary standard scale has been developed for measuring colour intensity in water samples. When a water is rated as having a colour of five units, it means: the colour of this water is equal in intensity to the colour of distilled water containing 5 milligrams of platinum as potassium chloroplatinate per liter. Highly coloured water is objectionable for most process work in the industrial field because excessive colour causes stains. While colour is not a factor of great concern in relation to household applications, excessive colour lacks appeal from an aesthetic standpoint in a potable water. Further, it can cause staining. The Aesthetic Objective (AO) for colour in drinking water is $< = 15$ true colour units.

The provision of treated water at or below the AO will encourage rapid notification by consumers should problems leading to the formation of colour arise in the distribution system. In general, colour is reduced or removed from water through the use of coagulation, settling and filtration techniques. Aluminium sulphate is the most widely used coagulant for this purpose. Superchlorination, activated carbon filters and potassium permanganate have been used with varying degrees of success in removing colour. Table 6.5 summarises water treatment methods currently used.

Table 6.5. Waste-water treatment methods.

Objective of treatment	Method or technology
	Odour
Rotten egg smell	Manganese green-sand filter up to 6 ppm H_2S with pH not lower than 6.7
	Over 6 ppm H_2S constant chlorination by filtration/dechlorination
	Open aeration followed by oxidising, catalyst filter
	The water should be tested at the source for H_2S determination as the gas escapes rapidly
Petroleum	Locate and eliminate seepage. Activated carbon will adsorb oil and gasoline (most hydrocarbons) on a short term basis. Air-strip with (40:1 air/water ratio) followed by 2 ft^3 carbon units in series
Aromatic, fishy, earthy or woody smell	Activated carbon type filter or
	Cartridge-activated carbon filter for drinking and cooking
Sharp metallic smell	Water softener can remove 0.5 ppm or iron (Fe) for every grain/gal. of hardness up to 10 ppm at minimum pH of 6.7 (unaerated water)
	Over 10 ppm Fe: chlorination with sufficient retention tank time for full oxidation followed by filtration and dechlorination
	Pressure aeration plus filtration for up to 20 ppm Fe
	Appearance
Rust	Up to 10 ppm iron removed by manganese greensand filter if pH is 6.7 or higher
	Manganese-treated pumicite catalyst filter if pH is 6.8 or higher and oxygen is 15 % of total iron content
	Downflow water softened with good backwash, up to 10 ppm, use calcite filter followed be downflow water softener
Black staining	Manganese greensand or manganese zeolite-type catalyst-filter to limit of 6 ppm or 15 ppm, respectively (combined Fe and Mn), with pH not lower than 6.7 value
	Process used for iron removal usually will handle manganese
	Manganese punicite medium catalyst-filter with ultra-filtration-type membrane element
	For whole-house system, remove by absorption via special macroporous Type 1 anion exchange resin regenerated with NaCl up to 3 ppm
	Above 3 ppm, constant chlorination with full retention time, followed by filtration and/or dechlorination
Gelatinous slime	Destroy iron bacteria with a solution of hydrochloric acid, then constant chlorination, followed by activated carbon filtration or calcite filter.
	Potassium permanganate chemical feed followed by MnZ/anthracite filter
Hydrocarbon sheen	Same as Petroleum
Murky	For mud, clay, sediment—use a calcite or pumicite filter, up to 50 ppm
	For sand, grit or clay—use a hydrocyclone, sand trap and/or install new well screen
	Taste
Salinity	There is no commercial residential treatment for sodium over 1800 ppm
	Deionise drinking water only with disposable mixed bed-anion/cation resin
	Reverse osmosis for drinking and cooking water only

(Contd...)

Objective of treatment	Method or technology
	Home distillation system for drinking water.
Medicine	Single faucet activated carbon filter or whole-house tank-type activated absorption filter
Chemical tastes (other)	Pesticides-herbicides: Activated carbon filter will absorb limited amount. Must continue to monitor the product water closely

SOME GENERAL COMMENTS

So there we have it—a broad overview of a complex subject that spans both technical and legal arenas. Much of the discussions have focused on drinking water, but from this point forward we will depart from the subject and only address this in passing. Recognise that there are a large number of technologies that are applied to treating water. The combination of technologies needed for a water treatment application depend on what we are ultimately trying to achieve in terms of final water quality.

Although the term pollution control has fallen out of favour today and what has become fashionable is Pollution Prevention, the fact remains that what we are doing is removing unwanted contaminants from water, whether it be to meet drinking water purposes or to meet a discharge standard to a local (non-potable) water body. The contaminants may be caused by man or they simply exist from nature. Either way, we are applying technologies aimed at removing these constituents and ultimately these concentrated forms of pollutants require disposal. In this regard, physical methods alone are quite limited, because they represent a non-destructive form of treatment. Their objective is both to remove suspended contaminants and to concentrate them within the limitations of the technology or hardware. From that point on, further concentration is required in order reconstitute the collected contaminants in a form that can be readily handled for ultimate disposal and or destruction. This is known as dewatering. But as noted above, water often contains much more than just suspended matter.

For newcomers to this subject, there is a section of general questions for thinking and discussing among your colleagues. These will help reinforce some of the general concepts and principles covered in this first chapter, and help you to prepare for the more technical discussions that follow.

List of Abbreviations Used in this Chapter:

Bats	Best available technologies
CCL	Contaminant candidate list
CWA	Clean Water Act
DNAPLs	Dense nonaqueous phase liquids
DWEL	Drinking water equivalent level
GAC	Granular activated carbon
MCLG	Maximum contaminant levels
NDWAC	National Drinking Water Advisory Council
NPDWR	National Primary Drinking Water Regulations
NSDWRs	National Secondary Drinking Water Regulations
PAC	Powdered Activated Carbon
RFD	Reference dose
RO	Reverse osmosis
SCADA	Supervisory Control and Data Acquisition
SDWA	Safe Drinking Water Act
TT	Treatment techniques

USEPA   United States Environmental Protection Agency
UV   Ultra violet

Example 6.1: If the per capita contribution of suspended solids and BOD is 100 gm and 60 gm, find the population equivalents of:

(i) A combined system serving 12000 persons and having 80 gm of per capita daily BOD; and (ii) 2,00,000 liters daily of industrial waste-water containing 1500 mg/l of suspended solids.

Solution:

(i) In the first case, population equivalent is based on BOD criterion.

Total daily BOD load = $12000 \times 80 = 96 \times 10^4$ gram.

Standard per captia BOD = 60 gram.

Therefore,   population equivalent $= \dfrac{96 \times 10^4}{60} = 16000$

(ii) In the second case, population equivalent is based on the criterion of suspended solids.

Total daily load of suspended solids $= 2,00,000 \times 1500 \times 10^{-3}$

$= 3,00,000$ gram.

Standard per capita suspended solids = 100 gram.

Therefore,   population equivalent $= \dfrac{3,00,000}{100} = 3000$

Example 6.2: A 500 ml aqueous salt solution has 125 mg of salt dissolved in it. Express the concentration of this solution in terms of: (i) mg/l; (ii) ppm; (iii) gpg; (iv) per cent; and (v) lb/mil gal.

Solution:

(i) 125 mg/500 ml $\times$ 1000 ml/l = 250 mg/l.

(ii) 250 mg/l = 250 ppm.

(iii) 250 mg/l $\times$ 1 gpg/17.1 mg/l = 14.6 gpg.

(iv) The fact that 500 ml of water has a mass of 500 gram gives

per cent = 0.125 g/500 g $\times$ 100 = 0.025 per cent

Or divide 250 mg/l by 10000 to get 0.025 per cent.

(v) 250 mg/l $\times$ 8.34 = 20.90 lb/mil gal.

Example 6.3: How many pounds of chlorine gas should be dissolved in 8 mil gal of water to result in a concentration of 0.2 mg/l?

Solution:

0.2 mg/l $\times$ 8.34 = 1.67 lb/mil gal

and

1.67 lb/mil gal $\times$ 8 mil gal $\approx$ 13 lb.

Chapter 7

Introduction to Separation Processes and Mass Transfer

INTRODUCTION

A variety of water treatment processes involve the transfer of material from one phase to another to bring about treatment. Typical examples include aeration (e.g., the transfer of oxygen from the atmosphere to water), adsorption (e.g., the transfer of dissolved organic constituents in water to a solid material) and sedimentation (e.g., the removal of particulate matter from a liquid to form a solid precipitate). The purpose of this chapter is to introduce the principles of transferring mass from one phase to another, commonly known as separation processes. Topics discussed in this chapter include: (i) an introduction to separation processes used in water treatment; (ii) an introduction to mass transfer and diffusion; (iii) calculation of diffusion coefficients; (iv) models for mass transfer at an interface; (v) correlations for mass transfer at an interface; and (vi) two-film theory for mass transfer across an interface.

SEPARATION PROCESSES USED IN WATER TREATMENT

The removal of dissolved gases and volatile contaminants, the dissolution of gas in aqueous solutions, the adsorption of chemical species in activated carbon and the precipitation of calcium carbonate in pipes and channels are examples of the transfer of material from one phase to another. Such mass transfer operations, classified as separation processes, are commonly employed in water treatment. Common separation processes used for water treatment are presented in Table 7.1. To introduce the subject of separation processes, the following topics are discussed in this section: (i) separation process terminology; (ii) separation mechanisms; (iii) the minimum energy required for separation; (iv) solid-liquid operations; (v) gas-liquid operations; (vi) contact modes for describing phase equilibrium and mass transfer; and (vii) the minimum amount of extracting phase needed for separation.

Separation Process Terminology

The terminology utilised in this chapter to describe separation processes and mass transfer is presented in Table 7.2. Much of the terminology used to describe separation processes is derived from the field of chemical engineering.

Table 7.1. Separation process used in water treatment and other environmental systems.

Term	Phase	Process (es)
Absorption	Gas $\longrightarrow$ liquid	Aeration, O_2 transfer, SO_2 scrubbing, chlorination, chlorine dioxide and ammonia addition, ozonation
Gas stripping	Liquid $\longrightarrow$ gas	H_2S removal from groundwater, removal of NH_3 and volatile organic chemicals, soil-vapour extraction
Evaporation	Liquid $\longrightarrow$ gas	Drying of sludges
Leaching	Solid $\longrightarrow$ liquid	Landfill leachate, dissolution of non-aqueous phase liquids (NAPLs) in groundwater
Adsorption	Gas $\longrightarrow$ solid Liquid $\longrightarrow$ solid	Inorganics and organics removal using activated carbon and activated alumina
Desorption	Solid $\longrightarrow$ liquid Solid $\longrightarrow$ gas	Reactivation of activated carbon, release of toxics from sediments
Sublimation	Solid $\longrightarrow$ gas	Volatile organic chemical release
Ion exchange	Liquid $\longrightarrow$ solid	Demineralisation of water
Humidification	Liquid $\longrightarrow$ gas	Evaporative loss of heat
Precipitation	Liquid $\longrightarrow$ solid	Softening and dissolved species removal
Membrane filtration	Liquid $\longrightarrow$ solid	Particle removal
Reverse osmosis	Liquid $\longrightarrow$ liquid	Dissolved species removal
Sedimentation, flotation	Liquid $\longrightarrow$ solid	Particle removal
Filtration	Liquid $\longrightarrow$ solid	Particle removal
Sludge dewatering, thickening	Liquid $\longrightarrow$ liquid	Removal of water from sludges

Separation Mechanisms

A variety of separation mechanisms are involved in the processes listed in Table 7.1. Precipitation occurs when the concentration of soluble compounds exceeds the solubility limit. Sedimentation occurs in response to differences in density between particles and a liquid and membrane filtration occurs by the physical straining of particles. In the separation processes considered in this chapter, the driving force for mass transfer is a concentration gradient. When a concentration gradient is present between two phases in contact with each other or between two locations within a single phase, matter will flow from the region of higher concentration to the region of lower concentration at a rate that is proportional to the difference between the two concentrations, as given by the following equation:

$$J_A = k_f \ (\Delta C_A) \qquad \qquad ... (7.1)$$

where,

J_A = mass flux of component A, g/m^2·s

k_f = fluid-phase (liquid or gas) mass transfer coefficient, m/s

ΔC_A = difference in concentration of component A, mg/l

Because of its importance, Eq. 7.1, which describes the rate of mass transfer due to a concentration gradient, is investigated thoroughly in this chapter.

Natural water contains a broad spectrum of inorganic and organic constituents with diverse origins. Dissolved minerals, for example, enter the water source by natural weathering processes of erosion and dissolution. Synthetic organic chemicals, on the other hand, can end up in water supplies by municipal and industrial discharge, pesticide and herbicide usage or poor residual management practice (such as disposal of solvents on the ground). These constituents are converted from a concentrated to a dilute state as they are mixed in water. Water treatment reverses the process of dilution, at the cost of expending thermodynamic work or energy.

Table 7.2. Separation process terminology.

Term	Definition
Absorption	Mass transfer process involving the addition of gases from air into water (e.g, aeration involving addition of oxygen to water)
Adsorbate, sorbate	Substance being removed at phase interface
Adsorbent, sorbent	Phase onto which adsorbate accumulates
Adsorption	Accumulation of solutes at the surface of a solid or a liquid in contact with another phase, resulting in an increased concentration of molecules from that medium in the immediate vicinity of the surface. Adsorption occurs on the surface and in the macropores and mesopores of a solid such as GAC
Batch operation	Contained system with no flow in or out. Typically, the two phases are brought together, allowed to approach equilibrium and then separated. This operation yields the same operating line as a continuous co-current operation
Co-current flow	Contacting phase (e.g., liquid and gas, water and powdered activated carbon) flows are in the same direction, operated in either stages or continuous contact mode
Continuous contact	Continuously changing phase concentrations as a function of position (e.g., in a column operation filled with packing, ion exchange resin or adsorbent)
Continuous operation	Contained system with continuous flow through system
Countercurrent flow	Contacting phase (e.g., liquid and gas) flows are in opposite directions, operated in either stages or continuous contact mode
Cross flow	Contacting phase (e.g., liquid and gas) flows are perpendicular to each ohter, operated in either stages or continuous contact mode
Desorption	Mass transfer process involving removal of volatile substances from water into air (e.g., air stripping)
Diffusion	Random thermal motion of molecules or small particles through which molecules and small particles intermingle
Diffusion coefficient	Parameter that represents the diffusion rate of a solute in a solvent at a specific temperature. Frequently used synonymously with diffusivity
Diffusivity	Property of a substance indicative of diffusion rate. Frequently used synonymously with diffusion coefficient
Extracting phase	Phase to which compounds are transferred in water treatment (e.g., gas phase for stripping, liquid phase for absorption, solid phase for adsorption or ion exchange)
Fluid-fluid operations	Fluid is in contact with another fluid (e.g., air and water in a stripping tower), operated as fixed or fluidised beds
Fluid-solid operations Imbibition	Fluid is in contact with a solid (e.g., packing), operated as fixed or fluidised beds. A special case of fluid-solid absorption whereby the solute can be completely dissolved in either the fluid or the solid phase
Mass transfer	Separation or transport of comopnents from one phase to another (e.g., liquid to gas, gas to liquid, liquid to solid)

(Contd...)

Term	Definition
Mass transfer zone	When two phases are brought together in a continuous contact mode, a concentration gradient forms in both phase. In principle, this gradient is very long, but the length is often defined as a high percentage of the influent concentration (e.g., 95 per cent) to a low percentage of the influent concentration (e.g., 5 per cent). Accordingly, a high mass transfer rate yields transfer zone
Staged operation	Two or more process stages in series. Stages are usually completely mixed and the concentrations are uniform in both phase within each stage
Solute	A dissolved substance
Solvent	Liquid into which other compounds (solutes) can be dissolved
Sorption	A generalised term for the many phenomena commonly included under the terms adsorption and absorption when the nature of the phenomenon involved in a particular case is unknown or indefinite
Stripping	See desorption

The minimum amount of work required to concentrate or remove dissolved constituents from water is given by the change in Gibbs free energy at constant temperature T and pressure P for a reversible process:

$$W_{min} = -\Delta G_{min} \qquad \ldots (7.2)$$

where,

W_{min} = minimum amount of work, kJ/mole

ΔG_{min} = free-energy change that is required to take the solute from the pure form (i.e., pure liquid or precipitate) to the dilute concentration, kJ/mole

The free-energy change ΔG_{min} may be calculated from the following expression:

$$\Delta G_{min} = n_i \Delta G_{f,pure}^o + RT \sum_{i=1}^{N} n_i \ln\{i\} - W_{min} \qquad \ldots (7.3)$$

where,

n_i = mole fraction of constituent i in the system, dimensionless

$\Delta G_{f,pure}^o$ = free-energy change of formation of pure form per mole of i at standard conditions (i.e., 298 K and 1 atm of pressure), kJ/mole

R = universal gas constant, 8.3144×10^{-3} kJ/mole·K

N = total number of components, unitless

T = absolute temperature, K $(273 + °C)$

$\{i\}$ = activity of ionic species, mole/l

The activity of a constituent, for dilute solutions is equal to the molar concentration of i. In general, the more dilute a constituent, the more energy that is required to separate it into its pure state or residual form.

The free energy required to form a saturated solution of an organic chemical from the liquid form of the organic chemical is equal to the free-energy change required to form a separate pure liquid phase of the chemical (assuming no water dissolves in the organic phase). At equilibrium; $\Delta G_{min} = 0$ and $\Delta G_{f,pure}^o$ can be calculated from the solubility of the chemical. In this case, the activity of the chemical is equal to its molar concentration (assuming the activity coefficient is one).

$$\Delta G_{min} = 0 = \Delta G^\circ_{f,pure} + RT \ln [i]_s \qquad \qquad ... (7.4)$$

where,

$[i]_s$ = the solubility of organic chemical i, mole/l

Solving (Eq. 7.4) for $\Delta G^\circ_{f,pure}$ yields

$$\Delta G^\circ_{f,pure} = -RT \ln [i]_s \qquad \qquad ... (7.5)$$

The minimum energy is the reversible energy requirement. To achieve this minimum, it is necessary to operate with essentially no concentration gradient to eliminate entropy losses. Often, mass transfer processes are operated far from a reversible situation by providing large concentration driving forces, which reduces the process size; consequently, the actual energy that is required for separation is far greater than ΔG_{min}. For example, W_{min} is about 3600 kJ/m^3 to desalinate water and current reverse-osmosis processes use about 36000 kJ/m^3.

Solid-Liquid Operations

In water treatment, the two phases involved in separation processes are either gas-liquid or liquid-solid. For liquid-solid operations, the principal processes are membrane and granular filtration, flotation, sedimentation, precipitation, ion exchange, sorption, adsorption and imbibition. During solute uptake by aquifer material in groundwater, it is difficult to differentiate between absorption (imbibition) and adsorption; consequently, this phenomenon is termed sorption (see Table 7.2). The common liquid-solid operations that are used in water treatment were reported previously in Table 7.1.

Gas-Liquid Operations

For gas-liquid operations, solutes are transferred from liquid to gas or from gas to liquid by desorption (stripping) or absorption (scrubbing), respectively. The major feature that distinguishes absorption from adsorption is that in absorption the solute is dissolved in both phases. The common gas-liquid operations for water treatment are presented in Table 7.1.

Contact Modes

There are two major methods for bringing two phases into contact: batch operation and continuous operation, as defined in Table 7.2. Continuous system may be operated with or without discrete stages.

Batch operation

A batch operation is a contained system with no flow in or out, as defined in Table 7.2. An example of a batch operation is an adsorption equilibrium isotherm conducted by adding powdered activated carbon (PAC) to a bottle.

Continuous operation

A continuous operation involves flow within the system as defined in Table 7.2. Three flow patterns are possible with continuous operations, which are defined in Table 7.2: (i) co-current; (ii) countercurrent; and (iii) cross-flow.

Co-current operation

A co-current operation consists of a system in which the two contacting phases flow in the same direction. In water treatment, there are natural draft cocurrent stripping devices in which the falling water creates a natural draft of air to flow through the device.

Countercurrent operation

A countercurrent operation consists of a system in which the two contacting phases flow in opposite directions. Countercurrent operation represents the preferred mode of operation for many air-stripping processes, such as packed towers.

Cross-flow operation

In a cross-flow operation, the two contacting phases flow perpendicular to one another. Sedimentation is an example of a cross-flow process and is widely used in water treatment, but cross-flow operations are typically impractical for most water treatment processes.

Staged operation

In a staged operation, the process is operated as a series of stages, as defined in Table 7.2. The flow pattern in a staged operation can be co-current, cross-flow or countercurrent. An example of a countercurrent staged operation is a low-profile air stripper.

Operating Diagrams

The impact of the concentration gradient on the rate of mass transfer between two phases can be evaluated graphically using a concept called operating diagrams or McCabe-Thiele diagrams. Operating diagrams are drawn by plotting the solute concentration in the extracting phase (e.g., air for gas transfer, activated carbon for adsorption) as a function of the solute concentration in the aqueous phase. The operating diagram consists of two lines: (i) an equilibrium line; and (ii) operating line. Operating diagrams can be used to determine the minimum amount of the extracting phase needed for treatment and to examine graphically the trade-off between the size of the mass transfer contacting device and the quantity of extracting phase needed [e.g., air water ratio for stripping or powdered activated carbon (PAC) required for adsorption].

Equilibrium line

The equilibrium line is derived from two-phase equilibrium relationships and gives the solute concentration in the extracting phase that exists when the extracting and aqueous phases are in equilibrium with each other. Examples of two-phase equilibrium relationships are Henry's law for air stripping and the Freundlich isotherm for adsorption.

Operating line

The operating line is derived from a mass balance on the contacting device, relating the solute concentration in each phase initially to the solute concentration in each phase after contact has begun. An example using a batch reactor, in which powdered activated carbon (PAC) is added to a vessel containing a solution of water and an organic solute, is shown in Fig. 7.1. Initially, there is no solute adsorbed onto the PAC. The mass balance for this system is as follows:

$$\begin{pmatrix} \text{Mass present} \\ \text{initially in solution} \end{pmatrix} = \begin{pmatrix} \text{Mass} \\ \text{adsorbed} \end{pmatrix} + \begin{pmatrix} \text{mass remaining} \\ \text{in solution after} \\ \text{adsorption} \end{pmatrix} \qquad ... (7.6)$$

$$VC_0 = Mq + VC \qquad \qquad ... (7.7)$$

where,

      V = volume of liquid in vessel, l
      C_0 = initial concentration of solute in vessel, mg/l
      M = mass of carbon, g
      q = concentration of solute adsorbed to the activated carbon at any time, mg/g
      C = concentration of the solute in the water after adsorption, mg/l

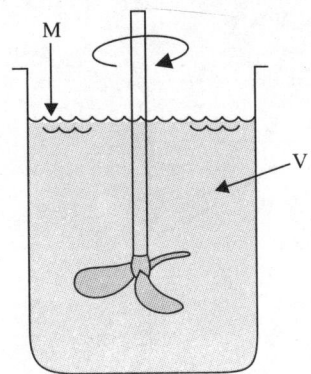

Fig. 7.1. Batch contactor for powdered activated carbon.

Equation 7.6 can be rearranged as follows:

$$q = \frac{V}{M}(C_0 - C) \qquad \qquad ... (7.8)$$

The operating line, which is the solute concentration in the extracting phase as a function of the concentration in the aqueous phase at any point in time after contact has started, is defined by Eq. 7.8. When the PAC is first added to the vessel, there is no solute on the PAC. As time proceeds, the solute becomes adsorbed onto the PAC and q and C at a particular time are related to one another by the operating line. It should be noted that although adsorption in a batch reactor proceeds toward equilibrium over the passage of time, the operating line does not identify the time progression of the process, but only relates the dependent variables q and C.

The operating diagram for the relationship described in Eq. 7.8 is shown in Fig. 7.2. Equation 7.8 is the equation of a straight line with a slope of $-V/M$ and several operating lines with different values for V/M have been shown. The equilibrium line is shown in Fig. 7.2 as a dashed line.

Driving force

The driving force for mass transfer, as shown in Fig. 7.2, is the difference between the actual solute concentration in solution and the concentration in solution that would be in equilibrium with the extracting phase. Initially, the solute is entirely in the aqueous phase and the solute is transferred rapidly to the PAC. As time progresses, the concentration on the PAC increases and the concentration in the aqueous phase decreases, which slows the rate of mass transfer. After a very long time, the solute concentration in the water is in equilibrium with the concentration on the PAC and bulk mass transfer ceases. Thus, the concentration gradient or driving force, is defined as the difference between the actual and equilibrium concentration C_e in the aqueous phase.

Because the equilibrium concentration is identified by the equilibrium line and the actual concentration (determined by mass balance) is identified by the operating line, the horizontal distance between these lines describes the concentration gradient. Equilibrium occurs and mass transfer ceases when the operating line and equilibrium line intersect.

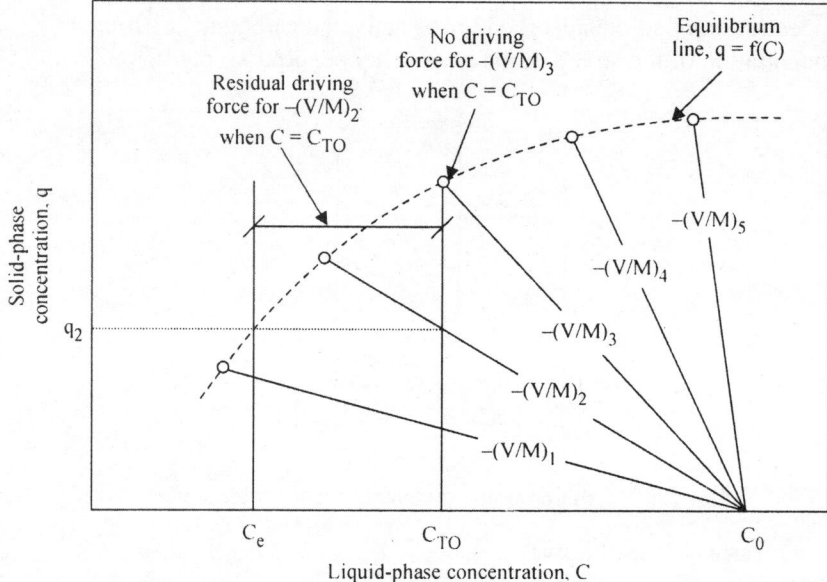

Fig. 7.2. Operating lines for a constant initial concentration C_0 and different adsorbent doses, V/M (equilibrium line is also plotted for reference).

Analysis using operating diagrams

The operating diagram can be used to determine the minimum amount of extracting phase required for treatment, which is an initial indicator of the feasibility of a process. For example, if millions of tonnes of activated carbon are required to treat a given water, then adsorption with activated carbon is not a feasible treatment option and no further analysis is necessary. If a separation process appears to be feasible based on the amount of extractant, then more detailed design and economic calculations are warranted.

An operating line analysis for an adsorption process is shown in Fig. 7.2. For a given volume of water, the quantity of PAC required can be defined by the V/M ratio, with greater values of V/M (greater slope of the operating line) corresponding to smaller amounts of PAC. If the treatment objective is the concentration shown as C_{TO} in Fig. 7.2, the minimum amount of PAC required can be determined from the operating line with the slope of $(V/M)_3$, which is the operating line that intersects the equilibrium line at the value of C_{TO}. Operating lines with greater slope, $(V/M)_4$ and $(V/M)_5$, intersect the equilibrium line at a concentration higher than C_{TO} and therefore would be unable to meet the treatment objective.

The operating diagram also qualitatively demonstrates the trade-off between the quantity of the extracting phase and the size of the contacting device. For the operating line identified as $(V/M)_3$, the driving force (horizontal distance between the equilibrium and operating lines) becomes infinitesimally small as equilibrium is approached. The small driving force results in a slow rate of mass transfer, requiring an exceedingly long time to reach the treatment objective. In a flow-through system treating a specified water flowrate, a long time corresponds to a long detention time and hence a very large contactor. The operating lines labelled as $(V/M)_1$ and $(V/M)_2$, have lower slopes, which correspond to greater quantities of carbon, but have larger concentration gradients when the actual concentration (operating line) reaches the treatment objective, resulting in shorter contact times. Thus, for the operating lines shown, the line labelled $(V/M)_3$ would use the most carbon but have the smallest contactor, the line labelled $(V/M)_2$ would have an intermediate carbon usage rate and contactor size, the line labelled $(V/M)_3$ would use the minimum amount of carbon but have a large (theoretically, infinitely large) contactor and the lines labelled $(V/M)_4$ and $(V/M)_5$ would be unable to meet the treatment objective.

Plug flow co-current continuous operation

Although Eq. 7.7 is for a batch operation, it is equally valid for a plug flow co-current continuous operation. For example, if a quantity of PAC per time, M, is added to water with a flow rate, Q, the mass balance is the same as presented previously in Eq. 7.6.

$$\begin{pmatrix} \text{Mass present} \\ \text{initially in solution} \end{pmatrix} = \begin{pmatrix} \text{mass} \\ \text{adsorbed} \end{pmatrix} + \begin{pmatrix} \text{mass remaining} \\ \text{in solution after} \\ \text{adsorption} \end{pmatrix}$$

$$QC_0 = \dot{M}q + QC \qquad \qquad ...(7.9)$$

$$q = \frac{Q}{\dot{M}}(C_0 - C) \qquad \qquad ...(7.10)$$

where,

Q = flow rate, l/s

C_0 = initial concentration of solute in the solution, mg/l

$\dot{M}$ = mass of PAC added per time, g/s

C = concentration of the solute in the water at any time, mg/l

q = concentration of solute adsorbed to the activated carbon at any time, mg/g

The PAC dosage in the plug flow system $\dot{M}/Q$ is identical to the PAC dosage in the batch reactor M/V and Eqs. 7.8 and 7.10 are essentially identical.

MASS TRANSFER AND DIFFUSION

In the previous section it was noted that the rate of mass transfer is the product of a mass transfer coefficient and a driving force (see Eq. 7.1). Although the mass transfer coefficient is often determined using empirical correlations, the correlations are based on one of several models of mass transfer and the models in turn are based on the physico-chemical process of molecular diffusion. As a result, an understanding of molecular diffusion is a necessary part of an understanding of mass transfer. The concept of mass flux, which can be used to explain several important concepts regarding diffusion, including Fick's first law, the Stokes-Einstein equation and Brownian velocity, is defined in this section. Fick's first law is the fundamental expression for the rate of mass transfer due to diffusion. The Stokes-Einstein equation forms the basis for correlations for the calculation of diffusivity and Brownian motion is important because it is used to determine the size of particles that settle and the flocculation and filtration rate of small particles.

Concept of Mass Flux

In mass transfer operations, mass flux is used to describe mass transfer:

$$J_A = \frac{m}{at} \qquad \qquad \text{... (7.11)}$$

where,

J_A = mass of solute A transferred across an interface, mg/m²·s
m = mass of compound, mg
a = area perpendicular to mass flux, m²
t = time during which mass is transferred, s

As shown in Eq. 7.11, the mass flux is expressed on a per-area basis; consequently, it is an intensive variable because it does not depend on the area perpendicular to the mass flux. (Intensive properties do not depend on the size of the system, e.g., concentration or temperature.)

Fick's First Law

Fick's first law of diffusion is used to describe the mass transport rate by diffusion. The negative sign indicates that solutes diffuse from regions of high concentration to regions of low concentration.

$$J_A = -D_{AB} \frac{dC_A}{dz} \qquad \qquad \text{... (7.12)}$$

where,

J_A = mass flux of component A in direction of concentration gradient, mg/m²·s
D_{AB} = diffusion coefficient of component A in solvent B, m²/s
C_A = concentration of component A, mg/l
z = distance in direction of mass transfer (or in direction of concentration gradient), m

Fick's first law is used to describe the rate of diffusion from a relative point of view; in other words, the mass transfer is superimposed on top of or in addition to, mass transfer from other processes such as advection. The governing equations for unit processes are often developed by writing mass balance expressions around a control volume using a fixed point of view (stationary frame of reference). It is useful, therefore, to examine the difference between the mathematical expression for diffusion that is defined from a relative reference frame (flux = J) and the expression for diffusion defined from a stationary reference frame (flux = N). The general expression for the molar flux of component A in solvent B can be written as

$$N_A = -D_{AB} \frac{dC_A}{dz} + x_A (N_A + N_B) \qquad \qquad \text{... (713)}$$

where,

N_A = molar flux of component A relative to stationary frame of reference, mole/m²·s
N_B = molar flux of solvent B relative to stationary frame of reference, mole/m²·s
$N_A + N_B$ = total molar flux, mole/m² ·s
D_{AB} = diffusion coefficient of component A in solvent B, m²/s
C_A = molar concentration of component A, mole/l
z = position in direction of flow and diffusion flux (or in direction of concentration gradient), m
x_A = mole fraction of A in solution, mole/mole

Molar flux in absence of advection

For the special case of no advective flow of the solvent ($N_B = 0$), Eq. 7.13 can be rearranged to yield the expression

$$N_A = \frac{1}{1-x_A}\left(-D_{AB}\frac{dC_A}{dz}\right) \qquad \dots (7.14)$$

where,

N_A = molar flux of component A relative to stationary frame of reference, mole/m^2 ·s

Here, N_B can be zero when dissolution or evaporation of the solute occurs and the solvent is stagnant.

For many environmental applications, x_A is very small. For example, the aqueous solubilities of chloroform and oxygen are about 9.3 g/l and 9.3 g/l at 20°C, respectively; consequently, the largest mole fractions that can be found in water are 0.0014 and 5.23 × 10^{-6}, respectively. However, for highly miscible solvents in water or VOCs in gases, it is advisable to examine whether the $1/(1 - x_A)$ factor is important. Thus, the following expression is satisfactory for solving most mass transfer problems in the absence of advection:

$$N_A = -D_{AB}\frac{dC_A}{dz} \qquad \dots (7.15)$$

where,

N_A = molar flux of component A relative to stationary frame of reference, mole/m^2 ·s

D_{AB} = diffusion coefficient of component A in solvent B, m^2/s

C_A = molar concentration of component A, mole/l

z = distance in direction of mass transfer (or in direction of concentration gradient), m

Equation 7.15 is also valid when the sum of the fluxes N_A and N_B are equal to zero, as in a case where the diffusion of species A is countered by the diffusion of B (equal molar counterdiffusion).

As defined by Fick's law, J_A is the flux of component A with respect to the centroid of the diffusing mass of solute, as shown in Eq. 7.12 but is equal to Eq. 7.15 when x_A is small. The concentration can be converted from molar units to mass units by multiplying by the molecular weight. Concentrated solutions can cause fluid flow by diffusion and Eq. 7.14 must be used in such cases. The key point is that J_A is the flux of component A with respect to the center of solute mass and it is only equal to the solute flux for a stationary frame of reference, N_A, when considering dilute, non-flowing systems.

Molar flux in the presence of advection

The mass transfer of component A due strictly to advection (in the absence of diffusion) may be written as:

$$\frac{Q_L C_A}{a} = v(C_A) \qquad \dots (7.16)$$

where,

Q_L = flow rate of fluid, m^3/s

a = cross-sectional area perpendicular to direction of flow, m^2

C_A = concentration of component A, mole/l

v = velocity in direction of concentration gradient, where $v = Q_L/a$, Q_L being the liquid flow rate   and a is the cross-sectional area, m/s

Consequently, Eq. 7.13 may be re-written for a dilute solution with advection to the expression

$$N_A = -D_{AB} \frac{dC_A}{dz} + v(C_A) \qquad \qquad ... (7.17)$$

where,

N_A = molar flux of component A relative to stationary frame of reference, mole/m^2 ·s

D_{AB} = diffusion coefficient of component A in solvent B, m^2/s

C_A = concentration of component A, mole/l

v = velocity in direction of concentration gradient, m/s

Stokes-Einstein Equation and Brownian Motion

In Einstein's simplified explanation of Brownian motion and diffusion, Fick's law is derived based on random motion of particles. Einstein's simplified explanation of Brownian motion and diffusion provides an answer to the following important question that leads to the development of correlations for diffusion coefficients: How is the rate at which molecules diffuse related to fluid-particle-solute collisions? A derivation of the theoretical basis calculating diffusion coefficients is provided below.

Stokes-Einstein equation

A theoretical value for the diffusion coefficient can be derived from considering the momentum exchange between fluid and solute molecules or small particles. As shown in Fig. 7.3, for a perfect hit and an elastic collision, the solvent molecule hits the solute and moves in the opposite direction after the collision. For one dimension (in the direction of the concentration gradient), the average momentum change and time can then be calculated using the following equations:

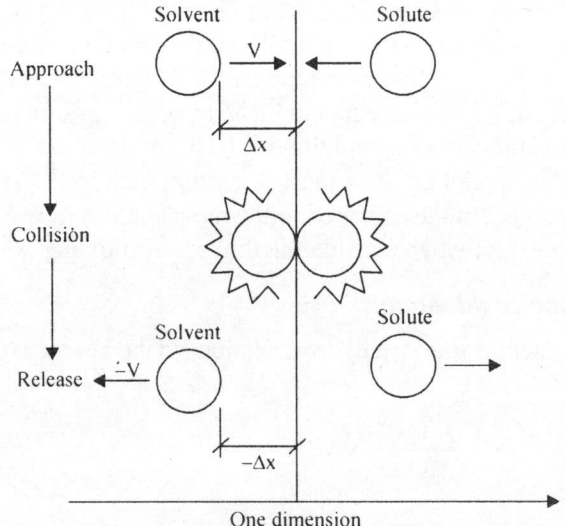

Fig. 7.3. Schematic of fluid molecule collision with solute molecule.

$$F_c = m \frac{v - (-v)}{\Delta t} = m \frac{2v}{\Delta t} \qquad \qquad ... (7.18)$$

where,

  F_c = average collision force exerted due to momentum change, N
  m = mass of solute molecule, kg
  v = average velocity of fluid molecule in one dimension, m/s
  Δt = time interval between collisions, s

$$\Delta t = \frac{\Delta x}{v} + \frac{-\Delta x}{-v} = \frac{2\Delta x}{v} \qquad \qquad ... (7.19)$$

where,

  Δt = time interval, s
  Δx = average distance in one dimension that solute or particle moves before striking another
       solvent molecule in time interval Δt, m

Substituting Eq. 7.19 into Eq. 7.18 yields

$$F_c = m\frac{v^2}{\Delta x} \qquad \qquad ... (7.20)$$

Based on the kinetic theory of gases, the kinetic energy of gas molecules in one direction is given by

$$KE = \frac{1}{2}mv^2 = kT \qquad \qquad ... (7.21)$$

where,

  KE = kinetic energy, N·m
  k = Boltzmann's constant, 1.3804×10^{-20} g·m^2/s^2·K
  T = absolute temperature, K (273 + °C)

Consequently, by solving Eq. 7.21 for mv^2 and substituting into Eq. 7.20, the average collision force can be related to kT:

$$F_c = \frac{mv^2}{\Delta x} = \frac{2kT}{\Delta x} \qquad \qquad ... (7.22)$$

For a liquid, the viscous force that is exerted is given by Stokes' law:

$$F_v = 6\pi R\mu_l \frac{\Delta x}{\Delta t} \qquad \qquad ... (7.23)$$

where,

  F_v = viscous force, N
  R = particle radius, m
  μ_l = viscosity of liquid, kg/m·s

Equating the viscous force (Eq. 7.23) to the collision force (Eq. 7.22) yields

$$6\pi R\mu_l \frac{\Delta x}{\Delta t} = \frac{2\,kT}{\Delta x} \qquad \qquad ... (7.24)$$

The diffusion coefficient which is equal to the square of the average distance that the solute moves in one direction before it collides with another solvent molecule divided by 2 times the average time between collisions, was derived from the kinetic theory of gases.

Rearranging Eq. 7.24 to obtain the diffusion coefficient yields the Stokes-Einstein equation:

$$D_l = \frac{\Delta x^2}{2\Delta t} = \frac{kT}{6\pi R \mu_l} \qquad \text{... (7.25)}$$

where,

D_l = liquid-phase diffusion coefficient, cm^2/s

Equation 7.25 was derived from the kinetic theory of gases and does not strictly apply to liquids. Nonetheless, Eq. 7.25 provides a basis for liquid diffusivity correlations. For example, the diffusivity correlations should have absolute temperature in the numerator and viscosity and something related to molecular size in the denominator. Further, Eq. 7.25 can be used to obtain a good prediction of the liquid diffusivity for large molecules.

For gases, the viscous force is not the same as it is for liquids because the fluid does not appear continuous for small particles or solutes. The size of a molecule or atom is of the order of 0.1 to 1 nm and the average distance that a solute molecule travels before it collides with another molecule is of the order of 100 nm in a gas at 1 atm and room temperature. These collisions cause drag or shear. This effect is considered using the Cunningham correction factor, which is shown in the following equation. For gases, the viscous force is equal to the following:

$$F_v = 6\pi R \mu_g \left(\frac{1}{C_c}\right) \frac{\Delta x}{\Delta t} \qquad \text{... (7.26)}$$

where,

F_v = viscous force for gases, N
R = particle radius, m
μ_g = viscosity of gas, kg/m·s
C_c = Cunningham correction factor
Δx = average distance in one dimension that solute or particle moves before striking another solvent molecule in time interval Δt, m
Δt = time interval, s

The Cunningham correction factor is determined from the expression

$$C_c = 1 + \frac{\lambda}{R}\left[1.257 + 0.4 \exp\left(-\frac{R}{\lambda}\right)\right] \qquad \text{... (7.27)}$$

where,

λ = mean free path, m

When λ/R is large, the following expression is obtained:

$$D_g = \frac{kT}{6\pi R \mu_g}\left(1 + \frac{\lambda}{R}\right) \qquad \text{... (7.28)}$$

where,

D_g = gas-phase diffusion coefficient, cm^2/s

The Cunningham correction factor gives rise to the collision function that appears in gas-phase correlations for gas-phase diffusivity.

Brownian velocity

By definition, the Brownian velocity is the average distance that the solute molecule or particle moves in one direction before it collides with another solvent molecule divided by the average time between collisions.

Thus, the Brownian velocity in liquid is given by the expression

$$v_B = \frac{\Delta x}{\Delta t} = \frac{2D_t}{\Delta x} = \frac{2kT}{\Delta x 6\pi R\mu_l} \qquad \qquad ... (7.29)$$

where,

v_B = Brownian velocity of molecules, m/s

Upon examination of Eq. 7.29, the Browning velocity depends on distance and this should not be surprising because Brownian motion is motion that is caused by random motion and the average distance that particles or solute molecules move can be determined from the expression.

$$(x) = \sqrt{2D_l t} = \sqrt{\frac{2kT}{6\pi R\mu_l} t} \qquad \qquad ... (7.30)$$

where,

(x) = average distance that particle or solute molecule moves as a result of Brownian motion, m

t = average time that particle or solute molecule moves as a result of Brownian motion, s

CALCULATION OF DIFFUSION COEFFICIENTS

To calculate the mass transfer rate, the diffusion coefficient or diffusivity, must be determined. Diffusion coefficients have been determined experimentally for some solutes and are available in reference texts, such as Robinson and Stokes for ionic species. When experimentally determined values are unavailable, diffusion coefficients can be calculated using a variety of correlations.

The methodology used to determine diffusion coefficients is presented in this section. Models used to determine diffusion coefficients are presented in Table 7.3.

Table 7.3. Models used for estimating molecular diffusivity in liquids.

Application	Model	Comments
All types of molecules of particles	Stokes-Einstein equation. $$D_l = \frac{kT}{3\pi\mu_l d_M}$$	Diffusivities are slightly larger than measured values for large spherical molecules (MW > 1000 Da) or particles in liquids. Diffusivity for small molecules with MW in order of 100 Da is underestimated
For large molecules	Polson equation[a] $D_l = 2.74 \times 10^{-5}(MW)^{-1/3}$	Polson; useful when MW > 1000 Da
For none-lectrolytes and small molecules	Hayduk-Laudie correlation[a] $$D_l = \frac{13.26 \times 10^{-5}}{(\mu_w)^{1.14}(V_b)^{0.589}}$$	Hayduk and Laudie (1974); convert units for μ_w (g/CM · S) and Table 7.5 for V_b (cm^3/mole)
For electrolytes in absence of electric field	Nernst-Haskell equation: $$D_{AB}^{\circ} = \frac{RT}{F^2}\left[\frac{1/n^+ + 1/n^-}{1/\lambda_+^{\circ} + 1/\lambda_-^{\circ}}\right]$$	Adapted from Reid; use Table 7.6 for λ values

[a] D_l is expressed in units of cm^2/s.

Typical values of solute diffusion coefficients in gases and liquids are as follows:

Liquids: ~ 10^{-6} to 10^{-5} cm²/s (10^{-10} to 10^{-9} m²/s).

Gases: ~ 10^{-2} to 10^{-1} cm²/s (10^{-6} to 10^{-5} m²/s).

Liquid-Phase Diffusion Coeffcients for Non-electrolytes

The diffusion coefficients of some common solutes found in water treatment are presented in Table 7.4. Methods for calculating diffusion coefficients for large and small molecules in water are as follows.

Table 7.4. Molecular diffusion coefficients in water at 25°C.

Constituent	D_l, cm²/s	Constituent	D_l, cm²/s
Dissolved gases		Strong electrolytes	
O_2	2.35×10^{-5}	NaCl	0.8×10^{-5}
CO_2	2.0×10^{-5}	$CaCl_2$	1.33×10^{-5}
SO_2	1.7×10^{-5}	$MgCl_2$	1.25×10^{-5}
Cl_2	1.4×10^{-5}	$NaHCO_3$	0.63×10^{-5}
NH_3	2.0×10^{-5}	Na_2SO_4	1.2×10^{-5}
Non-electrolytes		KCl	1.0×10^{-5}
Vinyl chloride	1.1×10^{-5}	Weak electrolytes	
Methane	1.8×10^{-5}	Formic acid	1.4×10^{-5}
Carbon tetrachloride	0.81×10^{-5}	Ethanol	1.3×10^{-5}
Tetrachloroethylene	0.75×10^{-5}	Glucose	0.7×10^{-5}
Trichloroethylene	0.84×10^{-5}	Lactose	0.5×10^{-5}
Chloromethane	1.3×10^{-5}	Urea	1.4×10^{-5}
1, 1, 1,-Trichloroethane	0.80×10^{-5}	Humic acid[a]	0.1×10^{-5}
Benzene	0.89×10^{-5}		
Chloroform	0.92×10^{-5}		
1, 2-Dichloroethane	0.91×10^{-5}		
Bromoform	0.88×10^{-5}		

[a] 5 nm in diameter.

Large molecules

The Stokes-Einstein equation yields diffusion coefficients that are slightly larger than measured values for large spherical molecules [molecular weight (MW) > 1000 Da] or particles in liquids, which is reasonable given the simplicity of the development of the Stokes-Einstein equation. Conversely, the Stokes-Einstein equation underestimates the diffusivity for small molecules with MW of the order of 100 Da. Substituting the parameters for water (at 20°C) into the Stokes-Einstein equation results in the expression

$$D_1 = \frac{kT}{3\pi\mu_1 d_M} = \frac{(1.3804 \times 10^{-16}\, g \cdot cm^2/s^2 \cdot K)\,(293\ K)}{3\pi(1.05 \times 10^{-2}\, g/cm \cdot s)d_M}$$

$$= \frac{4.08 \times 10^{-13}\, cm^3/s}{d_M} \qquad \ldots (7.31)$$

where,

D_l = liquid-phase diffusion cm²/s

k = Boltzmann constant, 1.3804×10^{-16} g·cm²/s²·K

T = absolute temperature, K (273 + °C)

μ_l = liquid viscosity at 20°C, 1.05×10^{-2} g/cm·s

d_M = molecular diameter, cm

Equation 7.31 is only valid for spherical molecules. If the molecular density is taken to be 2.9 g/cm³, then a molecule with a MW of 50000 Da has a diameter of 3.8 nm. Using this density and substituting it into Eq. 7.31 results in the following expression at 20°C:

$$D_l = 4.0 \times 10^{-5}(MW)^{-1/3} \qquad \qquad \dots (7.32)$$

where,

D_l = liquid-phase diffusion coefficient, cm²/s

MW = molecular weight, Da or g/mole

Polson obtained the expression given below by correlating experimental data for globular proteins:

$$D_l = 2.74 \times 10^{-5} (MW)^{-1/3} \qquad \qquad \dots (7.33)$$

where,

D_l = liquid-phase diffusion coefficient, cm²/s

MW = molecular weight, Da or g/mole

It is remarkable that Eqs. 7.32 and 7.33 are similar given that Eq. 7.32 is a theoretical result derived based on the Stokes-Einstein equation and Eq. 7.33 is an empirical expression.

Small molecules

The diffusivities of small uncharged molecules in water can be calculated using the Hayduk-Laudie correlation which is derived from the Wilke-Chang correlation. Both aqueous and non-aqueous solutions are considered in the Wilke-Chang correlation, whereas the Hayduk-Laudie correlation was derived using only water-phase data. Thus, the Hayduk-Laudie correlation is, in general, more suitable than the Wilke-Chang correlation for calculating diffusivities in aqueous solutions.

The Hayduk-Laudie correlation is given by the expression:

$$D_l = \frac{13.26 \times 10^{-5}}{(\mu_w)^{1.14}(V_b)^{0.589}} \qquad \qquad \dots (7.34)$$

where,

D_l = liquid-phase diffusion coefficient of solute, cm²/s

μ_w = viscosity of water, cP (1 kg/m·s = 1000 cP)

V_b = molar volume of solute at normal boiling point, cm³/mole

The molar volume at the normal boiling point, V_b, can be estimated using the LeBas method. The atomic volumes for different elements, mixtures and functional groups for use in calculation of molar volume at the normal boiling point via the LeBas method are presented in Table 7.5. Contributions of the various functional groups are added together along with deductions for certain ring structures. Calculations of molar volume with the LeBas method is illustrated in Example 7.1.

Table 7.5. Atomic volumes for use in computing molar volumes at normal boiling point with LeBas method.

Element mixture or functional group	Atomic volume, cm³/mole	Circumstance
Air	29.9	
Antimony	34.2	
Arsenic	30.5	
Bismuth	48.0	
Bromine	27.0	
Carbon	14.8	
Chlorine	21.6	Terminal as in R–Cl
	24.6	Medial as in R–CHCl–R
Chromium	27.4	
Fluorine	8.7	
Germanium	34.5	
Hydrogen	3.7	In organic compound
	7.15	In hydrogen molecule
Iodine	37.0	
Lead	46.5–50.1	
Mercury	19.0	
Nitrogen	15.6	
	10.5	In primary amines
	12.0	In secondary amines
Oxygen	7.4	Doubly bond, as carbonyl oxygen
	7.4	In aldehydes or ketones
	9.1	In methyl esters
	9.9	In methyl ethers
	11.0	In higher ethers and esters
	12.0	In acids
	8.3	In union with S, P or N
Phosphorus	27.0	
Silicon	32.0	
Sulphur	25.6	
Tin	42.3	
Titanium	35.7	
Vanadium	32.0	
Water	18.8	
Zinc	20.4	
Ring deductions	6.0	Three-membered ring
	8.5	Four-membered ring
	11.5	Five-membered ring
	15	Six-membered ring
	30	Naphthalene ring
	47.5	Anthracene ring

Example 7.1: Calculation of diffusion coefficients for non-electrolytes in water: Calculate the diffusion coefficients of the following contaminants found in a groundwater at 13°C: (1) tetrachloroethene (MW 165.8) and (2) benzene (MW 78.11) and then compare the values. Use Table 7.5 to find the contributions of the various functional groups to the molar volume. At T = 13°C.

$$\mu_W = 1.202 \times 10^{-3} \text{ kg/m·s} = 1.202 \text{ cP}$$
$$r_W = 0.99973 \text{ g/cm}^3$$

Solution:

1. Estimate the molar volume at the boiling point.

(a) Tetrachloroethene: Use the information in Table 7.5 and add the contributions of the various functional groups as shown below.

(i) Write the chemical formula for tetrachloroethene:

$$C_2Cl_4$$

(ii) Calculate the contribution of each atom to the molar volume using the data in Table 7.5:

$$2C = 2(14.8) = 29.6 \text{ cm}^3/\text{mole} \qquad 4Cl = 4(21.6) = 86.4 \text{ cm}^3/\text{mole}$$

(iii) Calculate the molar volume by adding the contributions of each atom:

$$V_b = 29.6 + 86.4 = 116 \text{ cm}^3/\text{mole}$$

(b) Benzene:

(i) Write chemical formula for benzene:

$$C_6H_6$$

(ii) Calculate the contribution of each atom to the molar volume using Table 7.5:

$$6C = 6(14.8) = 88.8 \text{ cm}^3/\text{mole} \qquad 6H = 6(3.7) = 22.2 \text{ cm}^3/\text{mole}$$

Six-member ring $= -15 \text{ cm}^3/\text{mole}$

(iii) Calculate the molar volume by adding the contributions of each atom:

$$V_b = 88.8 + 22.2 - 15 = 96 \text{ cm}^3/\text{mole}$$

2. Calculate the liquid-phase diffusion coeffcients for each compound using 7.34.

(a) Tetrachloroethene:

$$D_l = \frac{13.26 \times 10^{-5}}{(1.202 \text{ cP})^{1.14}(116 \text{ cm}^3/\text{mole})^{0.589}} = 6.54 \times 10^{-6} \text{ cm}^2/s$$

(b) Benzene:

$$D_l = \frac{13.26 \times 10^{-5}}{(1.202 \text{ cP})^{1.14}(96 \text{ cm}^3/\text{mole})^{0.589}} = 7.31 \times 10^{-6} \text{ cm}^2/s$$

Diffusivity of oxygen

The liquid-phase diffusivity of oxygen in water can be determined from a correlation that was obtained from a best fit of literature values.

$$D_{l,O_2} = 10^{(A+B/T)}(1.0 \times 10^{-9}) \qquad \qquad \dots (7.35)$$

where,

$\quad D_{l,O_2}$ = liquid-phase diffusivity of oxygen, m²/s

$\quad$ A $\quad$ = fitting parameter, 3.15, unitless

$\quad$ B $\quad$ = fitting parameter, −831.0, unitless

$\quad$ T $\quad$ = absolute temperature, K (273 + °C)

Liquid-Phase Diffusion Coefficients for Electrolytes

The liquid-phase diffusion coefficients of electrolytes in the absence of an electric field can be calculated using the Nernst-Haskell equation:

$$D_i^\circ = \frac{RT}{F^2}\left(\frac{1/n^+ + 1/n^-}{1/\lambda_+^\circ + 1/\lambda_-^\circ}\right) \qquad \qquad ...(7.36)$$

where,

D_i° = liquid-phase diffusivity at infinite dilution, cm²/s

R = universal gas constant, 8.314J/mole·K

T = absolute temperature, K (273 + °C)

n^+ = cation valence

n^- = anion valence

F = Faraday's constant, 96500 C/eq

λ_+° = limiting positive ionic conductance, (A/cm²) (cm/V) (cm³/eq)

λ_-° = limiting negative ionic conductance, (A/cm²) (cm/V) (cm³/eq)

The diffusion coefficient is related to the equivalent conductance because it is related to the current (coulombs per second) that would result from an electric field. Also, electroneutrality requires that the counterion must also migrate in the same direction. Consequently, the equivalent conductance for both the cation and anion is required to estimate the diffusion coefficient.

The equivalent conductance is the current through a unit area that results from applying an electric field for a given electrolyte concentration. The following review of the units of electromotive energy is presented to better understand the equivalent conductance:

$$\text{One volt} = \frac{\text{Work}}{\text{Charge}} = \frac{\text{Joules}}{\text{Coulomb}} \qquad 1V = \frac{J}{C} = \frac{N \cdot m}{C}$$

$$\text{Electric field strength} = V/cm \qquad 1 \text{ electron} = 1.6 \times 10^{-19} C$$

$$\text{Ampere} = \text{Coulombs per second} \qquad A = C/s$$

The current flow would be related to the equivalent conductance as given by the equation.

$$\text{Ampere/cm}^2 = \text{Electric field strength} \times \lambda_i \times C_i$$

$$...(7.37)$$

where,

λ_i = equivalent conductance of electrolyte i, (A/cm²) (cm/V) (cm³/eq)

C_i = concentration of electrolyte i, eq/cm³

As shown in Table 7.6, small ions have high equivalent conductances because they migrate through water very rapidly in response to an imposed electric field. Consequently, their diffusivity is higher than the value for large ions (Example 7.2).

Example 7.2: Calculating diffusion coefficients of ions. Estimate the diffusivity of HCl in a dilute aqueous solution at 25°C and pH 7. Use the information in Table 7.6 to find the limiting ionic conductances for the ions.

Solution:

1. Find the limiting ionic conductances. From Table 7.6.

$$\lambda_+^\circ = 349.8 \ (A/cm^2)(cm/V)(cm^3/eq)$$

$$\lambda_-^\circ = 76.3 \ (A/cm^2)(cm/V)(cm^3/eq)$$

2. Calculate the diffusion coefficients at infinite dilution. Use Eq. 7.36 and the limiting ionic conductances from step 1:

$$D_1^\circ = \left(\frac{RT}{F^2}\right) \times \left(\frac{1/n^+ + 1/n^-}{1/\lambda_+^\circ + 1/\lambda_-^\circ}\right)$$

$$= \frac{(8.314 \ J/mole \cdot K)(298 \ K)}{(96500)^2 \ C^2/eq^2}$$

$$\times \left\{ \frac{\left(\frac{1}{1} + \frac{1}{1}\right)\frac{mole}{eq}}{\left(\frac{1}{349.8} + \frac{1}{76.3}\right)\left[\left(\frac{cm^2}{A}\right) \times \left(\frac{A}{C/s}\right)\right]\left[\left(\frac{V}{cm}\right) \times \left(\frac{J/C}{V}\right)\right]\left[\frac{eq}{cm^3}\right]} \right\}$$

$$= 3.33 \times 10^{-5} \ cm^2/s$$

Table 7.6. Limiting ionic conductances in water at 25°C [(A/cm²)(cm/V)(cm³/eq)].

Cation	Formula	λ_i°	Cation	Formula	λ°
Hydrogen	H^+	349.8	Hydroxide	OH^-	197.6
Lithium	Li^+	38.7	Chloride	Cl^-	76.3
Sodium	Na^+	50.1	Bromide	Br^-	78.3
Potassium	K^+	73.5	Iodide	I^-	76.8
Ammonium	NH_4^+	73.4	Nitrate	NO_3^-	71.4
Silver	Ag^+	61.9	Perchlorate	ClO_4^-	68.0
Thallium	Tl^+	74.7	Bicarbonate	HCO_3^-	44.5
Magnesium	$(1/2)mg^{2+}$	53.1	Formate	HCO_2^-	54.6
Calcium	$(1/2)Ca^{2+}$	59.5	Acetate	$CH_3CO_2^-$	40.9
Strontium	$(1/2)Sr^{2+}$	50.5	Chloroacetate	$ClCH_2CO_2^-$	39.8
Barium	$(1/2)Ba^{2+}$	63.6	Cyanoacetate	$CNCH_2CO_2^-$	41.8
Copper	$(1/2)Cu^{2+}$	54	Propionate	$CH_3CH_2CO_2^-$	35.8
Zinc	$(1/2)Zn^{2+}$	53	Butyrate	$CH_3(CH_2)_2CO_2^-$	32.6
Lanthanum	$(1/3)La^{3+}$	69.5	Benzoic acid	$C_6H_52CO_2^-$	32.3
			Sulphate	$(1/2)SO_4^{2-}$	80

The diffusivity of an organic compound in the gas phase can be calculated using the Wilke-Lee modification of the Hirschfelder-Bird-Spotz correlation:

$$D_g = \frac{(1.084 - 0.249\sqrt{1/M_A + 1/M_B})(T^{1.5})\sqrt{1/M_A + 1/M_B}}{P_1(r_{AB})^2 f(kT/\varepsilon_{AB})} \quad \text{... (7.38)}$$

where,

D_g	=	gas-phase diffusivity of organic compound A in stagnant gas B, cm²/s
T	=	absolute temperature, K (273 + °C)
M_A, M_B	=	molecular weights of A and B, respectively, Da or g/mole
P_1	=	absolute pressure, N/m²
r_{AB}	=	molecular separation at collision, equal to $(r_A + r_B)/2$, nm
r_A	=	molecular separation at collision for component A, nm
r_B	=	molecular separation at collision for stagnant gas B, nm
ε_{AB}	=	energy of molecular attraction, equal to $\sqrt{\varepsilon_A \varepsilon_B}$, ergs (1 erg = 10^{-7}J)
ε_A	=	energy of molecular attraction for component A, ergs (1 erg = 10^{-7}J)
ε_B	=	energy of molecular attraction for stagnant gas B, ergs (1 erg = 10^{-7}J)
k	=	Boltzmann constant, 1.3804×10^{-16} g·cm²/s²·K
f	=	(kT/ε_{AB}) = collision function

The collision function originates from the Cunningham correction factor, already discussed in the begining of this chapter See. 7.2. The values of r_A and ε_A can be estimated for each component from the following equations:

$$r_A = 1.18 V_{b,A}^{1/3} \text{ (in nm for } V_{b,A} \text{ in l/mole)} \qquad \text{... (7.39)}$$

$$\frac{\varepsilon_A}{k} = 1.21 T_{b,A} \qquad \text{... (7.40)}$$

where,

r_A = molecular separation at collision for component A, nm
$V_{b,A}$ = molar volume of component A at normal boiling point, l/mole
$T_{b,A}$ = normal boiling point of component A, K

The diffusivity of a substance when the stagnant gas B is air can be calculated by assuming that air behaves like a single substance with respect to molecular collisions. The required parameters for air are

$$r_B = 0.3711 \text{ nm} \qquad \text{... (7.41)}$$

$$\frac{\varepsilon_B}{k} = 78.6 \qquad \text{..(7.42)}$$

$$\frac{\varepsilon_{AB}}{k} = \sqrt{\frac{\varepsilon_A}{k} \times \frac{\varepsilon_B}{k}} \qquad \text{... (7.43)}$$

and

$$f\left(\frac{kT}{\varepsilon_{AB}}\right) = 10^{\xi}$$

$$\xi = \begin{pmatrix} -0.14329 - 0.48343(ee) + 0.1939(ee)^2 + 0.13612(ee)^3 \\ -0.20578(ee)^4 + 0.083899(ee)^5 - 0.01149(ee)^6 \end{pmatrix} \qquad \text{... (7.44)}$$

$$ee = \log_{10}\left(\frac{kT}{\varepsilon_{AB}}\right) \qquad \text{... (7.45)}$$

MODELS FOR MASS TRANSFER AT AN INTERFACE

Mass transfer at an interface is an important concept in water treatment and is involved in these common treatment processes: air stripping, aeration, chlorination, adsorption, ion exchange, membrane processes and ozonation. To gain an understanding of mass transfer at an interface, the concept of mass flux is introduced first and then three theories for mass transfer at an interface are described.

Mass Transfer at Interfaces

The mass per time that is transferred from one phase to another depends on the mass transfer coefficient, the driving force for extraction and the surface area available for mass transfer per separation vessel volume. The driving force for extraction is caused by contacting enough extracting phase (e.g., air or activated carbon) that does not contain the contaminant with water and the minimum quantity and driving force for extraction are already discussed in the begining of this chapter. The surface area available per vessel volume is discussed at the end of this section. As given by the following expression, the mass flux is related to the mass transfer coefficient and the driving force as shown in Fig. 7.4:

$$N_A = k_f (C_b - C_s) \qquad \qquad ... (7.46)$$

where,

N_A = mass flux of solute A to interface, mg/m$^2 \cdot$s
k_f = fluid-phase (liquid or gas) mass transfer coefficient, m/s
C_s = concentration of solute A at interface, mg/l
C_b = concentration of solute A in bulk solution, mg/l

The mass transfer coefficient depends on the diffusivity and the mass transfer boundary layer thickness δ, as shown in Fig. 7.4. As shown in Fig. 7.4, the direction of the flux depends on the concentration gradient. Three prevailing theories are used to predict the mass transfer coefficient: (i) the film theory; (ii) the surface renewal or penetration theory; and (iii) the boundary layer theory. The three prevailing theories can be used to demonstrate the theoretical origin of mass transfer correlations. Many investigators have taken the theoretical forms of the mass transfer correlations and modified them to fit data that were collected for a variety of geometries (e.g., sphere, cylinder, plate) and flow regimes (e.g., laminar, transition and turbulent). Several of the common correlations used to determine mass transfer coefficients are given in Fig. 7.5.

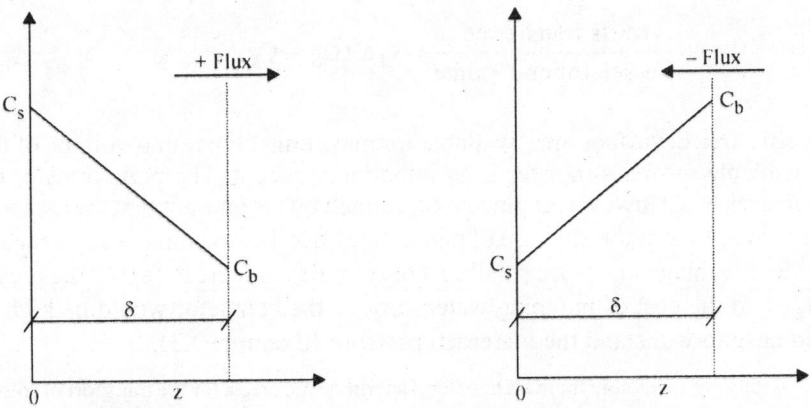

Fig. 7.4. Hypothetical fluxes at interface at steady state.

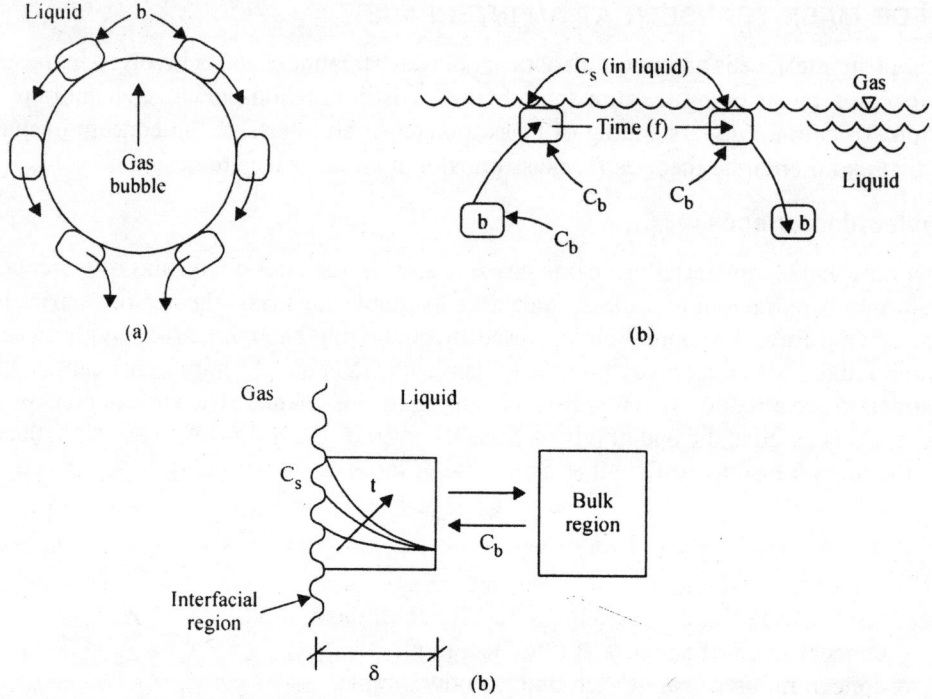

Fig. 7.5. Schematic of penetration and surface renewal theory for gas transfer: (a) gas bubble rising through liquid in which liquid packets b are in contact with the bubble for a time period t equal to the time it takes for the bubble to rise one diameter; (b) liquid packets b in turbulent eddy rising to liquid surface, as in an open channel and remaining in contact with gas for a time period t; and (c) increase in concentration of a dissolved gas at liquid surface interface with time for both cases (a) and (b).

Surface area available for mass transfer

To calculate the mass that is transferred to the extracting phase, the flux must be multiplied by the surface area available per contactor vessel volume as given by the expression

$$\frac{\text{Mass transferred}}{\text{Vessel volume} \times \text{time}} = k_f a \, (C_b - C_s) \qquad \qquad \text{... (7.47)}$$

where,

a = specific area or surface area available for mass transfer per unit volume of the vessel, m^2/m^3

The area a available for mass transfer is an important concept. The mass transfer rate can increase linearly with increases in a. However, engineers are limited by the amount that the area a can be increased because increases in a come at the expense of higher headloss. For example, in a packed bed of activated carbon it would be advantageous to use small carbon granules to increase a, but the pressure drop would become too large and the cost of pumping water through the contactor would be high. In addition, the contactor would have to withstand the increased pressure (Example 7.3).

Example 7.3: Calculating area available for mass transfer: Determine the area a for the transport of solute to PAC particles. The PAC concentration is 0.1 per cent by weight and the particle diameter is 44 μm. The density of the particles is 1.7 g/cm^3. Assume the density of water is 1 g/cm^3 and the surface of the PAC is like that of a smooth sphere.

Solution:

1. Calculate the concentration of particles in the water:

$$C_p = \frac{\text{Mass of particle}}{\text{Mass of water}} \times \frac{\text{Mass of water}}{\text{Volume of water}} = \frac{0.1\,g}{100\,g} \times \frac{1\,g}{cm^3} = 1000\,g/m^3$$

2. Calculate the particle area available for mass transfer:

$$a = \frac{\text{Surface area of particle}}{\text{Volume of particle}} \times \frac{\text{Volume of particle}}{\text{Mass of particle}} \times \frac{\text{Mass of particle}}{\text{Volume of water}}$$

$$= \left(\frac{4\pi R^2}{(4/3)\pi R^3}\right)\left(\frac{1}{\rho_p}\right) C_p$$

where, ρ_p = particle density, g/cm³

$$a = \left(\frac{3}{R}\right)\left(\frac{1}{\rho_p}\right) C_p = \left(\frac{3}{22 \times 10^{-6}\,m}\right)\left(\frac{1}{1.7\,g/cm^3}\right)\left(\frac{m^3}{10^6\,cm^3}\right)\left(\frac{1000\,g}{m^3\,water}\right)$$

$$= 80\ m^2/m^3\ water$$

Film theory

The film theory is the most straightforward of the theories that explain mass transfer at an interface. The system is considered to be composed of a well-mixed bulk solution (either gas or liquid), a stagnant film layer and an interface to another phase (for instance, a solid surface), as shown in Fig. 7.4. As a result of the solution being well-mixed, solutes are transported continually to the edge of the stagnant film layer and no concentration gradients exist in the bulk solution. Mass transfer to the interface occurs when the concentration at the interface to the other phase is different than the concentration in the bulk solution, causing a concentration gradient across the film layer. Because this layer is quiescent, the sole mechanism for transport across this layer is molecular diffusion. Processes that occur at the actual interface (such as a chemical reaction or adsorption to the surface) air assumed to occur much faster than the rate of diffusion and as a result, the rate of mass transfer is described by Fick's first law for diffusion across the film layer:

$$N_A = \text{Flux to interface} = -D_f\frac{dC}{dz} = -\frac{D_f}{\delta}(C_s - C_b) = k_f(C_b - C_s) \qquad \ldots (7.48)$$

where,

$\quad N_A$ = mass flux, mg/m²·s

$\quad D_f$ = fluid-phase diffusion coefficient of solute A, m²/s

$\quad k_f$ = fluid-phase mass transfer coefficient of solute A, m/s

$\quad \delta$ = film thickness, which is shown in Fig. 7.4, m

$\quad C$ = concentration, mg/l

$\quad z$ = distance in direction of mass transfer (or in direction of concentration gradient), m

In the film theory, the mass transfer coefficient is explicitly related to the film thickness, as shown in the expressions

$$k_f = \frac{Df}{\delta} \qquad \ldots (7.49)$$

$$\frac{k_f \delta}{D_f} = 1 = Sh \qquad \qquad ... (7.50)$$

where,

Sh = Sherwood number, dimensionless

The film thickness will vary from 10 to 100 μm for liquids and from 0.1 to 1 cm for stagnant gases. Unfortunately, there is no way to calculate the film thickness based on fluid mixing; consequently, the film theory cannot be used to determine the local mass transfer coefficient. Nevertheless, the film theory is used frequently to develop a conceptual view of mass transfer across an interface and to illustrate the importance of diffusion in controlling the rate of mass transfer.

Penetration and surface renewal theory

According to the penetration theory and the surface renew theory, packets of water move up to the gas-water interface and transfer solute either from the gas to the water or from the water to the gas and then the packets of fluid return to the bulk solution. The transport of fluid packets to and from the surface is shown in Figs. 7.5a and 7.5b. When a packet of water moves to the interface, the concentration of dissolved gases increases during aeration as shown on the bottom of Fig. 7.5c.

The essential difference between the penetration and surface renewal theory is that for penetration theory a fixed residence time at the surface is assumed. For surface renewal theory, it is assumed that there is an equal probability for the fluid elements to move to and from the surface up to a certain residence time. It is likely that surface renewal theory is more plausible because fluid turbulence is thought to be responsible for transporting the fluid element to and from the surface. Consequently, the time that fluid packets remain on the surface is random and all residence times are equally plausible. The surface renewal theory can be used to derive the theoretical basis for predicting mass transfer coefficients.

Boundary layer theory

When fluid passes a solid flat plate, a velocity gradient forms because the fluid velocity is assumed to be zero at the surface (no slip condition). This condition allows the concentration gradient to extend further into solution than it does in the penetration theory because the fluid is not able to move freely to the solid surface. The concentration gradient that is shown in Fig. 7.6 is for the case where solute is transported from the surface of the plate to the bulk solution and laminar flow conditions exist. Based on the mass balance for laminar flow past a flat plate, the following theoretical mass transfer correlation can be derived:

$$\frac{K_{f(avg)} L}{D_f} = 0.664 \, Re^{1/2} \, Sc^{1/3} \qquad \qquad ... (7.51)$$

where,

$k_{f(avg)}$ = Average fluid-phase mass transfer coefficient, m/s
L = Length of channel, m
D_f = Fluid-phase diffusion coefficient, m^2/s
Re = Reynolds number, dimensionless
Sc = Schmidt number, dimensionless

The Reynolds and Schmidt numbers for flow past a flat plate are defined as follows:

$$Re = \frac{\rho_f vL}{\mu_f} \qquad \ldots (7.52)$$

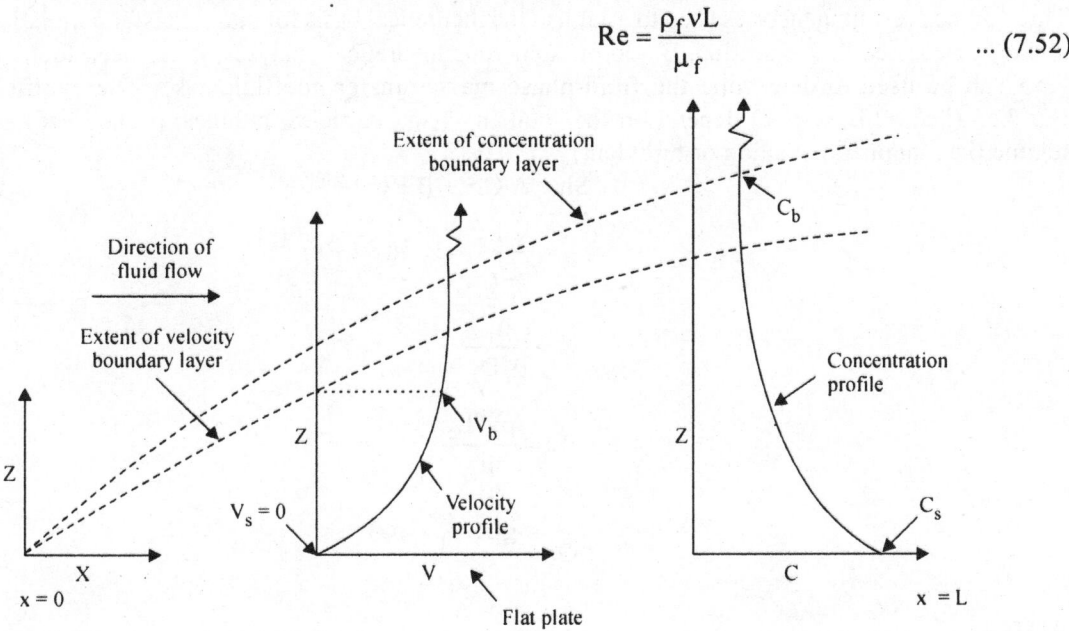

Fig. 7.6. Boundary layer theory diagram showing velocity and concentration profiles for laminar flow across flat plate.

$$Sc = \frac{\mu_f}{\rho_f D_f} \qquad \ldots (7.53)$$

where,

P_f = fluid-phase density, kg/m^3

v = velocity above boundary layer, m/s

μ_f = fluid-phase viscosity, kg/m·s

An identical result to Eq. 7.50 is found for isolated spheres that is used later in a correlation that was proposed by Gnielinski. However, Re is defined differently for a sphere:

$$Re = \frac{\rho_f vD}{\mu_f} \qquad \ldots (7.54)$$

where,

D = diameter of sphere, m

CORRELATIONS FOR MASS TRANSFER AT AN INTERFACE

Numerous mass transfer correlations have been developed to estimate mass transfer coefficients for various geometries and flow regimes. Topics considered in this section include: (i) the theoretical basis for mass transfer correlations; (ii) common forms of mass transfer correlations; and (iii) the relationship between mass transfer coefficients and diffusing species.

Theoretical Basis

Three of the prevailing theories used to establish the theoretical basis for mass transfer correlations were already discusssed in this section. Based on numerous theoretical analyses, it has been found that Eq. 7.55 can be used to determine the fluid-phase mass transfer coefficient k_f. The coefficients in Eq. 7.55 (i.e., A, B, a, b, c) depend on the geometry (e.g., particles, bubbles, packed bed) and flow regime (i.e., laminar, transition or turbulent):

$$Sh = A \, Gr^c + B \, Re^a \, Sc^b \qquad \qquad \text{... (7.55)}$$

$$= \frac{k_f L_c}{D_f} = f(Gr, Sc, Re) \qquad \qquad \text{... (7.56)}$$

$$Sc = \frac{\mu_f}{\rho_f D_f}$$

$$Re = \frac{\rho_f v L_c}{\mu_f} \qquad \qquad \text{... (7.57)}$$

$$Gr = \frac{gL_c^3 (\rho_p - \rho_f)}{\rho_f \upsilon^2} \qquad \qquad \text{... (7.58)}$$

where,

Sh	=	Sherwood number, dimensionless
Sc	=	Schmidt number, equal to ratio of momentum diffusivity to mass diffusivity, dimensionless
Gr	=	Grashof number, equal to ratio of buoyant forces to viscous forces, dimensionless
Re	=	Reynolds number, equal to ratio of inertial forces to viscous forces, dimensionless
A, B, a, b, c	=	constants, unitless
k_f	=	fluid-phase mass transfer coefficient, m/s
L_c	=	characteristic length, m
D_f	=	liquid-phase diffusion coefficient, m²/s
μ_f	=	fluid-phase viscosity, kg/m·s
P_f	=	fluid-phase density, kg/m³
ρ_p	=	particle density, kg/m³
v	=	velocity, m/s
υ	=	fluid-phase kinematic viscosity, equal to μ_f/ρ_f m²/s
g	=	acceleration due to gravity, 9.81 m/s²

The first term on the right side of Eq. 7.55, A Gr^c, accounts for molecular diffusion conditions under which there is no advection. The Grashof number accounts for mass transfer due to natural advection, which can be caused by density differences at the interface. A familiar example from heat transfer is a space heater in which air flows past the heater as a result of lighter, heated air that rises. A similar phenomenon is observed for salts dissolved in water because the solution at the salt interface is usually denser and fluid flow will occur at the interface without forced advection. However, in many cases, the buoyant force is not important in environmental problems because the solutions are very dilute. In such a case, the first term is simply a constant and Eq. 7.55 becomes

$$Sh = A + B \ Re^a Sc^b \qquad \qquad ... (7.59)$$

The second term on the right side of Eqs. 7.55 and 7.59, $B \ Re^a \ Sc^b$, accounts for mass transfer that is enhanced by advection. Many investigators have developed the theoretical bases for Eq. 7.55 for various geometries and flow regimes and have also developed mass transfer correlations by fitting data to Eq. 7.59 for various geometries and flow regimes. The forms of mass transfer correlations for different geometry and flow conditions are discussed briefly in the next section. In spite of the plethora of papers on mass transfer, mass transfer correlations for complex geometry and flow conditions do not exist for every situation. For such cases, it is possible to estimate the mass transfer coefficient for component A if the mass transfer for another component, B, is known. This relationship is discussed at the end of this section.

Common Forms of Mass Transfer Correlations

A collection of mass transfer coefficient correlations for a variety of transfer situations are shown in Table 7.7. Some of these correlations are obtained by analogy to heat transfer, others by measurement or theoretical approximation. Correlations presented in Table 7.7 include: (i) Gnielinski correlation for packed beds, which can be used for calculating mass transfer coefficients in gas and liquid-phase adsorption; (ii) Onda correlation for absorption and air stripping, which can be used to determine the mass transfer coefficient in packed-tower air stripping; and (iii) the Gilliland correlation for gases and liquids in pipes. (***See table 7.7 at the last of this Chapter)

Relationship Between Mass Transfer Coefficients and Diffusing Species

Despite the availability of mass transfer correlations for diverse situations, there are many cases for which mass transfer correlations do not exist. Under such circumstances, the mass transfer coefficient of one solute can be used to estimate the mass transfer coefficient of another. For example, if the mass transfer coefficient for oxygen is known, the mass transfer coefficient for other compounds can be calculated. The procedure is convenient because the mass transfer coefficient for dissolved oxygen is relatively easy to measure and many correlations are available for oxygen transfer.

Mass transfer correlations depend on the exponent of the Schmidt number (i.e., the b that appears in Eq. 7.59, Sc^b) and this dependency allows for estimation of the mass transfer coefficient of one compound from the mass transfer coefficient of another compound. The two situations that arise are: (i) the fluid moves freely to the surface and surface renewal occurs; and (ii) the fluid cannot move freely to the surface because the interface is a solid and a boundary layer forms. The surface renewal case yields a dependence on the Sherwood number as shown here:

$$Sh = \frac{k_f L_c}{D_f} \ \alpha \ Sc^{0.5} = \left(\frac{\mu_f}{\rho_f D_f} \right)^{0.5} \qquad \qquad ... (7.60)$$

where,

k_f = fluid-phase mass transfer coefficient, m/s

L_c = characteristic length, m

D_f = fluid-phase diffusion coefficient, m²/s

Sc = Schmidt number, dimensionless

Table 7.7. Common mass transfer correlations.

Description	Empirical correlation	Nomenclature
Gnielinski correlation[a] for packed beds; used for calculating external liquid phase adsorbate mass transfer coefficient in packed beads	$$k_f = \frac{[1+1.5(1-\varepsilon)]D_f}{d_p}(2+0.644Re^{1/2}Sc^{1/3})$$ $$Re = \frac{\rho_f \Phi d_p V_L}{\varepsilon \mu_f}$$ $$Sc = \frac{\mu_f}{\rho_f D_f}$$ Constraints: Pe = Re.Sc > 500 $0.7 < Sc < 10^4$, $Re < 2 \times 10^4$ $0.26 < \varepsilon < 0.935$	K_f = fluid-phase mass transfer coefficient, m/s D_f = fluid-phase diffusion coefficient, m²/s ε = bed void fraction, dimensionless d_p = fluid-phase density, kg/m³ Φ = sphericity, equal to ratio of surface area of sphere/volume of sphere to surface area of particle/volume of particle, dimensionless V_L = liquid superficial velocity, m/s μ_f = fluid-phase viscosity, kg/m · s Re = Reynolds number, dimensionless Sc = Schmidt number, dimensionless[b] Pe = Peclet number, dimensionless
Onda correlation[c] for absorption or stripping in packed tower; used for calculating liquid and gas-phase mass transfer coefficient in packed air absorber or stripper.	$$k_l = 0.0051\left(\frac{L_m}{a_w\mu_l}\right)^{2/3}\left(\frac{\mu_l}{\rho_l D_l}\right)^{-0.5}(a_t d_p)^{0.4}\left(\frac{\rho_l}{\mu_l g}\right)^{-1/3}$$ $$k_g = 5.23(a_t D_g)\left(\frac{G_m}{a_t\mu_g}\right)^{0.7}\left(\frac{\mu_g}{\rho_g D_g}\right)^{1/3}(a_t d_p)^{-2}$$ $$a_w = a_t\left\{1-\exp\left[-1.45\left(\frac{\sigma_c}{\sigma}\right)^{0.75}(Re)^{0.1}(Fr)^{-0.05}(We)^{0.2}\right]\right\}$$ $$Re = \frac{L_m}{a_t\mu_l} \quad Fr = \frac{(L_m)^2 a_t}{(\rho_l)^2 g} \quad We = \frac{(L_m)^2}{\rho_l a_t \sigma}$$ Constraints: $d_p < 0.0508$ m (2 inch) $0.8 < L_m < 43$ kg/m²·s	k_l = liquid-phase mass transfer rate coefficient. m/s L_m = liquid mass loading rate, kg/m² · s a_w = wetted surface area of packing, m⁻¹ a_t = specific surface area of packing, m⁻¹ D_l = liquid-phase diffusivity, m²/s dp = nominal packing diameter, m ρ_l = liquid-phase density, kg/m³ μ_l = liquid-phase viscosity, kg/m·s g = acceleration of gravity, 9.81 m/s² k_g = gas-phase mass transfer coefficient, m/s μ_g = gas-phase viscosity, kg/m·s D_g = gas-phase diffusion coefficient, m²/s ρ_g = gas-phase density, kg/m³

(Contd...)

Description	Empirical correlation	Nomenclature
	$(1.1 < L_m < 63 \text{ gpm/ft}^2)$	G_m = gas mass loading rate, kg/m^2·s
	$0.014 < G_m < 1.7 \text{ kg/m}^2 \cdot \text{s}$	σ_c = critical surface tension of packing, kg/s^2
	$(2.21 < G_m < 268 \text{ cfm/ft}^2)$	σ = surface tension of water, kg/s^2
		Re = Reynolds number, dimensionless
		Fr = Froude number, dimensionless
		We = Weber number, dimensionless
Gilliland correlation[d] for gases or liquids in pipes; used for calculating gas or liquid-phase mass transfer coefficient in pipes	$\text{Sh} = \dfrac{k_f D}{D_g (\text{or} D_l)} = 0.023 \text{Re}^{0.83} \text{Sc}^{0.33}$	Sh = Sherwood number, dimensionless
		K_f = fluid-phase mass transfer coefficient, m/s
	$\text{Re} = \dfrac{\rho_g D V_s}{\mu_g} \left(\text{or} \dfrac{\rho_l D U_L}{\mu_l} \right)$	D = diameter of pipe, m
		D_l = liquid-phase diffusivity, m^2/s
		D_g = gas-phase diffusivity, m^2/s
	$\text{Sc} = \dfrac{\mu_g}{\rho_g D_g} \left(\text{or} \dfrac{\mu_l}{\rho_l D_l} \right)$	Re = Reynolds number, dimensionless
		Sc = Schmidt number, dimensionless
	For turbulent flow;	ρ_g = gas density, kg/m^3
	Re > 2100	μ_g = gas viscosity, kg/m·s
	$0.6 < \text{Sc} < 3000$	ρ_l = liquid density, kg/m^3
		μ_l = liquid viscosity, kg/m·s
		V_s = superficial gas velocity, m/s
		U_L = liquid superficial velocity, m/s

[a] Adapted from Gnielinski

[b] Typical values for the Schmidt number for gas and liquid phases are 0.7 and 1000, respectively.

[c] Adapted from Onda.

[d] Adapted from Gilliland and Sherwood and Linton and Sherwood.

Thus, from Eq. 7.60, it can be seen that k_f depends on the diffusivity according to the expression:

$$k_f \propto D_f^{0.5} \qquad \qquad ...(7.61)$$

Furthermore, if $k_{f,A}$ is known, then $k_{f,B}$ can be determined from the relationship.

$$\frac{k_{f,B}}{k_{f,A}} = \left(\frac{D_{f,B}}{D_{f,A}}\right)^{0.5} \qquad \qquad ...(7.62)$$

where,

$k_{f,B}$ = fluid-phase mass transfer coefficient for B, m/s
$k_{f,A}$ = fluid-phase mass transfer coefficient for A, m/s
$D_{f,B}$ = fluid-phase diffusion coefficient of B, m²/s
$D_{f,A}$ = fluid-phase diffusion coefficient of A, m²/s

As discussed earlier in this section the boundary layer theory yields a dependence of Sh as shown here:

$$Sh = \frac{k_f L}{D_f} \propto Sc^{1/3} = \left(\frac{\mu_f}{\rho_f D_f}\right)^{1/3} \qquad \qquad ...(7.63)$$

From Eq. 7.63, k_f depends on the diffusivity according to the expression

$$K_f \propto D_f^{2/3} \qquad \qquad ...(7.64)$$

If $k_{f,A}$ is known, then $k_{f,B}$ can be determined from the following expression

$$\frac{k_{f,B}}{k_{f,A}} = \left(\frac{D_{f,B}}{D_{f,A}}\right)^{2/3} \qquad \qquad ...(7.65)$$

The dependency of the ratio of the mass transfer coefficients for constituents A and B in the ratio of the diffusivities of A and B for some common situations that are encountered in water treatment are given in Table 7.8.

Table 7.8. Dependency of ratio of mass transfer coefficients for compounds A and B on ratio of diffusivities of A and B for some common situations in water treatment.

Situation	$\frac{k_{f,B}}{k_{f,A}} = \left(\frac{D_{f,B}}{D_{f,A}}\right)^{n}$	Comment
Transport from fluid to solid or from solid to liquid	$\frac{k_{f,B}}{k_{f,A}} = \left(\frac{D_{f,B}}{D_{f,A}}\right)^{2/3}$	A boundary layer forms because the velocity at the solid-fluid interface is zero
Mass transfer resistance in water at air-water interface	$\frac{k_{l,B}}{k_{l,A}} = \left(\frac{D_{l,B}}{D_{l,A}}\right)^{0.5}$	For an air-water interface, there is no velocity gradient for the water at the boundary because the viscosity of air is 50 times lower than that for water at 1 atm
Mass transfer resistance in air at air–water interface	$\frac{k_{g,B}}{k_{g,A}} = \left(\frac{D_{g,B}}{D_{g,A}}\right)^{2/3}$	At the air-water interface, a boundary layer can form for air-phase mass transport because the viscosity of air is approximately 50 times lower at 1 atm than the water viscosity and a gas-phase velocity gradient appears. The only exception would be at very high air Reynolds numbers and the dependency would tend toward 1/2

Situation	$$\frac{k_{f,B}}{k_{f,A}} = \left(\frac{D_{f,B}}{D_{f,A}}\right)^{n}$$	Comment

where,

$k_{f,B}$ = fluid-phase mass transfer coefficient for B, m/s

$k_{f,A}$ = fluid-phase mass transfer coefficient for A, m/s

$D_{f,B}$ = fluid-phase diffusion coefficient for B, m²/s

$D_{f,A}$ = fluid-phase diffusion coefficient for A, m²/s

n   = exponent used to describe the relationship between the ratio of the mass transfer coefficients of compounds A and B to the ratio of diffusion coefficients of compounds A and B

$k_{l,A}$ = liquid-phase mass transfer coefficient for A, m/s

$k_{l,B}$ = liquid-phase mass transfer coefficient for B, m/s

$D_{l,A}$ = liquid diffusion coefficient for A, m²/s

$D_{l,B}$ = liquid diffusion coefficient for B, m²/s

$k_{g,A}$ = gas-phase mass transfer coefficient for A, m/s

$k_{g,B}$ = gas-phase mass transfer coefficient for B, m/s

$D_{g,A}$ = gas-phase diffusion coefficient for A, m²/s

$D_{g,B}$ = gas-phase diffusion coefficient for B, m²/s

TWO-FILM THEORY FOR MASS TRANSFER ACROSS A GAS-LIQUID INTERFACE

The various models and correlations that have been developed to describe the transport across a single interface have been introduced and discussed in this chapter. In this section, the methods for describing the transport of solute across a gas-liquid interface are considered. The two-film theory can be used to describe mass transfer across an interface. In applying the two-film theory, two situations are considered: (i) without chemical reaction; and (ii) with chemical reaction.

Without Chemical Reaction

The driving force for mass transfer between one phase and another results from the displacement of the system from equilibrium. The two-film theory is a model that is used to describe the interaction of two films (one gas and one liquid) at the gas-liquid interface. The two situations where mass transfer occurs between air and water at steady state are shown in Fig. 7.7. The situation for stripping where mass is transferred from the water to the air is shown in Fig. 7.7a and the situation for absorption in which mass is transferred from the air to the water is shown in Fig. 7.7b. A detailed explanation is only provided for stripping because the mechanisms and assumptions for mass transfer are essentially the same for both cases and the only difference is that mass is transferred in the opposite direction.

Conditions in bulk solution

The two-film theory is used to describe the mass transfer rate for: (i) the air stripping of VOCs such as methane, trichloroethane and tetrachloroethane and other gases such as hydrogen sulphide; and (ii) the absorption of gases such as oxygen, carbon dioxide, nitrogen or ozone. The following discussion will address the stripping of a volatile component A from water. As shown in Fig. 7.7a, the concentration of A in the bulk water is larger than the concentration of A at the air-water interface. Consequently, A diffuses from the bulk solution, where its liquid concentration is C_b, to the air-water interface where its liquid concentration is C_s. The difference between C_b and C_s is the driving force for stripping in the liquid phase. There is a discontinuity in the concentration at the air-water interface because A partitions in air at a different concentration based on equilibrium or Henry's law. Similarly, the concentration of A in the air at the air-water interface, y_s, is larger than the concentration of A in the bulk air, y_b and it diffuses from the air-water interface to the bulk air. The difference between y_s and y_b is the driving force for stripping in the gas phase.

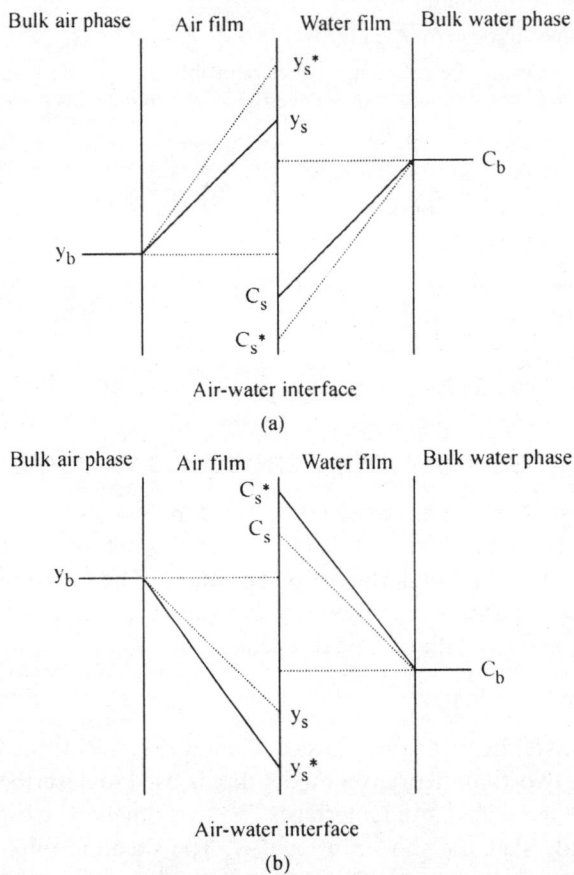

Fig. 7.7. Two film theory: mass transfer driving gradients that occur for (a) stripping and (b) absorption.

Conditions at interface

Local equilibrium occurs at the air-water interface because random molecular movement (on a local scale of nanometers in water and thousands of nanometers on the air side) causes constituent A to dissolve in the aqueous phase and volatilise into the air at a rate more rapidly than diffusion to or away from the air-water interface. Accordingly, local equilibrium may be assumed and Henry's law can be used to relate y_s to C_s:

$$y_s = HC_s \qquad \qquad \qquad ... (7.66)$$

where,

y_s = gas-phase concentration of A at air-water interface, mg/l
H = Henry's law constant, L of water/l of air, dimensionless
C_s = liquid-phase concentration of A at air-water interface, mg/l

For a dilute solution, the flux of A through the gas phase film must be equal to the flux through the liquid phase film. Thus:

$$N_A = k_l (C_b - C_s) = k_g (y_s - y_b) \qquad \qquad ... (7.67)$$

where,

N_A = flux of A across air-water interface, mg/m^2.s
k_l = liquid-phase mass transfer coefficient for rate at which contaminant A is transferred from bulk aqueous phase to air water interface, m/s
C_b = liquid-phase concentration of A in bulk solution, mg/l
C_s = liquid-phase concentration of A at air-water interface, mg/l
k_g = gas-phase mass transfer coefficient for rate at which contaminant A is transferred from air- water interface to bulk gas phase, m/s
y_s = gas-phase concentration of A at air-water interface, mg/l
y_b = gas-phase concentration of A in bulk solution, mg/l

Both k_l and k_g are sometimes referred to as local mass transfer coefficients for the liquid and gas phases because they depend upon the conditions at or near the air-water interface in their particular phase. The flux cannot be determined directly from Eq. 7.67 because the interfacial concentrations y_s and C_s are not known and cannot be measured easily. Consequently, it is necessary to define another flux equation in terms of hypothetical concentrations that are easy to determine. If it is hypothesised that all the resistance to mass transfer is on the liquid side, then there is no concentration gradient on the gas side and a hypothetical concentration, C_s^* can be defined as shown in Fig. 7.7a:

$$Y_b = HC_s^* \qquad \qquad ... (7.68)$$

where,

C_s^* = liquid-phase concentration of A that is in equilibrium with bulk air concentration, mg/l.

Alternatively, if it is hypothesised that all the resistance to mass transfer is on the gas side, then there is no concentration gradient on the liquid side and a hypothetical concentration y_s^* can be defined as shown in Fig. 7.7b:

$$y_s^* = HC_b \qquad \qquad ... (7.69)$$

where,

y_s^* = gas-phase concentration of A that is in equilibrium with bulk water concentration, mg/l.

Overall mass transfer relationship

For stripping operations, mass balances are normally written on the liquid side and it is convenient to calculate the mass transfer rate using the hypothetical concentration C_s^* and an overall mass transfer coefficient K_L, as shown in the equation

$$N_A = K_L (C_b - C_s^*) \qquad \qquad ... (7.70)$$

where,

N_A = mass flux of A across air-water interface, mg/m^2·s
K_L = overall mass transfer coefficient, m/s

C_b = liquid-phase concentration of A in bulk solution, mg/l

C_s^* = liquid-phase concentration of A at air-water interface assuming no concentration gradient in air phase, mg/l.

In all cases, the hypothetical, gas-side and liquid-side mass fluxes that are given in Eqs. 7.67 and 7.70 are all equal to one another:

$$N_A = k_l = (C_b - C_s) = k_g (y_s - y_b) = K_L (C_b - C_s^*) \qquad ...(7.71)$$

Equation 7.71 relates K_L to k_l and k_g and accounts for mass transfer resistances on both the gas and liquid sides of the interface, which is known as the two-film theory. The overall mass transfer coefficient can be related to the local mass transfer coefficients starting with the relationship

$$C_b - C_s^* = (C_b - C_s) + (C_s - C_s^*) \qquad ...(7.72)$$

Substituting $C_s = y_s/H$ and $C_s^* = Y_b/H$ into Eq. 7.72 and combining with Eq. 7.71 yields

$$\frac{N_A}{K_L} = \frac{N_A}{k_l} + \frac{N_A}{Hk_g} \qquad ...(7.73)$$

or

$$\frac{1}{K_L} = \frac{1}{k_l} + \frac{1}{Hk_g} \qquad ...(7.74)$$

Thus, according to the two-film theory, the mass flux across the interface, can be calculated using the expression

$$N_A = K_L (C_b - y_b/ H) \qquad ...(7.75)$$

Equation 7.75 is convenient to use in mass balances because the driving force for stripping $(C_b - y_b/H)$ involves concentrations that are easy to measure. The overall mass transfer coefficient can be estimated from the local mass transfer coefficients and the local mass transfer coefficients can be determined from correlations.

Determining phase that controls mass transfer rate

Evaluating which phase controls the mass transfer rate is important in optimising the design and operation of aeration and air-stripping processes. For example, when the liquid-phase resistance controls the mass transfer rate, increasing the mixing of the air will have little impact on the removal efficiency. From Eq. 7.74, the overall resistance to mass transfer is equal to the sum of the resistance in the liquid and gas phases and can be re-written as

$$R_T = R_L + R_G/H \qquad ...(7.76)$$

where,

R_T = overall resistance to mass transfer, equal to $1/K_L$, s/m

R_L = liquid-phase resistance to mass transfer, equal to $1/k_l$, s/m

R_G = gas-phase resistance to mass transfer, equal to $1/k_g$, s/m

To evaluate which phase controls the rate of mass transfer, Eq. 7.76 can be rearranged as follows:

$$\frac{R_L}{R_T} = \frac{1/k_l}{1/k_l + 1/Hk_g} = \frac{1}{1 + \dfrac{1}{(k_g/k_l)H}} \qquad ...(7.77)$$

$$\frac{R_L}{R_T}(\text{per cent}) = \frac{100}{1 + \dfrac{1}{(k_g/k_l)H}} \qquad \ldots (7.78)$$

Reported values of the ratio k_g/k_l range are: (i) 40 to 200 for surface aerators; (ii) 5 to 50 for packed towers; and (iii) 2.2 to 3.6 for diffused bubble aeration. Assuming $k_g/k_l = 100$, Eq. 7.78 can be used to determine which phase controls the rate of mass transfer across the interface. Generally, the liquid phase controls the rate of mass transfer for compounds with H values greater than 0.05. The gas phase controls mass transfer of compounds with H values less than 0.002. For compounds with H values between 0.002 and 0.05 either the liquid or the gas phase may control the mass transfer rate. Accordingly, the general trend is that the phase that controls the mass transfer rate is the phase that is less preferred by the solute.

Application of two-film theory

When designing aeration and stripping processes, the rate of mass transfer is often expressed on a volumetric basis rather than an interfacial area basis. The flux term is converted to a volumetric basis by multiplying by the surface area available for mass transfer per contactor vessel volume, a, as already defined in the preveious section. Equation 7.74 can be expressed in terms of a volumetric mass transfer rate by dividing by the area a:

$$\frac{1}{K_L a} = \frac{1}{k_l a} + \frac{1}{Hk_g a} \qquad \ldots (7.79)$$

where,

   K_L = overall liquid mass transfer coefficient, m/s
   a   = surface area available for mass transfer per volume of contactor, m^2/m^3
   k_l = liquid-phase mass transfer coefficient, m/s
   k_g = gas-phase mass transfer coefficient, m/s

Although Eq. 7.79 is based on the liquid-side mass transfer resistance, it includes the resistance to mass transfer in the gas phase and is an exact representation of the mass flux across the air-water interface.

A similar relationship can be developed for an overall mass transfer coefficient on the gas side, which is useful for absorption operations. The mass flux may be obtained from the gas-phase mass balance by multiplying the overall gas-phase mass transfer coefficient K_G by the hypothetical driving force $(Y_b - y_s^*)$. The final equations are given as:

$$N_A = K_G (Y_b - y_s^*) \qquad \ldots (7.80)$$

$$\frac{1}{K_G} = \frac{1}{k_l} + \frac{1}{k_g} \qquad \ldots (7.81)$$

$$\frac{1}{K_G a} = \frac{1}{k_l a} + \frac{1}{k_g a} \qquad \ldots (7.82)$$

where,

   K_G = overall gas-phase mass transfer coefficient, m/s

H = Henry's law constant L of water/l of air, dimensionless

k_l = liquid-phase mass transfer coefficient, m/s

k_g = gas-phase mass transfer coefficient, m/s

With Chemical Reaction

If a chemical reaction occurs after a solute enters the water at the air–water interface, the mass transfer rate of absorption may be faster than the rate of transfer by diffusion alone. The reason for the increase is that the reaction occurs within the mass transfer boundary layer. The reaction causes a much sharper concentration gradient; and, as predicted by Fick's first law, the larger concentration gradient causes a faster mass transfer rate. There are two possible situations: (i) as shown in Fig. 7.8a, some solute is left after the mass transfer boundary layer; and (ii) as shown in Fig.7.8b, there is no solute left after the mass transfer boundary layer because the chemical reaction proceeds rapidly.

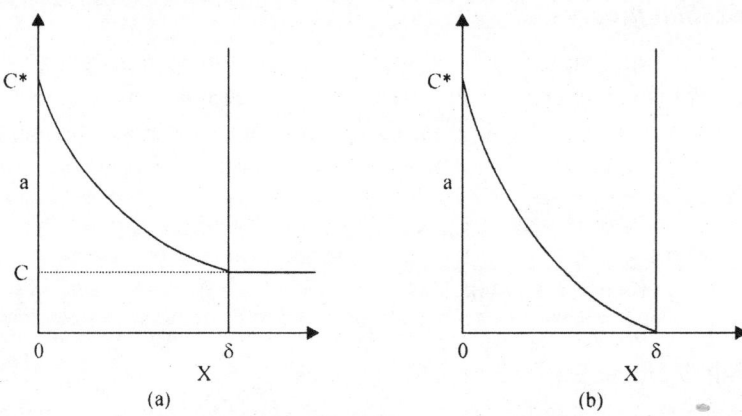

Fig. 7.8. Concentration profiles in the mass transfer boundary layer for two situations where a chemical reaction is occurring: (a) some solute is left after the mass transfer boundary layer; and (b) no solute is left after the mass transfer boundary layer.

Several gases commonly used in water treatment applications, notably chlorine, sulphur dioxide, carbon dioxide and ozone, undergo hydrolysis reactions or rapid chemical reactions with other solutes and water. The increase in rate of mass transfer must be considered in the process design of absorption equipment. The magnitude of this increase depends on the type of chemical reaction (e.g., reversible, irreversible, series), reaction order, reaction rate constants, concentration of reactants and the solute diffusion coefficients.

While innumerable reaction combinations are possible, chemical reactions for gases in water used in water treatment fall mainly in two categories: first-order, irreversible reactions and rapid or instantaneous reversible reactions.

First-order, irreversible reactions

Consider a first-order, irreversible reaction in the water phase with rate constant k_l, as illustrated by the nearly irreversible reaction of SO_2 in water:

$$SO_2(g) + H_2O \xrightarrow{k_l} HSO_3^- + H^+ \qquad \qquad ... (7.83)$$

where,

k_1 = first-order reaction rate constant, s^{-1}

Assuming that mass transfer resistances in the gas phase are negligible, it has been shown that the gas transfer flux at steady state is given by the expression

$$N_A = k_1 \left(C^* - \frac{C}{\cosh(Ha)} \right) \frac{Ha}{\tanh(Ha)} \qquad ... (7.84)$$

where,

N_A = gas transfer flux, $mg/m^2 \cdot s$
k_1 = liquid-phase mass transfer coefficient, m/s
C^* = liquid-phase concentration in equilibrium with bulk gas concentration, mg/L
C = concentration in bulk water, mg/l
Ha = Hatta number, dimensionless
cosh = hyperbolic cosine
tanh = hyperbolic tangent

The Hatta number is given by the expression

$$Ha = \frac{\sqrt{D_t k_1}}{k_1} \qquad ... (7.85)$$

where,

D_t = molecular diffusion coefficient of solute in water, m^2/s
k_1 = first-order reaction rate constant, s^{-1}

When a single gas such as pure chlorine or carbon dioxide is absorbed, mass transfer resistances in the gas phase are negligible and Eq. 7.84 can be used to predict the enhancement of the mass transfer rate due to reaction. The mass transfer coefficient k_1 shown in Eq. 7.84 can be estimated using a mass transfer correlation and the Hatta number can be calculated from the diffusivity and first-order rate constant of the species.

The second term that appears on the right-hand side of Eq. 7.84 defines the driving force. If Ha > 5, the reaction occurs entirely in the water near the interface and the bulk water concentration of the dissolving solute is zero, as shown in Fig. 7.8b. In this case, Eq. 7.84 simplifies to the following expression [note that tanh (5) = 1]:

$$N_A = k_1 \, Ha \, C^* \qquad ... (7.86)$$

The flux in the absence of reaction for a zero bulk solution concentration would be:

$$N_A = k_1 \, C^* \qquad ... (7.87)$$

By comparing Eqs. 7.86 and 7.87, it is seen that the rate of mass transfer is enhanced by a factor that is equal to the Hatta number, Ha:

$$E_{flux} = \frac{\text{Flux with reaction}}{\text{Flux without reaction}} = \frac{(Ha)k_1}{k_1} = Ha \qquad ... (7.88)$$

where,

E_{flux} = enhancement in mass transfer due to reaction
Enhancements of Ha = 1000 are possible for fast radical reactions.

Relationship between Overall Mass Transfer Coefficients and Diffusing Species

To relate the mass transfer coefficients of one compound to another when mass transfer resistances exist on both the water and gas side of the interface, Eq. 7.74 or 7.79 is combined in several ways depending on which mass transfer coefficients are known. In one approach the overall mass transfer coefficient of B can be determined from the gas and liquid-side mass transfer coefficients for compound A. The following equation can be derived by combining Eq. 7.79 with Eqs. 7.62 and 7.65:

$$\frac{1}{K_{L,B}a} = \left[\frac{1}{k_{l,A}a(D_{l,B}/D_{l,A})^n} + \frac{1}{H_B k_{g,A}a(D_{g,B}/D_{g,A})^m} \right] \qquad \ldots (7.89)$$

where,

$K_{L,B}$ = overall mass transfer rate of B, s^{-1}
a = area available for mass transfer per volume of container, m^2/m^3
$k_{l,A}$ = liquid-phase mass transfer coefficient of compound A, m/s
$D_{l,B}$ = liquid-phase diffusivity of compound B, m^2/s
$D_{l,A}$ = liquid-phase diffusivity of compound A, m^2/s
n = empirical exponent
H_B = Henry's constant for compound B, L of water/l of air, dimensionless
$k_{g,A}$ = gas-phase mass transfer coefficient of compound A, m/s
$D_{g,A}$ = gas-phase diffusivity of compound A, m^2/s
$D_{g,B}$ = gas-phase diffusivity of compound B, m^2/s
m = empirical exponent

As already discussed and shown in Table 7.8, n and m have values between ½ for no boundary layer (no velocity gradient at the interface) and ⅔ for a boundary layer (velocity gradient at the interface). Normally, the gas side has a velocity gradient and the water side has no velocity gradient and that n and m are ½ and ⅔, respectively, as shown in the equation:

$$K_{L,B}a = \left[\frac{1}{k_{l,A}a(D_{l,B}/D_{l,A})^{1/2}} + \frac{1}{H_B k_{g,A}a(D_{g,B}/D_{g,A})^{2/3}} \right]^{-1} \qquad \ldots (7.90)$$

In another approach, the ratio of k_g/k_l is assumed to be a constant for all compounds. This ratio is relatively constant for a given device and does not depend on the compound.

Therefore, the following simplification can be made:

$$\frac{K_{g,i}a}{K_{l,i}a} = \frac{K_g}{K_l} \qquad \ldots (7.91)$$

where,

$k_{g,i}$ = gas-phase mass transfer coefficient of compound i, m/s
$k_{l,i}$ = liquid-phase mass transfer coefficient of compound i, m/s
a = surface area available for mass transfer per volume of separation vessel, m^2/m^3
k_g/k_l = ratio of gas-phase mass transfer coefficient to liquid-phase mass transfer coefficient, which tends to be constant for a given separation device

Values of the ratio k_g/k_l were presented in the above discussion on determination of the phase that controls the mass transfer rate.

Re-write Eq. 7.79 in terms of a compound, i:

$$\frac{1}{K_{L,i}a} = \frac{1}{K_{l,i}a} + \frac{1}{H_i k_{g,i}a} \qquad \text{... (7.92)}$$

Multiply both sides of Eq. 7.92 by $k_{l,i}a$ and solve for $k_{L,i}a$:

$$K_{L,i}a = k_{l,i}a \left[1 + \frac{1}{H_i(k_{g,i}a / k_{l,i}a)} \right]^{-1} \qquad \text{... (7.93)}$$

where,

$K_{L,i}$ = overall mass transfer coefficient for compound i, m/s
H_i = Henry's constant for compound i, L of water/L of air, dimensionless

Substituting Eq. 7.91 into Eq. 7.93 results in the expression

$$K_{L,i}a = k_{l,i}a \left[1 + \frac{1}{H_i(k_g/k_l)} \right]^{-1} \qquad \text{... (7.94)}$$

At this point, it is convenient to choose a reference compound, such as oxygen, that has all resistance to mass transfer on the liquid side (and is easy to measure) because the overall mass transfer coefficient is equal to the liquid-phase mass transfer coefficient for such a compound:

$$\frac{1}{K_{L,O_2}a} = \frac{1}{k_{l,O_2}a} + \frac{1}{H_{O_2} k_{g,O_2}a} = \frac{1}{k_{l,O_2}a} \qquad \text{... (7.95)}$$

$$K_{L,O_2}a = k_{l,O_2}a \qquad \text{... (7.96)}$$

where,

K_{L,O_2} = overall mass transfer coefficient for oxygen, s^{-1}
k_{l,O_2} = liquid-phase mass transfer rate constant for oxygen, s^{-1}
a = area available for mass transfer per volume of container, m^2/m^3
H_{O_2} = Henry's law constant for oxygen, L of water/l of air, dimensionless
k_{g,O_2} = gas-phase mass transfer coefficient for oxygen, m/s

The mass transfer coefficient of a given compound, i, can be related to the mass transfer coefficient for oxygen by dividing Eq. 7.94 by Eq. 7.96:

$$\frac{K_{L,i}a}{K_{L,O_2}a} = \frac{k_{l,i}a}{k_{l,O_2}a} \left[1 + \frac{1}{H_i(k_g/k_l)} \right]^{-1} \qquad \text{... (7.97)}$$

where,

$K_{L,i}$ = overall mass transfer coefficient for component i, s^{-1}
$k_{l,i}$ = liquid-phase mass transfer coefficient for component i, s^{-1}
H_i = Henry's law constant for component i, L of water/L of air, dimensionless
a = area available for mass transfer per volume in vessel, m^2/m^3
k_l = liquid-phase mass transfer coefficient, m/s
k_g = gas-phase mass transfer coefficient, m/s

k_g/k_l = ratio of gas-phase to liquid-phase mass transfer coefficients, which tends to be relatively constant for a given device and does not depend on the compound

The ratio of the mass transfer coefficients $k_{L,i}a / k_{L,O_2}a$ in Eq, 7.97 can be determined from Table 7.8 for the case where the mass transfer resistance is in the water at the air-water interface:

$$\frac{k_{L,i}a}{k_{L,O_2}a} = \frac{k_{l,i}}{k_{l,O_2}} = \left(\frac{D_{l,i}}{D_{l,O_2}}\right)^{1/2} \qquad \ldots (7.98)$$

where,

$D_{l,i}$ = liquid-phase diffusivity of compound i, m^2/s

D_{l,O_2} = liquid-phase diffusivity of oxygen, m^2/s

The final expression for determining the mass transfer coefficient of any compound i using oxygen as a reference compound results from substituting Eq. 7.98 into Eq. 7.97:

$$K_{L,i}a = K_{L,O_2}a\left(\frac{D_{l,i}}{D_{l,O_2}}\right)^{1/2}\left[1+\frac{1}{H_i(k_g/k_l)}\right]^{-1} \qquad \ldots (7.99)$$

Chapter 8

Aeration and Gas Transfer

INTRODUCTION

Aeration and other gas transfer operations serve a multitude of purposes in water, waste-water and biological treatment. Aeration occupies a significant place in waste-water quality management and is an important factor in the purification of polluted water. Gas transfer is a physical phenomena in which gas molecules are exchanged between a liquid and a gas at a gas-liquid interface. This physical phenomenon of gas molecules exchanged between the liquid and gas at the liquid-gas interface may also be accompanied by biological, biochemical, biophysical and chemical action. These results are often the primary purpose of the gas transfer operation and methods of achieving the desired results may vary. Principal objectives of aeration, however, usually add or remove gases or volatile substances to water or carry out both objectives simultaneously. In the biological process, aerators function to transfer the required oxygen and include sufficient mixing to maintain uniform dispersed oxygen throughout the basin and keep biological solids in suspension in aerobic basins and the activated sludge process. For high-rate organic loadings, the power required may be determined by oxygen transfer requirements rather than mixing.

Aeration, aside from agitation purposes, may have the following functions:

1. Add oxygen in natural or waste-water treatment and disposal for promotion of biochemical and chemical processes.
2. Add oxygen to groundwater to oxidise dissolved iron and manganese.
3. Remove carbon dioxide and reduce corrosion and interference with lime-soda softening.
4. Remove hydrogen sulphide, odours, tastes and decrease metal corrosion and concrete and cement deterioration.
5. Lessen interference with chlorination.
6. Remove methane.
7. Remove volatile oils and odour and taste-producing substances produced by algae and other micro-organisms.

Air stripping and aeration are two water treatment unit processes that utilise the principles of mass transfer to accomplish specific water treatment objectives. Both of these water treatment unit processes bring air and water into intimate contact to transfer volatile substances from the water (e.g., hydrogen sulphide, carbon dioxide, volatile organic compounds) into the air or from the air (e.g., carbon dioxide, oxygen) into the water. The mass transfer process involving the removal of volatile substances from water into the air is known as desorption. Air stripping is one of the most common desorption processes used in water treatment. The addition of gases from air into water is the mass transfer process known as absorption. Aeration involving the addition of oxygen to water is a commonly used absorption process.

An understanding of the principles of the underlying mass transfer processes, including how to calculate diffusion coefficients and the basis for mass transfer correlations, is necessary to design air strippers and aerators effectively. In this chapter, the focus is on the application of the aforementioned mass transfer principles to water treatment unit processes. Specific topics considered in this chapter include: (i) an introduction to air stripping and aeration; (ii) gas-liquid equilibria (Henry's law); (iii) the classification of air stripping and aeration systems; (iv) the fundamentals of packed tower air stripping; (v) analysis and design for packed tower air stripping; (vi) an analysis of low-profile air strippers; (vii) an analysis of spray aerator; and (viii) other air stripping and aeration processes.

AIR STRIPPING AND AERATION

Water treatment objectives that can be achieved through the gas-liquid mass transfer process are summarised in Table 8.1. A number of the objectives listed in Table 8.1 can be achieved using air stripping and aeration, the two unit processes discussed in the following sections of this chapter. In both air stripping and aeration, air-water contactors are used to increase the contact opportunities between the gas and liquid phases. By increasing the air-water contact opportunities, the desorption or adsorption mass transfer process is accelerated above the rate that would occur naturally, meaning volatile substances move more rapidly from the water into the air or gases that are not as soluble move more rapidly from the air into the water. The increase in contact opportunities between the two phases occurs through increasing the air-water interface in the air-water contactor by increasing the air-water ratio.

Table 8.1. Application of air/water mass transfer in water treatment.

Examples	Water treatment objectives
Adsorption	
O_2	Oxidation of Fe^{2+}, Mn^{2+}, S^{2-}; lake destratification
O_3	Disinfection, colour removal, oxidation of selected organic compounds
Cl_2	Disinfection; oxidation of Fe^{2+}, Mn^{2+}, H_2S
ClO_2	Disinfection
CO_2	pH control
SO_2	Dechlorination
NH_3	Chloramine formation for disinfection
Desorption	
CO_2	Corrosion control
O_2	Corrosion control
H_2S	Odour control
NH_3	Nutrient removal
Volatile organics (e.g., $CHCl_3$)	Taste and odour control, removal of potential carcinogens

Bringing About Air-Water Contact

Over the years, a number of methods have been developed to bring about effective air-water contact. Packed towers or slat countercurrent flow towers, known as gas-phase contactors, have a continuous gas phase and a discontinuous water phase and are typically used to remove (or strip) gases or volatile chemicals from water. Air-water contactors such as basins with diffused aeration, also called bubble columns, are known as flooded contactors. In flooded contactors the water phase is continuous and the gas phase is discontinuous because the air is present as discontinuous bubbles. Flooded contactors are typically used to add gases (e.g., O_2, CO_2, O_3) into water.

One confusing concept with air stripping and aeration is that aerators can be used to accomplish air-water contact in both air stripping and aeration processes. In general, aerators are a relatively simple method for increasing the air-water ratio by (i) spraying water into the air; and (ii) introducing air into the water through surface turbines or submerged nozzles and diffusers (bubble columns). Thus, aerators allow both of the mass transfer processes, desorption and absorption, to occur in a relatively cost-effective manner. However, because backmixing can occur in aeration systems, a high degree of removal may be difficult to achieve.

Air Stripping

Two major types of air-water contactors are used for air stripping: (i) towers; and (ii) aerators. The two principal factors that control the selection of the type of air-water contactor for stripping are: (i) the desired degree removal of the compound; and (ii) the Henry's constant of the compound. Towers are used when either a high degree of removal is desired or the compound has a high affinity for water (is not very volatile so it has a low Henry's constant), as shown on Fig. 8.1. Aerators are used when either the desired degree of removal is not very high or the gas has a low affinity for water (is very volatile so it has a high Henry's constant). When removal of less than 90 per cent are required, both spray and diffused aeration systems, including mechanical aeration, may be economically attractive.

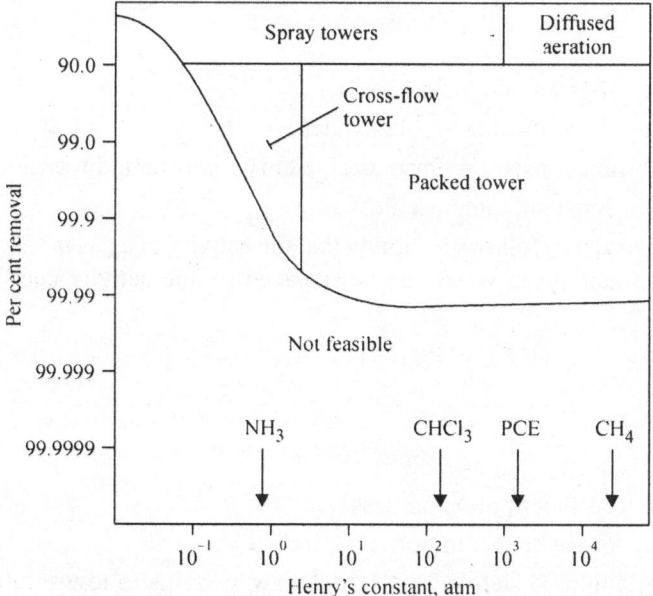

Fig. 8.1. Schematic diagram for selection of feasible aeration process for control of volatile compounds.

Aeration

Aeration is used to increase the oxygen content in the water by adding air into water through: (i) diffusers in a pipe, channel or process basin; (ii) cascading water over stacked trays; and (iii) surface turbines and wheels that mix air into water at the top of basins. Oxygenation can also be accomplished using pure oxygen.

Gas-Liquid Equilibria

When gas-free water is exposed to air, compounds such as oxygen and nitrogen will diffuse from the air into the water until the concentration of these gases in the water reaches equilibrium with the gases in the air. Conversely, if water in deep wells is brought to the ground surface, dissolved gases such as methane or carbon dioxide will be released to the air because their concentrations in groundwater typically exceed equilibrium conditions with air. The eruption of a carbonated beverage after it is opened is a more familiar example of carbon dioxide release after a pressure change. In each case, the driving force for mass transfer is the difference between the existing and equilibrium concentrations in the two phases.

Henry's Law

For most aeration and air stripping applications in water treatment, Henry's law can be used to describe equilibrium partitioning of a gas or organic contaminant between air and water. For a closed vessel that contains both water and air and component A in equilibrium at a constant temperature, as represented by Eq. 8.1, Henry's law is given in Eq. 8.2:

$$A_{aq} \in A_{gas} \qquad\qquad \text{... (8.1)}$$

$$\frac{A_{gas}}{A_{aq}} = K_{eq} = H \qquad\qquad \text{... (8.2)}$$

where,

A_{gas} = activity of gas in air, mole/l air

A_{aq} = activity of gas in water, mole/l water

K_{eq} = equilibrium constant defined as H, Henry's constant, dimensionless

H = Henry's constant, dimensionless

Equation 8.2 can be rewritten as follows by noting that the activity of a gas in air can be approximated by the partial pressure and activity in water can be replaced by the activity coefficient times the molar, concentration:

$$P_A = H_A^c \gamma_A [A] \qquad\qquad \text{... (8.3)}$$

where,

P_A = partial pressure of A, atm

H_A^c = Henry's constant, atm/(mole/l)

γ_A = activity coefficient of A, unitless

$[A]$ = aqueous-phase concentration of A, mole/l

The equilibrium condition, as defined by Henry's law, pertains to low solute concentrations where the solute molecules are only surrounded by water molecules. For such cases, the equilibrium constant does not depend on concentration. However, at higher concentrations, the equilibrium constant given in

Eqs. 8.2 and 8.3 is not constant and is a function of the solute concentration. For example, Eq. 8.3 is generally valid for concentrations of A less than 0.01 mole/l, but it has been shown in some cases to be valid for concentrations as high as 0.1 mole/l. The presence of air does not affect the Henry's law constant for volatile organic chemicals, (VOCs) or gases because the constituents of interest have low concentrations in air. Thus, for most natural fresh water (ionic strength less than 0.01 M) the activity coefficient is assumed to be equal to 1.0 and Eq. 8.3 is written as

$$P_A = H_A^c[A] \qquad \qquad ...(8.4)$$

where, the terms are as defined above.

Others Forms of Henry's Law

While thermodynamic equilibrium constants such as Henry's law do not have units, the inherent units of H_A^c in Eq. 8.3 are atm/(mole/l) or atm l/mole because the standard conditions for pressure and concentration are given in atmosphere and mole per liter, respectively. However, as shown in Fig. 8.2, the Henry's law constant is commonly reported in one of three units depending on whether: (i) concentration or partial pressure of A is used for characterising the gas phase; and (ii) concentration or mole fraction is used for characterising the aqueous phase. The three types of units commonly used for Henry's law constants are: (i) mass or mole of compound A in air per liter of air divided by mass or mole of compound A in water per liter of water; (ii) atm of component A divided by the dimensionless mole fraction of A in the liquid phase; and (iii) the partial pressure of A in atm divided by mass or mole of compound A in water per liter of water. Because the reported units of Henry's law constant vary, it is necessary to convert from one system of units to another. Various unit conversions that can be used to convert from one system of units to another are shown in Table 8.2.

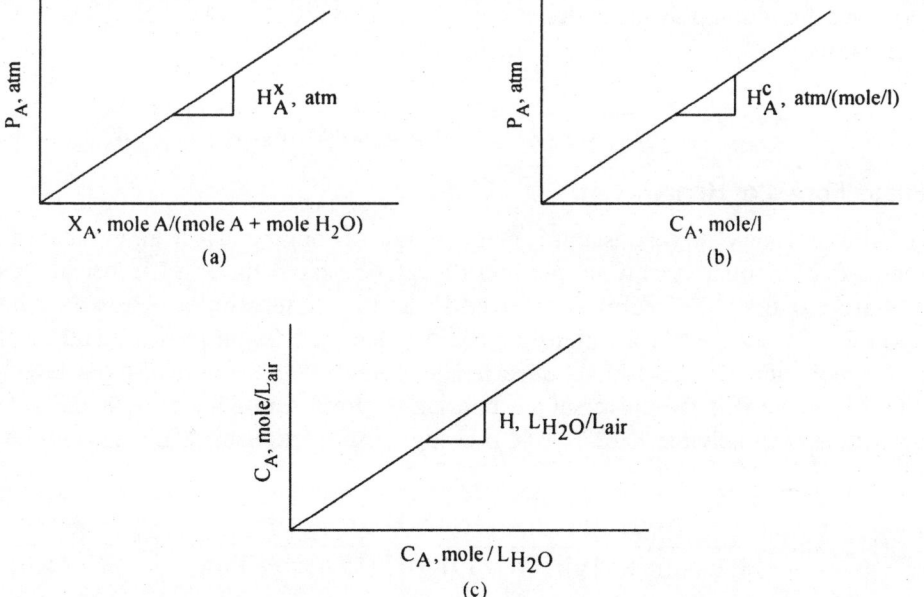

Fig. 8.2. Common units for reporting Henry's law constant: (a) partial pressure of A versus mole fraction (H has units of atm); (b) partial pressure of A versus molar concentration [H has units of atm/(mole/L)]; and (c) molar concentration of A in air versus molar concentration in water (H has dimensionless units of l H_2O/l air).

Table 8.2. Unit conversions for Henry's law constants.

$$H\left(\frac{L_{H_2O}}{L_{Air}}\right) = \frac{H_A^c\,[atm/(mole/l)]}{RT}$$

$$H\left(\frac{L_{H_2O}}{L_{Air}}\right) = \frac{H_A^x\,(atm)}{RT \times (55.6\ mole\ H_2O/l\ H_2O)}$$

$$H_A^c\,[atm/(mole/l)] = \frac{H(atm)}{(55.6\ mole\ H_2O/l\ H_2O)}$$

$$H_A^x\,(atm) = H(L_{H_2O}/L_{Air}) \times RT \times (55.6\ mole\ H_2O/l\ H_2O)$$

For example, if the units of the Henry's law constant is atmospheres (atm), then the following equilibrium expression may be used:

$$P_A = H_A^x x_A \qquad \qquad ...(8.5)$$

where,

P_A = partial pressure of A, atm

H_A^x = Henry's constant based on mole fraction, atm

x_A = mole fraction of A in water, dimensionless

= mole A/ (mole A + mole H_2O)

mole H_2O = (1000 g/l)/(18 g/mole) = 55.6 mole/l

It should also be noted that the partial pressure of A is equal to the mole fraction of A in air times the total pressure. For a dilute solution, the moles of A, typically below 0.01 mole/l, is negligible as compared to the moles of water (55.6 mole/l). Thus Eq. 8.5 can be written as

$$P_A\ ;\ H_A^x\,\frac{mole\ A}{mole\ water} = H_A^x\,\frac{mole\ A/l}{55.6\ mole\ water/l} = H_A^x\,\frac{C_A}{55.6} \qquad ...(8.6)$$

Dimensionless Forms of Henry's Law

The preferred unit for Henry's law constants is mass or mole of compound A in air per liter of air divided by mass or mole of compound A in water per liter of water because this unit is especially useful when performing mass balances. This unit is referred to as the dimensionless Henry's constant. The dimensionless Henry's constant H is obtained by noting that at 1.0 atm pressure and °C the volume occupied by 1.0 mole of air is 22.414 l. At other temperatures, using the universal gas law, 1.0 mole of air is equal to RT, where R is the universal gas constant expressed in units of 0.08205 atm/(mole/l)·K and the temperature is in kelvins, K (273 + °C). Using these relationships, Eq. 8.6 can be written as follows:

$$H = (H_A^x, atm)\left\{\frac{1}{[R, atm/(mole/l\ air)\cdot K]\,(T, K)}\right\}\left[\frac{1}{(55.6\ mole/l\ water)}\right] = \frac{H_A^x}{RT(55.6)} \qquad ...(8.7)$$

If the Henry's law constant has units of atm/(mole/l), then it can be converted to dimensionless units using the expression

$$P_A = H_A^c C_A \qquad \qquad ...(8.8)$$

where,

P_A = partial pressure of A, atm

H_A^c = Henry's constant based on concentration, atm/ (mole/l)

C_A = concentration of A, mole/l

Accordingly, following the approach used in developing Eq. 8.7, the dimensionless Henry's constant H is related to Henry's constant based on concentration, H_A^c, by the expression:

$$H = \left(\frac{H_A^c, atm}{mole/l}\right)\left\{\frac{1}{[R, atm/(mole/l\ air) \cdot K]\ (T,\ K)}\right\} = \frac{H_A^c}{RT} \qquad ... (8.9)$$

$$= \frac{H_A^c}{RT}$$

The use of Eqs. 8.7 and 8.9 is illustrated in Example 8.1.

Example 8.1: What is the dimensionless Henry's law constant for a compound that has a value of 500 atm? What is the Henry's law constant in atmospheres and atm/(mole/l) for a compound that has a dimensionless Henry's law constant of 0.001? Assume the temperature is 25°C.

Solution:

1. Determine the dimensionless Henry's law constant. Inserting the Henry's law constant of 500 atm into Eq. 8.7 results in the expression:

$$H = \frac{H_A^x}{RT(55.6)} = \frac{500\ atm}{[0.082054\ atm/(mole/l)\cdot K][(273 + 25)\ K](55.6\ mole/l)}$$

$$= 0.368$$

2. Determine Henry's law constant in atmospheres. Rearranging Eq. 8.7 and solving for H using a dimensionless Henry's constant of 0.001, the following result is obtained:

$$H_A^x = RT \times 55.6H$$

$$= [0.082054\ atm/(mole/l)\cdot K][273 + 25)\ K](55.6\ mole/l)(0.001)$$

$$= 1.36\ atm$$

3. Determine Henry's law constant in atm/(mole/l). Rearranging Eq. 8.9,

$$H_A^c = RTH = [0.082054\ atm/(mole/l)\cdot K][(273 + 25)\ K](0.001)$$

$$= 2.44 \times 10^{-2}\ atm/(mole/l)$$

Estimation of Henry's Constant from Vapour Pressure and Solubility

When the Henry's law constant for a component of interest is not readily available, it can be estimated if the vapour pressure and aqueous solubility of the compound are known. There are two situations where it is possible to estimate the Henry's constant of component A: (i) when component A is soluble in the aqueous phase; and (ii) when component A is slightly soluble in the aqueous phase. Using this approach, the estimated values are typically within ±50 to 100 per cent of the experimentally reported values and should, therefore, only be used when measured values of the constants are not available. For slightly soluble compounds in water, the Henry's law constant may be estimated using the following expression:

$$H_A^c = \frac{P_{v,A}}{C_{s,A}} \qquad \qquad \text{... (8.10)}$$

where,

$P_{v,A}$ = vapour pressure of A, atm

$C_{s,A}$ = aqueous solubility of A, mole/l

For soluble compounds, the Henry's law constant may be estimated as follows:

$$H_A^x = P_{v,A} \qquad \qquad \text{... (8.11)}$$

Factors Influencing Henry's Constant

Temperature, pressure, ionic strength, surfactants and solution pH (for ionisable species such as NH_3 and CO_2) can influence the equilibrium partitioning between air and water. The impact of total system pressure on H is negligible because other components in air have limited solubility in water.

Effect of temperature

Henry's constants for several organic compounds (at different temperatures) and gases (at 20°C) are listed in Table 8.3. Assuming that the standard enthalpy change (ΔH_{dis}°) for the dissolution of a component in water is constant over the temperature range of interest, the change in H with temperature can be estimated using the van't Hoff equation:

$$H_{T_2} = H_{T_1} \exp\left[\frac{-\Delta H_{dis}^\circ}{R}\left(\frac{1}{T_2} - \frac{1}{T_1}\right)\right] \qquad \qquad \text{... (8.12)}$$

where,

H_{T_2} = dimensionless Henry's law constant at temperature T_2

H_{T_1} = dimensionless Henry's law constant at temperature T_1

ΔH_{dis}° = standard enthalpy change of dissolution in water (kcal/kmole)

R = universal gas constant (1.987 kcal/kmole.K)

T_1, T_2 = absolute temperatures, K (273 + °C)

Equation 8.12 can be simplified to the following expression and K_C and ΔH_{dis}° values for selected compounds are reported in Table 8.3:

$$H = K_C \exp\left(-\frac{\Delta H_{dis}^\circ}{RT}\right) \qquad \qquad \text{... (8.13)}$$

Table 8.3. Dimensionless Henry's law constants at 20°C and temperature dependence for gases in water.

Compound	$\Delta H_{dis}^{\circ\ a}$	K_C	H
Ammonia	8.63	1526	0.0006
Carbon dioxide	4.77	4013	1.1

(Contd ...)

Compound	ΔH_{dis}° ^{a}	K_C	H
Chlorine	4.01	420	0.43
Chlorine dioxide	6.75	4300	0.04
Hydrogen sulphide	4.26	567	0.38
Methane	3.55	12402	28.41
Oxygen	3.34	9627	32.15
Ozone	5.80	83848	3.74
Sulphur dioxide	5.53	358	0.03
Carbon tetrachloride	9.32	85,80,096	0.96
Tetrachloroethylene	9.88	1,79,26,362	0.82
Trichloroethylene	7.85	2,90,732	0.41
Benzene	8.47	357678	0.18
Chloroform	9.21	9,40,789	0.13

$^a\Delta H_{dis}^{\circ}$ in units of kcal/kmole, $\times 10^3$.

Another common method of expressing the temperature dependence of H is to define K_C and $\Delta H_{dis}^{\circ}/R$ as fitting parameters K_A and K_B, respectively, using the equation

$$H = \exp\left(K_A - \frac{K_B}{T} \right) \qquad \text{... (8.14)}$$

Values of K_A and K_B for several compounds valid for temperatures ranging from 283 to 303 K Eq. 8.2.

Example 8.2: Henry's law constant and temperature effect.

Calculate the Henry's law constant at 5 and 20°C for trichloroethylene (TCE) using Eq. 8.13 and Table 8.3.

Solution:

1. At 5°C: Using Eq. 8.13 and the constants provided in Table 8.3, the Henry's law constant for TCE at 20°C can be estimated as follows:

$$K_c = 2,90,732$$

$$\Delta H_{dis}^{\circ} = 7.85 \times 10^3 \text{ kcal/kmole}$$

$$H = K_C \exp\left(-\frac{\Delta H_{dis}^{\circ}}{RT} \right) = 2,90,732 \exp\left\{ -\frac{(7.85 \times 10^3 \text{ kcal/kmole})}{\{(1.987 \text{ kcal/kmole})[(273 + 5) \text{ K}]\}} \right\}$$

$$= 0.20$$

2. At 20°C:

$$H = K_C \exp\left(-\frac{\Delta H_{dis}^{\circ}}{RT} \right) = 2,90,732 \exp\left\{ -\frac{(7.85 \times 10^3 \text{ kcal/kmole})}{\{(1.987 \text{ kcal/kmole})[(273 + 20) \text{ K}]\}} \right\}$$

$$= 0.41$$

Ionic strength

Gases or synthetic organic chemicals (SOCs) have a higher apparent Henry's law constant (H_{app}) when the dissolved solids are high, as shown in Eq. 8.3, which can be used to calculate H_{app}:

$$H_{app} = \frac{(P_A/C_A)}{RT} = \gamma_A H \qquad \qquad ... (8.15)$$

where,

H_{app} = apparent Henry's law constant, dimensionless
P_A = partial pressure of A, atm
C_A = concentration of A, mole/l
γ_A = activity coefficient of A
H = dimensionless Henry's constant

The activity coefficient γ_A is a function of ionic strength and can be calculated using the following empirical equation:

$$\log \gamma_A = K_s \times I \qquad \qquad ... (8.16)$$

where,

K_s = Setschenow or 'salting-out', constant, l/mole
I = ionic strength of water, mole/l.

Values of K_s need to be determined by experimental methods because there is no general theory for predicting them. Significant increases in volatility and the apparent Henry's constant are observed only for high-ionic-strength water such as seawater.

Effect of surfactants

Surfactants can impact the volatility of compounds. In most natural water, surfactant concentrations are relatively low; consequently, surfactants do not affect the design of most aeration devices. However, when surfactants are present in relatively high concentrations, the volatility of other compounds may be lowered by several mechanisms. The dominant mechanism is collection of surfactants at the air-water interface, decreasing the mole fraction of the volatile compound at the interfacial area, thereby lowering the apparent Henry's law constant. For example, the solubility of oxygen in water can be lowered by 30 to 50 per cent due to the presence of surfactants.

Another surfactant effect for hydrophobic organic compounds is the incorporation of dissolved organic compounds into micelles in solution. Above the critical micelle concentration, the formation of additional micelles will decrease the concentration of the organic compound at the air-water interface and lower the compound's volatility.

Impact of pH

The pH does not affect the Henry's constant directly, but it does affect the distribution of species between ionised and unionised forms, which influences the overall gas-liquid distribution of the compound because only the unionised species are volatile.

Uncharged weak acids such as H_2CO_3, HCN or H_2S cannot be stripped at pH values significantly above their pK_a value. For example, if hydrogen sulphide is to be stripped, then the following equilibrium condition applies (note, the second ionisation constant can be neglected because $pK_{a2} > 18$):

$$H_2S \ \epsilon \ \ HS^- + H^+ \qquad \qquad ... (8.17)$$

The first ionisation constant is given by:

$$K_a = 7.94 \times 10^{-8} = \frac{[HS^-][H^+]}{[H_2S]} (25°C) \qquad \qquad ... (8.18)$$

$$pK_a = (-\log 7.94 \times 10^{-8}) = 7.1$$

The corresponding Henry's law constant is given below:

$$H_{app} = \frac{(P_{H_2S}/C_{T,S})}{RT} = \frac{H}{(1+K_a/[H^+])} \qquad \dots (8.19)$$

If the pH is equal to 5.1, two units lower than the pK_a, then hydrogen sulphide is only 1 per cent ionised and the apparent Henry's constant is essentially the same as the H value. If the pH is two units higher than the pK_a value, then hydrogen sulphide is 99 per cent ionised and the apparent Henry's constant is 1 per cent of the H value.

CLASSIFICATION OF AIR STRIPPING AND AERATION SYSTEMS

Gas-liquid contactors are classified as either gas-phase contactors or flooded contactors and then further classified into four sub-groups based on the method that is used to either remove gas from water or add gas to water. Air stripping and aeration systems use gas-liquid contactors to either remove gases from water or add air (or oxygen) to water. The four basic types of air-water contact systems that are discussed in this section are: (i) droplet or thin-film air-water contactors; (ii) diffusion or bubble aerators; (iii) aspirator-type aerators; and (iv) mechanical aerators. Characteristics and typical applications of air-water contact systems that fall into one of these four groups are summarised in Table 8.4. Some of these systems may be used to contact water with gases other than air and while these uses are listed in Table 8.4.

Table 8.4. Characteristics of gas-liquid contacting systems.

Type of contacting device	Process description	Method of gas introduction	Typical applications	Oxygen transfer rate, kg O_2/kWh	Number of transfer units (NTU)	Hydraulic head required, m (ft)	Loading factor
Spray aerator	Water to be treated is sprayed through nozzles to form disperse droplets; typically a fountain configuration. Nozzle diameters usually range from 2.5 to 4 cm (1–1.6 inch) to minimise clogging	Natural aeration through convection	H_2S, CO_2 and marginal VOC removal; taste and odour control, oxygenation	–	0.5–0.7	1.5–7.6 (5–25)	Surface area of 0.10–0.30 m^2·s/l
Spray tower	Water to be treated is sprayed downward through nozzles to form disperse droplets in a tower configuration; air-water ratio is controlled; typically countercurrent flow	Forced-draft aeration	H_2S, CO_2 and VOC removal; taste and odour control	–	1–1.5	1.5–7.6 (5–25)	Surface area of 0.10–0.30 m^2·s/l

(Contd ...)

Type of contacting device	Process description	Method of gas introduction	Typical applications	Oxygen transfer rate, kg O_2/kWh	Number of transfer units (NTU)	Hydraulic head required, m (ft)	Loading factor
Packed tower	Water to be treated is sprayed onto high-surface area packing to produce a thin-film flow; process configuration typically countercurrent	Forced-draft aeration	H_2S, CO_2 and VOC removal; taste and odour control	–	1–4	3–12 (10–40)	
Cascade	Water to be treated flows over the side of sequential pans, creating a waterfall effect to promote droplet-type aeration	Aeration primarily by natural convection	CO_2 removal, taste and odour control, aesthetic value, oxygenation	–	0.5–0.7	0.9–3 (3–10)	
Multiple tray	Water to be treated trickles by gravity through trays containing media [layers 0.1–0.15 m (4–6 inch) deep] to produce thin-film flow. Typical media used include coarse stone or coke [50–150 mm (2–6 inch) in diameter] or wood salts	Natural or forced-draft aeration	H_2S, CO_2 removal, taste and odour control	–	<1	1.5–3 (5–10)	0.007– 0.014 m/s (10 to 20 gpm/ft^2)
Low profile (sieve tray)	Water flows from entry at the top of the tower horizontally across series of perforated trays. Large air flow rates are used, causing frothing upon air-water contact, which provides large surface area for mass transfer. Units are typically less than 3 m (10 ft) high.	Air introduced at bottom of tower	VOC removal	–	–	–	Water flow rates less than 0.065 m^3/s (1000 gpm)
Diffuser	Fine bubbles are supplied through porous diffusers submerged in the water to be treated; tank depth is typically restricted to 4.5 m (15 ft)	Compressed air or ozone	Fe and Mn removal, CO_2 removal, taste and odour control, oxygenation, ozonation	0.5	0.5–1.5	–	0.1–1 L air/L water

(Contd ...)

Type of contacting device	Process description	Method of gas introduction	Typical applications	Oxygen transfer rate, kg O_2/kWh	Number of transfer units (NTU)	Hydraulic head required, m (ft)	Loading factor
Dispersed air	Compressed air is supplied through a stationary sparger orifice-type dispersion apparatus located directly below a submerged high-speed turbine	Compressed air or ozone	Ozonation especially when high concentrations of Fe and Mn are present due to clogging of porous diffusers	1.5	1–2		
Hydraulic aspirator	A gas stream is educted into the liquid stream with a venturi-type device	Compressed ozone, CO_2, Cl_2	Ozonation, CO_2 addition, Cl_2 disinfection	1.5–3.5	–	3–6 (10–20)	
Mechanical aspirator	A hollow-blade impeller rotates at a speed sufficient to aspirate and discharge a gas stream into the water	Compressed air or ozone	Ozonation, CO_2 addition	0.7			
Mechanical aerator	Surface aerators (brush or turbine types) and aeration pumps are primary types of mechanical aerators	Mechanical agitation of water into surrounding air	O_2 absorption, VOC removal when < 90% required	1.5–4.5 (turbine) 2.5 (brush)			

Droplet or Thin-Film Air-Water Contactors

Droplet or thin-film air-water contactors are gas-phase contactors designed to produce small droplets of water or thin films, which promote rapid mass transfer.

Droplet air-water contactors

Contactors that use droplets are spray devices, such as towers and the fountain spray aerator shown in Fig. 8.3. Spray devices are designed to provide the desired droplet size for the desired contact time with the gas phase, which is typically air. Spray aerators are an efficient method of gas transfer; however, for efficient operation spray aerators should be placed in large basins or reservoirs in favourable climatic conditions.

Thin-film air-water contactors

Thin films of water are created in cascade and multiple-tray aerators and packed columns and towers. Air-water contact occurs when water flows by gravity over the surfaces of packing materials that are placed in trays, columns or towers and between trays. The thin liquid film that forms as water flows downward is disrupted continuously by the irregular surfaces of the packing material, maximising the exposure of the water to the atmosphere and encouraging air-water mixing.

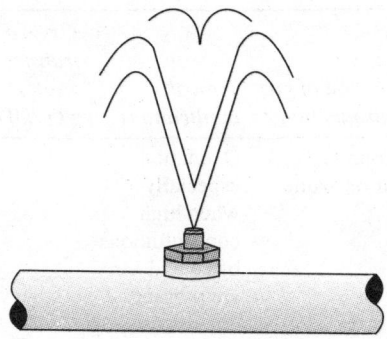

Fig. 8.3. Schematic of fountain spray aerator.

Cascade aerators

Cascade aerators are commonly used for treating groundwater and may be located at the groundwater source or reservoir. Cascade aerators are also called step aerators as water flows downward in a thin film over a series of steps or baffles, sometimes constructed of concrete. Cascade aerators are generally less efficient than other types of thin-film aerators because water flow is less turbulent, resulting in less air-water mixing as air contact is due to natural convection.

Multiple-tray aerators

There are several types of multiple-tray aerators, all based on the same design concept of stacked trays, where water is distributed over the top tray with a spray nozzle or special distribution trough and then flows from the upper tray over the tray sides into lower trays. Tray-type aerators may be either natural draft type such as coke tray aerators or forced draft type such as wood slat aerators.

Of all the types of multiple-tray aerators, wood slat towers are the most efficient. The slat towers are either forced or induced draft and are enclosed in a wood, fibreglass or metal shell, as shown in Fig. 8.4. The slats are generally stacked and centered vertically and the horizontal spaces between the slats are staggered so that the water trickling down one tray strikes the middle of the slat of the next tray.

Tray aerators are typically designed for natural draft, as shown in Fig. 8.5 and typically the tray is filled with packing material such as coke. Coke tray aerators provide somewhat more turbulence in air-water contact because the large surface area of the coke provides a large air-water contact area. Tray aerators are built with splash skirts to reduce the water loss and icing of the protective retaining screens.

A type of multiple-tray aerator that has recently grown in use for removal of VOCs from contaminated water is the low-profile air stripper, also called the sieve tray column, as shown on Fig. 8.6a. Because the water flows horizontally across each tray, the desired removal efficiency can be obtained by increasing the length or width of the trays instead of the height.

Packed towers

Packed towers are circular or square towers that are filled with an irregular shaped inert packing material, as shown in Fig. 8.7. Packing material is available in a wide variety of sizes and shapes depending on the manufacturer. Operationally, water is pumped to the top of the tower and into a liquid distributor where it is dispersed as uniformly as possible across the packing surface and then it flows by gravity through the packing material and is collected at the bottom of the tower. Airflow may be in the same direction as the water (co-current), in the opposite direction as the water (countercurrent) or across the water (cross-flow). For countercurrent operation, a blower is used to introduce fresh air into the bottom of the tower and the air flows countercurrent to the water up through the void spaces between the wetted packing material as shown in Fig. 8.7.

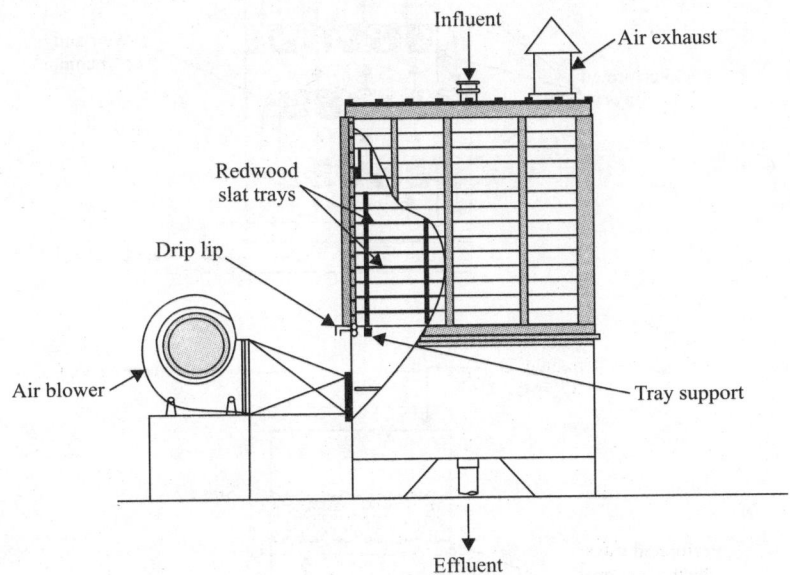

Fig. 8.4. Forced-draft multiple-slat tower cascade aerator.

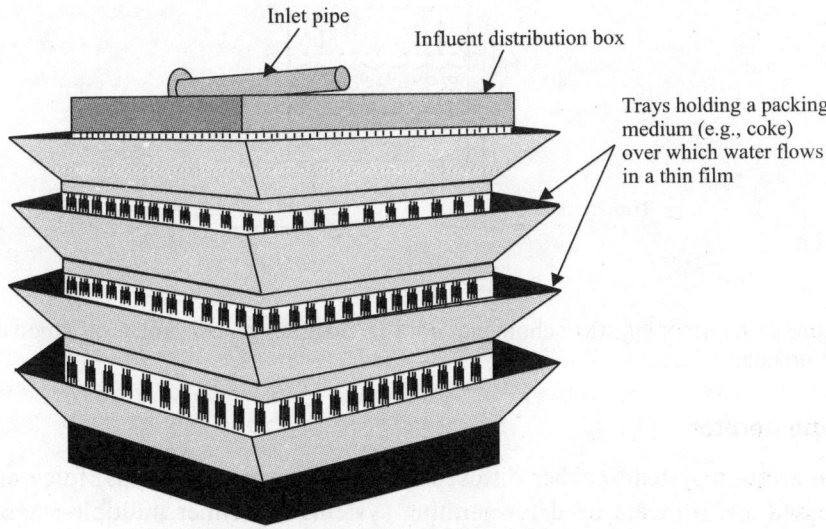

Fig. 8.5. Natural draft coke tray aerator.

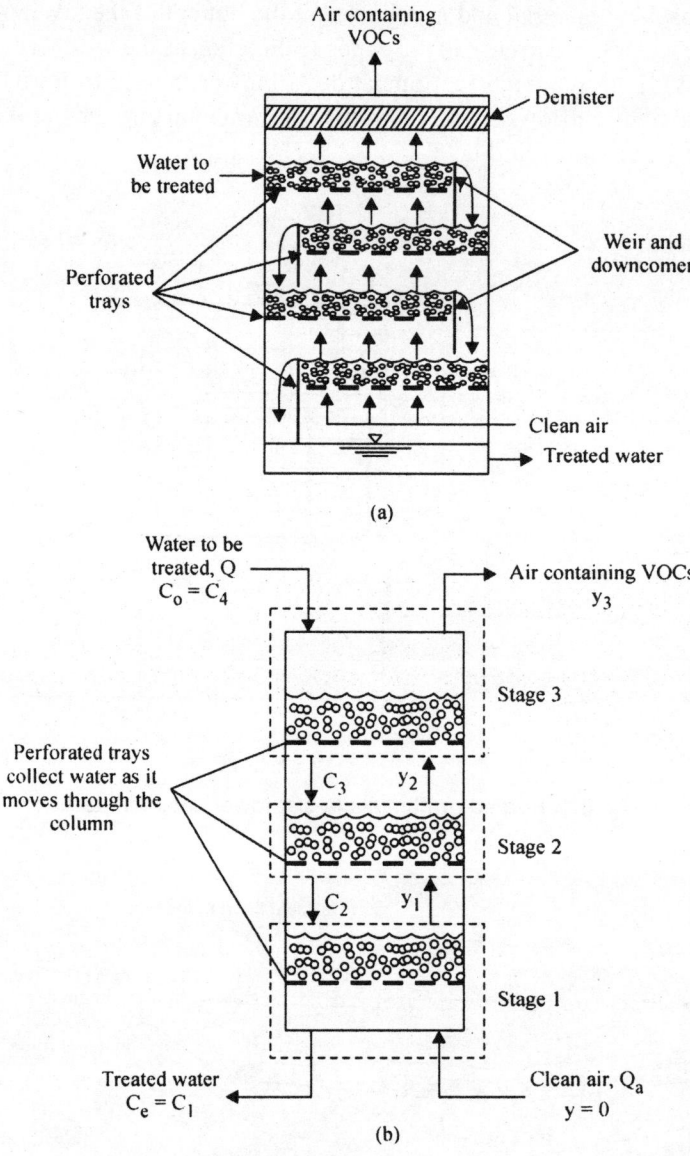

Fig. 8.6. Low-profile air stripping: (a) schematic; and (b) diagram of low-profile air stripping as multistage, countercurrent process.

Diffusion-Type Aerator

Diffusion-type aeration systems (either diffused-air or dispersed-air systems) force air into the water using compressed air. Blowers used for aerating systems are either multiple-stage or single-stage centrifugal or rotary positive displacement. Rotary blowers are often used for small installations, where water depth varies significantly.

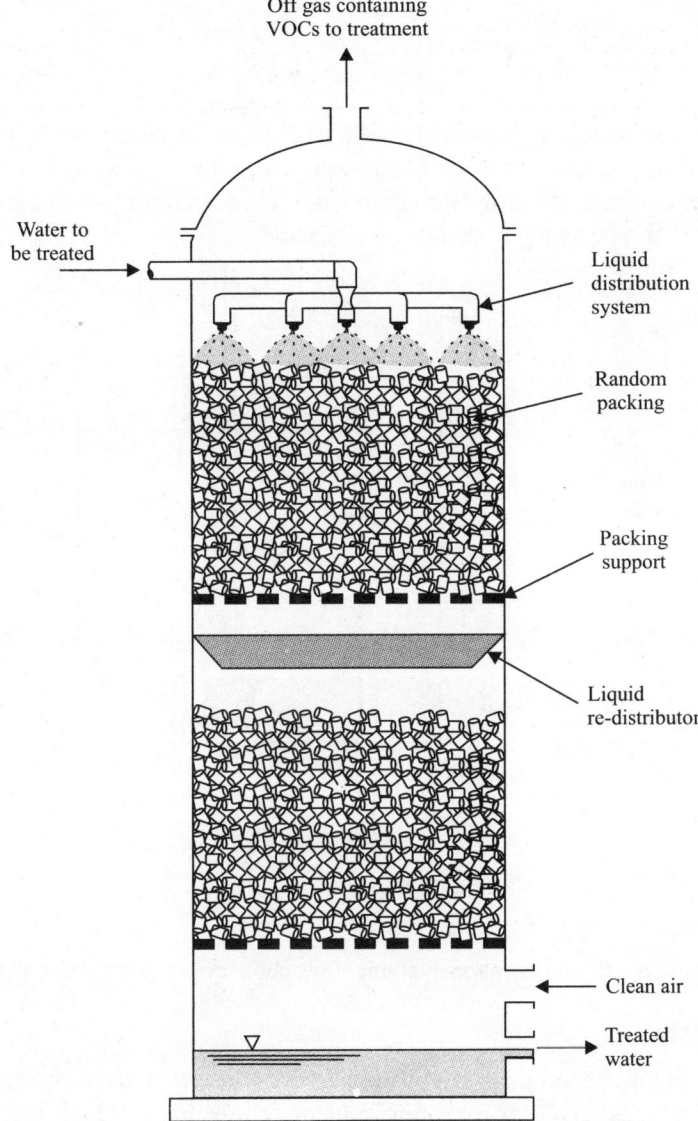

Fig. 8.7. Schematic of a countercurrent packed tower.

Diffused air

Compressed air is generally introduced through porous membranes, porous plates or tubes or wound fibre or metallic filaments at the bottom of a basin or tank, as illustrated by the bubble column in Fig. 8.8a. Diffused-air systems generally require filters to screen out particulates in the air because the air is forced to flow through very fine pores, which can easily plug.

Dispersed air

Mechanical mixers and a stationary orifice-type sparger air dispersion system are used to force air into water in dispersed-air systems, as shown in Fig. 8.8b. The mechanical mixer in the contactor aids in the air-water mixing and therefore, the gas transfer efficiencies are generally much better than in simple diffused-air systems. The air dispersion outlet is generally located a small distance above the tank bottom to reduce the pressure requirements of the air compressor. Dispersed-air systems usually do not require air filtration as the orifice-type spargers do not readily plug.

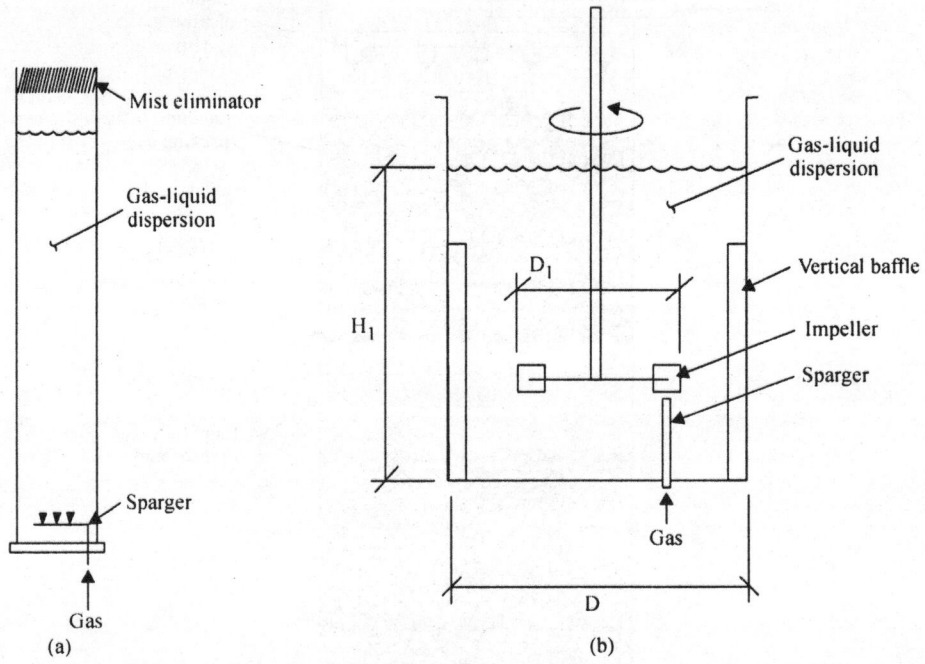

Fig. 8.8. Common diffused aeration systems; (a) bubble column; and (b) agitated vessel.

Aspirator-Type Aerator

Air aspiration is commonly accomplished with either hydraulic aspirators or mechanical devices. A typical hydraulic aspirator is a type of hydraulic eductor or injector in which pressurised water flows through a throat similar to a venturi tube to create a low-pressure condition that in turn draws atmospheric air or gas into the water.

A second type of aspirator is a mechanical aspirator that consists of a hollow-blade impeller that rotates at a speed sufficient to aspirate and discharge atmospheric air into the water. Hydraulic aspirators are mostly small in size, but a single unit under favourable conditions and operating with atmospheric air produces twice the oxygen transfer rate of ordinary mechanical aspirators.

Mechanical Aerators

Surface aerators and aeration pumps are the two basic types of mechanical aerators. Surface aerators may be of the turbine type or the brush type. Aeration pumps consist of a turbine mixer with a draft tube.

Advantages and Disadvantages of Various Air Stripping and Aeration Systems

Each type of air stripping and aeration system offers process and cost advantages for a specific mass transfer problem. Low-profile air strippers versus packed towers and diffusion-type aerators versus droplet and thin-film aerators are compared in the following section.

Low-profile air strippers versus packed towers

The advantages of using a low-profile air stripper over a conventional packed tower stripper are that: (i) the low-profile air stripper is smaller and more compact; and (ii) periodic maintenance is easier to perform on the low-profile air stripper. The disadvantage is that for a given removal the low-profile air stripper requires a significantly higher airflow than a conventional countercurrent packed tower. Consequently, the operational costs can be greater for the low-profile air stripper. In addition, low-profile air strippers are limited to lower water flow rates than countercurrent packed towers.

Diffusion-type aerators versus droplet and thin-film aerators

The advantages of diffusion-type aerators include: (i) negligible head loss for the process water system; and (ii) less space requirements than for the droplet and thin-film aerators. Diffused-air systems may be extremely efficient for reservoir management.

Selection of Appropriate Equipment

Selection of the appropriate equipment is based on relative transfer efficiencies, available hydraulic head, ease of maintenance and cost considerations (capital and operating costs). Transfer efficiency and some cost considerations are considered below.

Transfer efficiency

An important consideration in selecting a process for a specific application is the upper limit of transfer efficiency that can be economically achieved by the process. Many of the processes described in Table 8.4 have limited transfer efficiency. Commercially available cascade aerators cannot achieve greater than 50 or 60 per cent removal of chloroform, a relatively volatile organic contaminant. Cascade and multiple-tray aerators encounter corrosion and algae and slime growth problems, particularly if the process water contains hydrogen sulphide. Chlorination and copper sulphate treatment of the process water may help control these problems, but that is an additional operational issue with which to contend. In general, cascade aerators are about half as efficient as multiple-tray aerators.

Cost considerations

For many applications of air-water mass transfer in water treatment, such as stripping of volatile contaminants or addition of a reactant gas, one or more of the four types of air-water contacting systems described above may be used. When more than one system may be appropriate to address the air-water mass transfer issue, capital and operating costs generally are the deciding factors.

Capital cost

The capital cost of each of the processes discussed is closely related to the mass transfer efficiency of the process. In general, the lower the transfer efficiency, the larger the facility required for achieving a certain removal.

Operating cost

Operating cost is primarily a function of the hydraulics of the process and the method of gas dispersion. Equipment complexity or heavy maintenance is typically not a major consideration because most mass transfer equipment is relatively simple, although it should be noted that fouling by chemical precipitation and/or biological growth must be controlled. When adding a reactive gas such as oxygen, the operating cost must also include chemical costs. The principal cost may be the chemical itself and/or generation of the compound and to a lesser degree the feeding equipment and gas transfer contact tank (if any).

FUNDAMENTALS OF PACKED TOWER AIR STRIPPING

Packed towers are either cylindrical columns or rectangular towers containing packing that disrupts the flow of liquid, thus producing and renewing the air-water interface. A schematic of a countercurrent cylindrical packed tower is shown in Fig. 8.7 and the operation was described in previous section. Packed towers have high liquid interfacial areas and void volumes greater than 90 per cent, which minimises air pressure drop through the tower. An important part of the packed tower system that is not shown in Fig. 8.7 is a demistor, which eliminates entrained water drops and aerosols in the off-gas of a packed tower. Aerosols must be eliminated because they can be displeasing to local communities from an aesthetic point of view and they can freeze in cold climates, causing icing problems.

The packing material in a tower is important to the efficient transfer of volatile contaminants from the water to the air as it provides a large air-water interfacial area. Various types of packing shapes and materials are available commercially, as shown in Fig. 8.9. The packing can be structured packing or individual pieces that are randomly placed in the tower.

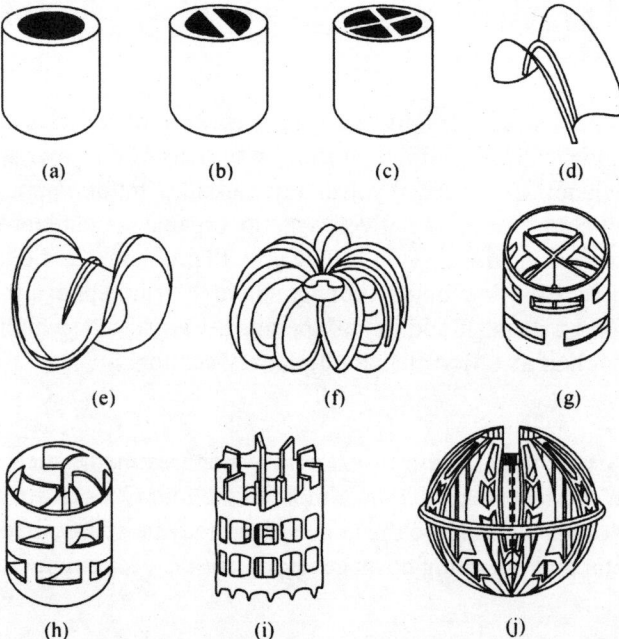

Fig. 8.9. Typical examples of packing materials used in air stripping towers: (a) Raschig ring; (b) lessing ring; (c) partition ring; (d) intalox saddle; (d) berl saddle; (f) tellerette; (g) pall ring; (h) pall ring; (i) nor Pac ring; and (j) tri-pac.

ANALYSIS OF LOW-PROFILE AIR STRIPPERS

Over the past 25 years, low-profile air stripping has become increasingly common, particularly in developing countries. Unit compactness is a key advantage of low-profile air strippers compared to packed towers. Design guidelines for low-profile air strippers, including a comparison with countercurrent packed tower air strippers, are discussed below.

Description

A schematic of a low-profile air stripper, which consists of a stack of sieve trays with contaminated water entering the top of the unit and exiting the bottom as treated water and clean air entering the bottom of the unit and exiting at the top containing VOCs, is shown in Fig. 8.6. Low-profile air strippers operate as a countercurrent process with water entering at the top of the unit and flowing across each sieve tray, as shown in Fig. 8.6a. Inlet and outlet channels or downcomers are placed at the ends of each tray to allow the water to flow from tray to tray. Fresh air flows upward from a blower positioned beneath the bottom tray through perforated holes into a water layer on each tray. Large air flow rates are typically used, causing very small bubbles or frothing to form upon air contact with the water. The frothing provides a high air-water surface area for mass transfer to occur. Low-profile air stripping can be described conceptually as a countercurrent, staged operation, as demonstrated in Fig. 8.6b.

Both packed towers and low-profile air strippers are capable of providing greater than 99 per cent removal of most VOCs. There are numerous advantages and disadvantages of low-profile air strippers when compared to packed towers.

Advantages

1. Unit compactness: Because the water flows horizontally across each tray, augmenting the length or width of the trays, instead of the height of the unit, will increase the removal efficiency. A typical low-profile air stripper is less than 3 m (10 ft) tall, whereas packed towers are often on the order of 8 m (26 ft) in height. There are many situations when architectural or height restrictions require use of a compact, low-profile air stripper even when cost analysis favours a packed tower.
2. Fouling: Low-profile air strippers are less susceptible to fouling by inorganics because there is no packing. Low-profile air strippers are also much easier to disassemble and clean, compared to packed towers, as the trays are stackable and can be easily removed for cleaning.

Disadvantages

1. Narrow range of air flow rates: A low-profile air stripper must operate under a narrow range of air flow rates. If the air flow rate is too high, a flooding condition results. If the air flow rate is too low, water will flow through the holes in the sieve trays, a condition known as weeping. Because of the importance of operating the low-profile air stripper under proper conditions, it is necessary that the manufacturer design the sieve tray columns to assure proper performance. Because the air flow rate is finely tuned by the manufacturer, it is not possible to adjust the air flow rate downward should the amount of water treated decrease. In contrast, the air flow rate for a packed tower can be more readily adjusted should a shift in water flow rate occur.
2. Higher air-to-water ratios: The air-to-water ratio required for a low-profilPe air stripper is on the order of 100 to 900, compared to a typical air-to-water ratio of 30 for a packed tower. The higher air flow rate for low-profile air strippers is an important consideration, especially when off-gas treatment is required. The higher air-to-water ratio for low-profile air strippers will

result in higher costs to operate the blower due to a higher pressure drop and higher costs to treat the off-gas.

3. Foaming: If the water has a tendency to foam, then packed tower aeration must be used.

Design Approach

The diffused aeration approach is not applicable because of the frothing that occurs in low-profile air stripping. An empirical Fickian approach to mass transfer was applied to low-profile air stripping and it was shown that the mass transfer rate constants for low-profile air stripping of TCE and PERC compare favourably to mass transfer rate constants of VOCs in packed tower aeration and mechanical aeration.

The following methods are available for determining the size of a low-profile air stripper:

1. Analytical equations.
2. Manufacturer-supplied software.
3. McCabe–Thiele graphical method.

A description of the recommended method for preliminary sizing of a low-profile air stripper from the design manual follows:

1. Determine the minimum and maximum volume of water to be air stripped, the minimum temperature of the water and the maximum concentration of VOCs in the untreated water to be air stripped.
2. Determine the desired concentration (per cent removed) of the VOCs in the treated water.
3. Calculate the theoretical number of sieve trays needed to remove the VOCs to the desired concentration.
4. Estimate the tray efficiency and the number of actual trays needed.
5. Estimate the size (cross-sectional area) of the perforated plate section of each tray.
6. Estimate the pressure drop through the air stripper.
7. Estimate the size of the air blower motor (in kilowatts).

Example 8.3: Low-profile air stripping.

Determine the actual number of trays needed for a low-profile air stripper compared to the theoretical number of trays for the following conditions. The influent concentration of PERC is 15 mg/l and the treatment objective is 0.005 mg/l. The water flow rate is 0.003 m³/s (48 gpm) and the water temperature is 15°C. The air flow rate is 0.7 m³/s (1500 cfm). The dimensionless Henry's constant of PERC at 15°C is equal to 0.569.

Solution:

1. Calculate the theoretical number of sieve trays required to remove the compound to the desired concentration.

    a. Determine air-to-water ratio and the stripping factor:

$$\frac{Q_a}{Q} = \frac{(0.7 \ m^3/s)}{(0.003 \ m^3/s)} = 233 \qquad S = \frac{Q_a}{Q} \times H = 233 \times 0.569 = 133$$

   b. Determine the number of theoretical trays:

$$N_{th} = \frac{\ln[1 + (C_0/C_{T0})(S-1)]}{\ln(S)} - 1$$

$$= \frac{\ln[1 + (15/0.005)(133-1)]}{\ln(133)} - 1 = 1.64$$

c. The appropriate number of theoretical trays is thus equal to 2.

2. Determine the number of actual trays. Use a tray efficiency of 0.5, which is within the appropriate range of 0.4 to 0.6:

$$N_{act} = \frac{N_{th}}{Eff_{tray}} = \frac{2}{0.5} = 4$$

ANALYSIS OF SPRAY AERATORS

Spray towers and spray fountains are the two main types of spray aerators. A fixed grid of nozzles is used to either spray water in towers (spray towers) or spray water vertically into the air from the water surface (spray fountains), as shown in Fig. 8.3. The primary type of spray aerator that is used in water treatment is a fountain spray aerator, which is popular because existing reservoirs and large basins may be readily retrofit with fountain spray aerators. When used in reservoirs and large basins, spray aerators are used to strip taste and odour-causing compounds from raw water stored in reservoirs, oxygenate groundwater to remove iron and manganese and strip VOCs.

Description

Air-water contact occurs by spraying fine water droplets from pressurised nozzles into the air, which creates a large air-water surface for mass transfer. Three types of pressurised spray nozzles are typically used in water treatment applications: (i) hollow cone; (ii) full cone; and (iii) fan spray (see Fig. 8.10). Full-cone nozzles create a uniform pattern of droplets over the entire angle of spray, while hollow-cone nozzles create a circular pattern of droplets, primarily around the circumference of the angle of spray. Although hollow-cone nozzles do not distribute droplets as well as full-cone nozzles and have a larger pressure drop requirement, hollow-cones are generally preferred over full cones because they create smaller diameter drops and have a larger nozzle orifice. Hollow-cone spray nozzles are also prone to plugging and may require strainers upstream of the nozzle to discourage nozzle plugging.

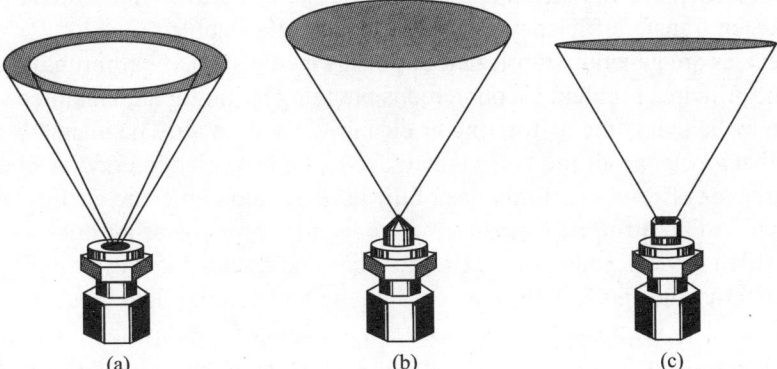

(a)   (b)   (c)

Fig. 8.10. Common spray nozzles: (a) hollow cone; (b) full cone; and (c) fan spray.

Design Approach

Contaminant removal occurs during the time the water droplet is in contact with the air, so the basis for spray aeration design equations is the mass transfer from the droplet across the air-water interface.

OTHER AIR STRIPPING AND AERATION PROCESSES

Other types of air stripping and aeration processes, such as spray towers, diffused aerators and mechanical aerators are introduced and discussed briefly in this section.

Spray Towers

There are a variety of configurations in which spraying can be used. Some configurations are analogous to packed tower designs and some are more complex designs, which are typically used for air pollution control such as cyclone scrubbers and Venturi scrubbers. Historically, spray systems have been used in water treatment for aeration, degasification of well water and odour removal.

Only a few studies of spray towers have been conducted on mass transfer in clean-water systems. Based on these studies, it has been found that spray systems are limited with respect to the removals that can be achieved and a substantial portion of removal in a spray system may occur at the nozzle. Typically, one to three transfer units are reported as a maximum limit that can be achieved in spraying systems. The apparent limitation in per cent removal is the product of backmixing of air and spray disturbance due to wall or adjacent spray contact.

The residual NTU at zero height, in this case between 0.1 and 0.2 transfer units, is the result of the removal occurring at the nozzle. Some process designs may take advantage of the removal occurring at the nozzles by recycling flow or by incorporating several banks of nozzles. The aforementioned non-ideal effects have hindered development of a general empirical design model. With the data presently available, a spray tower cannot be designed for a precise removal. Rather, the design approach serves merely as a basis to estimate the approximate removal efficiency of a spray system.

Diffused Air (Aeration)

Air diffusers or injection aerators bubble compressed air into water through orifices, nozzles in air piping, diffuser plates or tubes or spargers. Diffused aeration equipment can be classified in two general types depending on bubble size generated.

Large bubble devices have the advantage of low maintenance over fine bubble devices. A lower adsorption and oxygen transfer efficiency results from large gas bubbles.

Fine bubble devices are generally fabricated of porous media such as carborundum, nylon or tightly wrapped saran. The principal problem encountered is plugging requiring high maintenance to keep units operative, which may be overcome by filtering or cleaning the input air. The main advantage over large bubble devices is that greater absorption is obtained due to the increased interfacial area of the relatively small bubbles. Variables affecting performance of diffused aeration units are air flow rate, liquid depth and tank width. Types of air diffusers include simple open pipes for coarse air bubbles to more efficient oxygen transfer baffled devices and porous ceramic tubes and domes for fine air bubbles. Diffusers are located at basin bottoms and spacing depends on the type and aeration level required. Greater oxygen transfer can be achieved by locating the air diffuser at a greater depth below the water surface. Optimum balance between oxygen transfer and mixing is usually achieved at an 8–16 ft. diffuser depth.

Efficiencies show how most diffused air systems in the lower range of mechanical aeration devices tested. Fine bubble devices are usually higher than large bubble diffusers and are competitive with the low end of submerged turbine aeration. Ascending bubbles in waste-water acquire smaller terminal velocities than would drops falling freely in air through the same distance. This increases exposure time. Spiral and cross-current flow lengthens the travel path in the liquid.

Air diffusion is employed in water as well as waste-water treatment. The best known application of waste-water treatment is in the activated sludge process. Floating compressors are also used in raising oxygen content of receiving water overloaded with waste material that might become septic and destroy water stratification. Figure 8.11a shows construction of domed air diffuser and typical diffused-air aeration system is shown in Fig. 8.11b. The most commonly used diffuser system consists of a matrix of perforated tubes (or membranes) or porous plates arranged near the bottom of the tank to provide maximum gas to water contact.

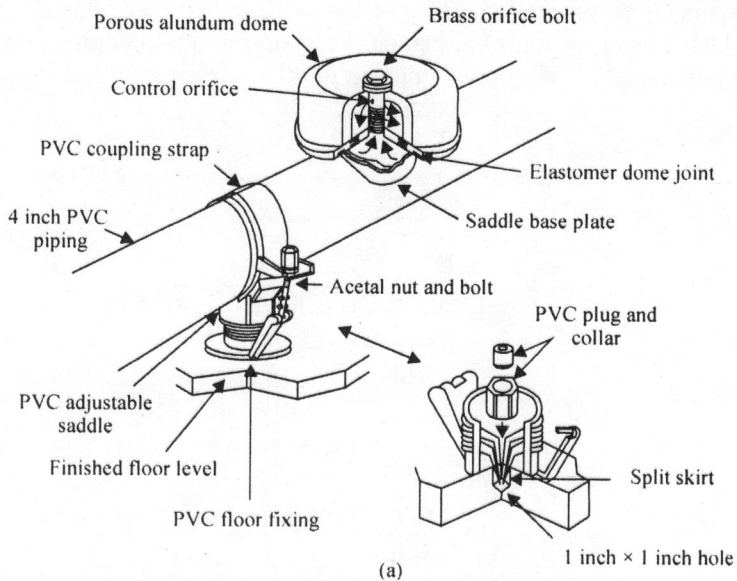

(a)

Fig. 8.11. (a) Detailed drawing showing construction of domed air diffuser, piping and basic floor attachments.

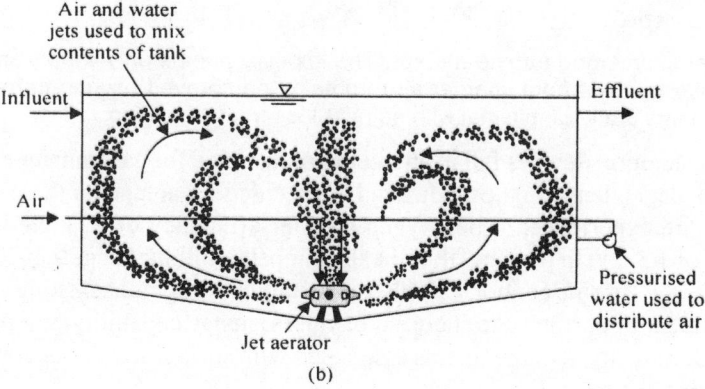

(b)

Fig. 8.11. (b) Typical example of a diffused-air aeration system.

Submerged Turbine

Mechanical aerators of this type are widely employed in waste-water treatment. The design objectives of a good submerged turbine aeration system is to provide by mechanical and fluid action sufficient shear to create a fine bubble distribution and maximise the air retention in the system. Here the mechanical function is important to keep the activated floc in mobile, useful suspension, as well as injection aeration for oxygen gas transfer.

While there may be a number of systems possible to achieve the aforementioned goals, the usual arrangement currently seen consists of a radial flow impeller located above an orifice sparge ring or an open air pipe. Air rising from the pipe is dispersed by the impeller and distributed throughout the liquid as shown in Fig. 8.12. Balance between air and impeller flow is important since in cases where air rates are too high, the gas may overcome the pumping action of the impeller and this action results in oxygen transfer efficiency loss. Lower air rates yield good dispersion in a wide range of gas rates while they independently maintain adequate mixing. Submerged turbines are fixed unit devices.

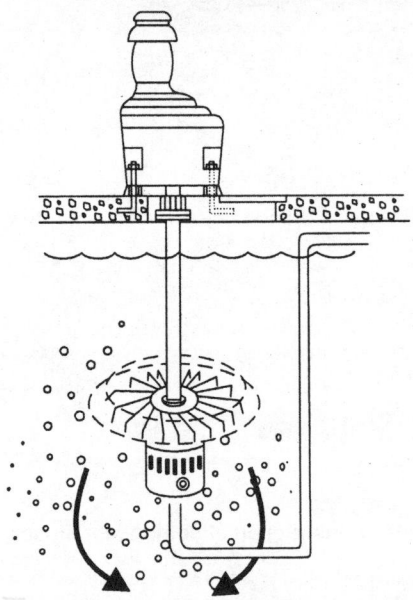

Fig. 8.12. Schematic of submerged turbine aerator. The impeller pumps only water. Shearing action in the turbine produces smaller bubbles from sparger air to obtain an improved oxygen transfer efficiency. Icing, spray and misting problems are kept minimal with no heat loss from splashing.

Submerged turbine aeration devices fall in an intermediate range for gas transfer efficiencies. Oxygen transfer can be varied independently of the mixing and is a decided advantage for these devices, particularly where wide loadings are experienced. The oxygen transfer efficiency of a single impeller submerged turbine is in the range of 1.5–2.0 lb oxygen/HP-hr.; a dual impeller turbine range 2.5–3.0 lb. oxygen/HP-hr. oxygen transfer efficiency. Turbine aerators can also be installed to augment existing diffused air systems. This offers such plants an opportunity to increase oxygen transfer capability at a minimum additional capital investment. Additionally, icing problems associated with surface aerator operation in cold weather do not exist with turbine aerators.

Surface Aeration

Surface aerators have found increasing application in activated sludge plants and aerated lagoons for waste-water treatment. The surface aerator is a device which brings to the surface the waste in water for contact with air.

A number of designs of surface aerators are in use. The bladed or paddle-surface aerator pumps liquid from beneath the blades and sprays the liquid across the water surface. The brush aerator utilises a rotating steel brush which sprays liquid from rotating blades with mixing achieved by an induced velocity below the rotating element. A draft tube is employed in some designs. Surface aerators are usually float-mounted. These designs can be broken into two classes, namely as to whether the pumping device operates at the liquid surface or substantially below the surface. Figure 8.13 shows the surface aeration process and variables.

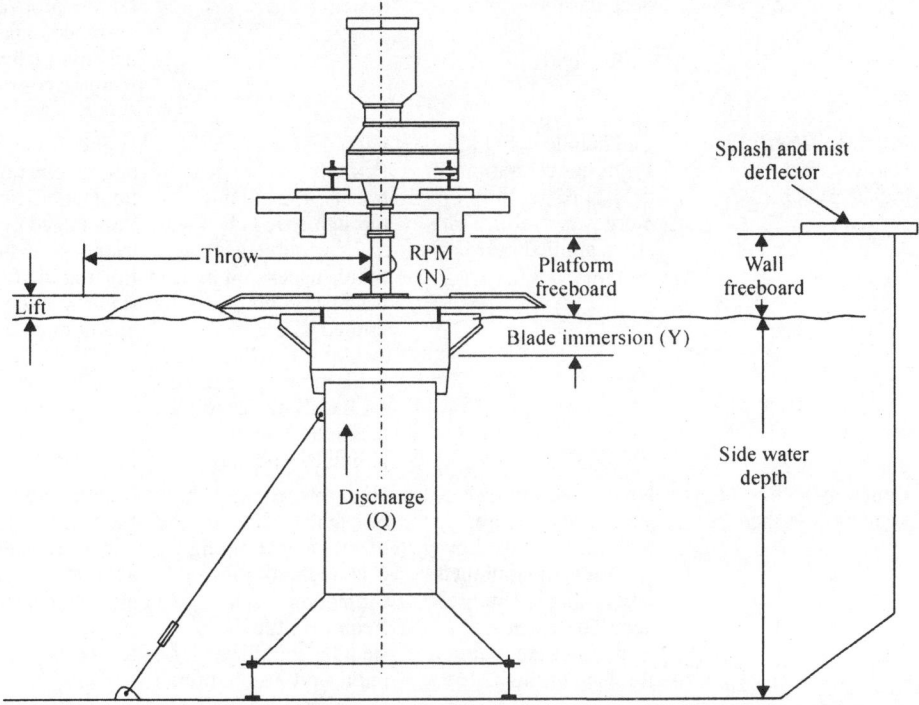

Fig. 8.13. Surface aeration process and variables.

The oxygen transfer occurs directly to the waste while it is being sprayed through the air. Additionally, oxygen transfer takes place with entrained air at the impeller and in the area around the aerator resulting from splashing liquid impinging into the liquid body.

Surface aerators generally provide higher efficiencies than other devices, oxygenation efficiencies of low-speed surface aerators range from 3.0–3.5 lb oxygen/HP-hr. These units can be controlled for varying oxygen demand requirements by making use of submergence adjustment, cycle timers and speed control with variable speed motors. High or motor-speed surface aerators are essentially axial flow pumps. Pumping action occurs using a marine-type propeller on the end of the motor shaft. As the propeller rotates, water drawn up through the draft tube is discharged at a high rate against deflector plates, producing horizontal liquid sprays. In most cases, adequate mixing occurs. Additional improvement, such as lower impellers in the case of deep basins, may be implemented. Combined units where a surface aerator is placed on the same shaft with a submerged turbine device have also been used. These units are adaptable to existing diffused air systems for increased oxygen transfer. Efficiencies will vary depending on the design combination. Table 8.5 shows mechanical aeration systems comparison.

Table 8.5. Mechanical aeration systems comparison.

System	Oxygen transfer	Solids suspension	Advantages	Disadvantages
Diffused aeration	Transfer depends on bubble size; fine bubble diffusers higher than large bubble diffusers	Basin design requires special attention with diffuser location critical for good suspension. Not good for use in deep basins	Quiet operation. Flexibility through variable gas rates enables the tailoring of oxygen transfer to system loads	Fine bubble diffusers are subject to plugging problems, thus often requiring air filtration. Diffused air systems require properly designed long, narrow basins which may increase construction costs
Submerged turbine aerator	Intermediate efficiency below surface but higher than diffused air	Can handle very high solids concentrations in deep tanks of 20 ft. or more water level. Full HP is applied near basin bottom	High degree of flexibility in oxygen transfer is available through speed changes and a variable gas rate. Solid suspension in deep basins is good. Not limited by area available. Can be used in high rate systems or for basins designed for maximum use of premium land	Higher installed HP due to lower oxygen transfer efficiencies. Submerged piping cannot be installed on floating platforms for shallow lagoon operation
Surface aerators	Generally highest of systems described	High flows produce good suspension of biological solids. Lower impellers or combined units using submerged aeration design techniques are required for deep basins	High oxygen transfer efficiency. Elimination of submerged piping. Can be float-mounted for lagoon systems. Requires little or no standby equipment for multi-unit installations. Sub-mergence adjustment allows changes in oxygen transfer to suit varying loads in the system	Requires sufficient area for proper aeration. High-rate systems applied in deep basins may have insufficient surface area to utilise full power
Combined surface and submerged turbine aerators	Generally design-dependent on split between surface aerator and submerged. turbine	Generally good and adjustable for use in deep basins through proper design	Aerator speed and gas rate changes are flexible. Advantage of turbine features for good solids suspension in deep basins. Partial oxygen transfer can be maintained with blowers shut down	Submerged piping required. Combined aeration generally requires a higher installed HP

Chapter 9

Screening, Sedimentation, Clarification, Flotation and Coagulation

INTRODUCTION

Natural forms of pollutants have always been present in surface waters. Long before the dawn of civilisation, many of the impurities were washed from the air, eroded from land surfaces, or leached from the soil and ultimately found their way into surface water. With few exceptions, natural purification processes were able to remove or otherwise render these materials harmless. Indeed, without these self-cleaning processes, the water-dependent life on earth could not have developed as it did.

The self-purification mechanisms of natural water systems include physical, chemical and biological processes. The speed and completeness with which these processes occur depend on many variables that are system-specific. Hydraulic characteristics such as volume, rate and turbulence of flow, physical characteristics of bottom and bank material, variations in sunlight and temperature, as well as the chemical nature of the natural water, are all system variables that have an influence on the natural purification processes. In natural waters, these system variables are set by nature and can seldom be altered.

The same physical, chemical and biological processes that serve to purify natural water systems also work in engineered systems. In water and waste-water treatment plants, the rate and extent of these processes are managed by controlling the system variables. A thorough knowledge of the natural purification processes is thus essential to the understanding of both the assimilative capacity of surface waters and the operation of engineered systems.

Thus the major physical processes involved in purification of water are screening, sedimentation, clarification, flotation and coagulation which are discussed in this chapter.

SCREENING

Screening devices are used to remove coarse solids from waste-water. Coarse solids consist of sticks, rags, boards and other large objects that often and, inexplicably, find their way into waste-water collection systems. Because the primary purpose of screens is to protect pumps and other mechanical equipment and to prevent clogging of valves and other appurtenances in the waste-water plant, screening is normally the first operation performed on the incoming waste-water.

Waste-water screens are classified as fine or coarse, depending on their construction. Coarse screens usually consist of vertical bars spaced 1 or more centimeters apart and inclined away from the incoming flow. Solids retained by the bars are usually removed by manual raking in small plants, while mechanically cleaned units are used in larger plants. Fine screens usually consist of woven-wire cloth or perforated plates mounted on a rotating disk or drum partially submerged in the flow or on a travelling belt. Fine screens should be mechanically cleaned on a continual basis. Typical screening devices are shown in Fig. 9.1.

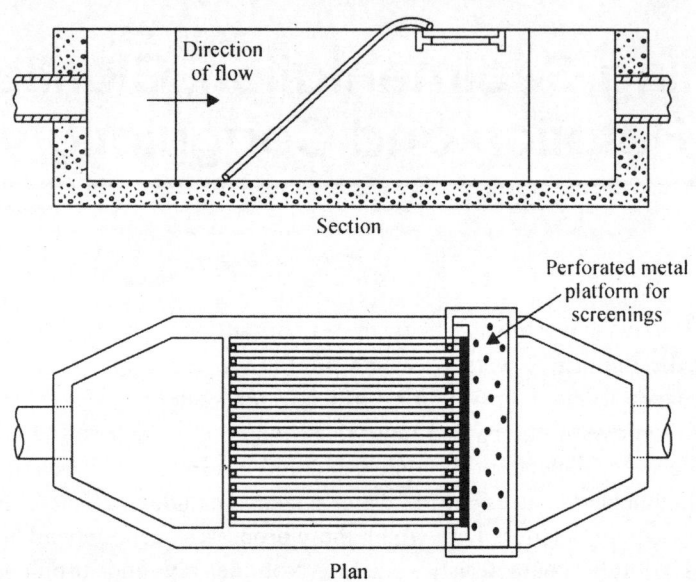

Fig. 9.1. Screening devices used in waste-water treatment manually cleaned bar rack.

Screening devices are contained in rectangular channels that receive the flow from the collection system. Manually cleaned devices should be readily accessible for cleaning, and mechanically cleaned systems should be enclosed in suitable housing. Proper ventilation must be provided to prevent accumulation of explosive gases. A straight channel section should be provided a few meters ahead of the screen to ensure good distribution of flow across the screen. Hydraulically, flow velocity should not exceed 1.0 m/s (3.3 ft/s) in the channel, with 0.3 m/s (1 ft/s) considered good design. Head loss across the screen will depend on the degree of clogging. Clean bars and screens result in a head loss of less than 0.1 m. Provisions should be made for a head loss of up to 0.3 m for manually cleaned or for manually operated, mechanically cleaned screens.

The quantity of solids removed by screening depends primarily on screen-opening size. The quantity of screenings removed from a typical municipal waste-water as a function of the screen size is illustrated in (Fig. 9.2). Screened solids are coated with organic material of a very objectionable nature and should be promptly disposed of to prevent a health hazard and/or nuisance condition. Disposal in a sanitary landfill, grinding and returning to the waste-water flow and incineration are the most common disposal practices.

Suspended solid matter in a fluid media represents a heterogeneous system that from a very general viewpoint is a fundamental fluid dynamics problem that virtually all chemical engineers are familiar with. It is so frequently encountered in numerous industry applications that considering only the problem of water treatment simply does not do the subject justice. Having noted this, the principles covered in this chapter are general enough to be applied to the class of unit operations aimed at separation, handling or processing heterogeneous or two-phase systems, including those dealing with gas-solid suspensions.

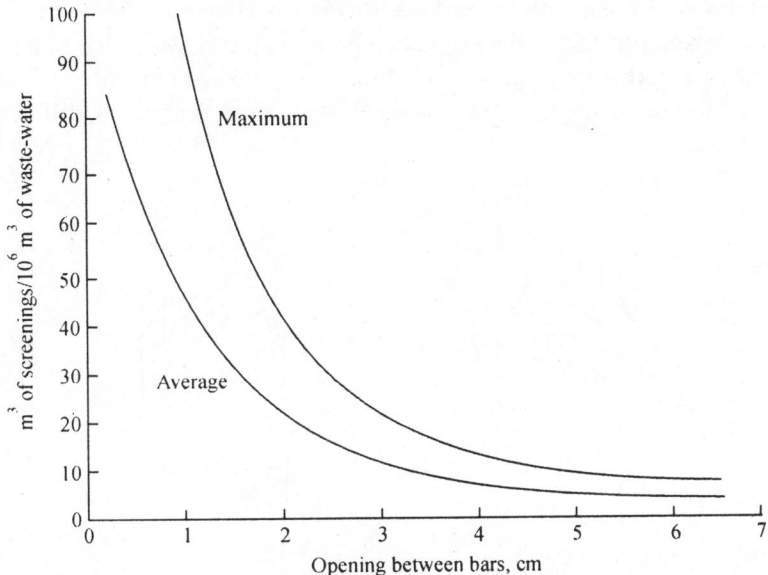

Fig. 9.2. Quantity of screening from municipal waste-water as a function of bar spacing using mechanically cleaned bar screens.

Examples of operations where heterogeneous systems are encountered include sedimentation of dust in chambers and cyclone separators, separation of suspensions in settlers, separation of liquid mixtures by settling and centrifuging, hydraulic and pneumatic transport, hydraulic and air classification, flotation, mixing by air and others. Each of these operations involves the simultaneous flow of gas and solid or liquid and solid phases. The widespread and successful application of these hydrodynamic processes to a large number of industrial problems is based on our ability to take advantage of one or a combination of five primary forces: gravity, centrifugal, buoyancy, pressure and electric. Gravity is the controlling force for separations achieved in settlers; centrifugal force is applied to cyclone separators, dryers and mixers; and pressure forces are employed in sprayers, pneumatic transport and filters. Electrical forces are employed in special techniques, such as precipitators. Buoyancy is related to gravity and takes advantage of density differences.

BEHAVIOUR OF A SINGLE PARTICLE IN A SUSPENSION

During the motion of viscous flow over a stationary body or particle, certain resistances arise. To overcome these resistances or drag and to provide more uniform fluid motion, a certain amount of energy must be expended. The developed drag force and, consequently, the energy required to overcome it, depend largely on the flow regime and the geometry of the solid body. Laminar flow conditions prevail when the fluid medium flows at low velocities over small bodies or when the fluid has a relatively high viscosity. Flow around a single body is illustrated in Fig. 9.3. As shown in Fig 9.3a, when the flow is laminar a well-defined boundary layer forms around the body and the fluid conforms to a streamline motion. The loss of energy in this situation is due primarily to fiction drag. If the fluid's average velocity is increased sufficiently, the influence of inertia forces becomes more pronounced and the flow becomes turbulent. Under the influence of inertia forces, the fluid adheres to the particle surface, forming only a very thin boundary layer and generating a turbulent wake, as shown in Fig. 9.3b. The pressure in the wake is significantly lower than that at the stagnation point on the leeward side of the particle. Hence, a net force, referred to as the pressure drag, acts in a direction opposite to that of the fluid's motion. Above a certain value of the Reynolds number, the role of pressure drag becomes significant and the friction drag can be ignored.

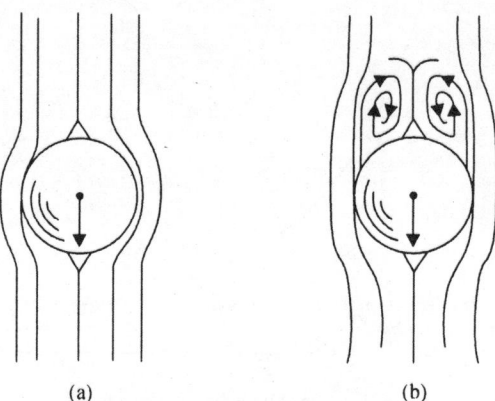

(a) (b)

Fig. 9.3. Flow around a single particle.

We shall begin discussions by analysing a dilute system that can be described as a low concentration of non-interacting solid particles carried along by a water stream. In this system, the solid particles are far enough removed from one another to be treated as individual entities. That is, each particle individually contributes to the overall character of the flow. Let's consider the dynamics of motion of a solid spherical particle immersed in water independent of the nature of the forces responsible for its displacement. A moving particle immersed in water experiences forces caused by the action of the fluid. These forces are the same regardless whether the particle is moving through the fluid or whether the water is moving over the particle's surface. For our purposes, assume the water to be in motion with respect to a stationary sphere. The fluid shock acting against the sphere's surface produces an additional pressure, P. This pressure is responsible for a force, R (called the drag force) acting in the direction of fluid motion. Now consider an infinitesimal element of the sphere's surface, dF, having a slope, α, with respect to the normal of the direction of motion (Fig. 9.4). The pressure resulting from the shock of the fluid against the element produces a force, dF, in the normal direction. This force is equal to the product of the surface area and the additional pressure, PdF_0. The component acting in the direction of flow, dR, is equal to dτcosα. Hence, the force, R, acting over the entire surface of the sphere will be:

$$R = \int PdF_0\cos\alpha = \int PdF = PF \qquad \qquad ...(9.1)$$

where dF is the projection of dF_0 on the plane normal to the flow. The term F refers to a characteristic area of the particle, either the surface area or the maximum cross-sectional area perpendicular to the direction of flow. The pressure P represents the ratio of resistance force to unit surface area (R/F) and it depends on several factors, namely the diameter of the sphere (d), its velocity (u), the fluid density (ρ) and the fluid viscosity (μ); i.e., $P = f(d, u, \rho, \mu)$.

Applying dimensional analysis, the following dimensionless groups are identified:

$$Eu = \phi(Re) \qquad \ldots (9.2)$$

where Eu is the dimensionless Euler number, defined as $P/u^2\rho$ and Re is the Reynolds number (Re = du ρ/μ). By substituting for density using the ratio of specific gravity to the gravitational acceleration, an expression similar to the well-known Darcy-Weisbach expression is obtained:

$$R/F = C_D(u^2/2g)y \qquad \ldots (9.3)$$

where C_D is the drag coefficient, which is a dimensionless parameter that is related to the Reynolds number. The relationship between C_D and Re for flow around a smooth sphere is given by the plot shown in (Fig. 9.5). As shown in this plot, there are three regions that can be approximated by expressions for straight lines. These three regions are the Stokes law region, Newton's law region, and the intermediate region. Refer to the sidebar discussion for these expressions. By substituting the expressions for the drag coefficient into equation (9.3), we obtain a convenient set of expressions that will enable us to calculate settling velocities. There are plenty of empirical correlations in the literature for the drag coefficient for different geometry objects, but practice is to use a simple sphere and then account for geometry effects by means of a correction or shape factor. We will cover this further on. For now, let's examine the phenomenon of particle settling more closely.

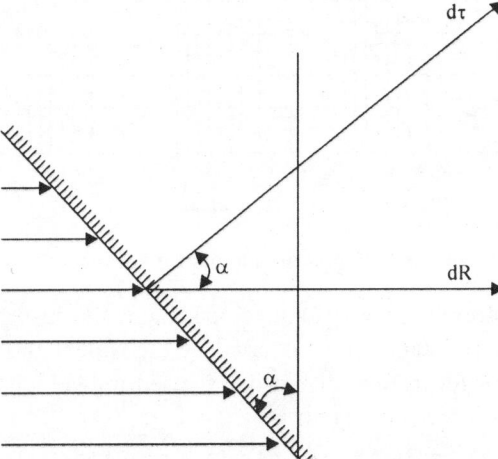

Fig 9.4. Infinitesimal element of a sphere's surface inclined at angle α from the direction of flow.

SETTLEMENT OF PARTICLES

If a particle at rest (with mass 'm' and weight 'mg') begins to fall under the influence of gravity, its velocity is increased initially over a period of time. The particle is subjected to the resistance of the surrounding water through which it descends. This resistance increases with particle velocity until the accelerating and resisting forces are equal. From this point, the solid particle continues to fall at a constant maximum velocity, referred to as the terminal velocity, u_t. In the settling velocity, the force responsible for moving a spherical particle of diameter 'd' can be expressed by the difference between its weight and the buoyant force acting on the particle.

The buoyant force is proportional to the mass of fluid displaced by the particle, that is, as the particle falls through the surrounding water, it displaces a volume of fluid equivalent to its own weight:

$$u = (\pi d^3/6)\, g\, (\rho_p - \rho) \qquad \qquad ...(9.4)$$

where,

ρ_p = density of the solid particle

ρ = density of the fluid

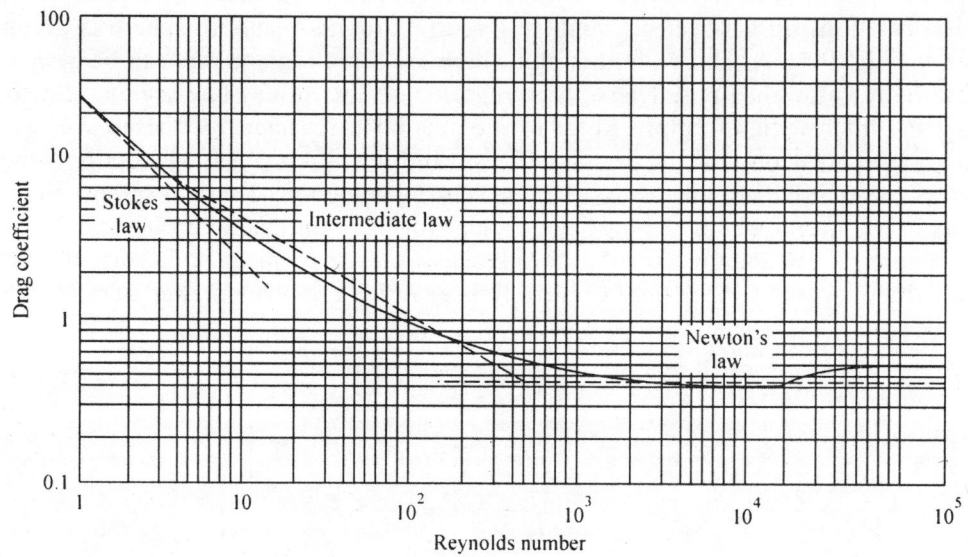

Fig. 9.5. Drag coefficients for spheres.

We now have all the information necessary to develop some working expressions for particle settling. Look back at equation 9.3 (the resistance force exerted by the water) and the expressions for the drag coefficient. The important factor for us to realise is that the settling velocity of a particle is that velocity when accelerating the resisting forces are equal:

$$(\pi d^3/6)\, g\, (\rho_p - \rho) = C_D\, (\pi d^2/4)\, (\rho u_s{}^2/2) \qquad \qquad ...(9.5)$$

From this point on, it's some very simple algebra. One can solve equation 9.5 for u_s and then substitute an expression in for C_D for each of the three flow regimes (laminar, turbulent and intermediate). One can work through the details, but the working expressions are summarised for one in the sidebar discussion on this page. Remember that to apply these equations one need to know the flow regime and so you need to make some assumptions when applying any one of these expressions.

One point we can make is that the expressions can be further developed to given us an idea of the maximum size particles that will settle out in the first two regimes of settling. If we take Stokes law for example, the maximum size particle whose velocity follows Stokes' law can be found by substituting $\mu Re/d_p\rho$ for the settling velocity into the first sidebar equation and then setting $Re = 2$ (the limiting Reynolds number value for the flow regime). This then gives us the following useful expression:

$$d_{max} \approx 1.56[\mu^2/(\rho\, (\rho_p - \rho))]^{1/3} \qquad \qquad ...(9.6)$$

The minimum size particles that do not follow Stokes' law occurs at $Re \approx 10^{-4}$. The settling velocity in this lower bound regime is less than that computed by the Stokes' law expression and generally an empirical correction factor is applied to account for particle slippage. This correction factor, which is applied by dividing the value into the Stokes' law calculated u_s value is: $K = 1 + A \lambda/d$, where λ is defined as the mean free path of a fluid molecule and constant A varies between 1.4 and 20 (as a point of reference, A = 1.5 for air). But this is a correction factor we will likely never have to consider in a conventional water treatment assignment. A more convenient set of expressions for settling velocity can be derived by expressing the three settling regime equations in terms of dimensionless groups. We won't get tangled in the derivations, although they are reasonably straightforward, but rather just list these expressions. The relationships are based on the dimensionless Archimedes number, defined as:

$$Ar = [d^3 \rho^2 g/\mu^2][\rho_p - \rho)/\rho] \qquad \ldots (9.7)$$

Note that the Archimedes number is a dimensionless group that describes the physical properties of the heterogeneous system. It can be related to the Reynolds number (and hence the settling velocity, u_s) for each settling regime as follows: For the Stoke's settling regime:

$$Re = Ar/18 \qquad \ldots (9.8)$$

Note that the upper limiting or critical value of the Archimedes number for this range occurs at $Re = 2$ and hence $Ar_{cr,1} = 18 \times 2 = 36$. This means that the laminar settling regime corresponds to $Ar < 36$. For the intermediate settling regime, where $2 < Re < 500$, we have the following expression:

$$Re = 0.152 \, Ar^{0.715} \qquad \ldots (9.9)$$

For the critical value $Re = 500$, the limiting value of Ar for the intermediate settling regime is $Ar_{cr,2} = 83000$. In other words, the intermediate settling regime corresponds to $36 < Ar < 83000$.

For the Newton's law (turbulent settling regime) region, where $Ar > 83000$, the expression of interest is:

$$Re = 1.74 \, Ar^{1/2} \qquad \ldots (9.10)$$

The usefulness of these relationships lies in the recognition that by evaluating the Archimedes number, we can establish the theoretical settling range for the particles we are trying to separate out of a waste-water stream. This very often gives us a starting point for evaluating the settling characteristics of suspended solids for dilute systems. Note that from the definition of the Reynolds number, we can readily determine the settling velocity of the particles from the application of the above expressions ($u_s = \mu Re/d_p \rho$). The following is an interpolation formula that can be applied over all three settling regimes:

$$Re = Ar/[18 + 0.575 \, Ar^{1/2}] \qquad \ldots (9.11)$$

For low values of Ar, the second term in the denominator may be neglected and equation 9.11 simplifies to equation 9.8; at high Ar values, we may neglect the first term in the denominator and the expression simplifies to equation 9.10, which corresponds to the Newton's law range.

The settling velocity of a nonspherical particle is less than that of a spherical one. A good approximation can be made by multiplying the settling velocity, u_s, of spherical particles by a correction factor, ψ, called the sphericity factor. The sphericity or shape factor is defined as the area of a sphere divided by the area of the non-spherical particle having the same volume:

$$u'_s = \psi u_s \qquad \ldots (9.12)$$

The factor $\psi < 1$ must be determined experimentally for particles of interest. Typical values are $\psi = 0.77$ for particles of rounded shape; $\psi = 0.66$ for particles of angular shape; $\psi = 0.43$ for particles of a flaky geometry. The above analysis applies only to the free settling velocities of single particles and does not account for particle-particle interactions. Hence, the application of these formulae only applies to very dilute systems. At high particle concentrations, mutual interference in the motion of particles exists and the rate of settling is considerably less than that computed by the given expressions. In the latter case, the particle is settling through a suspension of particles in a fluid, rather than through a simple fluid medium. The above provides us with a theoretical staring basis for particle settling. Let's now take a closer look at some of the standard hardware.

GRAVITY SEDIMENTATION

Thickeners and Clarifiers

Sedimentation involves the removal of suspended solid particles from a liquid stream by gravitational settling. This unit operation is divided into thickening, i.e., increasing the concentration of the feed stream, and clarification, removal of solids from a relatively dilute stream. A thickener is a sedimentation machine that operates according to the principle of gravity settling. Compared to other types of liquid/ solid separation devices, a thickener's principal advantages are:

1. Simplicity of design and economy of operation.
2. Its capacity to handle extremely large flow volumes.
3. Versatility, as it can operate equally well as a concentrator or as a clarifier.

In a batch-operating mode, a thickener normally consists of a standard vessel filled with a suspension. After settling, the clear liquid is decanted and the sediment removed periodically. The operation of a continuous thickener is also relatively simple. Fig. 9.6 illustrates a cross-sectional view of a standard thickener. A drive mechanism powers a rotating rake mechanism. Feed enters the apparatus through a feed well designed to dissipate the velocity and stabilise the density currents of the incoming stream. Separation occurs when the heavy particles settle to the bottom of the tank.

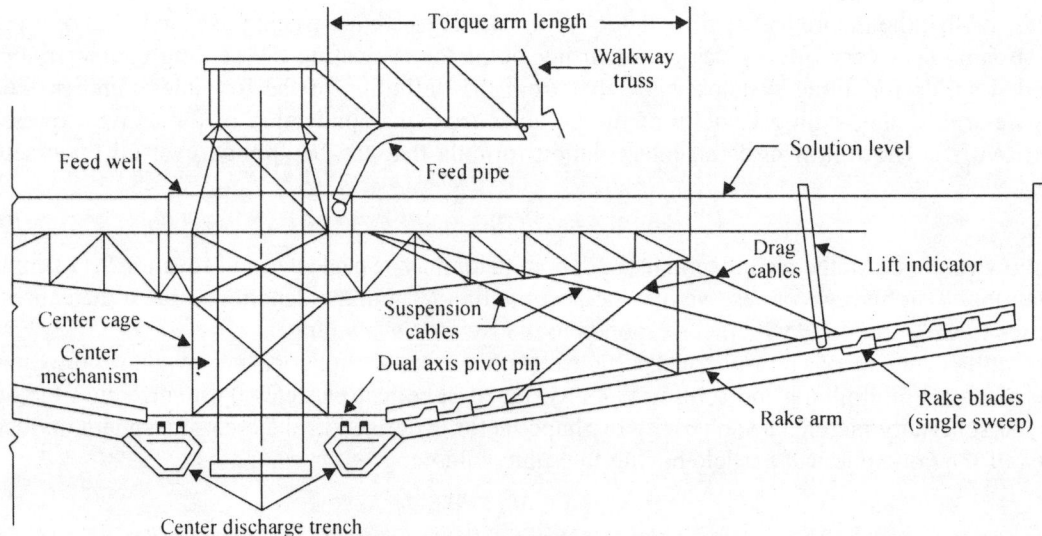

Fig. 9.6. Cross-sectional view of a thickener.

Some processes add flocculants to the feed stream to enhance particle agglomeration to promote faster or more effective settling. The clarified liquid overflows the tank and is sent to the next stage of a process. The underflow solids are withdrawn from an underflow cone by gravity discharge or pumping.

Thickeners can be operated in a countercurrent fashion. Applications are aimed at the recovery of soluble material from settleable solids by means of continuous countercurrent decantation (CCD). The basic scheme involves streams of liquid and thickened sludge moving countercurrently through a series of thickeners. The thickened stream of solids is depleted of soluble constituents as the solution becomes enriched. In each successive stage, a concentrated slurry is mixed with a solution containing fewer solubles than the liquor in the slurry and then is fed to the thickener. As the solids settle, they are removed and sent to the next stage. The overflow solution, which is richer in the soluble constituent, is sent to the preceding unit. Solids are charged to the system in the first-stage thickener, from which the final concentrated solution is withdrawn. Wash water or virgin solution is added to the last stage and washed solids are removed in the underflow of this thickener. The flow scheme for a three-stage CCD system is illustrated in Fig. 9.7. The feed stream, F, is mixed with overflow Q_2 (from thickener 2) before entering stage 1. The overflow of concentrated solution, Q_1, is withdrawn from the first stage. The underflow from the first stage, U_1, is mixed with third-stage overflow, Q_3 and fed to the second stage. Similarly, the second-stage underflow, U_2, is mixed with wash water and fed to thickener 3.

The washed solids are removed from the third stage as the final underflow, U_3. Continuous clarifiers handle a variety of process wastes, domestic sewage and other dilute suspensions.

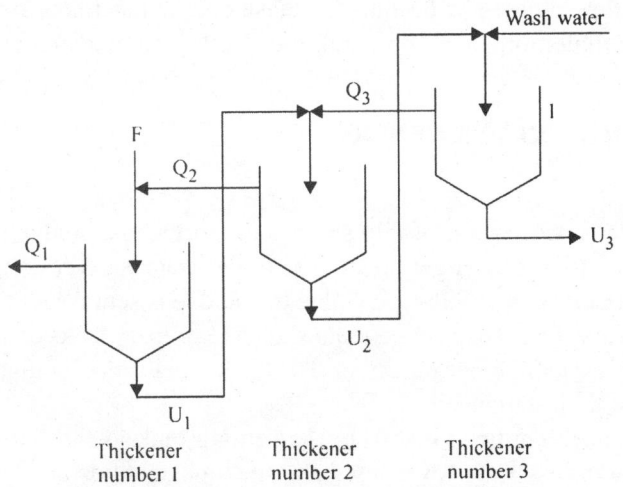

Fig. 9.7. Flow scheme for three-stage CDD.

They resemble thickeners in that they are sedimentation tanks or basins whose sludge removal is controlled by a mechanical sludge-raking mechanism. They differ from thickeners in that the amount of solids and weight of thickened sludge are considerably lower. In the cylindrical clarifier (Fig. 9.8), sedimentation machine, the feed enters up through the hollow central column or shaft, referred to as a siphon feed system. The feed enters the central feed well through slots or ports located near the top of the hollow shaft. Siphon feed arrangements greatly reduces the feed stream velocity as it enters the basin proper. This tends to minimise undesirable cross currents in the settling region of the vessel.

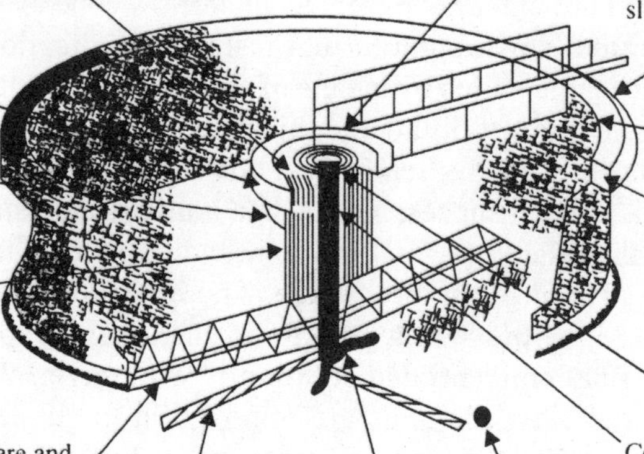

Fig. 9.8. A cylindrical clarifier and its features.

Most cylindrical units are equipped with peripheral weirs; however, some designs include radial weirs to reduce the exit velocity and minimise weir loadings. The unit shown also is equipped with adjustable rotating overflow pipes. The designs for these unique machines are pretty much universal and nearly identical configurations and comparable operating parameters can be found throughout the world.

IMPORTANT COMMENTS ON THICKENING

Gravity Thickening

This process involves the concentration of thin sludges to more dense sludge in special circular tanks designed for this purpose. Its use is largely restricted to the watery excess sludge from the activated sludge process, and in large plants of this type where the sludge is sent direct to digesters instead of to the primary tanks. It may also be used to concentrate sludge to primary tanks or a mixture of primary and excess activated sludge prior to high rate digestion. The thickening tank is equipped with slowly moving vertical paddles built like a picket fence.

Sludge is usually pumped continuously from the settling tank to the thickener which has a low overflow rate so that the excess water overflows and the sludge solids concentrate in the bottom. A blanket of sludge is maintained by controlled removal which may be continuous at a low rate. A sludge with a solids content of ten per cent or more can be produced by this method. This means that with an original sludge of two per cent, about four-fifths of the water has been removed and one of the objectives in sludge treatment has been attained.

Flotation Thickening

Flotation thickening units are becoming increasingly popular at sewage treatment plants, especially for handling waste activated sludges. With activated sludge they have the advantage over gravity thickening tanks of offering higher solids concentrations and lower initial cost for the equipment.

Dissolved Air-Pressure Flotation

The objective of flotation-thickening is to attach a minute air bubble to suspended solids and cause the solids to separate from the water in an upward direction. This is due to the fact that the solid particles have a specific gravity lower than water when the bubble is attached. Dissolved air flotation depends on the formation of small diameter bubbles resulting from air released from solution after being pressurised to 40 to 60 psi. Since the solubility of air increases with pressure, substantial quantities of air can be dissolved. In current flotation practice, two general approaches to pressurisation are used: (i) air charging and pressurisation of recycled clarified effluent or some other flow used for dilution, with subsequent addition to the feed sludge; and (ii) air charging and pressurisation of the combined dilution liquid and feed sludge. Air in excess of the decreased solubility, resulting from the release of the pressurised flow into a chamber at near atmospheric pressures, comes out of solution to form the minute air bubbles. Sludge solids are floated by the air bubbles that attach themselves to and are enmeshed in the floc particles. The degree of adhesion depends on surface properties of the solids. When released into the separation area of the thickening tank, the buoyed solids rise under hindered conditions analogous to those in gravity settling and can be called hindered separation or flotation. The upward moving particles form a sludge blanket on the surface of the flotation thickener.

Parameters: The primary variables for flotation thickening are: (i) pressure; (ii) recycle ratio; (iii) feed solids concentration; (iv) detention period; (v) air-to-solids ratio; (vi) type and quality of sludge; (vii) solids and hydraulic loading rates; and (viii) use of chemical aids. Similar to gravity sedimentation, the type and quality of sludge to be floated affects the unit performance. Flotation thickening is most applicable to activated sludges but higher float concentrations can be achieved by combining primary with activated sludge. Equal or greater concentrations may be achieved by combining sludges in gravity thickening units.

Centrifugation

Centrifugation has been demonstrated to be capable of thickening a variety of waste-water sludges. Centrifuges are a compact, simple, flexible, self contained unit and the capital cost is relatively low. They have the disadvantages of high maintenance and power costs and often a poor, solids-capture efficiency if chemicals are not used.

LAMELLA CLARIFIER

Cross-flow lamellar clarification is a technology used in industrial environments to remove oils and solids from residual water. It takes advantage of the natural tendency of oils to float and the decantation principle for suspended solids that are denser than water. The originality of this process is the combination of natural flotation and clarification techniques in one system. The strip decanter performs as well as conventional clarifiers, but is more compact and occupies a smaller area. The process is mainly used for dealing with oils and grease present in residual liquids emanating from industrial activities in the · petrochemical, chemical, mechanical, metallurgical and food-processing sectors. Depending on the specific design criteria (available space, quantity and quality of water to be treated, structural and hydraulic constraints, budget limits and need for mobility), the equipment can be housed in a circular or rectangular reservoir, or can consist of mechanical components attached to a below-grade concrete tank. If the oils are partially or completely emulsified, the cross-flow lamellar clarifier can be equipped with a coalescer, which uses a physical process to trigger separation of the oil and water phases. The coalescer is filled with various elements (rings, plastics, honeycombs, other appropriate materials) which maximise the potential contact surface. The accretion of microglobules to these elements leads to phase separation. System performance depends on the specific nature of the effluent to be treated and varies according to the type of industry. Depending on the particular situation, removal of free oils and greases and of suspended solids, varies from 90 to 99 per cent. With no chemical amendment (i.e., demulsifying agent), 20–40 per cent removal of emulsified oils and greases can be achieved. The addition of an agent enables the process to achieve 50–99 per cent removal, depending on the application.

This technology is designed to handle effluents containing a maximum of 10000 mg/l of grease and 3000 mg/l of solids. Hydraulic load is not a limiting factor. The cross-flow lamellar clarifier consists of the following basic units: primary screening chamber, separator plate cell, sludge silo, oil and grease storage chamber. First, water is pretreated in the primary screening chamber to remove part of the floating oil and grease and to allow sedimentation of large solid particles (> 500 mm). Then the effluent feeds through the plate cell where separation of phases is accomplished as follows: oils are deflected upwards by the plates to form a film on the surface of the water, sludges settle to the bottom and the purified water flows horizontally to the reservoir outlet. Before leaving the system, water passes through a calibrated opening which controls the unit's hydraulic load. Sludges are recovered in a conical extraction silo which aids their compaction and provides easier handling. Dryness rates of 1 to 5 per cent are achieved depending on sludge type. The sludge is kept apart from water to be treated, so as not to draw it back into the process flow. Floating oils are recovered from the water surface and channelled into a storage reservoir located beneath the sludge silo.

Equipment installation and start-up take less than a week. Running the system requires no energy input except for the effluent pump. Maintenance is limited to monthly testing, and cleaning of the separator plates once every six months. If needed, various products (polymers; demulsifiers and coagulants) can be added to the process to improve its performance. Running the system requires no special safety measures. If required, the equipment can be designed to provide a safe environment for treating effluents containing volatile compounds with risk of explosion.

Various parameters are taken into account in designing a cross-flow lamellar clarifier. These include:

1. Hydraulic load.
2. Suspended matter load.
3. Hydrocarbon load.
4. Desired performance.
5. Space available for clarifier's installation.
6. Operating costs are low.

SEDIMENTATION PROCESS IN GREATER DETAIL

To examine sedimentation in greater detail, let us examine the events occurring in a small-scale experiment conducted batchwise, as illustrated in Fig 9.9. Particles in a narrow size range will settle with about the same velocity. When this occurs, a demarcation line is observed between the supernatant clear liquid (zone A) and the slurry (zone B) as the process continues. The velocity at which this demarcation line descends through the column indicates the progress of the sedimentation process.

The particles near the bottom of the cylinder pile up, forming a concentrated sludge (zone D), whose height increases as the particles settle from zone B. As the upper interface approaches the sludge buildup on the bottom of the container, the slurry appears more uniform as a heavy sludge (zone D); the settling zone B disappears; and the process from then on consists only of the continuation of the slow compaction of the solids in zone D. By measuring the interface height and solids concentrations in the dilute and concentrated suspensions, a graphic representation of the sedimentation rates can be prepared as shown in Fig. 9.10.

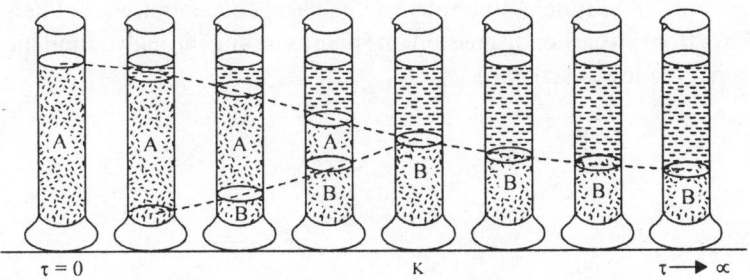

Fig. 9.9. Batch sedimentation in glass cylinder test.

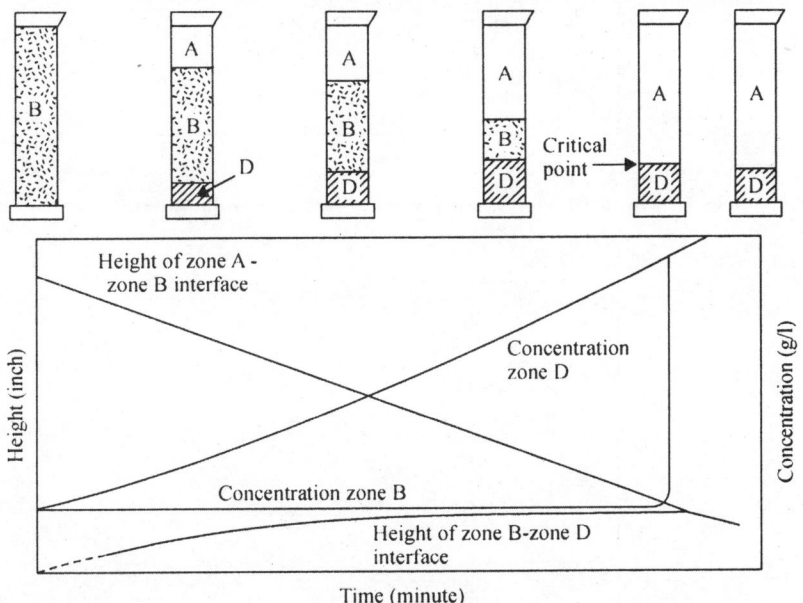

Fig. 9.10. Plot of interface height and solids concentration versus time.

The plot shows the difference in interface height plotted against time, which is proportional to the rate of settling as well as to concentration. Examining these data in more detail as a plot of sediment height, Z, versus time, t, in Fig. 9.11, we note that $Z \propto t$, meaning that the sedimentation rate is and continues to be, constant. Then the sedimentation rate of the heavy sludge decreases with time, which corresponds to the curve on the graph after point K. Naturally, the higher the concentration of the initial suspension, the slower the sedimentation process. Observations show that the solids concentration in the dilute phase is constant up to the point of complete disappearance of phase A. This is illustrated by the plot in Fig. 9.11 and corresponds to a constant rate of sedimentation in the phase.

Note, however, that the concentration in phase B changes with height Z and time t (as shown by Fig. 9.11) and hence, each curve in Fig. 9.12 represents the distribution of concentrations at any Oven moment. The initial concentration is C_1, which remains in the dilute phase during the process. After a sufficient period of time, the concentration increases to C_2, but in zone D. Obviously, if the concentration of the feed suspension is too high, no dilute phase will exist, even during the initial period of sedimentation. Hence, there is no constant sedimentation rate. In this case, concentration, not height, will change with time only. As follows from an earlier discussion on spheres falling through a fluid medium, sedimentation is faster in liquids having low viscosities.

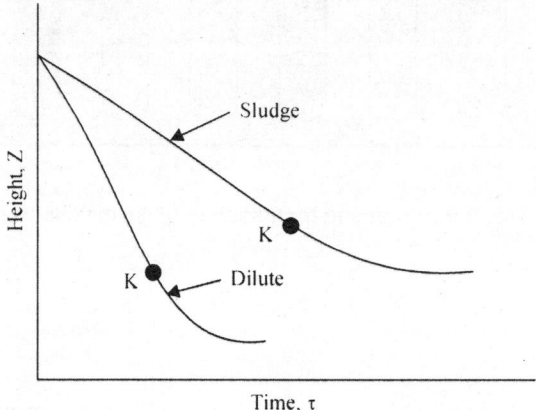

Fig. 9.11. Plot showing the kinetics of sedimentation.

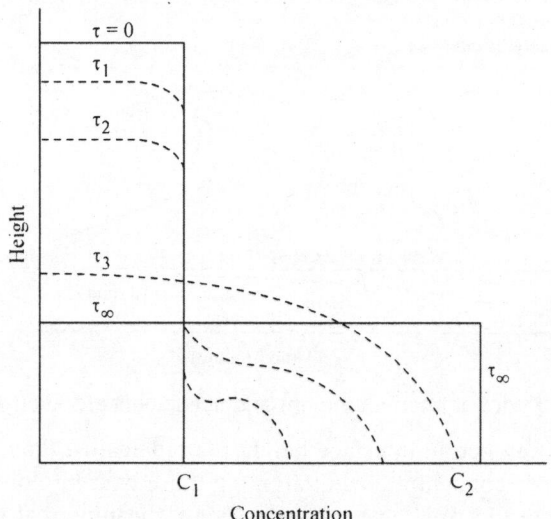

Fig. 9.12. Plot enabling derivation of sedimentation rates.

Hence, sedimentation rates are higher at elevated temperatures. In addition to temperature, an increase in the process rate may be realised by increasing particle sizes through the use of coagulation or agglomeration. In the case of colloidal suspensions, this is achieved by the addition of an electrolyte. Instead of using the concentration of the initial suspension to describe the process, we introduce a void fraction for the suspension. The void fraction is the ratio of the liquid volume, V_f, filling the space among the particles, to the total volume which is the sum of the liquid volume and the actual volume of the solid particles, V_p:

$$\varepsilon = V_f/(V_p + V_f) \qquad \qquad ...(9.13)$$

As particles settle, forming a thickened zone, the void fraction, ε, decreases. At the total settling of the slurry, the void fraction is at a minimum, its value depending on the shape of the particles. For example, the minimum void fraction of spherical particles is $\varepsilon_{min} = 0.2\ 15$; for small crystals, $\varepsilon_{min} = 0.4$. For most systems, the void fraction of a thickened sludge is approximately $\varepsilon_{min} = 0.6$; however, values should be determined experimentally for the specific system.

As the sludge compacts, its void fraction and height, X, decrease. If the initial void fraction, ε_0, is known when the sludge height is X_0, an average void fraction, ε, can be estimated assuming that the height of the sludge decreases to X. For a vertical cylinder of cross-sectioned area F, the initial volume of the sediment is FX_0 and hence, the volume of solid particles in the sediment is $(1 -\varepsilon)X_0F$. Similarly, the volume of solid particles for voidage ε is $(1 - \varepsilon)XF$. Consequently,

$$(1 -\varepsilon_0)X_0F = (1 - \varepsilon)XF \qquad \qquad ...(9.14)$$

Hence:

$$\varepsilon = (X_0/X)\ (1 -\varepsilon_0) \qquad \qquad ...(9.15)$$

For a unit volume of slurry, its weight, γ, is the sum of the weights of the solid particles, $\gamma_p(1 - \varepsilon)$ and of the liquid, $\gamma_f\varepsilon$, where γ_f = specific weight of liquid and γ_p = specific weight of particles:

$$\gamma = \gamma_p(1 - \varepsilon) + \gamma_f\varepsilon \qquad \qquad ...(9.16)$$

or $$\varepsilon = (\gamma_p - \gamma)/(\gamma_p - \gamma_f) \qquad \qquad ...(9.17)$$

This expression can be used to compute the void fraction from experimentally determined values of the specific weights.

Let's now direct our attention to the sedimentation process in the zone of constant settling velocity, i.e., in the dilute phase. To simplify the analysis w assume spherical particles of the same size. The process may be simplified further by viewing sedimentation as fixed particles in an upward-moving stream of viscous liquid, whose average velocity is u_f. Due to the viscosity of the liquid, a certain velocity gradient exists relative to the distance from the surface of spherical particle, du/dx. This velocity also depends on the average distance among particles, which is determined at any moment by a void fraction, ε, of the slurry and the particle diameter, d. The average velocity of the liquid may be presented in this case as $u_f = f\ (d, \varepsilon, du/dx)$. And rewritten on the basis of dimensional analysis in the following form:

$$u_f = K_1\ d\ (du/dx)\phi_2(\varepsilon) \qquad \qquad ...(9.18)$$

where,

K_1 = constant

$\phi_1(\varepsilon)$ = dimensionless function of the void fraction

The resistance to liquid flow around particles may be presented by an equation similar to the viscosity equation but with considering the void fraction. Recall that the shear stress is expressed by the ratio of the drag force, R, to the active surface, $K_2\pi d^2$. The total sphere surface is πd^2 and K_2 is the coefficient accounting for that part of the surface responsible for resistance. Considering the influence of void fraction as a function $\phi_2(\varepsilon)$, we obtain:

$$(R/K_2)\pi d^2 = \mu(du/dx)\phi_1(\varepsilon) \qquad \qquad ...(9.19)$$

Dividing equation 9.19 by equation 9.18, we obtain;

$$R = (K_2/K_1)\pi d\mu u_f[\phi_2(\epsilon)/\phi_1(\epsilon)] \qquad \ldots (9.20)$$

For very dilute suspensions, in which the void fraction does not influence the sedimentation process, the function $\phi_2(\epsilon)/\phi_1(\epsilon) = \phi(\epsilon)$ reduces to unity. It is known also that in dilute suspensions the sedimentation of small particles follows Stokes' law:

$$R_\alpha = 3\,\pi d\mu u_f \qquad \ldots (9.21)$$

Equating equations 9.20 and 9.21, we find that (K_2/K_1) is equal to 3. Hence, the resistance of the liquid relative to a spherical particle in the sedimentation process is

$$R = 3\,\pi d\mu u_f/\phi(\epsilon) \qquad \ldots (9.22)$$

This resistance is balanced by the gravity force acting on a particle:

$$W = (\pi d^3/6)(\gamma_p - \gamma) \qquad \ldots (9.23)$$

where γ_p is the actual specific weight of a particle and γ is the average specific weight of the sludge, which depends on void fraction ϵ. Using equation 9.17, we replace $(\gamma_p - \gamma)$ with $(\gamma_p - \gamma_f)\epsilon$, where γ_f is the specific weight of the liquid:

$$W = (\pi d^3/6)\,(\gamma_p - \gamma_f)\epsilon \qquad \ldots (9.24)$$

By comparing the gravity force acting on the particle (equation 9.24) with the resistance to liquid flow (equation 9.22), we obtain the average liquid velocity relative to the particles:

$$u_f = [d^2\,(\gamma_p - \gamma_f)/18\mu]\epsilon\,\phi(\epsilon) \qquad \ldots (9.25)$$

In practice, however, the liquid velocity relative to fixed particles, u_f, is not very useful. Instead, the velocity of settling relative to the walls of an apparatus, $u_f - u$, is of practical importance. The volume of the solid phase moving downward should be equal to that of liquid moving upward. This means that volume rates of these phases must be equal. Consider a column of slurry having a unit cross section and imagine the liquid and solid phases to have a well defined interface. The column of solid phase will have a base $1 - \epsilon$ and the liquid column phase will have a base ϵ. Hence, the volumetric rate of the solid column will be $(1 - \epsilon)u$ and that of the liquid column will be $(u_f - u)\epsilon$. Because these flowrates are equal to each other, we obtain

$$(1 - \epsilon)u = (u_f - u)\epsilon \qquad \ldots (9.26)$$

Therefore, the settling velocity of the solid phase relative to the wall of an apparatus, depending on the average liquid velocity relative to the sludge with void fraction ϵ, will be

$$u = u_f\epsilon \qquad \ldots (9.27)$$

Substituting this expression into equation 9.25, we obtain the actual settling velocity:

$$u = [d^2\,(\gamma_p - \gamma_f)18\mu]\epsilon^2\,\phi(\epsilon) \qquad \ldots (9.28)$$

Note that the term in parentheses expresses the velocity of free failing, according to Stokes' law:

$$u_p = d^2\,(\gamma_p - \gamma_f)/18\mu \qquad \ldots (9.29)$$

or

$$u = u_p\epsilon^2\,\phi(\epsilon) \qquad \ldots (9.30)$$

For a very dilute suspension, i.e., $\epsilon = 1$ and $\phi(\epsilon) = 1$, the settling velocity will be equal to the free-fall velocity. As no valid theoretical expression for the function $\phi(\epsilon)$ is available, common practice is to rely on experimental data. Note that a unit, volume of thickened sludge contains ϵ volume of liquid and $(1 - \epsilon)$

volume of solid phase, i.e., a unit volume of particles of sludge contains $\varepsilon/(1 - \varepsilon)$ volume of liquid. Denoting σ as the ratio of particle surface area to volume, we obtain the hydraulic radius as the ratio of this volume, $\varepsilon/(1 - \varepsilon)$, to the surface, σ, when both values are related to the same volume of particles:

$$r_h = \varepsilon/[(1 - \varepsilon)\sigma] \qquad \ldots (9.31)$$

For spherical particles, σ is equal to the ratio of the surface area, πd^2, to the volume $\pi d^3/6$, i.e., $\sigma = 6/d$. Hence,

$$r_h = \varepsilon/[(1 - \varepsilon)6] \qquad \ldots (9.32)$$

For a specified void fraction, the diameter of the sphere is a measure of the distance between sludge particles ($u_f = f(d, \varepsilon, du/dx)$). However, it is more practical to introduce the hydraulic radius and instead of $\phi_1(\varepsilon)$ and $\phi_2(\varepsilon)$, according to equation 9.30, we assume the following value:

$$\phi(\varepsilon) = [\varepsilon/(1 - \varepsilon)]\theta(\varepsilon) \qquad \ldots (9.33)$$

where $\theta(\varepsilon)$ is the new experimental function of the void fraction. Hence, the settling velocity equation may be rewritten in the following form:

$$u = u_p [\varepsilon/(1 - \varepsilon)]\theta(\varepsilon) \qquad \ldots (9.34)$$

By representing the velocity in this manner, we can anticipate a small change in the function $\theta(\varepsilon)$ because the influence of the flow pattern is, to a large extent, accounted for in the hydraulic radius. Laboratory and pilot scale testing have shown that the function $\phi(\varepsilon)$ may be presented by the following empirical equation:

$$\phi(\varepsilon) = 10^{-1.82 (1 - \varepsilon)} \qquad \ldots (9.35)$$

Multiplying equation 9.35 by $(1 - \varepsilon)/\varepsilon$, we obtain the function $\theta(\varepsilon)$. For $\varepsilon \leq 0.7$, i.e., for thickened sludges, this function is practically constant and equal to $\theta(\varepsilon) = 0.123$. The settling velocity of spherical particles is therefore:

$$u = u_p[\varepsilon^2 \times 10^{-1.82(1- \varepsilon)}] \qquad \ldots (9.36)$$

For more thickened sludges:

$$u = [0.123 \, \varepsilon^3/(1 - \varepsilon)]u_p \qquad \ldots (9.37)$$

Thus far, the analysis has been based on independently settling spherical particles. To relate to the design of the unit operations, we now must consider the kinetics of non-spherical particle settling and the sedimentation of flocculent particles. In contrast to single-particle settling, such systems form a certain structural unity similar to tissue. The sludge is compacted under the action of gravity force, i.e., the void fraction decreases and the liquid is squeezed out from the pore structure.

The formation of a regular sediment from a flocculent may be achieved by the addition of electrolytes, as described earlier. The general characteristic of normal settling of non-spherical particles (as well as flocculent ones) is that the sediment carries along with it a portion of the liquid by trapping it between particle cavities. This trapped volume of liquid flows downward with the sludge and is proportional to the volume of the sludge.

That is, it can be expressed as $a(1 - \varepsilon)$, where a is a coefficient and $(1 - \varepsilon)$ is the volume of particles. Consequently, a portion of the liquid remains in a layer above the sludge, and a portion is carried along with the sludge corresponding to the modified void fraction:

$$\varepsilon' = \varepsilon - a(1- \varepsilon) \qquad \ldots (9.38)$$

This is the difference of the total relative liquid volume and liquid moving together with particles. Substituting ε' for ε in equation 9.37, we obtain the settling velocity at $\varepsilon \le 0.7$:

$$u = 0.123(1 + a)^2 u_p \left[(\varepsilon - a/(1 + a))\right]^3/(1 - \varepsilon) \qquad \text{.... (9.39)}$$

Denoting $a/(1 + a) = \beta$, the above expression can be written as follows:

$$u = [0.123 u_p(\varepsilon - \beta)^3]/[(1 - \beta)^2(1 - \varepsilon)] \qquad \text{... (9.40)}$$

Similarly equation 9.36 for slurries with non-spherical particles is

$$u = u_p[(\varepsilon - \beta)^2/(1 - \beta)]10^{-1.82(1 - \varepsilon)/(1 - \beta)} \qquad \text{... (9.41)}$$

Parameter β is equal to the ratio of the liquid volume entrained and the sum of the volumes of this liquid and particles. Values of β are determined experimentally from measured settling velocities. In general, the smaller the effective particle size, the more liquid is entrained by the same mass of solids phase. For example, particles of carborundum with $d = 12.2 \, \mu m$ have $\beta = 0.268$; $d = 9.6 \, \mu m$, $\beta = 0.288$; and $d = 4.6 \, \mu m$, $\beta = 0.35$.

AN ALTERNATIVE ANALYSIS—THE FORGOTTEN METHOD OF DIMENSIONAL ANALYSIS

The dimensional analysis is known as 'lost art'—because it is usually not heavily emphasised in engineering education today. However, for well over 100 years it has provided simply a wealth of practical design correlations that are still relied upon in virtually all aspects of chemical engineering, ranging from classes of problems dealing with heat and mass transfer, reaction kinetics, momentum exchanges in flow dynamics. Much of sedimentation theory, and indeed the basis for more sophisticated analyses, are based on relating dimensionless groups that have been correlated with experimental observation. The following discussion will walk you through the approach of dimensional analysis as applied to the general theory of sedimentation, providing us with some expressions that will give us more than a head start in analysing specific separation problems in water treatment.

To accomplish this, let's take a few steps back and start by examining the forces acting on a single particle settling through a continuous fluid medium. These forces are gravity, G, buoyant or Archimedes forces, A, centrifugal field, C and an electrical field, Q. The system diagram for defining this is shown in Fig. 9.13. Geometrically summing all the forces, we obtain:

$$P = G + A + C + Q \qquad \text{... (9.42)}$$

If force P is greater than zero, the particle will be in motion relative to the continuous phase at a certain velocity, w. At the beginning of the particle's motion, a resistance force develops in the continuous phase, R, directed at the opposite side of the particle motion. At low particle velocity (relative to the continuous phase), fluid layers running against the particle are moved apart smoothly in front of it and then come together smoothly behind the particle (Fig. 9.14). The fluid layer does not intermix (a system analogous to laminar fluid flow in smoothly bent pipes). The particles of fluid nearest the solid surface will take the same time to pass the body as those at some distance away.

Because the liquid layers move at different velocities relative to each other, planes of slip exist between them and from Newton's law, the forces of viscous friction arise. Consequently, the resistance force depends on the viscosity of the medium as is determined by the viscosity coefficient, μ. At higher particle velocities (or higher medium velocities relative to the particle) the flow around the object is broken, forming swirling fluid patches (Fig. 19.14b). The formation of vortices is influenced by the relative flow velocity, the shape of the particle and the smoothness of the object's surface. The higher the velocity, the more complicated the particle shape; and/or the greater the roughness, the more intense the vortex formation. Eventually, this leads to the generation of eddies along the downstream surface of the particle (Fig. 9.14c).

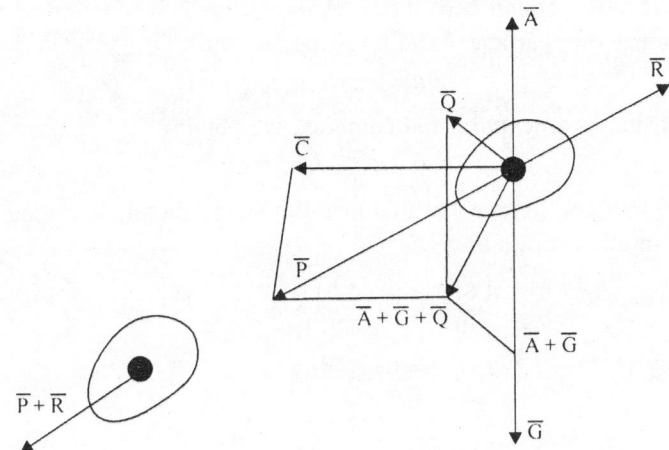

Fig 9.13. System of forces acting on a single particle.

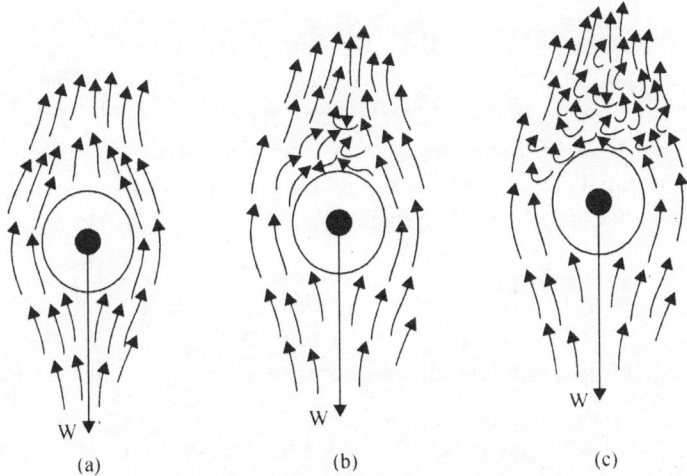

Fig. 9.14. Flow around a particle: (a) laminar; (b) transional; (c) turbulent.

The eddies are entrained by the flow and, at a certain distance from the particle, they disappear while being replaced by new eddies. Due to eddy formation and their breaking away from the particle, a low-pressure zone forms at the front of the particle. Hence, as described earlier, a pressure gradient is formed between the front and rear of the particle. This gradient is responsible primarily for the resistance to particle motion in the medium. The amount of this resistance depends on the energy expended toward eddy formation: the more intensive this formation, the greater the energy consumption and, hence, the greater the resistance force. The inertia forces generated by eddies play an important role. They are characterised by the mass and velocity of the fluid relative to the particle.

The total resistance is the sum of friction and eddy resistances. Both factors act simultaneously, but their contribution in the total resistance depends on the conditions of the flow in the vicinity of the particle. Hence, for the most general case the resistance force is a function of velocity, w, density, ρ, viscosity, μ, the linear size of a particle, ℓ and its shape, ψ. Thus,

$$R = (w, \rho, \mu, \ell, \psi) \qquad \qquad ...(9.43)$$

Assuming this relationship as an exponential complex, we obtain

$$R = A' \, w^x \rho^y \mu^z \ell^\alpha \qquad \qquad ...(9.44)$$

where A' is dimensionless coefficient that includes the shape factor, ψ noting the dimensions of all parameters appearing in this expression:

$$[R] = LMT^{-2}$$
$$[w] = LT^{-1}$$
$$[\rho] = L^{-3}M$$
$$[\ell] = L$$
$$[\mu] = L^{-1}MT^{-1}$$

where L, M and T are the principal unit measures-length, mass and time. Expressing this in terms of its dimensions:

$$LMT^{-2} = (LT^{-1})^x (L^{-3}M)^y L^z (L^{-1}MT^{-1})^\alpha$$

or

$$LMT^{-2} = L^{x-3y+z-\alpha} \, M^{y+\alpha} T^{-x-\alpha} \qquad \qquad ...(9.45)$$

For the dimensions on the LHS of this expression to satisfy the RHS, the exponents on the principal units of measure, must be equal. Thus, we have the following system of three equations (corresponding to the number of values with independent dimensions):

$$x - 3y + z - \alpha = 1$$
$$y + \alpha = 1$$
$$-x - \alpha = -2$$

Because this system of equations cannot be solved (there are fewer equations than variables), we express all exponents in terms of α:

$$x = 2 - \alpha; \; y = 1 - \alpha; \; z = 2 - \alpha$$

We may now write:

$$R = A'w^{2-\alpha} \, \rho^{1-\alpha} \, \ell^{1-\alpha} \mu^\alpha \qquad \qquad ...(9.46)$$

Equation 9.46 is a general expression that may be applied to the treatment of experimental data to evaluate exponent α. This, however, is a cumbersome approach that can be avoided by rewriting the equation in dimensionless form. Equation 9.42 shows that there are n = 5 dimensional values and the number of values with independent measures is m = 3 (m, kg, sec.).

Hence, the number of dimensionless groups according to the π-theorem is $\pi = 5 - 3 = 2$. As the particle moves through the fluid, one of the dimensionless complexes is obviously the Reynolds number: $Re = w\ell\rho/\mu$. Thus, we may write:

$$R = A' \, (w\ell\rho/\mu)^{-\alpha} \rho w^2 \ell 2 \qquad \qquad ...(9.47)$$

As one of two possible dimensionless numbers is now known, the second one can be obtained by dividing both sides of the equation through by the remaining values:

$$Eu = R/(\rho w^2 \ell^2) \qquad \qquad ... (9.48)$$

The result is a modified Euler number. You can prove to yourself that the pressure drop over the particle can be obtained by accounting for the projected area of the particle through particle size, ℓ, in the denominator. Thus, by application of dimensional analysis to the force balance expression, a relationship between the dimensionless complexes of the Euler and Reynolds numbers, we obtain:

$$Eu = A'Re^{-\alpha} \qquad \qquad ... (9.49)$$

Coefficient A' and exponent α must be evaluated experimentally. Experiments have shown that A' and α are themselves functions of the Reynolds number. Equation 9.47 shows that the resistance force increases with increasing velocity. If the force field (e.g., gravity) has the same potential at all points, a dynamic equilibrium between forces P and R develops shortly after the particle motion begins. As described earlier, at some distance from its start the particle falls at a constant velocity. If the acting force depends on the particle location in space, in a centrifugal field, for example, it will move with uniformly variable speed until it travels outside of the boundary of the field's action or runs into an obstacle such as a vessel wall. We now shall define the relationship between particle motion and the acting factors. Moving under the action of force P, with acceleration a_g at infinitesimal distance $\delta \ell$, the particle with mass m_p performs work $m_p a_g \delta \ell$.

This work is spent on overcoming the resistance force and the displacement of fluid mass, m_c, in a volume equal to the volume of the particle, V, at the same distance, but in opposite direction and with the same acceleration, a_g:

$$m_p a_g \delta \ell = m_c a_g \delta \ell + R \, \delta \ell \qquad \qquad ... (9.50)$$

Dividing through by $\delta \ell$ and expressing the masses of the particle and medium in terms of their volumes and densities, we obtain

$$V(\rho_p - \rho_c)a_g = R \qquad \qquad ... (9.51)$$

Consider again the simple motion of a sphere. In this case, the equivalent diameter of a sphere, d_{eq}, is equal to its geometric diameter, d. Equating the above expressions and replacing ℓ by d (and denoting the Euler, umber, Eu, by Υ), we obtain an expression for the resistance force:

$$R = \Upsilon \rho_c w^2 d^2 \qquad \qquad ... (9.52)$$

where,

$$\Upsilon = A'Re^{-\alpha} \qquad \qquad ... (9.53)$$

In sedimentation, the Eule, number is often referred to as the resistance number. Multiplying and dividing the RHS of equation 9.52 by $\pi/8$, we obtain:

$$R = C_D F \rho_c w^2/2 \qquad \qquad ... (9.54)$$

where,

$C_D = (8/\pi)\Upsilon$ = drag coefficient
$F \quad = \pi d^2/4$ = the cross-sectional area of the spherical particle

Equation 9.54 is Newton's resistance law. Substituting Equation 9.53 into the definition for C_D, we obtain:

$$C_D = BRe^{-\alpha} \qquad \qquad ... (9.55)$$

where,

$B = 8 A'/\pi$ since $\Upsilon = f(Re)$ and $C_D = f_1(Re)$.

The Reynolds number for a sphere is

$$Re = wd\rho_c/\mu = wd/\nu \qquad \text{... (9.56)}$$

Substituting the resistance force into equation 9.51 and expressing F and V in terms of d, the basic equation of sedimentation theory is obtained:

$$d^3(\rho_p - \rho_c)a_g = \tfrac{3}{4} C_D w^2 d^2 \qquad \text{... (9.57)}$$

or

$$d^3(\rho_f/\rho_c)a_g = \tfrac{3}{4} C_D w^2 d^2 \qquad \text{... (9.58)}$$

where $\rho_f = \rho_p - \rho_c$ is the effective density of the system. In separation calculations for heterogeneous systems, the important parameter is settling velocity:

$$w = \{(4/3) \times (d/C_D)(\rho_f/\rho_c)a_g\}^{\frac{1}{2}} \qquad \text{... (9.59)}$$

Application of the above formulae is difficult because the drag coefficient, C_D, is a function of velocity and particle geometry. A generalised calculation procedure for settling under the influence of gravity was developed many years ago. The method, however, also may be applied to settling due to the influence of any force field, provided the relationship between particle acceleration and the co-ordinates of field is defined. The procedure is based on expressing the basic equation of sedimentation (Eq. 9.58) in terms of a relationship of criteria. For this, both sides of equation 9.58 are 2 divided by v^2 and the RHS multiplied and divided by the acceleration due to gravity, g:

$$(gd^3/v^2)(\rho_f/\rho_c)(a_g/g) = \tfrac{3}{4} C_D w^2 d^2/v^2 \qquad \text{... (9.60)}$$

The LHS of this expression contains a dimensionless group known as the Galileo number, defined as $Ga = gd^3/v^2$.

Multiplying by a simplex composed of densities results in the Archimedes number:

$$Ar = Ga (\rho_f/\rho_c) \qquad \text{... (9.61)}$$

And introducing the ratio of accelerations, $K_s = a_g/g$, where K_s indicates the relative strength of acceleration, a_g, with respect to the gravitational acceleration g. This is known as the separation number. The LHS of equation 9.60 contains a Reynolds number group raised to the second power and the drag coefficient. Hence, the equation may be written entirely in terms of dimensionless numbers:

$$Ar K_s = \tfrac{3}{4} C_D Re^2 \qquad \text{... (9.62)}$$

The Archimedes number contains parameters that characterise the properties of the heterogeneous system and the criterion establishing the type of settling. The criterion of separation essentially establishes the separating capacity of a sedimentation machine. The product of these criteria is:

$$S_1 = Ar K_s \qquad \text{... (9.63)}$$

This product contains information on the properties of the suspension and characterises the settling process as a whole. Substituting equation 9.62 into 9.63 gives:

$$S_1 = \tfrac{3}{4} C_D Re^2 \qquad \text{... (9.64)}$$

This expression represents the first form of the general dimensionless equation of sedimentation theory. As the desired value is the velocity of the particle, equation 9.64 is solved for the Reynolds number:

$$Re = [(4/3)S_1/C_D]^{\frac{1}{2}} \qquad \text{... (9.65)}$$

To determine the size of a particle having a velocity, w, in the gravitational field, both sides of equation 9.63 are multiplied by the complex Re/ArK_s:

$$(Re^3/Ar(1/K_s)) = (4/3)\ Re/C_D \qquad \ldots (9.66)$$

The dimensionless complex Re^3/Ar is expressed simply as:

$$\zeta = Re^3/Ar = (w^3/gv)(\rho_f/\rho_c) \qquad \ldots (9.67)$$

Denoting $S_2 = \zeta/K_s$, then equation 9.62 may be rewritten as

$$S_2 = (4/3)Re/C_D \qquad \ldots (9.68)$$

This is the second form of the dimensionless equation for sedimentation. The Reynolds number may also be calculated from this equation:

$$Re = \tfrac{3}{4}\,C_D S_2 \qquad \ldots (9.69)$$

As with S_1, the Reynolds number is the dependent variable and S_2 is the determining one. For settling under the influence of gravity, we note that $a_g = g$, $K_s = 1$ and hence, $S_1 = Ar$ and $S_2 = \zeta$. Therefore, the general dimensionless equations for sedimentation are applicable in any force field. They need be transformed only into the appropriate dimensionless groups describing the type of force field influencing the process. Again, for gravity settling, $Ar = \tfrac{3}{4}\,C_D Re^2$ and $\zeta = 4Re/3C_D$. The dimensionless numbers of sedimentation, S_1 and S_2, as well as C_D and Υ are all functions of Re. The parameter Υ must be determined experimentally. Equation 9.53 can be written as a straight line when expressed in terms of its logarithms:

$$\log(\Upsilon) = \log(A') - \alpha \log(Re) \qquad \ldots (6.70)$$

Coefficient A' and exponent α can be evaluated readily from data on Re and γ. The dimensionless groups are presented on a single plot in Fig 9.15. The plot of the function $C_D = f_1(Re)$ is constructed from three separate sections. These sections of the curve correspond to the three regimes of flow. The laminar regime is expressed by a section of straight line having a slope $\beta = 135°$ with respect to the x-axis. This section corresponds to the critical Reynolds number, $Re'_{cr} < 0.2$. This means that the exponent a in equation 9.53 is equal to 1. At this α value, the continuous-phase density term, ρ_c, in equation 9.46 vanishes.

Therefore, the inertia forces have an insignificant influence on the sedimentation process in this regime. Theoretically, their influence is equal to zero. In contrast, the forces of viscous friction are at a maximum. Evaluating the coefficient B in equation 9.55 for $\alpha = 1$ results in a value of 24.

Hence, we have derived the expression for the drag coefficient of a sphere, $C_D = 24/Re$.

The first critical values of the dimensionless sedimentation numbers, S_1 and S_2, are obtained by substituting for the critical Reynolds number value, $Re'_{cr} = 0.2$, into the above expressions:

$$S'_{1cr} < 3.6 \qquad \ldots (9.71)$$

$$S'_{2cr} < 0.0022 \qquad \ldots (9.72)$$

Substituting the expression for C_D, we again obtain the settling velocity of an isolated particle in laminar flow:

$$w = (d^2/18v)(\rho_f/\rho_c)a_g \qquad \ldots (9.73)$$

Changing kinematic viscosity, v, to dynamic viscosity, the velocity of particle sedimentation in the laminar regime is:

$$w = (d^2/18\mu)\rho_f a_g \qquad \ldots (9.74)$$

From equations 9.64 and 9.60 at $S_1 \leq 3.6$, we obtain the first critical value of the particle diameter:

$$d'_{cr} \leq 1.53\{v^2\,\rho_c/a_g\,\rho_f\}^{1/3} \qquad\qquad \dots (9.75)$$

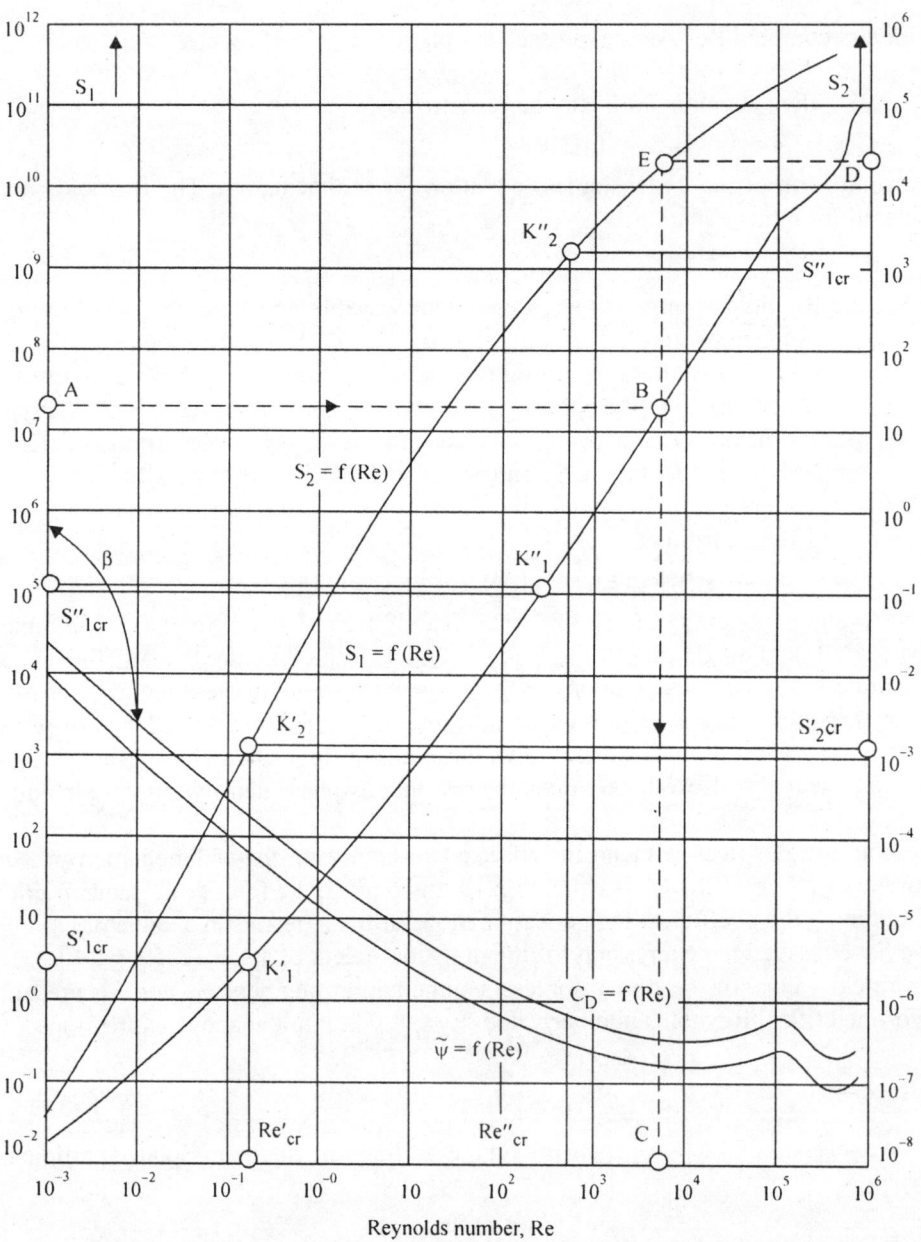

Fig. 9.15. Dimensionless sedimentation theory plot.

In applying this equation it is possible to determine the maximum size particle in laminar flow, taking into account the given conditions of sedimentation (ρ_c, ρ_p, μ and a_g). However, this equation does not determine what the flow regime is when $d > d'_{cr}$.

The turbulent regime for C_D is characterised by the section of line almost parallel to the x-axis (at the $Re''_{cr} > 500$). In this case, the exponent α is equal to zero. Consequently, viscosity vanishes from equation 9.46. This indicates that the friction forces are negligible in comparison to inertia forces. Recall that the resistance coefficient is nearly constant at a value of 0.44. Substituting for the critical Reynolds number, $Re'_{cr} \geq 500$, into equations 9.65 and 9.68, the second critical values of the sedimentation numbers are obtained:

$$S''_{1cr} \geq 82500$$

$$S''_{2cr} > 1515$$

And substituting $C_D = 0.44$ into equation 9.59,

$$w = 1.75[d\ (\rho_f/\rho_c)a_g]^{1/2} \quad m \qquad \qquad \ldots (9.76)$$

At $S''_{1cr} > 82500$, we obtain the second critical value of particle size:

$$d''_{cr} = 43.5\{v^2\ \rho_c/a_g\ \rho_f\}^{1/2} \qquad \qquad \ldots (9.77)$$

Those particles with sizes $d > d''_{cr}$ at a given set of conditions (v, ρ_c, ρ_p and a_g) will settle only in the turbulent flow regime. For particles with sizes $d'_{cr} < d$, d''_{cr} will settle only when the flow around the object is in the transitional regime. Recall that the transitional zone occurs in the Reynolds number range of 0.2 to 500. The sedimentation numbers corresponding to this zone are: $3.6 < S_1 < 82500$; and $0.0022 < S_2 < 1515$.

The slope of the curve in the transitional zone changes from 135° to 180°. It shows that the exponent in changes as follows: $0 \leq \alpha \leq 1$. This means that the friction and inertia forces are commensurable in the process of sedimentation. Several empirical formulae have been proposed for estimating the resistance coefficient in the transition zone. One such correlation is

$$C_D = 13/(Re)^{1/2} \qquad \qquad \ldots (9.78)$$

Introducing this to equation 9.59 produces:

$$w_s = 0.22d[(a_g\ \rho_f)^2/\mu\ \rho_c]^{1/3} \qquad \qquad \ldots. (9.79)$$

When we consider many particles settling, the density of the fluid phase effectively becomes the bulk density of the slurry, i.e., the ratio of the total mass of fluid plus solids divided by the total volume. The viscosity of the slurry is considerably higher than that of the fluid alone because of the interference of boundary layers around interacting solid particles and the increase of form drag caused by particles. The viscosity of a slurry is often a function of the rate of shear of its previous history as it affects clustering of particles, and of the shape and roughness of the particles. Each of these factors contributes to a thicker boundary layer.

Experimental measurements of viscosity almost always are recommended when dealing with slurries and extrapolations should be made with caution. Most theoretically based expressions for liquid viscosity are not appropriate for practical calculations or require actual measurements to evaluate constants. For non-clustering particles, a reasonable correlation may be based on the ratio of the effective bulk viscosity, μ_B, to the viscosity of the liquid. This ratio is expressed as a function of the volume fraction of liquid x' in the slurry for a reasonable range of compositions:

$$ty\mu_B/\mu = (1/x')10^{1.82(1-x)} \qquad \qquad \ldots (9.80)$$

A correction factor, R_c, incorporating both viscosity and density effects can be developed for a given slurry, which provides a more convenient expression based on the following equation:

$$u_t = (\rho_p - \rho)gd^2/18\mu \qquad \ldots (9.81)$$

as

$$v_H = R_c (\rho_p - \rho)gd^2/18\mu \qquad \ldots (9.82)$$

where V_H is the terminal velocity in hindered settling.

Measurements of the effective viscosity as a function of composition may be fitted to equation 9.80 or presented in graphic form as in Fig 9.16. The correction factor, R_c may also be determined by accounting for the volume fraction, η_v, of particles through the Andress formula:

$$R_c = (1 - \eta_v)^2/(1 + 2.5\,\eta_v + 7.35\,\eta_v^2) \qquad \ldots (9.83)$$

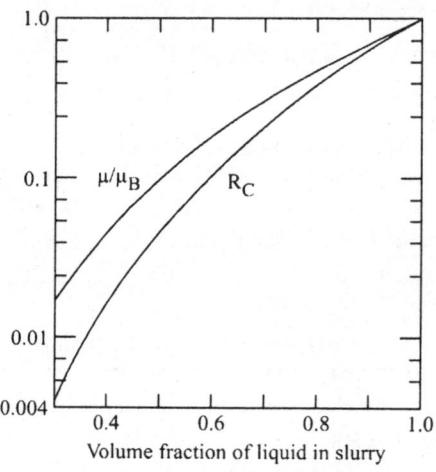

Fig. 9.16. Settling factor for hindered settling.

In summarising sedimentation principles, we note that the particle settling velocity is the principal design parameter that establishes equipment sizes and allowable loadings for separating heterogeneous systems. However, design calculations are not straightforward because prediction of the settling velocity requires knowledge of the flow regime in the vicinity of the particles. Therefore, the following generalised design method is recommended.

From known values of d, ρ_p, ρ_c, μ and a_g, compute the first sedimentation dimensionless number (equation 9.63). From the plot given in Fig. 9.17, obtain the corresponding Reynolds number, Re and evaluate the theoretical settling velocity. If the flow regime is laminar, the settling velocity may be calculated directly from equation 9.74 and the regime checked by computing the Reynolds number for the flow around an individual particle. After determining w, determine the appropriate shape factor, ψ (either from literature values or measurements) and correction factor, R_c. The design settling velocity then will be:

$$w_0 = R_c\,\psi\,w \qquad \ldots (9.84)$$

Sedimentation equipment is designed to perform two operations: to clarify the liquid overflow by removal of suspended solids and to thicken sludge or underflow by removal of liquid. It is the cross section of the apparatus that controls the time needed for settling a preselected size range of particles out of the liquid for a given liquid feed rate and solids loading. The area also establishes the clarification capacity. The depth of the thickener establishes the time allowed for sedimentation (i.e., the solid's residence time) for a given feed rate and is important in determining the thickening capacity. The clarification capacity is established by the settling velocity of the suspended solids. Sedimentation tests are almost always recommended when scaling up for large settler capacities. By means of material balances, the total amount of fluid is equal to the sum of the fluid in the clear overflow plus the fluid in the compacted sludge removed from the bottom of the thickener. The average vertical velocity of fluid at any height through the thickener is the volumetric rate passing upward at that level divided by the unit's cross section. Note that if the particle settling velocity is less than the upward fluid velocity, particles will be entrained out in the overflow, resulting in poor clarification. For those size particles whose settling velocity approximately equals that of the upward fluid velocity, particles remain in a balanced suspension, i.e., they neither rise nor fall and the concentration of solids in the clarification zone increases. This eventually results in a reduction of the settling velocity until the point where particles are entrained out in the overflow.

The thickener must be designed so that the settling velocity of particles is significantly greater than the upward fluid velocity, to minimise any increase in the solids concentration in the clarification zone.

Solids concentration varies over the thickener's height, and at the lower levels where the solution is dense, settling becomes retarded. In this region the upward fluid velocity can exceed the particle settling velocity irrespective of whether this condition exists in the upper zone or not. Figure 9.17 illustrates this situation, where curve II denotes a higher feed rate.

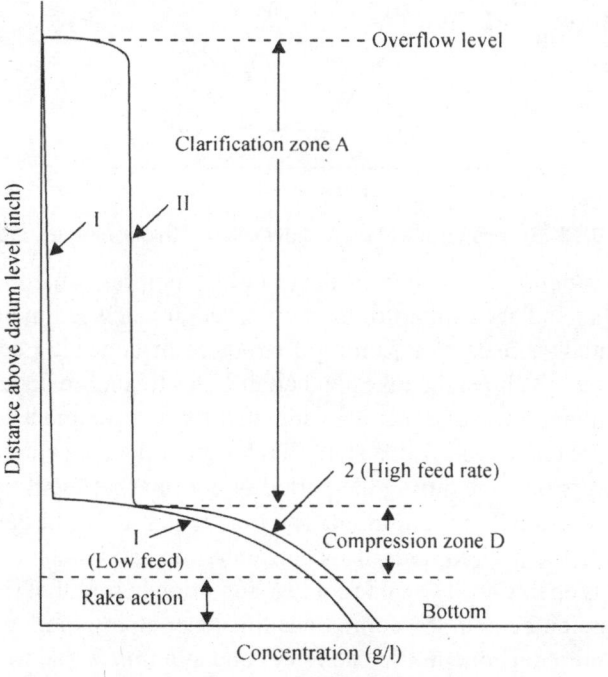

Fig. 9.17. Plot of concentration versus height in a continuous sedimentation device. Curve (1) - low feedrate; Curve (2) - high feed rate.

A proper design must therefore be based on an evaluation of the settling rates at different concentrations as compared to the vertical velocity of the fluid. If the feed rate exceeds the maximum of the design, particulates are unable to settle out of the normal clarification zone. Hence, there is an increase in the solids concentration, resulting in hindered settling. The result is a corresponding decrease in the sedimentation rate below that observed for the feed slurry. The feed rate corresponding to the condition of just failing to initiate hindered settling represents the limiting clarification capacity of the system. That is, it is the maximum feed rate at which the suspended solids can attain the compression zone. The proper cross-sectional area can be estimated from calculations for different concentrations and checkced by batch sedimentation tests on slurries of increasing concentrations. You will find some problems in the section on Questions for Thinking and Discussing that illustrate the need to check the thickener's calculated area against concentrations at various points in the vessel (including both the clarification and thickening zones). Figure 9.18 shows the effect of varying the underflow rate on the thickening capacity. In this example, the depth of the thickening zone (compression zone) increases as the underflow rate decreases; hence, the underflow solids concentration increases, based on a constant rate of feed.

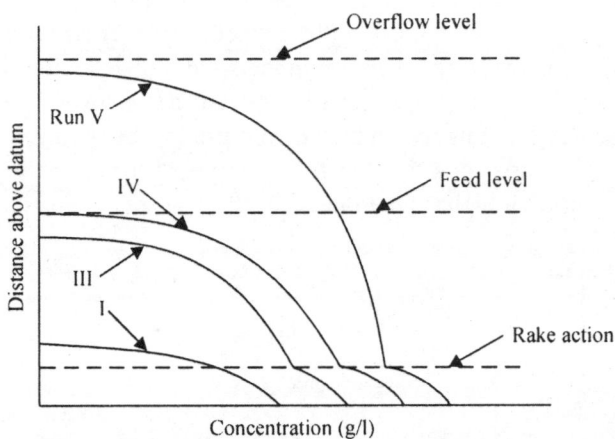

Fig. 9.18. Shows effect of underflow rate on thickening capacity.

The curves of concentration as a function of depth in the compression zone are essentially vertical displacements of each other and are similar to those observed in batch sedimentation. When the sludge rakes operate, they essentially break up a semirigid structure of concentrated sludge. Generally, this action extends to several inches above the rakes and contributes to a more concentrated underflow. The required height of the compression zone may be estimated from experiments on batch sedimentation. The first batch test should be conducted with a slurry having an initial concentration equivalent to that of the top layer of the compression zone during the period of constant rate settling.

This is referred to as the critical concentration. The time required for the sample slurry to pass from the critical concentration to the desired underflow concentration can be taken as the retention time for the solids in the continuous operation. The underlying assumption here is that the solids concentration at the bottom of the compression zone in the continuous thickener at any time is the same as the average concentration of the compression zone in the batch unit and at a time equal to the retention time of the solids in the continuous thickener. Hence, it is assumed that the concentration at the bottom of the thickener is an implicit function of the thickening time. The retention time is obtained from a batch test by observing the height of the compression zone as a function of time. The slope of the compression curve is described by the following expression:

$$-dZ/dt = k(Z - Z_\alpha) \qquad \qquad \text{... (9.85)}$$

where Z, Z_α are the heights of compression at times t and infinity, respectively and k is a constant that depends on the specific sedimentation system. Integrating this expression gives:

$$\ln (Z - Z_\alpha) = -kt + \ln (Z_c - Z_\alpha) \qquad \qquad \text{... (9.86)}$$

where Z_c is the height of the compression zone at its critical concentration. This expression is the equation of a straight line and normally is plotted as $\log[(Z - Z_\alpha)/(Z_0 - Z_\alpha)]$ versus time, where Z_0 is the initial slurry concentration. If batch tests are performed with an initial slurry concentration below that of the critical, the average concentration of the compression zone will exceed the critical value because it will consist of sludge layers compressed over varying time lengths. A method for estimating the required time to pass from the critical solids content to any specified underflow concentration can be done as follows:

1. Extrapolate the compression curve to the critical point or zero time.
2. Locate the time when the upper interface (between the supernatant liquid and slurry) is at height Z'_0, halfway between the initial height, Z_0 and the extrapolated zero-time compression zone height, Z'_0. This time represents the period in which all the solids were at the critical dilution and went into compression. The retention time is computed as $t - t_c$, where t is the time when the solids reach the specified underflow concentration. The procedure is illustrated in Fig. 9.19. It is recommended that you determine the required volume for the compression zone to be based on estimates of the time each layer has been in compression. The volume for the compression zone is the sum of the volume occupied by the solids plus the volume of the entrapped fluid. This may be expressed as:

$$V = Q(\Delta t)/\rho_s + \int \{m_t Q/m_s \, \rho_t\}dt \qquad \qquad \text{... (9.87)}$$

where Q = solids mass feed per unit time; $\Delta t = t - t_c$ = retention time; m_t = mass of liquid in the compression zone; m_s = mass of solids in the compression zone.

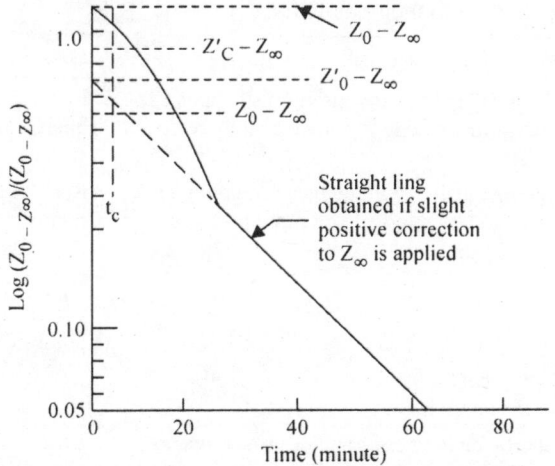

Fig. 9.19. Extrapolation of sedimentation data to estimate time for critical concentration.

This expression is based on our earlier assumption that the time required to thicken the sludge is independent of the interface height of the compression zone. An approximate solution to this expression can be obtained if we assume m_l/m_s to be constant, i.e., an average mass ratio in the thickening zone from top to bottom. Then,

$$V = Q \, \Delta t[1/\rho_s + (1/\rho_l)(\{m_l/m_s\}_{avg}] \qquad \qquad ... (9.88)$$

More reliable results can be obtained by assuming average conditions over divided parts of the compression zone. That is, the above expression can be applied to divisions of the compression zone and the total volume obtained by the sum of these calculations. Try some of the problems in the section on Questions for Thinking and Discussing to strengthen your understanding of the principles covered.

Factors Affecting Sedimentation

1. Detention time. An adequate detention time is important for complete sedimentation. The shorter the detention time, the less the settling, higher the turbidity and vice versa.
2. Velocity. Higher velocities cause the scouring (resuspension) of the settled floc, which may rise to the surface and cause high effluent turbidity. Very low velocity would not distribute the solids in the basin properly.
3. Surface turbulence. The greater the turbulence, the less is the rate of sedimentation and vice versa.
4. Short circuits. Short-circuiting causes a short detention time, which results in an inefficient sedimentation.
5. Temperature. The higher the temperature, the faster is the sedimentation and vice versa. At higher temperatures water is lighter and settling is faster.
6. Dimensions. Proper dimensions of the basins are also important—particularly depth, which is generally 15 feet or more.
7. Inlets and outlets. Inlets and outlets of the horizontal flow basins should be properly located to allow proper mixing of the water to prevent the short circuits.

Table 9.1 shows the sedimentation problems and their solutions.

Table 9.1. Sedimentation problems and their solutions.

Problems	Possible causes	Possible solutions
Poor sedimentation of a normal floc.	Sludge blanket is too high. This happens when fine sludge is not adequately removed	Increase withdrawal of sludge until blanket is at least 5 feet below the water surface
	Too much sludge in the basin. It could be due to less removal and more build-up	Remove sludge until all heavy sludge is removed. Watch sludge density of discharged sludge. Removal should cease when sludge becomes thin and watery
	Detention time is too short	Check detention time and adjust, if possible
	Not enough time for complete sedimentation	
	Short-circuiting of flocculated water; occurs in summer and winter due to stratification	Check baffling and mixing mechanisms and increase the mixing

(Contd...)

Problems	Possible causes	Possible solutions
Light feathery-flaky floc scouring into settled water	Inadequate flocculation due to excessive lime usage	Reduce lime dose until sludge is denser
Solids are gritty and like white sand	Calcium carbonate precipitate happens during winter in waters with carbonate hardness and lime treatment in coagulation and flocculation phase	Reduce lime dose and increase alum dose until floc becomes lighter
Scrapers of sedimentation basin are stopping	Buildup of heavy solids in basin due to inefficient sludge removal	Open sludge discharge valve; check sludge density; discharge until thin. Set timers to increase frequency and interval of discharge of sludge
	Too much sand and silt in influent water	Check presettled water turbidity and its nature. If turbidity is due to silt, then increase sludge removal
	Some obstructing object such as a tool, piece of wood or brick in basin	Drain basin and look for obstructing object
Clouds of floc are in the peripheral part of a solid contact basin	If basin is a walker type, the tangential gates are not adjusted properly. If gates are too tight for a flow, they cause a higher velocity to the coagulated water. Floc cloud hits basin wall and moves upward to cause clouds.	Open gates until water level in coagulation zone is about two inches higher than that of the flocculation zone. Consult operational manual for proper adjustment of gates
	In case of an accelator	Slow down the central mixer speed. There might be insufficient sludge removal from outer sludge hoppers. Increase sludge removal from outer hopers
Slurry concentration is too low in solid contact basin	In case of an accelerator-like basin	Reduce sludge disctiarge from outer hoppers. Scrappers will bring in finer floc toward center for recirculation
	In case of a walker-like basin	Increase discharge frequency and reduce discharge interval to get rid of heavy solids and retain lighter solids for recirculation. If possible, increase speed of recirculation
In solid contact basin slurry concentration is too high	Opposite of previously defined problem	Do opposite of previously defined problem. The more the sludge discharge of the fine floc, the less is the slurry concentration and vice versa
After heavy rain in summer, water does not flocculate and settle in solid contact basin	High turbidity and low hardness	This water can be treated at high pH with normal dose of coagulant (7–10 mg/l alum). May need small dose of a polymer. Sludge should be removed efficiently from basins. Sludge is heavy due to silt. Run jar test using a coagulant and a polymer; determine their optimum doses for best results and use the combination

(Contd ...)

Problems	Possible causes	Possible solutions
There is poor sedimentation (in solid contact basin) of cold water with low turbidity.	Low pH	High pH of the water will help in flocculation and sedimentation by providing OH⁻ ions to react with the coagulant. Add small dose of lime
	Insufficient detention time. In winter, low temperature will increase the density of the water, which slows down sedimentation process	Increase detention time by reducing flow through the basin. (Put on more basins)
	Insufficient coagulant dose. Colder water needs higher dose of a coagulant for proper sedimentation	Run jar test; determine required chemical doses and apply them

CLARIFIERS

There are commercially available clarifier simulation software for comprehensive (2D) analyses of waste-water treatment processes in circular and rectangular clarifiers. With these software, you can predict processes like:

1. Distribution of sludge in the settler.
2. Flow streamlines in the settler.
3. Vertical and horizontal flow velocities.
4. Vertical and horizontal flow velocities.
5. Sludge concentration in the effluent.
6. Return sludge concentration.
7. Total mass of sludge in the settler.

Different processes like eddy turbulence, bottom current, stagnation of flows, and storm-water events can be simulated, using either laminar or turbulent flow model for simulation. All processes are displayed in real-time graphical mode (history, contour graph, surface, etc.); you can also record them to data files. Thanks to innovative sparse matrix technology, calculation process is fast and stable: a large number of layers in vertical and horizontal directions can be used, as well as a small time step.

Mechanical Clarification Process and The Chemistry of Clarification

The process of clarification is all about is removing suspended solids from water. Important concepts that we have eluded to, but may be not spelled out so clearly up to now are:

1. Stable solids suspensions in water—The mechanisms involved in keeping solids suspended in water.
2. Chemical treatments—How organic polymers and inorganic coagulants work to counteract solids stabilisation mechanisms and enhance removal of solids from water.
3. The function of clarification unit operations—How these units work and how chemical treatment enhances their performance.

A term that we should get into our vocabulary is 'subsidence'. This term essentially means settling. While a degree of clarification can be accomplished by subsidence, most industrial processes require better quality water than can be obtained from settling only. Most of the suspended matter in water would settle, given enough time, but in most cases the amount of time required would not be practical.

As we have shown from our derivations of expressions describing the classical theory of sedimentation, settling characteristics depend upon the :

1. Weight of the particle.
2. Shape of the particle.
3. Size of the particle.
4. Viscosity and/or frictional resistance of the water, which is a function of temperature.

The settling rates of various size particles at 50°F (10°C) is illustrated in Table 9.2.

Table 9.2. Some settling rates for different particles (assumed spherical) and sizes.

Particle diameter (mm)	Particle type	Time to settle one foot
10.0	Gravel	0.3 sec.
1.0	Coarse sand	3.0 sec.
0.1	Fine sand	38.0 sec.
0.01	Silt	33.0 minutes
0.001	Bacteria	35.0 hours
0.0001	Clay particles	230 days
0.00001	Colloidal particles	65 years

Look closely at the settling times in Table 9.1 the times span from a fraction of a second to almost a lifetime! A great deal of the suspended matter found in waste-water fall into the colloidal suspension range, so obviously we cannot rely on gravitational force alone to separate out the pollutants.

COAGULATION

The term coagulation refers to the first step in complete clarification. It is the neutralisation of the electrostatic charges on colloidal particles. Because most of the smaller suspended solids in surface waters carry a negative electrostatic charge, the natural repulsion of these similar charges causes the particles to remain dispersed almost indefinitely. To allow these small suspended solids to agglomerate, the negative electrostatic charges must be neutralised. This is accomplished by using inorganic coagulants, which are water soluble inorganic compounds, organic cationic polymers or polyelectrolytes. The most common and widely used inorganic coagulants are:

1. Alum-aluminium sulphate-$Al_2(SO_4)_3$.
2. Ferric sulphate-$Fe_2(SO_4)_3$.
3. Ferric chloride-$FeCl_3$.
4. Sodium aluminate-$Na_2Al_2O_4$.

Inorganic salts of metals work by two mechanisms in water clarification. The positive charge of the metals serves to neutralise the negative charges on the turbidity particles. The metal salts also form insoluble metal hydroxides which are gelatinous and tend to agglomerate the neutralised particles. The most common coagulation reactions are as follows:

$$Al_2(SO_4)_3 + 3Ca(HCO_3)_2 = 2Al(OH)_3 + 3CaSO_4 + 6CO_2$$
$$Al_2(SO_4)_3 + 3Na_2CO_3 + 3H_2O = 2Al(OH)_3 + 3Na_2SO_4 + 3CO_2$$
$$Al_2(SO_4)_3 + 6NaOH = 2Al(OH)_3 + 3Na_2SO_4$$
$$Al_2(SO_4)_3 \ C(NH_4)_2SO_4 + 3Ca(HCO_3) = 2Al(OH)_3 + (NH_4)_2SO_4 + 3CaSO_4 + 6CO_2$$

$$Al_2(SO_4)_3\ K_2SO_4 + 3Ca(HCO_3)_2 = 2Al(OH)_3 + K_2SO_4 + 3CaSO_4 + 6CO_2$$

$$Na_2Al_2O_4 + Ca(HCO_3)_2 + H_2O = 2Al(OH)_3 + CaCO_3 + Na_2CO_2$$

$$Fe(SO_4)_3 + 3Ca(OH)_2 = 2Fe(OH)_3 + 3CaSO_4$$

$$4Fe(OH)_2 + O_2 + 2H_2O = 4Fe(OH)_3$$

$$Fe_2(SO_4)_3 + 3Ca(HCO_3) = 2Fe(OH)_3 + 3CaSO_4 + 6CO_2$$

The effectiveness of inorganic coagulants is dependent upon water chemistry and in particular—pH and alkalinity. Their addition usually alters that chemistry. Table 9.3 illustrates the effect of the addition of 1 ppm of the various inorganic coagulants on alkalinity and solids concentration.

Table 9.3. Coagulant, acid and sulphate - 1 ppm equivalents.

1 ppm formula or chemical	ppm Alkalinity reduction	ppm SO_4 increase	ppm Na_2SO_4 increase	ppm CO_2 increase	ppm Total solids increase
$Al_2(SO_4)_3 \cdot 18H_2O$	0.45	0.45	0.64	0.40	0.16
$Al_2(SO_4)_3 \cdot (NH_4)2SO_4 \cdot 24H_2O$	0.33	0.44	0.63	0.29	0.27
$Al_2(SO_4)_3 \cdot K_2SO_4 \cdot 24H_2O$	0.32	0.43	0.60	0.28	0.30
$FeSO_4 \cdot 7H_2O$	0.36	0.36	0.61	0.31	0.13
$FeSO_4 \cdot 7H_2O + (SCl)_2$	0.54	0.36	0.51	0.48	0.18
$Fe_2(SO_4)_3$	0.76	0.76	1.07	0.64	0.27
H_2SO_4 - 96%	1.00	1.00	1.42	0.88	0.36
H_2SO_4 - 93.2% (66°Be)	0.96	0.95	1.36	0.84	0.34
H_2SO_4 -77.7% (66° Be)	0.79	0.79	1.13	0.70	0.28
$NaSO_4$	–	0.64	0.95	–	1.00
$Na_2Al_2O_4$	Increase 0.54	–	–	Reduces 0.47	0.90

Aluminium salts are most effective as coagulants when the pH range is between 5.5 and 8.0 pH. Because they react with the alkalinity in the water, it may be necessary to add additional alkalinity (called buffering) in the form of lime or soda ash. Use the values in Table 9.4 to guide you. Iron salts, on the other hand, are most effective as coagulants at higher pH ranges (between 8 and 10 pH). Iron salts also depress alkalinity and pH levels; therefore, additional alkalinity must be added. Sodium aluminate increases the alkalinity of water, so care must be taken not to exceed pH and alkalinity guidelines. As is evident from the reactions discussed above, a working knowledge of the alkalinity relationships of water is mandatory. By using inorganic coagulants we can wind up producing a voluminous, low-solids content sludge is the difficult to dewater and dries very slowly. The properties of the sludge to be generated and estimated quantities needs to be carefully determined, in part from pilot-scale and bench testing prior to the design and construction of a plant.

Polymers are often described as long chains with molecular weights of 1000 or less to 50,00,000 or more. Along the chain or backbone of the molecule are numerous charged sites. In primary coagulants, these sites are positively charged. The sites are available for adsorption onto the negatively charged particles in the water. To accomplish optimum polymer dispersion and polymer/particle contact, initial mixing intensity is critical. The mixing must be rapid and thorough, polymers used for charge neutralisation cannot be overdiluted or over-mixed. The farther upstream in the system these polymers can be added, to better their performance. Because most polymers are viscous, they must be properly diluted before they are added to the influent water. Special mixers such as static mixers, mixing tees and specially designed chemical dilution and feed systems are all aids in polymer dilution. Static or motionless mixers in particular are popular for this application.

Table 9.4. Recommended alkali and lime 1 ppm equivalents.

Chemical 1ppm	Formula (1 ppm)	Alkalinity increase (1 ppm)	Free CO_2 reduction (1 ppm)	Hardness as $CaCO_3$ increases
Sodium bicarbonate	$NaHCO_3$	0.60	–	–
Soda ash (56% Na_2, 99.16% Na_2CO_3)	Na_2CO_3	0.94	0.41	–
Caustic soda (76% Na_2O, 98.06% Na_2CO_3)	$NaOH$	1.22	1.09	–
Quicklime (90% CaO)	CaO	1.61	1.41	1.61
Hydrated lime (93 % $Ca(OH)_2$)	$Ca(OH)_2$	1.26	1.10	1.36

Once the negative charges of the suspended solids are neutralised, flocculation begins. Flocculation can be thought of as the second step of the coagulation process. Charge reduction increases the occurrence of particle-particle collisions, promoting particle agglomeration. Portions of the polymer molecules not absorbed protrude for some distance into the solution and are available to react with adjacent particles, promoting flocculation. Bridging of neutralised particles can also occur when two or more turbidity particles with a polymer chain attached come together. It is important to remember that during this step, when particles are colliding and forming larger aggregates, mixing energy should be great enough to cause particle collisions but not so great as to break up these aggregates as they are formed. In some cases flocculation aids are employed to promote faster and better flocculation. These flocculation aids are normally high molecular weight anionic polymers. Flocculation aids are normally necessary for primary coagulants and water sources that form very small particles upon coagulation. A good example of this is water that is low in turbidity but high in colour (colloidal suspension).

The is colour removal is perhaps the most difficult impurity to remove from waters. In surface water colour is associated with dissolved or colloidal suspensions of decayed vegetation and other colloidal suspensions. The composition of this material is largely tannins and lignins, the components that hold together the cellulose cells in vegetation. In addition to their undesirable appearance in drinking water, these organics can cause serious problems in downstream water purification processes. For examples:

1. Expensive demineraliser resins can be irreversibly fouled by these materials.
2. Some of these organics have chelated trace metals, such as iron and manganese within their structure, which can cause serious deposition problems in a cooling system.

There are many ways of optimising colour removal in a clarifier. The three most common methods are:

1. Prechlorination (before the clarifier) significantly improves the removal of organics as well as reducing the coagulant demand.

2. The proper selection of polymers for coagulation has a significant impact on organic removal.
3. Colour removal is affected by pH. Generally, organics are less soluble at low pH.

EQUIPMENT OPTIONS

Although we have discussed the major hardware, it is still worthwhile reviewing these in relation to the major classes of clarifier processes. The major categories of this process are:

1. ·Conventional.
2. Upflow.
3. Solids-contact.
4. Sludge-blanket.

Conventional clarification is the simplest form of the process. It relies on the use of a large tank or horizontal basin for sedimentation of flocculated solids. Figure 9.20 provides a sketch of the basic configuration. The basin normally contains separate chambers for rapid mix and settling. The first two steps critical in achieving good clarification. An initial period of turbulent mixing is needed for contact between the coagulant and suspended solids. This is followed by a period of gentle stirring which helps to increase particle collisions and floc size. Retention times are typically between 3 and 5 minutes , 15 to 30 minutes for flocculation and 4 to 6 hours for settling. Coagulants are added to the waste-water in the rapid mix chamber or sometimes immediately upstream. The water passes through the mix chambers and enters the settling basin. Refer again to Fig. 9.20, which is a classical large-tank clarifier. The water passes out to the circumference, while the flocculated particles settle to the bottom. Accumulated sludge are scraped into a sludge collection basin for removal and disposal. The clean water flows over a weir and is held in a tank which is referred to as a clearwell. A rectangular version of a conventional clarifier is illustrated in Fig. 9.21. This unit is referred to as a horizontal basin clarifier.

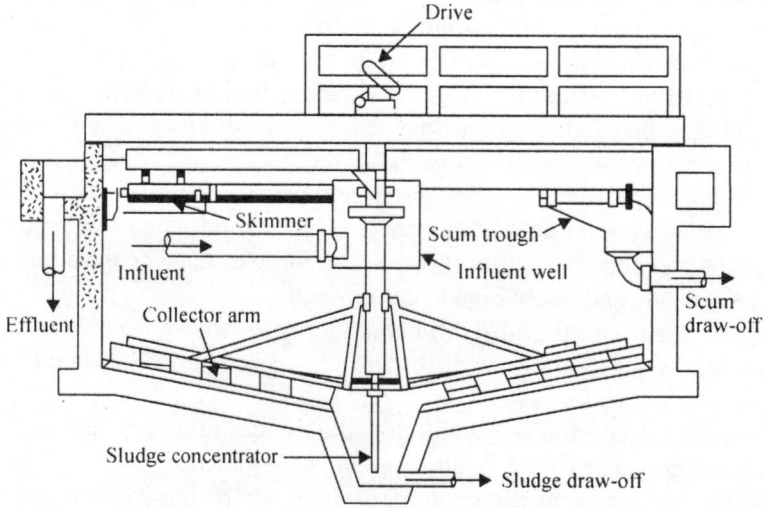

Fig. 9.20. Sketch of a large tank or circular clarifier.

It is often advantageous to employ a zone of high solids contact to achieve a better quality effluent. This is accomplished in an upflow clarifier, so called because the water flows upward through the clarifier as the solids settle to the bottom. Most upflow clarifiers are either solids-contact or sludge-blanket type clarifiers, which differ somewhat in theory of operation. Cross-sections of these two types of units are illustrated in Figs. 9.22 and 9.23. Both units have an inverted cone within the clarifier. Inside the cone is a zone of rapid mixing and a zone of high solids concentration. The coagulant is added either in the rapid mix zone or somewhere upstream of the clarifier.

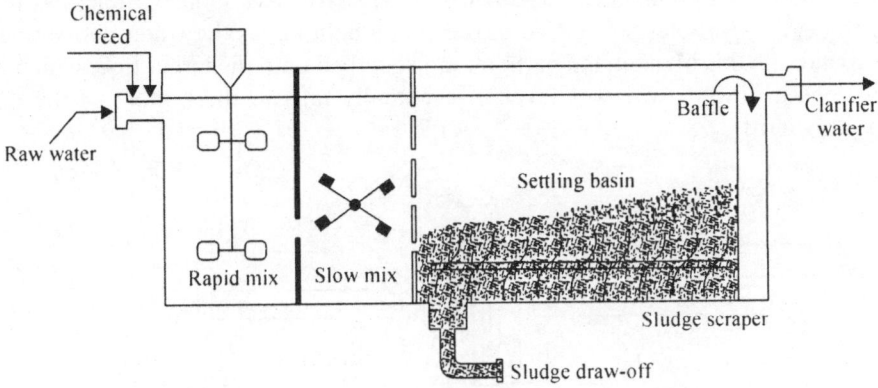

Fig. 9.21. Sketch of a horizontal basin clarifier.

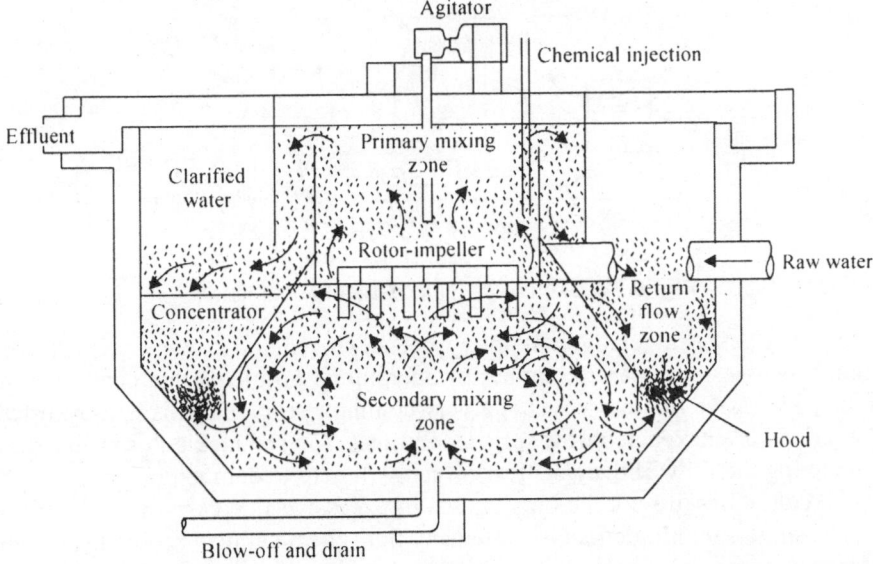

Fig. 9.22. Sketch of a solid-contact clarifier.

In the solids-contact clarifier, raw water is drawn into the primary mixing zone, where initial coagulation and flocculation take place. The secondary mixing zone is used to produce a large number of particle collisions so that smaller particles are entrained in the larger floc. Water passes out of the inverted cone into the settling zone, where solids settle to the bottom and clarified water flows over the weir. Solids are drawn back into the primary mixing zone, causing recirculation of the large floc. The concentration of solids in the mixing zones is controlled by occasional or continuous blowdown of sludge.

The sludge-blanket clarifier (Fig. 9.23) goes one step further, by passing the water up from the bottom of the clarifier through a blanket of suspended solids that acts as a filter. The inverted cone within the clarifier produces an increasing cross-sectional area from the bottom of the clarifier to the top. Thus, the upward velocity of the water decreases as it approaches the top. At some point, the upward velocity of the water exactly balances the downward velocity of a solid particle and the particle is suspended, with heavier particles suspended closer to the bottom. As the water containing flocculated solids passes up through this blanket, the particles are absorbed onto the larger floc, which increases the floc size and drops it down to a lower level. It eventually falls to the bottom of the clarifier to be recirculated or drawn off.

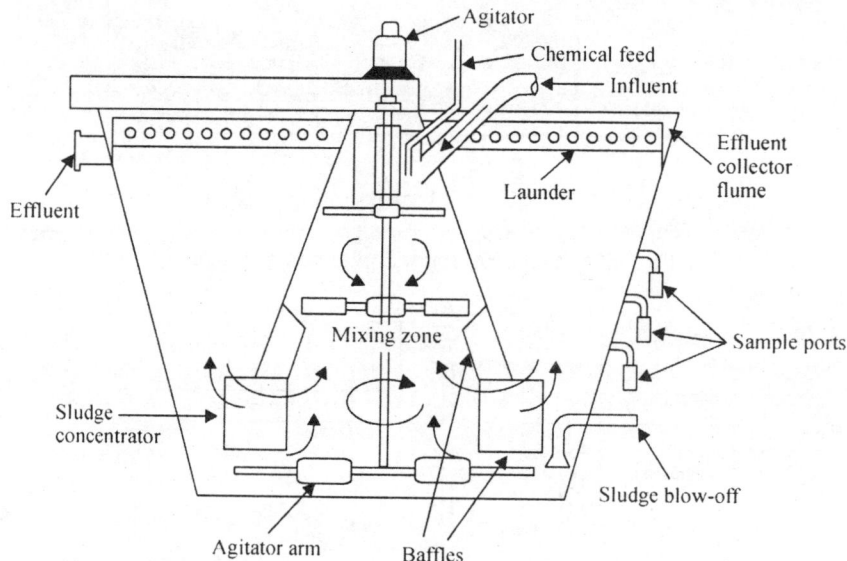

Fig. 9.23. Sketch of a sludge-blanket clarifier.

Although these processes seem relatively simple, especially in relation to many chemical manufacturing operations or unit processes, there are a number of operational problems that can make the life of an operator miserable. Excessive floc carryover is a very common problem. This is most often associated with hydraulic overload or unexpected flow surge conditions. You can tackle this problem by relying on equalise flow (metering the flow of the clarifier), which will help to dampen out surges. Unfortunately, hydraulic overload conditions are not the only causes of excessive floc carryover. Other reasons many be thermal currents, short-circuiting effects, low density floc, chemical feed problems. Another common operator problem is simply no floc in the centerwell. This can result from underfeeding of chemicals or a loss of the sludge bed recirculation.

One has to investigate and apply trial and error field tests to resolve some of these problems. When new equipment are installed, it is wise to spend time during a shake-down and start-up period to explore the operational limitations of the process and train operators on how to handle these types of problems.

RECTANGULAR SEDIMENTATION TANKS

The process concept for sedimentation tanks has hardly changed over the past 80 years. Dimensioning these vessels according to existing guidelines guarantees safe operation. With ever tightening legislation, however, the question of expansion or upgrading of existing sewage treatment plants arises. Expansion is an expensive solution and impossible if the available space is scarce so that a new, construction has to be built. The basis of upgrading consists in changed process concepts which are able to exploit the unused potential of existing tanks. An essential prerequisite for upgrading plants is the question whether the settling volume for activated sludge is sufficient. Beside the clarified water discharge, the feeding method for the sludge/water mix and the skimmer system have an essential influence on the separation efficiency in tanks. The inlet height is approximately $\frac{2}{3}$ to $\frac{1}{3}$ of the tank depth and the skimming direction in the counter flow. This concept is based on the empirical knowledge of normally minor turbulence in the tank. However, changed process concepts with a bottom-near inlet and concurrent skimming are able to minimise such turbulences to such an extent that the sludge load and thus the separation efficiency can be increased.

It is important to recognise that the process-engineering installations in rectangular sedimentation tanks have a great influence on the performance of this final treatment stage. Of particular importance is the design of the inlet section as turbulences are generated there by mixing with the waste-water inflow which may have an intense influence on the sedimentation process. The density of flows has a strong influence on the separation efficiency. The density flow sinks to the tank bottom during inflow and passes to the tank end. The rising density flow initiates back flow of the clarified water on the surface. To increase the separation efficiency of tanks, the density flow should be minimised or the engineering process modified in a way that the density flow will be integrated with the sedimentation process.

Inlet dimensions are important sizes. The density flow can be substantially influenced by the inlet structure by minimising the potential and kinetic energy of the waste-water stream with a suitable feed design. The inlet should have bottom-near feed openings to have an as small intermixing zone as possible between the activated sludge/water mix and the tank content. The velocity gradient in the inlet section should be small to avoid floc disturbance by shearing forces. The inlet section sludge scraping also influences the separation efficiency, most notably it impacts on the degree of thickening in the sedimentation tank. For minimisation of the turbulences in the secondary clarification tank and a consequently improved separation efficiency and for sludge thickening, practice has shown that scraping the sludge in the direction of the density flow works best. The scraping velocity should be low in order to prevent re-suspension of the activated sludge flocs. To increase the degree thickening and to minimise the volume flow of the return sludge, a minimum sludge residence time in the tank must be provided. Although a sludge hopper for thickening the activated sludge is not necessary, a sludge hopper at the tank end tends to increase the surface load of the tank.

Further information on the operation and design basic for rectangular settling basins can be obtained from references 1 through 3.

AIR FLOATATION SYSTEMS

Air floatation is one of the oldest methods for the removal of solids, oil and grease and fibrous materials from waste-water. Suspended solids and oil and grease removals as high as 99 per cent + can be attained with these processes. Air floatation is simply the production of microscopic air bubbles, which enhance the natural tendency of some materials to float by carrying waste-water contaminants to the surface of the tank for removal by mechanical skimming. Many commercially available units are packaged rectangular steel tank floatation systems; shipped completely assembled and ready for simple piping and wiring on site. Models typically range from 10 to over 1000 square feet of effective floatation area for raw waste-water flows to over 1000 gallons per minute. Complete systems often include chemical treatment processes. A dissolved air floatation (DAF) system can produce clean water in wash operations where reduction of oil and grease down to 2 mg/l is achievable in certain applications. In addition to municipal and heavy industry applications, DAF has found a home with commercial vehicle washing, industrial laundries, food processing. Vehicle wash applications need not be confined to automobiles and trucks, but can extend to buses, tank cars and many other types of vehicles. There is a broad and varied market for DAF. Excessive oil and grease (especially emulsified oils), plus high levels of suspended solids and metals are good candidates for our DAF systems. Another example is an industrial laundry, where there is the need for good waste treatment systems. A DAF system can be used in a variety of food processing applications, including Vegetable Oils, Animal and Seafood Fats, Red Meat Butchering, Poultry processing and Kitchen and Equipment Washing. Oil Drilling on and offshore is under pressure from the USEPA. The removal of the oils from the dirty water which surfaces when drilling wells cannot be dumped. DAF technology has proven very effective in this industry. Some waste stream problems found in shipyards and aboard ships are oily bilge water and solids containing copper and other heavy metals used in marine paints. DAF is a proven method in these applications as well. In metal finishing operations DAF will remove cadmium, chromium, lead, zinc and other toxic heavy metals from a waste stream. Finally, when combined with other treatment processes DAF can be applied upstream of large water treatment systems to handle contaminated water before it mingles with the rest of the waste stream.

When the primary target is oil removal, we should distinguish between the forms of oil. There are two forms of oil that we find in waste-water. Free oil is oil that will separate naturally and float to the surface. Emulsified oil is oil that is held in suspension by a chemical substance (Detergents - Surfactants) or electrical energy. When making an evaluation, free oil will normally separate by gravity and float to the surface in approximately 30 minutes. Emulsified oil is held in a molecular structure called a micelle and will not separate on its own. Hence, there is the need for a more sophisticated method of treating suspensions containing emulsified oils.

A good way to see how DAF technology works is to fill a glass full of water from the tap and observing the tiny, almost microscopic bubbles in the water. Dissolved-air floatation uses the same principle in order to introduce tiny bubbles into water. These bubbles form because the water inside the pipes, which is at high pressure, had somehow dissolved enough air that the water becomes supersaturated with air when the pressure drops before the water falls into your glass. As a result, the excess air precipitates out in the form of tiny bubbles. These bubbles are much smaller than we produce by other means of dispersing air in water.

The floatation process was developed in the mining and coal processing industries as a way of separating suspended solids from a medium such as water. As noted above, the floatation process has found uses in other fields, such as waste-water treatment. The process introduces fine air bubbles into the mixture, so that the air bubbles attach to the particles and lift them to the surface.

Dissolved-air floatation uses a particular way of introducing the air bubbles into the floatation tank. A dissolved-air floatation machine dissolves air into the water to be treated by passing the water through a pressurising pump, introducing air, and holding the air-water mixture at high pressure long enough for the water to become saturated with air at the high pressure. Typical pressures are 20–75 psig. After saturating the water with air at high pressure, the water passes through a pressure relief nozzle, after which air precipitates as tiny bubbles. This process for creating air bubbles has two advantages over other processes. Dissolved-air floatation typically produces bubbles in the 40–70 micron range, whereas in normal foam fractionation, a bubble of 500 microns is considered small. The smaller bubbles have much more surface area for their volume than do the larger bubbles. A particular volume of air has

10 times the surface area when distributed as 50 micron bubbles as it does when distributed as 500 micron bubbles. Looking at this fact another way, you need 10 times the air flow with 500 micron bubbles as you need with 50 micron bubbles in order to achieve the same air-water interfacial area.

The second and probably more important, advantage of producing bubbles by precipitation is that the process provides a more positive attachment between air bubbles and the particles or globules that you want to remove. Particles and globules in the water act as nucleation sites for the precipitation process; the precipitating air seeks out these sites to begin bubble formation. This is better than relying on chance encounters between waste particles and large bubbles introduced by some other means.

A typical DAF process is not simply a physical separation technique. One must consider the entire treatment process, which is based on chemical coagulation, clarification and rapid sand filtration. This process train is widely accepted and is applicable to the treatment of coloured and turbid surface water for municipal and industrial applications. Normally the clarification stage employs DAF. The suspended solid matter in the chemically treated water is separated by introducing a recycle stream containing small bubbles which floats this material to the surface of the tank. This is achieved by recycling a portion of the clarified flow back to the DAF unit. The recycle flow is pumped to higher pressure and is then mixed with compressed air. The flow passes through a tank where the air dissolves to saturation at the higher pressure. When the pressure is released at the clarifier, the dissolved air precipitates as a cloud of micro bubbles which attach to the particulate matter causing it to rise to the surface.

The DAF process is particularly well suited for the removal of floc formed in the treatment of low alkalinity, low turbidity, coloured water. This type of floc tends to be very fragile and voluminous, making traditional gravity sedimentation inefficient. Floatation processes do not require large, heavy floc in order to achieve efficient solids removal. This results in lower chemical dosages and reduced time required for flocculation. The compressive forces applied to the sludge by the buoyant bubble/floc agglomerates result in greatly reduced volumes of waste-water from the clarification process. This enhances the efficiency of chemical use and reduces the volume of residuals to be treated. An equally important benefit of the technology is the efficiency of the clarification process. Since the performance of the filtration process is directly affected by the amount of solids in the clarified water, the high degree of solids removal achieved in the DAF results in an overall increase in system performance. In order to meet stringent standards for turbidity removal in potable water applications, this high performance is essential.

In recapping, DAF is the process of removing suspended solids, oils and other contaminants via the use of bubble floatation. Air is dissolved into the water, then mixed with the waste-stream and released from solution while in intimate contact with the contaminants. Air bubbles form, saturated with air, mix with the waste-water influent and are injected into the DAF separation chamber. The dissolved air then comes out of solution, producing literally millions of microscopic bubbles. These bubbles attach themselves to the particulate matter and float then to the surface where they are mechanically skimmed and removed from the tank. Most systems are versatile enough to remove not only finely divided suspended solids, but fats, oils and grease (FOG). Typical wastes handled include various suspended solids, food/animal production/processing wastes, industrial wastes, hydrocarbon oils/emulsions and many others. Clarification rates as high as 97 per cent or more are achievable. The basic flow scheme for a DAF system is illustrated in Fig. 9.24.

The conventional DAF saturation design relies on a recycle pump combined with a saturation vessel and air compressor to dissolve air into the water. This type of system is effective, however it has the drawbacks of being labour intensive, is expensive and can destabilise its point of equilibrium, creating 'burps' due to incorrect, loss or creeping of EQ set-point in the saturation vessel. Such designs are slow to recover and can upset the floatation process. Air transfer efficiency is roughly 9 per cent with 80 per cent entrainment. This operational methodology can result in an increase in chemical use, labour costs, downtime, effluent loadings, production schedules due to the EQ loss. To overcome these shortcomings, some equipment suppliers have devised operational and control schemes that are best categorised as pollution prevention techniques.

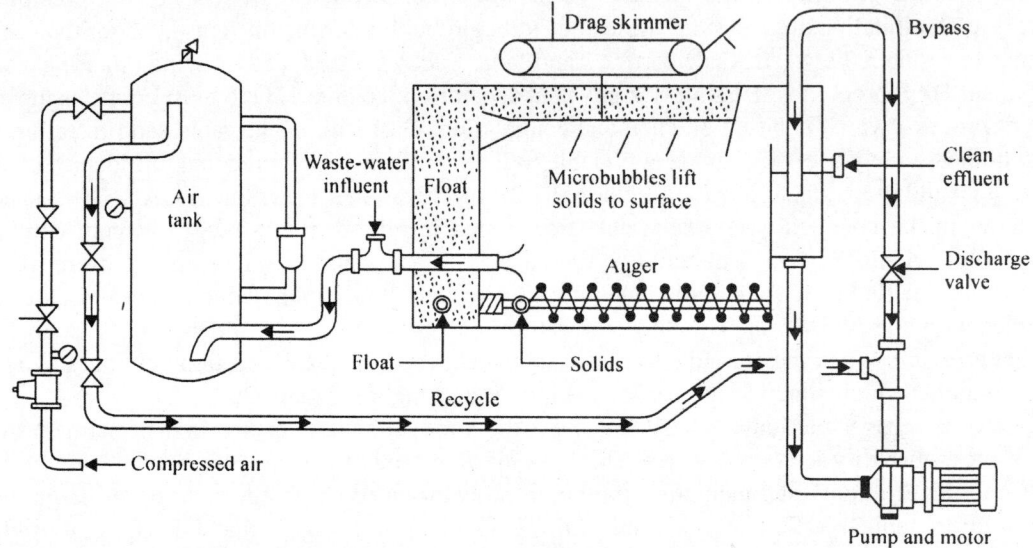

Fig. 9.24. Conventional DAF process scheme.

A variation for one vendor is shown in Fig. 9.25. The design and control of the system takes into consideration the following parameters: flow rate, water temperature, waste characteristics, chemical pre-treatment options, solids loading, hydraulic loading, the air to solids ratio. Units are designed on the basis of peak flow rate expected. Chemical pre-treatment is often used to improve the performance of contaminant removal. The use of chemical flocculants is based on system efficiency, the specific DAF application and cost. Commonly used chemicals include trivalent metallic salts of iron, such as $FeCl_2$ or $FeSO_4$ or aluminium, such as $AlSO_4$. Organic and inorganic polymers (cationic or anionic) are generally used to enhance the DAF process.

The most commonly used inorganic polymers are the polyacrylamides. Chemical flocculant concentrations employed normally range from 100 to 500 mg/liter. The waste-water pH may require adjustment between 4.5 and 5.5 for the ferric compounds or between 5.5 and 6.5 for the aluminium compounds using an acid such as H_2SO_4 or a base such as NaOH. In many applications, the DAF effluent requires additional pH adjustment, normally with NaOH to assure that the effluent pH is within the limits specified by the POTW. The pH range of the effluent from a DAF is typically between 6 and 9.

The mechanism by which flocculants work and enhance DAF is as follows. Attachment of most bubbles to solid particles can be effected through surface energies while others are trapped by solids or by hydrous oxide flocs as the floc spreads out of the water column. Colloidal solids are normally too small to allow the formation of sufficient air-particle bonding. They must first be coagulated by a chemical such as the aluminium or iron compounds mentioned above. The solids are essentially absorbed by the hydrous metal oxide floc generated by these compounds. Often, a coagulant aid is needed in combination with the flocculant to agglomerate the hydrous oxide floc, to increase particle size and improve the rate of floatation. Mechanical/chemical emulsions can also be broken through the application of pH and polymer reactions.

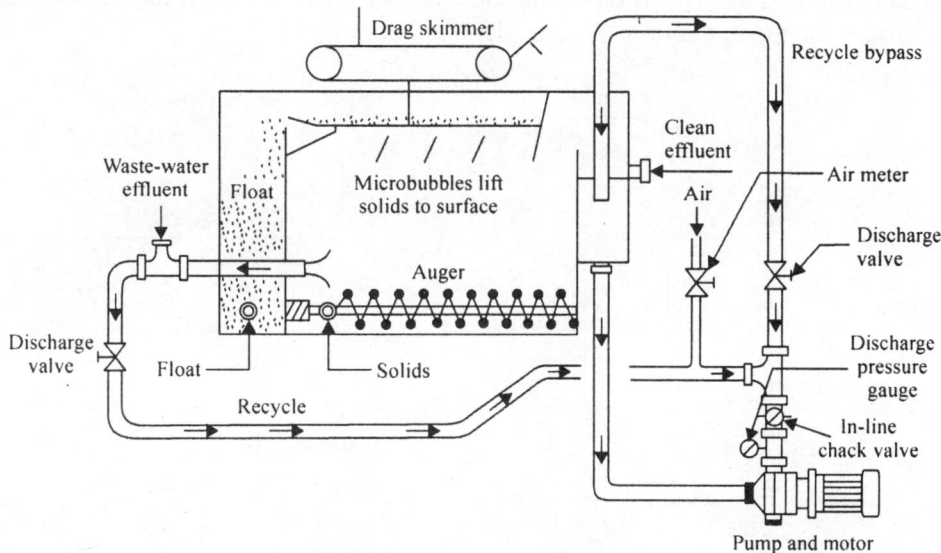

Fig. 9.25. Variation of DAF.

The material that we recover from the surface of the DAF is referred to as the 'float'. The float often contains 2 to 10 per cent solids. These solids will need to be dewatered before ultimately finding a home for them. In general, for many applications, air floatation is the system of choice. Microbubbles are produced by inducing air into a vortex as the floc is formed. This controlled induction of air allows the micro-bubbles to permanently attach to the floc, resulting in the clarifier. In fully-integrated systems, the clarifier has a built-in sludge-holding section. Figure 9.26 illustrate a continuous floatation system for treating waste-water containing free-floating or emulsified oils and a typical process flow sheet, respectively.

SEPARATION USING COALESCERS

A coalescer achieves separation of an oily phase from water on the basis of density differences between the two fluids. These systems obviously work best with non-emulsified oils. Applications historically have been in the oil and gas industry, and hence the most famous oil/water separator is the API separator (API being the abbreviation for the American Petroleum Institute). Modern-day designs are more sophisticated than the early, simple separators of a few decades ago that were introduced by the petroleum industry. Commercial systems are comprised of cylindrical vessels, rectangular vessels, above and underground installations. Figure 9.27 shows an underground system Tank Direct.

In this diagram the key features are: A—Diffusion baffle: this serves four roles. First to dissipate the velocity head, thereby improving the overall hydraulic characteristics of the separator. Next, to direct incoming flow downward and outward maximising the use of the separator volume. Third, to reduce flow turbulence and to distribute the flow evenly over the separator's cross-sectional area. Finally, to isolate inlet turbulence from the rest of the separator. B—Internal chambers: In the sediment chamber, heavy solids settle out, and concentrated sludges of oil rise to the surface. As the oily water passes through the parallel corrugated plate coalescer (an inclined arrangement of parallel corrugated plates) the oil rises and coalesces into sheets on the underside of the plate. The oil creeps up the plate surface and breaks loose at the top in the form of large globules.

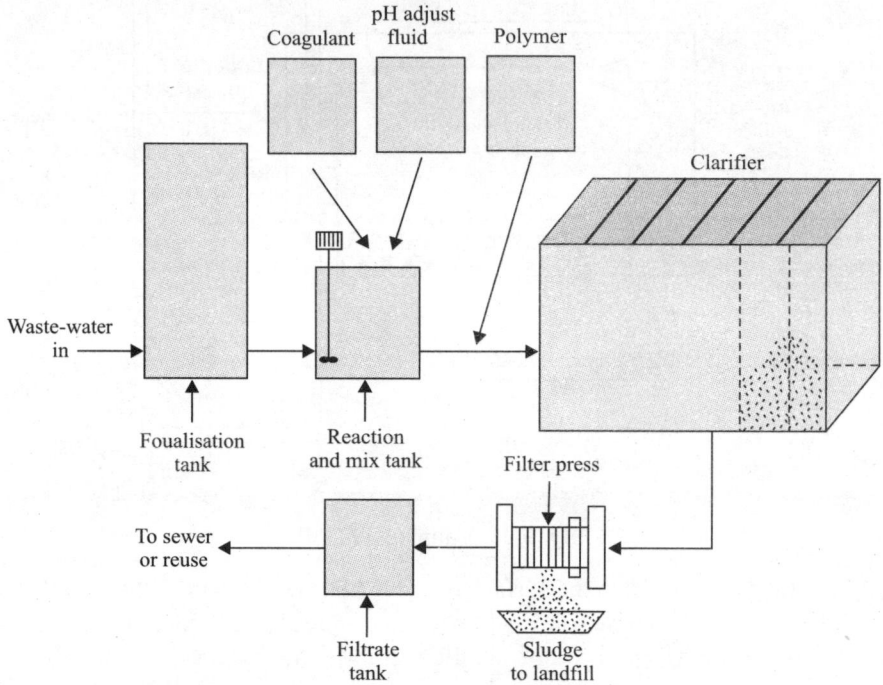

Fig. 9.26. Continuous floatation system flow sheet.

The globules then rise rapidly to the surface of the separation chamber where the separated oil accumulates. The effluent flows downward to the outlet downcomer, where it is discharged by gravity displacement from the lower region of the separator. C—PetroScreen. This part of the design focuses on improved separation efficiency. It is essentially a polypropylene coalescer, a bundle of oleophilic (i.e., oil extracting) fiber. The oleophilic fiber are layered from coarse to fine and are encapsulated within a solid framework. These are used to intercept droplets of oil that are too small to be removed by the parallel corrugated plate coalescer section. D— Monitoring systems: Various monitoring and control instruments are included in a system for level sensing and pumpout control purposes.

There are variations of coalescer designs, each achieving different degrees of separation depending upon the application and properties of the influent. One fairly popular design used primarily in the oil and gas industry is a so-called deoiler cyclone vessel and cyclone system. These systems take advantage of the hydroclone (or cyclone) principle of separation.

More correctly stated, the cyclonic action results in an increase in the oil droplet size, enabling a more efficient separation of the phases. Deoiling cyclones have steep droplet-cut size curves, like the one illustrated in Fig. 9.28.

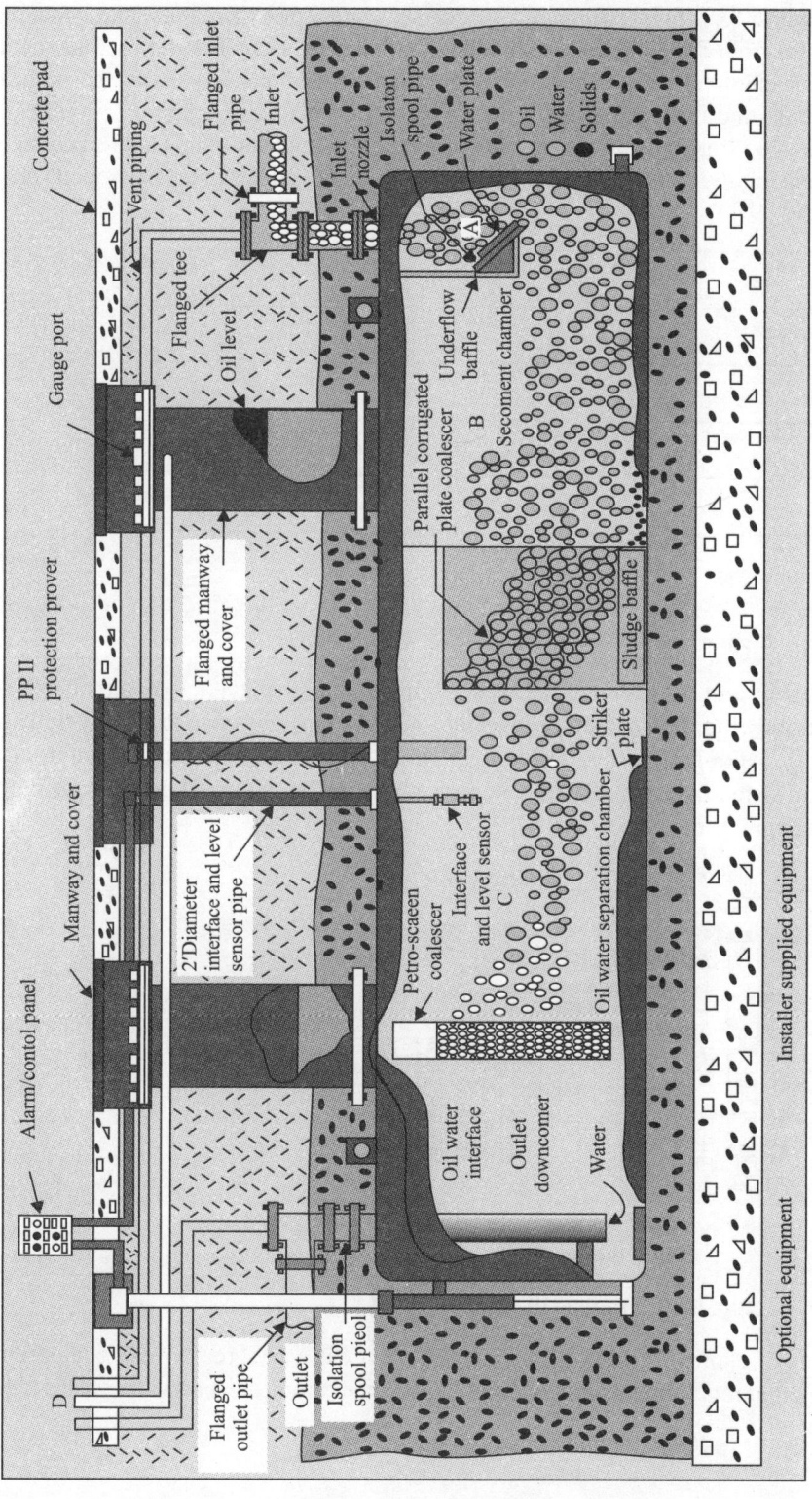

The typical performance curve for a high efficiency cyclone separator (or hydroclone) shows that a small increase in the droplet size from 5 to 10 microns typically increases the separation from 15 per cent to over 90 per cent. The inlet chamber of a conventional deoiling hydroclone is usually the largest chamber in the coalescer vessel, and has a residence time up to about 20 seconds. Some commercial units employ specially designed low-intensity pre-coalescing internals and inlet vanes that take advantage of the residence time by optimising the flow distribution. This enhances the coalescence of droplets and enables a pre-coalescing stage.

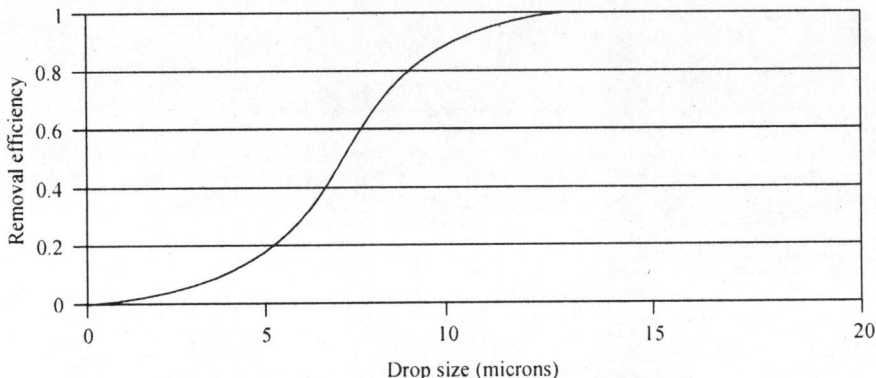

Fig. 9.28. Cut size curve for a deoiling hydroclone.

An example of a pre-coalescer unit is illustrated in Fig. 9.29. In general, the technology largely relies on many years of experience and empiricism. Before investing in a system, it is wise to run batch tests and perhaps pilot tests using vendor facilities. Units may range in size, complexity of internals, configuration (e.g., rectangular, slant rib designs that borrow concepts from the classical Lamella separator, to cylindrical, the classical design case).

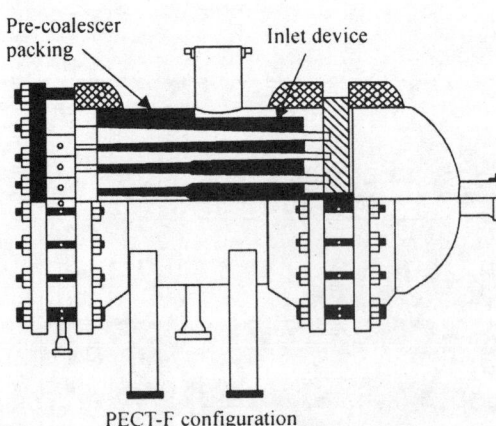

PECT-F configuration

Fig. 9.29. Example of a pre-coalescer.

There are varying degrees of claims for removal efficiency. Cost can vary greatly with coalescers, depending not only on throughput requirements and add-on controls and monitors, but with construction as well. Because of strict environmental regulations tied to underground and above ground tanks, double wall vessel construction is often needed. Double wall vessels are normally constructed in several ways.

One common construction is achieved by wrapping a secondary steel wall completely around the primary vessel. Each double wall vessel is constructed using the same basic fabrication techniques as used on single wall vessels, the area between the vessel walls, known as the interstice, can be monitored by a leak detection system installed in the monitor tube, located on the vessel head. Other variations of vessel construction exist and careful consideration to the advantages, disadvantages and impacts on cost must be considered.

Example 9.1: Design a primary settling tank of rectangular shape for a town having a population of 50000 with a water supply of 180 liters per capita per day.

Solution: Assuming that 80 per cent of water supplied to the city is converted into sewage.

Total sewage flow = 0.8 × 50000 × 180 = 7200 × 10^3 liters/day. Let us assume a detention period of 2 hours

$$\text{Therefore,} \qquad \text{Capacity required} = \frac{7200}{24} \times 2 = 600 \text{ m}^3 \qquad \qquad \text{... (1)}$$

Again, let us assume an overflow rate of 30 $m^3/d/m^2$ for average flow.

$$\text{Surface area} \qquad = 7200/30 = 240 \text{ m}^2 \qquad \qquad \text{... (2)}$$

Therefore,   Effective depth   = 600/240 = 2.5 m.   ... (3)

Again,   B × L   = 240 m²

Taking   L = 4 B,

B (4B) = 240

From which,   B = 7.46 m.

Keep   B = 7.5 m and L = 30 m.

Provide 4 m for inlet and outlet arrangements.

Therefore,   Total length = 30 + 4 = 34 m.

Also, provide 1 m depth for sludge accumulation and 0.5 m as free board. Hence total depth = 2.5 + 1 + 0.5 = 4 m.

Hence, the dimensions of the tank will be 34 m × 7.5 m × 4 m.

Example 9.2: Find out the quantity of alum required to treat 9 million liters of water per day. The dosage of alum is 14 mg per liter. Also work out the amount of CO_2 released per liter of water treated.

Solution:

$$\text{Quantity of alum per day} = \frac{14 \times 9 \times 10^6}{10^6} = 126 \text{ kg.}$$

The chemical reaction is as follows:

$Al_2(SO_4)_3.18H_2O + 3Ca(HCO_3)_2 = 2Al(OH)_3 + 3CaSO_4 + 18H_2O + 6CO_2$

Now,   Molecular wt. of alum = (2 × 26.97) + (3 × 32.066) + (36 × 1.008) + (30 × 16)

= (53.94 + 96.198 + 36.288 + 480)

= 666.426, say 666.

Molecular wt. of CO_2 = (1 × 12.01) + (2 × 16)

= 44.01, say 44.

Thus, 666 mg of alum will release 6 × 44 mg of CO_2

Therefore, 14 mg of alum will release $= \dfrac{14 \times 6 \times 44}{666}$

$$= 5.55 \text{ mg of } CO_2$$

Example 9.3: A water treatment plant consumes ferrous sulphate and lime as coagulant at the rate of 10 mg of ferrous sulphate per liter of water. Find out the quantities of ferrous sulphate and lime required to treat 9 million liters of water.

Soltuion:

The chemical reactions involved are as follows:

$$FeSO_4, 7H_2O + Ca(OH)_2 = Fe(OH)_2 + CaSO_4 + 7H_2O$$

$$Ca(OH)_2 = CaO + H_2O$$

Now,   Molecular wt. of $FeSO_4, 7H_2O = (1 \times 55.85) + (1 \times 32.066) + (11 \times 16) + (14 \times 1.008)$

$$= 278.028, \text{ say } 278.$$

Molecular wt. of CaO $= (40.08 + 16)$

$$= 56.08, \text{ say } 56.$$

Quantity of ferrous sulphate $= \dfrac{10 \times 9 \times 10^6}{10^6} = 90 \text{ kg.}$

Also, 278 kg of ferrous sulphate will react with 56 kg of lime.

Therefore, Quantity of lime required corresponding to 90 kg of ferous sulphate $= \dfrac{50 \times 90}{278}$

$$= 18.13 \text{ kg.}$$

Example 9.4: What is the minimum settling velocity, in feet per hour, of suspended particles that can be completely removed in a settling tank that has an overflow rate of 700 gpd/ft^2?

Solution:

$$\frac{700 \text{ gal}}{ft^2 \cdot d} \times \frac{1 \text{ ft}^3}{7.5 \text{ gal}} \times \frac{1 \text{ d}}{24 \text{ hrs.}} = 3.9 \text{ ft/hr.}$$

Only a fraction of particles that settle slower than 3.9 ft/hr. will be removed; the slower the settling velocity, the smaller is the percentage removed. For example, only 20 per cent of particles that settle at a velocity of $0.2 \times 3.9 = 0.78$ ft/hr. will be captured in the sludge layer.

Chapter 10

Filtration

INTRODUCTION

Filtration is a fundamental unit operation that, separates suspended particle matter from water. Although industrial applications of this operation vary significantly, all filtration equipment operate by passing the solution or suspension through a porous membrane or medium, upon which the solid particles are retained on the medium's surface or within the pores of the medium, while the fluid, referred to as the filtrate, passes through.

In a very general sense, the operation is performed for one or both of the following reasons. It can be used for the recovery of valuable products (either the suspended solids or the fluid) or it may be applied to purify the liquid stream, thereby improving product quality or both. Examples of various processes that rely on filtration include adsorption, chromatography, operations involving the flow of suspensions through packed columns, ion exchange and various reactor engineering applications. In petroleum engineering, filtration principles are applied to the displacement of oil with gas (i.e., liquid-liquid separations), in the separation of water and miscible solvents (including solutions of surface-active agents) and in reservoir flow applications. In hydrology, interest is in the movement of trace pollutants in water systems, the purification of water for drinking and irrigation and to prevent saltwater encroachment into freshwater reservoirs. In soil physics, applications are in the movement of water, nutrients and pollutants into plants. In biophysics, the subject of flow through a porous media touches upon life processes such as the flow of fluids in the lungs and the kidney. Although there are numerous industry-specific applications of filtration, water treatment has historically and continues to be the largest general application of this unit operation.

The objective of this chapter is to provide an overview of filtration terminology and basic engineering principles, as well as calculation methods that describe the filtration process in a generalised way. The basis equations describing the generalised process of filtration have been around for nearly 100 years and with few refinements, continue to be applied to modern design practices.

TERMINOLOGY AND GOVERNING EQUATIONS

There are essentially four important physical parameters that characterise a filter media and are used as a basis for relating the characteristics of the material to the system flow dynamics. These are porosity, permeability, tortuosity and connectivity.

We may begin by describing any porous medium as a solid matter containing many holes or pores, which collectively constitute an array of tortuous passages. Refer to Fig. 10.1 for an example. The number of holes or pores is sufficiently great that a volume average is needed to estimate pertinent properties. Pores that occupy a definite fraction of the bulk volume constitute a complex network of voids. The manner in which holes or pores are embedded, the extent of their interconnection and their location, size and shape characterise the porous medium. The term porosity refers to the fraction of the medium that contains the voids. When a fluid is passed over the medium, the fraction of the medium (i.e., the pores) that contributes to the flow is referred to as the effective porosity of the media. In a general sense, porous media are classified as either unconsolidated and consolidated and/or as ordered and random. Examples of unconsolidated media are sand, glass beads, catalyst pellets, column packing materials, soil, gravel and packing such as charcoal.

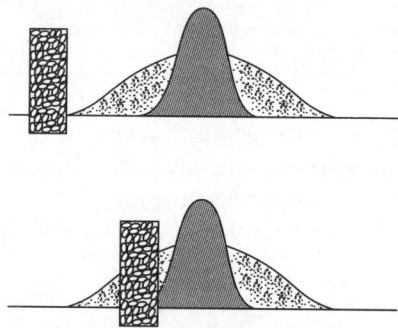

Fig. 10.1. Distribution of pores.

Examples of consolidated media are most of the naturally occurring rocks, such as sandstones and limestones. Materials such as concrete, cement, bricks, paper and cloth are manmade consolidated media.

Ordered media are regular packings of various types of materials, such as spheres, column packings and wood. Random media have no particular correlating factor. Porous media can be further categorised in terms of geometrical or structural properties as they relate to the matrix that affects flow and in terms of the flow properties that describe the matrix from the standpoint of the contained fluid. Geometrical or structural properties are best represented by average properties, from which these average structural properties are related to flow properties.

Pore Structure

A microscopic description characterises the structure of the pores. The objective of a pore-structure analysis is to provide a description that relates to the macroscopic or bulk flow properties. The major bulk properties that need to be correlated with pore description or characterisation are the four basic parameters: porosity, permeability, tortuosity and connectivity. In studying different samples of the same medium, it becomes apparent that the number of pore sizes, shapes, orientations and interconnections are enormous. Due to this complexity, pore-structure description is most often a statistical distribution of apparent pore sizes. This distribution is apparent because to convert measurements to pore sizes one must resort to models that provide average or model pore sizes. A common approach to defining a characteristic pore size distribution is to model the porous medium as a bundle of straight cylindrical or rectangular capillaries (Fig. 10.2). The diameters of the model capillaries are defined on the basis of a convenient distribution function.

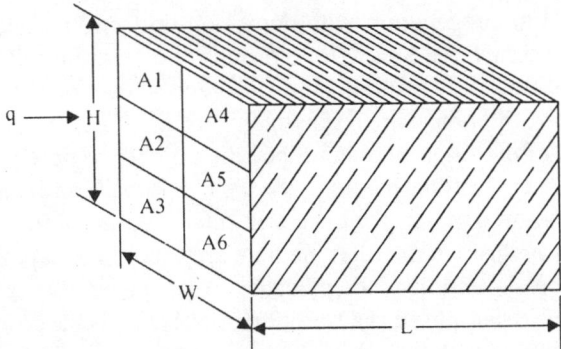

Fig. 10.2. Simplified capillary view of flow through pores.

Pore structure for unconsolidated media is inferred from a particle size distribution, the geometry of the particles and the packing arrangement of particles. The theory of packing is reasonably well established for symmetrical geometries such as spheres and cylinders. Information on particle size, geometry and packing theory allows us to develop relationships between pore size distributions and particle size distributions. A macroscopic description is based on average or bulk properties at sizes much larger than a single pore. In characterising a porous medium macroscopically, one must deal with the scale of description. The scale used depends on the manner and size in which we wish to model the porous medium. A simplified approach is to assume the medium to be ideal; meaning homogeneous, uniform and isotropic. From an industrial viewpoint, the objective of the unit operation of filtration is the separation of suspended solid particles from a process fluid stream which is accomplished by passing the suspension through a porous medium that is referred to as a filter medium. In forcing the fluid through the voids of the filter medium, fluid alone flows, but the solid particles are retained on the surface and in the medium's pores. The fluid discharging from the medium is called the filtrate. The operation may be performed with either incompressible fluids (liquids) or slightly to highly compressible fluids (gases). The physical mechanisms controlling filtration, although similar, vary with the degree of fluid compressibility. Although there are marked similarities in the particle capture mechanisms between the two fluid types, design methodologies for filtration equipment vary markedly. This reference volume concentrates only on process liquid handling (i.e., incompressible fluid flow and processing).

Difference Between Incompressible and Compressible Filuids

The dependency of liquid volume on pressure may be expressed in terms of the coefficient of compressibility. The coefficient is constant over a wide range of pressures for a particular material, but is different for each substance and for the solid and liquid states of the same material. For liquids, volume decreases linearly with pressure. For gases volume is observed to be inversely proportional to pressure. If water in its liquid state is subjected to a pressure change from 1 to 2 atm, then less than a 10^{-3} per cent reduction in volume occurs (the compressibility coefficient is very small). However, when the same pressure differential is applied to water vapour, a volume reduction in excess of 2 occurs.

FILTRATION DYNAMICS

When a suspension of solids passes through a porous media, the solid particles are collected on the feed side of the plate while the filtrate is forced through the media and carried away on the leeward side. A filter medium is, by nature, inhomogeneous, with pores non-uniform in size, irregular in geometry and unevenly distributed over the surface. Since flow through the medium takes place through the pores only, the micro-rate of liquid flow may result in large differences over the filter surface. This implies that the top layers of the generated filter cake are inhomogeneous and, furthermore, are established based on the structure and properties of the filter medium. Since the number of pore passages in the cake is large in comparison to the number in the filter medium, the cake's primary structure depends strongly on the structure of the initial layers. As a result, the cake and filter medium influence each other. Pores with passages extending all the way through the filter medium are capable of capturing solid particles that are smaller than the narrowest cross-section of the passage. This is generally attributed to the phenomenon of particle bridging or, in some cases, physical adsorption. Adsorption is the grouping together of molecules on the surface of a solid or liquid; such 'groupings' are the result of attractive forces between molecules. Activated carbons are highly porous; they contain mazes of interconnecting channels. An imbalance of molecular forces in the walls attracts many substances; these are physically held (adsorbed) by the carbon surfaces. After much use, the carbon may be regenerated and used again. Depending on the particular filtration technique, different filter media can be employed. Examples of common media are sand, diatomite, coal, cotton or wool fabrics, metallic wire cloth, porous plates of quartz, chamotte, sintered glass, metal powder and powdered ebonite. The average pore size and configuration (including tortuosity and connectivity) are established from the size and form of individual elements from which the medium is manufactured. On the average, pore sizes are greater for larger medium elements. In addition, pore configuration tends to be more uniform with more uniform medium elements, the fabrication method of the filter medium also affects average pore size and form. For example, pore characteristics are altered when fibrous media are first pressed together.

Pore characteristics also depend on the properties of fibers in woven fabrics, as well as on the exact methods of sintering glass and metal powders. Some filter media, such as cloths (especially fibrous layers), undergo considerable compression when subjected to typical pressures employed in industrial filtration operations. Other filter media, such as ceramic, sintered plates of glass and metal powders, are stable under the same operating conditions. In addition, pore characteristics are greatly influenced by the separation process occurring within the pore passages, as this leads to a decrease in effective pore size and consequently an increase in flow resistance. This results from particle penetration into the pores of the filter medium. The separation of solid particles from a liquid via filtration is a complicated process. For practical reasons filter medium openings are designed to be larger than the average size of the particles to be filtered. The filter medium chosen should be capable of retaining solids by adsorption. Furthermore, interparticle cohesive forces should be large enough to induce particle flocculation around the pore openings.

Process Classification

There are two major types of filtration: 'cake' and 'filter-medium' filtration. In the former, solid particulates generate a cake on the surface of the filter medium. In filter-medium filtration (also referred to as clarification), solid particulates become entrapped within the complex pore structure of the filter medium. The filter medium for the latter case consists of cartridges or granular media. Among the most common examples of granular materials are sand or anthracite coal.

When specifying filtration equipment for an intended application one must first account for the parameters governing the application and then select the filtration equipment best suited for the job. There are two important parameters that must be considered, namely the method to be used for forcing liquid through the medium and the material that will constitute the filter medium. When the resistance opposing fluid flow is small, gravity force effects fluid transport through a porous filter medium. Such a device is simply called a gravity filter.

When gravity is insufficient to induce flow, the pressure of the atmosphere is allowed to act on one side of the filtering medium, while a negative or suction pressure is applied on the discharge side. This type of filtering device is referred to as a vacuum filter. The application of vacuum filters is typically limited to 15 psi pressure, although there are applications where this value can be exceeded. (Note: filtration is often used in combination with clarification).

Clarification

Clarifiers are designed to efficiently remove undissolved substances from waste-water; removal is dependent upon density differences and is often enhanced by chemical means. Clarifiers are tank-like structures that may be either circular or rectangular in shape. When waste-water enter these treatment areas denser undissolved substances settle out, others rise to the surface. A scraper (rake) moves across the bottom of the clarifier; settled matter (sludge) is moved to a collection area. A skimmer moves across the water's surface collecting floating material.

If still greater force is required, a positive pressure in excess of atmospheric can be applied to the suspension by a pump. This motive force may be in the form of compressed air introduced in a montejus or the suspension may be directly forced through a pump acting against the filter medium (as in the case of a filter press) or centrifugal force may be used to drive the suspension through a filter medium as is done in screen centrifuges.

In all of these cases, the process of filtration may be characterised as a hydrodynamic process in which the fluid's volumetric rate is directly proportional to the existing pressure gradient across the filter medium and inversely proportional to the flow resistance imposed by the connectivity, tortuosity and size of the medium's pores and generated filter cake. The pressure gradient constitutes the driving force responsible for the flow of the suspension.

Regardless of how the pressure gradient is generated, the driving force increases proportionally. However, in most cases, the rate of filtration increases more slowly than the rate at which the pressure gradient rises. The reason for this is that as the gradient rises, the pores of filter medium and cake are compressed and consequently the resistance to flow increases. For highly compressible cakes, both driving force and resistance increase nearly proportionally and any rise in the pressure drop has a minor effect on the filtration rate.

Dynamics of Cake Formation

Filtration operations are capable of handling suspensions of varying characteristics ranging from granular, incompressible, free-filtering materials to slime-like compositions, as well as finely divided colloidal suspensions in which the cakes are incompressible. These latter materials tend to contaminate or foul the filter medium. The interaction between the particles in suspension and the filter medium determines to a large extent the specific mechanisms responsible for filtration.

In practice, cake filtration is used more often than filter-medium filtration. Upon achieving a certain thickness, the cake must be removed from the medium. This can be accomplished by the use of various mechanical devices or by reversing the flow of filtrate back through the medium (hence, the name backflushing).

To prevent the formation of muddy filtrate at the beginning of the subsequent filtration cycle, a thin layer of residual particles is sometimes deposited onto the filter medium. For the same reason, the filtration cycle is initiated with a low, but gradually increasing pressure gradient at an approximately constant flowrate. The process is then operated at a constant pressure gradient while experiencing a gradual decrease in process rate.

The structure of the cake formed and, consequently, its resistance to liquid flow depends on the properties of the solid particles and the liquid phase suspension, as well as on the conditions of filtration. Cake structure is first established by hydrodynamic factors (cake porosity, mean particle size, distribution and particle specific surface area and sphericity). It is also strongly influenced by some factors that can conditionally be denoted as physico-chemical. These factors are:

1. The rate of coagulation or peptisation of solid particles.
2. The presence of tar and colloidal impurities clogging the pores.
3. The influence of electrokinetic potentials at the interphase in the presence of ions, which decreases the effective pore cross-section.
4. The presence of solvate shells on the solid particles (this action is manifested at particle contact during cake formation).

Due to the combining effects of hydrodynamic and physico-chemical factors, the study of cake structure and resistance is extremely complex and any mathematical description based on theoretical considerations is at best only descriptive.

The influence of physico-chemical factors is closely related to surface phenomena at the solid-liquid boundary. It is especially manifested by the presence of small particles in the suspension. Large particle sizes result in an increase in the relative influence of hydrodynamic factors, while smaller sizes contribute to a more dramatic influence from physico-chemical factors. No reliable methods exist to predict when the influence of physico-chemical factors may be neglected. However, as a general rule, for rough evaluations their influence may be assumed to be most pronounced in the particle size range of 15–20 µm.

In specifying and designing filtration equipment, attention must be given to those methods that minimise high cake resistance. This resistance is responsible for losses in filtration capacity, which in turn impact on operating time and removal efficiency. One option for achieving a required filtration capacity is the use of a large number of filter modules. Increasing the physical size of equipment is feasible only within certain limitations as dictated by design considerations, allowable operating conditions and economic constraints. A more flexible option from an operational viewpoint is the implementation of process-oriented enhancements that intensify particle separation. This can be achieved by two different methods. In the first method, the suspension to be separated is pre-treated to obtain a cake with minimal resistance. This involves the addition of filter aids, flocculants or electrolytes to the suspension. In the second method, the period during which suspensions are formed provides the opportunity to alter suspension properties or conditions that are more favourable to low-resistance cakes. For example, employing pure initial substances or performing a pre-filtration operation under milder conditions tends to minimise the formation of tar and colloids. Similar results may be achieved through temperature control, by limiting the duration of certain operations immediately before filtering such as crystallisation or by controlling the rates and sequence of adding reagents.

Filtration Conditions

Two significant operating parameters influence the process of filtration: the pressure differential across the filtering plate and the temperature of the suspension. Most cakes may be considered compressible and, in general, their rate of compressibility increases with decreasing particle size. The temperature of the suspension influences the liquid-phase viscosity, which subsequently affects the ability of the filtrate to flow through the pores of the cake and the filter medium. In addition, the filtration process can be affected by particle inhomogeneity and the ability of the particles to undergo deformation when subjected to pressure and settling characteristics due to the influence of gravity. Particle size inhomogeneity influences the geometry of the cake structure not only at the moment of its formation, but also during the filtration process. During filtration, small particles retained on the outer layers of the cake are often entrained by the liquid flow and transported to layers closer to the filter medium or even into the pores themselves. This results in an increase in the resistance across the filter medium and the cake that is formed. Particles that undergo deformation when subjected to transient or high pressures are usually responsible for the phenomenon known as pore clogging. The addition of coagulating and peptising agents can greatly improve filterability. These are additives which can drastically alter the cake properties and, subsequently lower flow resistance and ultimately increase the filtration rate and the efficiency of separation. Filter aids may be used to prevent the penetration of fine particles into the pores of a filter plate when processing low concentration suspensions. Filter aids build up a porous, permeable, rigid lattice structure that retains solid particles on the filter medium surface, while permitting liquid to pass through. They are often employed as pre-coats with the primary aim of protecting the filter medium. They may also be mixed with a suspension of diatomaceous silica type earth (>90 per cent silica content). Cellulose and asbestos fiber pulps were typically employed for many years as well.

Washing and Dewatering

When contaminated, polluted or valuable suspension liquors are present, it becomes necessary to wash the filter cake to effect clean separation of solids from the mother liquor or to recover the mother liquor from the solids. The operation known as dewatering involves forcing a clean fluid through the cake to recover residual liquid retained in the pores, directly after filtering or washing. Dewatering is a complex process on a microscale, because it involves the hydrodynamics of two-phase flow. Although washing and dewatering are performed on a cake with an initially well defined pore structure, the flows become greatly distorted and complex due to changing cake characteristics. The cake structure undergoes compression and disintegration during both operations, thus resulting in a dramatic alteration of the pore structure.

WASTE-WATER TREATMENT APPLICATIONS

Subsequent chapters address the application of filtration techniques to waste-water treatment in some detail. For now, only some general comments and terminology are introduced as part of this introductory chapter. In a very general sense, there are two types of waste-water flows—municipal and industrial. Although municipal waste-water vary in composition, there are ranges of properties that enable filtration equipment to be readily selected and specified. This is not always the situation when treating industrial waste-water streams. The compositions and properties of industrial waste-water vary significantly and even within specific industry sectors; these flows can be dramatically different. This is important to realise because although filtration is a physical process, it depends upon and is integrally a part of chemical treatment processes such as pre-conditioning, buffering and filter aid conditioning. These chemical treatment methods must be properly specified along with the filtration equipment itself in order to ensure that a properly designed filtration system is being applied.

Filtration equipment selection can be complex not only because of the wide variations in suspension properties, but also because of the sensitivities of suspension and cake properties to different process conditions and to the variety of filtering equipment available. Generalities in selection criteria are, therefore, few; however, there are some guidelines applicable to certain classes of filtration applications. One example is the choice of a filter whose flow orientation is in the same direction as gravity when handling polydispersed suspensions. Such an arrangement is more favourable than an upflow design, since larger particles will tend to settle first on the filter medium, thus preventing pores from clogging within the medium structure.

A further recommendation, depending on the application, is not to increase the pressure difference for the purpose of increasing the filtration rate. The cake may, for example, be highly compressible; thus, increased pressure would result in significant increases in the specific cake resistance. We may generalise the selection process to the extent of applying three rules to all filtration problems:

1. The objectives of a filtration operation should be defined.
2. Physical and/or chemical pre-treatment options should be evaluated for the intended application based on their availability, cost, ease of implementation and ability to provide optimum filterability.
3. Final filtration equipment selection should be based on the ability to meet all objectives of the application within economic constraints.

In applying these general criteria, one should focus on the intended application. In waste-water treatment applications, filtration can be applied at various stages. It can be applied as a pre-treatment method, in which case the objective is often to remove coarse, gritty materials from the waste-stream. This is a pre-conditioning step for waste-water which will undergo further chemical and physical treatment downstream.

Filtration may also serve as the preparatory step for the operation following it. The latter stages may be drying or incineration of solids, concentration or direct use of the filtrate. Filtration equipment must be selected on the basis of their ability to deliver the best feed material to the next step. Dry, thin, porous, flaky cakes are best suited for drying where grinding operations are not employed. In such cases, the cake will not ball up and quick drying can be achieved. A clear, concentrated filtrate often aids downstream treatment, whereby the filter can be operated to increase the efficiency of the downstream equipment without affecting its own efficiency.

Filtration may also be applied as a part of the final stages of treatment in the process. This is most commonly referred to as a polishing operation. Indeed, filtration may be applied both as pre-treatment and polishing stages and even as an intermediate stage in the waste-water treatment process. Filtration equipment selection depends upon the specific operation that the equipment must perform.

Proper pH control can result in clarification that might otherwise not be feasible, since an increase in alkalinity or acidity may change soft, slimy solids into firm, free-filtering ones. In some cases pre-coats are employed, not because of the danger of filter cloth clogging, but to allow the use of a coarser filter medium, such as metallic cloth.

Equipment Selection Methodology

Equipment selection is seldom based on rigorous equations or elaborate mathematical models. Where equations are used, they function as a directional guide in evaluating data or process arrangements. Projected results are derived most reliably from actual plant operational data and experience where duplication is desired; from standards set up where there are few variations from plant to plant, so that results can be anticipated with an acceptable degree of confidence (as in municipal water filtration) or from pilot or laboratory tests of the actual material to be handled. Pilot plant runs are typically designed for short durations and to closely duplicate actual operations.

Proper selection of equipment may be based on experiments performed in the manufacturer's laboratory, although this is not always feasible. Sometimes the material to be handled cannot readily be shipped; its physical or chemical conditions change during the time lag between shipping and testing or special conditions must be maintained during filtration that cannot be readily duplicated, such as refrigeration, solvent washing and inert gas use. A filter manufacturer's laboratory has the advantage of having numerous types of filters and apparatus available with experienced filtration engineers to evaluate results during and after test runs.

The use of pilot-plant filter assemblies is both common and a classical approach to design methodology development. These combine the filter with pumps, receivers, mixers, etc., in a single compact unit and may be rented at a nominal fee from filter manufacturers, who supply operating instructions and sometimes an operator. Preliminary tests are often run at the filter manufacturer's laboratory. Rough tests indicate what filter type to try in the pilot plant.

Comparative calculations of specific capacities of different filters or their specific filter areas should be made as part of the evaluation. Such calculations may be performed on the basis of experimental data obtained without using basic filtration equations. In designing a new filtration unit after equipment selection, calculations should be made to determine the specific capacity or specific filtration area. Basic filtration equations may be used for this purpose, with preliminary experimental constants evaluated. These constants contain information on the specific cake resistance and the resistance of the filter medium.

The basic equations of filtration cannot always be used without introducing corresponding corrections. This arises from the fact that these equations describe the filtration process partially for ideal conditions when the influence of distorting factors is eliminated. Among these factors are the instability of the cake resistance during operation and the variable resistance of the filter medium, as well as the settling characteristics of solids. In these relationships, it is necessary to use statistically averaged values of both resistances and to introduce corrections to account for particle settling and other factors. In selecting filtration methods and evaluating constants in the process equations, the principles of similarity modelling are relied on heavily.

Within the subject of filtration, a distinction is made between micro and macromodelling. The first one is related to modelling cake formation. The cake is assumed to have a well defined structure, in which the hydrodynamic and physico-chemical processes take place. Macromodelling presents few difficulties, because the models are process-oriented (i.e., they are specific to the particular operation or specific equipment). If distorting side effects are not important, the filtration process may be designed according to existing empirical correlations. In practice, filtration, washing and dewatering often deviate substantially from theory. This occurs because of the distorting influences of filter features and the unaccounted for properties of the suspension and cake.

Existing statistical methods permit prediction of macroscopic results of the processes without complete description of the microscopic phenomena. They are helpful in establishing the hydrodynamic relations of liquid flow through porous bodies, the evaluation of filtration quality with pore clogging, description of particle distributions and in obtaining geometrical parameters of random layers of solid particles.

SAND FILTRATION

When people think of sand filtration, they automatically relate to municipal water treatment facilities. In general that's the arena that this classical filter has ruled, but it most certainly has found applications in pure industrial settings, oftentimes for niche applications where suspended solids and organic matter persist in process water. Sand filtration is almost never applied as the primary treatment method. Most often it is a pre-treatment or final stage, but sometimes intermediate stage of water treatment and is most often used along with other filtration technologies, carbon adsorption, sedimentation and clarification, disinfection and biological methods. The term sand filtration is somewhat misleading and stems from older municipal waste-water treatment methods. While there is a class of filtration equipment that relies principally on sand as the filter media, it is more common to employ multiple media in filtration methods and equipment, with sand being the predominant media. In this regard, the terms sand filtration and granular media filtration are considered interchangeable in our discussions. The design, operation and maintenance of these systems are very straightforward and indeed may be viewed as the least complex or simplest filtration practices that exist.

WASTE-WATER TREATMENT PLANT OPERATIONS

Before getting into the subject of sand filtration, we should first attempt to put the technologies of municipal waste-water treatment into some perspective. Waste-water treatment plants can be divided into two major types: biological and physical/chemical. Biological plants are more commonly used to treat domestic or combined domestic and industrial waste-water from a municipality. They use basically the same processes that would occur naturally in the receiving water, but give them a place to happen under controlled conditions, so that the cleansing reactions are completed before the water is discharged into the environment.

Physical/chemical plants are more often used to treat industrial waste-water directly, because they often contain pollutants which cannot be removed efficiently by micro-organisms although industries that deal with biodegradable materials, such as food processing, dairies, breweries and even paper, plastics and petrochemicals, may use biological treatment. Biological plants generally use some physical and chemical processes also.

A physical process usually treats suspended, rather than dissolved pollutants. It may be a passive process, such as simply allowing suspended pollutants to settle out or float to the top naturally—depending on whether they are more or less dense than water. Or the process may be aided mechanically, such as by gently stirring the water to cause more small particles to bump into each other and stick together, forming larger particles which will settle or rise faster—a process known as flocculation. Chemical flocculants may also be added to produce larger particles. To aid floatation processes, dissolved air under pressure may be added to cause the formation of tiny bubbles which will attach to particles.

Filtration through a medium such as sand as a final treatment stage can result in a very clear water. In contrast—ultrafiltration, nanofiltration and reverse osmosis (RO) are processes which force water through membranes and can remove colloidal material (very fine, electrically charged particles, which will not settle) and even some dissolved matter. Absorption (adsorption, technically) on activated charcoal is a physical process which can remove dissolved chemicals. Air or steam stripping can be used to remove pollutants that are gasses or low-boiling liquids from water and the vapours which are removed in this way are also often passed through beds of activated charcoal to prevent air pollution. These last processes are used mostly in industrial treatment plants, though activated charcoal is common in municipal plants, as well, for odour control.

Examples of chemical treatment processes, in an industrial environment, would be:

1. Converting a dissolved metal into a solid, settleable form by precipitation with an alkaline material like sodium or calcium hydroxide. Dissolved iron or aluminium salts or organic coagulant aids like polyelectrolytes can be added to help flocculate and settle (or float) the precipitated metal.

2. Converting highly toxic cyanides used in mining and metal finishing industries into harmless carbon dioxide and nitrogen by oxidising them with chlorine.

3. Destroying organic chemicals by oxidising them using ozone or hydrogen peroxide, either alone or in combination with catalysts (chemicals which speed up reactions) and/or ultraviolet light.

In municipal treatment plants, chemical treatment—in the form of aluminium or iron salts— is often used for removal of phosphorus by precipitation. Chlorine or ozone (or ultraviolet light) may be used for disinfection, that is, killing harmful micro-organisms before the final discharge of the waste-water. Sulphur dioxide or sulphite solutions can be used to neutralise (reduce) excess chlorine, which is toxic to aquatic life. Chemical coagulants are also used extensively in sludge treatment to thicken the solids and promote the removal of water. A conventional treatment plant is comprised of a series of individual unit processes, with the output (or effluent) of one process becoming the input (influent) of the next process. The first stages will usually be made up of physical processes that take out easily removable pollutants. After this, the remaining pollutants are generally treated further by biological or chemical processes. These may convert dissolved or colloidal impurities into a solid or gaseous form, so that they can be removed physically or convert them into dissolved materials which remain in the water, but are not considered as undesirable as the original pollutants. The solids (residuals or sludges) which result from these processes form a side stream which also has to be treated for disposal.

Common Processes

Common processes found at a municipal treatment plant include:

Preliminary treatment

Preliminary treatment to remove large or hard solids that might clog or damage other equipment. These might include grinders (comminuters), bar screens and grit channels. The first chops up rags and trash; the second simply catches large objects, which can be raked off; the third allows heavier materials, like sand and stones, to settle out, so that they will not cause abrasive wear on downstream equipment. Grit channels also remove larger food particles (i.e., garbage).

Primary settling basins

In primary settling basins the water flows slowly for up to a few hours, to allow organic suspended matter to settle out or float to the surface. Most of this material has a density not much different from that of water, so it needs to be given enough time to separate. Settling tanks can be rectangular or circular. In either type, the tank needs to be designed with some type of scrapers at the bottom to collect the settled sludge and direct it to a pit from which it can be pumped for further treatment—and skimmers at the surface, to collect the material that floats to the top (which is given the rather inglorious name of 'scum'). Figure 10.3 shows the operation of a typical primary settling tank.

Secondary treatment

In secondary treatment usually biological, tries to remove the remaining dissolved or colloidal organic matter. Generally, the biodegradation of the pollutants is allowed to take place in a location where plenty of air can be supplied to the micro-organisms. This promotes formation of the less offensive, oxidised products. Engineers try to design the capacity of the treatment units so that enough of the impurities will be removed to prevent significant oxygen demand in the receiving water after discharge.

There are two major types of biological treatment processes: attached growth and suspended growth.

In an attached growth process, the micro-organisms grow on a surface, such as rock or plastic. Examples include open trickling filters, where the water is distributed over rocks and trickles down to underdrains, with air being supplied through vent pipes; enclosed biotowers, which are similar, but more likely to use shaped, plastic media instead of rocks; and so-called rotating biological contacters or RBC's, which consist of large, partially submerged discs which rotate continuously, so that the micro-organisms growing on the disc's surface are repeatedly being exposed alternately to the waste-water and to the air. The most common type of suspended growth process is the so-called activated sludge system. This type of system consists of two parts, an aeration tank and a settling tank or clarifier. The aeration tank contains a 'sludge' which is what could be best described as a 'mixed microbial culture', containing mostly bacteria, as well as protozoa, fungi, algae, etc. This sludge is constantly mixed and aerated either by compressed air bubblers located along the bottom or by mechanical aerators on the surface. The waste-water to be treated enters the tank and mixes with the culture, which uses the organic compounds for growth— producing more micro-organisms—and for respiration, which results mostly in the formation of carbon dioxide and water. The process can also be set up to provide biological removal of the nutrients nitrogen and phosphorus. Refer to Fig. 10.3 for a simplified process flow sheet.

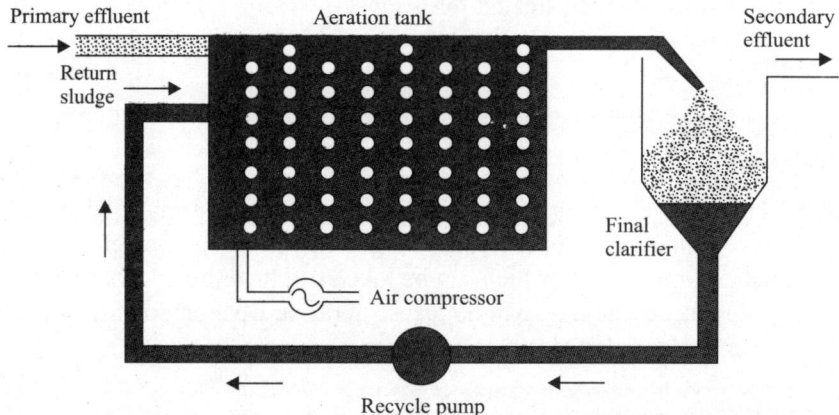

Fig. 10.3. Simplified process flow sheet of activated sludge process.

After sufficient aeration time to reach the required level of treatment, the sludge is carried by the flow into the settling tank or clarifier, which is often of the circular design. (An important condition for the success of this process is the formation of a type of culture which will flocculate naturally, producing a settling sludge and a reasonably clear upper or supernatant layer. If the sludge does not behave this way, a lot of solids will be remain in the water leaving the clarifier and the quality of the effluent waste-water will be poor.) The sludge collected at the bottom of the clarifier is then recycled to the aeration tank to consume more organic material. The term 'activated' sludge is used, because by the time the sludge is returned to the aeration tank, the micro-organisms have been in an environment depleted of 'food' for some time and are in a 'hungry' or activated condition, eager to get busy biodegrading some more wastes. Since the amount of micro-organisms or biomass, increases as a result of this process, some must be removed on a regular basis for further treatment and disposal, adding to the solids produced in primary treatment. The type of activated sludge system that has just been described is a continuous flow process. There is a variation in which the entire activated sludge process take place in a single tank, but at different times. Steps include filling, aerating, settling, drawing off supernatant, etc. A system like this, called a sequencing batch reactor, can provide more flexibility and control over the treatment, including nutrient removal and is amenable to computer control.

Nutrient removal

Nutrient removal refers to the treatment of the waste-water to take out nitrogen or phosphorus, which can cause nuisance growth of algae or weeds in the receiving water. Nitrogen is found in domestic waste-water mostly in the form of ammonia and organic nitrogen. These can be converted to nitrate nitrogen by bacteria, if the plant is designed to provide enough oxygen and a long enough 'sludge age' to develop these slow-growing types of organisms. The nitrate which is produced may be discharged; it is still usable as a plant nutrient, but it is much less toxic than ammonia. If more complete removal of nitrogen is required, a biological process can be set up which reduces the nitrate to nitrogen gas (and some nitrous oxide). There are also physical/chemical processes which can remove nitrogen, especially ammonia; they are not as economical for domestic waste-water, but might be suited for an industrial location where no other biological processes are in use. (These methods include alkaline air stripping, ion exchange and 'breakpoint' chlorination).

Phosphorus removal

Phosphorus removal is most commonly done by chemical precipitation with iron or aluminium compounds, such as ferric chloride or alum (aluminium sulphate). The solids which are produced can be settled along with other sludges, depending on where in the treatment train the process takes place. 'Lime' or calcium hydroxide, also works, but makes the water very alkaline, which has to be corrected and produces more sludge. There is also a biological process for phosphorus removal, which depends on designing an activated sludge system in such a way as to promote the development of certain types of bacteria which have the ability to accumulate excess phosphorus within their cells. These methods mainly convert dissolved phosphorus into particulate form. For treatment plants which are required to discharge only very low concentrations of total phosphorus, it is common to have a sand filter as a final stage, to remove most of the suspended solids which may contain phosphorus.

Disinfection

Disinfection, usually the final process before discharge, is the destruction of harmful (pathogenic) micro-organisms, i.e., disease-causing germs. The object is not to kill every living micro-organism in the water—which would be sterilisation—but to reduce the number of harmful ones to levels appropriate for the intended use of the receiving water.

The most commonly used disinfectant is chlorine, which can be supplied in the form of a liquefied gas which has to be dissolved in water or in the form of an alkaline solution called sodium hypochlorite, which is the same compound as common household chlorine bleach. Chlorine is quite effective against most bacteria, but a rather high dose is needed to kill viruses, protozoa and other forms of pathogen. Chlorine has several problems associated with its use, among them: (i) that it reacts with organic matter to form toxic and carcinogenic chlorinated organics, such as chloroform; (ii) chlorine is very toxic to aquatic organisms in the receiving water—the USEPA recommends no more than 0.011 pans per million (mg/l); and (iii) it is hazardous to store and handle. Hypochlorite is safer, but still produces problems: (i) and (ii). Problem (ii) can be dealt with by adding sulphur dioxide (liquefied gas) or sodium sulphite or bisulphite (solutions) to neutralise the chlorine. The products are nearly harmless chloride and sulphate ions. This may also help somewhat with problem: (i). A more powerful disinfectant is ozone, an unstable form of oxygen containing three atoms per molecule, rather than the two found in the ordinary oxygen gas which makes up about 21 per cent of the atmosphere. Ozone is too unstable to store and has to be made as it is used. It is produced by passing an electrical discharge through air, which is then bubbled through the water. While chlorine can be dosed at a high enough concentration so that some of it remains in the water for a considerable time, ozone is consumed very rapidly and leaves no residual. It may also produce some chemical by-products, but probably not as harmful as those produced by chlorine. The other commonly used method of disinfection is ultraviolet light. The water is passed through banks of cylindrical, quartz-jacketed fluorescent bulbs. Anything which can absorb the light, such as fouling or scale formation on the bulbs, surfaces or suspended matter in the water, can interfere with the effectiveness of the disinfection. Some dissolved materials, such as iron and some organic compounds, can also absorb some of the light. Ultraviolet disinfection is becoming more popular because of the increasing complications associated with the use of chlorine.

Sludge from primary settling basins, called primary or 'raw' sludge, is a noxious, smelly, gray-black, viscous liquid or semi-solid. It contains very high concentrations of bacteria and other micro-organisms, many of them pathogenic, as well as large amounts of biodegradable organic material. Because of the high concentrations, any dissolved oxygen will be consumed rapidly and the odourous and toxic products of anaerobic biodegradation (putrefaction) will be produced. The greasy floatable skimmings from primary treatment are another portion of this putrescible solid waste stream. In addition to the primary sludge, waste-water plants with secondary treatment will produce a 'secondary sludge', consisting largely of micro-organisms which have grown as a result of consuming the organic wastes. While not quite so objectionable, due to the biodegradation which has already taken place, it is still very high in pathogens and contains much material which will decay and produce odours if not treated further. Ultimately, the sludge must all be disposed of. The way in which this is done depends on the quality of the sludge and determines how it needs to be treated. The most desirable final fate for these solids would be for beneficial use in agriculture, since the material has organic matter to act as a soil conditioner, as well as a some fertiliser value. This requires the highest quality 'biosolids', free of contamination with toxic metals or industrial organic compounds and low in pathogens. At a somewhat lower quality, it can be used for similar purposes on non-agricultural land and for land reclamation (e.g., strip mines). Poorer quality sludge can be disposed of by landfilling or incineration.

One commonly used method of sludge treatment, called digestion, is biological. Since the material is loaded with bacteria and organic matter; why not let the bacteria eat the biodegradable material? Digestion can be either aerobic or anaerobic. Aerobic digestion requires supplying oxygen to the sludge; it is similar to the activated sludge process, except no external 'food' is provided. In anaerobic digestion, the sludge is fed into an air-free vessel; the digestion produces a gas which is mostly a mixture of methane and carbon dioxide. The gas has a fuel value and can be burned to provide heat to the digester tank and even to run electric generators. Some localities have compressed the gas and used it to power vehicles. Digestion can reduce the amount of organic matter by about 30 to 70 per cent, greatly decrease the number of pathogens and produce a liquid with an inoffensive, 'earthy' odour. This makes the sludge safer to dispose of on land, since the odour does not attract as many scavenging pests, such as flies, rodents, gulls, etc., which spread pathogens from the disposal site to other areas— and there are fewer pathogens to be spread.

A liquid sludge, which might contain 3 to 6 per cent dry weight of solids, can be dewatered to form a drier sludge cake of may be 15 to 25 per cent solids, which can be hauled as a solid rather than having to be handled as a liquid. Equipment used to dewater sludge includes centrifuges, vacuum filters and belt presses or plate-and-frame presses. Chemical coagulants are commonly added to help form larger aggregates of solids and release the water. Further processes such as composting and heat drying can produce a drier product with lower pathogen levels. Another approach involves treatment with lime (calcium oxide), which kills pathogens due to its highly alkaline nature as well as the heat that is generated as it reacts with the water in the sludge; this also results in a drier product. A final disposal method which eliminates all of the pathogens and greatly reduces the volume of the sludge is incineration. This is not considered a beneficial use, however and is becoming less popular due to public concerns over air emissions.

Sludges from physical-chemical treatment of industrial waste streams containing heavy metals and non-biodegradable toxic organic compounds often must be handled as hazardous wastes. Some of these will end up in hazardous waste landfills or may be chemically treated for detoxification—or even for recovery of some components for recycling. Recalcitrant organic compounds can be destroyed by carefully controlled high-temperature incineration or by other innovative processes, such as high-temperature hydrogen reduction.

GRANULAR MEDIA FILTRATION

Granular media filtration is used for treating aqueous waste streams. The filter media consists of a bed of granular particles (typically sand or sand with anthracite or coal). The anthracite has adsorptive characteristics and hence can be beneficial in removing some biological and chemical contaminants in the waste-water. This material may also be substituted for activated charcoal.

The bed is contained within a basin and is supported by an underdrain system which allows the filtered liquid to be drawn off while retaining the filter media in place. As water containing suspended solids passes through the bed of filter medium, the particles become trapped on top of and within, the bed. The filtration rate is reduced at a constant pressure unless an increase in the amount of pressure is applied to force the waste-water through the filter bed. In order to prevent plugging of the upper surface and uppermost depth of the bed, the filter is backflushed at high velocity to dislodge the filtered particles. The backwash water contains high concentrations of solids and is sent to further treatment steps within the waste-water treatment plant.

The filter application is typically applied to handling streams containing less than 100 to 200 mg/liter suspended solids, depending on the required effluent level. Increased-suspended solids loading reduces the need for frequent backwashing. The suspended solids concentration of the filtered liquid depends on the particle size distribution, but typically, granular media filters are capable of producing a filtered liquid with a suspended solids concentration as low as 1 to as high as 10 mg/liter. Large flow variations will affect the effluent's quality.

Granular media filters are usually preceded by sedimentation in order to reduce the suspended solids load on the filter. Granular media filtration can also be installed ahead of biological or activated carbon treatment units to reduce the suspended solids load and in the case of activated carbon to minimise plugging of the carbon columns. Granular media filtration is only marginally effective in treating colloidal size particles in suspensions. Usually these particles can be made larger by flocculation although this will reduce run lengths. In cases where it is not possible to flocculate such particles (as in the case of many oil/water emulsions), other techniques such as ultrafiltration must be considered. Figure 10.4 illustrates a common sand filter that most people are familiar with in swimming pool applications. Such systems rely on very fine sand media that can typically remove suspended particles about 0.5 μm in size. Filtration is an effective means of removing low levels of solids from wastes provided the solids concentration does not vary greatly and the filter is backwashed at appropriate intervals during the filtration cycle. The operation can be easily integrated with other treatment steps and further, is well suited to mobile treatment systems as well as on-site or fixed installations. In short, sand filtration technologies, although simple, are quite versatile in meeting treatment challenges.

A typical multi-media sand filtration unit is shown in Fig. 10.5. In this configuration a coarse layer of media is used to reduce the contaminant loading to the final layer. This allows multimedia filters to use finer media. Such units generally remove suspended solids down to about 15 μm and they require large volumes of water to properly remove contaminant that is trapped deep within the bed. Often manufacturers of these types of systems claim 90 per cent removal of 0.5 μm particles and larger. This can be a misleading statement as quite often only about 5 per cent of the 0.5 μm particles will be removed. Grouping the 0.5 μm particles with much larger particles allows the claim to be met by removing a few large volume particles from the tower sump, even though the vast majority of fine particles remain to foul heat exchange equipment.

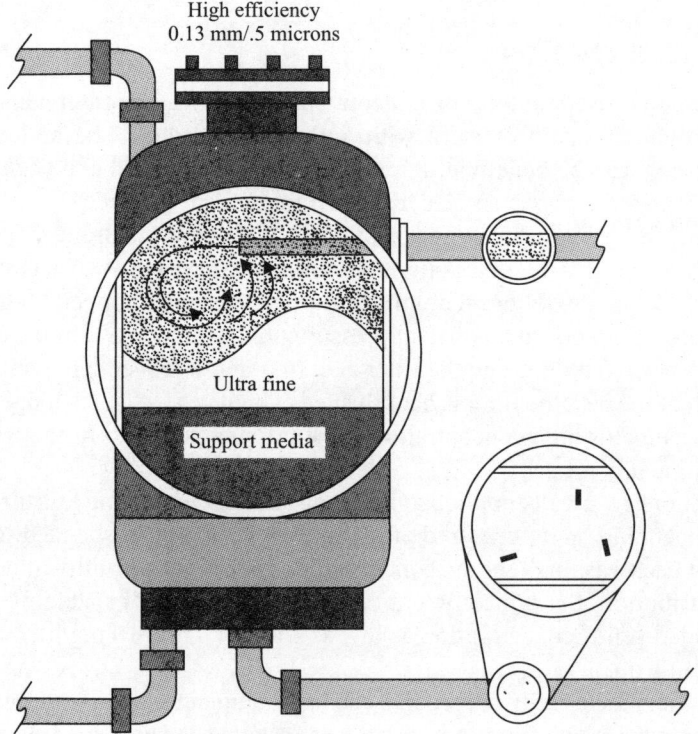

High efficiency
0.13 mm/.5 microns

Ultra fine

Support media

Fig. 10.4. A simple sand filtration unit.

A typical physical-chemical treatment system incorporates three 'dual' medial (sand anthracite) filters connected in parallel in its treatment train. The major maintenance consideration with granular medial filtration is the handling of the backwash. The backwash will generally contain a high concentration of contaminants and require subsequent treatment.

In this application, the operations of precipitation and flocculation play important roles. Precipitation is a physico-chemical process whereby some or all, of a substance in solution is transformed into a solid phase. It is based on alteration of the chemical equilibrium relationships affecting the solubility of inorganic species. Removal of metals as hydroxides and sulphides is the most common precipitation application in waste-water treatment. Lime or sodium sulphide is added to the waste-water in a rapid mixing tank along with flocculating agents. The waste-water flows to a flocculation chamber in which adequate mixing and retention time is provided for agglomeration of precipitate particles. Agglomerated particles are then separated from the liquid phase by settling in a sedimentation chamber and/or by other physical processes such as filtration.

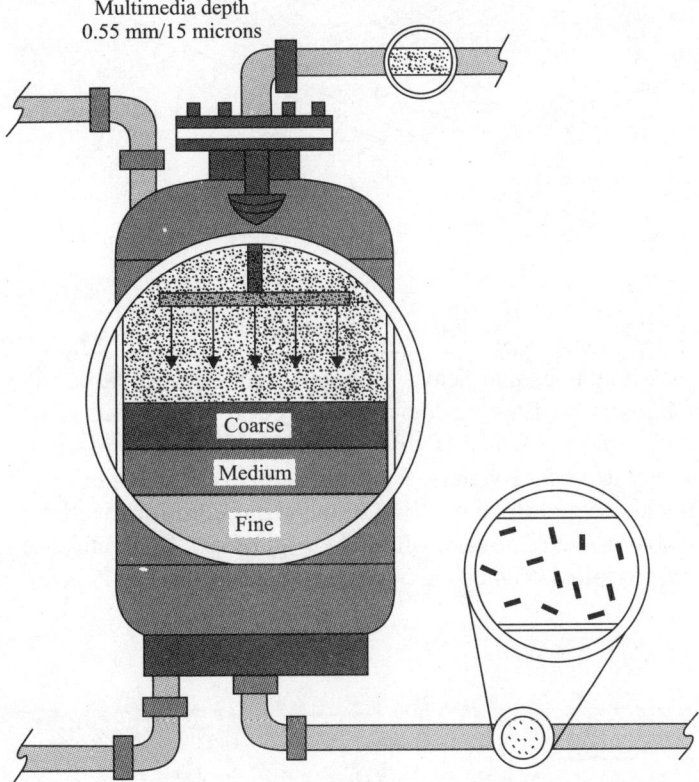

Fig. 10.5. Multimedia sand filter.

Precipitation is often applied to the removal of most metals from waste-water including zinc, cadmium, chromium, copper, fluoride, lead, manganese and mercury. Also, certain anionic species can be removed by precipitation, such as phosphate, sulphate and fluoride. Note that in some cases, organic compounds may form organometallic complexes with metals, which could inhibit precipitation. Cyanide and other ions in the waste-water may also complex with metals, making treatment by precipitation less efficient. A cutaway view of a rapid sand filter that is most often used in a municipal treatment plant is illustrated in Fig. 10.6. The design features of this filter have been relied upon for more than 60 years in municipal applications.

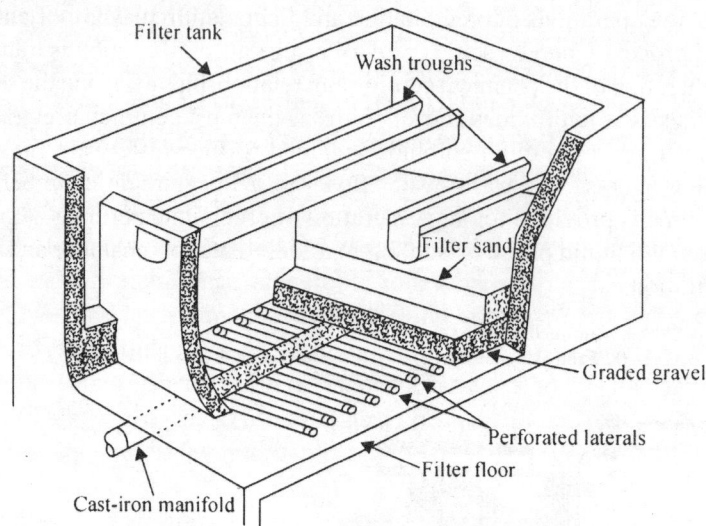

Fig. 10.6. Cutaway view of a rapid sand filter.

SAND FILTERS

A typical sand filter system consists of two or three chambers or basins. The first is the sedimentation chamber, which removes floatables and heavy sediments. The second is the filtration chamber, which removes additional pollutants by filtering the runoff through a sand bed. The third is the discharge chamber. The treated filtrate normally is then discharged through an underdrain system either to a storm drainage system or directly to surface waters. Sand filters are able to achieve high removal efficiencies for sediment, biochemical oxygen demand (BOD) and fecal coliform bacteria. Total metal removal, however, is moderate and nutrient removal is often low. Figure 10.7 illustrates one type of configuration. Typically, sand filters begin to experience clogging problems within 3 to 5 years.

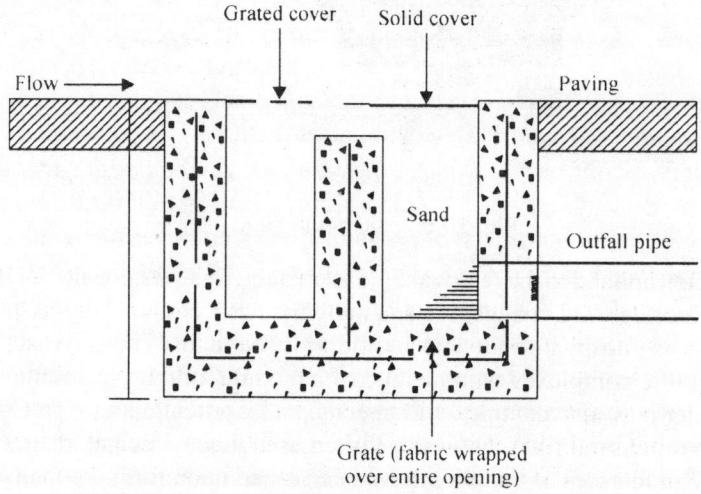

Fig. 10.7. Example of a sand filter configuration.

Accumulated trash, paper, debris should be removed every six months or as needed. Corrective maintenance of the filtration chamber includes removal and replacement of the top layers of sand and gravel as they become clogged. Table 10.1 provides some typical removal efficiencies for specific pollutants.

Table 10.1. Typical removal efficiencies.

Pollutant	Per cent removal	Pollutant	Per cent removal
Fecal coliform	76	Total organic carbon (TOC)	48
Biochemical oxygen demand (BOD)	70	Total nitrogen (TN)	21
Total suspended solids (TSS)	70	Iron, lead, zinc	45

Precipitation, Flocculation and Agglomeration

The process of flocculation is applicable to aqueous waste streams where particles must be agglomerated into larger more settleable particles prior to sedimentation or other types of treatment. Highly viscous waste streams will inhibit the settling of solids. In addition to being used to treat waste streams, precipitation can also be used as an *in situ* process to treat aqueous wastes in surface impoundments. In an *in situ* application, lime and flocculants are added directly to the lagoon and mixing, flocculation and sedimentation are allowed to occur within the lagoon.

Precipitation and flocculation can be integrated into more complex treatment systems. The performance and reliability of these processes depends greatly on the variability of the composition of the waste being treated. Chemical addition must be determined using laboratory tests and must be adjusted with compositional changes of the waste being treated or poor performance will result.

Precipitation is non-selective in that compounds other than those targeted may be removed. Both precipitation and flocculation are non-destructive and generate a large volume of sludge which must be disposed of. Coagulation, flocculation, sedimentation and filtration, are typically followed by chlorination in municipal waste-water treatment processes.

Coagulation involves the addition of chemicals to alter the physical state of dissolved and suspended solids. This facilitates their removal by sedimentation and filtration. The most common primary coagulants are alum ferric sulphate and ferric chloride. Additional chemicals that may be added to enhance coagulation include activate silica, a complex silicate made from sodium silicate and charged organic molecules called polyelectrolytes, which include large-molecular-weight polyacrylamides, dimethyl-diallylammonium chloride, polyamines and starch. These chemicals ensure the aggregation of the suspended solids during the next treatment step—flocculation. Sometimes polyelectrolytes (usually polyacrylamides) are also added after flocculation and sedimentation as an aid to the filtration step.

Coagulation may also remove dissolved organic and inorganic compounds. The hydrolysing metal salts may react with the organic matter to form a precipitate or they may form aluminium hydroxide or ferric hydroxide floc particles on which the organic molecules adsorb. The organic substances are then removed by sedimentation and filtration or filtration alone if direct filtration or in-line filtration is used. Adsorption and precipitation also removes inorganic substances.

The process of sedimentation involves the separation from water, by gravitational settling of suspended particles that are heavier than water. The resulting effluent is then subject to rapid filtration to separate out solids that are still suspended in the water. Rapid filters typically consist of 24 to 36 inches of 0.5 to 1 mm diameter sand and/or anthracite. Particles are removed as water is filtered through the media at rates of 1 to 6 gallons/minute/square foot. Rapid filtration is effective in removing most particles that remain after sedimentation. The substances that are removed by coagulation, sedimentation and filtration accumulate in sludge which must be properly disposed of.

Coagulation, flocculation, sedimentation and filtration will remove many contaminants. Perhaps most important is the reduction of turbidity. This treatment yields water of good clarity and enhances disinfection efficiency. If particles are not removed, they harbour bacteria and make final disinfection more difficult.

Hydraulic Performances

The hydraulic performances required of the sand with slow filters are inferior to those for rapid filters. In the case of slow filters, one can use fine sand, since the average filtration velocity that is usually necessary lies in the range 2 to 5 m/day.

In slow filtration, much of the effect is obtained by the formation of a filtration layer, including the substances that are extracted from the water. At the early stages of the operation, these substances contain micro-organisms able to effect, beyond the filtration, biochemical degradation of the organic matter. This effect also depends on the total surface of the grains forming the filter material. The probability of contact between the undesirable constituents of the water and the surface of the filter medium increases in proportion to the size of the total surface of the grains.

The actual diameter of the sands used during slow filtration typically lies between 0.15 and 0.35 mm. It is not necessary to use a ganged sand. The minimum thickness of the layer necessary for slow filtration is 0.3 to 0.4 m and the most efficient filtration thickness typically is at 2 to 3 cm.

The actual requirements for the sand in slow filtration are chemical in nature. Purity and the absence of undesirable matters are more important than grain-size distribution in the filtration process. On the other hand, the performance of rapid filters requires sands with quite a higher precise grain size. In the case of rapid filtration, the need for hydraulic performances is greater than in slow nitration. This means that the grain-size distribution of the medium is of prime concern in the latter case.

In waste-water treatment plants, the purity of the sand media used must be examined regularly. In addition, both the head loss of the filter beds and an analysis of the wash water during the operation of washing the filters must be checked regularly. Special attention must also be granted to the formation of agglomerates. The presence of agglomerates is indicative of insufficient washing and the possible formation of undesirable microbiological development zones within the filter bed.

Primary mechanisms that control the operation of sand filtration

The primary mechanisms that control the operation of sand filtration are:
1. Straining.
2. Settling.
3. Centrifugal action.
4. Diffusion.
5. Mass attraction or the effect of van der Waals forces.
6. Electrostatic attraction.

Straining action consists of intercepting particles that are larger than the free interstices left between the filtering sand grains. Assuming spherical grains, an evaluation of the interstitial size is made on the basis of the grains diameter (specific diameter), taking into account the degree of non-homogeneity of the grains.

Porosity constitutes a important criterion in a description based on straining. Porosity is determined by the formula V_L/V_C, in which V_C is the total or apparent volume limitated by the filter wall and V_L is the free volume between the particles. The porosity of a filter layer changes as a function of the operation time of the filters. The grains become thicker because of the adherence of material removed from the water, whether by straining or by some other fixative mechanism of particles on the filtering sand. Simultaneously the interstices between the grains diminish in size. This effect assists the filtration process, in particular for slow sand filters, where a deposit is formed as a skin or layer of slime that has settled on the bed making up the active filter. Biochemical transformations occur in this layer as well, which are necessary to make slow filters efficient as filters with biological activity.

Filtration occurs correctly only after build-up of the sand mass. This formation includes a 'swelling' of the grains and, thus, of the total mass volume, with a corresponding reduction in porosity. The increases and swellings are a result of the formation of deposits clinging to the empty zones between grains.

The porosity of a filter mass is an important factor. This property is best defined by experiment. A general rule of thumb is that for masses with the effective size greater than 0.4–0.5 mm and a specific maximum diameter below 1.2 mm the porosity is generally between 40 and 55 per cent of the total volume of the filter mass. Layers with spherical grains are less porous than those with angular material.

The second important mechanism in filtration is that of settling. From Stoke's law of laminar particle settling, the settling velocity of a particle is given by:

$$U = \frac{1}{18} \frac{g}{v} \frac{\Delta \rho}{\rho} D^2 \qquad\qquad \text{... (10.1)}$$

where,

ρ	=	volumetric mass density of the water
$\rho, \Delta\rho$	=	Volumetric mass density of the particles in suspension
D	=	diameter of the particles
g	=	9.81 m/s^2
v	=	kinematic viscosity (e.g., 10^{-4} m/s at 20°C)

In sedimentation zones the flow conditions are laminar. A place is available for the settling of sludges contained in the water to be filtered.

Although the total inner surface that is available for the formation of deposits in a filter sand bed is important, only a part of this is available in the laminar flow zones that promote the formation of deposits. Usually material with a volumetric mass slightly higher than that of water is eliminated by sedimentation during filtration. Such matter could be, for example, organic granules or particles of low density. In contrast, colloidal material of inorganic origin—sludge or clay, for instance—with a diameter of 1–10 μm is only partially eliminated by this process, in which case the settling velocities in regard to the free surface become insufficient for sedimentation. The trajectory followed by water in a filter mass it is not linear. Water is forced to follow the outlines of the grains that delineate the interstices. These changes in direction are also imposed on particles in suspension being transported by the water. This effect leads to the evacuation of particles in the dead flow zones. Centrifugal action is obtained by inertial force during flow, so the particles with the highest volumetric mass are rejected preferentially.

Diffusion filtration is another contributor to the process of sand filtration. Diffusion in this case is that of Brownian motion obtained by thermal agitation forces. This compliments the mechanism in sand filtration. Diffusion increases the contact probability between the particles themselves as well as between the latter and the filter mass. This effect occurs both in water in motion and in stagnant water and is quite important in the mechanisms of agglomeration of particles (e.g., flocculation).

The next mechanism to consider is the mass attraction between particles which is due to van der Waals forces. These are universal forces contributing to the transport and fixation mechanism of matter. The greater the inner surface of the filters, the higher is the probability of attractive action. van der Waals forces imply short molecular distances and generally play minor role in the filtration process. Moreover, they decrease very quickly when the distance between supports and particles increases. Nevertheless, the indirect effects, which are able to provoke an agglomeration of particles and, thus, a kind of flocculation, are not to be neglected and may become predominant in the case of flocculation-filtration or more generally in the case of filtration by flocculation. Electrostatic and electrokinetic effects are also factors contributing to the filtration process. Filter sand has a negative electrostatic charge. Microsand in suspension presents an electrophoretic mobility. The value of the electrophoretic mobility or of the corresponding zeta potential, depends on the pH of the surrounding medium. Usually a coagulation aid is used to condition the surface of microsand. In filtration without using coagulant aids, other mechanisms may condition the mass more or less successfully. For instance, the formation of deposits of organic matter can modify the electrical properties of the filtering sand surfaces. These modifications promote the fixation of particles by electrokinetic and electrostatic processes, especially coagulation. Also, the addition of a neutral or indifferent electrolyte tends to reduce the surface potential of the filtering sand by compression of the double electric layer. This is based on the principles of electrostatic coagulation.

The sand, as the carrier of a negative charge spread over the surface of the filter according to the model of the double layer, will be able to fix the electropositive particles more exhaustively. This has a favourable effect on the efficiency of filtration of precipitated carbonates or of flocs of iron or aluminium hydroxide-oxide. Optimal adherence is obtained at the isoelectric point of the filtrated material. In contrast, organic colloidal particle carriers of a negative charge such as bacteria are repulsed by the electrostatic mechanism in a filter with a fresh filter mass. In this case, the negative charges of the sand itself appear unchanged. With a filter that is conditioned in advance, there are sufficient positively charged sites to make it possible to obtain an electrochemical fixation of the negative colloids.

BED REGENERATION

In addition to washing the bed, a degraded mass containing agglomerates or fermentation zones (referred to as mud balls) can be regenerated by specific treatment techniques. Among the regeneration techniques that are usually used are sodium chloride, regeneration through application of chlorine and treatment with potassium permanganate, hydrogen peroxide or caustic soda. Cleaning methods based on the use of caustic soda are aimed at eliminating thin clay, hydrocarbons and gelatinous aggregates that form in filtration basins. After the filter has been carefully washed with air and water or only with water, according to its specific operating scheme, a quantity of caustic soda is spread over a water layer approximately 30 cm thick above the filter bed. The solution is then diffused in the mass by slow infiltration. After about 6 to 12 hours, the filter is washed very carefully. Sodium chloride is used specifically for rapid filters. The cleaning solution is spread in solution form in a thin layer of water above the freshly washed sand bed. After 2 or 3 hours of stagnation, slow infiltration in the mass is achieved by opening an outlet valve for the filtered water. The brine is then allowed to work for about a 24 hour period. The filter is placed back into service after a thorough washing. Sodium chloride works on proteinic agglomerates, which are bacterial in origin.

The use of potassium permanganate ($KMnO_4$) is applied to filters clogged with algae. A concentrated solution containing potassium permanganate is spread at an effective concentration over the surface of the filters to obtain, a characteristic pink-purple colour on the top of the mass and allowed to infiltrate the bed for a 24 hour period. After this operation, the filter is carefully washed once again.

Hydrogen peroxide is typically used in the range of 10 to 100 ppm. The cleaning method is similar to that used for permanganate. The addition of phosphates or polyphosphates makes it easier to remove ferruginous deposits. This method can be used *in situ* for surging the isolation sands of the wells. Adjunction of a reductor as bisulphite can be useful to create anaerobic conditions for the elimination of nematodes and their eggs when a filter has been infected. Hydrochloric acid solution is applied to the recurrent cleaning of rapid filters for sand, iron and manganese removal. This operation has the advantage of causing the formation of chlorine *in situ* which acts as a disinfectant. Instantaneous cleaning of a filtering sand bed can be accomplished by the use of chlorine. A water layer is typically used as a dispersion medium. Further infiltration of the solution is obtained by percolation into the bed. The action goes on for several hours, after which the filter is washed. Chlorine is used from concentrated solutions of sodium hypochlorite. An alternative method involves the application of dioxide. This method has the advantage of arresting the formation of agglomerates of biological origin by permanent treatment of the filter wash water with chlorine.

Flocculation and Filtration Together

The sand filtration process is normally comprised of a clarification chain including other unit operations which precede filtration in the treatment sequence and cannot be conceived of completely independent of the filtration stage. The conventional treatment scheme consists of coagulation-flocculation-settling followed by filtration. When the preceding process, in this case flocculation and/or settling, becomes insufficient, subsequent rapid filtration can be used to ensure a high quality of the effluent treated. However, this action is achieved at the expense of the evolution of filter head loss. Problems in washing and cleanliness of the mass may arise. Filtration is often viewed as serving as a coagulant flocculator. This is referred to as flocculation-filtration. The presence of thin, highly electronegative colloids (e.g., activated carbons) introduced in the form of powder in the settling phase may be a problem for the quality of the settled effluent. The carbon particles, which are smaller than 50 μm, penetrate deeply into the sand filter beds. They may rapidly provoke leakage of rapid filters. The same holds for small colloids other than activated carbon. Activated silica, which may have a favourable or an unfavourable effect on filtration, is composed of ionised micella formed by polysilicic acid-sodium polysilicate. This become negatively charged colloidal micella. The behaviour of activated silicas depends on the conditions of neutralisation and the grade of the silicate used in the preparation of the material. Activated silica is a coagulant aid that contributes to coalescence of the particles. Hence, it brings about an improvement in the quality of settled or filtrated water, depending on the point at which it is introduced. Pre-conditioning of the sand surface of filters by adding polyelectrolytes is an alternative use of sand filters as a coagulator-flocculator. In the treatment of drinking water the method depends on the limitations of these products in foodstuffs. The addition of polyphosphates to a water being subjected to coagulation usually has a negative effect; specifically the breaking of the agglomeration velocity of the particles during flocculation will occur in sand filtration. The addition of polyphosphates simultaneously with phosphates can be of value in controlling corrosion. This sometimes makes it possible to avoid serious calcium carbonate precipitation at the surface of filter grains when handling alkaline water. The application concerns very rapidly incrusting water while maintaining high hardness in solution. The addition of polyphosphates involves deeper penetration of matter into the filter mass. Hence, the breaking of flocculation obtained by the action of polyphosphates enables the thinner matters to penetrate the filters more deeply. These products favour the 'in-depth effects' of the filter beds. Their use necessitates carefully checking that they are harmless from a hygienic point of view. The depth penetration of material in coagulation-filtration is almost opposite to the concept of using the filter as a screen. Precipitation initiated by germs plays a significant role. Empirical relations are normally relied on in the design of filters as a function of the penetration in depth of coagulated material. The concentration of those residual matters in filtered water (C_f) depends on several factors: the linear infiltration rate (v_f), the effective size of the filter medium (ES), the porosity of the filter medium (ε), the final loss of head of the filter bed (Δh), the depth of penetration of the coagulated matter (l), the concentration of the particles in suspension in the water to be filtered (C_0) and the water height (H). The following generalised relation is often found among the filtration engineer's notes.

$$C_f = f\left(v_f \times (ES) \times \varepsilon^4 \times \frac{\Delta h}{l} \times C_0 \times H \right) \qquad \qquad ... (10.2)$$

It should be noted that the total loss of head of a filter bed is in inverse ratio to the depth of penetration of the matter in suspension. In a normal waste-water treatment plant, the water is brought onto a series of rapid sand filters and the impurities are removed by coagulation-flocculation-filtration. Backwashing is typically performed in the counterflow mode, using air and water. One type of common filter is illustrated in Fig. 10.8, consisting of closed horizontal pressurised filters.

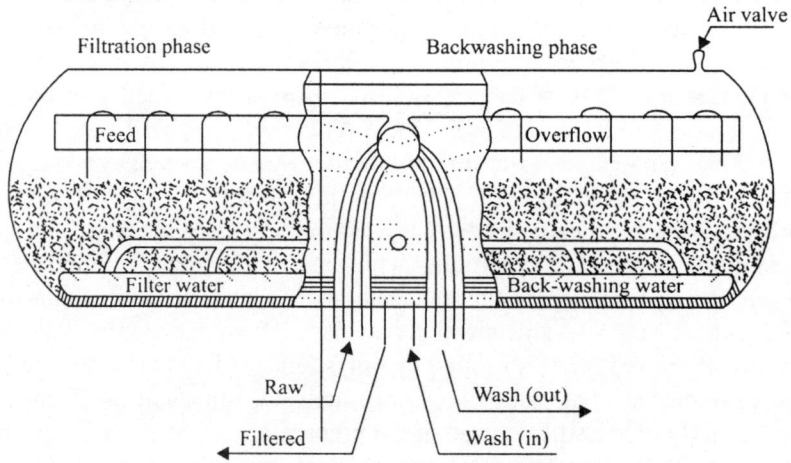

Fig. 10.8. Cross-section of a typical filtration unit.

SLOW SAND FILTRATION

Slow sand filtration involves removing material in suspension and/or dissolved in water by percolation at slow speed. In principle, a slow filter comprises a certain volume of areal surface, with or without construction of artificial containment, in which filtration sand is placed at a sufficient depth to allow free flow of water through the bed. When the available head loss reaches a limit of approximately 1 m, the filter must be pulled out of service, drained and cleaned. The thickness of the usual sand layer is approximately of 1 to 1.50 m, but the formation of biochemically active deposits and clogging of the filter beds takes place in the few topmost centimeters of the bed. The filter mass is pored onto gravels of increasing permeability with each layer having a thickness of approximately 10 to 25 cm. The lower-permeability layer can reach a total thickness of 50 to 60 cm. So-called gravels 18 to 36 cm in size are used and their dimensions are gradually diminished to sizes of 10 to 12 cm or less for the upper support layer. The sand filter must be cleaned by removal of a few centimeters of the clogged layer. This layer is washed in a separate installation. The removal of the sand can be done manually or by mechanical means. The removed sand may not be replaced entirely by fresh sand. Placing pre-conditioned and washed sand is recommended as this takes into account the biochemical aspects involved in slow filtration. An alternative to manual or mechanical removal involves cleaning using a hydraulic system. Sometimes slow filtration is used without previous coagulation. This is generally practiced with water that does not contain much suspended matter. If the water is loaded (periodically or permanently) with clay particles in suspension, pre-treatment by coagulation-flocculation is necessary. Previous adequate oxidation of the water, in this case pre-ozonisation producing biodegradable and metabolisable organic derivatives issuing from dissolved substances, can be favourable because of the biochemical activity in slow filters. There are several disadvantages to the use of slow filters. They may require a significant surface area and volume and may therefore involve high investment costs. They are also not flexible— mainly during the winter, when the open surface of the water can freeze. During the summer, if the filters are placed in the open air, algae may develop, leading to rapid clogging during a generally critical period of use. Algae often cause taste and odour problems, in the filter effluent. Additional construction costs to cover slow filters are often necessary.

RAPID SAND FILTRATION

Rapid filtration is performed either in open gravitational flow filters or in closed pressure filters. Rapid pressure filters have the advantage of being able to be inserted in the pumping system, thus allowing use of a higher effective loading. Note that pressure filters are not subject to development of negative pressure in a lower layer of the filter. These filters generally support higher speeds, as the available pressure allows a more rapid flow through the porous medium made up by the filter sand. Pressure filtration is generally less efficient than the rapid open type with free-flow filtration. Pressure filters have the following disadvantages. The injection of reagents is complicated and it is more complicated to check the efficiency of backwashing. Work on the filter mass is difficult considering the assembly and disassembly required. Also, the risk of breakthrough by suction increases. Another disadvantages is that pressure filters need a longer filtration cycle, due to a high loss of head available to overcome clogging of the filter bed.

Another option is to use open filters, which are generally constructed in concrete. They are normally rectangular in configuration. The filter mass is posed on a filter bottom, provided with its own drainage system, including bores that are needed for the flow of filtered water as well as for countercurrent washing with water or air.

There are several types of washing bottoms. One type consists of porous plates which directly support the filter sand, generally without a layer of support gravel. Even if the system has the advantage of being of simple construction, it nevertheless suffers from incrustation. This is the case for softened water or water containing manganese. Porous filters bottoms are also subject to erosion or disintegration upon the filtration of aggressive water.

The filter bottom is often comprised of pipes provided with perforations that are turned toward the underpart of the filter bottom and embedded in gravel. The lower layers are made up of gravel of approximate diameter 35–40 mm, decreasing up to 3 mm. The filter sand layer, located above this gravel layer, serves as a support and equalisation zone. Several systems of filter bottoms comprise perforated self-supporting bottoms or false bottoms laid on a supporting basement layer. The former constitutes a series of glazed tiles, which includes bores above which are a series of gravels in successive layers.

All these systems are surpassed to some extent by filter bottoms in concrete provided with strainers. The choice of strainers should in part be based on the dimensions of the slits that make it possible to stop the filter sand, which is selected as a function of the filtration goal. Obstruction or clogging occurs only rarely and strainers are sometimes used.

Strainers may be of the type with an end that continues under the filter bottom. These do promote the formation of an air space for backwashing with air. If this air space is not formed, it can be replaced by a system of pipes that provide for an equal distribution of the washing fluids.

Pressure filters are worth noting. These are usually set up in the form of steel cylinders positioned vertically. Another variation consists of using horizontal filtration groups. This has the drawback that the surface loading is variable in the different layers of the filter bed; moreover, it increases with greater penetration in the filter bed (the infiltration velocity is lowest at the level of the horizontal diameter of the cylinder). The filter bottom usually consists of a number of screens or mesh sieves that decrease in size from top to bottom or, as an alternative, perforated plates supporting gravel similar to that used in the filter bottoms of an open filter system.

Filter mass washing can influence the quality of water being filtered. Changes may be consequent to fermentation, agglomeration or formation of preferential channels liable to occur if backwashing is inadequate.

Backwashing requires locating a source that will supply the necessary flow and pressure of wash water. This water can be provided either by a reservoir at a higher location or by a pumping station that pumps treated water. Sometimes an automated system is employed with washing by priming of a partial siphon pumping out the treated water stored in the filter itself. An example is shown in Fig. 10.9. The wash water must have sufficient pressure to assure the necessary flow. Washing of the filter sands is accomplished followed by washing with water and in most cases including a short intermediate phase of simultaneous washing with air and water. Due to greater homogenisation of the filter layer and more efficient washing, the formation of fermentation areas and agglomerates in the filter mass of treatment plants for surface water (mud balls) is diminished. The formation of a superficial crust on the filter sand is avoided by washing with air. After washing with air, water flow is gradually superimposed on the air flow. This operational phase ends at the same time that the wash air is terminated, to avoid the filter mass being blown away. The wash water contains materials that eventually require treatment in a sludge treatment plant. Their concentration varies as a function of the washing cycle. Accounting for the superficial load in filtration, velocity of the wash water and length of the filtration cycle, it may be assumed that the water used for washing will not attain 5 per cent of the total production.

For new installations the first washing cycles result in the removal of fine sand as well as all the other materials usually undesirable in the filter mass, such as particles of bitumen on the inner surface of the water inlet or other residuals from the crushing or straining devices of the filter media.

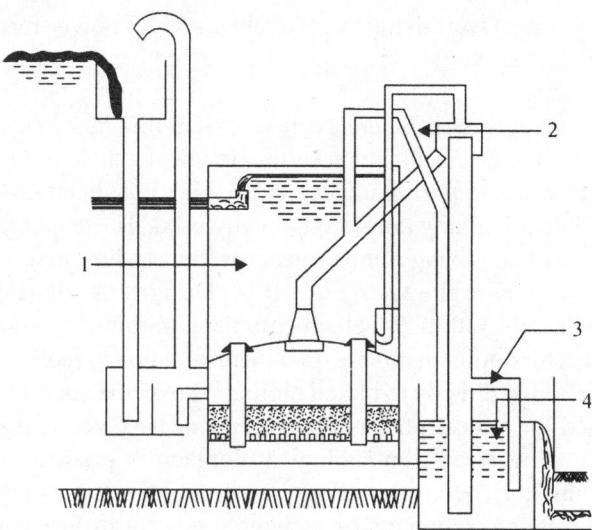

Fig. 10.9. Automatic backwashing filter with a partial siphon system: 1 filtered water (reserve); 2 partial siphoning; 3 initiation; and 4 restitution.

Consequently it is normal that at the beginning of operation of a filter sand installation, dark coloured deposits appear at the surface of the filter mass. In the long term they have no consequence and disappear after a few filtration and wash cycles. If, after several weeks of filtration, these phenomena have not disappeared, it will be necessary to examine the filter sand. The elimination of fine sand must stop after 1 or 2 months of activity. If this sand continues to be carried away after the first several dozen washings it is necessary to re-examine the hydraulic criteria of the washing conditions the granulometry of the filter mass and the filter's resistance to shear and abrasion.

CHEMICAL MIXING AND SOLIDS CONTACT PROCESSES

Chemical mixing and flocculation or solids contact are important mechanical steps in the overall coagulation process. Application of the processes to waste-water generally follows standard practices and employs basic equipment. Chemical mixing thoroughly disperses coagulants or their hydrolysis products so the maximum possible portion of influent colloidal and fine supracolloidal solids are absorbed and destabilised. Flocculation or solids contact processes increase the natural rate of contacts between particles. This makes it possible, within reasonable detention periods, for destabilised colloidal and fine supracolloidal solids to aggregate into particles large enough for effective separation by gravity processes or media filtration.

These processes depend on fluid shear for coagulant dispersal and for promoting particles contacts. Shear is most commonly introduced by mechanical mixing equipment. In certain solids contact processes shear results from fluid passage upward through a blanket of previously settled particles. Some designs have utilised shear resulting from energy losses in pumps or at ports and baffles.

Chemical Mixing

Chemical mixing facilities, should be designed to provide a thorough and complete dispersal of chemical throughout the waste-water being treated to insure uniform exposure to pollutants which are to be removed. The intensity and duration of mixing of coagulants with waste-water must be controlled to avoid overmixing or undermixing. Overmixing excessively disperses newly-formed floc and may rupture existing waste-water solids. Excessive floc dispersal retards effective flocculation and may significantly increase the flocculation period needed to obtain good settling properties. The rupture of incoming waste-water solids may result in less efficient removals of pollutants associated with those solids. Undermixing inadequately disperses coagulants resulting in uneven dosing. This in turn may reduce the efficiency of solids removal while requiring unnecessarily high coagulant dosages. In water treatment practice several types of chemical mixing units are typically used. These include high-speed mixers, in-line blenders and pumps and baffled mixing compartments or static in-line mixers (baffled piping sections). An example of a high-speed mixer is shown in Fig. 10.10. Designs usually call for a 10 to 30 seconds detention times and approximately 300 fps/ft velocity gradient. Variable speed mixers are recommended to allow varying requirements for optimum mixing. In mineral addition to biological waste-water treatment systems, coagulants may be added directly to mixed biological reactors such as aeration tanks or rotating biological contactors. Based on typical power inputs per unit tank volume, mechanical and diffused aeration equipment and rotating fixed-film biological contactors produce average shear intensities generally in the range suitable for chemical mixing. Localised maximum shear intensities vary widely depending on the speed of rotating equipment or on bubble size for diffused aeration.

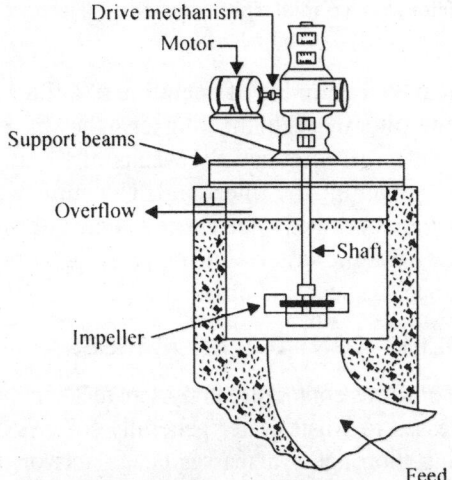

Fig. 10.10. Example of an impeller mixer.

The proper measure of flocculation effectiveness is the performance of subsequent solids separation units in terms of both effluent quality and operating requirements, such as filter backwash frequency. Effluent quality depends greatly on the reduction of residual primary size particles during flocculation, while operating requirements relate more to the floc volume applied to separation units.

Flocculation units should have multiple compartments and should be equipped with adjustable speed mechanical stirring devices to permit meeting changed conditions. In spite of simplicity and low maintenance, non-mechanical, baffled basins are undesirable because of inflexibility, high head losses and large space requirements. Mechanical flocculators may consist of rotary, horizontal-shaft reel units as shown in Fig. 10.11. Rotary vertical shaft turbine units as shown in Fig. 10.12 and other rotary or reciprocating equipment are other examples. Tapered flocculation may be obtained by varying reel or paddle size on horizontal common shaft units or by varying speed on units with separate shafts and drives. In applications other than coagulation with alum or iron salts, flocculation parameters may be quite different. Lime precipitates are granular and benefit little from prolonged flocculation.

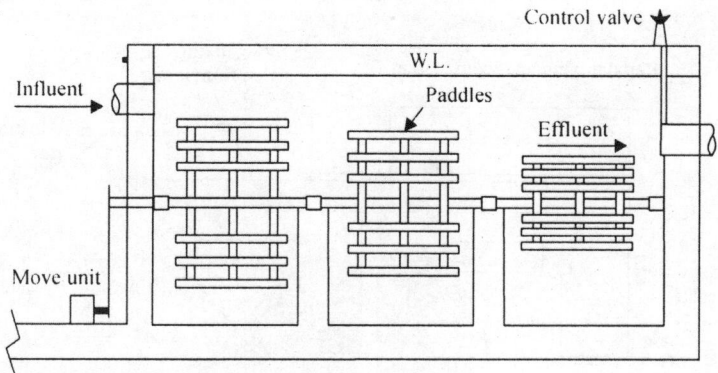

Fig. 10.11. Mechanical flocculation basin horizontal shaft-reel type.

Polymers which already have a long chain structure may provide a good floc at low mixing rates. Often the turbulence and detention in the clarifier inlet distribution is adequate.

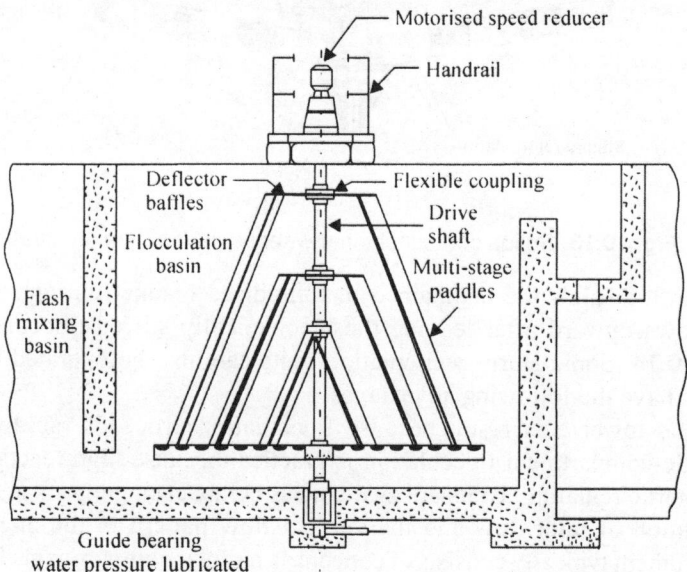

Fig. 10.12. Mechanical flocculator vertical shaft-paddle type.

Solids Contacting

Solids contact processes combine chemical mixing, flocculation and clarification in a single unit designed so that a large volume of previously formed floc is retained in the system. The floc volume may be as much as 100 times that in a 'flow-through' system. This greatly increases the rate of agglomeration from particle contacts and may also speed up chemical destabilisation reactions. Solids contact units are of two general types: slurry-recirculation and sludge-blanket. In the former, the high floc volume concentration is maintained by recirculation from the clarification to the flocculation zone, as illustrated in Fig. 10.13.

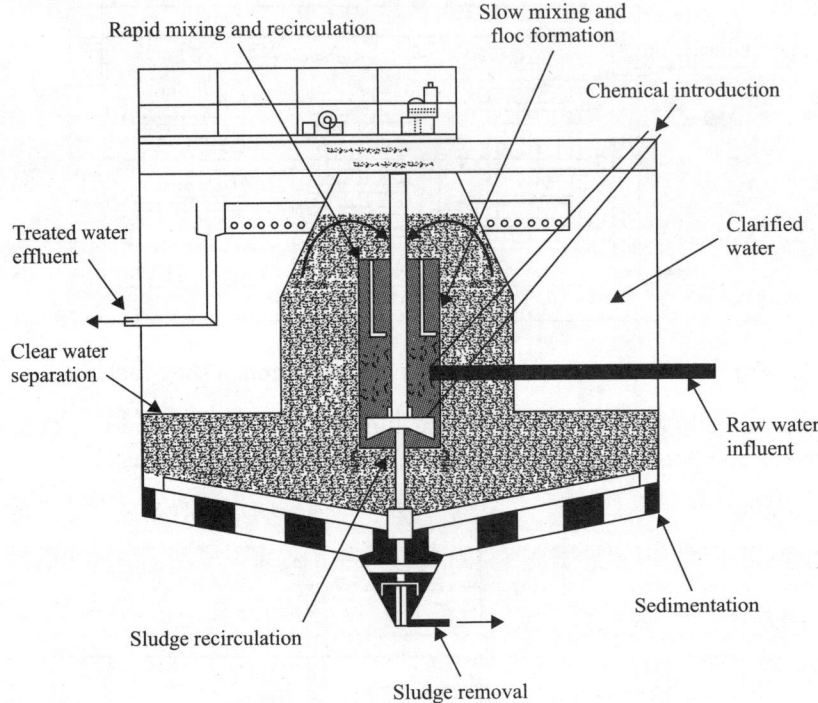

Fig. 10.13. Solids contact clarifier without sludge blanket filtration.

In the latter, the floc solids are maintained in a fluidised blanket through which the waste-water under treatment flows upward after leaving the mechanically stirred-flocculating compartment, as illustrated in Fig. 10.14. Some slurry-recirculation units can also be operated with a sludge blanket. Solids contact units have the following advantages.

Reduced size and lower cost result because flocculation proceeds rapidly at high floc volume concentration. Single-compartment flocculation is practical because high reaction rates and the slurry effects overcome short circulating. Units are available as compact single packages, eliminating separate units; even distribution of inlet flow and the vertical flow pattern in the clarifier improve clarifier performances. Equipment typically consists of concentric circular compartments for mixing, flocculation and settling. Velocity gradients in the mixing and flocculation compartments are developed by turbine pumping within the unit and by velocity dissipation at baffles. For ideal flexibility it is desirable to independently control the intensity of mixing and sludge scraper drive speed in the different compartments.

Operation of slurry-recirculation solids contact units is typically controlled by maintaining steady levels of solids in the reaction zone. Design features of solids contact clarifiers should include:

1. Rapid and complete mixing of chemicals, feedwater and slurry solids must be provided. This should be comparable to conventional flash mixing capability and should provide for variable control, usually by adjustment of recirculator speed.

2. Mechanical means for controlled circulation of the solids slurry must be provided with at least a 3:1 range of speeds. The maximum peripheral speed of mixer blades should not exceed 6 ft/sec.
3. Means should be provided for measuring and varying the slurry concentration in the contacting zone up to 50 per cent by volume.
4. Sludge discharge systems should allow for easy automation and variation of volumes discharged. Mechanical scraper tip speed should be less than 1 fpm with speed variation of 3:1.
5. Sludge-blanket levels must be kept a minimum of 5 feet below the water surface.
6. Effluent launders should be spaced so as to minimise the horizontal movement of clarified water.

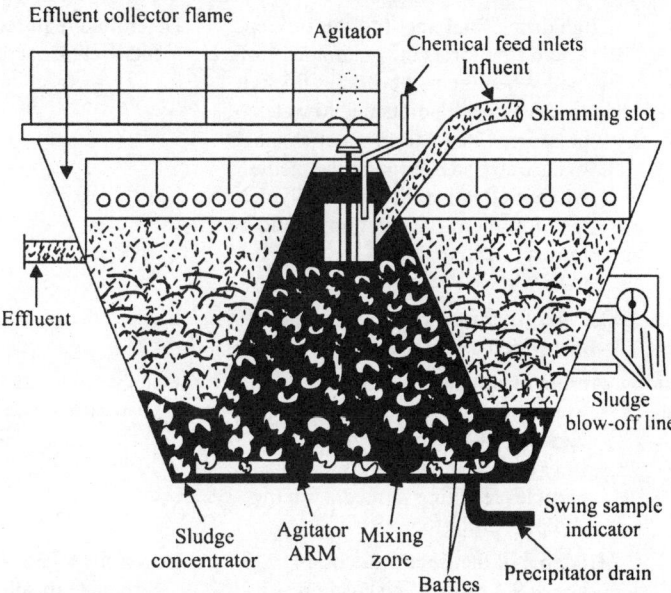

Fig. 10.14. Solids contact clarifier with sludge blanket filtration.

Further considerations include skimmers and weir overflow rates. Skimmers should be provided on all units since even secondary effluents contain some floatable solids and grease. Overflow rates and sludge scraper design should conform to the requirements of other clarification units.

Table 10.2. Filtration problems and their solutions.

Problems	Possible causes	Possible solutions
Mud balls are in a sand filter	High level of $CaCO_3$ in water. Calcium carbonate forms aggregates in the surface mat	Lower the pH (below 9.3) of the influent water to convert $CaCO_3$ into soluble calcium bicarbonate. Also add 0.5–0.75 mg/l sodium hexametaphosphate to the filter influent to keep calcium in solution
	Abrupt high rate of washing. Mud balls are formed by faulty distribution of wash water caused by abrupt and quick	Start the filter washing by surface agitation and gradually increasing the wash water flow

(Contd...)

Problems	Possible causes	Possible solutions
	opening of the wash water valve. It causes openings in the filter bed by pushing the media and letting some of the surface mat aggregates of $CaCO_3$ fall below the media's surface. They grow close to 1″ in size and can cause clogging of filter and high effluent turbidity	
Cracks are in surface mat of a sand filter	High dose of polymer to filter influent. Micro floc forms the gelatinous surface mat. When the mat becomes thick, it cracks at weak points due to water pressure. Cracks form water channels to let water pass through the media without effective filtration. Polymer feed for micro flocculation should be lowest effective dose—generally 0.25 to 0.75 mg/l.	Determine required polymer dose by jar test and decrease it as required
	Longer filter runs can cause a thicker mat which can crack at thinner places	Shorten the filter run adequately
There is jet action and sand boiling in a mixed media filter	Abrupt opening of the backwash water valve. Fast movement of backwash water in part of the filter bed can cause boiling of media particles resulting in media mixing and poor filtration	Open wash water valve slowly and partially for first few minutes of the washing
There is high turbidity in filter effluent of a conventional treatment plant when all other phases have no problems	High pH of the filter influent: pH above 9.4 causes fine calcium carbonate ($CaCO_3$) particles to go through filters to cause a high turbidity reading	Lower pH to 9.3 which will convert calcium carbonate to soluble calcium bicarbonate. Remember there is no insoluble calcium carbonate below pH 9.3.
	Overfeed of a polyphosphate: A dose of a polyphosphate such as sodium hexametaphosphate over 0.75 mg/l to the filter will cause excessive removal of calcium carbonate coating of the filter media which results in a high turbidity reading	Run a jar test; determine optimum dose of polyphosphate and apply. Generally, 0.5 mg/l dose is adequate
	Overfeed of a polymer. Overdose of a polymer to the filter influent will cause cracks and channels in the media for the turbidity particles to go through without filtration	Check polymer dose to the filter influent. Usually, a dose above 0.5 mg/l can cause a high turbidity reading. Run jar test; determine optimum polymer dose and apply
	Air bubbles in the turbidity measuring cell. There is a higher concentration of dissolved gases	Warm the sample to remove gases before taking the turbidity reading

(Contd...)

Problems	Possible causes	Possible solutions
	in the water, especially in winter, which start coming out as gas bubbles at room temperature	
	Scratched or smudged sample cell	Always use a scratch-free sample cell since scratches will also give a false high turbidity reading. Wipe the cell with soft tissue paper
Surface sweeps are being buried under the media	High calcium carbonate content. High calcium carbonate deposition on media particles will cause the swelling of media	Lower the pH below 9.3 and apply small amount of polyphosphate as previously discussed. The amount of swelling can be determined by washing a sample of media with 5 per cent hydrochloric acid. The difference of the dried sample weight or volume before and after washing, times 100, is the per cent of swelling in weight and volume, respectively. It may require the acid washing of the whole media to reduce its volume. This will also improve the filter performance
There is biofouling of a membrane	High bacterial count in feed water	Check bacterial count in pre-treated water and reduce it by proper pre-disinfection
	Membrane has been out of service too long. Some bacteria in and on a wet membrane will start to multiply if the membrane is out of service too long	Flush membrane with filtered water and sanitise it as recommended by the manufacturer
There is chemical scaling of the membrane	A high metallic (calcium, magnesium and iron) salt content. These metals will become concentrated in the feed water during filtration and precipitate	Use an antiscalant such as sodium hexametaphosphate or polyacrylate as recommended by the manufacturer
	pH of feed water is too high. This will cause more deposition of metallic salts	Lower the pH to dissolve the scaling substances

Example 10.1: Find the area of slow sand filters required for a town having a population of 15000 with an average rate of demand as 160 liters per head per day.

Solution:

Maximum daily demand $= (15000 \times 160 \times 1.50)$

$$= 36,00,000 \text{ liters.}$$

Assuming a rate of filtration as 150 liters per hour per m^2 of filter area,

$$\text{Area of filter required} = \frac{36,00,000}{150 \times 24} = 1000 \text{ m}^2.$$

Let the size of one unit be 16.00 m × 12.50 m. Then provide 6 such units of slow sand filters including one unit as standby. The units may be arranged in series with 3 units on either side.

Example 10.2: Find the area of rapid sand filters required for a town having a populatiion of 80000 with an average rate of demand as 200 liters per head per day.

Solution:

$$\text{Maximum daily demand} = (80000 \times 200 \times 1.50)$$
$$= 2,40,00,000 \text{ liters.}$$

Assuming a rate of filtration as 5000 liters per hour per m² of filter area,

$$\text{Area of filter required} = \frac{2,40,00,000}{5000 \times 24} = 200 \text{ m}^2.$$

Let the size of one unit be 8 m × 5 m. Then provide 6 such units of rapid sand filters including one unit as stand-by. The units may be arranged in series with 3 units on either side.

Example 10.3: A filter unit has a surface area of 50 ft². The flow rate through the filter is 0.25 mgd. Compute the filtration rate and the velocity at which the water flows through the filter bed.

Solution: First, convert the flow rate to gallons per minute, as follows:

$$2,50,000 \text{ gal/d} \times \frac{1 \text{ day}}{24 \text{ hours}} \times \frac{1 \text{ hour}}{60 \text{ minutes}} = 174 \frac{\text{gallons}}{\text{minutes}}$$

Then,

$$\text{Filtration rate} = \frac{174 \text{ gpm}}{50 \text{ ft}^2} = 3.5 \text{ gpm/ft}^2$$

$$\text{Velocity} = 3.5 \frac{\text{gal}}{\text{ft}^2 \cdot \text{minute}} \times \frac{1 \text{ ft}^3}{7.5 \text{ gal}} = 0.47 \text{ ft/minute}$$

Example 10.4: A square filter box is to be designed for a filtration rate of 2.8 l/m²·s. What are the required surface area and side dimension of the unit if the flow rate is 6 ml/d?

Solution: The flow rate in terms of liters per second is

$$6 \times 10^6 \text{ l/d} \times 1 \text{ d}/ 24 \text{ hr.} \times 1 \text{ hr.}/3600 \text{ s} = 69.4 \text{ l/s}$$

Since filtration rate = flow rate/area,

$$\text{Area} = \text{flow rate/filter rate}$$
$$= 69.4 \text{ l/s} \div 2.8 \text{ l/m}^2 \cdot \text{s} = 25 \text{ m}^2$$

and

$$\text{Side dimension} = \sqrt{25 \text{ m}^2} = 5 \text{ m}$$

Chapter 11

Flocculation, Adsorption, Desalination and Ion Exchange

INTRODUCTION

The treatment processes described in this chapter are, in a sense, both physical and chemical operations, because they affect both the physical and the chemical composition of the water subjected to them. There is flocculation, for example, when small, often colloidal, particles conjoin as a result of interparticle contact while the suspending fluid is being stirred or otherwise submitted to hydraulic shear; there is adsorption when particles of colloidal or molecular size are adsorbed on surfaces of activated carbon added to water as a powder or forming the granules of fixed or fluidised beds and exposing relatively immense interfacial areas to the carrying, suspending or filtering water; there is desalination when salt is separated physically from water by induced passage through a selectively permeable membrane and when water is separated physically from salt by thermal conversion to the vapour or solid phase or by passage through a permeable membrane and there is demineralisation by ion exchange when ions are transported from water to solid ion-exchange media in fixed or fluidised beds.

FLOCCULATION

In recent years flocculators or polyelectrolytes have widely been used. Flocculators are organic high molecular weight compounds comprising many inorganic groups. These groups undergo ionisation when dissolved in water. During the course of ionisation, the radical proper molecule becomes positively charged in some polyelectrolytes (cation exchange flocculators) and negatively charged in others (anion exchange flocculators). The two important flocculators are polyacrylamide (weak anion exchange flocculator) and BA-2 flocculator (cation exchange type). When polyacrylamide is introduced into water already containing coagulant flakes or precipitating suspension, the flakes consolidate and gain weight and so settle down. Polyacrylamide is available as 8 per cent solution in a gelatinous white mass and dissolves in water, provided that intensive stirring is ensured. It should be noted that polyacrylamide alone is not capable of inducing coagulation of colloids and formation of suspension. It is sorbed by the particles of the suspension, sews them and forms a net structure that helps in the formation of large particles or flocs. These flocculent aggregations are denser and heavier than individual particles and settle down quickly.

The BA-2 polyelectrolyte is employed without any coagulant, because it brings about coagulation of colloids itself. Most of the colloidal particles carry negative surface charge, but molecules of BA-2 are positively charged. Hence neutralisation (mutual) of charges takes place. As a result, a molecule of the flocculator is sorbed by the particles of the suspension being formed along with coarse aggregates and complexes that increases the rate of suspension settling in the coagulation tank. The BA-2 type flocculator is a slightly viscous 11 per cent concentration liquid. This type of flocculator can also be used successfully in treating coloured water, if taken in increased dosages.

A combination of mixing and stirring or agitation that produces aggregation is called flocculation, even though more specific terms such as coagulation and thickening may be preferred by physical chemists and chemical engineers to identify the origin of the particulates concerned. Thus, the purpose of waste-water flocculation is to form aggregates or flocs from the finely divided matter. Although not used routinely, the flocculation of waste-water by mechanical or agitation may be worthy of consideration when it is desired to: (i) increase the removal of suspended solids and BOD in primary settling facilities; (ii) condition waste-water containing certain industrial wastes; and (iii) improve the performance of secondary settling tanks following the activated-sludge process.

When used, flocculation can be accomplished: (i) in separate tanks or basins specifically designed for the purpose; (ii) in in-line facilities such as the conduits and pipes connecting the treatment units; and (iii) in combination flocculator-clarifiers. Paddles for mechanical agitation should have variable-speed drives permitting the adjustment of the top paddle speed downward to 30 per cent of the top value. Similarly, where air flocculation is employed, the air supply system should be adjustable so that the flocculation energy level can be varied throughout the tank. In both mechanical and air-agitation flocculation systems, it is common practice to taper the energy input so that the flocs initially formed will not be broken as they leave the flocculation facilities (whether separate or in-line).

Flocculation is aided by eddying motion of the fluid within the tanks and proceeds through the coalescence of fine particles at a rate that is a function of their concentration and of the natural ability of the particles to coalesce upon collision. As a general rule, therefore, coalescence of a suspension of solids becomes more complete as time elapses. For this reason, detention time is also a consideration in the design of sedimentation tanks. The mechanics of flocculation are such, however, that as the time of sedimentation increases, less and less coalescence of remaining particles occurs.

Mixing is the specific blending, mingling or co-mmingling of coagulating chemicals or materials with water or waste-water in order to create a more or less homogeneous single or multiple-phase system, whereas stirring describes the disturbing of the flow pattern of a fluid in a mechanically orderly way for the purpose of effecting a dynamic redistribution of particles. Random rather than orderly turbulence is distinguished by the term agitation.

Generally speaking, mixing is a brief operation seeking a quick response, often in advance of stirring or agitation, whereas stirring and agitation are more protracted operations normally aiming at the conjunction of suspended particles or flocs, but sometimes intended to break up large flocs in order to maintain particle numbers. Aggregation of particles is referred to as floc growth; break-up of flocs, as floc shear.

Mixing and Stirring Devices

The sources of power for flocculating devices are gravitational, pneumatic, or mechanical. Generally speaking, mechanical and pneumatic devices are relatively flexible in power input; gravitational devices, relatively inflexible. Because of this, gravitational devices are seldom included in large plants, even though they may possess quite useful features.

Baffled channels

Baffled channels are prime examples of gravitational mixing and stirring devices. Most other gravitational devices are relatively inefficient. Baffled channels differ from unobstructed open channels and for that matter, from pipelines in that shear gradients or turbulence are not merely functions of frictional resistance to flow. Velocity gradients are purposely intensified by induced changes in the direction of flow (Fig. 11.1). For baffled channels of capacity C, in which a loss of head h is incurred when the rate of flow is Q, the useful power input $P = Q\rho gh$, where ρg is the weight density of water. Accordingly the permissible channel loading at a given value of Gt_d is:

$$Q/C = \sqrt{Q\rho gh/\mu C}/(Gt_d) = \sqrt{Qgh/vC}/(Gt_d) \qquad \ldots (11.1)$$

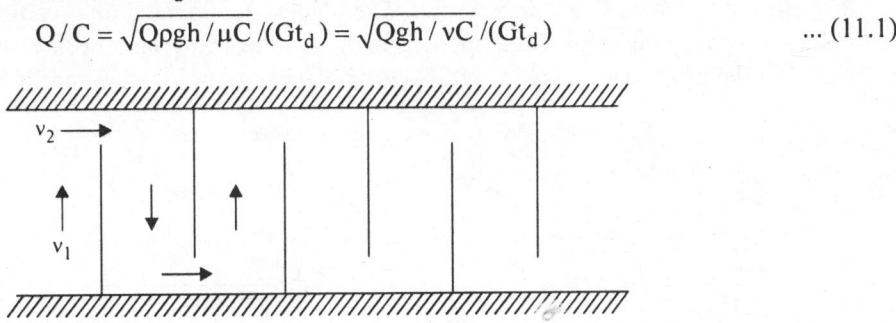

Fig. 11.1. Baffled channel (schematic diagram). Plan of round-the-end baffles or vertical section of over-and-under baffles.

Here g is the gravity constant and v the kinematic viscosity of the fluid. Each foot of lost head is $62.4 \times 1.547/550 = 0.175$ hp per mgd or $62.4 \times 1.547/737.6 = 0.131$ kw per mgd. In practice, head losses commonly lie between 0.5 and 2 ft, velocities vary from 0.5 to 1.5 fps and detention times run from 10 to 60 minutes. For $(n-1)$ equally spaced over-and-under or around-the-end baffles and for velocities v_1 and v_2 in the channels and baffle slots, respectively, the loss of head approaches $nv_1^2/2g + (n-1)v_2^2/2g$ in addition to normal channel friction. In arriving at this estimate the assumption is made that necessary velocities must be re-developed at each change in direction of flow (Fig. 11.1).

Pneumatic mixing and stirring

When air is injected or diffused into water after suitable compression, it normally expands isothermally. Accordingly, the work done by the air is $\int p \, dV$, where p is the absolute pressure intensity and V is the volume of air. Because pV is constant ($p_a V_a$, for example), $V_a \int_{vc}^{va} dV/V = p_a V_c \ln(V_a/V_c) = p_c V_a \ln(p_c/p_a)$, where the subscripts a and c denote free or atmospheric conditions and compressed conditions, respectively. If, for example, Q_a cu ft per minute of free air are injected into water from a diffuser situated h ft below the water surface, the power dissipated usefully by the rising air bubbles is essentially $P = (14.7 \times 144 \times 2.303/60) Q_a \log[(h+34)/34]$ ft-lb per second or:

$$P = 81.5 \, Q_a \log[(h+34)/34] \qquad \ldots (11.2)$$

Compressed air is diffused into treatment units also in aeration for gas exchange, in cleaning granular filters by air scour and in aerating and stirring activated-sludge units.

Mechanical mixing and stirring

The impellers employed in mechanical mixing and stirring generate both mass flow and turbulence. Three types are in common use: (i) paddles; (ii) turbines; and (iii) propellers.

Paddles consist of blades attached directly to vertical or horizontal shafts (Fig. 11.2c,d). The moving blades (rotors) may be complemented by stationary blades (stators) that oppose rotational movement of the entire mass of water within the treatment unit and help to suppress vortex formation. However, stators are not often used in water or waste-water treatment practice. Paddles are rotated at slow to moderate speeds of 2 to 15 rpm. The currents generated by them are both radial and tangential.

Turbines comprise flat or curved blades attached by a connecting radius arm to a vertical or horizontal shaft. Operating in the middle range of speeds (10 to 150 rpm), they generate much the same kind of currents as do paddles (Fig. 11.2a).

Propellers are shaped like ship's screws. The blades are mounted on a vertical or inclined shaft and generate strong axial currents. Their speed is high—150 to 1500 rpm or more—and they may be placed off center in the treatment unit (Fig. 11.2b). Propellers are employed primarily in flash mixers.

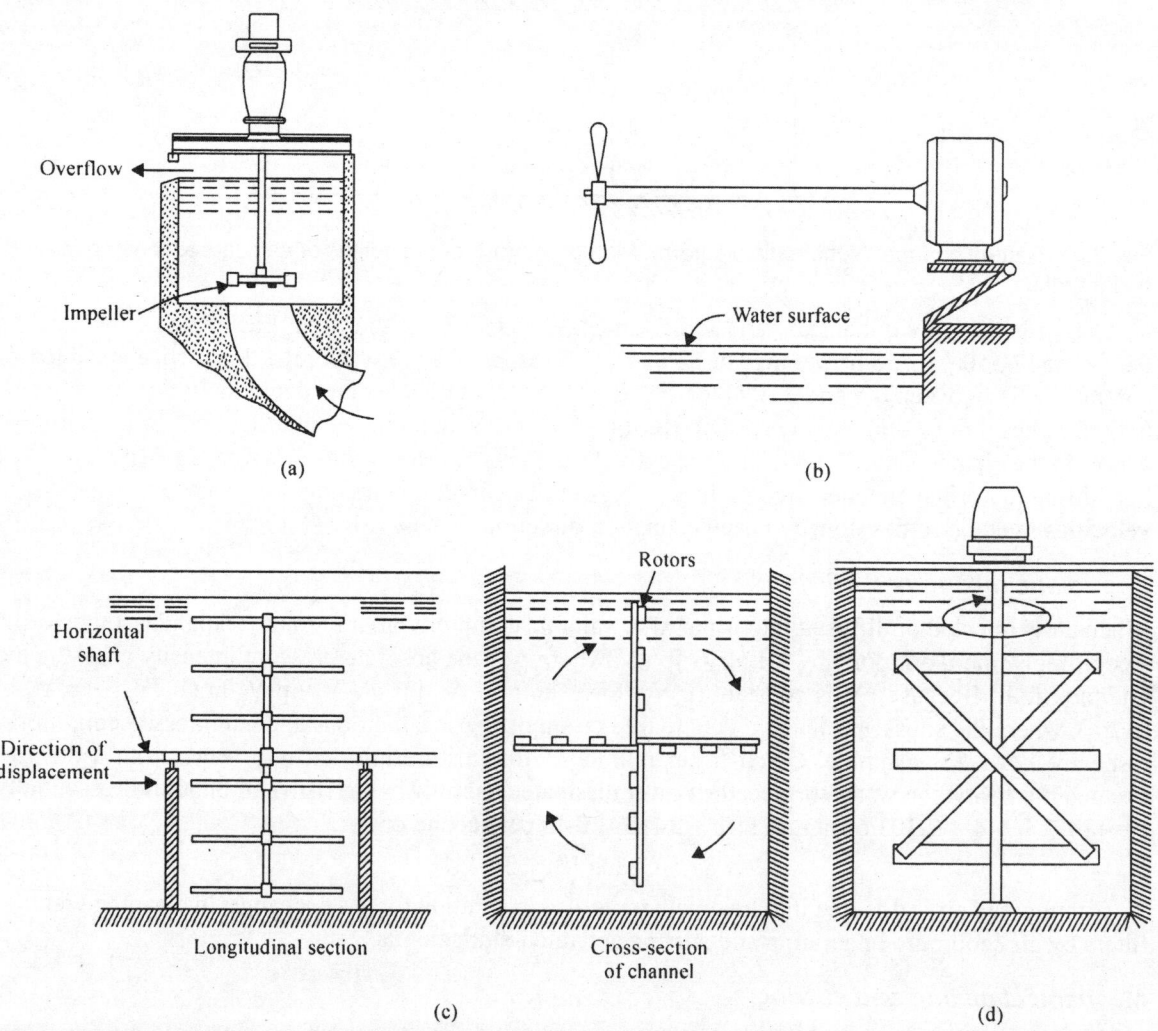

Fig. 11.2. Mixing and stirring impellers. (a) Flash mixer; (b) mixing propeller tilted into horizontal position; (c) paddle or blade mixer with horizontal shaft; and (d) paddle mixer with vertical shaft.

Paddle flocculators, which are most widely used in modern plants, may be arranged so that the flow is parallel to horizontal shafts installed lengthwise in parallel or sequential basins or at right angles to vertical shafts or horizontal shafts installed across the width of one or more tanks or compartments.

The useful power input for mixing and stirring is lowered by the rotational movement of water masses as a whole and by vortex formation. Shear gradients become smaller because the velocity differential between the impeller and the water is narrowed and power expended in changing water levels by vortex formation is not put to use for mixing purposes. Stators are useful adjuncts to all types of impellers.

The useful power input of an impeller is a function of the impelling force F_I and coefficient of drag C_D, that is, the drag force, $F_D = C_D F_I$ of the paddle, blade or propeller, the stator, if any and the tank wall; the relative velocity v of the impeller and the fluid and the area A of the impeller blade (Fig. 11.3). The useful power input[*] $P = F_D v = C_D F_I v$, where $F_I = \rho A v^2 / 2$. Therefore:

$$P = \tfrac{1}{2} C_D \, \rho A v^3 \qquad \qquad \text{... (11.3)}$$

If k is the ratio of the fluid velocity to the impeller velocity (v_i), the relative velocity of the blade $v = v_i - k v_i = (1 - k) v_i = 2\pi (1 - k) \, rn/60$. Here r is the effective radius arm of the blade and n is the number of revolutions per minute. It follows that the useful power expended by a single blade is:

$$P = 5.74 \times 10^{-4} \, C_D \, \rho [(1 - k) \, n]^3 \, r^3 A \qquad \qquad \text{... (11.4)}$$

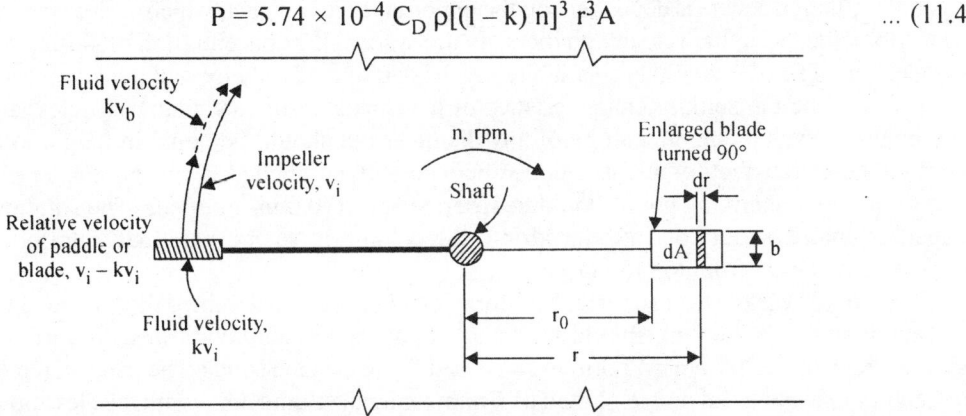

Fig. 11.3. Velocity and power relationships of mechanical mixers or stirrers (schematic diagram).

If the dimension of A in the direction of the radius arm is substantial, but the width, b, of the blade is constant, Fig. 11.3 shows that $r^3 A = \int_{r0}^{r} r^3 \, dA = b \int_{r0}^{r} r^3 \, dr = \tfrac{1}{4} b (r^4 - r_0^4)$ because $dA = b \, dr$. Moreover, if the impeller includes a series of blades, $r^3 A$ is replaced by $\Sigma \, r^3 A$. In these terms, therefore,

$$P = 1.44 \times 10^{-4} \, C_D \, \rho [(1 - k) \, n]^3 \, b \, \Sigma \, (r^4 - r_0^4) \qquad \qquad \text{... (11.5)}$$

In flocculation practice, peripheral speeds of paddles range from 0.3 to 3 fps, k is about 0.25 in the absence of stators and C_D is approximately 1.8 for flat plates. At n = 1 to 5 rpm or more, paddles 8 ft in diameter have Reynolds numbers of 7.57×10^4 to 3.79×10^5 and the power number is directly proportional to G^2. Accordingly, G becomes as meaningful as P. G and Gt values have been tapered from high to low for best results in specific cases. Wherever possible the power dissipation function P/C should be determined by measuring the actual torque and rotor speed.

---

[*] Because power input identifies only geometric and kinematic similarity, chemical engineers prefer to express mixing and stirring relationships by a dimensionless power number $P = 2.16 \times 10^5 P/(32\rho n^3 r^5) = 6.75 \times 10^3 P/(\rho n^3 r^5) = K R^p F^q$, where K, p and q are coefficients changing with the conditions of flow, P is the power input and R and F are the Reynolds and Froude numbers, respectively.

Flocculator performance

The performance of mixing and stirring devices for flocculation is a function of: (i) the substances to be removed; and (ii) the additives, if any, introduced to promote coagulation. Examples of common operational practices in successful chemical coagulation are, in the simplest cases, routine additions of standard amounts of chemicals; in more closely supervised cases, daily, hourly or even more frequent adjustments of coagulant dosages to the results obtained in jar tests.

However, if comparative performance is to be identified, analysis must turn in a different direction, at the hand of the fundamental equations of purification or process kinetics $dy/dt = k\phi(y)$, where dy/dt is the rate of performance, y the property, quality or manifestation under observation, t the time and k the rate constant, which may depend on a number of factors governing the process kinetics in one way or another.

Analysis of Flocculant Settling

Particles in relatively dilute solutions will not act as discrete particles but will coalesce during sedimentation. As coalescence or flocculation occurs, the mass of the particle increases and it settles faster. The extent to which flocculation occurs is dependent on the opportunity for contact, which varies with the overflow rate, the depth of the basin, the velocity gradients in the system, the concentration of particles and the range of particle sizes. The effects of these variables can be determined only by sedimentation tests.

To determine the settling characteristics of a suspension of flocculant particles, a settling column may be used. Such a column can be of any diameter but should be equal in height to the depth of the proposed tank. Satisfactory results can be obtained with a 6 inch (150 mm) diameter plastic tube about 10 ft (3 m) high. Sampling ports should be inserted at 2 ft (0.6 m) intervals. The solution containing the suspended matter should be introduced into the column in such a way that a uniform distribution of particle sizes occurs from top to bottom.

Care should be taken to ensure that a uniform temperature is maintained throughout the test to eliminate convection currents. Settling should take place under quiescent conditions. At various time intervals, samples are withdrawn from the ports and analysed for suspended solids. The per cent removal is computed for each sample analysed and is plotted as a number against time and depth, as elevations are plotted on a survey grid. Between the plotted points, curves of equal per cent removal are drawn. A settling column and the results of a sedimentation test are shown in Fig. 11.4. The resulting curves are shown, but the plotted numbers representing the individual samples have been omitted from the figure. Determination of the amount of material removed, using the curve given in Fig. 11.4.

ADSORPTION

Adsorption can be defined as the accumulation of substances at the interface between two phases. In water and waste-water treatment, the interface is between the liquid and solid surfaces that are artificially provided. The material removed from the liquid phase is called the adsorbate and the material providing the solid surfaces is called the adsorbent.

The adsorbent most commonly used in water and waste-water treatment is activated carbon. Activated carbon is manufactured from carbonaceous material such as wood, coal, petroleum residues, etc. A char is made by burning the material in the absence of air. The char is then oxidised at higher temperatures to create a very porous structure. This 'activation' step provides irregular channels and pores in the solid mass, resulting in a very large surface-area-per-mass ratio. Surface areas ranging from 500 to 1500 m^2/g have been reported, with all but a small fraction of the surface area being associated with the pores.

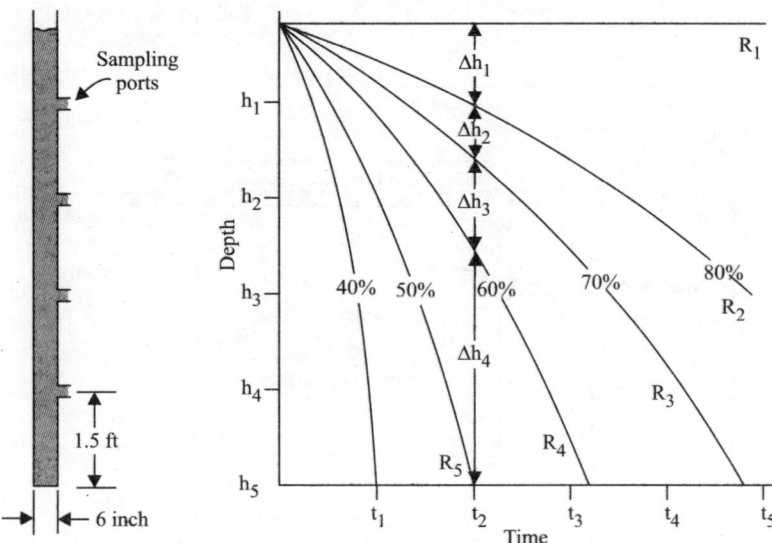

Fig. 11.4. Settling column and settling curves for flocculant particles.

Once formed, activated carbon is crushed into granules ranging from 0.1 to 2 mm in diameter or is pulverised to a very fine powder. Dissolved organic material adsorbs to both exterior and interior surfaces of the carbon. When these surfaces become covered, the carbon must be regenerated. Although adsorption properties and mechanisms are essentially the same, application techniques for granular activated carbon and powdered activated carbon are considerably different.

The contact system for granular activated carbon (GAC) consists of a cylindrical tank which contains a bed of the material (Fig. 11.5). The water is passed through the bed with sufficient residence time allowed for completion of the adsorption process. The system may be operated in either a fixed-bed or moving-bed mode. Fixed-bed systems are batch operations that are taken off the line when the adsorptive capacity of the carbon is used up.

Although fixed granular carbon beds can be cleaned in a place with superheated steam, the most common practice is to remove the carbon for cleaning in a furnace. The regeneration process is essentially the same as the original activation process. The adsorbed organics are first burned at about 800°C in the absence of oxygen. An oxidising agent, usually steam, is then applied at slightly higher temperatures to remove the residue and re-activate the carbon. In a moving-bed system, spent carbon is continuously removed from the bottom of the bed, with regenerated carbon being replaced at the top. Most modern applications use the moving-bed system with a countercurrent flow; that is, the water is introduced at the bottom of the bed and moves upward against the flow of carbon.

The major problem associated with granular-activated-carbon-contact systems is plugging of the bed by suspended solids in the water. Provisions may be made in the design of the vessel for back- washing the bed in a fashion similar to filter backwashing. Other designs avoid plugging by operating with the bed in a fluidised state. Sufficient upflow velocity is provided to maintain the bed at about 10 per cent expansion at all times so suspended solids in the influent can pass through. This mode of operation has an added advantage in that adsorbed organics increase the density of the carbon and the spent carbon migrates to the bottom of the fluidised bed for removal to the regeneration process.

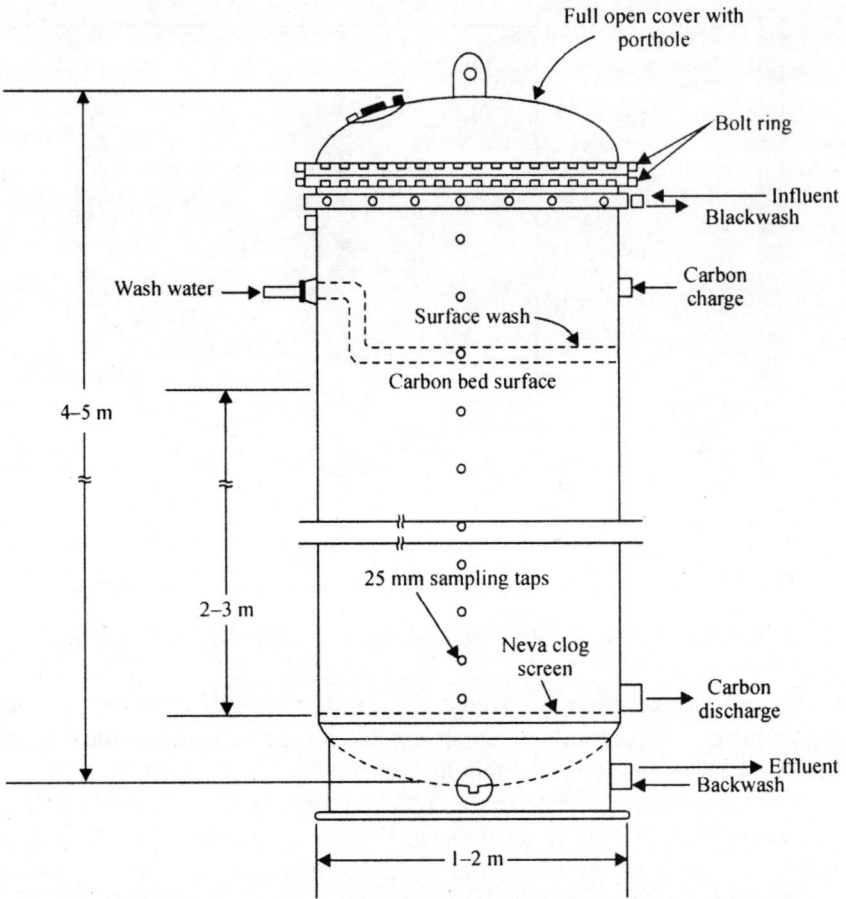

Fig. 11.5. Typical activated-carbon adsorption column.

Design of granular-activated-carbon systems is based on flow rates and contact times. Flow rates of 0.08 to 0.4 m³/m².min and contact times of 10 to 50 minutes, based on empty-tank cross-section and volume, are common practice. Downtime of up to 40 per cent should be included in the plant capacity, with 5 to 10 per cent make-up carbon being provided after each regeneration cycle.

Carbon columns can be arranged in parallel to increase the capacity and in series to increase the contact time. To approximate the countercurrent approach in a series of fixed-bed columns, water proceeds from the column which has been used the longest to the one in use for the shortest time.

Powdered activated carbon (PAC) cannot be used in a fixed-bed arrangement because of its small size and the subsequent high head loss that would result from passing water through it. Powdered activated carbon is contacted with the water in open vessels where it is maintained in suspension for the necessary contact time and then removed by conventional solids-removal processes. Flocculation equipment is sufficient for this purpose. Powdered activated carbon is much more difficult to regenerate than granular. Most systems employ a fluidised bed arrangement in which a mixture of steam and other hot gases holds the carbon in suspension while the regeneration processes occur. In some cases, sand is fluidised along with the carbon to help hold heat in the system.

In waste-water treatment, powdered activated carbon can be added to the aeration basin and removed with the biological solids in the secondary clarifier. In this case, both refractory and biodegradable organics are adsorbed. Biomass growth on the carbon surface utilises the biodegradable fraction. Removal efficiency for biodegradable organics may be improved by this process, but usually at the expense of refractory organic removal efficiency. Use of powdered activated carbon in secondary waste-water systems results in an inseparable mixture of biological solids and carbon. Thermal regeneration of the carbon also results in destruction of the biomass, eliminating the need for other sludge processing and disposal techniques, but increasing the size of the carbon regeneration system.

In current practices, most systems treating potable water use powdered activated carbon, while advanced waste-water systems use moving-bed, granular activated carbon. Better regeneration procedures would greatly enhance the use of powdered activated carbon in waste-water treatment, particularly in the secondary processes.

DESALTING

The removal of dissolved solids from water is described as desalting, desalination or salt-water conversion. The salinity of raw water varies with their origin. The degree of salinity is normally expressed in milligrams per liter of dissolved solids, chloride ion, Cl^- or common salt, $NaCl$. The following classification is based on the water source and its relative saltiness: (i) mildly saline water: brackish mixtures of saline and sweet water with salt concentrations of 1000 to 2000 mg per l of dissolved solids; (ii) moderately saline water: inland water with salt concentrations of 2000 to 10000 mg per l of dissolved solids; (iii) severely saline water: inland and coastal water with salt concentrations of 10000 to 30000 mg per l of dissolved solids; and (iv) sea-water: offshore water of the oceans and their seas with salt concentrations of 30000 to 36000 mg per l of dissolved solids, including 19000 ppm Cl, 10,600 ppm Na, 1270 ppm Mg, 880 ppm S, 400 ppm Ca, 380 ppm K, 65 ppm Br, 28 ppm C, 13 ppm Sr, 4.6 ppm B and others.

Inland seas that do not drain to river systems may be saltier than the ocean itself.

Desalting or demineralising processes share as a common objective the removal of salt from saline water. By contrast, the present section is concerned with separating water from saline water. In Table 11.1 the classification of saline-water conversion processes is shown.

Table 11.1. Classification of saline-water conversion processes.

Constituent removed		Phase to which transported	Process number and kind
Salt	→	Vapour	1. (None known)
			2. Electrodialysis
	→	Liquid	Osmionic
			Thermal diffusion
	→	Solid	3. Ion exchange
			Adsorption on carbon electrode
Water	→	Vapour	1. Distillation
	→	Liquid	2. Solvent extraction
			Reverse osmosis
			3. Freezing
	→	Solid	Contact freezing
			Adsorption

Distillation

Of the processes removing water from saline solutions, distillation is the oldest and in terms of established plants, the most productive. It differs from other processes by its passage of water through the vapour phase. Of interest is the fact that the potential of this operation is readily assessed in terms of the minimum work to be accomplished by the heat energy applied. If, for example, needed energy is derived from steam, as it usually is, the minimum work requirement is a function of the temperature of the steam. Because saturated steam can provide about 205 Btu per lb of work energy when it is condensed at atmospheric pressure and rejected at a temperature of about 70°F, the expected volume of product water is 100 gal when 47 lb of steam are put to use efficiently. The commercial value of atmospheric-pressure steam thereby equals the price of about 2.5 kWh of electrical energy. This figure is useful for first estimates of cost along with a realisation that, in addition to final costs, the efficiency of energy production and utilisation is the cost-controlling factor in distillation plants. Hence plant design is directed to: (i) tapping the most economic sources of heat energy; and (ii) exploiting the most efficient processes of heat transfer. A cost target is offered, moreover, by the alternative possibility of supplying water from a natural freshwater source based, in each instance, on delivery of the water at the plant under the required systems head.

Three types of evaporators are illustrated in Fig. 11.6: a multiple-effect evaporator, a multistage flash evaporator and a vapour-compression still. All of them re-capture the latent heat of vapourisation of water, in countercurrent circulation with steam.

Multiple-effect evaporators are relatively efficient, whereas single-effect evaporators are relatively inefficient. Each component unit of a multiple-effect evaporator is maintained in sequence at slightly lower pressure and temperature in order to permit the steam produced in one effect to become the source of heat in the next. Pound for pound, the amount of product water then approximates the number of effects (Fig. 11.6a).

Multistage flash evaporation, too, is accomplished at successively lower pressures and temperatures. The incoming water is warmed by the heat of condensation and only a small amount of heat energy is required to flash the pre-heated water in the reduced-pressure stage into steam (Fig. 11.6b).

The vapour-compression process relies on mechanical compression of the vapour to boost its temperature high enough, because $pv = RT$, to supply through its own condensation the heat necessary to evaporate the feed water (Fig. 11.6c). Once started, this process does not draw upon further heat energy, only upon mechanical energy. When electric motors are made the prime movers of compressors, only mechanical energy is supplied. However, internal combustion engines can assist in pre-heating the feed water in their cooling jackets and in exhaust-gas heat exchangers.

There is much scale formation and corrosion as well as erosion at the high temperatures at which stills must be operated.

Innovative Trends in Desalination

In the field of thermal desalination, efforts are directed towards utilising the low grade heat and the waste heat as energy input for desalination, lesser chemical treatment and the advantage of scale up to higher capacity as a cost reduction strategy. In membrane desalination, work is being carried out on better pre-treatment methods like use of ultrafiltration, energy reduction by use of energy recovery devices and higher membrane life by better quality membranes. Work is pursued on hybrid desalination for producing different quality of product water for process industries and for potable use at lower cost.

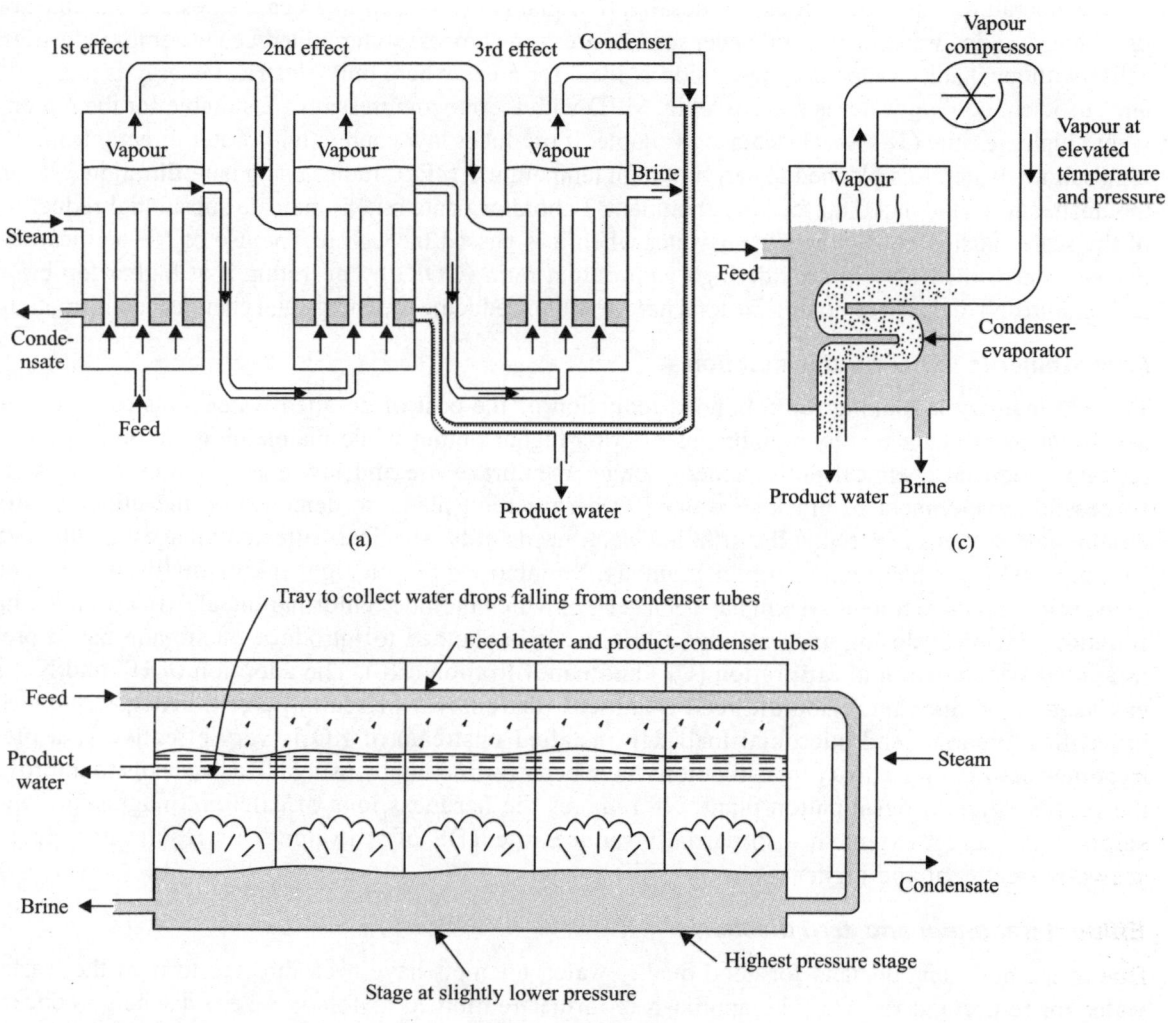

Fig. 11.6. Schematic arrangement of evaporation. (a) Multiple-effect evaporator; (ii) multistage flash evaporator; and (c) vapour-compression still.

Improved heat transfer

Basic studies carried out on a horizontal tube thin film (HTTF) boiling indicate that the bubble nucleation in the thin film on the tube takes place with rapid bubble growth. The application of forced convection due to liquid spray on the tube increases the convective contribution and results in early removal of bubble adhering to the surface which increases bubble frequency. The overall heat transfer coefficient in the range of 3–4 kW/m²K was obtained which is about three times the heat transfer coefficient as compared to submerged tube evaporator. High heat transfer coefficient implies low heat transfer area requirement and in turn low capital cost. The data collected on the fluid flow and heat transfer aspect of the boiling in a thin film were used in the design and installation of 1 m^3/d HTTF desalination unit for MED. Low temperature vapour compression desalination plants of 50–200 m^3/d capacities are suitable for providing drinking water in the rural/water scarcity areas and process water/boiler feed water for industries. MED with mechanical vapour compression is ideal for areas where only electrical energy is available and sufficient cooling water is not available. MED with thermocompression are suitable for the regions where high pressure (5–10 bar) steam is available. It produces low conductivity water directly from the high salinity water. It is planned to carryout high temperature MED studies using nanofiltration (NF) in the upstream for the make-up feed pre-treatment. Laboratory data on NF indicate substantial reduction of the scale causing constituents in seawater when it is passed through it. The use of NF permeate as make-up feed to MED will provide high gain output ratio (GOR) by operating it at higher top brine temperature. NF helps in removing the total hardness thus reducing the energy and chemical consumption.

Improvements in RO for desalination

The RO industry is looking for continued reduction of the cost of desalted water. This calls for the development of better quality membranes offering higher output while maintaining the optimum salt rejection, reduced chemical pre-treatment, longer membrane life and low energy requirement. After successful development of brackish water RO desalination plants to demonstrate the utility of RO desalination systems in meeting the drinking water needs of brackishness effected villages, a 100 m^3/d seawater RO plant has been set up in Trombay, Mumbai for producing drinking quality water. The conventional pre-treatment system has been set-up, which includes chlorination, clarification, media filtration, chemical dosing and cartridge filtration. It is planned to introduce membrane based pre-treatment system using ultrafiltration (UF) and nanofiltration (NF). The adoption of UF and NF is envisaged to reduce the elaborate feed treatment for removal of scaling constituents, suspended impurities, organics and microbial load. UF installed upstream of RO is very effective as a pre-treatment set-up. Preliminary investigations have been carried out by using NF as a means to improve the performance of desalination plant. NF reduces the hardness ions of calcium, magnesium and sulphate to a great extent. It also partially reduces the TDS of seawater. This results in reduced seawater treatment and higher recovery.

Effluent treatment and zero discharge

Due to an increasing demand for good quality water, attempts have been directed to treat the waste-water for re-use and recycle. The approach is further reinforced to follow a zero discharge concept wherever possible. R&D work in the field of thermal and membrane processes has been pursued for the treatment of waste-water and removing pollutants from the effluent stream for safe discharge into environment and recovery of significant fraction of the water for re-use. Selection of a process for treatment of a particular waste-water is based on product requirements, influent water characteristics and cost. Industrial waste-water often is the combined product of a number of different manufacturing processes in the complex. The membrane processes that are useful for waste-water treatment include: microfiltration, ultrafiltration, nanofiltration and reverse osmosis. The suspended solids in waste-water are successfully removed by microfiltration. Ultrafiltration is useful in separating macromolecules and the sub-micron particles including oil emulsion and very large molecules such as polymeric compound having a polymeric weight of 1000 and above. Nanofiltration is capable of separating molecules in the range of 300–1000 molecular weight. It also helps in selective separation of low molecular weight organics from salt solution. Reverse osmosis membranes have very small pore size (5–10 Å) suitable for removing ions and molecules. Laboratory scale studies are continuing on development of such membranes and their performance evaluation. As no two waste-water are exactly alike, it is necessary to carry out laboratory evaluations to determine the flux rate under different temperatures and pressures for individual waste-water samples. LTE and VC desalination is ideal for treating high salinity effluent and producing pure water for re-use.

Thus, the development work has generated capability in the country to design, fabricate, commission and operate small and large size desalination plants. Efforts are now directed towards reducing the cost of desalted water through technological innovations. In case of thermal processes, this calls for capital cost reduction through heat transfer enhancement and use of cheaper materials, low grade or waste heat utilisation and least chemical pre-treatment. Today, production of boiler quality water and high quality process water from sea water desalination is cheaper than that produced from conventional DM plant using raw water where the raw water contains 600 ppm or more salinity. In the case of membrane processes, attempts are continued towards the development of better membranes, least pre-treatment, longer membrane life and reduced energy consumption. Effluent treatment and water re-use through desalination route, as a step towards zero discharge appears promising. The technological innovations in desalination would lead to its large scale application and provide opportunities for the socioeconomic development of water scarcity areas and large coastal arid zones of the country.

ION EXCHANGE

A wide variety of dissolved solids, including hardness, can be removed by ion exchange. The discussion here will be limited to ion exchange for softening.

As practiced in water softening, ion exchange involved replacing calcium and magnesium in the water with another, non-hardness cation, usually sodium. This exchange takes place at a solids interface. Although the solid material does not directly enter into the reaction, it is a necessary and important part of the ion-exchange process. Early applications of ion exchange used zeolite, a naturally occurring sodium alumino-silicate material sometimes called greensand. Modern applications more often use a synthetic resin coated with the desirable exchange material. The synthetic resins have the advantage of a greater number of exchange sites and are more easily regenerated.

In equal quantities, calcium and magnesium are adsorbed more strongly to the medium than is sodium. As the hard water is contacted with the medium, the following generalised reaction occurs.

$$\begin{Bmatrix} Ca \\ Mg \end{Bmatrix} + [anion] + 2Na[R] \longrightarrow \begin{Bmatrix} Ca \\ Mg \end{Bmatrix} [R] + 2Na + [anion] \qquad \dots (11.6)$$

The reaction is virtually instantaneous and complete as long as exchange sites are available. The process is shown in Fig. 11.7.

When all of the exchange sites have been utilised, hardness begins to appear in the effluent. Referred to as breakthrough, this necessitates the regeneration of the medium by contacting it with a strong sodium-chloride solution. The strength of the solution overrides the selectivity of the adsorption site and calcium and magnesium are removed and replaced by the sodium.

$$\begin{Bmatrix} Ca \\ Mg \end{Bmatrix} [R] + 2NaCl \ (excess) \longrightarrow \begin{Bmatrix} Ca \\ Mg \end{Bmatrix} 2Cl + 2Na[R] \qquad \dots (11.7)$$

The system can again function as a softener according to Eq. (11.6).

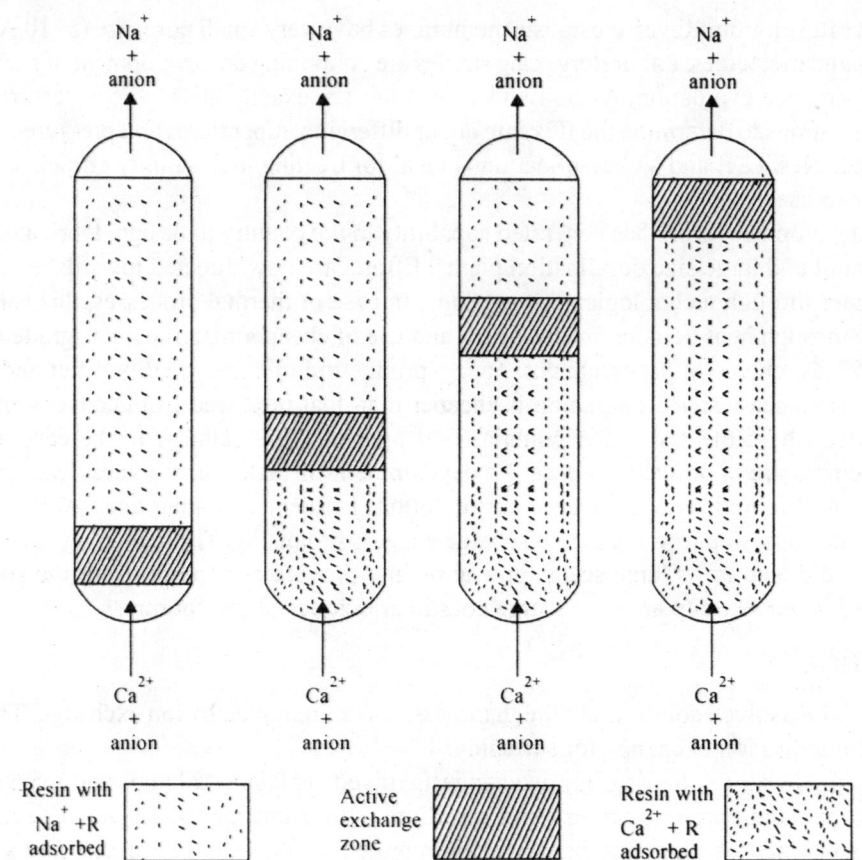

Fig. 11.7. Ion-exchange process.

The capacity and efficiency of ion exchange softeners vary with many factors, including type of solid medium, type of exchange material used for coating, quantity of regeneration materials and regeneration contact time. The overall quality of the water to be softened is also an important factor. Generally, the capacity of ion exchange materials ranges from 20 to 10 mequiv/g or about 15 to 100 kg/m³. Regeneration is accomplished using from 80 to 160 kg of sodium chloride per cubic meter of resin in 5 to 20 per cent solution at a flow rate of about 40 l/min·m².

The effluent from the regeneration cycle will contain the hardness accumulated during the softening cycle as well as excess sodium chloride. After regeneration, the medium should be flushed with softened water to remove the excess sodium chloride. These highly mineralised water constitute a waste stream that must be disposed of properly.

Ion exchange operations are usually conducted in enclosed structures containing the medium. Water is forced through the material under pressure at up to 0.4 m³/min·m². Single or multiple units may be used and the medium may be contained in either a fixed or a moving bed. Where continuous operation is necessary, multiple units or moving beds are used. Single-stage fixed beds can be used when the flow of treated water can be interrupted for regeneration. Most treatment-plant operations are of the continuous type, while home softeners are serviced intermittently.

Ion-exchange softening at water-treatment plants is becoming more common place as more efficient resins are developed and as the process is better understood by design engineers. Ion exchange produces a softer water than chemical precipitation and avoids the large quantity of sludges encountered in the lime-soda process. The physical and mechanical apparatus is much smaller and simpler to operate. There are several disadvantages, however. The water must be essentially free of turbidity and particulate matter or the resin will function as a filter and become plugged. Surfaces of the medium may act as an adsorbent for organic molecules and become coated. Iron and manganese precipitates can also foul the surfaces if oxidation occurs in or prior to, the ion-exchange unit. Softening of clear groundwater should be done immediately (before aeration occurs), while surface water should receive all necessary treatment, including filtration, prior to softening by ion exchange. The water should not be chlorinated prior to ion exchange softening.

Chapter 12

Membrane Separation Technologies

INTRODUCTION

A membrane is defined as an intervening phase separating two phases forming an active or passive barrier to the transport of matter. Membrane processes can be operated as: (i) dead-end filtration; and (ii) cross-flow filtration. Dead-end filtration refers to filtration at one end. A problem with these systems is frequent membrane clogging. Cross-flow filtration overcomes the problem of membrane clogging and is widely used in water and waste-water treatment.

There are five types of membrane processes, which are commonly used in water and waste-water treatment:

1. Electrodialysis.
2. Microfiltration.
5. Ultrafiltration.
6. Nanofiltration.
7. Reverse osmosis.

Through these processes dissolved substances and/or finely dispersed particles can be separated from liquids. All five technologies rely on membrane transport, the passage of solutes or solvents through thin, porous polymeric membranes.

The membrane itself is a polymeric coating or extrusion with inverted conical shaped pores. Membrane filters do not plug because the pore diameter is smaller at the top, which is the point of contact with the waste-water. Material passing through the membrane passes unimpeded through the membrane structure, therefore eliminating accumulation of material within the filter. Waste-water is pumped across the membrane surface at high flow rates. This parallel fluid flow eliminates the cake-like build-up typical of conventional filters such as bags and cartridges which must be frequently replaced. Some waste-water contaminants slowly accumulate on the membrane surface, forming a thin film, during normal operating conditions. This fouling process is normal and causes the filtration rate to slowly decrease with time. When membranes no longer produce clean water at the desired rate they are cleaned in place with soap and water and returned to service. Membranes can be repeatedly cleaned for years of productive, dependable service prior to replacement.

Most of these processes are oftentimes used with chemical mechanical polishing (CMP), which is fast becoming the established technology for planarising multilevel devices. This process requires large quantities of ultrapure water for rinsing slurry particles off the polished wafers. Treatment by this method is generally needed in order to maintain an acceptable level of total suspended solids (TSS) in industrial waste-water effluent. With large quantities of particles in the CMP waste-water stream, cross-flow filtration is the most economical method for TSS removal.

AN OVERVIEW OF MEMBRANE PROCESSES

Table 12.1 will give a basic appreciation for the technologies and provides a comparison of the factors that affect the performance of the five technologies.

Table 12.1. Factors impacting on the performance of membrane processes.

Technology	Driving force	Influencing factors			
		Size	Diffusivity	Ionic charge	Solubility
Microfiltration	Pressure	+++	−	−	−
Ultrafiltration	Pressure	+++	−	+	−
Nanofiltration	Pressure	+++	+	+	−
Reverse osmosis	Pressure	+	+++	+	+++
Electrodialysis	Electrical	+	+	+++	−

Table 12.2 provides a comparison of membrane structures. Between these two tables, one can get an idea of the operating conditions viz., membrane structural types, the driving forces involved in separation and the separation mechanisms.

Table 12.2. Comparison of membrane structures.

Technology	Structure	Driving force	Mechanism
Microfiltration	Symmetric microporous (0.02–10 μ)	Pressure, 1–5 atm	Sieving
Ultrafiltration	Asymmetric microporous (1–20 nm)	Pressure, 2–10 atm	Sieving
Nanofiltration	Asymmetric microporous (0.01–5 nm)	Pressure, 5–50 atm	Sieving
Reverse osmosis	Asymmetric with homogeneous skin and microporous support	Pressure, 10–100 atm	Solution diffusion
Electrodialysis	Electrostatically charged membranes (cation and anion)	Electrical potential	Electrostatic diffusion

Common types of membrane materials used are listed in Table 12.3. This gets us into the concept of geometry. There are three types of modules generally used, namely: tubular, spiral wound and hollow fiber. A comparison of the various geometries is given in Table 12.4.

Table 12.3. Summary of common membrane materials and their characteristics.

Technology	Membrane materials	Polar character
Microfiltration	Polypropylene (PP)	–
	Polyethylene (PE)	Non-polar
	Polycarbonate (PC)	Non-polar
	Ceramic (CC)	Non-polar
Ultrafiltration	Polysulphone (PSUF)	Non-polar
	Dynel	Non-polar
	Cellulose acetate (CA)	Non-polar
Nanofiltration	Polyvinylidene fluoride (PVDF)	Polar
	Cellulose acetate	Polar
Reverse osmosis	Polyamide	Polar
	Nylon	Polar
Electrodialysis	Styrene/Vinylpyridene	–
	Divinyl benzene	

Table 12.4. A comparison between module types.

Parameter		Module type	
	Tubular	spiral wound	Hollow fibre
Surface area, m^2/m^3	300	1000	15000
Inner dia./spacing, mm	20–50	4–20	0.5–2
Feed flow rate, liter/m^2-day	300–1000	300–1000	30–100
Production, m^3/m^3 of membrane per day	100–1000	300–1000	450–1500
Pretreatment requirements	Simple	Average	High
Extent of clogging	Little	Average	High
Mechanical Cleaning methods	Possible	Not possible	Not possible
Chemical cleaning methods	Possible	Possible	Possible

ELECTRODIALYSIS

The principle behind electrodialysis is that electrical potential gradients will make charged molecules diffuse in a given medium at rates far greater than attainable by chemical potentials between two liquids as in conventional dialysis. When a DC electric current is transmitted through a saline solution, the cations migrate toward the negative terminal or cathode and the anions toward the positive terminal, the anode. By adjusting the potential between the terminals or plates, the electric current and therefore, the flow of ions transported between the plates can be varied.

Electrodialysis can be applied to the continuous-flow type of operation needed in industry. Multi-membrane stacks can be built by alternately spacing anionic and cationic-selective membranes. Among the technical problems associated with the electrodialysis process, concentration polarisation is perhaps the most serious (discussed later). Other problems in practical applications include membrane scaling by inorganics in feed solutions as well as membrane fouling by organics. Efficient separation or pre-treatment in the influent streams can include activated carbon absorption to reduce or prevent such problems. Principal applications of electrodialysis include: (i) recovery of materials from liquid effluents, such as processes related to conservation, cleanup, concentration and separation of desirable fractions from undesirable ones; (ii) purification of water sources; and (iii) effluent water renovation for re-use or to meet point source disposal standards required to maintain suitable water quality in the receptor streams.

Concentration-polarisation is a problem which also exists in reverse osmosis systems, and is due to of a build-up in the concentration of ions on one side of the membrane and a decrease in concentration on the opposite side. This adversely affects the operation of membranes and can even damage or destroy them. Polarisation occurs when the movement of ions through the membrane is greater than the convective and diffusional movements of ions in the bulk solutions toward and away from the membrane. Along with a deleterious pH shift occurring at the membrane surface, polarisation may cause solution contamination and sharply decrease energy efficiency. Commercial electrodialyser designs incorporate various baffles or turbulence promoters and limit current densities to avoid these effects. Increased feed flow also assists mixing but requires additional power for pumping.

Treatment of brackish water in the production of potable supplies has been the largest application of electrodialysis. Costs associated with electrodialysis processes depend on such factors as the total dissolved solids (TDS) in the feed, the level of removal of TDS (per cent rejection) and the size of the plant.

A rough rule of thumb for the energy requirements for demineralising 1000 gallons of salt water by ED in large capacity plants (4 mgd) is 5 to 7 kWh per 1000 ppm of dissolved solids removed. Since the efficiency of electrodialytic demineralisation decreases rapidly with increasing feed concentrations, this process is best utilised for treatment of weakly saline (brackish) water containing less than 5000 ppm of total dissolved solids. In fact, for water at the low-concentration end of the brackish scale, ED may be the most cost-effective process of all. Electrodialysis is widely used in the United States in the dairy industry, namely in the desalting of cheese whey. Electrical requirements may vary from 5 to 14 kWh per pound of product solids. Another application of ED is the sweetening of prepared citrus juices. Other less extensive uses of electrodialysis in commercial operations include tertiary or advanced treatment of municipal sewage water and treatment of industrial waste-water such as metal-plating baths, metal-finishing rinse water, wood pulp wash water and glass-etching solutions. Potential applications of ED are many. A particular advantage of the electrodialysis process is its ability to produce solutions of high concentrations of soluble salts.

A combination of electrodialysis with conventional evaporation, for example, may be substantially cheaper than evaporation alone for the production of dry salt from saline solutions. Competing technologies include reverse osmosis and crystallisation.

Added control of the movement of the ions can be obtained by placing sheet-type membranes of cation or anion-exchange material between the outer plates, as shown diagrammatically in Fig. 12.1. These sheets of cation-selective resins and anion-selective resins permit the passage of the respective ions in the solution. Under an applied DC field, the cations and anions will collect on one side of each membrane through which they are transported and vacate the other side. Thus, if a NaCl solution is supplied to the central zone of the cell shown in Fig. 12.1, the Na$^+$ ions will migrate through membrane A, depleting the central zone (termed the diluting or product feed stream) of the salt ions. The two outer zones where the ions collect are commonly known as the concentrating or brine streams.

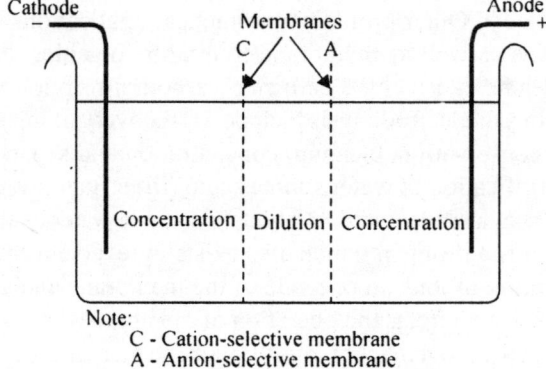

Fig. 12.1. Electrodialysis cell diagram.

Multi-membrane stacks can be built from alternately spacing anionic and cationic-selective membranes. Flow of solutions through specific compartments and appropriate recombination of transported ions permit desired enrichment of one stream and depletion of another. A schematic view of a typical stack based on the concept of alternating these concentrating and diluting compartments is shown in Fig. 12.2. The feed stream enters each compartment along the top of the figure and flows downward toward the lower exit ports and manifolds. However, as the ionised streams move tangentially along the membranes, cations are transported or attempt movement, toward the left and anions to the right, causing an alternate build-up and a depletion of ions in adjoining compartments. Thus, one resultant output stream is a diluted product water and the other is a concentrated stream of dissolved salt.

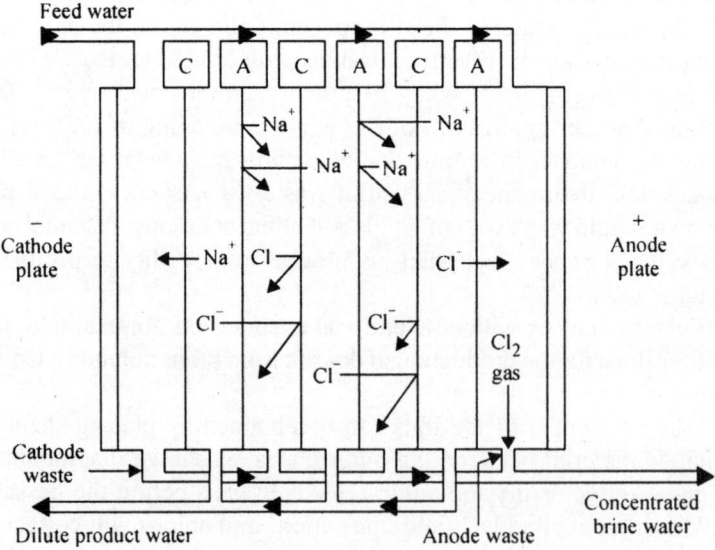

Fig. 12.2. Electrodialysis process diagram.

The process flow stream through a commercial demineraliser, incorporating two stacks in series demineralised water, is shown in Fig. 12.3. Several of the refinements required for continuous-flow operational systems are shown on this diagram, representing a two-stage demineraliser.

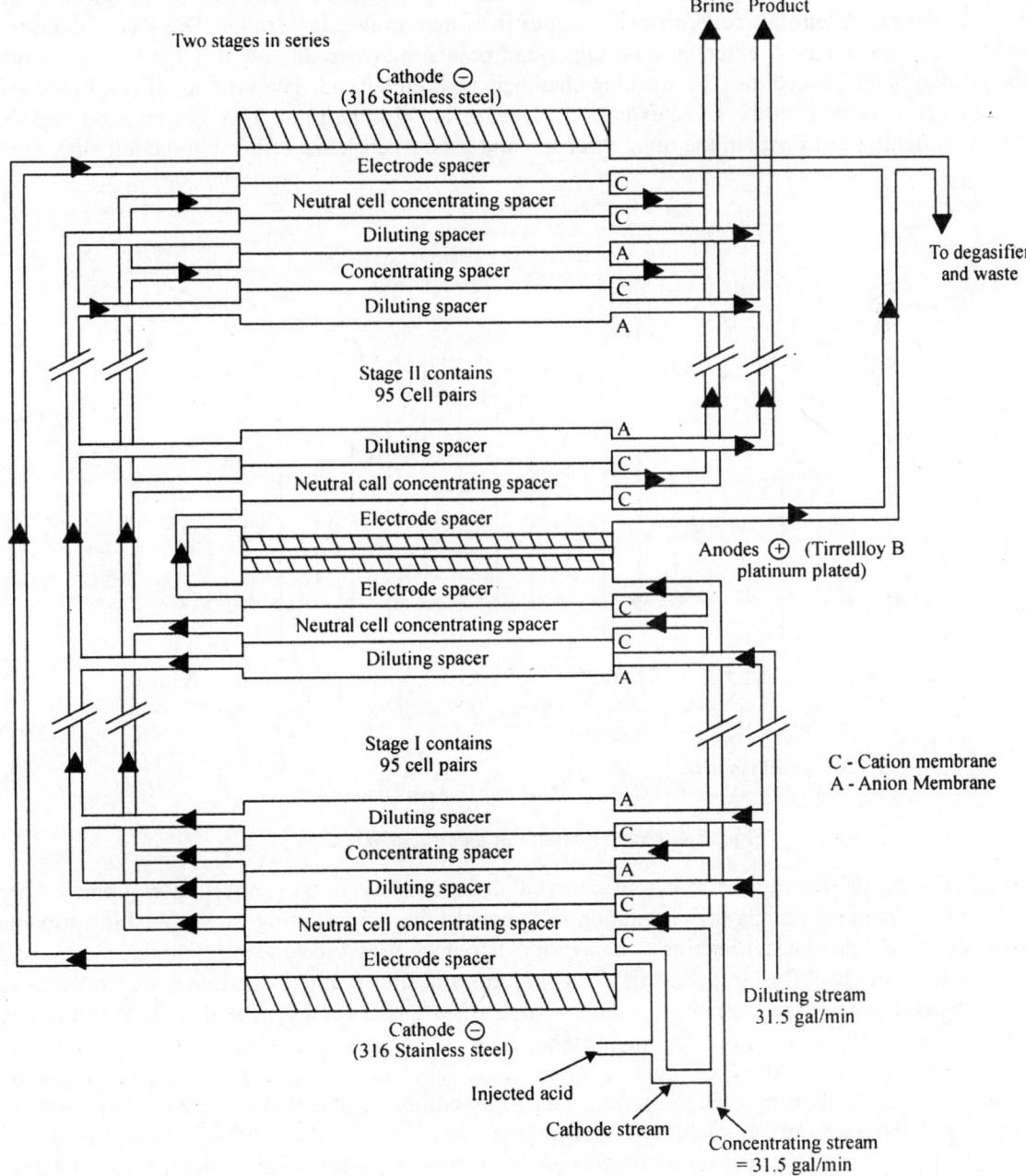

Fig. 12.3. Flow through electrodialysis stack.

Although ED is more complex than other membrane separation processes, the characteristic performance of a cell is, in principle, possible to calculate from a knowledge of ED cell geometry and the electrochemical properties of the membranes and the electrolyte solution.

Another kind of electrodialysis cell configuration, shown in Fig. 12.4, is a multiple electrodialysis system consisting of ten-unit cells, in series rather than manifolded in parallel. The feed solution is introduced at four points: It enters at both upper end points to sweep directly through both electrode chambers and is introduced into the working chambers near either end. The feed solution into the left side traverses depleted chambers and exits as depleted effluent at the right. The feed solution into the rightmost enriching cell flows in the other direction and exits as enriched effluent at the left side.

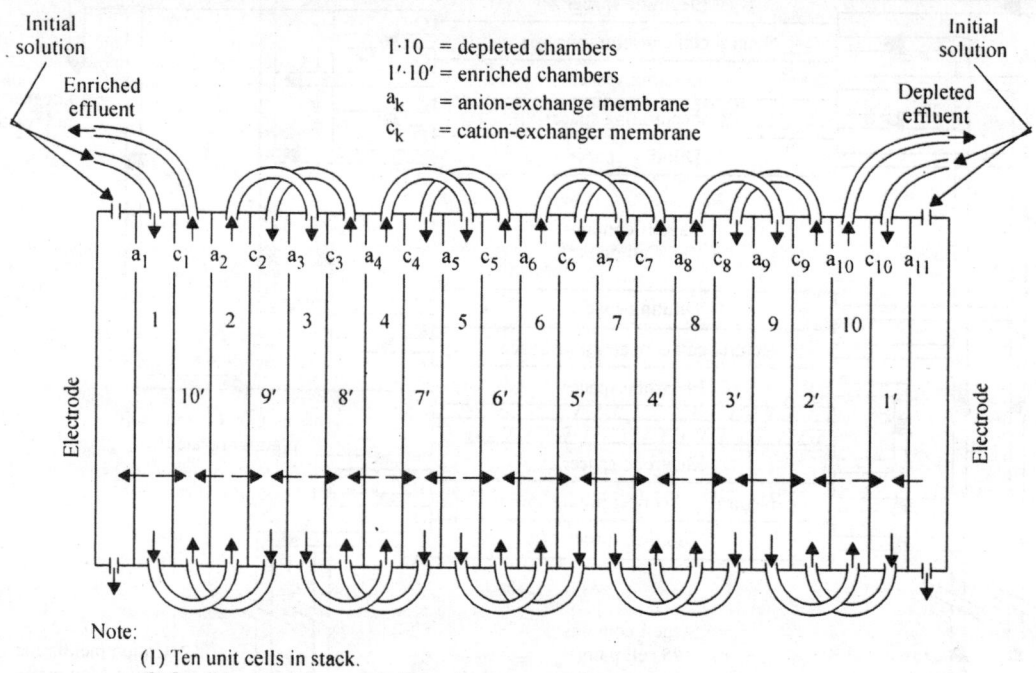

Note:
(1) Ten unit cells in stack.
(2) Small horizontal arrows indicate ion movement between compartments.

Fig. 12.4. Multiple-chamber electrodialysis unit.

In addition to the membrane stacks in electrodialysis units, various supporting equipments are essential. This includes pumps for circulation of concentrating and diluting flows; flushing streams for cathode and anode plates; injection systems for pH control; pressure concentration, pH alarms and control systems and backflushing controls; feed strainers and filters; and grounding systems. Because of the high pH of the cathode stream, substances, such as carbonates and hydroxides could precipitate on the cathode surface and adjoining membrane; often sulphuric acid is injected to maintain the stream at pH of 2 or less. Also, recirculating concentrate requires an acid addition to yield a low pH for stability along with additional substances such as sodium hexametaphosphate. A key point to remember is that the separation is achieved by removing ions by passing them through a semi-permeable membrane. The electric field applied across the membrane transports only ions. As noted the application of this technology is to desalting brackish water, to removing TDS from water and to the removal of certain heavy metals.

The issue of concentration polarisation results in an increase of the resistance of flow of ions across the membrane. The current must therefore be increased to overcome this resistance.

Major differences between ED and other processes are, first, the solute is transferred across the membrane against water in the other technologies discussed below, whereas only ionic species are removed by ED. As noted, two different membranes (anionic and cationic) are employed. Current consumption depends primarily on the TDS concentration. One should look at this very closely when comparing the operating cost benefits and tradeoffs of this technology to other options. Current efficiency can be calculated from the following formula:

$$E = FqN\xi/\eta I \qquad \qquad ... (12.1)$$

where,

q	=	the stream flow rate
F	=	Faraday's constant, 96540 amp-sec/g.eq
η	=	number of cells
ξ	=	removal efficiency, [(N–Ne)/N]
I	=	current, amp

ULTRAFILTRATION

Suspended materials and macromolecules can be separated from a waste stream using a membrane and pressure differential, called ultrafiltration. This method uses a lower pressure differential than reverse osmosis and doesn't rely on overcoming osmotic effects. It is useful for dilute solutions of large polymerised macromolecules where the separation is roughly proportional to the pore size in the membrane selected.

Ultrafiltration membranes are commercially fabricated in sheet, capillary and tubular forms. The liquid to be filtered is forced into the assemblage and dilute permeate passes perpendicularly through the membrane while concentrate passes out the end of the media. This technology is useful for the recovery and recycle of suspended solids and macromolecules. Excellent results have been achieved in textile finishing applications and other situations where neither entrained solids that could clog the filter nor dissolved ions that would pass through are present. Membrane life can be affected by temperature, pH, and fouling.

Ultrafiltration equipment are combined with other unit operations. The unique combination of unit operations depends on the waste-water characteristics and desired effluent quality and cost considerations. Like normal filtration, with ultrafiltration (UF), a feed emulsion is introduced into and pumped through a membrane unit; water and some dissolved low molecular weight materials pass through the membrane under an applied hydrostatic pressure. In contrast to ordinary filtration however, there is no build-up of retained materials on the membrane filter.

A variety of synthetic polymers, including polycarbonate resins, substituted olefins and polyelectrolyte complexes, are employed as ultrafiltration membranes. Many of these membranes can be handled dry, have superior organic solvent resistance and are less sensitive to temperature and pH than cellulose acetate, which is widely used in RO systems. In UF, molecular weight (MW) cut-off is used as a measure of rejection. However, shape, size and flexibility are also important parameters. For a given molecular weight, more rigid molecules are better rejected than flexible ones. Ionic strength and pH often help determine the shape and rigidness of large molecules. Operating temperatures for membranes can be correlated generally with molecular weight cut-off. For example, maximum operating temperatures for membranes with 5000 to 10000 MW cutoffs are about 65°C and for a 50000 to 80000 MW cut-off, maximum operating temperatures are in the range of 50°C.

The largest industrial use of ultrafiltration is the recovery of paint from water-soluble coat bases (primers) applied by the wet electrodeposition process (electrocoating) in auto and appliance factories. Many installations of this type are operating around the world. The recovery of proteins in cheese whey (a waste from cheese processing) for dairy applications is the second largest application, where a market for protein can be found (for example, feeding cattle and farm animals). Energy consumption at an installation processing 5,00,000 pounds per day of whey would be 0.1 kWh per pound of product. Another large-scale application is the concentration of waste-oil emulsions from machine shops, which are produced in association with cooling, lubrication, machining, rolling heavy metal operations and so on. Ultrafiltration of corrosive fluids such as concentrated acids and ester solution is also an important application. The chemical inertness and stability of ultrafilters make them particularly useful in the cleaning of these corrosive solutions. Uses include separation of colloids and emulsions and recovery of textile sizing chemicals. Biologically active particles and fractions may also be filtered from fluids using ultrafilters. This process is used extensively by beer and wine manufacturers to provide cold stabilisation and sterilisation of their products. It is also used in water pollution analysis to concentrate organisms from water samples. Food concentration applications can be applied to processing milk, egg white, animal blood, animal tissue, gelatine and glue, fish protein, vegetable extracts, juices and beverages, pectin solutions, sugar, starch, single-cell proteins and enzymes.

Figure 12.5 conceptually illustrates how ultrafiltration works. Water and some dissolved low molecular weight materials pass through the membrane under an applied hydrostatic pressure. Emulsified oil droplets and suspended particles are retained, concentrated and removed continuously as a fluid concentrate. The pore structure of the membrane acts as a filter, passing small solutes such as salts, while retaining larger emulsified and suspended matter. The pores of ultrafiltration membranes are much smaller than the particles rejected and particles cannot enter the membrane structure. As a result, the pores cannot become plugged. Pore structure and size (less than 0.005 microns) of ultrafiltration membranes are quite different from those of ordinary filters in which pore plugging results in drastically reduced filtration rates and requires frequent backflushing or some other regeneration step. In addition to pore size, another important consideration is the membrane capacity. This is termed flux and it is the volume of water permeated per unit membrane area per unit time. The standard units are gallons per day per square foot (gpd/ft^2) or cubic meters per day per square meter (m^3/day/m^2).

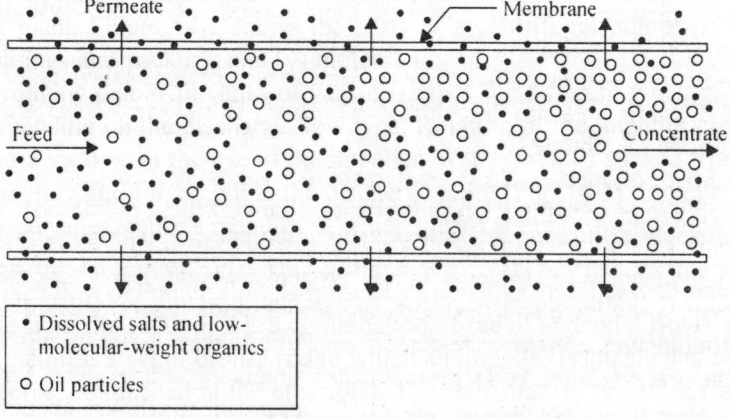

Fig. 12.5. Ultrafiltration basics.

Because membrane equipment, capital costs and operating costs increase with the membrane area required, it is highly desirable to maximise membrane flux. Ultrafiltration utilises membrane filters with small pore sizes ranging from 0.015μ to 8μ in order to collect small particles, to separate small particle sizes or to obtain particle-free solutions for a variety of applications. Membrane filters are characterised by a smallness and uniformity of pore size difficult to achieve with cellulosic filters. They are further characterised by thinness, strength, flexibility, low absorption and adsorption and a flat surface texture. These properties are useful for a variety of analytical procedures. In the analytical laboratory, ultrafiltration is especially useful for gravimetric analysis, optical microscopy and X-ray fluorescence studies.

All particles larger than the actual pore size of a membrane filter are captured by filtration on the membrane surface. This absolute surface retention makes it possible to determine the amount and type of particles in either liquids or gases—quantitatively by weight or qualitatively by analysis. Since there are no tortuous paths in the membrane to entrap particle sizes smaller than the pore size, particles can be separated into various size ranges by serial filtration through membranes with successively smaller pore sizes. Figure 12.6 shows pore size in relation to commonly known particle sizes.

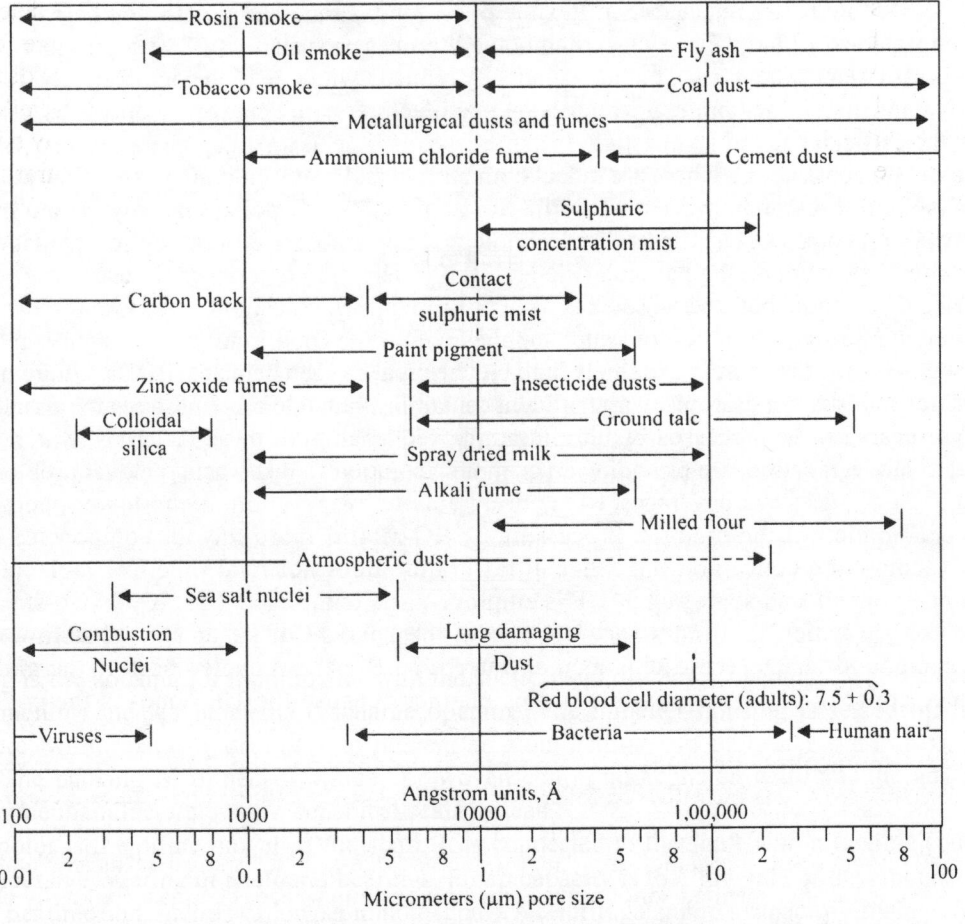

Fig. 12.6. Range of common particle sizes (diameter) over range of UF pore size.

Fluids and gases may be cleaned by passing them through a membrane filter with a pore size small enough to prevent passage of contaminants. This capability is especially useful in a variety of process industries which require cleaning or sterilisation of fluids and gases. The retention efficiency of membranes is dependent on particle size and concentration, pore size and length, porosity and flow rate. Large particles that are smaller than the pore size have sufficient inertial mass to be captured by inertial impaction. In liquids the same mechanisms are at work. Increased velocity, however, diminishes the effects of inertial impaction and diffusion. With interception being the primary retention mechanism, conditions are more favourable for fractionating particles in liquid suspension.

In contrast to reverse osmosis, where cellulose acetate has occupied a dominant position, a variety of synthetic polymers has been employed for ultrafiltration membranes. Many of these membranes can be handled dry, have superior organic solvent resistance and are less sensitive to temperature and pH than cellulose acetate. Polycarbonate resins, substituted olefins and polyelectrolyte complexes have been employed among other polymers to form ultrafiltration membranes. Preparation details for most of the membranes are proprietary. As noted earlier, molecular weight cut-off is used as a measure of rejection, however, shape, size and flexibility are also important parameters. For a given molecular weight, more rigid molecules are better rejected than flexible ones. Ionic strength and pH often help determine the shape and rigidness of large molecules. Membrane lifetimes are usually two years or more for treating clean streams (water processing), but are drastically reduced when treating comparatively dirty streams (e.g., oily emulsions). Membrane guarantees by manufacturers are determined only after pilot work is done on the particular stream in question. In some cases as little as a 90-day guarantee may be given for oil/water-waste applications. There are in fact both current and many emerging uses of ultrafilters in the areas of biological research, processing sterile fluids, air and water pollution analysis, and recovery of corrosive or non-corrosive chemicals. The technology is applicable to dewatering some sludges, but this use is highly dependent on the particular sludge itself. There are no commercial uses of UF for sludge dewatering at this time, but several sources have been found which claim that this represents a possible near-future application. Pollution of water supplies within the food industry is a significant problem, since many food wastes possess extremely high biochemical oxygen demand (BOD) requirement. In the potato starch industry, for example, waste effluent containing valuable proteins, free amino acids, organic acids and sugars can be processed by ultrafiltration. Reclamation of these materials, which are highly resistant to biodegradation, are providing an economic solution to this waste removal problem. For the concentration of juices and beverages, RO is preferable to evaporation due to lower operating costs and no degradation of the product. Processing by RO retains more flavour components than does heated, vacuum pan concentration. Since ultrafiltration does not retain the low molecular weight flavour components and some sugar, UF is employed as a complement to RO. A two-stage process may be used in which the first stage, UF, allows the passage of sugar and other low molecular weight compounds. This permeate is then dewatered by RO and recycled back to the main stream. High juice concentrations are possible in this manner because the UF removes the colloidal and suspended solids which would foul the RO and helps relieve some of the high hydraulic pressure due to high osmotic pressure of the juice. In a process as shown in Fig. 12.7, a citrus press liquor or multiphase suspension, is ultrafiltered following a coarse pre-filtration. The resultant clear permeate is processed through ion exchange and granular activated carbon adsorption units to remove low molecular weight contaminants and inorganic salts. The product is a natural citrus sugar solution suitable for re-use. The concentrated suspended solids are used in making animal feed. Pectins are a family of complex carbohydrates which are used to form gels with sugar and acid in the production of jellies, preserves, and other confections from fruit juices. The recovery of starch and other high molecular weight compounds from waste effluents is an important application for UF. The output from a 30-tonne-per-day starch plant would be about 4,32,000 gpd with a solids content of 0.5 per cent to 1.0 per cent and 9000 to 14000 mg/liter COD.

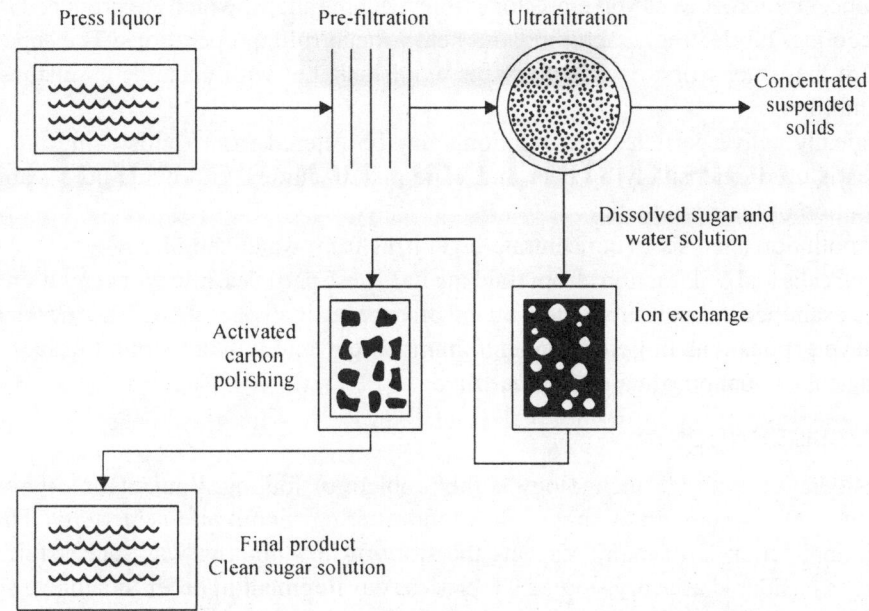

Fig. 12.7. Process flow scheme for sugar recovery from citrus press liquors.

Treatment systems which can both reduce the strength of this waste and recover valuable by-products, such as proteins, are an obvious advantage to this industry. Heat and acid coagulation, distillation, and freezing techniques are more costly and less efficient than UF in protein recovery. Reverse osmosis is also a competing process, but protein recovery by UF would lead to a somewhat higher purity. In the production of single-cell proteins as a food source, UF has several applications. For harvesting cells, UF can replace centrifugation in some applications since the efficiency of centrifugation decreases rapidly with particle size. Ultrafiltration is also well suited for recovering and concentrating the metabolic products of fermentation (enzymes, for example). In a related application, UF is also able to concentrate and desalt protein products, being more efficient than dialysis for this purpose. Moreover, a UF membrane module may be coupled with a fermenter so that toxic metabolites can be continuously removed from the system as fresh substrate is introduced. This permits the growth limitations of a batch fermenter to be relieved and permits a substantial increase in productivity. A membrane enzymatic reactor is similar to a membrane fermenter with the exception that no micro-organisms are present. Instead, enzyme-catalysed reactions take place and re-use of the enzyme is simplified. That is, purification problems and enzyme removal from end products can be eliminated.

Important Applications

1. Recovery of paint from water soluble cool bases (primers) applied by the wet electrodeposition process (electrocoating) in auto and appliance factories.

2. Recovery of proteins in cheese whey (a waste from cheese processing) for dairy applications. This is done if a market for protein can be found, in particular for feeding cattle and farm animals. In cheese whey processing, a typical unit might process 5,00,000 pounds a day of whey for 300 days a year.

3. The concentration of waste-oil emulsions from machine shops, which are produced in association with cooling, lubricating, machining and heavy metal rolling operations. The separation of the oil from the water works well with stable emulsions, but with unstable emulsions the oil will clog the filter.

4. Biologically active particles and fractions may be filtered from fluids using ultrafilters. This process is used extensively by beer and wine manufacturers to provide cold stabilisation and sterilisation of their products.

5. Water pollution analysis to concentrate organisms from water samples.

6. Filtering cells and cell fractions from fluid media. These particles, after concentration by filtration, may be examined through subsequent quantitative or qualitative analysis. The filtration techniques also have applications in fields related to immunology and implantation of tissues as well as in cytological evaluation of cerebrospinal fluid.

FOULING CONSIDERATIONS

A critical consideration with UF technology is the problem of fouling. Foulants interfere with UF by reducing product rates—sometimes drastically—and altering membrane selectivity. The story of a successful UF application is in many respects the story of how fouling was successfully controlled. Fouling must be considered at every step of UF process development in order to achieve success.

When we talk about this subject, the term foulant or foulant layer comes to the forefront. Foulant or fouling layer, are general terms for deposits on or in the membrane that adversely affect filtration. The term 'fouling' is often used indiscriminately in reference to any phenomenon that results in reduced product rates. 'Fouling' in this casual sense can involve several distinct phenomena. These phenomena can be desirable or undesirable, reversible or irreversible. Different technical terms apply to each of these possibilities.

Fouling is not always detrimental. The term dynamic membrane describes deposits that benefit the separation process by reducing the membrane's effective MWCO (molecular weight cut-off) so that a solute of interest is better retained. Concentration polarisation refers to the reversible build-up of solutes near the membrane surface. Concentration polarisation can lead to irreversible fouling by altering interactions between the solvent, solutes and membrane.

Cake layer formation builds on the membrane surface and extends outward into the feed channel. The constituents of the foulant layer may be smaller than the pores of the membrane. A gel layer can result from denaturation of some proteins. Internal pore fouling occurs inside the membrane. The size of the pore is reduced and pore flow is constricted. Internal pore fouling is usually difficult to clean.

Fouling can be characterised by mechanism and location. Membranes can foul in three places: on, above or within the membranes. The term agglomeration in the general sense, describes colloidal precipitates resulting from solute-solute attractions. Agglomerates can deposit on the membrane surface, reducing permeability. On the other hand, controlled aggregation of solutes can facilitate ultrafiltration.

Sorption or adsorption refers to deposition of foulants on the membrane surface resulting from electrochemical attractions. These attractions arise from non-covalent, intermolecular forces such as van der Waals forces and hydrogen bonding. Adsorption is associated with internal pore fouling, since most of the surface area of the membrane occurs internally. The high internal surface area of UF membranes is readily apparent from photomicrographs of cross-sections of UF membranes. The photomicrographs show sponge-like structures that suggest convoluted, tortuous pore pathways. Adsorption can lead to more extensive fouling. For instance, a protein might denature upon adsorbing to the surface of an ultrafilter. The denatured protein attracts other proteins, the process repeats and a deposit builds on the membrane surface.

UF membranes are often rated by molecular weight cut-off (MWCO); solutes above the MWCO are retained and those below the MWCO permeate through the membrane. MWCO can be determined by challenging a UF membrane with a polydisperse solute, such as dextran, in a cross-flow filtration experiment A retention profile or curve is determined by comparing the dextran molecular weight distribution in the feed to that in the permeate using size exclusion chromatography (SEC). MWCO is typically defined as the 90 per cent retention level or the molecular weight value on the ordinate where the retention curve crosses 90 per cent on the axis. MWCO ratings are relative. Membrane retentivity depends upon many factors, including the shape of the solute used, the fluid mechanics and the various interactions possible between the solvent, solute and membrane.

MWCO curves are useful for identifying membranes with appropriate selectivity for an intended separation. Predicting the best membrane MWCO is not always straightforward, however. A common assumption is that the membrane MWCO should closely match the molecular weight of the solute of interest. Remarkably, better UF performances are sometimes achieved with membranes having MWCO's significantly higher or lower than the molecular weight of the solute to be retained. For example, lower protein adsorption and flux loss are reported in the literature in the filtration of albumin with polyethersulphone membranes when membrane MWCO's were much larger and smaller than the molecular weight of albumin. How is this possible? Better performances with the low and high MWCO membranes are explained by considering the effects of fouling on the membranes. Higher product rates are sometimes realised with lower MWCO membranes because they exclude more potential foulants and internal pore fouling is reduced. Membranes with higher MWCO's will sometimes effectively separate smaller solutes because solutes aggregate into larger entities or because foulant forms an effective dynamic membrane. The dynamic membrane reduces the effective MWCO of the ultrafilter so that the solute is retained. The larger pores suffer less flow restriction due to adsorption and the greater hydraulic permeability of the larger pores yields high product rates.

A useful analytical tool for predicting and diagnosing foulings Fourier transform infrared spectroscopy (FTIR). FTIR can reveal important information that is useful for predicting and measuring foulants. The FTIR is a standard laboratory instrument for chemical analysis and has been applied for many years in the field of membrane science. It has been successfully applied in identifying which solutes in complex mixtures may cause fouling. The ability to distinguish foulants is advantageous in applications where complex process streams predominate. Fit with an attenuated total reflectance (ATR) accessory, FTIR allows us to look quickly and easily at the chemistry of the foulant layer and membrane surface. The ATR technique can also provide quantitative estimates of fouling.

FTIR can be used to screen membranes for fouling tendencies prior to the first ultrafiltration experiment. Screening can be done by means of a simple static adsorption test. Membranes showing greater static adsorption are expected to foul more during ultrafiltration and are disfavoured. Figure 12.8 illustrates the

FTIR results of a static adsorption test using a polysulphone ultrafilter as the substrate and a water extract of soya flour as the source of potential adsorbates.

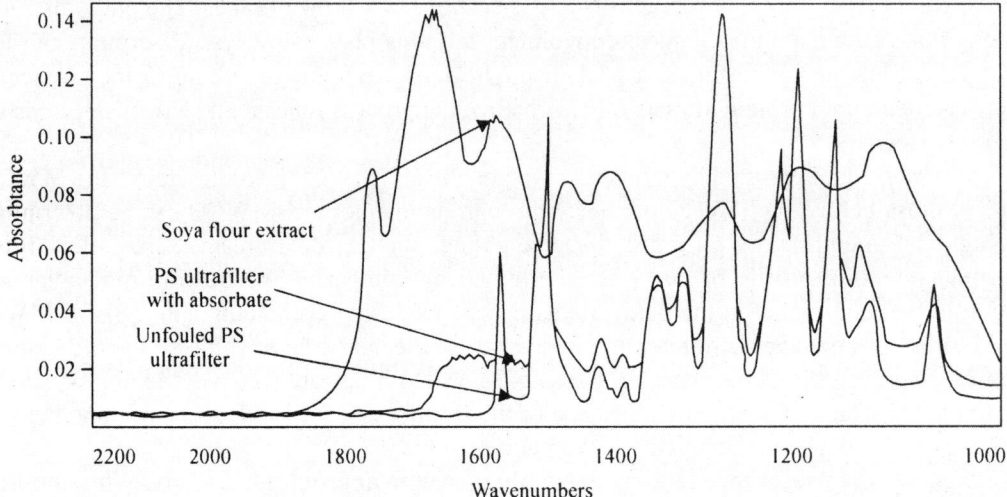

Fig. 12.8. Example of FTIR analysis of polysulphone (PS) ultrafilter static adsorption test.

FTIR can be used to diagnose fouling as well as to predict it. The techniques are similar. Among the diagnostic possibilities, one can:

1. Chemically identify the foulant(s) by searching spectral libraries.
2. Estimate the thickness of foulant layer by comparing the relative size of peaks due to the membrane and foulant.
3. Evaluate the effectiveness of various cleaners by measuring the disappearance of foulant peaks.
4. Surmise internal pore fouling if foulant peaks persist after the surface of the membrane has been thoroughly cleansed.

MICROFILTRATION AND NANOFILTRATION

In the case of microfiltration, a more porous membrane is used than in the other membrane separation technologies, thus yielding a relatively higher flux. It is mainly useful in removing turbid causing materials and can replace conventional granular filtration processes. The most significant design parameters are:

1. Transmembrane pressure.
2. Tangential velocity.
3. Size and geometry of modules.
4. Recirculation factor.

The clean water flux across a membrane without any material being deposited follows Darcy's law:

$$J_w = \Delta P / \mu R_m \qquad \ldots (12.2)$$

The net pressure differential across a membrane, taking into consideration the osmotic pressure is given by $(\Delta P - \Delta II)$ and hence, the expression for the permeate flux is:

$$J_w = (\Delta P - \Delta II) / \mu (R_m + R_c) \qquad \ldots (12.3)$$

where,

J_w =   permeate flux, m/s

ΔP =   pressure difference, N/m^2

R_m =   internal membrane resistance, 1/m

R_c =   resistance due to deposit on the surface, 1/m

μ =   dynamic viscosity, N-sec/m^2

The resistance due to the deposit on the surface is given by the following relationship:

$$R_c = 180 \ (1 - \varepsilon)^2 \ \delta/(d_p^2 \varepsilon^3) \qquad\qquad ... (12.4)$$

where,

ε =   porosity of deposit

δ =   thickness of the deposit, m

d_p =   average diameter of the particles, m

Commercial systems for waste-water treatment are designed to be submerged into built on-site rectangular concrete tanks. These are pre-engineered modular membrane systems that typically use a membrane with a 0.2 micron nominal pore size. A vacuum pump draws water through the membrane fibers of sub-modules submerged in the open top filter tanks. The fibers are the same polypropylene material as those used in the conventional filtration process. A typical system operates under vacuum and utilises improved filter cake characteristics at low pressures with a maximum driving pressure in the league of 85–100 kPa.

There have been only a few studies have evaluated membrane microfiltration of secondary waste-water effluent. Microfiltration membranes might be used to achieve very low turbidy effluents with very little variance in treated water quality. Because bacteria and many other micro-organisms are also removed, such membrane disinfection might avoid the need for chlorine and subsequent dechlorination. Metal salts of iron or aluminium may also be added to enhance membrane performance. For example, iron or aluminium coagulants may be added to precipitate otherwise soluble species such as phosphorus and arsenic as well as improving the removal of viral particles. Coagulation of colloidal materials may also increase the effective size of particles applied to membranes and increase permeate flux by: (i) reducing foulant penetration into membrane pores; (ii) forming a more porous cake on the membrane surface; (iii) decreasing the accumulation of materials on the membrane due to particle size effects of particle transport; and (iv) improving the backflushing characteristics of the membrane.

Membrane microfiltration at the pilot-scale has produced a permeate of similar or better quality than that produced by conventional filtration. Good removal of particulate contaminants, including coliform bacteria, have been observed. In this regard, the process appears to be as effective as chlorination for the removal of coliforms from secondary waste effluent. A key advantage is the ability to filter and disinfect in a single step without the need for subsequent dechlorination. Preliminary results indicate that coagulation pre-treatment in conjunction with membrane microfiltration can be used to reduce phosphorus concentrations as well. There does not appear to be any advantage in running the microfiltration unit in a cross-flow mode and there may even be some disadvantages. The permeate quality and evolution of pressure drop obtained from the membrane operated in the dead-end mode is found similar or superior to that obtained under crossflow conditions.

Although membrane processes have been used successfully for many years in desalting brackish water and seawater, new kinds of membrane processes are now capable of treating water for a wide range of other uses. Some of these new, robust processes promise to do a better job meeting our current water treatment goals than such conventional processes as granular media filtration, carbon filtration and disinfection with chlorine. Engineers classify membranes in many different ways, including describing them by the driving forces used for separating materials (i.e., pressure, temperature, concentration and electrical potential), the mechanism of separation, the structure and chemical composition and the construction geometry. In water treatment, the membranes most widely used are broadly described as pressure driven. Each membrane process is best suited for a particular water treatment function. For example, microfiltration (MF) and ultrafiltration (UF), which are very low pressure processes, most effectively remove particles and micro-organisms. The reverse osmosis (RO) process most effectively desalts brackish water and seawater and removes natural organic matter and synthetic organic and inorganic chemicals. The nanofiltration (NF) process softens water by removing calcium and magnesium ions. These so-called nanofilters are also effective in removing the precursors to disinfection by-products that result from such oxidants as chlorine.

Nearly a decade ago, the use of low-pressure membranes such as MF and UF for disinfection and particle removal was only a concept being studied or was used only on a limited basis. The water community foresaw the possibility of providing primary disinfection without the use of chemicals. Moreover, *Cryptosporidium*, a waterborne enteric pathogen responsible for several disease outbreaks, was gradually showing resistance to traditional disinfectants such as chlorine. Thus, researchers believed that greater emphasis should be placed on removing organisms through physical means as opposed to chemical means.

Membrane processes also offer other advantages over conventional treatments. They reduce the number of unit processes in treatment systems for clarification and disinfection and increase the potential for process automation and plant compactness. Designers also thought membrane plants could be much smaller than conventional plants of the same capacity and given their modular configuration, could be easily expanded. Additionally, these plants would produce less sludge than conventional plants because they wouldn't use such chemicals as coagulants or polymers.

Today many of the projected benefits of MF and UF have been realised. These technologies provide effective disinfection for potable water supplies as they reduce the levels of *Giardia* and *Cryptosporidium*, as well as a variety of bacteria, below detectable levels. MF and UF plants are now in operation throughout the world.

The anticipated US EPA Disinfectant/Disinfection By-Product Rule will lower the maximum contaminant level (MCL) for trihalomethanes (THM) from 100 to 80 µg/l and set an MCL for haloacetic acids (HAA) at 60 µg/l. Thus, NF and RO membrane processes are receiving considerably more attention since they are efficient at removing the precursors to these by-products. RO and NF membranes, in addition to desalting brackish water and seawater, are being used to remove inorganic chemicals such as nitrates, as well as synthetic organic chemicals.

Membrane plant design begins with the selection of the membrane, which can be organic or inorganic in composition. Membrane manufacturers strive to formulate membranes that provide a desired permeate quality, are durable and resistant to fouling and can be produced at a competitive cost. Most commercial water treatment NF and RO membranes are made up of organic polymers and are asymmetric. The active layer responsible for the separation process is typically a few micrometers thick and is supported on a highly permeable layer that adds mechanical strength to the membrane. One type of asymmetric membrane for NF and RO systems, the thin-film composite (TFC), shows great promise for potable water treatment. These membranes generally consist of an ultrathin active layer coated onto a microporous layer that, in turn, is supported on a mechanically strong base. TFC membranes typically have higher water permeability and chemical resistance than symmetric membranes. MF and UF membranes are constructed in either asymmetric or symmetric configurations. A number of hydrophilic and hydrophobic, polymeric materials are used in manufacturing these membranes. These include cellulosic polymers, polypropylene, polysulphones and polyamides. The choice of material will influence contaminant rejection characteristics, durability and fouling potential.

Membrane systems consist of membrane elements or modules. For potable water treatment, NF and RO membrane modules are commonly fabricated in a spiral configuration. All important consideration of spiral elements is the design of the feed spacer, which promotes turbulence to reduce fouling. MF and UF membranes often use a hollow fibre geometry. This geometry does not require extensive pre-treatment because the fibers can be periodically backwashed. Flow in these hollow fiber systems can be either from the inner lumen of the membrane fiber to the outside (inside-out flow) or from the outside to the inside of the fibers (outside-in flow). Tubular NF membranes are now just entering the marketplace.

MF and UF systems can be designed to operate in various process configurations. A common configuration is one in which the feedwater is pumped with a cross-flow tangential to the membrane. The only pre-treatment usually provided is a crude pre-screening (usually 50 to 300 µm). The water that permeates the membrane is clean. The water that does not permeate is recirculated as concentrate and blended with additional feedwater just after the preliminary filter. To control the concentration of the solids in the recirculation loop, some of the concentrate is discharged at a specified rate.

MF and UF systems may also operate in a direct filtration configuration, with no cross-flow (or recirculation). This is often termed dead-end filtration. All of the pre-screened feedwater passes through the membrane. Therefore, there is 100 per cent recovery of this water, except for the small fraction of the water used to periodically backwash the system. MF and UF plants typically rely on either liquid or pneumatic backwashing systems. Most MF and UF water treatment plants use this direct flow configuration, since it saves considerably on energy by not requiring recirculation. There are also capital cost savings since there is no need to purchase recirculation pumps and associated piping.

RO and NF systems usually operate in a series of stages. In a three-stage system, the first stage consists of three pressure vessels, which usually contain four to eight membrane elements; the second stage has two pressure vessels and the final stage has one. In full-scale plants, elements are approximately 1000 mm long and have a diameter of 200 mm. Permeate is collected from each pressure vessel. The concentrate from the first stage serves as the feed to the second; concentrate from the second stage serves as the feed to the third. Consequently, each successive stage of the array increases the total system recovery. For many groundwater applications, the pre-treatment required for RO or NF consists in adding acid or antiscalant and then passing the feed through a cartridge filter. However, for surface water, more extensive pre-treatment is necessary. This may involve conventional treatment, MF, UF, slow sand filtration or in some cases, granular activated carbon adsorption.

One innovative process configuration for surface water and tertiary waste-water treatment involves the use of double-membrane systems, consisting of a low pressure and a high-pressure membrane in series. This treatment is effective for both microbial and chemical contaminant control. The first membrane (MF or UF) is used to help prevent fouling of the second, higher-pressure membrane system (RO or NF). Another system uses membranes designed to be immersed in a process tank and suction instead of pressure to draw water through the membrane hollow fiber lumen.

Pilot testing is often a key aspect of successful membrane plant design. One of the most important reasons for conducting a pilot study is to evaluate the influence of water quality on membrane fouling. It is critical to determine if the process is feasible for a specific water source, particularly for those that exhibit significant water quality changes on a seasonal basis. Other reasons for pilot tests include demonstrating regulatory compliance, identifying the most effective and appropriate processes, evaluating new membrane products and establishing design criteria for a full-scale plant. Designers can also evaluate pre-treatment options by conducting experiments using parallel treatment trains with varying pre-treatment processes but identical operating conditions. This avoids the confounding factors in data interpretation, such as changing source water quality. Designers can also verify and fine-tune chemical cleaning procedures for site-specific conditions. In pilot studies, designers are able to optimise the operating conditions of each individual stage of a multistage installation.

As membranes filter out the impurities from the water, the membranes themselves become fouled (or clogged) and less effective. The fouling of membranes has been one the primary impediments to their more widespread application in water treatment. Membrane systems operate in one of two modes: constant transmembrane water flux (flow rate per unit membrane area) with variable pressure or constant pressure with variable transmembrane water flux. The former is the more common. Membrane fouling occurs during an increase in transmembrane pressure to maintain a particular water flux or during a decrease in water flux when the system is operated at constant pressure. In general, membranes can be fouled by an accumulation of inorganic particles (for example, clays, iron, manganese and silica) and organic compounds (such as humic and fulvic acids, hydrophilic and hydrophobic materials and proteins). Bacteria can also adhere to the membranes and create a biofilm. Accurate tests to predict fouling still do not exist. However, researchers can conduct 'autopsies' on fouled membranes prior to chemical cleaning to analyse the nature and composition of the contamination. They can then adjust pre-treatment and chemical cleaning procedures. Fouling can be controlled by hydrodynamic and chemical methods, periodic backwashing and chemical cleaning. Other methods include improving pre-treatment and changing operating conditions. Membrane fouling rates are functions of the operating conditions such as water flux and recovery. Typically, reducing the operating flux and recovery will reduce fouling. Research shows that, for MF and UF, increasing the frequency of backwashing also decreases the rate of fouling. Because filtered or raw water is used for backwashing, the net recovery of direct-flow MF and UF systems decreases as the frequency of backwashing increases. Improved pre-treatment also reduces membrane fouling. Water treatment operators have decreased MF and UF fouling rates by using coagulation/flocculation/sedimentation and dissolved air flotation pre-treatment.

Chemical scaling is another form of fouling that occurs in NF and RO plants. The thermodynamic solubility of salts such as calcium carbonate and calcium and barium sulphate imposes an upper boundary on the system recovery. Thus, it is essential to operate systems at recoveries lower than this critical value to avoid chemical scaling, unless the water chemistry is adjusted to prevent precipitation. It is possible to increase system recovery by either adjusting the pH or adding an anti-scalant or both.

REVERSE OSMOSIS

When pure water and a salt solution are introduced on opposite sides of a semi-permeable membrane in a vented container, the pure water diffuses through the membrane and dilutes the salt solution. At equilibrium, the liquid level on the saline water side of the membrane will be above that on the freshwater side; this process is known as osmosis and is depicted in Fig. 12.9. The view on the left illustrates the commencement of osmosis and the center view presents conditions at equilibrium. The effective driving force responsible for the flow is osmotic pressure. This pressure has a magnitude dependent on membrane characteristics, water temperature and salt solution properties and concentration. By applying pressure to the saline water, the flow process through the membrane can be reversed. When the applied pressure on the salt solution is greater than the osmotic pressure, fresh water diffuses in the opposite direction through the membrane and pure solvent is extracted from the mixed solution; this process is termed reverse osmosis (RO). The fundamental difference between reverse osmosis and electrodialysis is that in reverse osmosis the solvent permeates the membrane, while in electrodialysis the solute moves through the membrane.

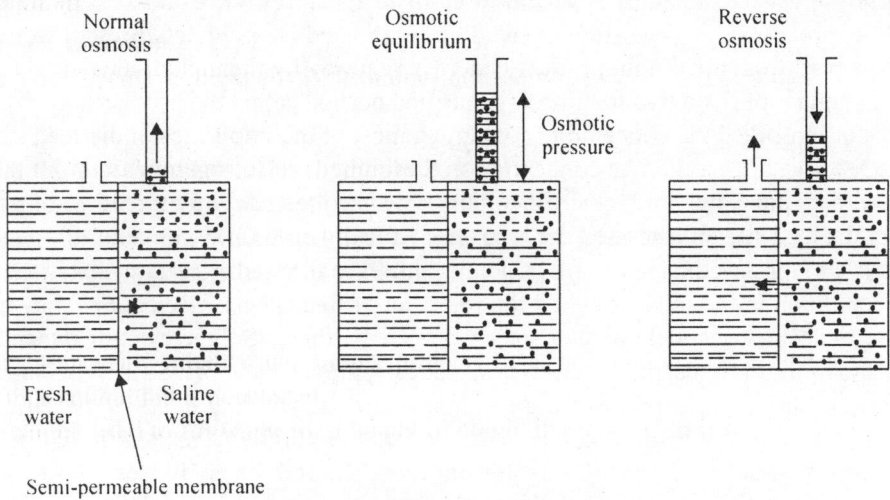

Fig. 12.9. Principle of reverse osmosis.

Reverse osmosis (RO) is a membrane treatment process used to separate dissolved solutes from water and describes any pressure-driven membrane that uses preferential diffusion for separation. A typical RO membrane is made of synthetic semi-permeable material, which is defined as a material that is permeable to some components in the feed stream and impermeable to other components and has an overall thickness of less than 1 mm. Water is pumped at high pressure across the surface of the membrane; causing a portion of the water to pass through the membrane, as shown schematically, in Fig. 12.10. Water passing through the membrane, called permeate, is relatively free of targeted dissolved solutes, while the remaining water, called concentrate (also commonly called retentate, reject water or brine), exits at the far end of the pressure vessel. The delineation of membrane processes, applications for RO, an historical perspective, a process description, process fundamentals and process design are presented in this chapter.

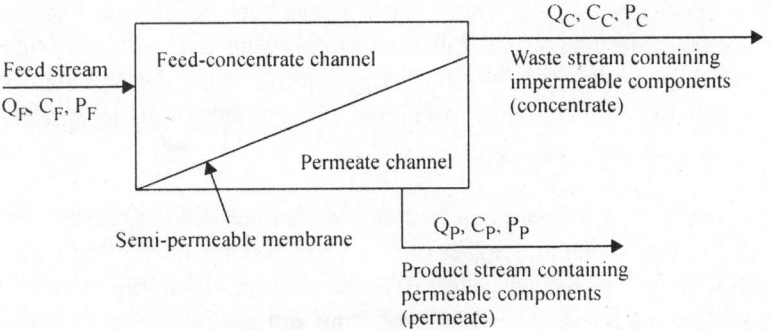

Fig. 12.10. Schematic of separation process through reverse osmosis membrane.

Reverse osmosis is a means for separating dissolved solids from water molecules in aqueous solutions as a result of the membranes being composed of special polymers which allow water molecules to pass through while holding back most other types of molecules; since true 'pores' do not exist in the membrane, suspended solids are also retained by superfiltration. In an actual reverse osmosis system operating in a continuous-flow process, feed water to be treated or desalinated is circulated through an input passage of the cell, separated from the output product water passageway by the membrane.

The feed stream is split into two fractions—a purified portion called the product water (or permeate) and a smaller portion called the concentrate containing most of the impurities in the feed stream. At the far end of the feed-water passage, the concentration (dewatered) reject stream exits from the cell. After permeating the membrane, the product (fresh-water) flow is collected. The percentage of product water obtained from the feed stream is termed the recovery, typically around 75 per cent.

The ratio (F–P)/F or the concentration of a solute species in the feed (F) minus that in the product (P) over the concentration in the feed, is called the rejection of that species. Rejections may be stated for particular ions, molecules or conglomerates such as TDS or hardness. Solids rejection depends on factors such as types and forms of solids, membrane types, recovery, pressure and pH. Suspended solids (typically defined as particles larger than 0.5 micron mean diameter and including colloids, bacteria, and algae) are rejected 100 per cent; that is, none can pass through the membrane. Weakly ionised dissolved solids (usually organics, but may include other materials such as silicates) undergo about a 90 per cent rejection at normal recoveries for certain membranes. Although pH can strongly influence the rejection, when the molecular weight of these solids is less than 100, rejection decreases appreciably. Ionised solids or salts, are rejected independent of molecular weight and at molecular weights considerably below 100. At 75 per cent recovery and pressures greater than 250 pounds per square inch, overall rejection of total dissolved solids (TDS) is about 90 per cent. Rejections vary with pressure because the actual salt flow through the membrane remains fairly constant, but the water permeation depends nearly linearly on pressure, affecting the ratio of concentrations. For example, rejection of sodium chloride can fall from 90 per cent at 300 pounds per square inch to 20 per cent at 50 pounds per square inch, indicating the need to operate at the highest pressures possible.

Cellulose acetate is a common membrane material, but others include nylon and aromatic polyamides. The mechanism at the membrane surface involves the influent water and impurities attempting to pass through the pressurised side, but only pure water and certain impurities soluble in the membrane emerge from the opposite side.

Various configurations of membranes with different surface-to-volume ratios and different flux capabilities (gallons per day per square foot or gpd/ft') have been developed. Each type of membrane is a flexible plastic filmno more than 4 to 6 mils thick, firmly supported. Basic designs include the plate and frame, the spiral-wound module (jellyroll configuration), the tubular and the newest of the process designs, the hollow-fine fiber. Fibers range from 25 to 250 microns (0.001 to 0.01 inch) in diameter, can withstand enormous pressure, are self-supporting and can be bundled very compactly within a containment pipe. While product flow per square foot of fiber surface is less than that for an equivalent area of flat membrane, the difference in surface area more than compensates for the reduced unit flux.

Major problems inherent in general applications of RO systems have to do with: (i) the presence of particulate and colloidal matter in feed water; (ii) precipitation of soluble salts; and (iii) physical and chemical make-up of the feed water. All RO membranes can become clogged, some more readily than others. This problem is most severe for spiral-wound and hollow-fiber modules, especially when sub-micron and colloidal particles enter the unit (larger particulate matter can be easily removed by standard filtration methods). A similar problem is the occurrence of concentration-polarisation, previously, discussed for ED processes. Concentration-polarisation is caused by an accumulation of solute on or near the membrane surface and results in lower flux and reduced salt rejection.

The degree of concentration that can be achieved by RO may be limited by the precipitation of soluble salts and the resultant scaling of membranes. The most troublesome precipitate is calcium sulphate. The addition of poly-phosphates to the influent will inhibit calcium sulphate scale formation, however and precipitation of many of the other salts, such as calcium carbonate, can be prevented by pre-treating the feed either with acid or zeolite softeners, depending on the membrane material.

Hydrolysis of cellulose acetate membranes is another operational problem and occurs whenever the feed is too acidic or alkaline; that is, the pH deviates beyond designed range limits. As may readily happen, whenever CO_2 passes through the membrane, the resultant permeate has a low pH. The operational solution is to remove the gas from the permeate by deaerators, by strong-base anion resins or a complementary system—for example, RO and ion exchange, in series. Aromatic polyamide or nylon membranes are much less sensitive to pH than cellulose acetate. Compounds such as phenols and free chlorine that are either soluble in the membrane or vice versa will be poorly rejected and may damage the membrane. Procedures to improve feed-water make-up and thus reduce such membrane damage include acid pre-treatment of the feed water, dechlorination, periodic cleaning or replacement of the membrane, sequestration of cations, coagulation and filtration of organics and use of alternative, more durable membrane materials.

Reverse osmosis process is applied—or undergoing evaluation for imminent application—to a number of water-upgrading needs including high-purity rinse water production for the electronics industry (semi-conductor manufacturing), potable municipal water supplied for newly-developed communities (for example, large coastal plants to upgrade brackish well water contaminated by seawater intrusion), boiler feed-water supplies, spent liquor processing for pulp and paper mills and treatment of acid mine drainage.

In desalting operations, distillation plants have provided the major portion of the world's capacity. As the world's requirements for treated water increase, however and water quality standards become more stringent, the membrane treatment processes in general and commercial RO processes in particular have been undergoing appreciable development. Important factors in the expansion of commercial RO applications are their favourably low power requirements and the realisation of continuous technical improvements in membranes which are used in RO systems. A general guideline in water benefication is that RO is most frequently considered for cases in which the TDS is greater than 2000 to 3000 ppm; ED generally applies when the TDS is less than 2000 to 3000 ppm. However, many exceptions exist, based on feed-water species and product requirements.

One of the most important applications of RO is in the reclamation of large volumes of municipal and industrial waste-water and the concentration of the solids for simplified disposal. The value of the reclaimed water offsets the cost of RO and dilute waste-water concentration leads to economies in any further required liquid waste treatment.

Unrestricted use of reclaimed waste-water for drinking water, however, requires careful examination. While practically a complete barrier to viruses, bacteria and other toxic entities that must be kept out of a potable supply, RO membranes could pose serious problems should any defect develop in their separation mechanism. Given the purity and clarity of RO-treated waste-water, however, it might be advantageous to use RO and then subject the product to well-established disinfection procedures.

One should remember that RO uses a semi-permeable membrane. As such, the membrane is permeable to only very light molecules like water. Under atmospheric conditions the fresh water flows into the solution which is called osmotic flow. But for purification purposes, this is no use and hence we employ the reverse of osmotic flow.

For this to happen, we need to apply external pressure in excess of osmotic pressure. The osmotic pressure is given by:

$$\Delta p = nRT \qquad \qquad ... (12.5)$$

Of course, one should be familiar with this equation (the ideal gas law), where 'n' is the molar concentration of solute, R is the universal gas law constant and T is absolute temperature in °K. The permeate flow can be calculated from:

$$J_w = A_m (\Delta P - \Delta p) \qquad \qquad ... (12.6)$$

in this expression, A_m is the membrane permeability coefficient.

It is useful to compare the merits of the various processes for seawater desalination. Although the comparison will be primarily qualitative, it should be helpful in providing a deeper insight into the strengths and weaknesses of process. Foremost among the aspects of comparison is the energy consumption of each process one consider. With the known 'process specification, it is theoretically possible to calculate the minimum work or energy needed for separation of pure water from brine. For the real process, however, the actual work required is likely to be many times the theoretically possible minimum. This is because the bulk of the work is required to keep the process going at a finite rate rather than to achieve the separation.

The minimum work needed is equal to the difference in free energy between the incoming feed (i.e., seawater or brackish water) and outgoing streams (i.e., product water and discharge brine). For the normal seawater (3.45 per cent salt) at a temperature of 25°C, for usual recoveries the minimum work has been calculated as equal to about 0.86 kWh/m^{-3}. Table 12.5 makes the desired comparison.

Table 12.5. Energy requirements of four industrial desalination processes.

Description	MSF	MEB	MEB/V C	RO
Possible unit size	60000	60000	24000	24000
Energy consumption (kWh/m^3) Electrical/mechanical	4–6	2–2.5	7–9	5–7
Energy consumption (kWh/m^3) Thermal	55–120	30–120	None	None
Electrical equivalent for thermal energy (kWh/m^{-3})	8–18	2.5–10	None	None
Total equivalent energy (kWh/m^{-3})	12–24	4.5–12.5	7–9	5–7

There are no major technical obstacles to desalination as a means of providing an unlimited supply of fresh water, but the high energy requirements of this process pose a major challenge. Theoretically, about 0.86 kWh of energy is needed to desalinate 1 m^3 of salt water (34500 ppm). This is equivalent to 3 kJ kg^{-1}. The present day desalination plants use 5 to 26 times as much as this theoretical minimum depending on the type of process used. Clearly, it is necessary to make desalination processes as energy-efficient as possible through improvements in technology and economies of scale.

Desalination as currently practiced is driven almost entirely by the combustion of fossil fuels. These fuels are in finite supply; they also pollute the air and contribute to global climate change. The whole character of human society in the 20th century in terms of its history, economics and politics has been shaped by energy obtained mostly from oil. Almost all oil produced to date is what is called conventional oil, which can be made to flow freely from wells (i.e. excluding oil from tar sands and shale).

While salinity or salty water, is generally used to describe and measure seawater or certain industrial wastes, we use the term total dissolved solids (TDS) to describe water high in various salt compounds and dissolved minerals. While one could have very high total dissolved solids and very low salinity from a chemistry standpoint, here we are talking about high TDS. Total dissolved solids (TDS) refers to the amount of dissolved solids (typically various compounds of salts, minerals and metals) in a given volume of water. It is expressed in parts per million (also known as milligrams per liter) and is determined by evaporating a small amount of water in the laboratory and weighing the remaining solids. Another way to approximately determine TDS is by measuring the conductivity of a water sample and converting the resistance in micromhos to TDS. TDS in municipally treated water in our area range from 90 ppm to over 1000 ppm. The most common range in city water is 200–400 ppm. The maximum contaminant level set by USEPA is 500 ppm. California sets its standard as 1000 ppm, probably due to the high number of groundwater sources in the state. The MCL is known as a Secondary Standard and in one sense, refers to the aesthetic quality of a given water. The higher the TDS, the less palatable the water is thought to be. Sea water ranges from 30000 to 40000 ppm. Generally, one wants a TDS of less than 500 for household use. In our experience, it appears that folks can tolerate for general household use, soft clean water with a TDS of up to 1500 ppm. When the levels start to exceed 1500 ppm, most people start to complain of dry skin, stiff laundry and corrosion of fixtures. White spotting and films on surfaces and fixtures is also common at these levels and can be very difficult or impossible to remove.

TDS affects taste also and water over 500–600 ppm can taste poor. When the levels top 1500 ppm, most people will report the water tastes very similar to weak alka-seltzer. TDS is removed by distillation, reverse-osmosis or electrodialysis. Depending on the water chemistry, reverse osmosis systems are the most popular, given their low cost and ease of use. Distillers work very well also and produce very high quality water, but require electricity and higher maintenance than reverse osmosis systems. For whole house treatment, commercial-sized reverse osmosis systems are usually the best approach.

PROCESS DESCRIPTION

Reverse osmosis is used to separate solutes from water, relying on the physical and chemical properties of the solutes for separation. A high-pressure feed stream is directed across the surface of a semi-permeable material and due to a pressure differential between the feed and permeate sides of the membrane, a portion of the feed stream passes through the membrane. The membrane can be a dense material (i.e., without pores or void spaces), as in the case of high-pressure RO membranes or porous material, as in the case of NF membranes. As water passes through the membrane, solutes are rejected and the feed stream becomes more concentrated. The permeate stream exits at nearly atmospheric pressure, while the concentrate remains at nearly the feed pressure. Reverse osmosis is a continuous separation process; that is, there is no periodic backwash cycle.

The smallest physical unit of production capacity in a membrane plant is called a membrane element. The membrane elements are enclosed in pressure vessels mounted on skids, which have piping connections for feed, permeate, and concentrate streams. A group of pressure vessels operated in parallel is called a stage. The concentrate from one stage can be fed to a subsequent stage to increase water recovery (a multistage system, sometimes called a brine-staged system) or the permeate from one stage can be fed to a second stage to increase solute removal (a two-pass system, also sometimes called a permeate-staged system). In multistaged systems, the number of pressure vessels decreases in each succeeding stage to maintain sufficient velocity in the feed channel as permeate is extracted from the feed water stream. A unit of production capacity, which may contain one or more stages, is called an array. Schematics of various arrays are shown in Fig. 12.11. The ratio of permeate to feed water flow (recovery) ranges from about 50 per cent for seawater RO systems to about 90 per cent for low-pressure NF systems. Several factors limit recovery, most notably osmotic pressure, concentration polarisation and the solubility of sparingly soluble salts. Two and three-stage systems are used to improve system recovery.

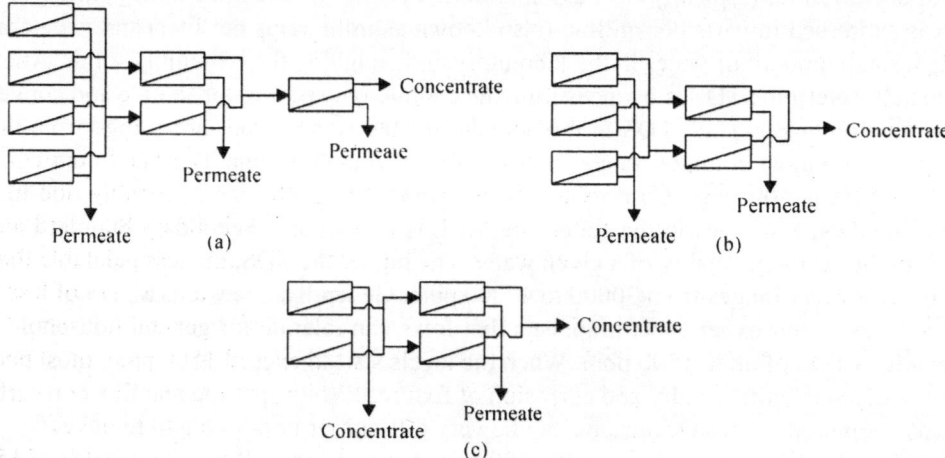

Fig. 12.11. Array configurations of reverse osmosis facilities: (a) 4 × 2 × 1 concentrate-staged array, (b) 3 × 2 concentrate-staged array; and (c) two-pass system.

Pre-treatment and Post-treatment

A schematic of a RO system with typical pre-treatment and post-treatment processes is shown in Fig. 12.12 and described below.

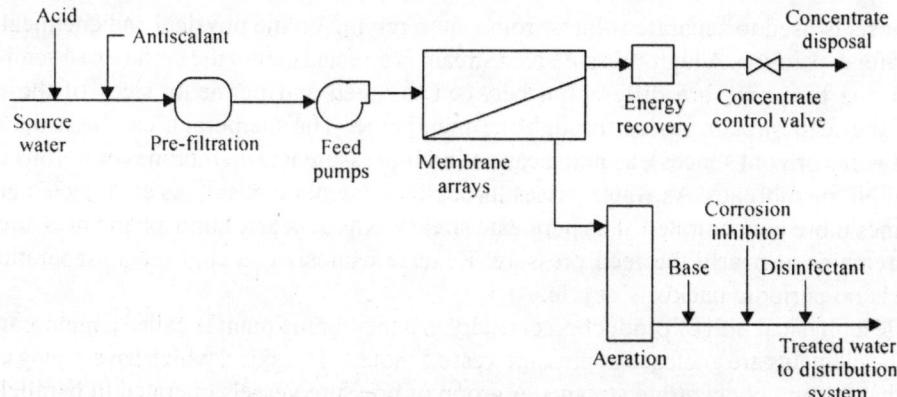

Fig. 12.12. Schematic of typical reverse osmosis facility.

Pre-treatment

Feed water pre-treatment is required in virtually all RO systems. When sparingly soluble salts are present, the purpose of the first pre-treatment process is to prevent scaling. Solutes are concentrated as water is removed from the feed stream and the resulting concentration can be higher than the solubility product of various salts. Without pre-treatment, these salts can precipitate onto the membrane surface and irreversibly damage the membrane. Scale control consists of pH adjustment and/or antiscalant addition. Adjusting the pH changes the solubility of precipitates and antiscalants interfere with crystal formation or slow the rate of precipitate formation.

The second pre-treatment process is pre-filtration to remove particles. Without a backwash cycle, particulate matter can clog the feed channels or accumulate on the membrane surface unless the concentration is low. As a minimum, cartridge filtration with a 5 µm strainer opening is employed for pre-filtration, although granular filtration or membrane filtration pre-treatment may be necessary for surface water sources. Disinfection is another typical pre-treatment step and is used to prevent biofouling. Some membrane materials are incompatible with disinfectants, so the disinfectant can only be applied in specific situations and must be matched to the specific membrane type to prevent premature membrane degradation. After scale control and pre-filtration, the feed-water is pressurised with feed pumps. The feed-water pressure ranges from 5 to 10 bar (73 to 145 psi) for NF membranes, from 10 to 30 bar (145 to 430 psi) for low-pressure and brackish water RO and from 55 to 85 bar (800 to 1200 psi) for seawater RO. The membrane arrays can consist of one or more stages that may be concentrate staged or permeate staged.

Post-treatment

Permeate typically requires post-treatment, which consists of removal of dissolved gases and alkalinity and pH adjustment. Membranes do not remove efficiently small, uncharged molecules, in particular dissolved gases. If hydrogen sulphide is present in the source groundwater, it must be stripped prior to distribution to consumers. If sulphides are removed in the stripping process, provisions must be made to scrub the sulphides from the stripping tower off-gas to prevent odour and corrosion problems. The stripping of carbon dioxide raises pH and reduces the amount of base needed to stabilise the water. Permeate is typically low in hardness and alkalinity and frequently has been adjusted to an acidic pH value to control scaling. Consequently, the permeate is corrosive to downstream equipment and piping. Alkalinity and pH adjustments are accomplished with various bases and corrosion inhibitors are used to control corrosion.

Concentrate Stream

The concentrate stream is under high pressure when it exits the final membrane element. This pressure is dissipated through the concentrate control valve. Some RO systems utilise energy recovery equipment on the concentrate line to minimise the energy lost through the concentrate control valve. Unlike cross-flow membrane filtration, the concentrate stream is not recycled to the head of the plant but is a waste stream that must be discarded. Concentrate disposal can be a significant issue in the design of RO facilities and the concentrate may require treatment before disposal. Methods for concentrate disposal include ocean, brackish river or estuary discharge; discharge to a municipal sewer and deep-well injection. Other concentrate disposal options, including evaporation ponds, infiltration basins and irrigation, are used by a small number of facilities.

Membrane Element Configuration

Reverse osmosis membrane elements are fabricated in either a spiral-wound configuration or a hollow-fine-fiber (HFF) configuration.

Spiral-wound modules

Spiral-wound modules are constructed of several elements in series. The basic construction of a spiral-wound element is shown in Fig. 12.13 and a photograph of typical elements is shown in Fig. 12.14. An envelope is formed by sealing two sheets of flat-sheet membrane material along three sides, with the active membrane layer facing out. A permeate carrier spacer material inside the envelope prevents the inside surfaces from touching each other and provides a flow path for the permeate inside the envelope. The open ends of several envelopes are attached to a perforated central tube known as a permeate collection tube. Feed-side mesh spacers are placed between the envelopes to provide a flow path and create turbulence in the feed-water. By rolling the membrane envelopes around the permeate collection tube, the exterior spacer forms a spirally shaped feed channel. This channel, exposed to element feed-water at one end and concentrate at the other end, is known as the feed-concentrate channel. Membrane feed-water passes through this channel and is exposed to the membrane surface. Spiral-wound elements are typically 1 m (40 inch) long and up to 0.3 m (12 inch) in diameter, although 0.2 m (8 inch) diameter elements are most widely used. Four to seven elements are arranged in series in a pressure vessel, with the permeate collection tubes of the spiral-wound elements coupled together. Some, manufacturers produce elements that are 1.5 m (60 inch) long to reduce the number of permeate tube interconnections within a pressure vessel.

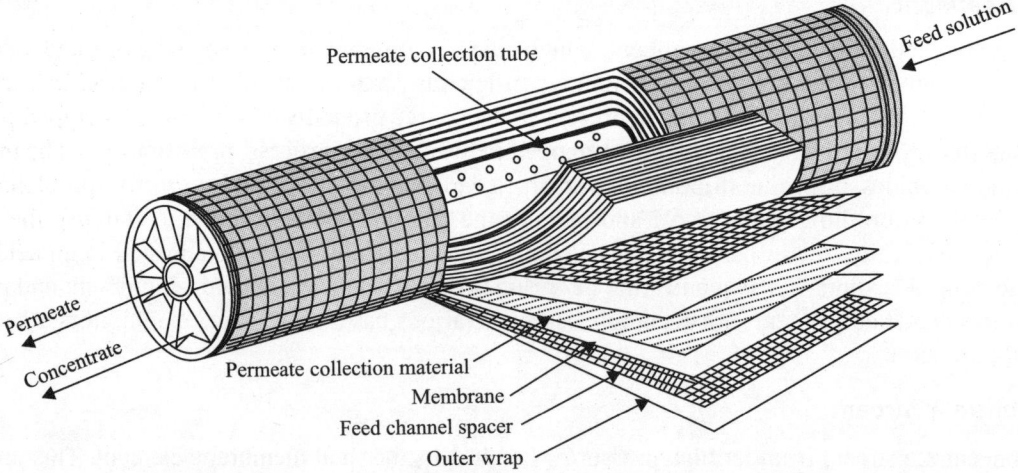

Fig. 12.13. Construction of spiral-wound membrane element.

During operation, the pressurised feed-water enters one side of the pressure vessel and encounters the first membrane element. As the water flows tangentially across the membrane surface, a portion of the water passes through the membrane surface and into the membrane envelope and flows spirally toward the permeate collection tube. The remaining feed-water, now concentrated, flows to the next element in series and the process is repeated until the concentrate exits the pressure vessel. Individual spiral-wound membrane elements have a permeate recovery of 5 to 15 per cent per element. Head loss develops as feed-water flows through the feed channels and spacers, which reduces the driving force for flow through the membrane surface. This feed-side head loss across a membrane element is low, typically less than 0.5 bar (psi) per element.

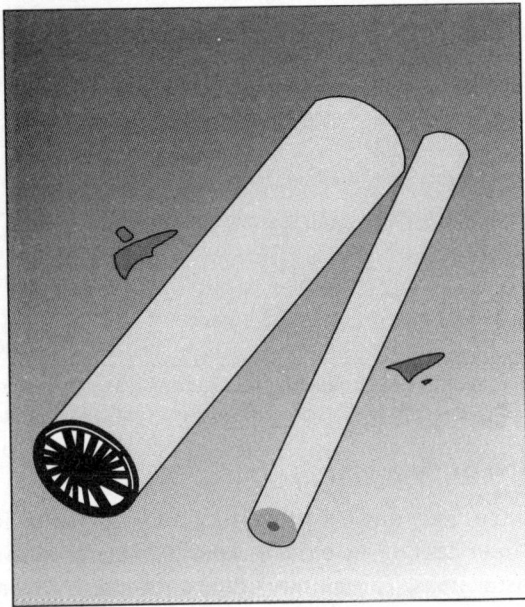

Fig. 12.14. Photograph of spiral-wound membrane elements.

Hollow-fine-fiber modules

The HFF configuration is similar to the hollow fibers used in membrane filtration. Feed-water passes over the outside of the fiber and is forced through the wall of the fiber and the permeate is collected in the lumen (or inner annulus) of the fiber. The original manufacturer of HFF membranes was DuPont, who manufactured fiber with an outside diameter (OD) of 0.085 mm (about the thickness of human hair) and inside diameter of 0.042 mm, considerably thinner than the hollow fibers used in membrane filtration, which have an OD of 1 to 2 mm (about the thickness of pencil lead).

The active surface of the membrane is on the outside surface of the fiber and is 0.1 to 1 μm thick. In a typical HFF module, the feed enters one end of the module and the concentrated brine exits from the opposite end. The fibers are folded and suspended lengthwise in the module, with the open ends of a set of fibers exposed at each end of the module. The fiber bundles are wound helically around a center tube. A single module can contain several hundred thousand fibers. Product water recovery per element is 30 per cent.

APPLICATIONS OF REVERSE OSMOSIS

Uses for RO in water treatment as well as alternative processes are listed in Table 12.6. These objectives encompass the desalination of ocean or brackish water, softening, natural organic matter (NOM) removal, and specific contaminant removal.

Table 12.6. Reverse osmosis objective and alternative processes.

Process objective	Membrane process name	Alternative processes
Ocean or seawater desalination	High-pressure RO, seawater RO	Multistage flash distillation (MSF), multieffect distillation (MED), vapour compression distillation (VCD)
Brackish water desalination	RO, low-pressure RO, NF	Multistage flash distillation,[a] multieffect distillation,[a] vapour compression,[a] electrodialysis, electrodialysis reversal
Softening	Membrane softening, NF	Lime softening, ion exchange
NOM removal for DBP control	NF	Enhanced coagulation/softening, GAC
Specific contaminant removal[b]	RO	Ion exchange, activated alumina, coagulation, lime softening, electrodialysis, electrodialysis reversal
High-purity process water	RO	Ion exchange, distillation

[a] MSF, MED and VCD are rarely competitive economically for brackish water desalination.
[b] Applicability of alternative processes depends on the specific contaminants to be removed and their concentration.

Desalination of Ocean Water or Seawater

The scarcity of freshwater sources may mean a strong future for the use of RO for desalination of ocean water or seawater. About 97.5 per cent of the earth's water is in the oceans and about 75 per cent of the world's population live in coastal areas. The salinity of the ocean ranges from about 34000 to 38000 mg/l as total dissolved solids (TDS), nearly two orders of magnitude higher than that of potable water (the WHO's guidance level for TDS is 1000 mg/l and the United States has a secondary standard of 500 mg/l). Although desalination of ocean water is an energy-intensive process, overall desalination costs are dropping and this process is becoming more competitive with other treatment options in areas where freshwater is scarce.

The Middle East is currently the most prominent market for desalination of ocean water. Virtually 100 per cent of the drinking water in Kuwait and Qatar and 40 to 60 per cent of the drinking water in Bahrain, Saudi Arabia and Malta is produced by desalination. Evaporative processes such as multistage flash (MSF) distillation and multieffect distillation (MED) are common in the Middle East, which has vast energy resources but little freshwater. Worldwide, 66 per cent of ocean water desalination is done with the MSF process, compared to 22 per cent using the RO process. When all brackish water sources are included, however, the distribution between evaporative and membrane processes changes dramatically. Multistage flash distillation and RO each contribute about 44 per cent of the total desalination capacity when all source water are considered.

Other Applications for Reverse Osmosis

Additional applications of RO include softening, NOM removal for controlling disinfection by-product (DBP) formation, desalination of brackish groundwater and specific contaminant removal. Nanofiltration or softening membranes are capable of removing 80 to 95 per cent of divalent ions like calcium and magnesium with low removal of low-molecular-weight (MW) monovalent ions such as sodium and chloride. By allowing passage of sodium and chloride, the osmotic pressure differential is minimised. Nanofiltration membranes can soften water without the voluminous sludge production of lime softening, although concentrate disposal can be a significant regulatory obstacle. Nanofiltration membranes that effectively remove hardness are also effective at removing NOM, making them an excellent treatment option for colour removal and DBP formation control because removing NOM and colour from water before disinfection with free chlorine typically reduces the formation of DBPs. Nanofiltration membranes have widespread use in Florida, where the groundwater is either brackish or very hard, highly coloured freshwater.

An additional use for RO is specific contaminant removal. The EPA has designated RO as a best available technology (BAT) for removal of numerous inorganic contaminants, including antimony, arsenic, barium, fluoride, nitrate, nitrite, and selenium and radionuclides, including beta-particle and photon emitters, alpha emitters and radium-226. Reverse osmosis has also been demonstrated to be effective for removing larger MW synthetic organics such as pesticides. Use of RO for specific contaminants, however, is rare because alternative technologies are frequently more cost effective and the disposal of the concentrate stream may present challenges.

PROCESS FUNDAMENTALS

The fundamentals of RO include the membrane material properties, the phenomenon of osmotic pressure, the mechanisms for water and solute permeation, the equations used to predict water and solute flux, the phenomenon of concentration polarisation and the causes and measurement of membrane fouling. These topics are addressed in this section.

Membrane Structure, Material Chemistry and Rejection Capabilities

An understanding of the mechanisms that control RO begins with an understanding of the membrane. Important properties include the physical structure, chemistry and rejection capabilities of the membranes.

Membrane structure

The resistance to flow through a membrane is inversely proportional to thickness. To achieve any appreciable water flux, the active membrane layer must be extremely thin, which in RO and NF membranes ranges from about 0.1 to 2 µm. Material this thin lacks structural integrity, so these membranes are comprised of several layers, with a thin active layer providing separation capabilities and thicker, more porous layers providing structural integrity. Multilayer membranes are fabricated in two ways. As previously mentioned, asymmetric membranes are formed from a single material that develops into active and support layers during the casting process (in other words, the membranes are chemically homogeneous but physically heterogeneous). Thin film composite membranes are composed of two or more materials cast on top of one another. An advantage of thin-film membranes is that separation and structural properties can be optimised independently using appropriate materials for each function.

The active layer of RO membranes must selectively allow water to pass through the material while rejecting dissolved solutes that may have dimensions similar to water molecules. Separation of small ions cannot be accomplished if they are convectively carried with liquid water. Thus, RO membranes are fabricated of a dense material, meaning a permeable but not porous material with no void spaces through which liquid water travels. Water and solutes dissolve into the solid membrane material, diffuse through the solid and reliquefy on the permeate side of the membrane. Low-pressure RO or NF membranes may have void spaces large enough for the convective flow of liquid water through the membrane.

Membrane material

Membrane performance is strongly affected by the physical and chemical properties of the material. The ideal membrane material is one that can produce a high flux without clogging or fouling and is physically durable, chemically stable, non-biodegradable, chemically resistant and inexpensive. The materials most widely used in RO are cellulosic derivatives and polyamide derivatives.

Cellulose acetate membranes

Membranes composed of cellulose acetate (CA) are typically of asymmetric construction. Cellulose acetate is hydrophilic, which helps to maintain high flux values and to minimise fouling. The structural properties of CA are not ideal, however and the material is not tolerant of temperatures above 30°C, tends to hydrolyse when the pH value is below 3 or above 8, is susceptible to biological degradation and degrades with free-chlorine concentrations above 1 mg/l, depending on the concentration and duration of contact. In addition, membrane compaction due to the high operating pressure and asymmetric construction causes a reduction of flux over time.

Polyamide membranes

Polyamide (PA) membranes are chemically and physically more stable than CA membranes, generally immune to bacterial degradation, stable over a pH range of 3 to 11 and do not hydrolyse in water. Under similar pressure and temperature conditions, PA membranes can produce higher water flux and higher salt rejection than CA membranes. However, PA membranes are more hydrophobic and susceptible to fouling than CA membranes and are not tolerant of free chlorine in any concentration. Any residual oxidant like chlorine in the feed will cause rapid deterioration of the membrane. For most applications, dechlorination is required if the feed-water is chlorinated and can be done with sodium bisulphite, sulphur dioxide or activated carbon. Sensors and instrumentation must be provided to monitor the feed- water for oxidants that may damage the material and shut down the system if any are detected. Some PA membranes have a rougher surface than CA membranes, which can increase susceptibility to biological and particulate fouling. Polyamide membranes are typically of thin-film construction. The PAs are used for the active layer and durable materials such as polyether-sulphone are used for the support material. The support layer is essentially a standard UF membrane and provides little resistance to flow.

Rejection capabilities

The rejection capabilities of RO and NF membranes are designated with either a per cent salt rejection or a molecular weight cut-off (MWCO) value. Salt rejection is typically used for RO membranes:

$$Rej = 1 - \frac{C_P}{C_F} \qquad \qquad ...(12.7)$$

where,

Rej = rejection, dimensionless (expressed as a fraction)
C_P = concentration in permeate, mole/l
C_F = concentration in feed water, mole/l

Rejection can be calculated for bulk parameters such as TDS or conductivity. For membrane rating, however, rejection of specific salts is specified. Sodium chloride rejection is normally specified for high-pressure RO membranes, whereas $MgSO_4$ rejection is often specified for NF or low-pressure RO membranes. Nanofiltration membranes can also be characterised by MWCO. The MWCO of NF membranes is typically determined by passage of solutes such as sodium chloride and magnesium sulphate. The MWCO of NF membranes is typically 1000 Daltons (Da), also known as atomic mass units (amu) or less.

Osmosis is the flow of solvent through a semi-permeable membrane, from a dilute solution into a concentrated one. Osmosis reduces the flux through a RO membrane by inducing a driving force for flow in the opposite direction.

The physico-chemical foundation for osmosis is rooted in the thermodynamics of diffusion.

Diffusion

Consider a vessel with a removable partition that is filled with two solutions to exactly the same level, as shown in Fig. 12.15a. The left side is filled with a concentrated salt solution, the right with pure water and the partition is gently removed without disturbing the solutions. Initially, the contents are in a non-equilibrium state and the salt will eventually diffuse through the water until the concentration is the same throughout the vessel. With salt ions diffusing from left to right across the plane originally occupied by the partition, conservation of mass requires a flux of water molecules in the opposite direction. Without a flux of water molecules from right to left, mass accumulates on the right side of the container, which is unthinkable with a continuous water surface. Equilibrium requires mass transport in both directions.

In Fig. 12.15b, the top of the vessel has been closed and fitted with manometer tubes and the removable partition has been replaced with a semi-permeable membrane. The semi-permeable membrane allows the flow of water but prevents the flow of salt. Filling the chambers with salt solution and pure water again creates a thermodynamically unstable system, which must be equilibrated by diffusion. Because the membrane prevents the flux of salt, however, mass accumulates in the left chamber, causing the water level in the left manometer to rise and in the right manometer to drop. This flow of water from the pure side to the salt solution is osmosis. Water flux occurs despite the difference in hydrostatic pressure that develops due to the difference in manometer levels.

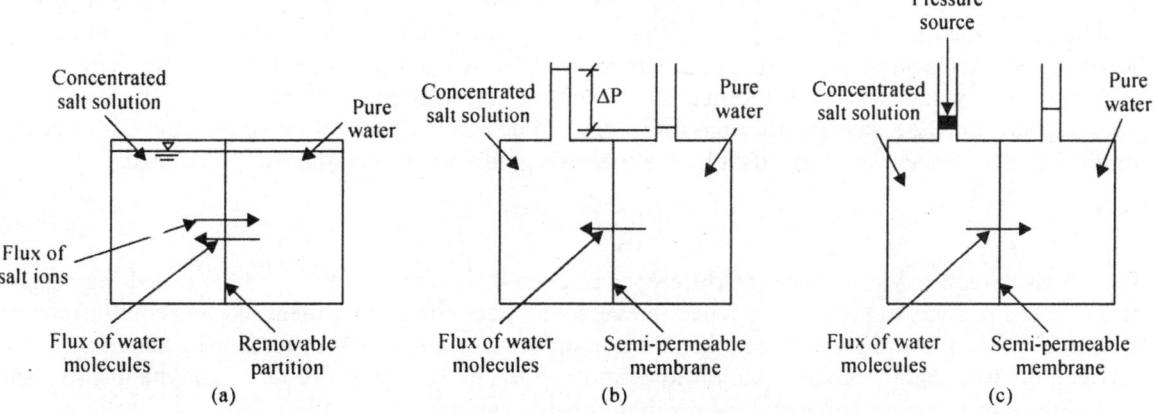

Fig. 12.15. Diffusion sketch for reverse osmosis: (a) diffusion; (b) osmosis; and (c) reverse osmosis.

The driving force for diffusion is typically described as a concentration gradient, although a more rigorous thermodynamic explanation is a gradient in Gibbs energy. When the vessels in Fig. 12.15 were filled with water and salt solutions, the two sides had different values of Gibbs energy due to differences in salt concentration. Equilibrium is defined thermodynamically when $\Delta G = 0$, so the gradient in Gibbs energy across the first vessel caused the simultaneous diffusion of salt ions and water molecules and the system was driven toward an equilibrium condition in which the Gibbs energy (and concentration) was equal throughout the system. In the second vessel, water stops flowing from right to left when the vessel reaches equilibrium but both pressure and concentration are unequal between the chambers. Equilibrium is achieved in the second vessel when the Gibbs energy is constant throughout and the Gibbs energy includes components to account for both the pressure and concentration differences.

Osmosis

To describe osmosis, a more general description of Gibbs energy is needed. The general form of the Gibbs function is:

$$\partial G = V \partial P - S \partial T + \sum_i \mu_i^\circ \partial n_i \qquad \qquad ... (12.8)$$

where,

G = Gibbs energy, J
V = volume, m^3
P = pressure, Pa
S = entropy, J/K
T = absolute temperature, K (273 + °C)
μ_i° = chemical potential of solute i, J/mole
n_i = amount of solute i in solution, moles

Chemical potential is defined as the change in Gibbs energy resulting from a change in the amount of component i when temperature and pressure are held constant:

$$\mu_i^\circ = \left. \frac{\partial G}{\partial n_i} \right|_{P,T} \qquad \qquad ... (12.9)$$

Therefore, the last term in Eq. 12.8 ($\mu_i^\circ \partial n_i$) describes the difference in Gibbs energy resulting from the difference in the amount of solute between the chambers (when volume is constant, the difference in amount is equivalent to the difference in concentration). Under constant-temperature conditions (i.e., $\partial T = 0$), Eq. 12.8 says equilibrium ($\partial G = 0$) will be achieved when the sum of the Gibbs energy gradient due to concentration is offset by the pressure gradient between the two chambers:

$$\partial G = 0 \qquad \text{when} \qquad V \partial P = -\sum_i \mu_i^\circ \partial n_i \qquad \qquad ... (12.10)$$

The pressure required to balance the difference in chemical potential of a solute is called the osmotic pressure and is given the symbol π. When the vessel in the second experiment reaches equilibrium, the difference in hydrostatic pressure between the manometers is equal to the difference in osmotic pressure between the two chambers. An equation for osmotic pressure can be derived thermodynamically using assumptions of incompressible and ideal solution behaviour:

$$\pi = \frac{-RT}{V_b} \ln x_W \qquad \qquad ... (12.11)$$

where,

π = osmotic pressure, bars
V_b = molar volume of pure water, l/mole
x_W = mole fraction of water, moles/mole
R = universal gas constant, 0.083145 l·bar/moles·K

In dilute solution (i.e., $x_W \cong 1$), Eq. 12.11 can be approximated by the Van't Hoff equation for osmotic pressure (Eq. 12.12), which is identical in form to the ideal gas law (PV = nRT):

$$\pi = \frac{n_S}{V}RT \qquad \text{or} \qquad \pi = CRT \qquad \qquad ... (12.12)$$

where,

n_S = total amount of all solutes in solution, moles
C = concentration of all solutes, moles/l
V = volume of solution, l

Equation 12.12 was derived assuming infinitely dilute solutions, which is often not the case in RO systems. To account for the assumption of diluteness, the non-ideal behaviour of concentrated solutions, and the compressibility of liquid at high pressure, a non-ideality coefficient (osmotic coefficient ϕ) must be incorporated into the equation. It should be noted that the thermodynamic equation for osmotic pressure (Eq. 12.11) contains no terms identifying the solute. Osmotic pressure is strictly a function of the concentration or mole fraction, of water in the system. Solutes reduce the mole fraction of water and the effect of multiple solutes is additive because they cumulatively reduce the mole fraction of water. Solutes that dissociate also have an additive effect on the mole fraction of water (for instance, addition of 1 mole of NaCl produces 2 moles of ions in solution, doubling the osmotic pressure compared to a solute that does not dissociate). Incorporating the osmotic coefficient and the dissociation of solutes into the van't Hoff equation yields

$$\pi = i\phi CRT \qquad \qquad ... (12.13)$$

where,

i = number of ions produced during dissociation of solute
ϕ = osmotic coefficient, unitless

If multiple solutes are added on an equal-mass basis, the solute with the lowest molecular weight produces the greatest osmotic pressure. The use of Eq. 12.13 is demonstrated in Example 12.1.

Example 12.1: Calculate the osmotic pressure of 100 mg/l solutions of the following solutes at a temperature of 20°C assuming an osmotic co-efficient of 1.0: (1) NaCl, (2) $SrSO_4$ and (3) glucose ($C_6H_{12}O_6$).

Solution:

1. Determine the osmotic pressure of NaCl using Eq. 12.13:

$$p = ifCRT$$

$$= \frac{(2 \text{ mole ion/mole NaCl})(1.0)(100 \text{ mg/l})(0.083145 \text{ L·bar/K·mole})(293 \text{ K})}{(10^3 \text{ mg/g})(58.4 \text{ g/mole})}$$

$$= 0.083 \text{ bar}$$

2. Determine the osmotic pressure for $SrSO_4$ using Eq. 12.13:

$$\pi = \frac{(2 \text{ mole ion/mole } SrSO_4)(1.0)(100 \text{ mg/l})(0.083145 \text{ L·bar/K·mole})(293 \text{ K})}{(10^3 \text{ mg/g})(183.6 \text{ g/mole})}$$

$$= 0.027 \text{ bar}$$

3. Determine the osmotic pressure for glucose (no dissociation, so i = 1) using Eq. 12.13:

$$\pi = \frac{(1.0)(100 \text{ mg/l})(0.083145 \text{ L·bar/K·mole})(293 \text{ K})}{(10^3 \text{ mg/g})(180 \text{ g/mole})}$$

$$= 0.014 \text{ bar.}$$

The osmotic pressure of a solution of sodium chloride, calculated with Eq. 12.13 and $\phi = 1$, is shown in Fig. 12.16 along with experimentally determined values. Over the range of salt concentrations of interest in seawater desalination, the osmotic coefficient for sodium chloride ranges from 0.93 to 1.03 and is shown as a function of solution concentration in Fig. 12.17. The deviation between measured and calculated values of osmotic pressure can be significantly greater for other solutes and higher concentrations, as shown for sucrose solutions in Fig. 12.18.

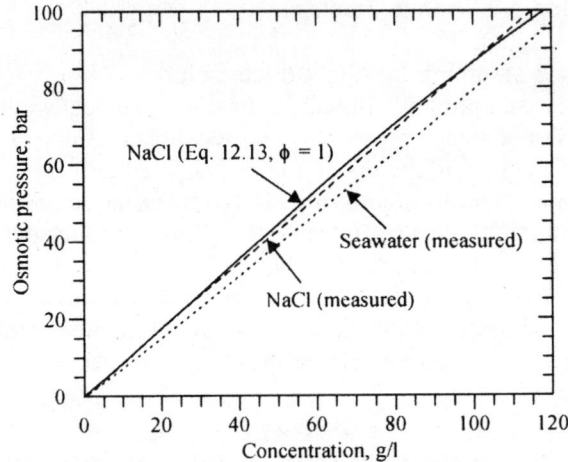

Fig. 12.16. Osmotic pressure of aqueous solutions of sodium chloride. Osmotic pressure calculated with Eq. 12.11 varies from osmotic pressure calculated using Eq. 12.13 by less than 0.5 per cent over the range of values shown.

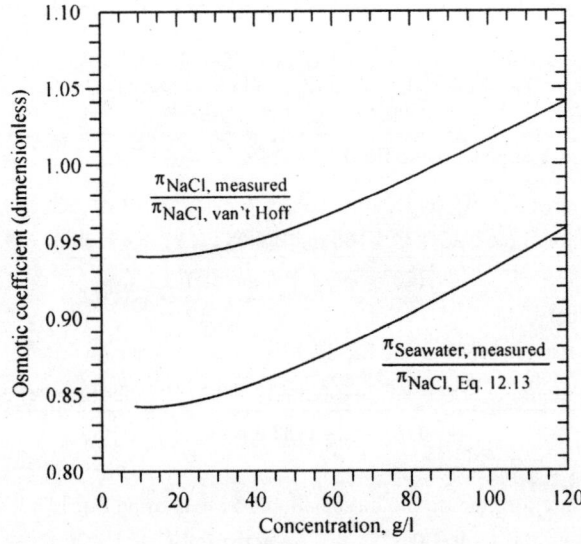

Fig. 12.17. Osmotic coefficients for sodium chloride and seawater (calculation of osmotic pressure for seawater with the van't Hoff equation is based on a concentration of NaCl equal to the TDS of the seawater).

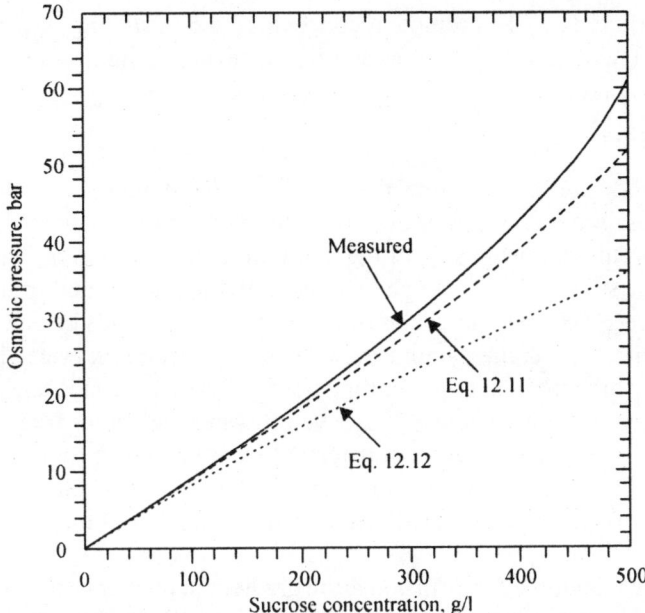

Fig. 12.18. Osmotic pressure of aqueous solution of sucrose.

Reported values for the osmotic pressure of seawater are about 10 per cent below measured values for sodium chloride, as shown in Fig. 12.16, due to the presence of compounds with a higher molecular weight than sodium chloride. The osmotic pressure for seawater can be calculated with Eq. 12.13 and an equivalent concentration of sodium chloride by using the osmotic coefficient for seawater shown in Fig. 12.17.

The flow of water through the semi-permeable membrane in Fig. 12.15b results from two opposing forces: (i) a gradient in chemical potential from right to left; and (ii) a gradient in pressure from left to right. These opposing forces are exploited in RO. Consider a new experiment using the apparatus in Fig. 12.15, which has been modified so that it is possible to exert an external force on the left side, as shown in Fig. 12.15c. Applying a force equivalent to the osmotic pressure places the system in thermodynamic equilibrium and no water flows. Applying a force in excess of the osmotic pressure places the system in non-equilibrium, with a pressure gradient exceeding the chemical potential gradient. In this case, the direction of liquid flow would be from left to right, that is, from the concentrated solution to the dilute solution. The process of causing water to flow from a concentrated solution to a dilute solution across a semi-permeable membrane by the application, of an external pressure in excess of the osmotic pressure is called reverse osmosis.

Water and Solute Flux

Several models have been developed to describe the flux of water and solutes through RO membranes. Unfortunately, some controversy remains about the mechanics of permeation. The main areas of controversy are whether RO membranes, have pores and the degree of coupling between water and solute flux. Two basic approaches have been used to describe RO. The first approach relies on fundamental thermodynamics and does not depend on a physical description of the membrane. The other approach uses physical and chemical descriptions of the membrane and feed solution to develop equations that predict both water and solute flux. To promote a conceptual understanding of permeation through RO membranes, a basic qualitative description of the solution-diffusion, pore flow and preferential sorption-capillary flow models are presented in the following sections.

Solution-diffusion model

The solution-diffusion model describes permeation through a dense membrane where the active layer is permeable but non-porous. Water and solutes dissolve into the solid membrane material, diffuse through the solid and reliquefy on the permeate side of the membrane. Dissolution of water and solutes into a solid material occurs if the solid is loose enough to allow individual water and solute molecules to travel along the interstices between polymer molecules of the membrane. Liquids behave as liquids because of attractive interactions with surrounding liquid molecules.. In particular, water has many anomalous properties (e.g., unusually high melting point, boiling point, viscosity, surface tension), which are due to hydrogen bonding between water molecules. Thus, even if water molecules travel along a defined path (which could be called a pore), they are not surrounded by other water molecules and therefore not present as a liquid phase. Diffusion occurs by movement of the water and solute molecules in the down-gradient direction of the driving force. Separation occurs when the flux of the water is different from the flux of the solutes.

As noted earlier, the driving force for diffusion is the gradient in the Gibbs energy. The proportionality between osmotic pressure and concentration is described by Eq. 12.13. Therefore, the driving force for any component can be written equivalently in terms of either pressure or concentration provided the mass transfer coefficient is reported in the proper units. For water, the driving force is expressed in terms of the net pressure gradient, that is, the applied pressure in excess of the osmotic pressure. Solute transport is expressed in terms of the concentration gradient and most models neglect the effect of applied pressure on solute transport. Flux through the membrane is determined by both solubility and diffusivity. Components of low solubility have a low driving force and components of low diffusivity have a low diffusion coefficient. The solution-diffusion model predicts that separation occurs because the solubility, diffusivity or both of the solutes are much lower than those of water, resulting in a lower solute concentration in the permeate than in the feed. The basic solution-diffusion model considers the water and solutes to be uncoupled; that is, the flux of each is completely independent of the other and there are no interactions between components. The model has also been extended to include some coupling between the water and solutes.

Pore flow models

The solution-diffusion model does not consider convective flow through the membrane. Other models consider RO membranes to have void spaces (pores) through which liquid water travels. The pore flow models consider water and solute fluxes to be coupled, meaning the solutes are convected through the membrane with the water. Thus, rejection occurs through mechanisms for membrane filtration, meaning the solute molecules are 'strained' at the entrance to the pores. Because solute and water molecules are similar in size, the rejection mechanism is not a physical sieving and must consider chemical effects such as electrostatic repulsion between the ions and membrane material. A limitation of pore flow models that consider only rejection at the membrane surface is that the pores of a RO membrane must be incredibly small to achieve rejection of small ions considerably smaller than 1 nm and approaching the size of water molecules. The basic assumptions of fluid mechanics must be re-considered for ions that are near the size of water molecules. The properties of water depend on attractive forces between adjacent molecules and fluid mechanics assumes that water is a continuum (the molecular mean free path is much smaller than the tube diameter) and for very narrow pores, these basic assumptions may not be valid.

Preferential sorption-capillary flow model

A third description of water and salt permeation through membranes is provided by the preferential sorption-capillary flow model, which assumes that the membrane has pores. Separation occurs when one component of the feed solution (either the solute or the water) is preferentially adsorbed to the pore walls and is transported through the membrane by surface diffusion. Membrane materials with a low dielectric constant, such as cellulose acetate, repel ions and preferentially adsorb water, forming a sorbed layer with a reduced concentration of salts. The sorbed layer moves through the membrane in response to the Gibbs energy gradient, leaving behind solution components that are repelled from the membrane surface. Separation is a function of the surface chemistry of the membrane and solutes, rather than pore dimensions, although the maximum pore dimension to effect good removal of solutes is two times the thickness of the adsorbed layer, as shown in Fig. 12.19.

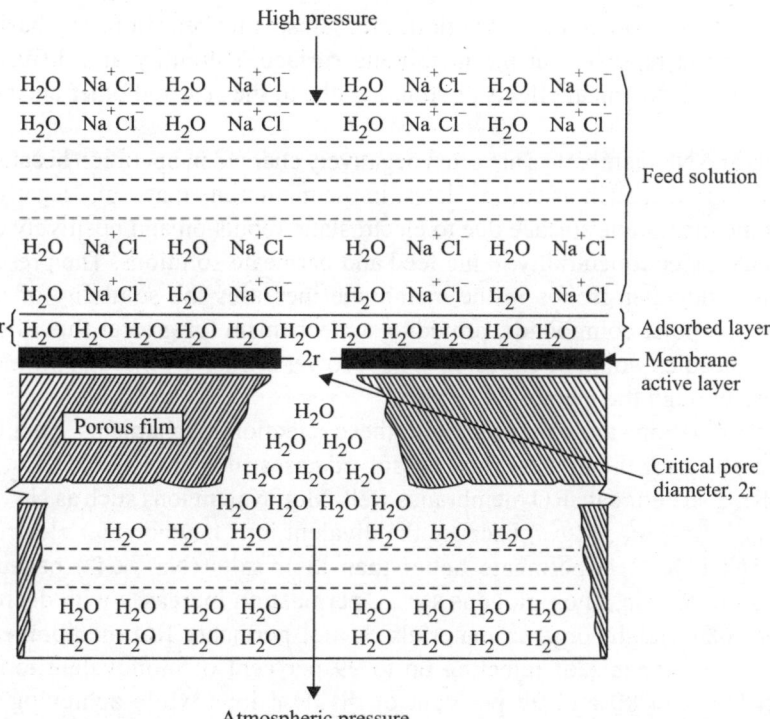

Fig. 12.19. Preferential-sorption capillary flow model. Ions are repelled from the membrane surface, resulting in an adsorbed layer of water. The adsorbed water flows through capillary pores in the membrane surface, and the repelled species are left in the feed solution. Good separation can be obtained if the pore diameter is less than 2 times the adsorbed layer thickness.

Coupling

Other models consider a combination of permeation mechanisms. The solution-diffusion-imperfection model assumes that water and solute permeate the membrane by both solution-diffusion and pore flow.

The permeation by solution-diffusion is uncoupled but the pore flow is completely coupled. The flux of water by solution-diffusion is proportional to the net applied pressure ($\Delta P - \Delta \pi$), the diffusion of solutes is proportional to the concentration gradient (ΔC) and pore flow is proportional to the applied pressure gradient (ΔP). To achieve high rejection, the pore flow must be a small fraction of the total flow.

In addition to coupling between water and solutes, coupling between solutes must be considered. Electroneutrality must be maintained in both the permeate and the concentrate streams. Thus, preferential transport of ions of one charge can influence the transport of ions of the opposite charge. For instance, negative rejection of hydrogen ions (the concentration of hydrogen ions in the permeate is higher than in the feed solution, manifested as a lower pH in the permeate) is sometimes observed in RO operations. This occurs because of higher flux of negatively charged ions, such as chloride, than the salt's co-ion sodium. Because hydrogen ions are more mobile than sodium ions, the flux of hydrogen ions increases to maintain electroneutrality in the permeate.

Rejection mechanisms

The membrane permeation models suggest various mechanisms for rejection. The basic mechanisms of rejection are electrostatic repulsion at the membrane surface, solubility and diffusivity through the membrane material due to chemical effects or straining due to the size and other chemical properties of molecules.

Reverse osmosis and NF membranes are often negatively charged in operation because of the presence of ionised functional groups, such as carboxylates, in the membrane material. Negatively charged ions may be rejected at the membrane surface due to electrostatic repulsion and positively charged ions may be rejected to maintain electroneutrality in the feed and permeate solutions. The presence of polar and hydrogen-bondable functional groups in the membrane increases the solubility of polar compounds such as water over non-polar compounds, providing a mechanism for greater flux of water through the membrane. Large molecules would be expected to have lower diffusivity through the membrane material or be unable to pass through the membrane at all.

Experimental observations are consistent with these rejection mechanisms. Small polar molecules such as water generally have the highest flux. Dissolved gases such as H_2S and CO_2, which are small, uncharged and polar, also permeate RO membranes well. Monovalent ions such as Na^+ and Cl^- permeate better than divalent ions (Ca^{2+}, Mg^{2+}) because the divalent ions have greater electrostatic repulsion. Acids and bases (HCl, NaOH) permeate better than their salts (Na^+, Cl^-) because of decreased electrostatic repulsion. Within a homologous series, permeation increases with decreasing molecular weight. High-molecular-weight organic materials do not permeate RO membranes at all. Reverse osmosis membranes are capable of rejecting up to 99 per cent of monovalent ions. Nanofiltration membranes reject between 80 and 99 per cent of divalent ions while achieving low rejection of monovalent ions.

Equations for Water and Solute Flux

Based on the models presented above, a variety of equations have been developed for the rate of water and solute mass transfer through an RO membrane. Ultimately, these models express flux as the product of a mass transfer coefficient and a driving force. The driving force for water flux through RO membranes is the net pressure differential or the difference between the applied and osmotic pressure differentials:

$$\Delta P_{NET} = \Delta P - \Delta \pi = (P_F - P_P) - (\pi_F - \pi_P) \qquad \ldots (12.14)$$

where,

ΔP_{NET} = net trans-membrane pressure, bars

Subscripts F and P refer to the feed and permeate, respectively.

The water flux through RO membranes is described by the expression

$$J_W = k_W (\Delta P - \Delta \pi) \qquad \qquad \dots (12.15)$$

where,

J_W = volumetric flux of water, $l/m^2 \cdot h$

k_W = mass transfer coefficient for water flux, $l/m^2 \cdot h \cdot bar$

Water flux is normally reported as a volumetric flux ($l/m^2 \cdot h$ or $gal/ft^2 \cdot d$) and the mass transfer coefficient is typically reported with units of $l/m^2 \cdot h \cdot bar$ or $gal/ft^2 \cdot d \cdot atm$. Equation 12.15 is valid at any point on the membrane surface between the feed water entrance and concentrate discharge in a membrane element, but it should be noted that both applied and osmotic pressures change continuously along the length of a spiral-wound element due to head loss and the changing solute concentration. As a result, overall flux must be determined by integrating Eq. 12.15 across the length of the membrane element, as will be demonstrated in the design section of this chapter.

The driving force for solute flux is the concentration gradient and the flux of solutes through RO membranes is expressed as

$$J_S = k_S (\Delta C) \qquad \qquad \dots (12.16)$$

where,

J_S = mass flux of solute, $mg/m^2 \cdot h$

k_S = mass transfer coefficient for solute flux, $l/m^2 \cdot h$ or m/h

ΔC = concentration gradient across membrane, mg/l

Solute flux is normally reported as a mass flux with units of $mg/m^2 \cdot h$ or $lb/ft^2 \cdot d$. Values of k_W and k_S are determined experimentally by membrane manufacturers. The flux of solutes through the membrane is equal to the flux of water multiplied by the solute concentration in the permeate as described by

$$J_S = C_P J_W \qquad \qquad \dots (12.17)$$

The ratio of permeate flow to feed water flow or recovery, is calculated as

$$r = \frac{Q_P}{Q_F} \qquad \qquad \dots (12.18)$$

where,

r = recovery, dimensionless

Using flow and mass balance principles, the solute concentration in the concentrate stream can be calculated from the recovery and solute rejection. The pertinent flow and mass balances using flow and concentration terminology as shown in Fig. 12.10 are:

$$\text{Flow balance:} \quad Q_F = Q_P + Q_C \qquad \qquad \dots (12.19)$$

$$\text{Mass balance:} \quad C_F Q_F = C_P Q_P + C_C Q_C \qquad \qquad \dots (12.20)$$

where,

Q = flow, m^3/s

C = concentration, mole/l or mg/l

Combining the mass and flow balances with Eq. 12.7 (rejection) and Eq. 12.18 (recovery) yields the following expression for the solute concentration in the concentrate stream:

$$C_C = C_F \left[\frac{1-(1-Rej)r}{1-r} \right] \qquad \dots (12.21)$$

where,

Rej = rejection (dimensionless, expressed as a fraction)
C_C = concentration in concentrate, moles/l or mg/l
C_F = concentration in feed water, moles/l or mg/l

Rejection is frequently close to 100 per cent, in which case Eq. 12.21 can be simplified as follows:

$$C_C = C_F \left(\frac{1}{1-r} \right) \qquad \dots (12.22)$$

As shown in Eqs. 12.9 and 12.10, water flux depends on the pressure gradient and solute flux depends on the concentration gradient. As feed-water solute concentration increases at constant pressure, the water flux decreases (because of higher $\Delta\pi$) and the solute flux increases (because of higher ΔC), which reduces rejection and causes a deterioration of permeate quality. As the feed water pressure increases, water flux increases but the solute flux is essentially constant. Therefore, as the water flux increases, the permeate solute concentration decreases and the rejection increases. These relationships are illustrated in Fig. 12.20.

Dependence of Flux on Temperature and Pressure

Membrane performance declines (water flux decreases, solute flux increases) due to fouling and membrane ageing. However, fluxes of water and solute also vary because of changes in feed water temperature, pressure, velocity and concentration. To evaluate the true decline in system performance due to fouling and ageing, water and solute flux data must be compared at standard conditions. Standard procedures have been established for normalising RO performance data. These procedures incorporate the use of temperature and pressure correction factors, evaluated at standard (subscript ST) and actual (subscript A) conditions:

$$J_{W,ST} = J_{W,A} \frac{TCF_{ST}}{TCF_A} \frac{PCF_{ST}}{PCF_A} \qquad \dots (12.23)$$

or

$$Q_{P,ST} = Q_{P,A} \frac{TCF_{ST}}{TCF_A} \frac{PCF_{ST}}{PCF_A} \qquad \dots (12.24)$$

where,

TCF = temperature correction factor (defined below), dimensionless
PCF = pressure correction factor (defined below), bar

Temperature affects the fluid viscosity and the membrane material. The relationship between membrane material, temperature and flux is specific to individual products, so TCF values should normally be obtained from membrane manufacturers, who determine values experimentally. If manufacturer TCF values are unavailable, the relationship between flux and fluid viscosity can be approximated by the following expression, which may be appropriate for membranes containing pores:

$$TCF = (1.03)^{T-25} \qquad \dots (12.25)$$

where, T = temperature, °C

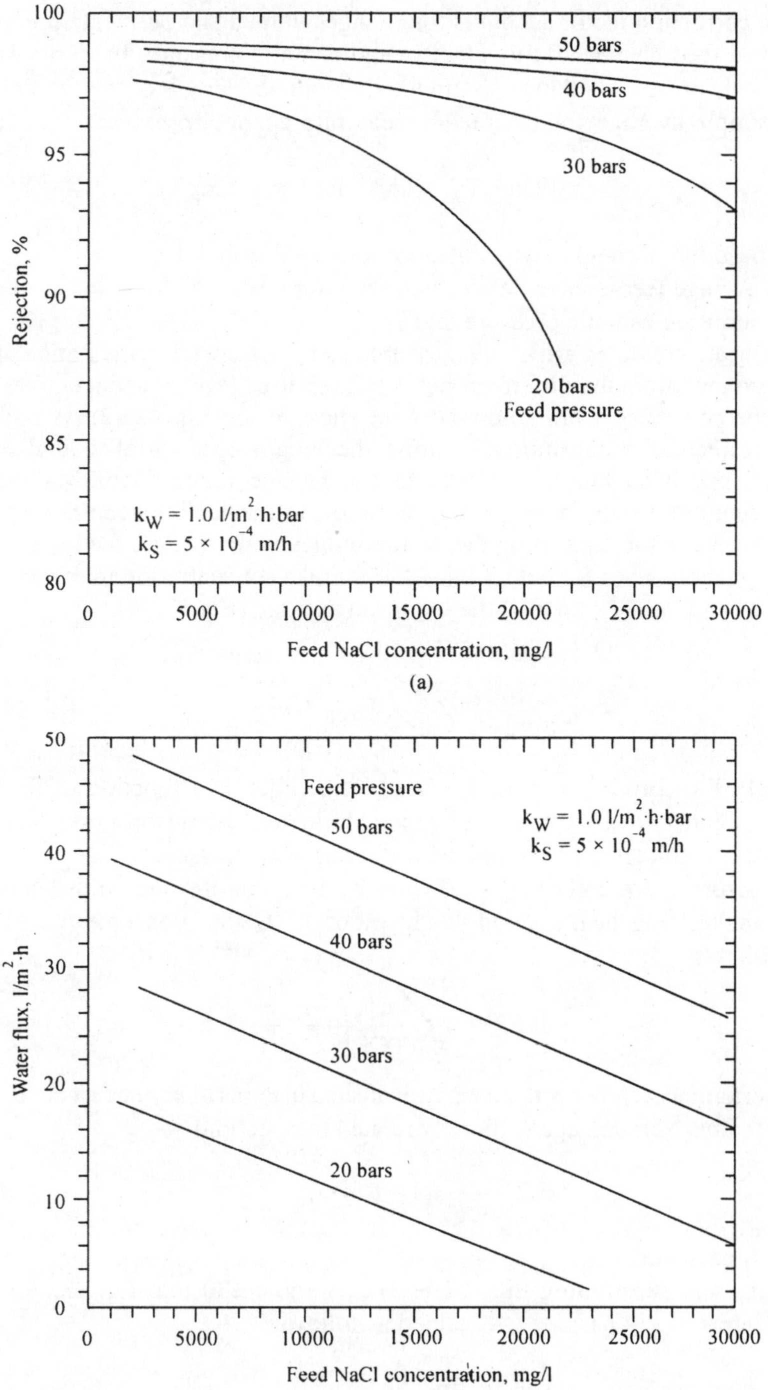

Fig. 12.20. Effect of feed water concentration and pressure on (a) per cent solute rejection; and (b) water flux.

The pressure correction factor accounts for changes in feed and permeate pressures, pressure drop (due to changes in feed channel fluid velocity) and osmotic pressure. In spiral-wound elements, the applied and osmotic pressures change continuously along the length of the feed-concentrate channel. ASTM uses a simple average for the feed-concentrate channel pressure in calculating the PCF as follows:

$$PCF = P_F - \tfrac{1}{2}h_L - P_P - \pi_{FC} + \pi_P \qquad ... (12.26)$$

where,

h_L = head loss through device feed-concentrate channel, bar

π_{FC} = average feed-concentrate osmotic pressure, bar

π_P = permeate osmotic pressure, bar

Feed and permeate pressures are easily measured using system instrumentation and the feed channel head loss is determined from the difference between measured feed and concentrate pressures. Osmotic pressures must be calculated from solute concentrations using Eq. 12.13. As with applied pressure, osmotic pressure increases continuously along the length of a spiral-wound element, but solute concentration is normally only measured in the feed and concentrate streams. Manufacturers use various procedures for determining the average concentration in the feed-concentrate channel and must be contacted for procedures for calculating the average concentration in the feed-concentrate channel. The two most common approaches for determining the average concentration in the feed channel are: (i) an arithmetic average (Eq. 12.27); and (ii) the log mean average (Eq. 12.28).

$$C_{FC} = \tfrac{1}{2}\,(C_F + C_C) \qquad ... (12.27)$$

$$C_{FC} = \frac{C_F}{r}\ln\left(\frac{1}{1-r}\right) \qquad ... (12.28)$$

Because the PCF incorporates terms for head loss, which is a function of feed flow and osmotic pressure, which is a function of solute concentration, the system design must establish standard conditions for feed flow rate and solute concentration as well as applied pressure.

Solute flux across the membrane is affected by temperature and solute concentration, so it is standardised by multiplying the measured flux by ratios of TCF and concentration at standard and actual conditions, as follows:

$$J_{S,ST} = J_{S,A}\,\frac{TCF_{ST}}{TCF_A}\,\frac{C_{FC,ST}}{C_{FC,A}} \qquad ... (12.29)$$

Membrane performance, however, is usually evaluated in terms of permeate concentration and rejection rather than solute flux. Salt passage is directly related to rejection:

$$SP = \frac{C_P}{C_F} = 1 - Rej \qquad ... (12.30)$$

where, SP = salt passage

By rearranging and substituting Eqs. 12.17, 12.23 and 12.30 into Eq. 12.29, standard membrane performance in terms of salt passage is obtained as follows:

$$SP_{ST} = SP_A\,\frac{PCF_A}{PCF_{ST}}\,\frac{C_{F,A}}{C_{F,ST}}\,\frac{C_{FC,ST}}{C_{FC,A}} \qquad ... (12.31)$$

Rearranging Eq. 12.31 in terms of rejection yields the expression

$$\text{Rej}_{ST} = 1 - (1 - \text{Rej}_A) \frac{\text{PCF}_A}{\text{PCF}_{ST}} \frac{C_{F,A}}{C_{F,ST}} \frac{C_{FC,ST}}{C_{FC,A}} \qquad \text{... (12.32)}$$

In multistage systems, it is necessary to standardise the water flux and recovery for each stage independently.

Concentration Polarisation

Concentration polarisation (CP) refers to the accumulation of solutes near the membrane surface, which has adverse effects on membrane performance. The flux of water through the membrane brings feed water (containing water and solute) to the membrane surface and as clean water flows through the membrane, the solutes accumulate at the membrane surface. In membrane filtration, particulate matter contacts the membrane and forms a cake layer. Because the rejection mechanisms for solutes are different, solutes stay in solution and form a boundary layer of higher concentration at the membrane surface. Thus, the concentration in the feed solution becomes polarised, with the concentration at the membrane surface higher than the concentration in the bulk feed water in the feed channel.

Concentration polarisation has several negative impacts on RO as follows:

1. Water flux is lower because the osmotic pressure gradient is higher due to the higher concentration of solutes at the membrane surface.
2. Rejection is lower due to an increase in solute transport across the membrane from an increase in the concentration gradient and a decrease in the water flux.
3. Solubility limits of solutes may be exceeded, leading to precipitation and scaling.

Equations for concentration polarisation can be derived from film theory and mass balances. In the membrane schematic shown in Fig. 12.21, feed water is travelling vertically on the left side of the membrane and water is passing through the membrane to the right. According to film theory, a boundary layer forms at the surface of the membrane. Water and solutes move one-dimensionally perpendicular to the membrane surface. During initial operation, the solute concentration at the membrane surface increases, causing a difference in concentration between the bulk feed water and membrane surface. During continuous operation, however, a steady-state condition is reached in which solutes no longer accumulate in the boundary layer and the concentration at the membrane surface reaches a constant value with respect to time. The solute flux toward the membrane surface is due to the convective flow of water, as described by the expression:

$$J_S = J_W C \qquad \text{... (12.33)}$$

A mass balance can be developed at the membrane surface as follows:

$$\text{Mass accumulation} = \text{Mass in} - \text{mass out} \qquad \text{... (12.34)}$$

With no accumulation of mass at steady-state, the solute flux toward the membrane surface must be balanced by fluxes of solute flowing away from the membrane (due to diffusion) and through the membrane (into the permeate) as follows:

$$\frac{dM}{dt} = 0 = J_W Ca - D_L \frac{dC}{dz} a - J_W C_P a \qquad \text{... (12.35)}$$

where,

M = mass of solute, g

t = time, second

D_L = diffusion coefficient for solute in water, m²/s
z = distance perpendicular to membrane surface, m
a = surface area of membrane, m²

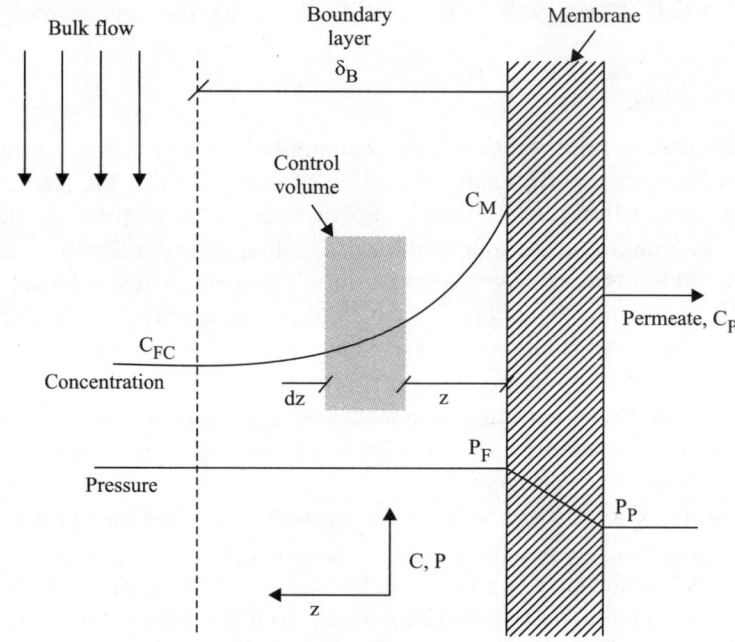

Fig. 12.21. Schematic of concentration polarisation.

Equation 12.35 applies not only at the membrane surface but also at any plane in the boundary layer because the net solute flux must be constant throughout the boundary layer to prevent the accumulation of solute anywhere within that layer (the last term in Eq. 12.35 represents the solute that must pass through the boundary layer and the membrane to end up in the permeate). Rearranging and integrating Eq. 12.35 across the thickness of the boundary layer with the boundary conditions $C(0) = C_M$ and $C(\delta_B)$ = C_{FC}, where C_{FC} is the concentration in the feed-concentrate channel and C_M is the concentration at the membrane surface, are done in the following equations:

$$D_L \cdot \int_{C_M}^{C_{FC}} \frac{dC}{C - C_P} = -J_W \int_0^{\delta_B} dz \qquad \ldots (12.36)$$

Integrating yields

$$\ln\left(\frac{C_M - C_P}{C_{FC} - C_P}\right) = \frac{J_W \delta_B}{D_L} \qquad \ldots (12.37)$$

$$\left(\frac{C_M - C_P}{C_{FC} - C_P}\right) = e^{J_W \delta_B / D_L} = e^{J_W / k_{CP}} \qquad \ldots (12.38)$$

where,

k_{CP} = concentration polarisation mass transfer coefficient, m/s

The solute concentration at the membrane surface is predicted by Eq. 12.38, which suggests that the concentration at the membrane can increase without limit as flux increases, but the concentration is constrained by solubility considerations. The upper limit to C_M has important implications for membrane performance.

The concentration polarisation mass transfer coefficient describes the diffusion of solutes away from the membrane surface. The following correlation between dimensionless numbers is often useful in mass transfer:

$$Sh = \alpha_1 (Re)^{\alpha_2} (Sc)^{\alpha_3} \qquad \qquad ... (12.39)$$

where,

Sh	=	Sherwood number, dimensionless
Re	=	Reynolds number, dimensionless
Sc	=	Schmidt number, dimensionless
$\alpha_1, \alpha_2, \alpha_3$	=	empirical parameters, dimensionless

Correlations for mass transfer coefficients depend on physical characteristics of the system and the flow conditions (e.g., laminar or turbulent). To promote turbulent conditions and minimise concentration polarisation in RO membrane elements, spiral-wound elements contain feed channel mesh spacers and maintain a high-velocity flow parallel to the membrane surface.

Although RO element feed channels can contain complex flow patterns, the Gilliland correlation for turbulent flow in pipes has to be used to calculate k_{CP} where the relationship between k_{CP} and the Sherwood number is given by

$$Sh = \frac{k_{CP} d_H}{D_L} \qquad \qquad ... (12.40)$$

where,

d_H = hydraulic diameter of feed channel, m

Substituting Eq. 12.40 into Eq. 12.39 and applying empirical values for $\alpha_1 = 0.023$, $\alpha_2 = 0.83$ and $\alpha_3 = 0.33$, the Gilliland correlation is defined as

$$k_{CP} = 0.023 \frac{D_L}{d_H} (Re)^{0.83} (Sc)^{0.33} \qquad \qquad ... (12.41)$$

$$Re = \frac{\rho v d_H}{\mu} \qquad \qquad ... (12.42)$$

$$Sc = \frac{\mu}{\rho D_L} \qquad \qquad ... (12.43)$$

where,

v = velocity in feed channel, m/s

ρ = feed water density, kg/m^3

μ = feed water dynamic viscosity, kg/m·s

The hydraulic diameter is defined as

$$d_H = \frac{4(\text{Flow cross-section})}{\text{Wetted perimeter}} \qquad \qquad ... (12.44)$$

For hollow-fiber membranes (circular cross-section), the hydraulic diameter is equal to the inside fiber diameter. Spiral-wound membranes can be approximated by flow through a slit, where the width is much larger than the feed channel height ($w \gg h$). As a result, the hydraulic diameter is twice the feed channel height, as shown in the equation:

$$d_H = \frac{4wh}{2w + 2h} \approx 2h \qquad \qquad \ldots (12.45)$$

where,

        h = feed channel height (feed-concentrate spacer thickness), m

        w = channel width, m

Concentration polarisation is expressed as the ratio of the membrane and feed-concentrate channel solute concentrations as follows:

$$\beta = \frac{C_M}{C_{FC}} \qquad \qquad \ldots (12.46)$$

where,

        β = concentration polarisation factor, dimensionless

Combining Eq. 12.38 with Eqs. 12.7 and 12.46 results in the following expression for the concentration polarisation factor, which is independent of concentration:

$$\frac{\beta - (1 - \text{Rej})}{\text{Rej}} = e^{J_W/k_{CP}} \qquad \qquad \ldots (12.47)$$

If rejection is high (greater than 99 per cent), Eq. 12.47 can be reasonably simplified as follows:

$$\beta = e^{J_W/k_{CP}} \qquad \qquad \ldots (12.48)$$

Concentration polarisation varies throughout the length of a membrane element; the parameters that change most significantly are the velocity in the feed channel (v) and the permeate flux (J_W). Variation in the concentration polarisation factor as a function of these parameters is shown in Fig. 12.22. As might be expected, concentration polarisation increases as the permeate flux increases and as the velocity in the feed channel decreases. The maximum concentration polarisation allowed for membrane elements is specified by manufacturers; $\beta = 1.2$ is a typical value. The importance of maintaining a high velocity in the feed-concentrate channel, particularly for membranes that achieve higher permeate flux; is clearly demonstrated in Fig. 12.22.

Fouling and Scaling

Nanofiltration and RO membranes are susceptible to fouling via a variety of mechanisms. The primary sources of fouling and scaling are particulate matter, precipitation of insoluble inorganic salts, oxidation of soluble metals and biological matter.

Particulate fouling

The primary source water for RO are groundwater and seawater, which typically have lower particle content than fresh surface water. Nevertheless, particulate fouling is a bigger concern in RO than in membrane filtration because the RO operational cycle does not include a backwashing step to remove accumulated solids (in fact, backwashing might cause the active layer of thin-film membranes to separate from the support layers). Virtually all RO systems require pre-treatment to minimise particulate fouling and fouling by residual particulate matter affects the cleaning frequency.

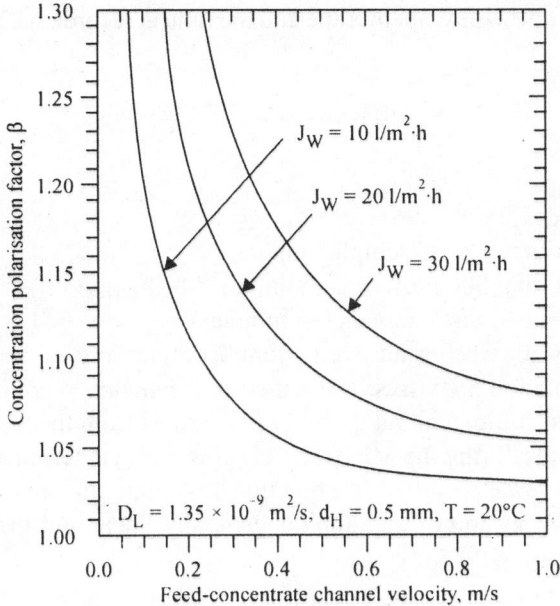

Fig. 12.22. Concentration polarisation factors as function of feed channel velocity and permeate flux.

Plugging and cake formation

Both inorganic and organic materials, including microbial constituents and biological debris, can cause particulate fouling, which includes plugging and cake formation. Plugging is the entrapment of large particles in the feed channels and piping. Hollow-fine-fiber membranes are reported to have more significant plugging problems because the high packing density of the fibers inside the pressure vessel results in very small spaces between the fibers. The mesh spacers in spiral-wound elements are sized to minimise plugging, but an excessive load of particulate matter may cause plugging anyway. Plugging is minimised by pre-filtration of the feed water and all RO membrane manufacturers recommend pre-filtration through 5 μm cartridge filters as a minimum pre-filtration step for protection of the membrane elements.

Particulate matter forming a cake on the membrane surface adds resistance to flow and affects system performance. Source water with excessive potential for particulate fouling require advanced pretreatment to lower the particulate concentration to an acceptable level. Coagulation and filtration (using sand, carbon or other filter media) are sometimes used for pretreatment as well as MF and UF.

Assessment of particle fouling

It is important to assess the fouling tendency prior to design and construction of an RO facility and to monitor the fouling tendency during operation. Empirical tests have been developed to assess particulate fouling, including the silt density index (SDI) and the modified fouling index (MFI). The SDI (ASTM, 2001b) is a timed filtration test using three time intervals through a gridded membrane filter with a mean pore size of 0.45 ± 0.02 μm and a diameter of 47 mm at a constant applied pressure of 2.07 bars (30 psi). The first interval is the duration necessary to collect 500 ml of permeate. Filtration continues through the second interval without recording volume until 15 minutes has elapsed (including the first time interval). Occasionally, a duration shorter than 15 minutes is used for water with high fouling tendency. At the end of 15 minutes, the third interval is started, during which an additional 500 ml aliquot of water is filtered through the now-dirty membrane and the time is recorded. The SDI is calculated from these time intervals:

$$SDI = \frac{100(1 - t_I / t_F)}{t_T}$$

where,

SDI = silt density index, min^{-1}

t_I = time to collect first 500 ml sample, minute

t_F = time to collect final 500 ml sample, minute

t_T = duration of first two test intervals (15 minutes)

The SDI has been criticised as being an overly simplistic measure of particulate fouling and not reflective of continuous membrane processes. Nevertheless, it remains in common use.

The MFI uses identical test equipment but different procedures from the SDI. The volume filtered is recorded at 30 seconds intervals during the MFI test. The flow rate is determined from volume and time data and the inverse of the flow rate is plotted as a function of volume filtered. An example of the plotted data is shown in Fig. 12.23. A portion of the graph is generally linear and the MFI is the slope of the graph in this region, that is,

$$\frac{\Delta t}{\Delta V} = \frac{1}{Q} = (MFI)V + b$$

where,

MFI = modified fouling index, s/l^2

V   = volume of permeate, liters, l

b   = intercept of linear portion of graph

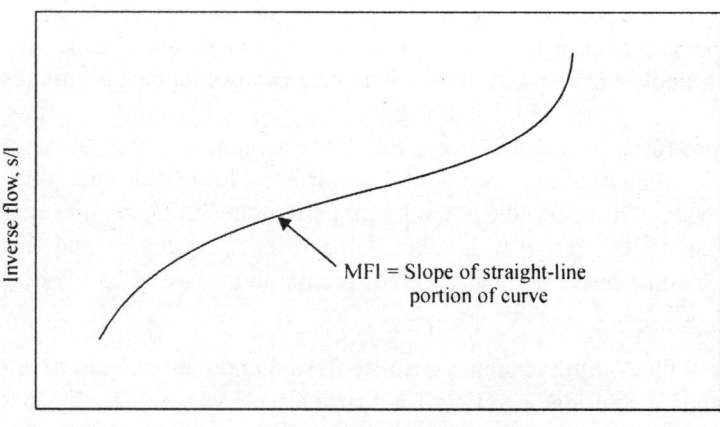

Fig. 12.23. Determination of modified fouling index (MFI).

Both the SDI and MFI use a 0.45 μm filter, so they only nominally measure fouling by material larger than that size. Recent research suggests that colloidal matter smaller than 0.45 μm may cause significant fouling of RO membranes. As a result, a revised MFI test that uses a 13 kDa UF membrane is being developed.

The SDI and MFI serve as rough guidelines for acceptable feed water quality and a maximum value of 4 to 5 is typically specified by manufacturers when the SDI and MFI are expressed in units of reciprocal minutes and s/l^2, respectively. Higher SDI or MFI values generally require pre-treatment to remove particulate matter. Lower values sometimes lead to longer intervals between chemical cleanings. However, manufacturers generally do not accept SDI or MFI values as indicative of the level of pre-treatment needed without conducting pilot tests and turbidity and particle count data are sometimes used to supplement SDI and MFI values.

Scaling or precipitation of inorganic salts

Inorganic scaling occurs when salts in solution are concentrated beyond their solubility limits and form precipitates. Because a common use of RO is desalination of brackish or saline water, the feed water typically has high concentrations of many inorganic ions. The concentration of important ions in seawater is shown in Table 12.7. If these ions are concentrated past the solubility product for sparing-soluble salts, precipitation occurs. The precipitation reaction for a typical insoluble salt is as follows:

$$CaSO_4(s) \rightleftarrows Ca^{2+} + SO_4^{2-} \qquad ... (12.51)$$

The solubility product is written as

$$K_{SP} = \{Ca^{2+}\}\{SO_4^{2-}\} \qquad ... (12.52)$$

where,

$\{Ca^{2+}\}$ = activity of calcium, mole/l
$\{SO_4^{2-}\}$ = activity of sulphate, mole/l

Table 12.7. Typical concentration of important solutes in seawater.

Salt	Concentration mg/l
Cations	
Sodium, Na^+	10800
Magnesium Mg^{2+}	1290
Calcium, Ca^{2+}	412
Potassium, K^+	399
Strontium, Sr^{2+}	7.9
Barium, Ba^{2+}	0.02
Anions	
Chloride, Cl^-	19400
Sulphate, SO_4^{2-}	2700
Total carbonate, CO_3^{2-}	142
Bromide, Br^-	67
Fluoride, F^-	1.3
Phosphate, HPO_4^{2-}	0.5
Total	35200

Several factors in RO operation affect ionic strength, which is used to calculate ion activity. The rate of ion or salt rejection is clearly important, as an ion with 99 per cent rejection will be concentrated more than one with 10 per cent rejection. The system recovery is also important because the more clean water withdrawn from a system, the higher the concentration of the rejected solutes. In fact, precipitation is one of the important factors that limit recovery in RO systems. Finally, the degree of concentration polarisation is important because precipitation occurs in the more concentrated zone near the membrane surface. The inorganic scale that forms on the membrane surface can reduce water permeability or permanently damage the membrane.

Avoiding scaling

In the absence of pre-treatment, precipitation must be avoided by minimising concentration polarisation, limiting salt rejection or limiting recovery. Concentration polarisation is minimised by promoting turbulence in the feed channels and maintaining minimum velocity conditions specified by equipment manufacturers. Limiting rejection is undesirable because it conflicts with process objectives. Limiting recovery, however, is often necessary to prevent precipitation. The highest recovery possible before any salts precipitate is the allowable recovery and the salt that precipitates at this condition is the limiting salt. Common limiting salts are listed in Table 12.8 and the most common scales encountered in water treatment applications are calcium carbonate ($CaCO_3$) and calcium sulphate ($CaSO_4$).

The allowable recovery without pre-treatment that can be achieved in RO is determined by performing solubility calculations for each of the possible limiting salts. The highest solute concentrations occur in the final membrane element immediately prior to the feed water exiting the system as the concentrate stream, so concentrate stream concentrations are used to evaluate solubility limits. In addition, the concentration in the concentrate steam must be adjusted for the level of concentration polarisation that is occurring. Incorporating the concentration polarisation factor defined in Eq. 12.46 with the expression for the solute concentration in the concentrate stream defined by Eq. 12.21 yields

$$C_M = \beta C_F \left[\frac{1 - (1 - Rej)r}{1 - r} \right] \qquad \dots (12.53)$$

Allowable recovery is determined by substituting the concentrations at the membrane into a solubility product calculation and solving for the recovery.

Table 12.8. Typical limiting salts and their solubility products.

Salt	Equation	Solubility product (pK_{sp} at 25°C)
Calcium carbonate (aragonite)	$CaCO_3(s) \rightleftarrows Ca^{2+} + CO_3^{2-}$	8.2
Calcium fluoride	$CaF_2(s) \rightleftarrows Ca^{2+} + 2F^-$	10.3
Calcium ortho-phosphate	$CaHPO_4(s) \rightleftarrows Ca^{2+} + HPO_4^{2-}$	6.6
Calcium sulphate (gypsum)	$CaSO_4(s) \rightleftarrows Ca^{2+} + SO_4^{2-}$	4.6
Strontium sulphate	$SrSO_4(s) \rightleftarrows Sr^{2+} + SO_4^{2-}$	6.2
Barium sulphate	$BaSO_4(s) \rightleftarrows Ba^{2+} + SO_4^{2-}$	9.7
Silica, amorphous	$SiO_2(s) + 2H_2O \rightleftarrows Si(OH)_4(aq)$	2.7

Activity coefficients cannot be ignored because: (i) the concentrate stream typically has a high ionic strength; and (ii) the limiting salts are composed primarily of divalent ions, whose activity deviates significantly from concentration at high ionic strength. The ionic strength is dependent on recovery and rejection, however, so the activity coefficients cannot be calculated until the recovery is determined. Because activity coefficients for ions are less than 1, ignoring ionic strength may yield a significantly lower value for allowable recovery than could actually be achieved. The assumption of 100 per cent rejection is often justified because divalent ions typically have rejection near 100 per cent. An assumption of 100 per cent rejection yields a slightly conservative value for allowable recovery because lower rejection will produce concentrate stream concentrations that are actually slightly lower. For NF and low-pressure RO systems that have divalent ion rejection significantly below 100 per cent, however, this assumption would be inappropriate. An additional complexity in limiting salt calculations is that carbonate and phosphate concentrations are dependent on pH. As can be imagined, accounting for ionic strength, recovery and pH in the calculations would result in equations that cannot be easily manipulated algebraically. Furthermore, the calculations must be repeated for each limiting salt in Table 12.8.

Pre-treatment to prevent scaling

Pre-treatment is necessary in virtually all RO systems to prevent scaling due to precipitation of sparingly soluble salts. Calcium carbonate precipitation is common and most systems require pre-treatment for this compound. In addition to the limiting salt calculations presented in the above example, calcium carbonate solubility can also be expressed in terms of the Langelier saturation index (LSI) and Stiff and Davis stability index (ASTM, 2001a, 2001f) and manufacturers' solubility programmes often report these values. Calcium carbonate precipitation can be prevented by adjusting the pH of the feed stream with acid to convert carbonate to bicarbonate and carbon dioxide. Sulphuric or hydrochloric acids are normally used, but using sulphuric acid can increase the sulphate concentration enough to cause precipitation of sulphate compounds. The pH of most RO feed water is adjusted to a pH value of 5.5 to 6.0. At this pH, most carbonate is in the form of carbon dioxide and passes through the membrane.

Use of antiscalants

Scaling of other limiting salts is commonly prevented with the addition of antiscalant chemicals. Antiscalants allow supersaturation without precipitation occurring by preventing crystal formation and growth. Sodium hexametaphosphate (SHMP) was commonly used as an antiscalant, but it has limited ability to extend the supersaturation range and adds phosphate compounds to the concentrate, which causes disposal problems. As a result, SHMP has been largely replaced with polymeric compounds. The degree of supersaturation allowed because of antiscalant addition depends on properties of the antiscalant, which are often proprietary and characteristics of specific equipment configurations. It is appropriate to rely on the recommendations of equipment and antiscalant manufacturers when determining appropriate antiscalant selection and doses necessary for a specific feed water analysis and design recovery.

Silica precipitates in an amorphous rather than crystalline form; thus, antiscalants that prevent crystal growth are ineffective for preventing silica precipitation. The precipitation of silica is kinetically slow, and a small degree of supersaturation can typically be accommodated in membrane systems. The presence of metals can increase silica precipitation and change its form, complicating the presence of silica in RO feed water. Recent advances and new antiscalant formulations are now available for both minimising silica precipitation and cleaning silica from membranes; these proprietary compounds have had varying degrees of success. When high silica concentrations are present, high-pH softening may be necessary to remove silica from the feed water to prevent precipitation on the membrane.

When antiscalants are ineffective

In cases where scaling is excessive and antiscalants are ineffective, additional pre-treatment measures, such as lime softening, may be required. The use of ion exchange as a pre-treatment to membrane processes is rarely cost effective on a municipal scale because the ions intended for removal by the RO process may quickly exhaust the ion exchange resins, resulting in high regeneration frequency. Ion exchange as pre-treatment to RO has been used effectively for the production of high-purity industrial process water, however.

A cost trade-off exists between methods of preventing scaling: operating at a lower recovery or the use of pre-treatment processes and chemicals. In some cases, it may be more cost effective to operate at a lower recovery to minimise pre-treatment costs. Pre-treatment and membrane equipment costs must be considered simultaneously and the design recovery set at the point that minimises overall system costs.

Metal oxide fouling

Groundwater used as the source water for RO and NF systems is often anaerobic. Iron and manganese, soluble compounds in their reduced states, can oxidise, precipitate and foul membranes if oxidants enter the feed water system. Iron fouling is more prevalent and can occur rapidly if any air enters the feed system. Fouling may be avoided by preventing oxidation or removing the iron or manganese after oxidation. If iron concentrations are low, precautions to prevent air from entering the feed system may be sufficient; anti-scalants often include additives to minimise fouling by low concentrations of iron. Pre-treatment to control iron might include oxidation with oxygen or chlorine followed by adequate mixing and hydraulic detention time and granular media or membrane filtration or greensand filtration in which oxidation and filtration take place simultaneously. When oxidants are used, precautions must be made to prevent them from reaching the membranes, particularly for polyamide membranes or other materials that are not oxidant resistant. Iron-fouling deposits are usually removable from RO membrane surfaces by commercially available cleaning agents and procedures.

An additional constituent present in many anaerobic groundwater is hydrogen sulphide. If air enters the feed water system, hydrogen sulphide can oxidise to colloidal sulphur, which can foul membranes. As with iron oxidation, precautions to prevent air from entering the feed system are important to prevent colloidal sulphur fouling. Sulphur deposits on membrane surfaces are typically irreversible.

Biological fouling

Biological fouling refers to the attachment or growth of micro-organisms or extracellular soluble material on the membrane surface or in the membrane element feed channels. Biological fouling is common in many RO systems and can have a variety of negative effects on performance, including loss of flux, reduced solute rejection, increased head loss through the membrane modules, contamination of the permeate, degradation of the membrane material and reduced membrane life. The primary source of microbial contamination is the feed water. Biological fouling is a significant problem in many RO systems.

Biological fouling is prevented by maintaining proper operating conditions, applying biocides and flushing membrane elements properly when not in use. Many RO and NF feed water (groundwater in many cases) have low microbial population. When operated properly, the shear in the feed channels helps to keep bacteria from accumulating or growing to unacceptable levels. When membrane trains are out of service, however, bacteria can quickly multiply. To avoid this problem, membranes should be flushed with permeate periodically or filled with an approved biocide if out of service for any significant period. Chlorine solutions can be used as a biocide for cellulose acetate membranes within recommended limits, but other chemical such as sodium bisulphite must be used with polymide membranes because of their susceptibility to degradation by chlorine.

The feed water to cellulose acetate membranes can be continuously chlorinated within limited concentrations to prevent biological growth, if necessary. Ultraviolet radiation, chloramination or chlorination followed by dechlorination can sometimes be used for polyamide membranes.

PROCESS DESIGN

During preliminary design of a RO system, the design engineer must perform the following activities:

1. Select the basic performance criteria: capacity, recovery, rejection and permeate solute concentration.
2. Evaluate alternatives for membrane equipment and operation, select the type of membrane element, and determine the array configuration (number of stages, number of passes, number of elements in a pressure vessel, number of vessels in each stage, feed pressure).
3. Select feed water pre-treatment requirements (methods to control fouling).
4. Select permeate post-treatment requirements.
5. Select concentrate treatment and disposal requirements.
6. Select ancillary membrane system features such as permeate backpressure control and interstage booster pumps.
7. Select equipment and procedures for process monitoring.

These elements of design are not independent of one another. For instance, recovery is often constrained by the solubility of limiting salts. As a result, selection of pre-treatment requirements, recovery, and array design must be done simultaneously and iteratively to determine the most economical design.

The basis for design information typically includes characteristics of the feed water (solute concentrations, turbidity, SDI and MFI values) from laboratory or historical data, required treated water quality (established by the client or regulatory limits) and required treated water capacity (established by demand requirements). The process design criteria for a hypothetical brackish water RO facility are shown in Table 12.9. Frequently, pilot testing is part of the design process.

Table 12.9. Typical design criteria for reverse osmosis facility.

Operating parameter	Units	Value
Feedwater pre-treatment		
Capacity	m^3/d	37900
Strainers		
Number	number	5
Nominal particle size rating	μm	5
Capacity, each	m^3/d	9480
Chemicals		
Sulphuric acid, maximum dose	mg/l	200
Scale inhibitor, maximum dose	mg/l	2
Feed pumps		
Number	no.	5
Capacity, each	m^3/d	9480
Pressure	bar	40

(Contd ...)

Operating parameter	Units	Value
Membrane system		
Feedwater flow rate	m^3/d	37900
Permeate flow rate	m^3/d	30300
Concentrate flow rate	m^3/d	7580
Recovery	%	80
Number of arrays	number	4
Capacity per array	m^3/d	9480
Array design criteria		
Membrane area per element	m^2	32.5
Elements per pressure vessel	no.	7
Number of stages per array	no.	2
Number of pressure vessels (stage 1)	no.	40
Stage 1 average permeate flux	$l/m^2 \cdot h$	21
Number of pressure vessels (stage 2)	no.	20
Stage 2 average permeate flux	$l/m^2 \cdot h$	17
Post-treatment[a]		
Caustic soda, maximum dose	mg/l	10
Corrosion inhibitor, maximum dose	mg/l	1
Chlorine, maximum dose	mg/l	2
Fluoride, maximum dose	mg/l	1
Concentrate disposal	Deep-well injection	

[a] Post-treatment may also include countercurrent packed tower for hydrogen sulphide or carbon dioxide removal.

Membrane array design involves determination of the quantity and quality of water produced by each membrane element in an array. This involves calculation of the flow, velocity, applied pressure, osmotic pressure, water flux and solute flux in each element, which leads to the determination of the number of stages, number of passes, number of elements in each pressure vessel and number of vessels in each stage. Membrane array design is a complex and iterative process using a large number of interrelated design parameters. Several important design parameters such as mass transfer coefficients are specific to individual products and available only from membrane manufacturers. Because of the complexity of the calculations and dependence on manufacturer information, array design is often done with design software provided by membrane manufacturers. Nevertheless, an understanding of the mechanics of the design procedure as described in the following paragraphs is important to interpreting the results from manufacturer design software.

Design calculations

The most common type of membrane element in use is the spiral-wound element. As described earlier, feed-water enters one end of the pressure vessel and flows through several spiral-wound elements in series. As, the water passes through each element, some water passes through the membrane into the permeate carrier channel, resulting in continuously changing conditions along the length of the membrane element. The net transmembrane pressure declines continuously across the length of a membrane element because of changes in both applied pressure (due to head loss in the feed channels) and osmotic pressure (due to concentration of salts). As, a result, fluxes of both water and solute are dependent on the position within a spiral-wound element and the design procedure must integrate along the length of the membrane element.

A differential slice of a membrane element is shown in Fig. 12.24. In this figure, the center plane represents the membrane surface, with the feed-concentrate channel above the membrane and the permeate channel below the membrane. The fluxes of water and solute are described by Eqs. 12.15 and 12.16, but the applied pressure differential, osmotic pressure differential and concentration differential depend on the location within the pressure vessel:

$$J_{W,Z} = k_W(\Delta P_Z - \Delta \pi_Z) = k_W\left[(P_{FC,Z} - P_{P,Z}) - (\pi_{M,Z} - \pi_{P,Z})\right] \qquad \dots (12.54)$$

$$J_{S,Z} = k_S(\Delta C_Z) = k_S(C_{M,Z} - C_{P,Z}) \qquad \dots (12.55)$$

where,

$C_{M,Z}$ = concentration at the membrane surface, $C_{M,Z} = B_Z C_{FC,Z}$, mole/l

$\pi_{M,Z}$ = osmotic pressure at the membrane surface, bar

Other terms are defined in Fig. 12.24.

The water and solute mass transfer coefficients (k_W and k_S) are dependent on the properties and configurations of specific membrane elements and cannot be generalised. These values are embedded in the manufacturer's design software and are typically not publicised but can be generated from pilot data if they cannot be obtained from the manufacturer.

Solute flux calculations are complicated by the presence of multiple solutes, which may have different value for the mass transfer coefficient. For instance, a low-pressure NF membrane has low rejection of monovalent ions but high rejection of divalent ions and the mass transfer coefficients would reflect this difference in rejection.

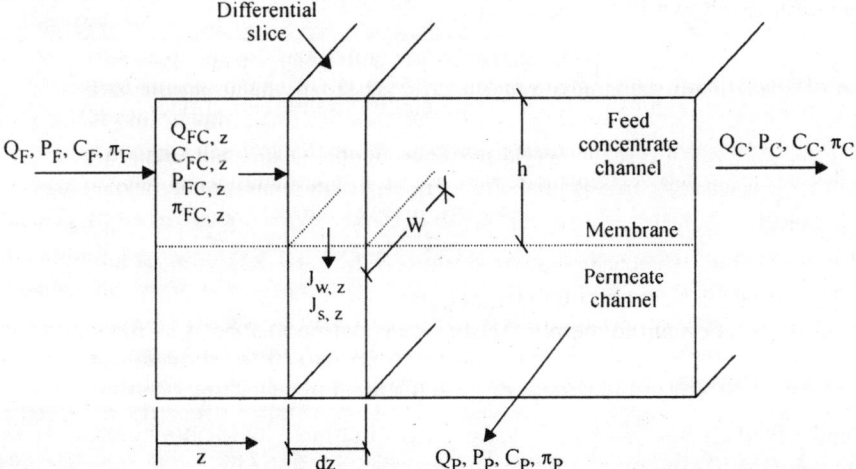

Fig. 12.24. Differential slice of spiral-wound membrane element. Because the feed flows axially* along the pressure vessel and the permeate flows spirally toward the center of the vessel, the feed and permeate flows are perpendicular to each other.

The permeate flow and mass solute flow through the membrane are equal to the flux times the membrane area in the differential element and the cumulative transfer of water and solute across the membrane is determined by integrating the flow between the feed end and the position z within the pressure vessel, as shown in the following:

$$Q_{P,Z} = \int_0^z J_{W,Z} w \, dz \qquad \ldots (12.56)$$

$$M_{S,Z} = \int_0^z J_{S,Z} w \, dz \qquad \ldots (12.57)$$

where,

w   = effective width .of feed-concentrate flow channel, m

$M_{S,Z}$ = mass of solute transferred, mg/s

The water flow rate (and velocity) in the feed-concentrate channel declines as permeate is produced and the flow rate at any point in the channel can be determined by subtracting the net permeate production up to that point from the feed-water flow rate as follows:

$$Q_{FC,Z} = Q_F - Q_{P,Z} \qquad \ldots (12.58)$$

Similarly, the solute concentration in the feed-concentrate channel can be determined by performing a mass balance on the solute as follows:

$$C_{FC,Z} = \frac{Q_F C_F - M_{S,Z}}{Q_{FC,Z}} \qquad \ldots (12.59)$$

Water and solute flux are affected by concentration polarisation and the concentration of solute at the membrane surface. Some manufacturers have developed relationships describing concentration polarisation for specific element designs and these relationships should be used if available. If no manufacturer information is available, the Gilliland correlation can be used to estimate the concentration polarisation factor. Because both flux and velocity are changing, β must be calculated using Eq. 12.47, but as a function of position, as shown in the equation:

$$\beta_Z = Rej(e^{-J_{W,Z}/k_{CP,Z}}) + (1 - Rej) \qquad \ldots (12.60)$$

The mass transfer coefficient k_{CP} depends on velocity in the feed-concentrate channel, which can be calculated from the expression

$$v_Z = \frac{Q_{FC,Z}}{hw} \qquad \ldots (12.61)$$

where,

h = height of feed-concentrate channel, m

The solute concentration at the membrane surface is defined by Eq. 12.46, using concentrations as a function of position.

$$C_{M,Z} = \beta_Z C_{FC,Z} \qquad \ldots (12.62)$$

Pressure in the feed channel drops due to head loss, but head loss is not constant across the length of the membrane element. Turbulent conditions are maintained, so head loss in the channel is proportional to the square of the velocity and the first power of length (consistent with the Darcy-Weisbach equation) as given by the expression:

$$h_L = \delta_{HL} v^2 L \qquad \ldots (12.63)$$

where,

h_L = head loss in feed-concentrate channel, bar

δ_{HL} = head loss coefficient, bar·s^2/m^3

v = water velocity in feed-concentrate channel, m/s

L = channel length, m

Finally, the permeate solute concentration can be calculated from the ratio of the solute and water fluxes per Eq. 12.17:

$$C_{P,Z} = \frac{J_{S,Z}}{J_{W,Z}} \qquad \qquad ...(12.64)$$

Additional design calculations, such as the calculation of osmotic pressure from concentration, have been presented earlier in this chapter.

Functional specifications

Because design criteria cannot be developed independently of manufacturer data, procurement of RO systems is often accomplished by means of a functional specification. By this method, an engineer develops the system requirements, designs the pre-treatment processes, designs the RO system support facilities and defines the basic requirements of the RO system. The functional specifications outline the operating requirements of the system, physical constraints of the system and warranty agreements between the manufacturer and the owner. Bid proposals are returned by the interested manufacturers that outline the particulars of the system being supplied, estimates of system product quality as a function of time, system capital costs and system operating costs as a function of time. The proposals are typically reviewed by the engineer to determine the optimum life cycle cost.

Pilot Testing

An important aspect of long-term RO operation is loss of performance due to compaction, fouling or degradation of the membrane. Unfortunately, fouling cannot be quantitatively predicted from water quality measurements and parameters such as SDI and MFI provide only a general indication of the severity of fouling.

Therefore, it is necessary to perform pilot testing for nearly all RO installations. Pilot testing is guided by membrane system selection and operating conditions developed during array design and serves to verify the array design criteria and identify pre-treatment requirements to prevent excessive fouling.

Commercial RO pilot plants

Reverse osmosis pilot plant systems are typically available from membrane manufacturers or consulting engineering firms. A typical commercially available skid-mount system, the skid unit contains six pressure vessels, each containing spiral-wound membrane elements in series. The pressure vessels can be operated as two independent systems, with each system containing three pressure vessels that can be piped as a 2 × 1 array, which allows membranes from two manufacturers to be tested simultaneously.

The pilot plant system is operated with a programmable logic controller (PLC). Chemicals are added to the feed water to prevent fouling of the membrane. Manufacturer-supplied specifications for pilot plant systems are usually provided so that the pilot unit can be properly operated. These specifications are usually obtained from the manufacturer and provide useful guidelines when planning and operating the pilot plant units.

Pilot test parameters

For most RO pilot studies, the following parameters should be recorded:

1. Date and time of sample analysis.
2. Flow rates (feed, concentrate and permeate).
3. Pressure (feed, concentrate and permeate).
4. Feed-water temperature.
5. Conductivity (online reading recommended).
6. Power consumption.
7. Chemical usage.
8. pH (feed, concentrate and permeate).

Pre-treatment

Pre-treatment is necessary to prevent scaling and fouling. The common pre-treatment strategies include the injection of acids and anti-scalants to prevent the precipitation of sparingly soluble salts and filtration to prevent plugging by particulate matter. Very clean source water may be able to operate with only cartridge filtration prior to the membrane units, but more advanced filtration methods, including coagulation, flocculation, sedimentation and granular filtration or membrane filtration, are required for source water containing significant quantities of particulate matter. Pre-treatment must be selected and designed in concert with the array design because the membrane element performance is dependent on the level of pre-treatment. Details of fouling and pre-treatment were presented earlier in this chapter.

Post-treatment

The permeate from a RO facility typically requires additional treatment. Feed-water pH adjustment prior to RO, along with extensive removal of divalent ions by the RO process, produces treated water with low pH, low alkalinity and low hardness, which are conditions that cause water to be corrosive. Anaerobic groundwater frequently contains hydrogen sulphide, which passes through the membrane and causes odour problems in the treated water. Finally, residual disinfection is always required for municipal water distribution.

Permeate-stability

A number of strategies can be used to increase the stability (reduce the corrosivity) of the water. When the feed-water is acidified for scale control, carbonate alkalinity in the raw water is converted to carbonic acid, which passes through the membrane. Thus, addition of a base such as caustic soda can restore both pH and alkalinity to acceptable levels. Without additional measures, however, such water will still be corrosive. Stability can be improved by adding hardness ions to the water, so base addition with chemicals containing calcium is sometimes preferred over caustic soda. Small systems sometimes can add an acceptable amount of hardness by passing the permeate through a bed of calcareous media such as dolomite or calcite. In lieu of adding hardness to the water, corrosion inhibitors may be effective. Another strategy for producing a stable finished water is to blend the permeate with a by-pass stream of raw water that meets all other water treatment requirements (such as filtration if a surface water source is used). Proper blending of raw and permeate water may produce a finished water with the desired pH, alkalinity and hardness. However, DBP precursor concentration in the raw water and the potential for DBP formation need to be evaluated when considering blending options.

Hydrogen sulphide

Anaerobic groundwater can contain hydrogen sulphide, a highly odourous compound that is not removed during RO. Hydrogen sulphide can be removed by oxidation or aeration. Oxidation to sulphate can be accomplished with oxidants such as chlorine, but large doses are needed (the stoichiometric chlorine requirement is about 9 times the hydrogen sulphide concentration on a mass basis and insufficient amounts can oxidise sulphide to elemental sulphur, which is equally undesirable).

Thus, hydrogen sulphide is commonly removed after the membrane process in an air-stripping process using counter-current packed towers. Odour control can be a significant issue when stripping water that contains sulphide.

It is necessary to consider all post-treatment goals simultaneously and select treatment options that achieve all objectives. For instance, air stripping to remove sulphide before base addition will strip carbon dioxide and increase the permeate pH; subsequent pH adjustment with caustic soda will not restore alkalinity because the carbonate will be gone. Alternatively, pH adjustment before stripping can prevent effective stripping because sulphide.

Disinfection

Chlorine is commonly used for disinfection. The RO process is effective at removing DBP precursors; thus, free chlorination can typically be practiced without forming significant quantities of DBPs. Blending with raw water for stability, however, may increase DBP formation when using free chlorine.

Disposal of Residuals

Disposal of the concentrate stream is frequently a challenge in RO plant design. The factors that contribute to this problem are identified in Table 12.10. In addition to the concentrate stream, RO plants must also dispose of spent cleaning solutions. Both of these residuals are discussed in this section.

Table 12.10. Factors affecting concentrate disposal.

Issue	Description
Volume	The waste stream volume from many water treatment processes is less than 5 per cent of the feed stream volume. In RO, the waste stream volume ranges from 15 to 50 per cent of the feed stream volume
Salinity/toxicity	The high salinity of the concentrate stream makes it toxic to many plants and animals, limiting options for land application or surface water discharge and rendering it unusable for recycling or re-use. Many concentrate streams are anaerobic, which can be toxic to fish without sufficient dilution. In addition, RO processes used for specific contaminant removal (i.e., arsenic, radium) may produce concentrate streams that can be classified as a hazardous material
Regulations	Concentrate is classified as an industrial waste by the US EPA. Concentrate disposal is regulated under several different federal, state and local laws and the interaction between these regulatory requirements can be complex. Regulatory considerations are often as important as cost and technical considerations for determining viable concentrate disposal options

Concentrate

Several surveys of concentrate disposal methods are available. The most common concentrate disposal options are: (i) discharge to a brackish surface water (include oceans, brackish rivers, or estuaries); (ii) discharge to a municipal sewer; and (iii) deep-well injection. In the United States, about half of all plants discharge concentrate to a surface water, a third discharge to a municipal sewer and about 10 per cent discharge to a deep well. Deep-well injection is most common in Florida. Other concentrate disposal options, including evaporation ponds, infiltration basins and irrigation, are used by a small number of facilities. Beneficial uses, such as a water supply for reconstructed brackish water wetlands, has been proposed in some areas. Concentrate disposal is an integral part of the design of RO facilities.

Cleaning solutions

Spent cleaning solutions from RO plants are frequently acidic or basic solutions and contain detergents or surfactants. In many cases, the cleaning solution volume is small compared to the concentrate stream and can be diluted into and disposed of with the concentrate. In some cases, treatment of the cleaning solution may be required prior to disposal, but treatment may consist only of pH neutralisation. Detergents and surfactants should be selected with disposal issues in mind.

Energy Recovery

Reverse osmosis is an energy-intensive process. A significant component of operating costs is electrical power for the feed pumps because of the high pressure necessary to operate RO membranes. Although pressure drops significantly as permeate passes through the membrane, the head loss through the feed channels is relatively small and the concentrate exits the final membrane element at 80 to 90 per cent of the feed pressure, with backpressure maintained by a concentrate control valve.

If concentrate is discharged to a deep well, a portion of this pressure can be used to drive the disposal process. If, however, the concentrate is discharged to a surface water, this pressure must be dissipated prior to discharge. Pressure in the concentrate stream dissipated across the concentrate control valve is wasted energy because it performs no useful work in the treatment system. Because the concentrate steam is both high energy and relatively high volume, the amount of waste energy is substantial.

Energy recovery devices are being used more frequently to reclaim the wasted energy in the concentrate stream. Several types of devices are available, including reverse-running turbines, Pelton wheels, pressure exchangers and electric motor drives. Typically, recovered energy from the residual pressure of the concentrate stream is used to pressurise the feed stream.

In all of the available devices, the concentrate stream spins a rotor, losing energy in the process and exits the energy recovery device at a significantly lower pressure. In the reverse-running turbine and pressure exchanger, the energy recovery device is in contact with both the feed and concentrate streams, with a single rotor transferring pressure from the concentrate to the feed stream. Pelton wheel devices use a rotor connected directly to the feed pump via an extended shaft and the energy recovered from the concentrate stream provides hydraulic assistance to the operation of the feed pumps. The main moving part is the Pelton wheel and shaft. Electric motor drives are more complex, utilising a hydraulic drive system connected to the pump motor.

More than 90 per cent of the energy expended to pressurise the concentrate stream can be recovered. Depending on the price for electricity, capital costs of energy recovery equipment may be recouped within 3 to 5 years. Energy recovery devices were first utilised on seawater RO systems because they operate at high pressure and low recovery, compounding the energy loss. Recent trends and improvements in energy recovery equipment and rising electricity prices suggest that energy recovery will be applied in more and more low-pressure systems.

In addition to providing pressure to the feed stream, another application is to use the energy recovery system to add pressure between stages. In normal operation, the second or later stages produce less permeate because of lower applied pressure (due to pressure drop in the first stage) and higher osmotic pressure (due to pressure drop in the first stage) and higher osmotic pressure (due to concentration of the feed stream in the first stage). The lower permeate flow and higher feed concentration also increase salt passage and degrade permeate quality. These effects are sometimes counteracted by installing booster pumps between stages, so that a higher feed pressure is available to offset the higher osmotic pressure. By using energy recovery devices to boost pressure between stages, the booster pumps can be eliminated, which offsets a portion of the capital cost of the energy control device.

Chapter 13

Disinfection

INTRODUCTION

The threat of microbiological contaminants in drinking water is eliminated by three complementary strategies: (i) preventing their access to the water source; (ii) employing water treatment to reduce their concentration in the water; and (iii) maximising the integrity of the distribution system for finished water. Early in the history of public drinking water systems, the emphasis was almost entirely on gaining access to a protected source. In recent years, greater emphasis has been directed toward providing effective water treatment to reduce microbiological contaminants. Today, there is increasing emphasis on employing both source protection and treatment to ensure that safe water is produced and on improving distribution system integrity to ensure that contamination does not occur during transport from the treatment plant to the consumer's tap. In the water treatment process, reducing microbiological contaminants is accomplished by two basic strategies, removing them from the water or inactivating them. Inactivated micro-organisms, although still present in the water, are no longer able to cause disease in the consumer. Processes that use inactivation as their strategy are traditionally referred to as disinfection, the focus of this chapter.

In water works practice, the term disinfection is used to refer to two activities: (i) primary disinfection—the inactivation of micro-organisms in the water; and (ii) secondary disinfection—maintaining a disinfectant residual in the treated water distribution system (residual maintenance). The characteristics that make a disinfectant the best choice for each of these purposes are not the same.

Primary disinfection is discussed in this chapter, along with the role disinfection plays in protecting the public, the strengths and weaknesses of inactivation versus removal, the kinetics of the disinfection process and some specific details about the design of disinfection facilities.

HISTORICAL PERSPECTIVE

The first continuous use of chlorination for disinfection of drinking water occurred in Middelkerke, Belgium, in 1902. Also, by the 1940s, disinfection with chlorine had become a world water treatment standard and even today, many water supplies are treated with chlorination alone. The presence of a free-chlorine residual in water at the tap was generally taken as a guarantee of microbiological safety by health officials and the public. Disinfection thus became established as the most important water treatment process.

During the last two decades of the twentieth century, events occurred that have also resulted in the questioning of the effectiveness of chlorination in controlling waterborne disease. In the 1980s, the protozoa *Giardia lamblia* was identified as an important waterborne pathogen. Because *G. lamblia* is more resistant to chlorine than other targets of disinfection, more stringent standards for reduction of pathogens were established. More recently, another protozoa, *Cryptosporidium parvum*, has also been identified as an important source of waterborne disease and is even more resistant to chlorine than *G. Lamblia*. In fact, chlorination has not proven to be a significant barrier for *C. parvum*.

The discovery of chlorination by-products and chlorine-resistant organisms is causing a re-evaluation of the use of chlorine as the primary disinfectant and a re-evaluation of the role of inactivation itself in the control of pathogens. For example, because methods are not available to determine if *C. parvum* oocysts found in water supplies will cause disease if ingested by a consumer, the Drinking Water Inspectorate in the United Kingdom recognises only removal, not inactivation, as a viable strategy for addressing the control of this pathogen.

New treatment processes have also come to the fore that show promise for the removal or inactivation of chlorine-resistant organisms and others as well. Membrane filtration processes, developed originally in the mid-1950s and later employed for sterilising laboratory solutions, juices and eventually brewed beverages, have now reached a stage in their development where they are commercially viable at large scale. Membranes are capable of removing pathogens much more effectively than traditional physical treatment processes like coagulation and granular media filtration. In fact, the removals that have been demonstrated using membranes are on the same order as the magnitude of inactivation of bacteria customarily achieved by chlorine. Disinfection with ultraviolet light (UV) is also effective for inactivating *Giardia* and *Cryptosporidium*. While chlorine remains the dominant drinking water disinfectant and disinfection (inactivation) remains the cornerstone of water treatment, this situation may change in the future.

METHODS OF DISINFECTION COMMONLY USED IN WATER TREATMENT

Five disinfection agents are commonly used in drinking water treatment today: (i) free chlorine; (ii) combined chlorine (chlorine combined with ammonia); (iii) ozone; (iv) chlorine dioxide; and (v) UV light. The first four agents are chemical oxidants whereas UV light involves the use of electromagnetic radiation. Of the five, by far the most common is free chlorine. Combined chlorine is also common, but its use is often limited to residual maintenance. Ozone is the strongest of the four oxidants and its use is becoming increasingly common, in part because of its stronger disinfecting properties and in part because it controls taste and odour compounds, specifically geosmin and methyl isoborneol. Information on each of these common disinfectants is summarised in Table 13.1.

Table 13.1. Characteristics of five most common disinfectants.

Issue	Disinfectant				
	Free chlorine	Combined chlorine	Chlorine dioxide	Ozone	Ultraviolet light
Effectiveness in disinfection					
Bacteria	Excellent	Good	Excellent	Excellent	Good
Viruses	Excellent	Fair	Excellent	Excellent	Fair
Protozoa	Fair to poor	Poor	Good	Good	Excellent
Endospores	Good to poor	Poor	Fair	Excellent	Fair

(Contd...)

Issue	Disinfectant				
	Free chlorine	*Combined chlorine*	*Chlorine dioxide*	*Ozone*	*Ultraviolet light*
Frequency of use as primary disinfectant	Most common	Common	Occasional	Common	Emerging use
Regulatory limit on residuals	4 mg/l	4 mg/l	0.8 mg/l	–	–
Formation of chemical by-products					
Regulated by-products	Forms 4 THMs[a] and 5 HAAs[b]	Trace of THMs and HAAs	Chlorite	Bromate	None
By-products that may be regulated in future	Several	Cyanogen halides, NDMA	Chlorate	Biodegradable organic carbon	None known
Typical application					
Dose, mg/l (kg/ml)	1–6	2–6	0.2–1.5	1–5	20–100 mJ/cm^2
Dose, lb/mg	8–50	17–50	2–13	8–42	–
Chemical source	Delivered: as liquid gas in tank cars, 1 tonne and 68 kg (150 lb) cylinders or as liquid bleach. Onsite generation from salt and water using electrolysis. Calcium hypochlorite powder is used for very small applications	Same sources for chlorine. Ammonia is delivered as aqua ammonia solution, liquid gas in cylinders or solid ammonium sulphate. Chlorine and ammonia are mixed in treatment process	ClO$_2$ is manufactured with an on-site generator from chlorine and chlorite. Same sources for chlorine. Chlorite as powder or stabilised liquid solution	Manufactured on-site using a corona discharge in dry air or pure oxygen. Oxygen is usually delivered as a liquid. Oxygen is also manufactured on-site in some large plants	Uses low-pressure or low-pressure high-intensity UV (254 nm) or medium-pressure UV (several wavelengths) lamps in the contactor itself
Typical contactor	In the past was added at beginning of plant and residual carried through. Increasingly engineered contactors are used	In the past was added at beginning of plant and residual carried through. Increasingly engineered contactors are used	In the past was added at beginning of plant and residual carried through. Increasingly engineered contactors are used	Has always been added in specially engineered contactors. These contactors are using more compartments and other techniques are being experimented with	Lamps are placed in gravity channels or in specially manufactured UV reactors. Because the contact time is so short, reactors must be tested for short circuiting

[a] THMs = trihalomethanes.
[b] HAAs = haloacetic acids.

Reactors Used for Chemical Oxidants

Before addressing the kinetics of disinfection discussed in the next section, it is useful to have some understanding of the contactors (reactors) that are typically used. All five disinfectants are best applied as a separate unit process. The three forms of chlorine are most often used in baffled, serpentine contact chambers or long pipelines when these are available. Both types of reactors can be designed to be highly efficient, closely approaching ideal plug flow. Ozone is generally introduced in bubble chambers where short circuiting can be a bigger problem. As a result, designs of ozone reactors often employ multiple chambers in series.

Reactors Used for Ultraviolet Light Disinfection

Ultraviolet light disinfection is often applied in proprietary reactors. Short circuiting is a special concern for UV reactors, particularly the proprietary reactors because their contact times are so short. Proprietary pressure vessels are particularly common where medium-pressure UV lamps are used because the high intensity of the UV lamps enables the delivery of a high UV dosage in a small space.

KINETICS OF DISINFECTION

For chemical disinfectants like chlorine, combined chlorine, chlorine dioxide and ozone, the specific mechanisms of micro-organism inactivation and the reaction kinetics are not well understood. The action of each chemical disinfecting agent depends on the properties of each micro-organism, the disinfectant and the water. As will be shown later, the reaction rates that have been observed can vary by as much as six orders of magnitude from one organism to the next, even for one reactant. Even for disinfection reactions where the reaction mechanism is well understood, for example, UV light, reaction rates vary by one and one-half orders of magnitude.

Nevertheless, there is one simple kinetic model that is widely used and there is enough commonality in the behaviour of all these reactions to allow the development of some phenomenological laws that are useful in modelling the course of all of these reactions. As these disinfection processes are physio-chemical processes, they are also subject to the rules of analysis. In the following discussion, the form of disinfection data resulting from laboratory experiments is examined by considering the shape of classical disinfection kinetic plots. Following this discussion, useful phenomenological kinetic models are discussed along with the merits of each.

Classical Disinfection Kinetics

Near the beginning of the twentieth century, Dr. Harriet Chick, a research assistant at the Lister Institute of Preventive Medicine in Chelsea, England, proposed that disinfection could be modelled as a pseudo-first-order reaction with respect to the concentration of the organisms. Chick demonstrated her concept by plotting organism survival versus time on a semi-log graph for disinfection data for a broad variety of disinfectants and organisms.

Chick worked with disinfectants like phenol, mercuric chloride and silver nitrate and organisms like (*Salmonella typhi, Salmonella paratyphi, Escherischia coli, Staphylococcus aureus, Yersinia pestis* and *Bacillus anthracis*). Over the subsequent years 'Chick's law' has been shown to be broadly applicable to disinfection data. Data from one of Chick's original experiments is shown in Fig. 13.1. The data are used to describe the inactivation of (*B. anthracis*) with a 5 per cent phenol solution at 33°C.

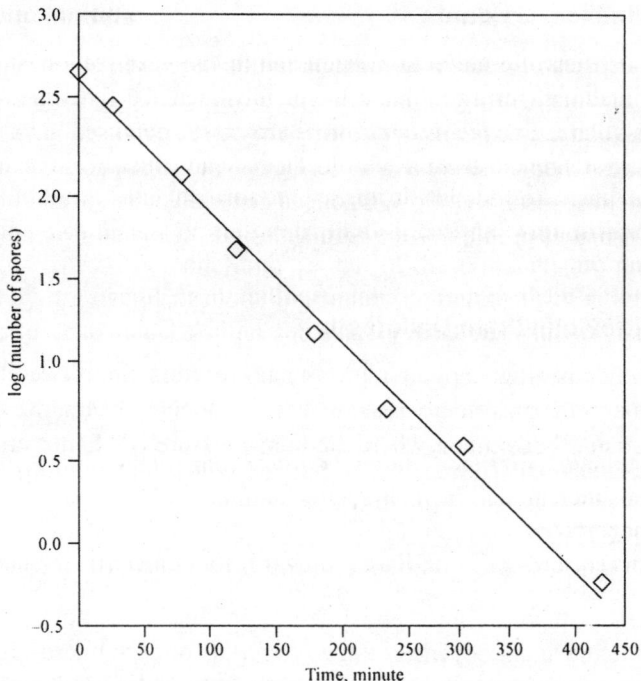

Fig. 13.1. Inactivation of *Bacillus anthracis* by phenol as originally plotted by Harriet Chick.

Application of Chick's concept met with immediate success and that success has continued through the years and across all the disciplines interested in disinfection. As a result, most disinfection data are first evaluated on a semi-log graph of survival, log(S), versus time t, where S = N/N₀.

Disinfection data do not always conform to Chick's linear semi-log plot. Two anomalies, accelerating rate and decelerating rate, are particularly common, as illustrated in Fig. 13.2. Reasons often cited in the literature for these particular curve shapes and the circumstances (organism, disinfectant and magnitude of disinfection) under which each type of curve is sometimes found are also given. It is important to note that most disinfection studies are conducted in the laboratory using a completely mixed batch reactor (CMBR). The reason a batch reactor is used is that the results obtained simulate the results that would be obtained with an ideal flow (i.e., without dispersion) plug flow reactor.

Chick's law

The most straightforward model of the disinfection process is given by Chick's law, in which the rate of the reaction is first order with respect to the concentration of the organisms being inactivated by the disinfectant. Chick's law takes the form:

$$\frac{dN}{dt} = -KN \qquad \qquad \dots (13.1)$$

where,

       dN/dt   = rate of change in number of organisms with time
       N   = concentration of organisms, organisms/volume
       K   = Chick's law rate constant, $time^{-1}$

After integration, Chick's law takes the form:

$$\ln\left(\frac{N}{N_0}\right) = -Kt$$

... (13.2)

Shape of semi-log plot of disinfection data	Reasons for shape	Examples
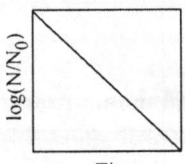	**Pseudo first order** The most common form of disinfection data Data fit Chick's law	With free chlorine: *E. coli*, Polio virus With ozone: Polio virus, *E. coli*, *G. lamblia*, *C. parvum* With UV: *C. parvum*, MS2 (<4 log), *G. lamblia* (<3 log)

(a)

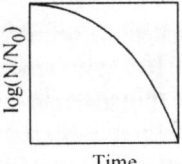	**Accelerating rate** Often observed at low disinfection doses Possible reasons: 1. Disinfectant must react with more than one critical site in organism. 2. Disinfectant must take time to diffuse to critical site. 3. Natural heterogeneity in resistance among organisms.	With combined chlorine: Most organisms at low inactivation With any disinfectant: Suspensions of aggregated virus particles or multicellular organisms With chlorine, combined chlorine or chlorine dioxide: *C. parvum*, endospores

(b)

(graph: log(N/N₀) vs. Time)	**Decelerating rate** Often observed after several logs of inactivation Possible reasons: 1. Decrease in germicidal properties of the disinfecting agent with time. 2. A resistance to the disinfectant increases with increasing exposure. 3. Natural heterogeneity in resistance among organisms. 4. Interference of particles with disinfection. 5. Organisms are in clumps that test as one unit but must be inactivated individually.	With combined chlorine: Most any organism at high removals With UV: Total coliform in secondary effluent, *G. lamblia* above 3 log removal

(c)

Fig. 13.2. Graphical forms of disinfection data.

When Chick first did her work, she actually plotted the log of the organism concentration directly against time, as shown in Fig. 13.1. Now that Eq. 13.2 has received broad recognition, it is more common to plot the log of the survival ratio versus time on a semi-log graph [$\log(N/N_0)$ vs. t]. When a log survival plot is used, however, significant errors in estimating the rate of disinfection can result from inaccurate measurement of the initial concentration of organisms, N_0. To improve the measurement of organism survival, several replicates of each point on the disinfection curve, including several measures of N_0, should be made. When the number of measured replicates is insufficient, a line fit through the data may not pass through zero [i.e., $\log(N/N_0)_{t=0} \neq 0$], which is not consistent with the definition of N_0. By definition, $N_0 = N_{t=0}$; thus, at t = 0, $\log(N/N_0) = 0$. Given the conditions cited above, it is often best to find the rate constant without requiring that the best fit pass through zero.

Another issue with Chick's law is the effect of the concentration of the disinfectant, as illustrated in Fig. 13.3, using data on the inactivation of polio virus with bromine. Note that there is a different slope for each concentration of bromine and using Eq. 13.2, the reaction has a different rate constant for each concentration. Thus, Chick's first-order concept is consistent with the data, but a better means for accounting for disinfectant concentration is necessary.

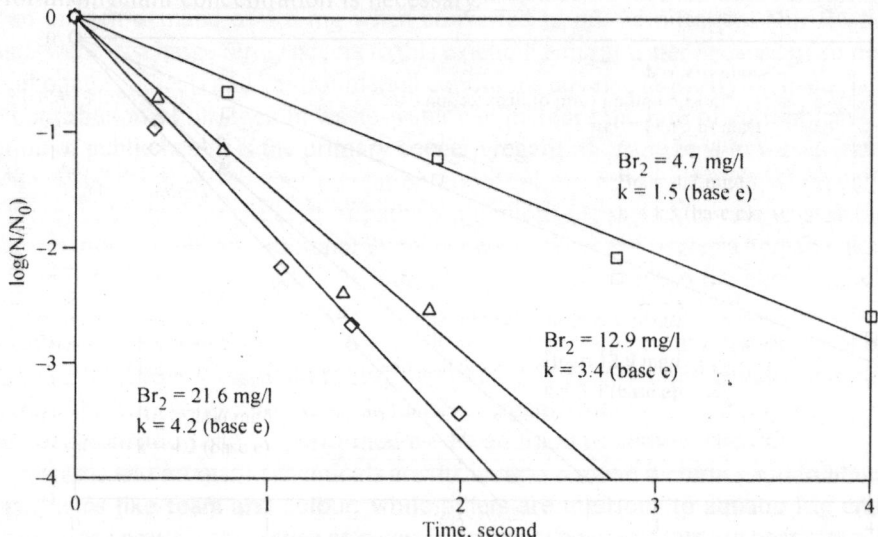

Fig. 13.3. Disinfection of Polio virus type I with three concentrations of bromine.

Chick-Watson model

Aware of the significance of the disinfectant concentration, Herbert Watson in the same year proposed an alternative approach. Watson proposed the following expression for a given level of inactivation by the disinfectant:

$$C^n t = \text{constant} \qquad \ldots (13.3)$$

where,

C	=	concentration of disinfectant, mg/l
n	=	empirical constant related to concentration, unitless
t	=	time required to achieve a constant percentage of inactivation (e.g., 99 per cent)
constant	=	value for given percentage of inactivation, unitless

Taking the log of Eq. 13.3 results in the linear equation:

$$n \log(C) + \log(t) = \log(\text{const}) \qquad \ldots (13.4)$$

Watson demonstrated the concept by plotting data points showing equal inactivation on a plot of $\log(C)$ versus $\log(t)$. The slope of the log-log plot, n, is often called the coefficient of dilution, which reflects the effect of diluting the disinfectant. When $n > 1$ the efficiency of the disinfectant decreases with dilution; if $n < 1$ time is more important than concentration; if $n = 1$ concentration and time are of equal importance. Such plots are still widely used today. As a matter of convention, Watson plots are generally constructed with data corresponding to a removal of 99 per cent.

An example of such a plot using data on the inactivation of polio virus with a number of disinfectants is shown in Fig. 13.4.

Chick's law and the Watson equation are often combined and referred to as the 'Chick-Watson model':

$$\frac{dN}{dt} = -\Lambda_{CW}C^nN = -K_{CW}N \qquad \qquad ... (13.5)$$

where,

dN/dt	=	rate of change in concentration of organisms with time
Λ_{CW}	=	coefficient of specific lethality, units vary with n
C	=	concentration of disinfectant, mg/l
n	=	disinfectant coefficient of dilution, unitless
N	=	concentration of organisms
K_{CW}	=	$\Lambda_{CW}C^n$

The dilution coefficient for all the data in Fig. 13.4 is approximately 1 and given the inaccuracies involved in collecting disinfection data, there is little evidence for a dilution coefficient other than unity. Therefore Eq. 13.5 can often be simplified to the form:

$$\ln\left(\frac{N}{N_0}\right) = -\Lambda_{CW}Ct \qquad \qquad ... (13.6)$$

The coefficient of specific lethality, Λ_{CW}, in Eq. 13.6, expressed in units of l/mg·min, can be determined by plotting log survival versus the product Ct on a semi-log graph.

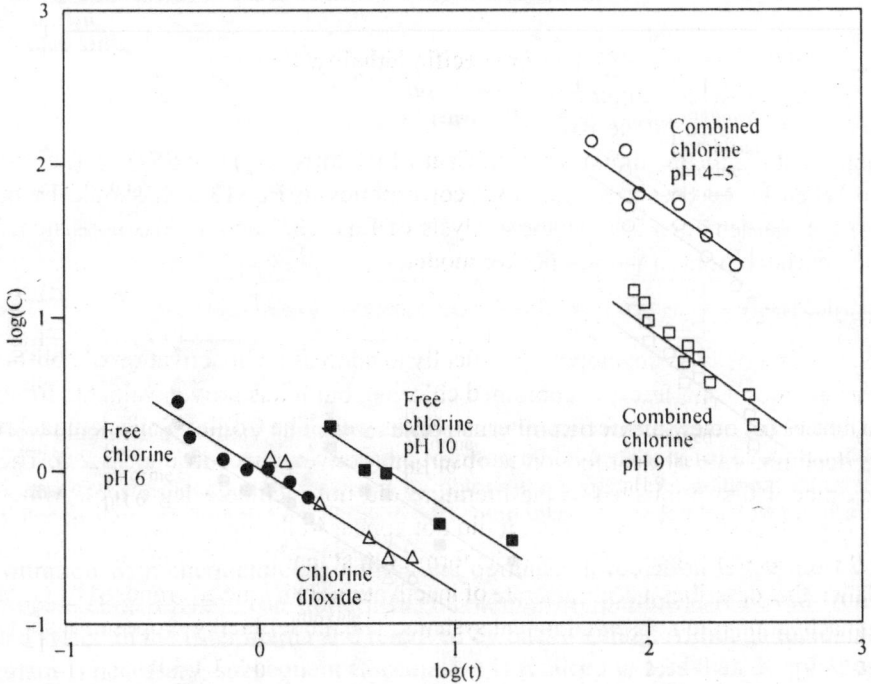

Fig. 13.4. Watson plot of requirements for 99 per cent inactivation of Polio virus type I.

Contemporary Kinetic Models

All of the disinfection models discussed above address disinfection data that can be described with a straight-line semi-log plot (Fig. 13.2a); however, it is common for disinfection reactions to exhibit a behaviour characteristic of the convex or concave rate curves (Figs. 13.2 b, c). As discussed earlier, only a limited understanding of the specific mechanisms for the various disinfection reactions is now available. Substantially different kinetics mechanisms may control the rate of inactivation of different micro-organisms with the same disinfectant or the same micro-organisms with different disinfectants. There is extensive literature on disinfection modelling; several of these models are presented in the following discussion because they are useful in modelling many common disinfection reactions. The models discussed below may be used to model disinfection data for reactions with accelerating and/or decelerating rates on a semi-log plot (Figs. 13.2b,c).

Rennecker-Marinas model

Some organisms do not exhibit significant inactivation until a certain Ct value has been exceeded. This inactivation response is observed, for example, when chemical disinfectants are applied to oocysts and endospores. The Marinas group at the University of Illinois has more recently addressed this situation by proposing the use of a lag coefficient, b for Eq. 13.6. The Rennecker-Marinas model can be summarised as follows:

$$\ln\left(\frac{N}{N_o}\right) = \begin{cases} 0 & \text{for } Ct < b & \qquad ...\,(13.7) \\ -\Lambda_{CW}(Ct - b) & \text{for } Ct \geq b & \qquad ...\,(13.8) \end{cases}$$

where,

Λ_{CW} = Chick-Watson coefficient of specific lethality, l/mg·min
b = lag coefficient, mg·min/l

Other terms are as defined previously.

The lag coefficient b is the maximum value of Ct at which $\ln(N/N_0) = \log(S_0) = 0$ (i.e., no inactivation has occurred). When b becomes zero, Eq. 13.8 corresponds to Eq. 13.6. It should be noted that the presentation of the mathematics used in the analysis of Eqs. 13.7 and 13.8 is consistent with but not identical to the approach used in the Rennecker model.

Collins-Selleck model

The Collins–Selleck model was developed specifically to address the inactivation of coliform organisms in domestic waste-water using free and combined chlorine, but it has proven valuable for modelling the behaviour of a number of other disinfection alternatives as well. The Collins-Selleck model is particularly useful when a declining rate of disinfection is observed (convex curve in Fig. 13.2c). The form of the Collins-Selleck model first published in the literature did not include a lag effect. Although the final formulation was published shortly after that, it did not appear in the peer-reviewed literature until some time later and there was some discussion about its form even at that time. Collins and Selleck began with a simple formulation that describes a declining rate of inactivation with time as proposed by Gard and adapted it to the lag in inactivation often observed in real systems. The model finally proposed is shown below:

$$\ln\left(\frac{N}{N_o}\right) = \begin{cases} 0 & \text{for } Ct < b & \qquad ...\,(13.9) \\ -\Lambda_{CS}[\ln(Ct) - \ln(b)] & \text{for } Ct \geq b & \qquad ...\,(13.10) \end{cases}$$

where,

Λ_{CS} = Collins-Selleck coefficient of specific lethality, unitless
b = lag coefficient, mg·min/l

Other terms are as defined previously.

There is a close parallel between the Collins-Selleck and Rennecker-Marinas models. Both have the same form, but in the Rennecker-Marinas model, the observed data are fit with a straight line when log survival is plotted versus Ct, whereas in the Collins-Selleck model, the data can be fit with a straight line when log survival is plotted versus log(Ct). Also each model has only two parameters, Λ_{CW} or Λ_{CS} and b. Based on a large number of tests, it has been found that most disinfection data can be fit to one of these two models.

Hom-Haas Model

Hom proposed an empirical model of survival versus disinfectant concentration that Haas subsequently refined. The integrated form of the Hom-Haas model, which has come to be commonly used, is a simple linear regression of survival, disinfectant concentration and time:

$$\ln\left(\frac{N}{N_0}\right) = -\kappa C^n t^m \qquad \qquad \text{... (13.11)}$$

where,

N/N_0 = survival of organisms
κ = die-off coefficient, variable
C = concentration of disinfectant, mg/l
t = time, minute
m, n = empirical constants, unitless

The Hom-Haas model has been examined extensively by Haas and found to be more robust than the Chick-Watson model and most other equations in fitting a variety of disinfection data. The Hom-Haas model can be used to describe reactions with either convex or concave shapes. The model has also been used to analyse bench-top disinfection data of a number of different organisms and to model the performance of disinfection systems where modelling of the hydrodynamics of the basin, the decay of the disinfectant and the disinfection process itself are all integrated into one model system.

Comparison of Disinfection Models

Some important characteristics of the disinfection models presented in this section are summarised in Table 13.2. The Chick-Watson model (Eq. 13.6) is not shown as it now becomes a special case of the Rennecker-Marinas model when b = 0. A selection of kinetic constants gathered from the literature are reported in Table 13.3. Constants are offered for the disinfection of total coliform from waste-water because this has long been a target organism in effluent re-use. Constants are presented for (E. coli) and poliovirus because these organisms have long been the targets of classical disinfection regulations. Constants are presented for Giardia and Cryptosporidium because the difficulty in inactivating these organisms is having profound effects on water treatment regulations. Constants are presented for Bacillus subtilis because its behaviour in disinfection is thought to be similar to Bacillus anthracis, a possible organism that may be used by terrorists.

Table 13.2. Comparison of disinfection models.

Disinfection model	Form of data	Plot as straight line on	Number of coefficients	Comment
Chick[a]: $$\ln\left(\frac{N}{N_0}\right) = -Kt$$ where K = Chick's rate constant, time^{-1}	Pseudo-first order	Semi-log graph: log (N/N$_0$) vs. t	1	Widely used in microbiology. Approximates a lot of disinfection data. Subject to anomalies (see Fig. 13.2)
Rennecker-Marinas: For Ct < b $$\ln\left(\frac{N}{N_0}\right) = 0$$ For Ct > b $$\ln\left(\frac{N}{N_0}\right) = -\Lambda_{CW}(Ct - b)$$ where b = lag coefficient, mg·min/l Λ_{CW} = coefficient of specific lethality, decades/(mg·min/l)	Pseudo-first order with lag	Semi-log graph: log (N/N$_0$) vs. Ct	2	Equation is consistent with 'Ct' concept. Can approximate most disinfection data. Performs poorly only if the disinfection reaction truly shows an accelerating rate from the start or if a decelerating rate of reaction is observed. When b = 0, this equation has the same form as the Chick-Watson equation when the dilution coefficient n = 1
Collins-Selleck: For Ct < b $$\ln\left(\frac{N}{N_0}\right) = 0$$ For Ct > b $$\ln\left(\frac{N}{N_0}\right) = -\Lambda_{CS}[\ln(Ct) - \ln(b)]$$ where b = lag coefficient, mg · min/l Λ_{CS} = log-based coefficient of specific lethality	Decelerating rate with lag	log-log graph: log(N/N$_0$) vs. log(Ct)	2	Equation is also consisten with 'Ct' concept. Can approximate most disinfection data. Performs poorly if only the accelerating phase is off interest or if several logs of first-order behaviour are observed
Hom-Haas: $$\ln\left(\frac{N}{N_0}\right) = -\kappa C^n t^m$$ where k = Hom-Haas rate constant, (mg/l)$^{-n}$(min)$^{-m}$ n = Hom-Haas coefficient of dilution, unitless	Takes any form	No plot: log (N/N$_0$) vs. t can be plotted for given value of C, but plot is not a straight line	3	Equation does the best job of fitting the widest variety for curve shapes. Particularly useful of modelling performance of one disinfectant-organism combination for which the disinfection kinetics have been well characterised. Comparing n and m from one disinfectant-

(Contd...)

Disinfection model	Form of data	Plot as straight line on	Number of coefficients	Comment
m = Hom-Hass time coefficient, unitless				organism combination with another is not always useful because the three constants sometimes vary in a complex way. Simplifies to the Chick-Watson equation when the time coefficient m = 1. Is consistent with the 'Ct' concept only when m = 1 and n = 1

[a] The Chick-Watson model is not shown because it is a special case of the Rennecker-Mariñas model when b = 0.

Table 13.3. Selected kinetic parameters (base e) based on data in the literature[a].

Organism	Disinfectant	Chick-Watson and Rennecker-Marinas		Collins-Selleck, Λ_{CS}
		Λ_{CW} l/mg·min or m^2/J	b, mg·min/l or J/m^2	
E. coli	Cl_2, pH 8.5, T = 2–5°C	3.75	0.2	–
	NH_2Cl	0.0375	10	–
	NH_2Cl	0.0327	–	–
	ClO_2	3.3	0.33	–
	O_3	8330	–	–
	UV	0.83	–	–
Total coliform (waste-water or waste-water seed)	HOCl	–	0.005	1.2
	OCl^-	–	0.1	1.9
	NH_2Cl	–	3.0	2.8
	ClO_2	–	0.9	2.2
	UV	–	4	26
Poliovirus	HOCl	0.2	–	–
	NH_2Cl	–	–	–
	ClO_2	0.47	28	–
	O_3	0.85	–	–
	UV	3	–	–
MS-2	Cl_2	3.4	–	–
	NH_2Cl	0.005	–	–
	ClO_2	–	–	–

(Contd...)

| Organism | Disinfectant | Chick-Watson and Rennecker-Marinas | | Collins-Selleck, Λ_{CS} |
		Λ_{CW}, l/mg·min or m²/J	b, mg·min/l or J/m²	
	O_3	–	–	–
	UV	0.96	–	–
Giardia	Cl_2, pH 7	–	68	3.8
	NH_2Cl	–	300	5
	ClO_2	0.21	–	–
	O_3	–	0.02	1.77
	O_3	1.9	–	–
	UV	38	–	–
C. parvum	Cl_2, pH 6	0.0013	375	–
	NH_2Cl	0.00077	5500	–
	ClO_2	0.083	35	–
	O_3	1.7	0.22	–
	O_3	0.83	–	–
	UV	25	–	–
B. subtilis	Cl_2, pH 6	0.0006	–	–
	NH_2Cl	0.00054	4560	–
	ClO_2	0.13	–	–
	O_3	2.12	4.91	–
	UV	0.004	170	–

ª Unless otherwise noted all kinetic parameters are given for 25°C.

Application of Models Used to Assess Disinfection

The true, detailed kinetics of most chemical disinfectants are exceedingly complex and they are influenced by the chemistry of the disinfectant as well as the nature of the susceptibility in the organism. Moreover, measuring disinfection effectiveness is difficult to do with great precision, partly because of the complexity of the chemical conditions but also due to the imprecise nature of most microbiological measurements. As a result, it is probably best to employ the simplest approach possible to describe the results of disinfection experiments. In order of increasing complexity, alternatives that might be considered are:

1. Ct tables. Numerical Ct (concentration × time) values are established to achieve a given degree of inactivation of a specific organism using a defined disinfectant under controlled conditions. Such an approach is consistent with all the models except the Hom-Haas model; however, when n and m are equal (see Eq. 13.11), this approach is also consistent with the Hom-Haas model. The US EPA uses this approach in regulating water disinfection. When required, different tables can be offered for a range of concentrations, as the US EPA did for the inactivation of *G. lamblia* with free chlorine.

2. Semi-log plots of survival versus Ct. The use of semi-log plots of log survival as a function of Ct is consistent with the Chick-Watson model and the Rennecker-Marinas model. In this approach,

it is assumed that the log survival values will plot as a linear function of time or the concentration-time product Ct on a semi-log plot and only one or two constants, Λ_{CW} and b, are required for application of the model. This approach is often successful when a modest degree of disinfection is required, a reduction of approximately 3 log inactivation, for example.

3. Log-log plots of survival and Ct product. The use of log-log plots is consistent with the Collins-Selleck model. This approach is useful for situations where a lag time is present (complex organisms, slow disinfectants, etc.) or where a declining rate of disinfection with time is observed. This approach is also useful when disinfection requirements are substantial, for example, 4 log reduction or more. In the Collins-Selleck model, it is assumed that the log survival will plot as a linear function against log (Ct) and that two constants, Λ_{CS} and b, are required for application of the model.

4. Correlating principal factors of influence. The Hom-Haas model is essentially a regression of disinfection performance against the two most important factors of influence, time and disinfectant concentration. In general, this model is the most powerful for use in characterising the results of a specific experiment.

Generally, as disinfection models becomes more complex, the precision with which they can be used to describe the results of a given disinfection experiment improves. However, comparing the constants of the simpler models provides better perspective on the performance of different disinfectants and on the resistance of different organisms. The ability to compare results is one of the reasons that Chick's law and the simplified Chick-Watson equation continue to be popular.

Review of Ct Approach to Disinfection

The product Ct has long been used as a basis for disinfection requirements. Such a practice is equally practical when the Collins-Selleck and Rennecker-Marinas models are used. The Ct product required for achieving a given level of disinfection for a specific micro-organism under defined conditions is a useful way of comparing alternate disinfectants and for comparing the resistance of a variety of pathogens. Indeed, the product Ct can be thought of as the dose of disinfectant.

The dose concept, analogous to Ct, is also applicable when UV light is used for disinfection. The product of the UV light intensity (mW/cm^2) and the time of exposure is used to compute the dose ($mW/cm^2 \times s = mJ/cm^2$). This product is often referred to as It (intensity × time). Modelling disinfection with UV light using It in place of Ct in Eq. 13.6 has also been successful. In the case of UV light, there is probably greater justification for this form because the mechanism of inactivation is not so much a function of light intensity but a function of exposure of the organism to a quantity of potentially damaging photons. The Ct concept also allows for the development of a broad overview of the relative effectiveness of different disinfectants and the resistance of different organisms.

The Ct required to produce a 99 per cent inactivation of several micro-organisms using the five disinfection techniques most often used in water treatment is illustrated in Fig. 13.5. Because of the difference in the behaviour from one organism and one disinfectant combination to the next, Ct and It products range over seven orders of magnitude. For example, the Ct product required to inactivate *C. parvum* must be three orders of magnitude higher with chloramines than with ozone. Comparing UV disinfection to disinfection with chemical oxidants, little similarity exists between the it values for one organism with the Ct values for the same organism. While the required UV doses vary over a range of two orders of magnitude, their variation is much less than that for other disinfectants.

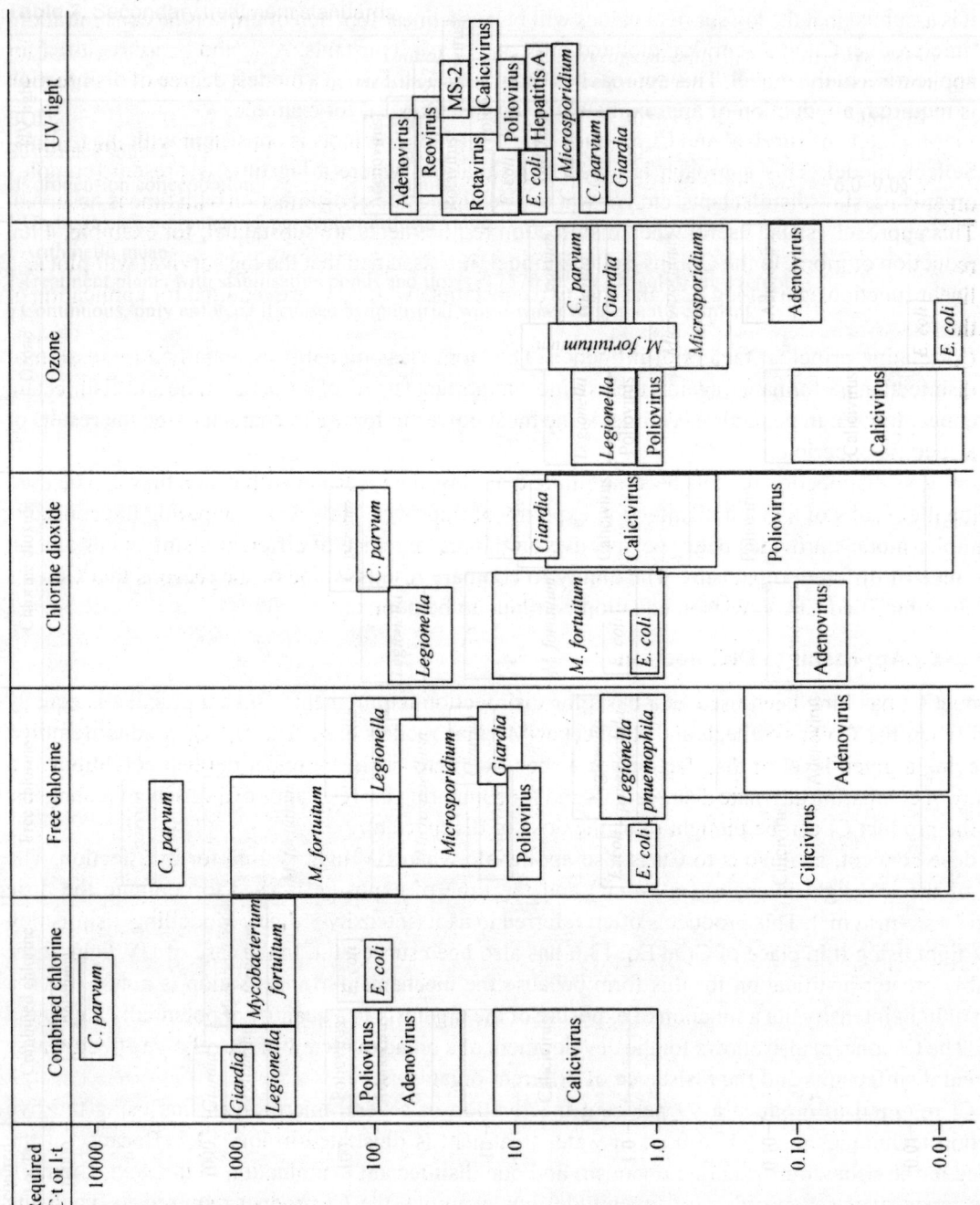

Fig. 13.5. Overview of disinfection requirements for 99 per cent inactivation.

The reduced variation may be a result of the fact that UV disinfection of all micro-organisms results from a similar protein dimerisation mechanism.

The US EPA began the practice of specifying Ct products that must be met as a way of regulating the control of pathogens in water treatment with the promulgation of the Surface Water Treatment Rule (US EPA, 1989a). A summary of the Ct and It values required to meet the primary disinfection requirements specified in the Surface Water Treatment Rule (US EPA, 1989a) and the proposed long-term disinfection rule (US EPA, 2004).

Declining Concentration of Chemical Disinfectant

Chick's experiments were conducted with stable disinfectant concentrations as the disinfectant was in excess. In the laboratory, researchers generally attempt to maintain the disinfectant concentration during the kinetic study so that disinfection rates can be measured with maximum precision. Given the complexities that exist in the microbiological world that can influence the outcome of such experiments, it is important to minimise variations in chemistry and physical conditions. With combined chlorine, a stable residual is usually a reasonable assumption in full-scale contactors as well. With free chlorine and chlorine dioxide it can also be reasonable for short contact times. For these same disinfectants at longer contact times or for ozone contactors at any contact time, residual decay must be accounted for.

Accounting for varying disinfection concentration can be addressed by dividing the problem into two parts: (i) modelling the decay of the disinfectant; and (ii) integrating that work into the model of the disinfection reaction itself. For all the common oxidising disinfectants (chlorine, combined chlorine, chlorine dioxide and ozone), it is often assumed that disinfectant decay can be modelled as first order, that is,

$$\frac{dC}{dt} = -k_d C \qquad \qquad ...(13.12)$$

where,

dC/dt = change in disinfectant concentration with time, mg/l·s or moles/l·s
k_d = first-order decay rate, s^{-1}
C = disinfectant concentration, mg/l or moles/l

The decay of these disinfectants is often characterised by two phases, an early phase of rapid decay followed by a later phase with slower decay. When two-phase decay occurs, a second-order model with a fast reaction step and a slow reaction step has been used successfully, but this model is rather difficult to use because it cannot be solved analytically. Another alternative is the parallel first-order decay model proposed by Haas and Karra, in which it is assumed that decay may proceed through two mechanisms, each first order but involving a different component of the chlorine residual:

$$\frac{dC}{dt} = -x k_{d1} C - (1-x) k_{d2} C \qquad \qquad ...(13.13)$$

where,

dC/dt = change in disinfectant concentration with time, mg/l·s or moles/l·s
x = fraction varying between 0 and 1, unitless
C = concentration of disinfectant, mg/l or moles/l
k_{d1}, k_{d2} = decay coefficients for two different mechanisms, s^{-1}

The first component, with an initial concentration of C_{0x}, is subject to first order decay with a faster rate constant, k_{d1} and the second component, with an initial concentration of $(1-x) C_0$, is subject to first-order decay with a slower rate constant, k_{d2}. As noted above, the value of x, by definition, is between zero and 1. When x = 0, the parallel first-order model becomes the simple first-order model; the same is true when x =1.

Finding an analytical solution in which the decay reaction and the disinfection reaction are integrated together adds to the complexity of the mathematics used to describe the disinfection process. Where analytical solutions are not available, it is possible to use computer models to simulate the two processes in parallel. Nevertheless, Haas and Joffe have developed analytical solutions for both the Chick-Watson model and the Hom-Haas model.

Influence of Temperature

The effect of temperature on the rate of a chemical reaction is described by the Arrhenius equation and is used here to describe the influence of temperature on the pseudo-first-order disinfection rate constant:

$$\ln(k_r) = \left(-\frac{E_a}{R}\right)\left(\frac{1}{T}\right) + \ln(A) \qquad \qquad \ldots (13.14)$$

where,

k_r = appropriate reaction rate constant, Λ_{CW}, Λ_{CS}, K or κ

E_a = activation energy, J/mole

R = universal gas constant, 8.314J/(mole·K)

T = reaction temperature, K (273 + °C)

A = collision frequency parameter

Once the rate is known at one temperature, the rate at another temperature can be determined if the activation energy E_a is known. In the disinfection literature, an empirical approach used is to specify θ in the following equation:

$$\frac{k_{r.T_1}}{k_{r.T_2}} = \theta^{T_1 - T_2} \qquad \qquad \ldots (13.14a)$$

where,

k_{r,T_1} = reaction rate constant at temperature 1

k_{r,T_2} = reaction rate constant at temperature 2

θ = empirical constant, dimensionless

T_1 = temperature corresponding to known rate constant $k_{r.T1}$ K (273 + °C)

T_2 = temperature corresponding to known rate constant $k_{r.T2}$ K (273 + °C)

Combining Eqs. 13.14 and Eq. 13.14a and solving for q, the following expression is obtained:

$$\theta = e^{E_a / RT_1 T_2} \qquad \qquad \ldots (13.15)$$

Because the product $T_1 T_2$ is somewhat insensitive to changes in temperature it is reasonable to assume θ is constant in empirical approach. Values of E_a from the literature are summarised in Table 13.4.

DISINFECTION KINETICS IN NON-IDEAL REACTORS

The disinfection kinetics described in previous section were based on studies conducted in CMBRs. While the results of batch kinetics provide insight into the behaviour of full-scale continuous-flow systems, such systems exhibit more complex non-ideal behaviour. Of particular importance is the impact of dispersion on the progress of the reaction.

Table 13.4. Activation energies for a variety of disinfection reactions.

Micro-organisms	Disinfectant	E_a, kJ/mol	$K_{25°C}/K_{5°C}$
C. parvum	HOCl	71.9	
C. parvum	HOCl	64.7	
		72[a]	6.4
C. parvum	ClO$_2$	67.5	
C. parvum	ClO$_2$	86.3	
		77[a]	8.0
C. parvum	NH$_2$Cl	75.6	
C. parvum	NH$_2$Cl	78.7	
C. parvum	NH$_2$Cl	59.2[b]	
		77[a]	8.0
C. parvum	O$_3$	102	
C. parvum	O$_3$	75.7	
C. parvum	O$_3$	81.2	
C. parvum	O$_3$	47.6	
		76[a]	7.8
C. muris	O$_3$	92.8	12
E. coli	O$_3$	37.1	2.7
G. lamblia	O$_3$	39.2	2.9
G. muris	O$_3$	70	6.6
N. gruberi	O$_3$	31.4	2.3
B. subtilis	O$_3$	46.8	3.6
B. subtilis	NH$_2$Cl	79.6	8.7

[a] Recommended value; and [b] old oocysts.

Influence of Dispersion

Three different approaches to modelling the performance of non-ideal reactors are: (i) the tanks-in-series (TIS) model; (ii) the dispersed-flow model (DFM); and (iii) the segregated-flow model (SFM). The TIS model is used to simulate the effects of dispersion on the residence time distribution (RTD) curve by examining the behaviour of a cascade of completely mixed flow reactors (CMFRs) in series. In the TIS model, dispersion is modelled using the number of reactors in series, n. A high value of n corresponds to low dispersion. The DFM is used to simulate the effects of dispersion on the RTD using models of axial dispersion in turbulent flow that would be expected in a channel or pipeline reactor. In the DFM, dispersion is described using the Peclet number (Pe) or the dispersion number (d, Pe = 1/d). A high value of Pe or a low value of d corresponds to low dispersion. The SFM is used to simulate the effects of non-ideal mixing by assuming the flow passes through a series of parallel plug flow reactors (PFRs) having detention times that, in sum, match the RTD of the real reactor. While the TIS model and the DFM incorporate assumptions about the nature of the RTD curve built in to their structure, an RTD curve must be provided for use in the SFM. In the TIS model, it is assumed that all the components that participate in the reaction are mixed completely throughout each reactor at all times. In the DFM, it is assumed that all components are mixed completely in the lateral direction but axial transport occurs by advection and dispersion.

When dispersion is low, both the TIS model and DFM produce similar results. In the SFM, it is assumed that the components that participate in the reaction are never completely blended in the reactor; rather the target reactant travels through the reactor in small cells or discrete elements that react with the bulk solution.

Application of the SFM Model

Disinfection processes are an ideal application of the SFM because micro-organisms actually do travel through the reactor as particles, separate from the fluid, but react with disinfectants in their environment as they pass through. If disinfection conditions are uniform throughout the reactor (e.g., the reactant residual or the intensity of inactivating irradiance is uniform throughout), the inactivation of each individual micro-organism is the same as it would be in a batch reactor after the same residence time. The RTD of a conservative tracer can reasonably be used to describe the RTD of the micro-organisms themselves. Thus, the disinfection process can be modelled in the following manner:

$$\frac{N_t}{N_0} = \sum_{t=0}^{t=\infty} E_t \left(\frac{N}{N_0}\right)_{Batch,\ t} \Delta t \qquad \qquad \dots (13.16)$$

where,

N_t, N_0	=	number of organisms at time t and 0, respectively
E_t	=	exit age distribution at time t
$(N/N_0)_{Batch,\ t}$	=	inactivation of micro-organisms that would be achieved in CMBR or PFR under same conditions after reaction time t
Δt	=	differential time step

The expression shown in Eq. 13.16 is analogous to the SFM method of analysis. Using the model presented in Eq. 13.16, the disinfection achieved in a real reactor can be estimated from the RTD curve and knowledge of the batch reaction in the following manner:

$$\frac{N}{N_0} = \int_{\theta=0}^{\theta=\infty} R[\theta]E[\theta]\ d\theta$$

where,

$R[\theta]$	=	inactivation of micro-organisms that would be achieved in CMBR or PFR under same conditions after reaction time equivalent to $\theta = (N/N_0)_{Batch,\ \theta}$
θ	=	normalised time (any observed time period divided by hydraulic residence time, $= t/\tau$), dimensionless
τ	=	hydraulic residence time, $= V/Q$
V	=	volume of reactor
Q	=	flow rate through reactor

Selleck first introduced the approach outlined above to modelling in the early 1970s. Trussell and Chao then employed this approach to demonstrate the influence of dispersion on chlorine contactor performance. Both Trussell and Chao worked on disinfection of coliform bacteria using combined chlorine and, in both studies, the disinfectant residual was assumed to be constant and uniform throughout the reactor.

Schieble introduced a similar approach to the modelling of UV disinfection in the US EPA disinfection design manual (US EPA, 1986). The approach is appropriate for UV disinfection if it is assumed that turbulent flow exists, no short circuiting occurs and each organism takes a path through the contactor such that its average exposure to UV light is equal to the average intensity of UV light in the reactor.

The promulgation of the US EPA's Surface Water Treatment Rule substantially increased disinfection requirements for drinking water in the United States (US EPA, 1989) and as a result, has stimulated further interest in methods of refining the rule's approach to specifying disinfection. Lawler and Singer re-introduced the concept again and later Haas demonstrated its application. Subsequently the SFM concept was incorporated in the integrated disinfection design framework, an effort to further optimise the design and operation of water disinfection systems.

When Dispersion is Important in Disinfection

Minimising dispersion and short circuiting in disinfection contactors is a widely accepted principle. To this end, California requires a minimum t_{modal} of 90 minutes and a minimum length-to-width ratio of 40:1 for baffled chlorine contactors in its regulations for reclaiming waste-water for non-restricted re-use. The US EPA limits the credit for disinfection contact time to the time it takes for the first 10 per cent of a tracer to pass through a disinfection contactor (t_{10}).

As a general rule, reducing dispersion is more important when disinfection goals are substantial. For example, dispersion is more important in the design of a contactor that must achieve 4 log of inactivation than it is in the design of a contactor that must achieve 1 log of reduction. The proceeding statement is true regardless of the organism under consideration or its specific reaction kinetics.

A thought experiment can be used illustrate the impact of dispersion. Assume a disinfection process is designed to achieve a 1 log reduction of a certain protozoa and a 4 log reduction of a particular virus. Further assume the reactor operates as designed and achieves exactly those objectives. A small by-pass pipe is installed and 1 per cent of the flow coming into the reactor is diverted so that it flows around the reactor and blends, without disinfection, with the treated water from the reactor. The result of the experiment is illustrated in Fig. 13.6. The small diversion as illustrated has almost no impact on the removal of protozoa but severely compromises the removal of the virus, exposing the consumer to virus levels nearly 100 times higher than the goal that was being sought.

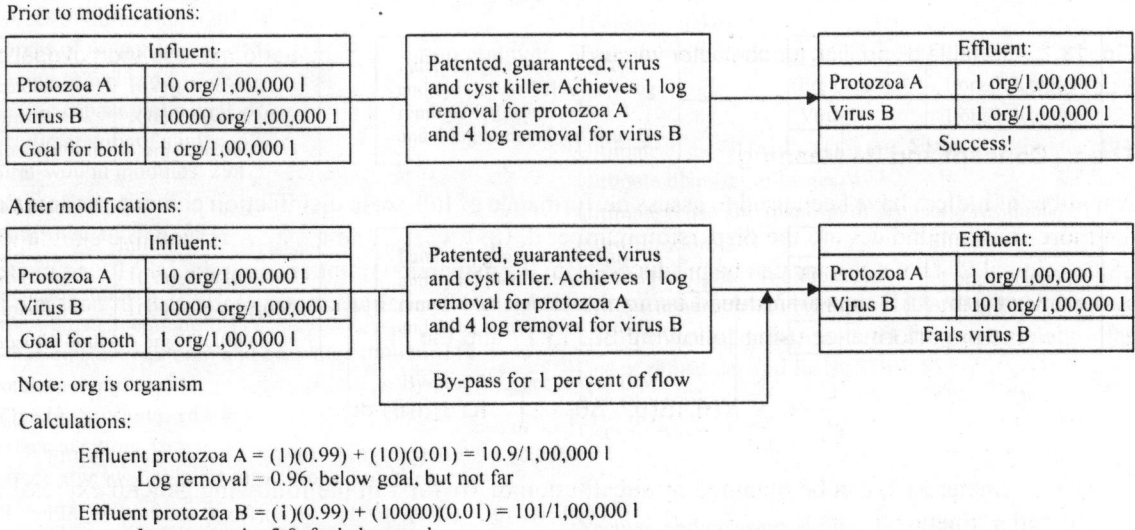

Prior to modifications:

Influent:			Patented, guaranteed, virus and cyst killer. Achieves 1 log removal for protozoa A and 4 log removal for virus B	Effluent:	
Protozoa A	10 org/1,00,000 l			Protozoa A	1 org/1,00,000 l
Virus B	10000 org/1,00,000 l			Virus B	1 org/1,00,000 l
Goal for both	1 org/1,00,000 l			Success!	

After modifications:

Influent:			Patented, guaranteed, virus and cyst killer. Achieves 1 log removal for protozoa A and 4 log removal for virus B	Effluent:	
Protozoa A	10 org/1,00,000 l			Protozoa A	1 org/1,00,000 l
Virus B	10000 org/1,00,000 l			Virus B	101 org/1,00,000 l
Goal for both	1 org/1,00,000 l			Fails virus B	

Note: org is organism

By-pass for 1 per cent of flow

Calculations:

Effluent protozoa A = (1)(0.99) + (10)(0.01) = 10.9/1,00,000 l
Log removal = 0.96, below goal, but not far

Effluent protozoa B = (1)(0.99) + (10000)(0.01) = 101/1,00,000 l
Log removal = 2.0; far below goal

Fig. 13.6. Thought experiment: Dispersion and short circuiting are more important when removal goals are high.

A dispersion model may be prepared to estimate the amount of dispersion that could be allowed without compromising plug flow performance more than 5 per cent (specifically without reducing the log removal more than 5 per cent). Assuming a simple first-order reaction, the plug flow contactor may be specified with a removal that ranges several orders of magnitude. As shown in Fig. 13.7, when the required removal is modest, so are the requirements for controlling dispersion. As the removal requirements increase to 3 log or more of removal, dispersion is limited. Removals higher than 3 log cannot be achieved if the dispersion is not reduced to the level specified or reactors designed to achieve these higher removals will have to operate at Ct values substantially greater than those required to meet the same treatment objective in a perfect plug flow reactor or in a batch reactor on the laboratory bench.

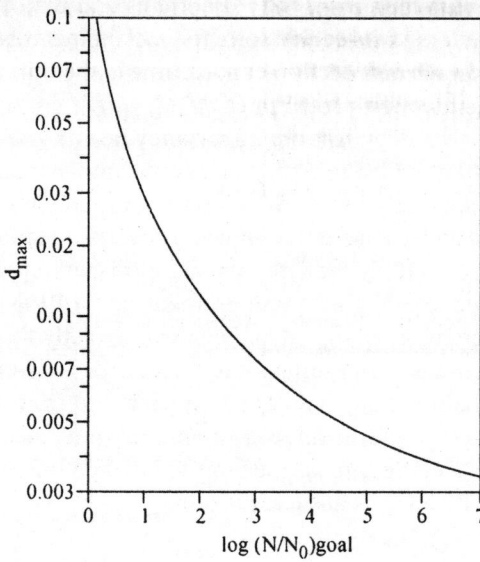

Fig. 13.7. Allowable dispersion for contactor versus inactivation goals. At d_{max}, performace is short of goal by 5 per cent.

The t_{10} Concept and its Meaning

A number of indices have been used to assess performance of full-scale disinfection contactors. Some of the more common indices are the dispersion number d, t_{10}/τ, t_{10}/t_{90} and t_{modal}. A reasonable simulation of the original RTD of a reactor can be produced from the dispersion number using the DFM for a closed system. Using an RTD curve produced using the DFM and available kinetic models, it is possible to estimate reactor performance using following Eq. 13.17 and the SFM.

$$\frac{\overline{C}}{C_0} = \sum_{i=1}^{N} R(\theta_i)E(\theta_i) \, \Delta\theta_i = \int_0^{\infty} R(\theta)E(\theta) \, d\theta \qquad \ldots (13.17)$$

The parameter $R(\theta)$ can be obtained by substitution of $\theta\overline{t}$ for τ in the following general expression for any order kinetics.

As a result, the dispersion number is perhaps the best measure of the suitability of a reactor for accomplishing disinfection. Nevertheless, regulators tend to prefer parameters like t_{10} or t_{modal} as these values are understood and determined more easily. Because of the US EPA's regulations, t_{10} deserves special attention where water treatment is concerned. To assess whether using the t_{10} value provides the same level of protection as controlling dispersion, Fig. 13.8 was constructed using Eq. 13.17 that would achieve 4 log of removal in a plug flow reactor. The performance estimated by the SFM for the reactor with dispersion (middle curve) is compared to the performance credit that would be allowed for the reactor based on the batch equation and the product Ct_{10} (bottom curve). The inactivations estimated by the SFM and by the product Ct_{10} both improve as dispersion is reduced. From the presentation on Fig. 13.8 it can be concluded that the US EPA's t_{10} approach is effective, but conservative.

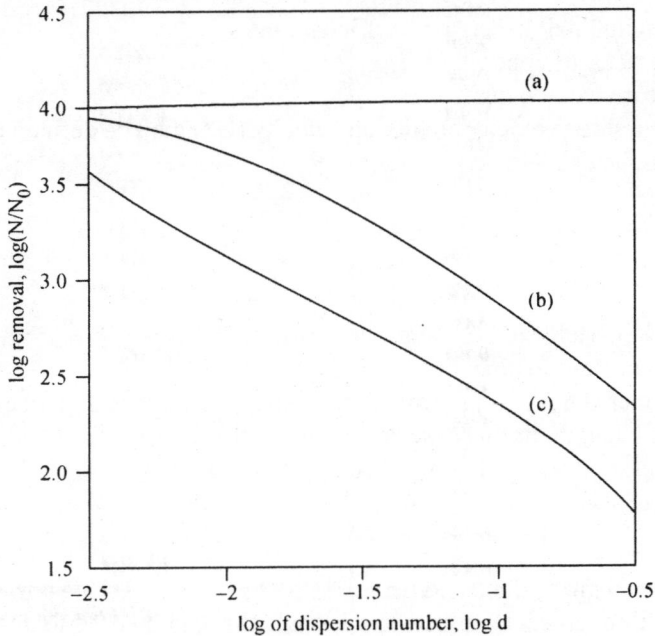

Fig. 13.8. Use of flow models to predict reactor disinfection performance: (a) ideal plug flow mode; (b) segregated-flow model (SFM) with dispersion, derived from residence time distribution (RTD) curve produced using closed-system dispersion flow model (DFM); and (c) predicted using t_{10} values derived from $E(\theta)$ curves based on closed-system DFM.

Designing Contactors that Exhibit Low Dispersion

Because disinfection requirements were less stringent in the past, disinfection was often not a special concern. Chlorine was added early in the treatment process and the chlorine residual carried throughout the plant. Following the THM rule in 1980 many utilities moved the point of chlorine addition to the end of the treatment process. Later, when the first Surface Water Treatment Rule came about, many utilities struggled to find a way to get more credit for contact time in their existing facilities, often by baffling them to increase t_{10}.

More and more, disinfection is being treated as a separate unit process, and it is this approach that will be discussed here. Engineered disinfection contactors are typically of three types: (i) serpentine basins; (ii) pipelines; and (iii) the over and underbaffled contactors often used in ozonation.

Design of pipeline contactors

For free chlorine, combined chlorine and chlorine dioxide the ideal contactor design is a long channel or pipeline with plug flow characteristics. Occasionally there is a long pipeline leaving the plant that has sufficient contact time to make it an attractive alternative. Axial dispersion in pipeline flow is the most straightforward case that will be considered. Taylor demonstrated that the longitudinal dispersion coefficient (D_L) can be described as:

$$D_L = 5.05 DU^* \qquad \qquad ...(13.18)$$

where,

D_L = longitudinal dispersion coefficient, m²/s
D = diameter of conduit, m
U^* = shear velocity, m/s

In the above formula the shear velocity or friction velocity (U^*) may be defined in terms of the velocity of flow and the friction factor:

$$U^* = \sqrt{\frac{f\ U^2}{8}} \qquad \qquad ...(13.18a)$$

where,

f = Darcy-Weisbach friction factor, unitless
U = velocity of flow in pipe, m/s

The dispersion number is defined in terms of the longitudinal dispersion coefficient, the velocity of flow and a characteristic length, in this case, the length of the pipe:

$$d = \frac{D_L}{UL} \qquad \qquad ...(13.19)$$

where,

d = dispersion number, dimensionless

Combining Eqs. 13.18 through 13.19 results in a formula that can be used to describe the dispersion of flow in a pipe:

$$d = 5.05 \left(\frac{D}{L}\right)\sqrt{\frac{f}{8}} \qquad \qquad ...(13.20)$$

Available data from laboratory experiments confirm Taylor's theory within a factor of 2. Generally somewhat more dispersion is found in field-scale measurements than would be predicted from the theory. For this reason, Sjenitzer gathered a great number of measurements, both in the laboratory and in the field and correlated them to produce the empirical expression:

$$d = 89500\ f^{3.6} \left(\frac{D}{L}\right)^{0.859} \qquad \qquad ...(13.21)$$

Using Sjenitzer's data, Trussell and Chao demonstrated that Eq. 13.21 provides a significantly better fit of the data than Eq. 13.20. Even Sjenitzer's equation, however, is only accurate for a long pipeline without bends, restrictions or other disturbances to flow. Generally, the flow in a pipeline with 30 minutes of contact time, a flow rate greater than 44 l/s (1 mgd) and a velocity greater than 0.6 m/s (2 ft/s) will be nearly ideal plug flow in behaviour.

Design of serpentine contact basins

A pipeline is convenient if it is already necessary for some other purpose, but long, baffled, serpentine basins are generally the most cost-effective means of achieving low dispersion (plug flow). Using such designs, it is not uncommon to achieve dispersion numbers less than 0.01 and t_{10}/τ of 80 per cent. An ideal basin would be long and narrow, similar to the contactor discussed in the previous section. Computational fluid dynamics should be used to optimise the design of any large disinfection contactor. In the following discussion, the design of serpentine basins to achieve a specified level of dispersion is addressed first and then, because of US regulatory requirements, designing these same facilities to meet a specified t_{10} will also be discussed.

Designing for a specified level of dispersion

To develop a better understanding of design criteria, it is useful to start with a more general form of Eq. 13.17 (the Taylor equation):

$$D_L = CR_h U^* \qquad \ldots (13.22)$$

where,

D_L = longitudinal dispersion coefficient, m^2/s
C = coefficient, unitless
R_h = hydraulic radius of channel, m
U^* = shear velocity, m/s

Here the coefficient C is a function of channel geometry and the Reynolds number. Elder applied Taylor's concept of dispersion to a logarithmic velocity profile and suggested that the coefficient C should have a value of approximately 5.9 for the circumstances in most chlorine contact chambers. For uniform flow in an open channel, the friction velocity can be defined as follows:

$$U^* = \frac{3.82nU}{R_h^{1/6}} \qquad \ldots (13.23)$$

where,

U = velocity of flow in channel, m/s
n = Manning coefficient, unitless

Combining Eqs. 13.23, 13.22 and 13.19, the following approximate formula for dispersion coefficient in a long open channel is obtained:

$$d = \frac{22.7nR_h^{5/6}}{L} \qquad \ldots (13.24)$$

Equation 13.24 may be re-written to describe dispersion using the channel volume and height and length aspect ratios:

$$d = 22.7 \frac{n}{\beta_L}\left(\frac{\beta_H}{2\beta_H + 1}\right)^{5/6}\left(\frac{\beta_H\beta_L}{V_{ch}}\right)^{1/18} \qquad \ldots (13.25)$$

where,

β_H = height aspect ratio or H/W (channel height/channel width)
β_L = length aspect ratio or L/W (channel length/channel width)
V_{ch} = channel volume, m^3

The dispersion values computed using Eq. 13.25 are not sensitive to the range of β_H values typical for concrete contact chambers (1 to 3). As a result, the following abbreviated form of Eq. 13.25 can be used satisfactorily:

$$d = \frac{0.14}{\beta_L} \qquad \qquad ...(13.26)$$

A plot of dispersion coefficients from field-scale tracer studies conducted on 17 different field-scale basins is illustrated in Fig. 13.9. Because the field tests were conducted in baffled, serpentine contactors, not long straight channels, none of the studies resulted in the performance predicted using Eq. 13.26. These basins include entrance effects, exit effects, 180° turns and other non-idealities and would not be expected to result in ideal performance. Nevertheless, the results shown in Fig. 13.9 are encouraging for two reasons: (i) confirmation of the implication of Eq. 13.26 that dispersion is inversely proportional to the length aspect ratio; and (ii) the basins fall short of ideal performance, as expected. Recognising this situation, a coefficient of ideality C_i was proposed such that

$$d = \frac{0.14 C_i}{\beta_L} \qquad \qquad ...(13.27)$$

where,

C_i = coefficient of ideality

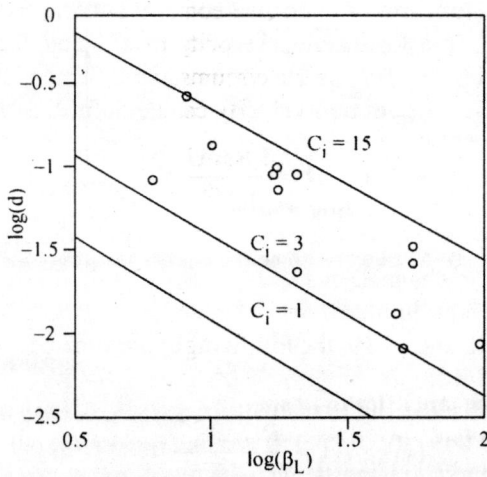

Fig. 13.9. Impact of contactor aspect ratio on dispersion.

Lines corresponding to C_i values between 3 and 15 are also displayed in Fig. 13.9 and all the data lie on or between them. Based on the data presented in Fig. 13.9, it appears that a good design should be able to equal or exceed the performance estimated by Eq. 13.27 with a C_i value of 3. A best-fit line corresponding to a C_i of 7.1 approximates the performance of a typical older reactor design.

Designing for a specified t_{10}/τ

Although the dispersion number is probably the most suitable means of assuring disinfection performance, a means of estimating t_{10}/τ must be used to be sure that the design will meet regulations. The impact of baffling rectangular contact tanks to improve hydraulic performance was evaluated in a recent AWWA

Research Foundation (AWWARF) study. A pilot contactor was baffled with nine different configurations having length-aspect ratios ranging from 4.8 to 98. In addition, tracer tests were conducted on a full-scale, 34 ML/d (9 mgd) contactor before ($\beta_L = 6.1$) and after ($\beta_L = 52$) modifications. Finally, an empirical correlation between t_{10} and β_L was developed and confirmed:

$$\left(\frac{t_{10}}{\tau} \right) = 0.198 \ln(\beta_L) - 0.002 \qquad \qquad \ldots (13.28)$$

All the data from the AWWARF tests are shown in Fig. 13.10 along with a line representing Eq. 13.28. Note the results from full-scale tests also lie close to model predictions.

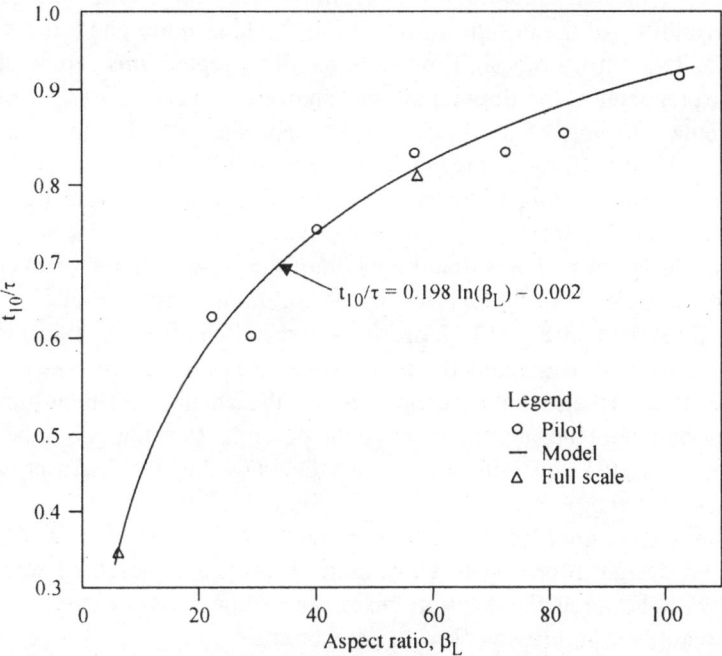

Fig. 13.10. Impact of contactor aspect ratio on t_{10}.

Although the design of an effective disinfection contact basin require attention to the length-aspect ratio, other design details are also important. Any design detail that causes disturbances in flow is undesirable. Unnecessary gates, ports or objects that constrict the flow lines are examples. Assuming these things are taken care of, however, special attention should be paid to three elements of design in every contactor: (i) inlet configuration; (ii) outlet configuration; and (iii) turns. Without proper attention to design, each of these is a likely cause of poor basin performance.

Basin inlets and outlets

Basin inlets are designed ordinarily as flow over a weir into the basin, flow through a pipe into the basin, or flow through a gate or gate valve into the basin. When pipes are used, it is always helpful to use a tee so that the flow is not directed down the basin. With any of these inlet configurations (including a pipe with a tee) it is also desirable to install a diffuser wall in between the inlet of the basin and the first pass.

Basin outlet configurations are analogous to inlets and have many of the same problems, although outlet effects are not ordinarily quite as important because outlets do not impart momentum to the basin flow. Often a diffuser wall is the best way to manage flow to outlets.

180° turns

To build a compact basin with the best possible length-aspect ratio, rectangular basins are baffled in a serpentine fashion. But the impact of baffling on the flow in a basin is not entirely benign. While increasing the tank's length-aspect ratio, the baffles also introduce flow separations at the 180° turns. Computational fluid dynamics (CFD) can be used to evaluate the flow in a chlorine contactor design including such flow separations and produce an estimate of the resulting tracer curve is illustrated in Fig. 13.11. Note that, although the overall t_{10} of the design shown in Fig. 13.11 is quite good, it also serves to illustrate the adverse impact of 180° turns on basin flow patterns. Flow separations can be observed at each turn and these impact the character of the flow for some distance down each pass. Based on some estimates, as much as 40 per cent of the volume in a baffled tank behaves as a dead zone. As a result, the increased dispersion that results from baffling can result in a decrease in the effective contact time (early tracer appearance and a great deal of tailing in the tracer curve). Most of the non-ideality in the tracer curve in Fig. 13.11 results from the 180° turns.

The most obvious design principle is that the width of the flow path must be kept constant around a turn. A number of methods have been devised for controlling the problem of flow separation at turns, and some of them, illustrated in Fig. 13.12 are hammerheads and fillets, turning vanes and diffuser walls. Turning vanes, hammerheads and fillets are used to reduce or eliminate the flow separation. Diffuser walls, in contrast, re-distribute the flow across the channel after the turn is complete. As a result, turning vanes, hammerheads and fillets have the potential to actually reduce the head loss due to the turn as well as to reduce the flow non-ideality introduced by the turn. Diffuser walls always increase head loss because they depend on head loss to re-distribute the flow.

None of the techniques to improve flow can be reliably designed by rules of thumb alone, although in some cases limited design information is available. Kawamura presented some useful criteria for designing diffuser walls between flocculation basins and sedimentation basin; for example, some of these guidelines also apply to improving flow in contactors:

1. Port openings should be uniformly distributed across the baffle wall.
2. A maximum number of ports should be provided so that flow is evenly distributed.
3. The size of the ports should be uniform in diameter.
4. Ports should be 75 mm or larger to avoid clogging.
5. Ports should be spaced with consideration to the structural integrity of the baffle. For wood baffles, this leads to 250 to 500 mm spacing.

Kawamura also recommends that ports be designed to cause a head loss of 0.3 to 0.9 mm. This criterion will be useful for many disinfection contact basins as well, but the narrow channels used in contact basins will result in the water column coming down the channel travelling at a higher velocity and with greater momentum. Whereas diffuser walls have the advantage that there are some design criteria available and they improve flow, they have the disadvantage that they increase the head loss.

In fact, head loss and construction are the two major limitations on designing baffled, serpentine basins. Many baffle and channel designs become so narrow that construction is difficult. Moreover the head loss from the 180° turn can become significant. Nevertheless, baffled contactors with length-aspect ratios as high as 100 and dispersion numbers below 0.01 are common.

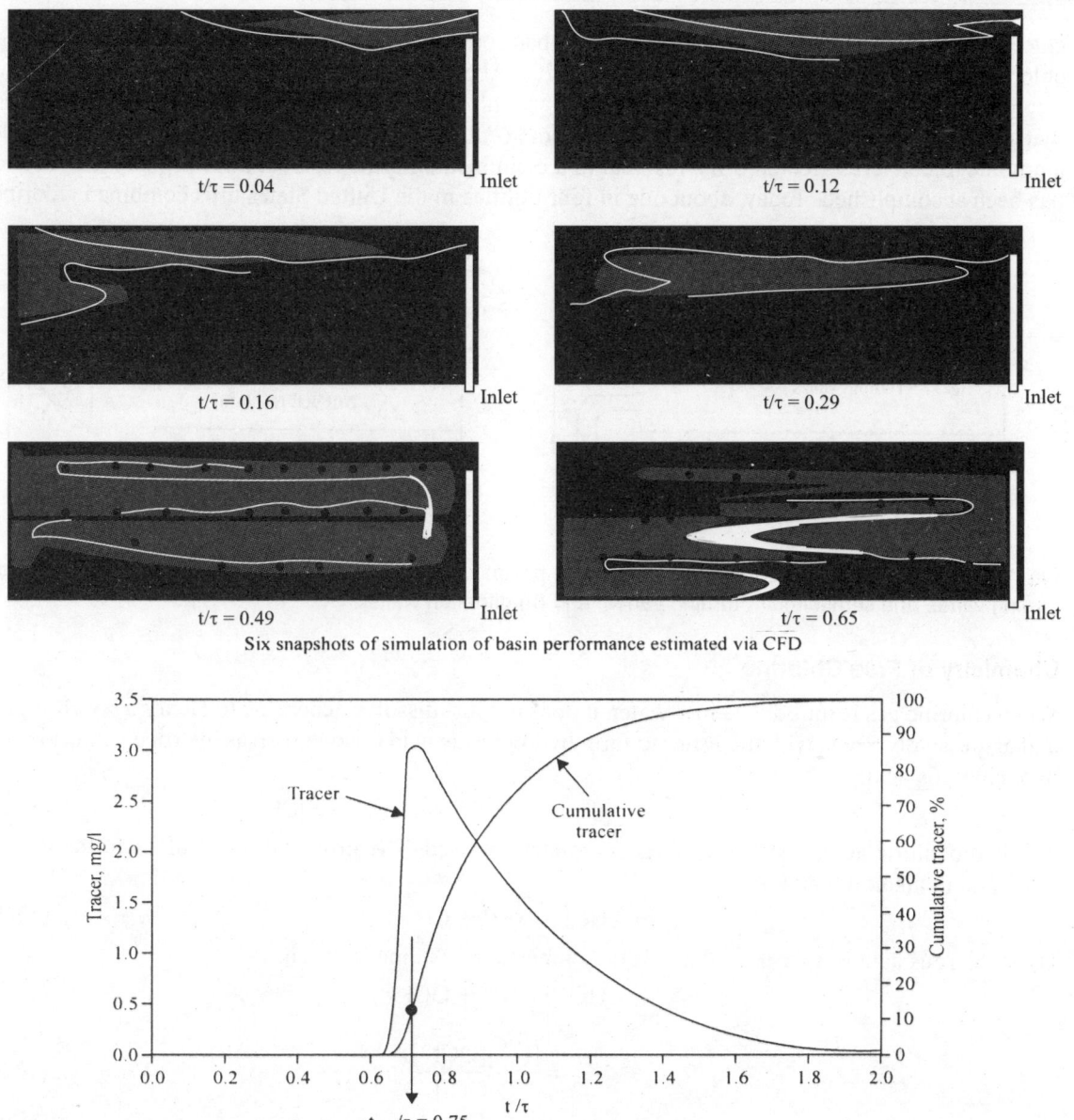

Fig. 13.11. Using computational fluid dynamics (CFD) to evaluate RTD of disinfection contactor (CFD by flow science for an optimised design for the Weber Basin Water Conservancy District in Utah; $\tau = 110$ min, $t_{10} = 83$ min).

DISINFECTION WITH FREE AND COMBINED CHLORINE

Until approximately World War II, free and combined chlorine (chlorine combined with ammonia) were both commonly used and viewed as effective disinfectants. In 1943, the US PHS demonstrated that free chlorine exhibits much more rapid kinetics in the disinfection of several bacteria. Since that time free chlorine has been understood to be more effective than combined chlorine. As a result, the use of combined chlorine declined between 1943 and the mid-1970s. About that time, it became widely recognised that free chlorine formed chemical by-products and that chloramines did so to a much lesser degree. Since that realisation, the use of chloramines has become much more popular particularly the addition of ammonia to convert a free-chlorine residual to a combined chlorine residual once primary disinfection has been accomplished. Today, about one in four utilities in the United States uses combined chlorine.

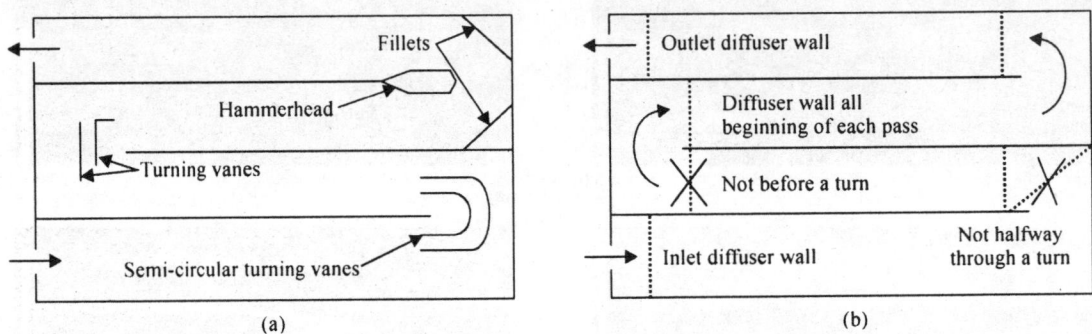

(a)   (b)

Fig. 13.12. Controlling flow separation in serpentine basins using various devices: (a) fillets, hammerhead and turning vanes and semi-circular turning vanes; and (b) diffusion walls.

Chemistry of Free Chlorine

When chlorine gas is introduced into water, it does not just dissolve according to Henry's law. It rapidly and aggressively reacts with the water to form hydrochloric acid (also known as hydrogen chloride) and hypochlorous acid:

$$Cl_2(g) + H_2O \longrightarrow HCl + HOCl \qquad \text{... (13.29)}$$

The hydrochloric acid, a strong acid, is completely ionised. The principal result of its formation is a reduction in alkalinity and pH:

$$HCl \longrightarrow H^+ + Cl^- \qquad \text{... (13.30)}$$

Hypochlorous acid is a weak acid and is only ionised in alkaline solution:

$$HOCl \rightleftarrows H^+ + OCl^- \qquad \text{... (13.31)}$$

$$K_a = \frac{[H^+][OCl^-]}{[HOCl]} \qquad \text{... (13.32)}$$

The distribution between HOCl and OCl⁻ is illustrated in Fig. 13.13 as a function of pH and temperature. Hypochlorous acid (HOCl) exhibits faster disinfection kinetics than does hypochlorite ion (OCl⁻) (Table 13.3). Consequently, where disinfection alone is concerned, a pH of 7 or slightly lower is desirable. As can be seen from Fig. 13.13, temperature has a small effect as well, warmer water causing hypochlorous acid to dissociate at somewhat lower pH.

Chlorine is relatively stable in pure water but reacts slowly with the organic matter naturally present in drinking water and rapidly with sunlight. Where sunlight is concerned, photons react with hypochlorite ion to produce oxygen, chlorite ion and chloride ion.

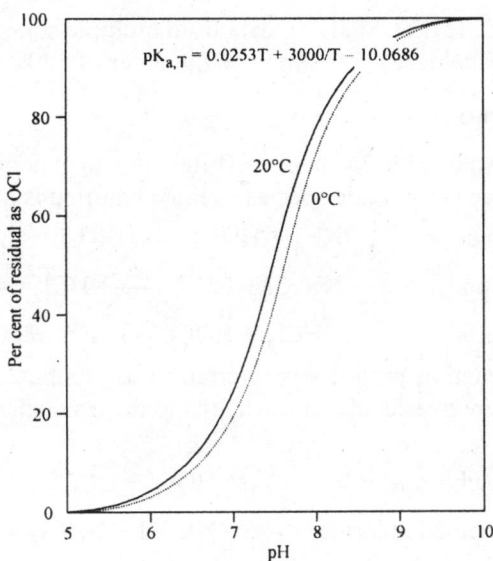

Fig. 13.13. Effect of temperature and pH on fraction of free chlorine present as hypochlorous acid.

The sensitivity of the reaction rate to pH, a consequence of the fact that the photolytic reaction is with hypochlorite ion and not hypochlorous acid, is illustrated in Fig. 13.14, constructed using the data of Nowell and Hoigne.

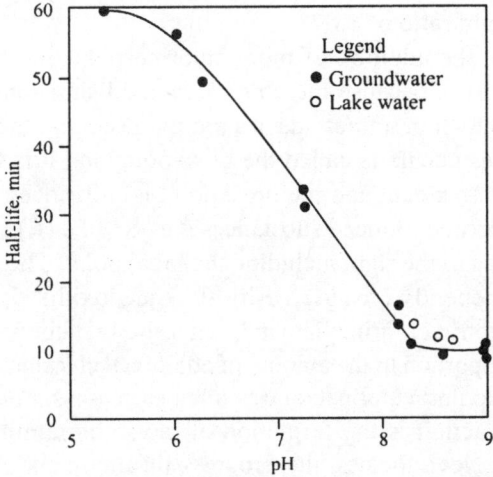

Fig. 13.14. Half-life of free chlorine residual in sunlight.

Although significant work has been done measuring the impact of natural organic matter on the decay of chlorine residuals, no universal relationships have evolved. Rather, the decay of free chlorine is often best modelled with the simple first-order reaction depicted in Eq. 13.12. Sometimes the process is modelled as two reactions operating in parallel, a fast reaction with rapidly reducible substances and a slower first-order reaction (Eq. 13.13). Studying data from multiple sources, Powell concluded that the activation energy for the chlorine decay reaction is on the order of 62 kJ/mole.

Chemistry of Combined Chlorine

When ammonia is present in water, chlorine behaves differently. In general, chlorine reacts successively with ammonia to form the three chloramine species as more chlorine is added.

1. Monochloramine formation $\quad NH_3 + HOCl \longrightarrow NH_2Cl + H_2O$ ... (13.33)

2. Dichloramine formation $\quad NH_2Cl + HOCl \longrightarrow NHCl_2 + H_2O$ ... (13.34)

3. Trichloramine formation $\quad NHCl_2 + HOCl \longrightarrow NCl_3 + H_2O$ ... (13.35)

The sum of these three reaction products is referred to as combined chlorine. The total chlorine residual is the sum of the combined residual and any free-chlorine residual. A summary of these definitions is provided below:

$$\text{Free chlorine} \quad = HOCl + OCl^- \quad \text{... (13.36)}$$

$$\text{Combined chlorine} \quad = NH_2Cl + NHCl_2 + NCl_3 \quad \text{... (13.37)}$$

$$\text{Total chlorine} \quad = \text{Free chlorine} + \text{combined chlorine} \quad \text{... (13.38)}$$

With all quantities expressed as milligrams per liter as Cl_2. When small amounts of chlorine are added to water, the reactions are much like the simple model just outlined. However, as the amount of chlorine added increases, the reactions become more complex. These reactions and their behaviour are partially illustrated by the three zones in Fig. 13.15. At first, as depicted in zone A, the total chlorine residual increases by approximately the amount of chlorine added until the mole ratio of chlorine to ammonia approaches 1 (a weight ratio of 5.07).

Beyond a molar ratio of 1, the addition of more chlorine reduces, rather than increases, the total chlorine residual (zone B). This is because the chlorine is oxidising some of the chloramine species. Eventually, essentially all of the chloramines species are oxidised and the point at which the complete oxidation of chloramine species occurs is called the breakpoint and it is the beginning of zone C. The exact location of the maximum residual and the breakpoint is influenced by the presence of dissolved organic matter, organic nitrogen and reduced substances e.g., S^{2-}, Fe (II), Mn (II). The presence of any of these will shift all three zones to the right including the breakpoint. The degree to which they shift the point of maximum residual depends on how easily they are oxidised. The shift in the breakpoint corresponds to their stoichiometric chlorine demand. After the breakpoint is reached, the free-chlorine residual increases directly in proportion to the amount of additional chlorine added. Prior to concerns about disinfection by-products, 'breakpoint' chlorination was often used as a simple means of ammonia removal.

In zone A, the principal reaction is the formation of monochloramine. This reaction is extremely rapid, with little interference. Nevertheless, the progress through zone A is somewhat influenced by reduced substances but is more profoundly influenced by pH. At low pH values, other reactions become more significant, including:

$$NH_2Cl + H^+ \rightleftharpoons NH_3Cl^+ \quad \text{... (13.39)}$$

$$NH_3Cl^+ + NH_2Cl \rightleftharpoons NHCl_2 + NH_4^+ \quad \text{... (13.40)}$$

Thus, when Palin examined the composition of residuals at pH 6, 7 and 8, he found no dichloramine in zone A at pH 8 but significant amounts at pH 6. The same holds true in zone B, except zone B, which is richer in chlorine, does show some evidence of dichloramine, even at pH 8. The principal reactions in zone B are the oxidation of ammonia to nitrogen gas and the oxidation of ammonia to nitrate ion. Between these, the conversion to nitrogen gas is the dominant reaction commonly observed:

$$3HOCl + 2NH_3 \longrightarrow N_2(g) + 3H_2O + 3HCl \qquad \text{(ammonia to nitrogen gas)} \qquad ... (13.41)$$

$$4HOCl + NH_3 \longrightarrow HNO_3 + H_2O + 4HCl \qquad \text{(ammonia to nitrate ion)} \qquad ... (13.42)$$

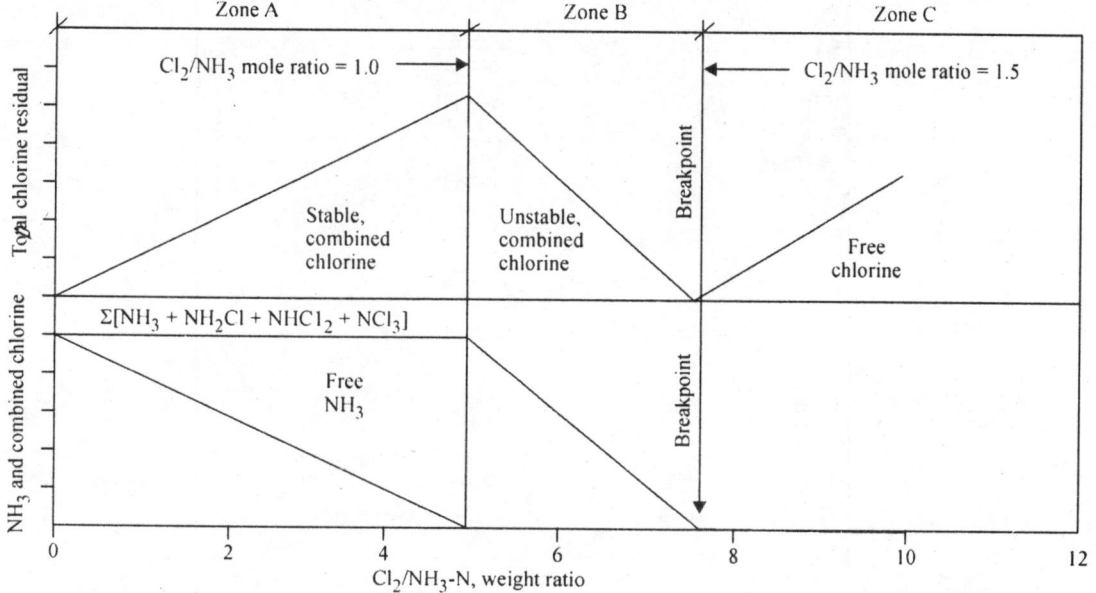

	Zone A	Zone B	Zone C
Time to metastable equilibrium	Seconds to a few minutes	10–60 min	10–60 min
Composition of metastable residual	Mostly monochloramine. Some dichloramine and traces of trichloramine at neutral or acid pH or at high Cl₂/NH₃ ratio	A mixture of monochloramine and dichloramine. Some free chlorine and traces of trichloramine at low pH	Mostly free chlorine. Trichloramine can be significant (aesthetically, but not as fraction of residual) at neutral pH, but especially in acid region

Fig. 13.15. Overview of chlorine breakpoint stoichiometry.

The previous discussions have presented the breakpoint reaction as an easily determined equilibrium or static condition. However, anyone who has attempted the laboratory work required to construct such a curve knows that the behaviour of the Cl_2–NH_3 system is actually quite dynamic and the breakpoint curve shown in Fig. 13.15 should be considered more of a metastable than an equilibrium state. As a result, the laboratory work required to support the construction of a breakpoint curve must be done with precise timing or it will not be reproducible, especially for Cl_2/NH_3 mole ratios above 1. Above this ratio the reaction proceeds rapidly until the metastable state is reached. Anywhere along the curve, the rate at which the reaction progresses is strongly influenced by the pH (Fig. 13.16), particularly in the vicinity of the breakpoint itself. Near the breakpoint, the reaction is at its maximum rate at a pH between 7 and 8.

The rate decreases rapidly at pH values outside that range. Facilities engineered to accomplish ammonia removal via the breakpoint reaction should be designed to accommodate the time for this reaction. Even in the optimum range, the reaction time can be significant.

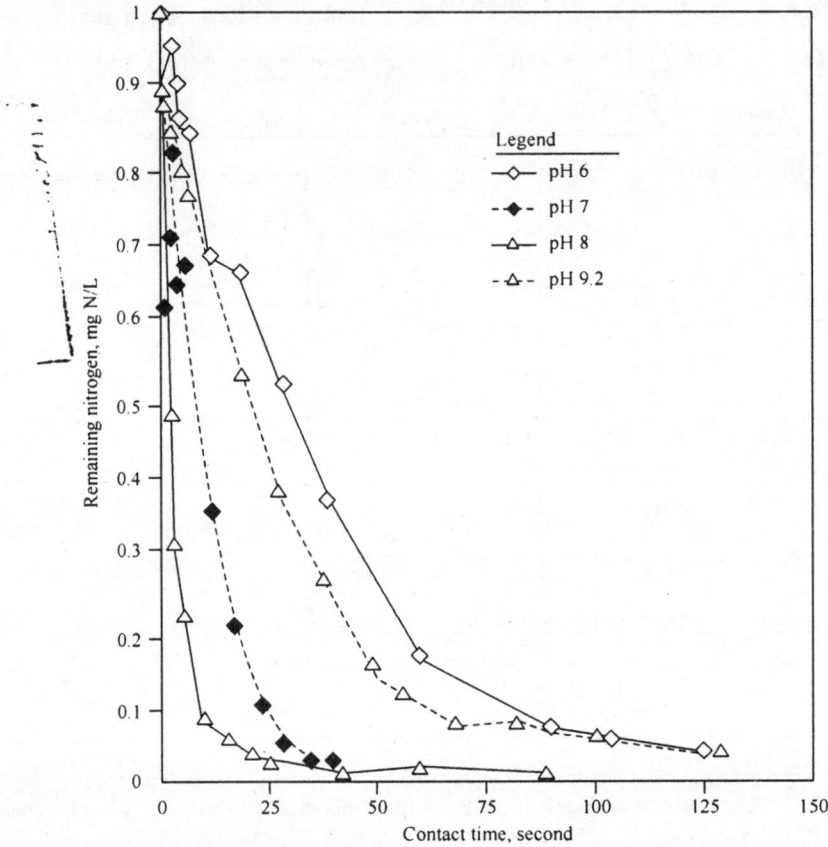

Fig. 13.16. Effect of pH on breakpoint chlorination.

Forms of Chlorine (Liquid, Gas, Hypochlorite, etc.,)

The forms of chlorine most often used in the treatment of drinking water are chlorine gas and sodium hypochlorite solution. Calcium hypochlorite powder is also used in some smaller systems. In the United States chlorine gas can be purchased in 68 kg (150 lb) cylinders, in 908 kg (1 tonne) cylinders (note that in Europe 1000 kg cylinders are used), by tank truck or in railroad tank cars of between 14.5 and 49.9 metric tonnes in capacity (16 and 55 American tonnes). Generally the cost of chlorine is lower when it is shipped in larger volumes, the cost delivered in 1 tonne cylinders being approximately half the cost delivered in 68 kg cylinders but nearly twice that when delivered by rail. As a result, some very large utilities purchase liquid chlorine by rail and re-package it for use at various sites.

Liquid Chlorine

The basic components in a chlorine facility are listed below:
    1. Storage of liquid chlorine gas.

2. Conduits for the transport of liquid chlorine.
3. Evaporation of liquid chlorine.
4. Conduits for the transport of chlorine gas under pressure.
5. Regulation of the chlorine feed rate.
6. Conduits for the transport of chlorine gas under vacuum.
7. Chlorine-to-water mass transfer.
8. Mixing of chlorine water with the process flow.
9. The chlorine contact facility.
10. Chlorine sampling and analysis.
11. Chlorination control system.

In small systems many of these elements are found in one device and other elements, like the control system, are very rudimentary. In large chlorination systems, each of these elements can sometimes present a separate, specific design challenge. Each of these elements requires different materials and different design considerations apply to each.

Design issues with liquid chlorine

The details of the design of systems for handling the delivery, storage and dosing of liquid chlorine are beyond the scope of this book. An overview of a variety of the more important issues is provided in Table 13.5. Chlorine is truly a hazardous material so it is important that care be taken in the design of these facilities.

Table 13.5. Overview of key design issues for chlorination systems.

Item	Description
Delivery	In cylinders 68 kg (150 lb) and 908 kg (1 tonne); in tank trucks 13600–18200 kg (15–20 tonnes); and in rail cars 14500–49900 kg (16 to 55 tonne)
Storage	In cylinders; in tank trucks; in rail cars or in custom tanks
Conduits for liquid	Use schedule 80 stainless steel, schedule 80 carbon steel or cast iron (do not use PVC). Should be chlorine seamlessly welded. Use cast-iron valves. Use pipe sizes recommended by White to avoid 'flashing'
Evaporation of liquid	Can use vapour pressure of container to feed up to 19 kg/d (40 lb/d) with 68 kg cylinder and up to 150 kg/d (330 lb/d) with 908 kg cylinders. Multiple cylinders are often used with automatic switchover. At feed rates above 680 kg/d (1.5 tonnes/d) a separate evaporator is recommended to convert liquid chlorine to chlorine gas
Conduits for chlorine gas under pressure	Use schedule 80 SS, schedule 80 carbon steel or cast-iron (do not use PVC). Should be seamlessly welded. Use cast-iron valves
Regulation of chlorine gas feed rate	This is accomplished by the chlorinator. Most chlorinators include four principal elements: (a) a pressure-reducing valve; (b) a rotometer; (c) a control valve; and (d) a vacuum regulator
Conduits for transport of chlorine gas under vacuum	Often constructed to Schedule 80 PVC or reinforced fiberglass pipe. Piping should be carefully sized to maintain pressure drop below 50–60 mm Hg
Chlorine-to-water mass transfer	Chlorine reacts vigorously with water to form hypochlorous and hydrochloric acids and as a result it is highly soluble. Chlorine-to-water mass transfer is normally accomplished via chlorine injector, a venturi-type device. The maximum solution strength downstream of the injector is approximately 3.5 g/l. The injector is also used to create the vacuum in the system

(Contd...)

Item	Description
Blending of chlorine water into process flow	Under normal conditions, blending must be accomplished before the chlorine residual monitoring point. With normal turbulent flow in a conduit, this requires travel down the conduit an axial distance of approximately 100 hydraulic radii. Blending can be expedited with devices normally used for rapid mixing or via flow structures that dissipate a lot of energy (e.g., a hydraulic jump or a fall over a sharp-crested weir). In small to medium-size plants (<200 ml/d or 50 mgd) the Water Champ has been found to be very effective. When ammonia is present, it is important that the chlorine be rapidly blended with the bulk flow. If this is not accomplished, disinfection is compromised and both chlorine and ammonia are lost in localised breakpoint reactions. In this case rapid mixing is required
Chlorine contact facility	Traditionally, the contact time in existing facilities (e.g., sedimentation basins, clearwells) has been used. Increasingly specially designed chlorine contact tanks are used. The most efficient designs, from the standpoint of dispersion, are long, straight pipelines and carefully designed, serpentine contact chambers. Most contact chambers are of concrete
Chlorine sampling and analysis	Reliable equipment for the sampling and analysis of free and total chlorine has been available for some time. There are now several devices available on the market
Chlorination control system	Historically control systems were manual, feed-forward, feedback and compound loop in design. Today control systems and methods of sampling and analysis have improved to where more complex designs are workable

Control of Gas Chlorination

Four methods have traditionally been used for controlling the feed rate of chlorine gas when it is being used for residual control in drinking water systems. Each is displayed in Fig. 13.17: (i) manual control; (ii) feed-forward control; (iii) residual feedback control; and (iv) compound loop control. Through the middle of the twentieth century manual control was the most common. A great deal of attention was required by the operator to ensure that a suitable residual was reliably provided, especially when the flow rate through the plant was adjusted. By the mid-1950s flow measurement and chlorine metering techniques improved to the point where feed-forward control began to appear. This important advance allowed automatic adjustment for flow but still required the operator to adjust for any changes in the water quality (chlorine demand) or any drifts in monitoring and feed rates.

By the mid-1960s direct residual control began to appear as well. In principle, the feedback method of control is more robust than feed-forward control, but residual measurement did not approach suitable levels of reliability and precision for two more decades. As a result, compound loop control evolved as a compromise. With this method, changes in flow could be accommodated via the flow signal and an additional control increment could be added via residual control for minor water quality changes. Properly maintained, compound loop control was the first system to provide reliable, continuous residual control. During the last decade of the twentieth century, computerised supervisory control systems had evolved to the point where these same inputs (flow and residual) could be combined with other measurements as well to provide improved reliability. None of these control systems, however, is an adequate substitute for vigilant attention from the operator.

Residual control system

With all of these alternatives, the sequence of events in the residual control system must be carefully designed and controlled. All the elements shown and labelled in the diagram as 'compound loop control' in Fig. 13.17 must be considered in designing the system and envisioning its method of control. The instructions of the supervisory control and data acquisition (SCADA) system must be conceived with a full understanding of the range of timing that can occur in all the other elements of the system.

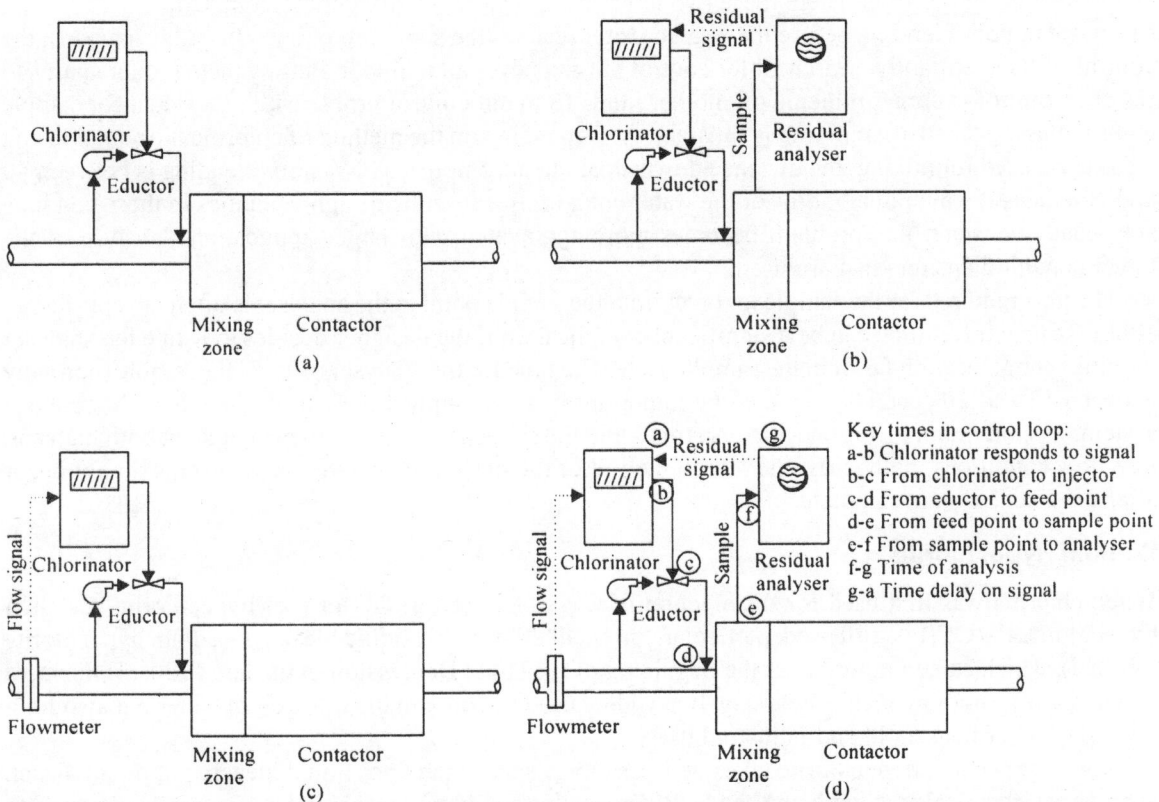

Fig. 13.17. Control of chlorine gas feed rate: (a) manual control; (b) feedback or residual control; (c) feed forward control; and (d) compound loop control.

The time required for the chlorinator to completely respond to an instruction from the SCADA must be considered. Normally this time is not too significant. The time required for the change in feed rate established by the chlorinator to be recognised at the eductor must be considered. This time to change feed rates is not normally too long either, but it can be too long when the chlorinator is located a long way from the injector and when the chlorine feed rate is very low. White suggests that this time be estimated by assuming that the change in pressure will be accommodated in a wave that travels about three times as fast as the gas flow in the line.

Next, the time for the water in the chlorine water line to travel from the eductor to the application point must be considered. This time is a function of the distance between the eductor and the application point and the flow rate (velocity) in the chlorine water line. Again, designs with long distances between the eductor and the application point can cause trouble. Ideally the chlorine is stored near the application point so that both the time in the vacuum line and the time in the chlorine line are minimised. When nearby storage is not possible, it is usually best to lengthen the vacuum line, not the chlorine water line, as a signal ordinarily travels much faster down a vacuum line.

Sampling point

Another important constraint is the time between the application point and the sampling point. There is an inherent design conflict where these two times are concerned. Putting these points too close together can result in poor blending before the treated water reaches the sampling point. When this happens, the control system constantly 'searches' for control but can never quite find it. Putting them too far apart can result in control-cycling problems of another kind. To avoid control problems, the residual for sample control must be taken after mixing is satisfactory. Depending on the method of chlorine introduction and the criteria used for mixing, the distance downstream to accomplish satisfactory blending is between 40 and 200 times the hydraulic radius of the water conduit. Because the design velocities in these conduits are usually constant, this problem becomes more aggravated with larger applications because of the larger conduit diameters that are used.

The time required for the sample to travel from the sample point to the analyser is often not considered either. Sample travel time can be a significant complication if the designer decides to locate the analyser in some central location far from the sampling site. The time for the analyser to assay the sample (normally between 15 and 20 seconds) can also be important in some applications. In designing such a control system, it is important to analyse all these times and the sequence in which they operate at both high and low-flow conditions, both early and late in the life of the design, to ensure that problems do not occur after the installation is complete.

Sodium Hypochlorite

When chlorine was first used for disinfection, it was often used in the form of hypochlorite. Calcium hypochlorite $Ca(OCl)_2$ is still used, particularly in small utilities, but liquid bleach or sodium hypochlorite $(NaOCl)$, which came into use near the beginning of the Great Depression in the late 1920s, is the most widely used form of hypochlorite today. It is widely used not only in disinfection of water but also for a myriad of other household and industrial uses.

High-test calcium hypochlorite takes up less storage space than does liquid bleach, but liquid bleach is an easier chemical to handle and feed. Whereas chlorine gas is prepared by an electrolytic process that breaks sodium chloride solution into chlorine gas and sodium hydroxide, ironically, sodium hypochlorite is generally prepared by mixing sodium hydroxide and chlorine gas together:

$$2NaOH + Cl_2 \longrightarrow NaOCl + NaCl + H_2O \qquad \text{... (13.43)}$$

On a weight basis, 1.128 kg of NaOH reacts with 1 kg of chlorine to produce 1.05 kg of NaOCl and 0.83 kg of NaCl. The process is complicated by the fact that the reaction generates a significant amount of heat. It is common practice to add an excess of NaOH because, as will be shown, hypochlorite is more stable at higher pH values. As a result, the density of one hypochlorite solution is not necessarily the same as another, even if both are of the same strength (per cent Cl_2). This density difference occurs because the final density depends on the amount of excess NaOH added during manufacture. Liquid bleach is usually delivered with a pH between 11 and 13. Hypochlorite can also be manufactured via onsite generation and this process is receiving more attention.

Stability of hypochlorite

Under some conditions, the strength of hypochlorite can significantly decline in just a few days. In fact, stability is one of the major issues that must be addressed in both designing and operating a hypochlorite facility. A utility should not undertake using hypochlorite unless it is prepared to dedicate time and energy to a regular programme of monitoring and controlling its decay. Of considerable significance is the fact that, when hypochlorite does decay, chlorate ion is one of the principal by-products of the reaction. The stability of hypochlorite is affected by the strength of the solution, the temperature at which it is stored, the pH and the contamination of heavy metals, which can catalyse its decay. Light is also a problem. As a general rule, the following actions accelerate the rate of decay:

1. Storing the hypochlorite at higher concentrations.
2. Storing the hypochlorite at lower pH.
3. Storing the hypochlorite at higher temperatures.
4. Storing the hypochlorite where it is exposed to sunlight.
5. The presence of certain heavy metals, notably copper and nickel.

Under these basic conditions the decomposition of hypochlorite ion to chlorate ion follows a disproportionation reaction, which exhibits second-order reaction kinetics and the following overall stoichiometry:

$$OCl^- + OCl^- \longrightarrow ClO_2^- + Cl^- \qquad \text{... (13.44)}$$

$$OCl^- + ClO_2^- \longrightarrow ClO_3^- + Cl^- \qquad \text{... (13.45)}$$

$$3OCl^- \longrightarrow ClO_3^- + 2Cl^- \qquad \text{... (13.46)}$$

The second reaction, as given by Eq. 13.45, is the faster of the two. As a result, the first reaction is the rate-limiting step in the consumption of hypochlorite ion. Bleach also decays via a slow reaction that forms oxygen and an acid forming reaction that also forms chlorate ion. These reactions are shown below:

$$OCl^- + OCl^- \longrightarrow O_2 + 2Cl^- \qquad \text{... (13.47)}$$

$$2HOCl + OCl^- \longrightarrow ClO_3^- + 2H^+ + 2Cl^- \qquad \text{... (13.48)}$$

Gordon and colleagues have shown that copper and nickel catalyse oxygen formation (see Eq. 13.47) and research at the Swiss Federal Institute for Environmental Science and Technology (EAWAG) has shown that a similar reaction occurs via proteolysis. The relationships between the three principal reactions that result in hypochlorite decay are displayed in Fig. 13.18.

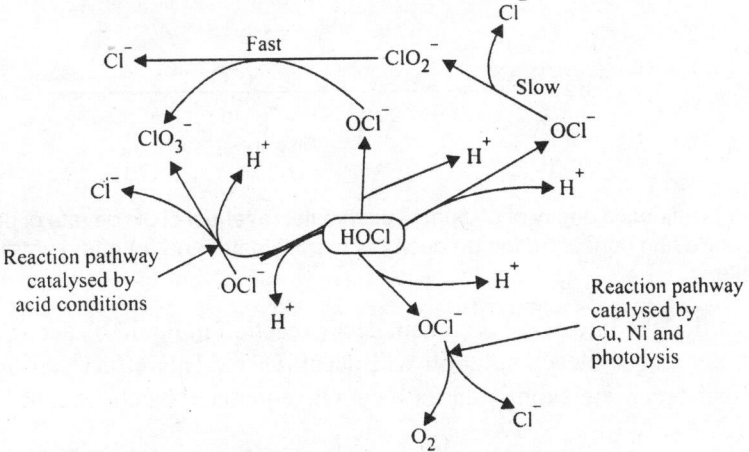

Fig. 13.18. Decay reactions of hypochlorite.

The pH at which a sodium hypochlorite solution is stored has important impacts on its rate of decay. Gordon examined this question and their results are displayed in Fig. 13.19a. The rate of decay stabilises to a very low level at pH 11 and above but increases rapidly below pH 10. Gordon and Colleagues present some evidence that a decay minimum occurs between pH 12 and 13. As liquid bleach is normally delivered at pH 12 or above, low-pH decay is normally not a problem with the undiluted product. Often it is delivered with enough excess hydroxide to allow a 50 per cent dilution without developing pH problems. Nevertheless, pH monitoring and control are cornerstones of a hypochlorite management programme.

Temperature is also an important consideration. Like most reactions, the decay of hypochlorite accelerates as the temperature increases. Gordon examined this question by comparing the rate of decay of three commercial bleaches, all at approximately the same temperature. In this work it was demonstrated that temperature is important and also that the rate of decay of most modern good-quality commercial bleaches is comparable.

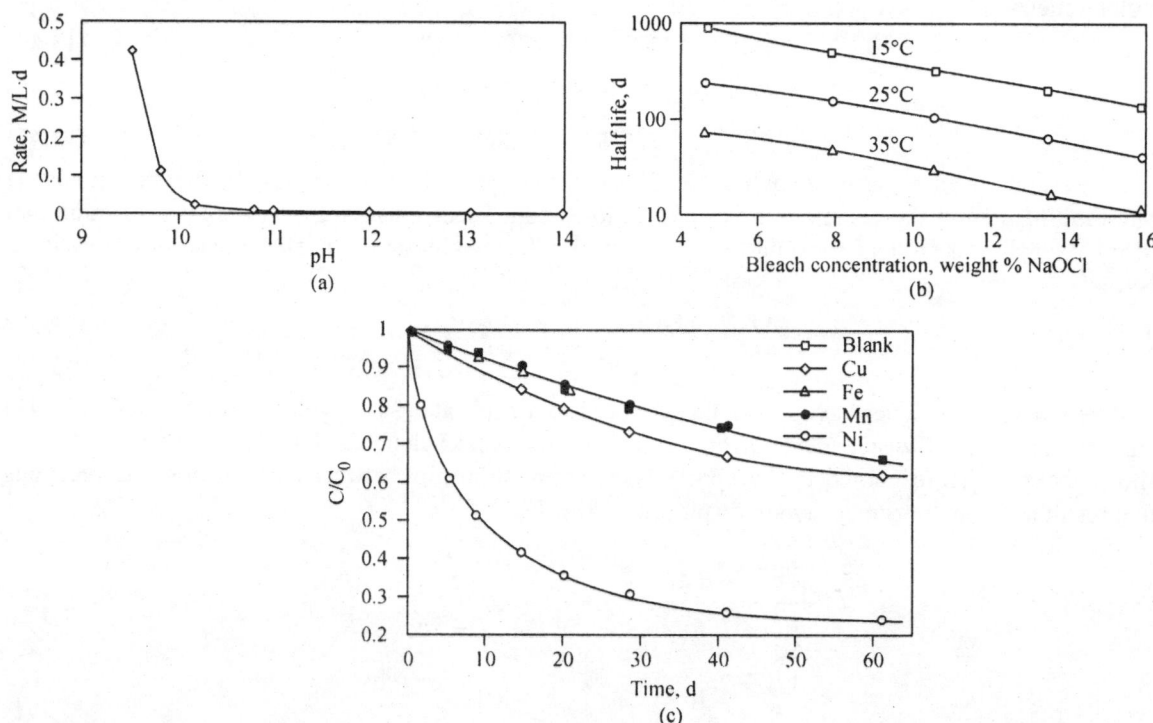

Fig. 13.19. Factors that influence decay of sodium hypochlorite: (a) effect of pH on rate of decay of hypochlorite; (b) effect of temperature and concentration on decay of hypochlorite; and (c) effect of trace metals on rate of decay of hypochlorite.

As mentioned earlier, the rate of the dominant decay reaction in liquid bleach (Eq. 13.46) is second order. As a result, a stronger bleach solution will decay faster. This effect can be illustrated by the solution of the second-order rate expression for a completely mixed batch reactor:

$$\frac{C}{C_0} = \frac{1}{1 - k_d C_0 t} \qquad \qquad \dots (13.49)$$

where,

C   = bleach concentration after time t, moles/l
C_0 = bleach concentration at time 0, moles/l
k_d = second-order decay coefficient, l/mole·s
t   = time, s

The effects of bleach strength and temperature are illustrated in Fig. 13.19b. Based on the data in this figure, diluting bleach delivered at a concentration of 15 per cent to a concentration of 7.5 per cent will increase its half-life from about 50 to about 140 days (at 25°C). If the 7.5 per cent bleach is also stored at 15°C instead of 25°C, the combined effect of dilution and temperature control will increase its half-life to more than 500 days.

Finally, since the work of Lister bleach technologists have understood that certain metals can catalyse the decomposition of bleach. In the mid-1950s rhodium, iridium, cobalt, copper, manganese, iron and nickel were issues. Today the principal concerns are copper and nickel and manganese has also been shown to exacerbate the destructive effect of nickel. Gordon conducted tests to examine the effect of a concentration of 1 mg/l of copper, iron, manganese and nickel, individually, on the decay of a 13.5 per cent bleach. These are illustrated in Fig. 13.19c. Gordon recommended that copper and nickel be kept below 0.1 mg/l. Bleaches are also filtered in an attempt to reduce metals contamination, but, with one exception, additional filtering of modern commercial bleaches showed only small improvements. It appears that many modern bleaches are produced in such a condition that additional filtering is of little benefit.

Formation of chlorite and chlorate ion

It was discovered that hypochlorite solutions containing significant concentrations of chlorate ion were responsible for introducing chlorate ions into drinking water. Of special significance in this regard is the fact that the principal reaction resulting in bleach decay also results in the production of both chlorate and chlorite ions (see Eqs. 13.45 and 13.44).

As mentioned earlier, the disassociation of chlorite to chloride and chlorate (Eq. 13.46) is much faster than the disproportionation of hypochlorite ion to chloride and chlorite (Eq. 13.44) and this minimises the formation of chlorite ion. As a result, the chlorite level in liquid bleach should normally be less than one-half per cent of the concentration of hypochlorite ion. Under these conditions, a dose of 1 mg/l results in a contribution of less than 5 μg/l of chlorite ion. Thus, even though chlorite is created in the process, it does not reach levels that pose a problem; it is estimated that chlorite stays below 0.5 per cent of the hypochlorite concentration. Unfortunately the same is not true where chlorate is concerned.

If all of the hypochlorite decomposition were the result of the second order decay reaction (Eqs. 13.44 and 13.47), the chlorate generated would be about 33 per cent of the hypochlorite decomposed (molar basis). But the other important contributions to hypochlorite decay (decomposition catalysed by light and by metals) normally lead to the pathway that produces oxygen and not chlorate. How significant is this effect? Gordon conducted an experiment to examine this question. The results of those tests, which included 12 tests with commercial bleaches, showed that the actual production of chlorate was slightly less, about 31 per cent (Fig. 13.20a). As a rule of thumb, it is conservative to assume that one-third of the bleach lost to decomposition ends up as chlorate ion.

Two surveys were also conducted to evaluate the contribution of chlorate ion to water systems using sodium hypochlorite for disinfection. The ratio of chlorate ion and hypochlorite ion in the bleaches being used as well as the concentration of chlorate ion in the drinking water system itself. A probability plot of the chlorate/hypochlorite ratio in the bleaches from both surveys is presented in Fig. 13.20b.

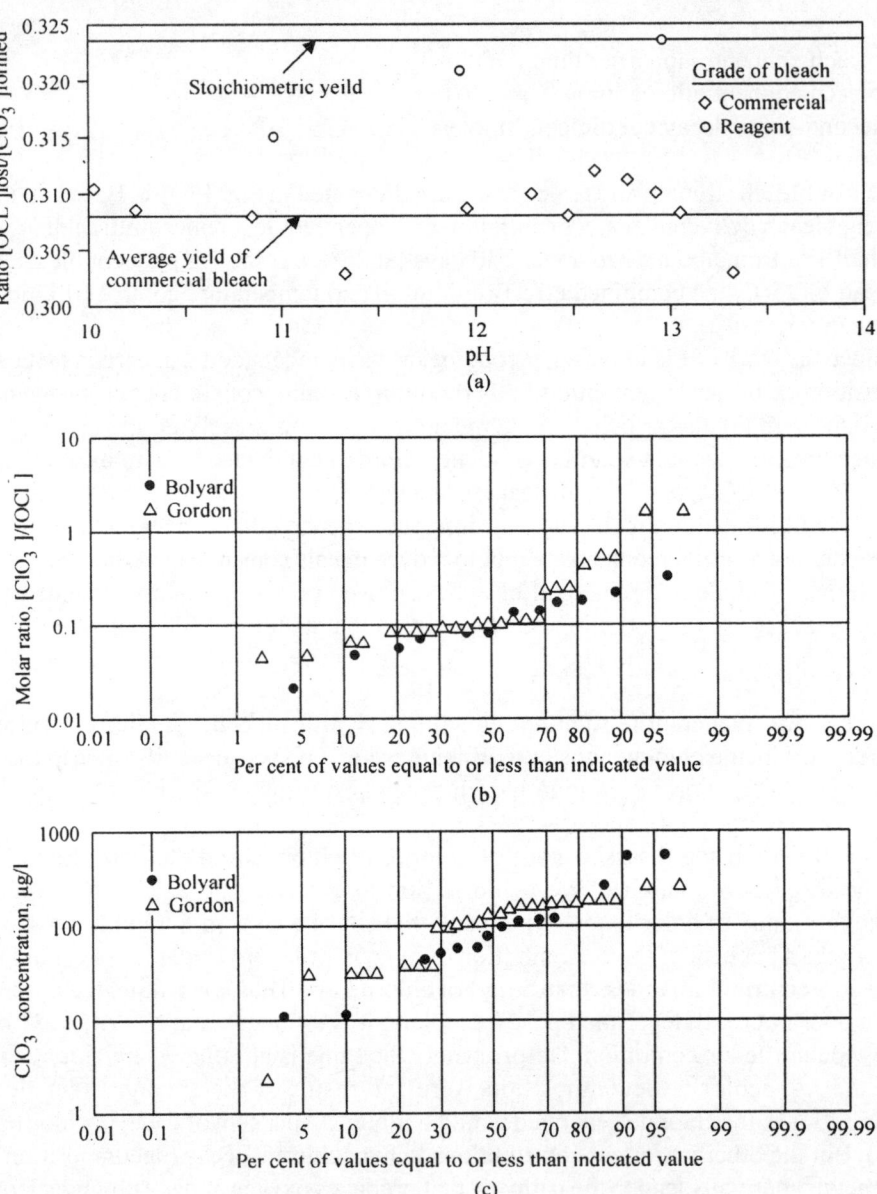

Fig. 13.20. Formation of and impacts of chlorate in hypochlorite feedstock: (a) chlorate formation during decomposition of reagent and commercial hypochlorite; (b) surveys of chlorate content in bleach; and (c) surveys of chlorate content in systems using bleach.

In both cases, the median was slightly less than 0.1 mole $[ClO_3^-]$/mole $[HOCl^-]$. On the other hand, levels greater than 1 mole $[ClO_3^-]$/mole $[HOCl^-]$ and as low as 0.02 mole $[ClO_3^-]$/mole $[HOCl^-]$ were observed, indicating that bleach manufacturing and storage practice result in substantial differences. At a ratio of 0.1, a chlorine dose of 3 mg/l would result in chlorate levels of approximately 0.1 mg/l. Thus the chlorate that is found in bleach under typical conditions of use should not be a significant issue. Many of the same considerations that affect the stability of bleach are also important in limiting its chlorate content as well. Nevertheless, surveys of chlorate in systems using hypochlorite did sometimes show the presence of significant chlorate (Fig. 13.20c), suggesting that utilities using hypochlorite should occasionally monitor for chlorate and consider modifying their practice if significant amounts are observed.

Feeding sodium hypochlorite

Experience with materials for the construction of large hypochlorite tanks has not been uniformly good. Early projects in Chicago had unsatisfactory experience with filament-wound fiberglass tanks and with underground concrete tanks with fiberglass lining. These tanks were replaced with hand lay-up fabricated fiberglass tanks using a vinyl resin binder and with plastic, continuous-weld, full-weight carbon steel tanks lined with a fiberglass reinforced polyester material at a thickness of 0.9 mm (35 mils). The latter gave acceptable performance. Properly fabricated fiberglass tanks or steel tanks with a rubber or polyvinyl chloride (PVC) lining give satisfactory service as well.

Hypochlorite is an extremely aggressive chemical and no equipment used to store or feed it can be expected to last indefinitely. Some particularly robust diaphragm and solenoid-metering pumps have been successfully used and this is the approach that is found in most plants. In very large plants (>380 ML/d or 100 mgd), White recommends metering the chemical by gravity from the storage tank through a Teflon-lined magnetic flowmeter and rate-modulating valve to the point of application.

Hypochlorite can be transported in schedule 80 PVC piping; except where sunlight is a problem, chlorinated polyvinyl chloride (CPVC) should be used. Ball valves and plug valves made of steel lined with PVC or propylene should be avoided. In general, precautions should also be taken for the potential for precipitation of calcium carbonate whenever the hypochlorite is mixed with carrying water or at the application point with the water being treated. The high specific gravity of hypochlorite solution must be overcome to accomplish effective mixing at the point of application. This can be accomplished by using a diffuser and carrying water (be cautious about the potential formation of $CaCO_3$) or by the use of a pumped jet mixer like that often used for coagulants. Mixing can also be accomplished by introducing the hypochlorite at a point of significant turbulence.

Ammonia

Ammonia can be supplied for water treatment applications in three forms: as a pure liquid (anhydrous ammonia), dissolved in water (aqueous ammonia) or as a dry ammonium salt, usually ammonium sulphate. Ammonia is not an expensive chemical, but the relative cost of these alternative forms of ammonia varies from one location and one application to another. Ironically anhydrous ammonia is only the least expensive alternative if it is purchased in lots of 20 tonnes (18.6 metric tonnes) or more, but 20 tonnes is enough to add 1 mg/l of ammonia to 18600 ML of water. In lower quantities, both ammonium sulphate and aqua ammonia are generally cheaper. For reasons of convenience, aqua ammonia is the form most commonly used. Exposure to high concentrations of ammonia vapour can be fatal. At concentrations of several 100 parts per million by volume (ppm_v), it causes throat and eye irritation and at higher concentrations it can cause convulsions or even rapid asphyxia. While not addressed in this discussion, appropriate precautions should be taken both in design and operation of ammonia facilities.

Storage and feeding of anhydrous ammonia

At normal temperatures and pressures anhydrous ammonia (>99 per cent NH_3) is a gas. However, it can be easily liquefied and is commonly stored and transported in liquid form in pressurised containers of the same size and same design as those used for chlorine (they are usually a different colour). At atmospheric pressure, liquid anhydrous ammonia has a density of 680 kg/m^3 (42.6 lb/ft^3 or 5.7 lb/gal), approximately two-thirds that of water. Anhydrous ammonia containers comply with International Code Council (ICC) regulations, which require a minimum working pressure of 1760 kPa (256 psig) with safety valves set to release at that pressure. Valves and fittings used for anhydrous ammonia are normally rated at 2070 kPa (300 psig). In the United States, bulk shipments of anhydrous ammonia are normally made in 23 and 73 metric-tonne (25 and 80 US tonne) rail tank cars, in 18 tonne (20 US tonne) tank trailers and cylinders the same size and design as those used to deliver 908 and 68 kg of anhydrous chlorine. It is common for vendors to provide storage tanks. Permanent (stationary) storage tanks for anhydrous ammonia can also be custom fabricated to any desired size. Such tanks must meet the same pressure restrictions as the shipping containers and are usually made of carbon steel. No copper, bronze or brass fittings should be used because ammonia attacks copper-based alloys. Storage tanks should be sheltered from the sun to prevent excessive pressure build-up. The vapour pressure of anhydrous ammonia at 10°C is slightly more than 611 kPa (89 psig). At 30°C it nearly doubles to 1183 kPa (172 psig). The formula below may be used to estimate the vapour pressure at temperatures between 0 and 40°C:

$$P_{v, NH_3} = 434.9 + 13.96T + 0.3645T^2 \qquad \qquad ... (13.50)$$

where,

P_{v, NH_3} = vapour pressure of anhydrous ammonia, kPa

T = temperature, °C

Anhydrous ammonia can be fed by two methods: direct feed and solution feed. In direct gas feed, the ammonia gas is directly introduced into the water to be treated. Unless the plant is very small, this method often suffers from poor distribution at the application point because of the low flow rate of ammonia gas. The solution feed method is analogous to the technology used to feed chlorine, except the vapour pressure of ammonia is higher. Precipitation of basic salts like $CaCO_3$ is often a problem in the vicinity of the application point.

Direct gas feed

Direct gas ammonia feeders are commercially available and differ only with respect to minor material changes from chlorinators in that they include an ammonia pressure-regulating valve, pressure gauges, a pressure relief valve, rotameters and a control valve with back-pressure regulator all in a modular cabinet. The ammoniator meters gaseous ammonia into the process stream under positive pressure. The high pressure in the storage tank is normally reduced to approximately 276 kPa (40 psi) using the pressure regulator. At this reduced pressure the ammonia flows through the rotameter where the gas flow can be read directly in mass/time units (In the United States the units are usually pounds per hour or pounds per day). Finally, the gas flows through the back-pressure valve, which maintains a constant back pressure on the system. This pressure is limited to a range of 101 to 122 kPa (15 to 18 psig). The back-pressure valve is used to keep the feed rate constant with changes in the pressure at the application site. Ammoniators should be housed separately from chlorination equipment.

A direct-feed ammonia application is illustrated In Fig. 13.21a. For completeness, an evaporator is included, although these are not always required. If ammonia feed rates are high enough, the anhydrous liquid would be withdrawn from the bottom of the storage tank and converted to gas in the evaporator prior to entering the ammoniator. The largest direct-feed ammoniators have a maximum feed capacity (determined by the rotameter rating) of 455 kg/d (1000 lb/d).

Solution feed of anhydrous ammonia

The design of these systems closely parallels the design of modern gas chlorination systems. An ammoniator and a gas diffuser are often used to feed the anhydrous ammonia solution (see Fig. 13.21a). A solution-feed ammoniator (see Fig. 13.21b) is typically recommended when higher feed rates or greater discharge pressures prohibit the use of direct-feed ammoniators. (It is important to remember that direct-feed ammoniators are limited by their back-pressure valve to a pressure of approximately 100 kPa). An important difference between these and chlorination systems, where design is concerned, is that the utility water for a solution ammoniation system must be softened to prevent the deposition of the hardness of the water in the system.

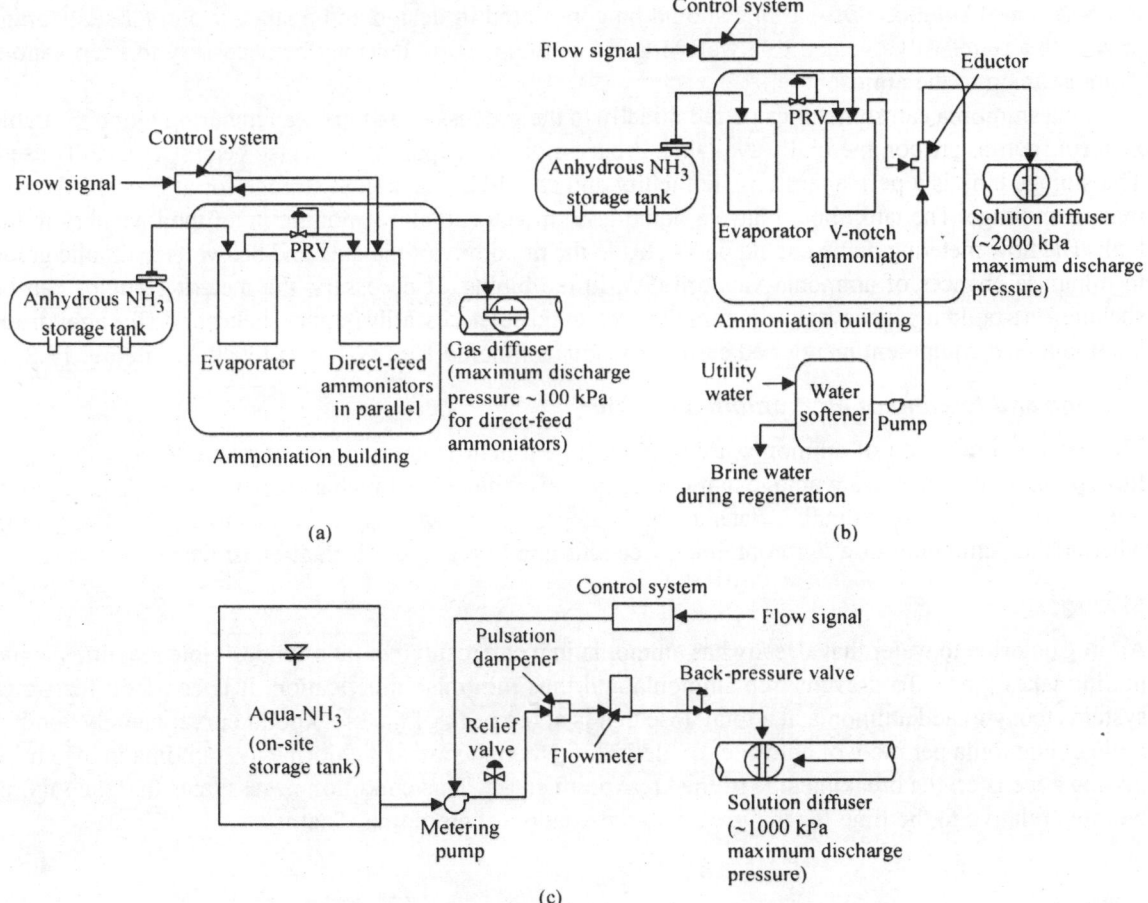

Fig. 13.21. Schematics of alternate ammonia feed systems; (a) direct feed of anhydrous ammonia; (b) solution feed of anhydrous ammonia; (c) aqua ammonia feed system.

Storage and feeding of aqua ammonia

Ammonia is very soluble in water. As an example, I volume of water will dissolve 1150 volumes of anhydrous ammonia at a temperature of 0°C and atmospheric pressure. As a consequence, ammonia is commercially available as an aqueous solution of between 20 and 30 per cent strength 'aqua ammonia'. It is usually dissolved in deionised or softened water and stored in low pressure tanks. The vapour pressure of 30 per cent aqua ammonia at 37.8°C (100°F), a temperature common in many parts of the world, is greater than 1 atm. To prevent off-gassing of ammonia in these locations, a slightly pressurised tank should be used. In contrast, the vapour pressure of 20 per cent aqua ammonia is less than 1 atm, permitting storage in a non-pressure tank with a minimum of off-gassing. Aqua ammonia is not commonly shipped long distances; hence the largest transport vessel in the United States is a 28300 l (7500 gal) tank trailer. There seems to be less standardisation for on-site aqua ammonia storage tanks, probably because low-pressure tanks are acceptable. Steel and fiberglass tanks are both used in water treatment applications.

Depending on the concentration of aqua ammonia, excessive temperatures can cause ammonia gas to come out of solution. Off-gassing should be considered in design and a slightly pressurised storage tank with a relief valve vented to a water trap or ammonia scrubber may be necessary to keep vapour from escaping to the atmosphere.

Aqua ammonia can sometimes be fed directly to the process stream using a metering pump. Suitable metering pumps are commercially available. Progressive cavity pumps have also been successfully used. The storage tank is a permanent on-site facility and should have enough storage for at least 10 days of maximum usage. The tank should have a liquid-level monitor to allow monitoring of the inventory in the tank. The flow metering pump should be located in the proximity of the tank and below its hydraulic grade to minimise chances of ammonia vapourisation in the piping. If necessary, the metering pumps can be sheltered in a building; however, the pumps themselves do not necessarily require shelter as do the anhydrous ammonia feed equipment mentioned earlier. An aqua ammonia feed system is illustrated in Fig. 13.21c.

Storage and feeding of ammonium sulphate

The most common salt of ammonia used in water treatment is ammonium sulphate, $(NH_4)_2SO_4$. This form of ammonia has the advantage that it does not raise the pH as much as the others do. As a result, it is easier to combine it with dilution water to obtain proper mixing. Mixing can be an important consideration when adding ammonia to water containing free chlorine to arrest the formation of DBPs.

Mixing

Adding chlorine to water that already has ammonia in it can result in a lot of undesirable reactions while mixing takes place. To prevent free ammonia and thus minimise nitrification, it is common for water systems today to add ammonia at a total dose that is at the peak of the breakpoint curve, namely about 1 mole of ammonia per mole of chlorine. By definition, then the ratio of chlorine to ammonia in the entire mixing zone is on the breaking side of the breakpoint curve. This condition necessitates that the mixing be rapid relative to the time for the irreversible oxidation of ammonia. That is,

$$t_{mix} = t_{rx} \qquad \qquad ... (13.51)$$

where,

t_{mix} = time required to obtain mixing to microscale, s

t_{rx} = half-life of breakpoint reaction, s

Although this circumstance is easily described in a qualitative way, it is quite difficult to characterise quantitatively because t_{rx} is a function of not only the pH but also the Cl_2/NH_3 ratio, namely the degree of mixing. When ammonia is added to a chlorinated water to arrest the formation of disinfection by-products, very good mixing is required to ensure that chemicals are efficiently used.

Managing combined chlorine (chloramine) residuals

Maintaining a combined chlorine residual involves some considerations that are not important when a free-chlorine residual is used. Chloramines have the advantage that their odour threshold is lower, that they are more effective in controlling growths on the pipe surfaces and that they are generally much more stable. It should be noted that combined chlorine residuals are subject to destruction by biological nitrification, especially if temperatures are warm and if ammonia is used in excess. Also there is recent evidence that the use of combined chlorine can result in the formation of low levels of 1-nitrosodimethylamine (NDMA), a suspected human carcinogen. Some of the conditions that aggravate NDMA formation, namely a high chlorine-to-ammonia ratio, are the same things that discourage nitrification.

DISINFECTION WITH CHLORINE DIOXIDE

When the regulation of the by-products of chlorination began, along with ozone, chlorine dioxide was a fairly high profile disinfection alternative. Chlorine dioxide is widely used in continental Europe, particularly Germany, Switzerland and France and produces almost no identifiable organic by-products, except a few aldehydes and ketones, produced at low levels. Chlorine dioxide was known to produce two inorganic by-products, chlorite and chlorate ion. As a result, most applications of chlorine dioxide were on low-TOC water that did not require a high dose to overcome oxidant demand. Late in the 1980s, concern about the toxicity of chlorite ion and chlorine dioxide itself reached a peak. Also based on field experience, it was found that the use of chlorine dioxide was sometimes responsible for a very undesirable 'cat urine' odour. As a precautionary measure, the State of California banned the use of chlorine dioxide for the disinfection of drinking water and several other states followed.

Eventually, when the disinfectant by-product rule was promulgated, an MCL of 0.8 mg/l was set for chlorite ion and a maximum disinfectant residual limit (MDRL) of 1 mg/l was set for chlorine dioxide. No MCL was placed on chlorate ion, but utilities were encouraged to be cautious about the production of chlorate and again as a precautionary measure, the State of California has set an action level of 0.8 mg/l. Methods for reducing the concentration of chlorite ion downstream of the use of chlorine dioxide have been demonstrated and it has been established that the cat urine odour only occurs when chlorite ion is exposed to a free-chlorine residual. As a result, it appears that chlorine dioxide may indeed play an important role in DBP control, particularly in systems using combined chlorine for residual maintenance and looking for a small boost in primary disinfection.

Generation of Chlorine Dioxide

The principal reactions that occur in most chlorine dioxide generators have been known for a long time. In industry, large-scale chlorine dioxide generators use chlorate as a feedstock, but for potable water applications chlorine dioxide is usually generated using a 25 per cent sodium chlorite solution. Although a sodium chlorite feedstock is a common starting point, a number of different approaches are used to convert the chlorite to chlorine dioxide. These include reactions with gaseous chlorine (Cl_2), aqueous chlorine (HOCl) or acid (usually hydrochloric acid, HCl). The reactions are:

$$2NaClO_2 + Cl_2(g) \longrightarrow 2ClO_2(g) + 2NaCl \qquad \text{... (13.52)}$$

$$2NaClO_2 + HOCl \longrightarrow 2ClO_2(g) + NaCl + NaOH \qquad \text{... (13.53)}$$

$$5NaClO_2 + 4HCl \longrightarrow 4ClO_2(g) + 5NaCl + 2H_2O \qquad \text{... (13.54)}$$

The stoichiometry of Eq. 13.52 requires 0.5 kg of chlorine and 1.34 kg of sodium chlorite to produce 1 kg of chlorine dioxide. Several of the alternative approaches used for the generation of chlorine dioxide are summarised in Table 13.6. The differences between Eqs. 13.52, 13.53 and 13.54 help to explain how generators can differ even though the same feedstock chemicals are used and why some should be pH controlled and others are less dependent on pH. In most generators, more than one reaction may be taking place. Chlorine dioxide generators are relatively simple mixing chambers. The reactors are frequently filled with media (Teflon chips, ceramic or Raschig rings) to generate hydraulic turbulence for mixing. A sample petcock valve on the discharge side of the generator is desirable to allow for monitoring of the generation process.

Table 13.6. Chlorine dioxide generation alternatives.

Generator type	Main reactions, reactants, by-products, key reactions, and chemistry notes	Special attributes
Acid-chlorite: (direct acid system)	$5NaClO_2 + 4HCl \rightarrow 4ClO_2(g) +$ $5NaCl + 2H_2O$ Low pH ClO_3^- also possible Slow reaction rates	Chemical feed pump interlocks required; production limit ~10–15 kg/d (25–30 lb/d); maximum yield is ~80 per cent of stoichiometric yield
Aqueous chlorine-chlorite: (Cl_2 gas ejectors with chemical pumps for liquids or booster pump for ejector water)	$Cl_2 + H_2O \rightarrow HOCl + HCl$ $HOCl + 2NaClO_2 \rightarrow ClO_2(g)$ $+ NaCl + NaOH$ Low pH ClO_3^- also possible Relatively slow reaction rates Excess Cl_2 or acid to neutralise NaOH	Production rates limited to ~450 kg/d (1000 lb/d); high conversion but yield only 80–92 per cent; more corrosive effluent due to low pH (~2.8–3.5); three chemical systems pump HCl, hypochlorite, chlorite and dilution water to reaction chamber
Recycled aqueous chlorine-chlorite: (saturated Cl_2 solution via a recycling loop prior to mixing with chlorite solution)	$2HOCl + 2NaClO_2 \rightarrow 2ClO_2 +$ $Cl_2 + 2NaOH$ Excess Cl_2 or HCl needed due to NaOH formed Concentration of ~3g/l required for maximum efficiency	Production rate limited to ~450 kg/d (1000 lb/d); yield of 92–98 per cent with ~10 per cent excess Cl_2 reported; highly corrosive to pumps; drawdown; calibration needed; maturation tank required after mixing
Gaseous chlorine-chlorite: (gaseous Cl_2 and 25 per cent solution of sodium chlorite; pulled by ejector into the reaction column)	$Cl_2(g) + 2NaClO_2 \rightarrow 2ClO_2(g)$ $+ 2NaCl$ Neutral pH Rapid reaction Potential scaling in reactor under vacuum due to hardness of feedstock	Production rates 2300–55000 kg/d (5000–1,20,000 lb/d); ejector-based, with no pumps; motive water is dilution water; near-neutral pH effluent; no excess Cl_2; turndown rated at 5–10X with yield of 95–99 per cent; less than 2 per cent excess Cl_2; highly calibrated flowmeters with minimum line pressure ~275 kPa (40 psig) needed

Sodium Chlorite

Sodium chlorite is used as a solution, normally with a concentration of approximately 25 per cent sodium chlorite or less. It is commercially available as a 38 or 25 per cent solution. The major safety concern for solutions of sodium chlorite is the unintentional and uncontrollable release of high levels of chlorine dioxide. Levels that approach an explosive mix can some times occur if the sodium chlorite is exposed to acid.

Another concern to be addressed with sodium chlorite is crystallisation. Like most salts, sodium chlorite solutions are prone to crystallisation at low temperatures and/or higher concentrations. When crystallisation occurs, it may obstruct flow in pipelines, valves and other equipment. Sodium chlorite is not stable as a powder. If dried, it is a fire hazard and can ignite when in contact with combustible materials. A sodium chlorite explosion may occur if too much water and inappropriate fire-fighting techniques are used to quench such a fire. Burning sodium chlorite will quickly generate enough heat to turn water to steam. At high temperatures, the breakdown products of sodium chlorite include oxygen. As a result, highly trained firefighters are required to extinguish closed containers or dry material that has been ignited. Stratification in holding tanks for sodium chlorite solutions may also occur and when it does, will adversely influence the chlorine dioxide yield in the generator. As stratification develops, the sodium chlorite solution being fed gradually changes from low to high density as the generator operates. In stratified tanks, excess chlorite will be fed to the generator as the bottom of the tank will have denser material and this material will have more chlorite than required. Similarly, the bulk tank will later yield chlorite that is too dilute. Although infrequent, such stratification is not readily apparent and may likely remain unnoticed by an operator unless the generator performance is evaluated frequently. Operators should be aware of the possibility of stratification and crystallisation during delivery conditions.

DISINFECTION WITH OZONE

The word ozone finds its origins in the Greek word ozein, which means 'to smell'. Ozone (O_3) is an allotrope of oxygen with three oxygen atoms. Ozone in air has a pungent odour that is noticeable to most persons at levels above 0.1 ppm_v. Because ozone is a strong oxidant, extended exposure to ozone-containing air is harmful.

Chemistry of Ozone

At high concentrations (>23 per cent) ozone is unstable (explosive) and under ambient conditions it undergoes rapid decay. Therefore, unlike chlorine gas, it cannot be stored inside pressurised vessels and transported to the water treatment plant. Once dissolved in water, ozone begins a process of decay that results in the formation of the hydroxyl radical. Where contaminants and microbes are concerned, ozone reacts in two ways: (i) by direct oxidation; and (ii) through the action of hydroxyl radicals (HO·) generated during its decomposition. The consensus is that action of ozone as a disinfectant is primarily dependent on its direct reactions; hence it is the residual of the ozone itself that is important.

Ozone Decay and Ozone Demand

The overall rate of ozone decay in water is generally consistent with first order kinetics, although, like chlorine, it can be modelled successfully through the use of a parallel first-order decay model, as shown in Eq.13.13. Although simple reactions serve as good phenomenological models for ozone decay, it is unlikely that they correctly characterise the actual mechanism of decay. From work done in this area, it appears more likely that ozone decay consists of a large number of nth-order reactions operating in parallel that, in sum, appear to be simple first order.

An introduction to ozone decay based on the models developed by Staehelin and Hoigne is provided in Fig. 13.22. The cyclic nature of the ozone decay process in pure water illustrated in Fig. 13.22a. The process must be initiated by a reaction between ozone and the hydroxide ion to form superoxide radicals

(O_2^-) and peroxide ions (HO_2^-), a slow process. As a result, decay is accelerated at higher pH. Once completed the decay reactions enter a cyclic process represented in the figure in the shape of a circle. The cyclic reactions are promoted by ozone itself—hence the decay of ozone. If the concentration of ozone is increased, the cycle is accelerated.

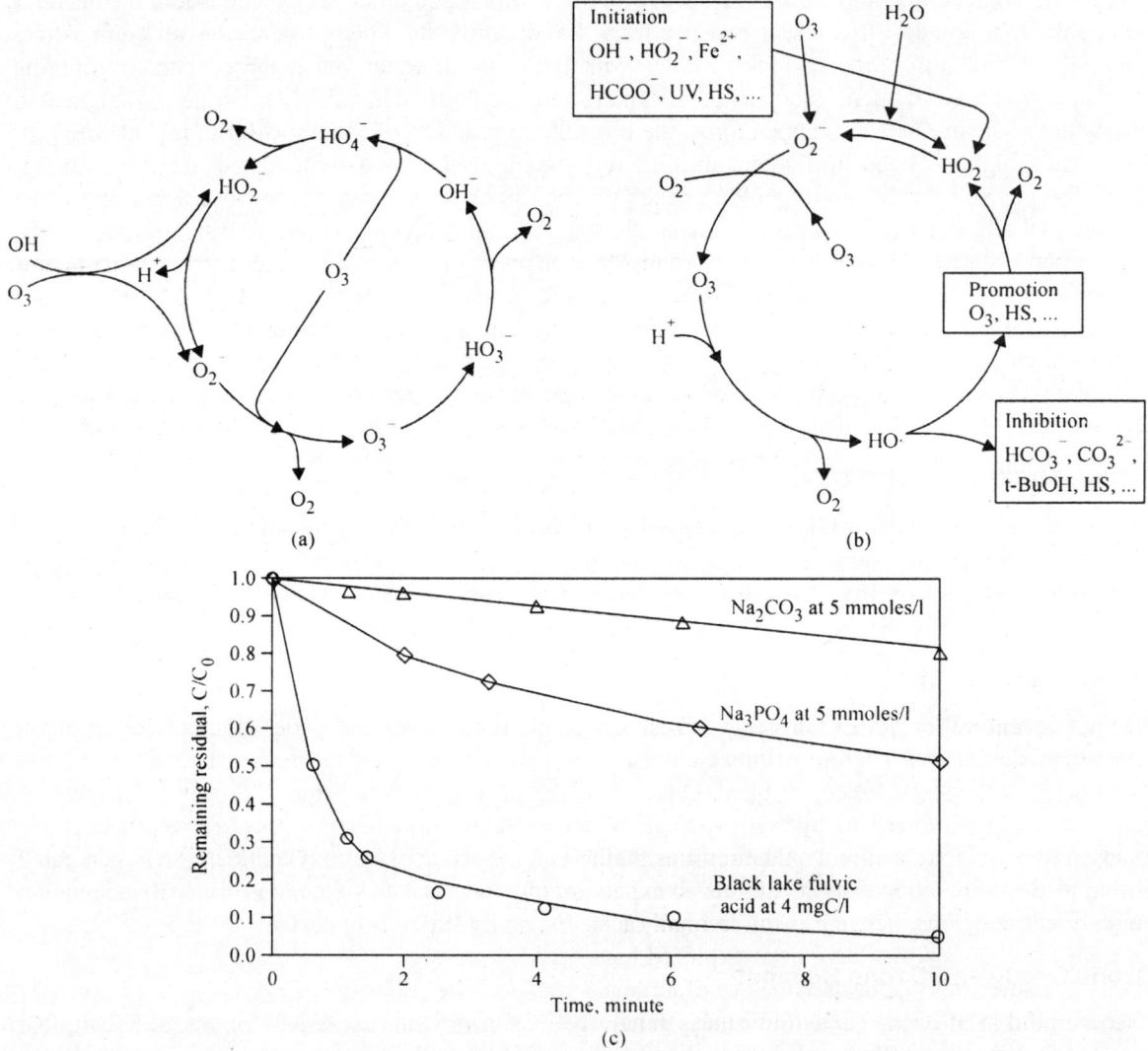

Fig. 13.22. Understanding ozone reaction pathways and decay of residual ozone in natural water: (a) influence of initiators, promotors and inhibitors; (b) the ozone decay wheel—reaction pathways in pure water; and (c) effect of fulvic acid and carbonate on ozone decay—all tests conducted at 20°C with GAC filtered, deionised tap water adjusted to pH 7 and $C_0 \sim 8$ mg/l.

In natural water, there are other 'initiators' besides hydroxide ion, as shown in Fig. 13.22b. Prominent among them are the ferrous ion and hydrogen peroxide. In natural water certain natural organic materials have also been shown to promote the cycle, accelerating decay. Finally, the continuation of the cycle depends on the action of the hydroxyl radical on the ozone residual. As a result, scavengers that react with the hydroxyl radical, removing it from the process, also slow the rate of decay. The carbonate and bicarbonate ions are important examples of such inhibitors. The data of Reckhow and co-workers, are shown in Fig. 13.22c to illustrate the action of fulvic acids as initiators and promoters and carbonate and bicarbonate ions as HO· traps or inhibitors. The factors that influence the stability of ozone residuals are summarised in Table 13.7.

Table 13.7. Factors that influence stability of aqueous ozone residuals.

Increases stability	Reduces stability
Low pH	High pH
High alkalinity	Low alkalinity
Low TOC	High TOC
Low temperature	High temperature

Conceptual Design

Including ozonation in a water treatment plant requires the addition of two components in the process treatment train: (i) a mass transfer device for dissolving the ozone into the water; and (ii) a contact chamber in which the disinfection reaction takes place. For several decades, the most common approach to ozonation has been to combine these two components by introducing the ozone into the water in large, deep basins using porous diffusers. Both mass transfer and the disinfection reaction then take place in these basins. Most often these contact basins are simulated at a pilot scale using tall bubble columns.

The conceptual design of any ozonation system requires a means for estimating mass transfer of ozone into the water, an understanding of the kinetics of ozone decay, an understanding of the disinfection kinetics and a means for estimating dispersion in the reactor. All of these components have been applied successfully to model the disinfection of *C. parvum* in bubble columns at pilot scale. Nevertheless, significant challenges remain in addressing all these issues in the variety of approaches that are considered in full-scale design.

Mass transfer of ozone into water

Because ozone is sparingly soluble, the facilities that are required to introduce it into water are usually expensive. As a result, the industry continues to experiment with new approaches. As mentioned earlier, both mass transfer and contact time are traditionally achieved via large, deep basins using porous diffusers. None of the innovations that have been explored have managed to displace this technology. The most common alternative approaches are the use of large basins with submerged turbines (also used for both mass transfer and contact time) and in-line mass transfer devices using venturi or static mixers. Submerged turbines are particularly useful when high manganese levels would adversely affect the performance of ozone diffusers. In-line mixers are particularly attractive when a low ozone dose and/or a low Ct product is required. From the process standpoint, in-line mixing is an attractive alternative for disinfection because the backmixing is minimised and plug flow is more closely approached. From the standpoint of cost of construction, in-line mixers have the advantage that the excavation depth is greatly reduced because depth is not required to support mass transfer. Finally, if a concentrated ozone/oxygen stream is used and a low ozone dose is required, mass transfer can be optimised fully (all the gases completely dissolved) and off-gas treatment may not be required. The principal drawback of in-line mixing is the need for additional pumping ahead of the injector and the complications of achieving adequate mass transfer over the wide range of flows often required in such systems.

Batch reactor testing

Batch testing is often conducted by bubbling ozone directly into a gas wash bottle containing the sample of interest. The ozone concentration is measured in the gas entering and exiting the bottle and the difference constitutes the ozone added to or consumed by the sample. For a number of reasons, the preferred technique is to prepare the ozonated water first and then add that to the sample of interest. In this case, the batch reactor might be a 1 or 2-L jar or a 0.5 to 1-L Teflon bag containing the water of interest. The concentrated ozone solution is prepared in a separate container by continuously bubbling ozone gas into a small volume of distilled-deionised (DI) water. At ambient temperature, the maximum ozone solution concentration may be about 15 mg/l. To prepare a more concentrated solution, the DI water can be chilled in an ice bath. At temperatures close to 1°C, the concentration of ozone in the stock solution can be as high as 40 mg/l. Aliquots of the ozone stock solution are then drawn and injected into the test water sample. The volume of each aliquot is calculated to deliver a pre-determined ozone dose to the test water sample. Water samples are then drawn from the test water at various times after the ozone is added and analysed for ozone residual concentration. This test is repeated using various ozone doses.

The profile of ozone residual concentration versus time can then be plotted. Two example ozone decay profiles in two water dosed with 1.0 mg/l ozone are shown in Fig. 13.23a. Both water have relatively high ozone demand, particularly water B. The profile of ozone decay in water A is typical of most moderate TOC, well-oxygenated surface water. The curve fit through the data points is that of a pseudo-first-order decay equation with a decay coefficient of 0.3 min^{-1}. The decay of ozone in some water does not always follow this uniform first-order decay model. Water B is an example of common ozone decay profiles where the ozone experiences an initial period of a high decay rate followed by a second period of slower decay. The curve fit through the data points for water B was accomplished using Eq. 13.13. Although this equation is based on the progress of two parallel first-order reactions, it should be viewed as a phenomenological model, not a mechanistic one. Based on experimental evidence, ozone and particularly the hydroxyl radical that is produced when it decays, participate simultaneously in many reactions of different orders at the same time.

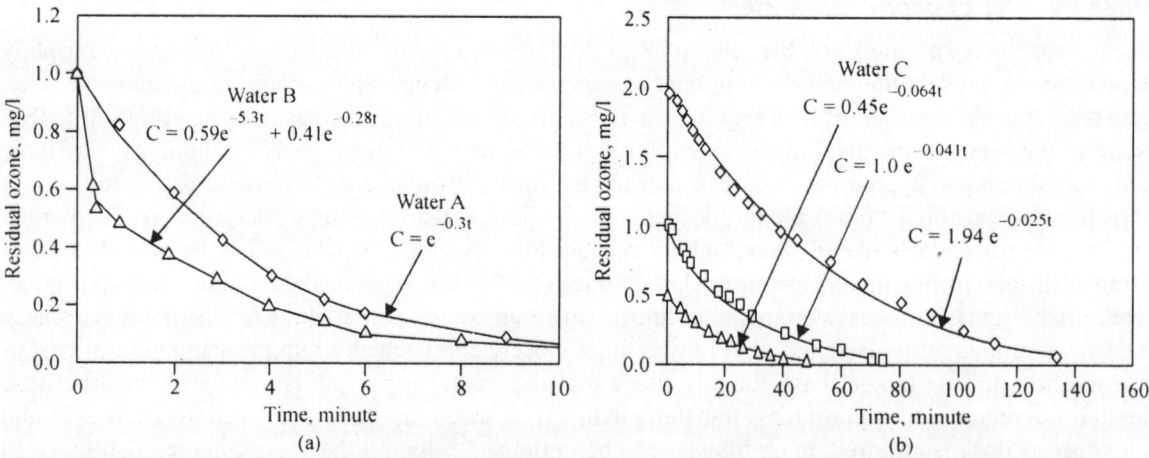

Fig. 13.23. Typical batch ozone decay curves for three different water: (a) water A, B; and (b) water C.

Another result of this complexity is that with ozone, as with chlorine, the rate of decay observed in a batch test like the one just described is also influenced by the ozone residual at the beginning of the decay period, as illustrated in Fig. 13.23b using the data from a pure mountain water supply. In general, these curves exhibit a low rate of decay; nevertheless they also show a rate of decay that was significantly higher when the residual at the beginning of the decay process (C_0) was increased. As with chlorine, the change in decay rate is approximately inversely proportional to C_0. As a result of these complexities, if only batch testing is conducted to determine the basis of design, a wide variety of test conditions must be evaluated to get an adequate database for design. Even with such data, a number of assumptions and approximations must be made during the process of design.

Continuous-flow reactor testing

At the bench or pilot scale, continuous-flow reactors are a much better choice for gathering information for design of ozonation facilities, especially for an over-under ozone contactor with ozone addition via diffusers. In full-scale designs of this type, the ozone is generally added in the first few compartments of the design and then the residual is allowed to decay as the water travels throughout the rest of the reactor. This approach to design and operation can be simulated by operating the small-scale continuous-flow unit so that it has the same detention time as the ozone addition compartments will have in the full-scale design. Once the reactor has reached steady-state operation, both the flow of water and the ozone dosing can be stopped and the decay of ozone residual can be observed as a function of time. The continuous operation simulates the ozone addition compartments and the decay curve can be used to estimate the residual in downstream compartments.

A continuous-flow set-up requires measuring the ozone gas flow rate, water flow rate, ozone concentration in the feed gas and ozone concentration in the off-gas.

The ozone dose to the reactor is then calculated as:

$$\text{Ozone dose, mg/l} = \frac{Q_g}{Q_l} \times (C_{g,in} - C_{g,out}) \qquad \qquad ... (13.55)$$

where,

$\quad Q_g \quad$ = gas flow rate, l/min
$\quad Q_l \quad$ = water flow rate l/min
$\quad C_{g,in} \quad$ = concentration of ozone in feed gas, mg/l
$\quad C_{g,out} \quad$ = concentration of ozone in off-gas, mg/l

For each ozone dose, the operating conditions are kept constant until steady-state conditions are reached. This stabilisation period can be between three and five times the average hydraulic contact time in the reactor. It is essential that the continuous reactors be operated with approximately the same detention time as the ozone addition compartments in the full-scale design. An RTD curve similar to the full-scale reactor is also highly desirable. Unfortunately, tall, narrow pilot columns with long aspect ratios are often used because they achieve more efficient ozone transfer. The use of tall columns is not a particularly good choice because they much more closely approach plug flow than full-scale designs. This test must also be conducted at various doses because it is important to understand the relationship between the ozone dose and the ozone residual in the water exiting the ozone addition section of the reactor. The ozone decay rate downstream of these compartments will vary with this residual as well.

Design of conventional ozone contactors

Several designs have been developed and implemented for the addition of ozone at a water treatment plant. The most common design is a multichamber over-under baffled contactor with ozone addition to the first one or two chambers via porous stone diffusers situated at the bottom of the chambers. Schematics of such a contactor are shown in Fig. 13.24. The water depth in the contactor is typically between 4.6 and 6 m (15 and 20 ft) to achieve high transfer efficiency of the added ozone.

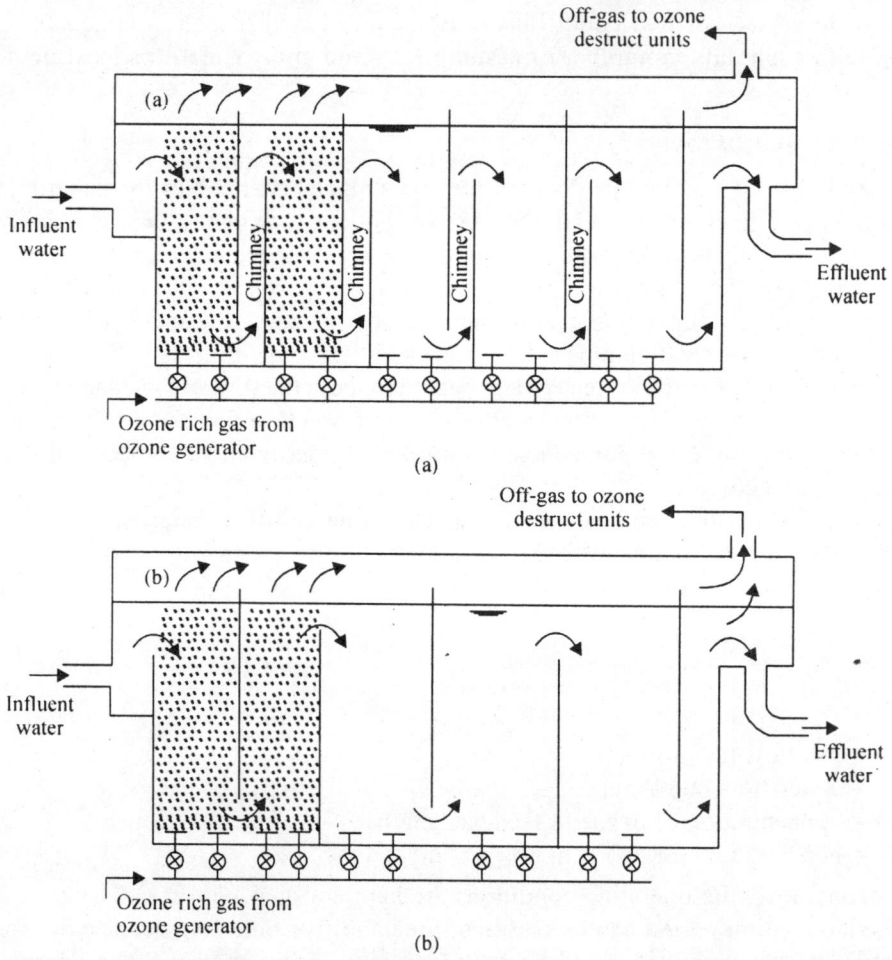

Fig. 13.24. Schematics cross-sectional views of two alternate designs for five-chamber, over-under ozone contact chamber: (a) with chimneys; and (b) without chimneys.

Water enters the first chamber from the top and exists from the bottom. This countercurrent flow configuration (between the water and the air) helps increase the overall ozone transfer efficiency in two ways: first, the contact time between the air bubbles and the water is greater in a countercurrent flow configuration than in a co-current flow configuration because the effective rise velocity of the air bubbles is equal to the algebraic sum of the rise velocity of the air bubble in still water and the water flow velocity. Second, the overall transfer efficiency in a gas-bubble diffusion process is somewhat higher in a countercurrent flow pattern than in a co-current pattern based on mathematical calculations.

To achieve countercurrent flow in subsequent chambers, the contactor is also designed with segments that return the flow back to the top. A design is shown in Fig. 13.24a, where the water exiting the bottom of the first chamber rises to the surface through a narrow chamber, commonly called a chimney, before it enters the top of the second chamber. The chimney design achieves countercurrent flow in all chambers where ozone is added. A design with no chimneys is shown in Fig. 13.24b. In this design, the flow configuration alternates from countercurrent to co-current as the water moves from one chamber to the next. While lower transfer efficiency may take place in the co-current chambers, experience has shown that the impact is minimal. The passage of the water through the narrow chimneys of the alternate design causes a significant flow separation as the water enters exits each down-flow contact chamber, resulting in high dispersion. In Fig. 13.25, schematic renderings of possible hydraulic flow patterns are showing a multichamber contactor where the water is forced through a narrow pathway. Chimneys between chambers are indicated in Fig. 13.25a. The design shown in Fig. 13.25b no longer has chimneys but still exhibits significant flow separation at the turns.

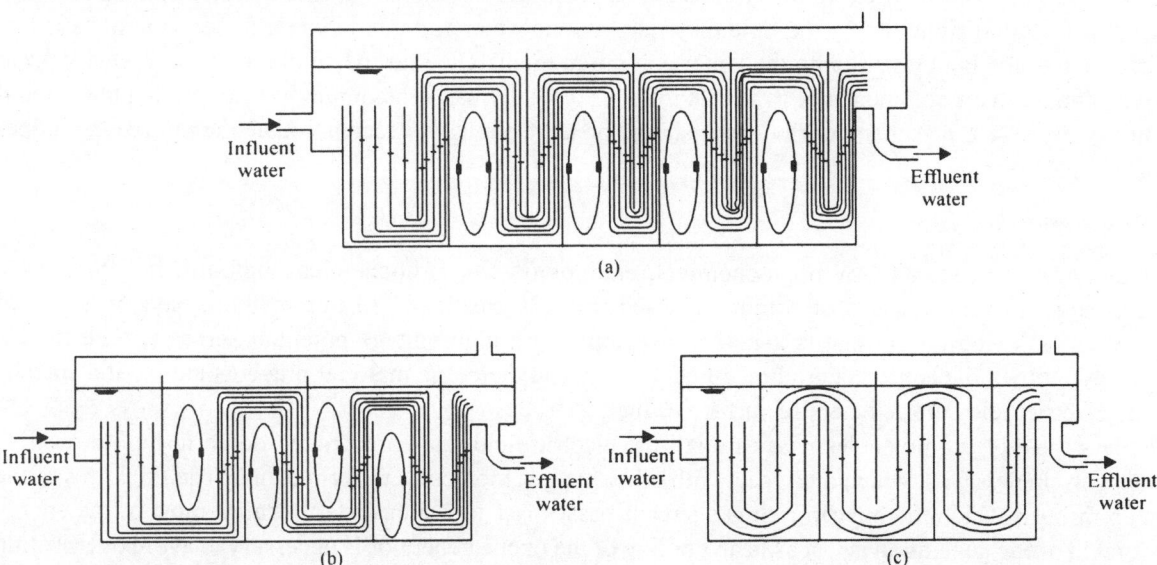

Fig. 13.25. Conceptual impact of ozone contactor design flow hydrodynamics: (a) with chimneys; (b) without chimneys; and (c) with uniform flow path.

The problem with the contactor design in Fig. 13.25b is that the openings through which the water flows between chambers are too narrow. The same principle that applies in the design of the serpentine chlorine contact chambers discussed previously applies here: the width of the flow path must be maintained. The flow path can be maintained by ensuring that the opening between two consecutive chambers is approximately the same width as the downstream chamber. The hydraulic flow pattern in a contactor designed with these considerations in mind is illustrated in Fig. 13.25c. It is noted that the hydraulic flow lines shown in Fig. 13.25 are only conceptual. A more accurate determination of the true hydraulic behaviour can be determined using computational fluid dynamic (CFD) modelling of the contactor.

Henry and Freeman conducted such modelling on various ozone contactor designs and determined that the contactor-baffling ratio (defined as the ratio of t_{10}/τ) is greatly impacted by the internal geometry of the contactor. The impact of the H/L ratio on the baffling ratio, where H is the water depth and L is the longitudinal width of the chamber, is shown in Fig. 13.26a. Increasing the H/L ratio from 2 to 4 increases the t_{10}/τ ratio from 0.55 to 0.65. The impact of the G/L ratio, where G is the depth of the flow path under the baffle, on the baffling ratio is illustrated in Fig. 13.26b. Increasing the G/L ratio from 0.2 to 1.0 increases the t_{10}/τ ratio from 0.45 to 0.65. Based on this work, a maximum t_{10}/τ ratio can be achieved with a H/L ratio of 4:1 and a G/L ratio of 1:1.

Porous stone diffusers are used in ozone contactors to produce fine bubbles, which greatly increases the overall ozone transfer efficiency from the gas phase to the water, especially when compared to the use of a perforated-pipe diffuser. While both types of diffusers are used, experience has shown that perforated-pipe diffusers produce an excessively large bubble size.

The cause of this problem is attributed to the way air exits the diffuser. When the diffuser is positioned horizontally, the air that exits on the underside of the diffuser seems to creep along the circumference of the diffuser before it rises into the water. As this creep occurs, the initial fine bubbles pick up more air and grow to large bubbles by the time they rise into the water column. Dome diffusers do not have this problem as the bubbles rise into the water column immediately after they exit the diffuser. Due to head loss limitations, a commercially available diffuser typically has a maximum gas flow rating that should not be exceeded. The chamber floor area should be large enough to accommodate the minimum number of diffusers.

Generation of Ozone

Ozone can be generated by photochemical, electrolytic and radiochemical methods, but the corona discharge method is the most commonly used in water treatment. In this method, oxygen is passed through an electric field that is generated by applying a high voltage potential across two electrodes separated by a dielectric material (see Fig. 13.27). The dielectric material prevents arcing and spreads the electric field across the entire surface of the electrode.

As the oxygen molecules pass through the electric field, they are broken down to highly reactive oxygen singlets $(O\cdot)$, which then react with other oxygen molecules to form ozone. The thickness of the gap through which the oxygen-rich gas stream passes is 1 to 3 mm wide. Because most of the energy used in ozone generation is lost as heat, cooling of the ozone generator is necessary to avoid overheating and subsequent decomposition of the ozone generated. Cooling is normally accomplished by passing a continuous stream of cooling water next to the ground electrode. Some of the key design factors that influence ozone generator performance are summarised in Table 13.8.

Table 13.8. Influence of increasing four key design factors on generator performance.

Design factor	Effect on ozone production
Frequency of applied current	Increase ozone production
Voltage of applied current	Increase ozone production
Gap between generator electrodes	Decrease ozone production
Dielectric constant of dielectric separating electrodes	Decreases ozone production

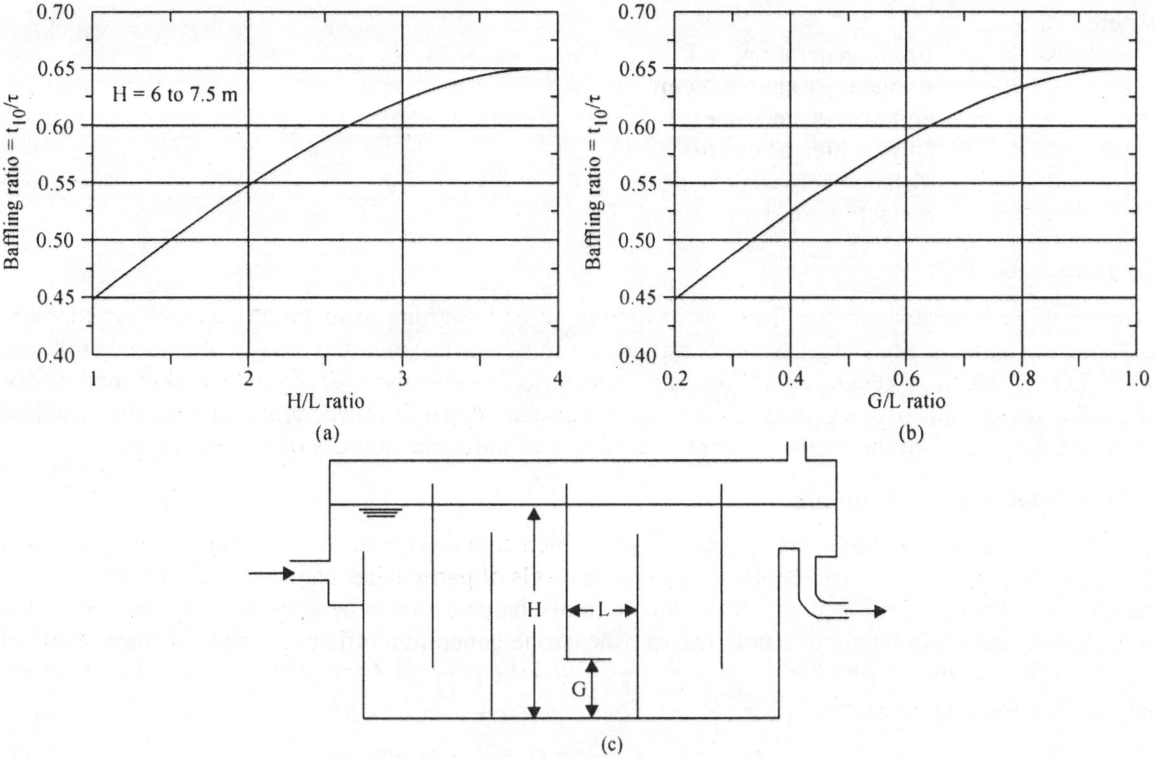

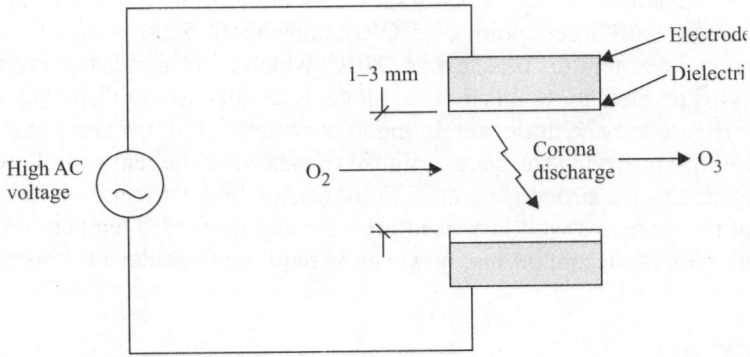

Fig. 13.26. Impact of internal contactor design on its baffling ratio: (a) impact of H/L ratio; (b) impact of G/L ratio; and (c) contactor schematic. Dimensions, H, G and L are defined in contactor schematic.

Fig. 13.27. Ozone generation by corona discharge.

The equation below, while not intended to be quantitative, provides a general idea of the significance of a number of the variables of importance to the design of a corona discharge ozone generator:

$$Q_{O_3} \; \alpha \; \left(f \frac{V^2 A}{d\varepsilon} \right) Q_{O_2} \qquad \ldots (13.56)$$

where,

Q_{O_3}	=	ozone generation, MT^{-1}
f	=	frequency of applied emf
V	=	emf across electrodes, V
A	=	surface area of electrodes, l^2
d	=	distance between electrodes, l
ε	=	dielectric constant

Oxygen Source

Ozone can be generated directly from the oxygen in air or from pure oxygen. Pure oxygen is generated on-site from ambient air at larger plants or provided through the use of liquid oxygen (commonly referred to as LOX), which is generated off-site and transported to the plant. The most suitable method for providing oxygen for ozone generation in a particular plant depends on economic factors, the principal ones being the scale of the facility and the availability of industrial sources of liquid oxygen.

Use of prepared, ambient air

The most accessible oxygen source is ambient air, which contains about 21 per cent oxygen by volume. Unfortunately, ambient air also contains significant levels of particulates and water vapour, which must be removed. Water vapour is detrimental to corona discharge ozone generators for two reasons: (i) the presence of water vapour significantly reduces the ozone generation efficiency; and (ii) trace levels of water can react with the nitrogen present in the air and the generated ozone to form nitric acid, which attacks the ozone generator itself:

$$O_3 + N_2 + O_2 + H_2O \xrightarrow{hv} 2HNO_3 \qquad \text{... (13.57)}$$

The moisture content of a gas is often defined by its dew point, which is the temperature to which the gas needs to be cooled to reach 100 per cent saturation. The lower the dew point of a gas, the lower is its moisture content. For example, an air stream with a dew point of 30°C contains about 28000 ppm_v of moisture. An air stream with a dew point of 5°C contains about 5000 ppm_v of water. The dew point specified for many ozone generators is as low as –80°C, which corresponds to a moisture content of less than 1.5 ppm_v. Drying ambient air to this moisture level is usually accomplished by a three-step process of compression, refrigeration and desiccant drying. A schematic of all the components of such a system is shown in Fig. 13.28. The drying steps are designed to maximise the removal of moisture from the gas stream before it is sent to the ozone generator. Compression and refrigeration help because the water vapour capacity of air decreases with increased pressure and decreased temperature, reducing the load on the desiccant system. Desiccant drying, however, is required to achieve the specifications for ozone generation.

Liquid oxygen delivery

Liquid oxygen is widely available as a commercial, industrial-grade chemical. Water treatment plants can purchase commercially available LOX, store it at the plant, and use it as the oxygen source for ozone generation. Liquid oxygen is delivered in trucks and stored in insulated pressurised tanks with capacities as high as 225 m^3 (60000 gal). Liquid oxygen is then drawn from the tank and piped to a vapouriser that warms and converts the oxygen to the gaseous form. Commercially available LOX is inherently low in contaminant content and water vapour as a result of its own manufacturing process. Therefore, minimal additional processing of the oxygen stream is required before it is introduced to the ozone generator.

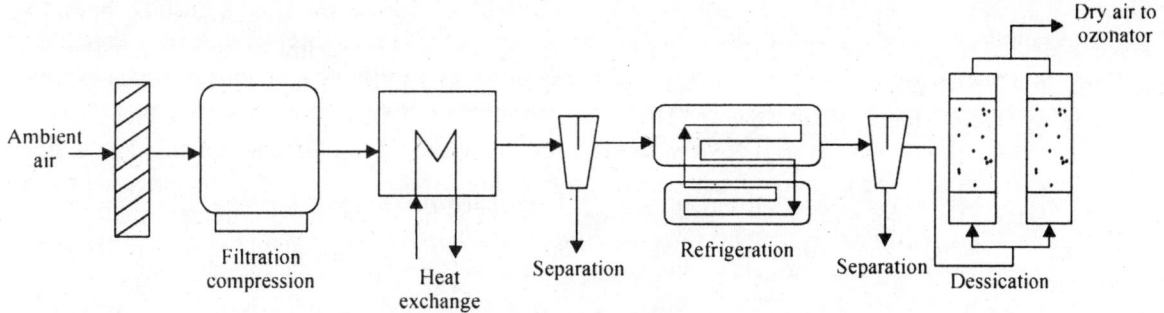

Fig. 13.28. Air preparation system for ozonation.

The use of LOX for ozone generation has several advantages over the use of ambient air, including: (i) simpler operation and maintenance because fewer processes are required; (ii) a smaller facility with lower capital cost; and (iii) a smaller number of ozone generators. The disadvantages of LOX include: (i) increased truck traffic caused by the need for regular LOX deliveries; and (ii) susceptibility to market pricing. Safety concerns associated with the storage of a large volume of concentrated oxygen must also be addressed.

On-site oxygen generation

Two types of on-site oxygen separation and concentration processes are used in water treatment plants that require oxygen: (i) pressure swing or vacuum swing adsorption (PSA or VSA) processes; and (ii) cryogenic oxygen generation processes. Generally, the economics of these processes improve as oxygen requirements increase. None are more economical than ambient air or LOX feed systems in small applications. The PSA systems are believed to be viable for systems requiring less than 30 tonnes/d (33 tonnes/d) of oxygen, while VSA systems are viable for systems requiring as much as 100 tonnes/d 110 tonnes/d). Only the largest applications typically utilise cryogenic oxygen generation. In fact, companies that sell LOX typically utilise cryogenic oxygen generation systems to generate LOX from ambient air. In direct municipal applications, cryogenic oxygen generation is commonly limited to large waste-water treatment plants that utilise large amounts of oxygen in pure-oxygen activated sludge processes. A large drinking water treatment plant rarely requires such large amounts of oxygen to warrant the use of cryogenic oxygen generation.

For example, a 2300 ML/d (600 mgd) water treatment plant having a design ozone dose of 3 mg/l requires about 6800 kg/d (15000 lb/d) of ozone. Assuming generators convert 8 per cent of the oxygen to ozone, this translates to approximately 85000 kg/d (94 tonne/d) of oxygen.

PSA and VSA systems

The PSA and VSA processes take advantage of the effect of gas pressure on the differences in the adsorption characteristics of the various constituents of ambient air on speciality adsorption resins. For the generation of oxygen, the affinity of the resin for nitrogen, water and carbon dioxide is higher than that for oxygen and increases with increased pressure. Therefore, the PSA or VSA system cycles between 'high' and 'low' pressures. During the high-pressure period, water moisture, carbon dioxide, nitrogen and any hydrocarbons present preferentially adsorb onto the resin while oxygen, now constituting about 90 to 95 per cent of the remaining gas passes through. Once the resin is saturated with the constituents removed, the system cycles to the low pressure, resulting in the desorption of the adsorbed material, which is then exhausted to the atmosphere before the cycle is repeated. For a PSA system, the high-pressure setting ranges from 200 to 400 kPa (30 to 60 psig) , while the low setting is atmospheric pressure. In a VSA system, the high-pressure setting is at 20 to 70 kPa (3 to 10 psig), while the low setting is achieved using a vacuum pump. The VSA systems are favoured over PSA systems because they utilise less energy. However, a VSA system requires additional equipment compared to a PSA system in the form of a vacuum pump as well as a downstream compressor to boost the pressure of the oxygen stream to the level required by the ozone generator. The need for an extra pump translates into higher capital cost and higher maintenance cost. The schematic layout of a typical PSA system is illustrated in Fig. 13.29.

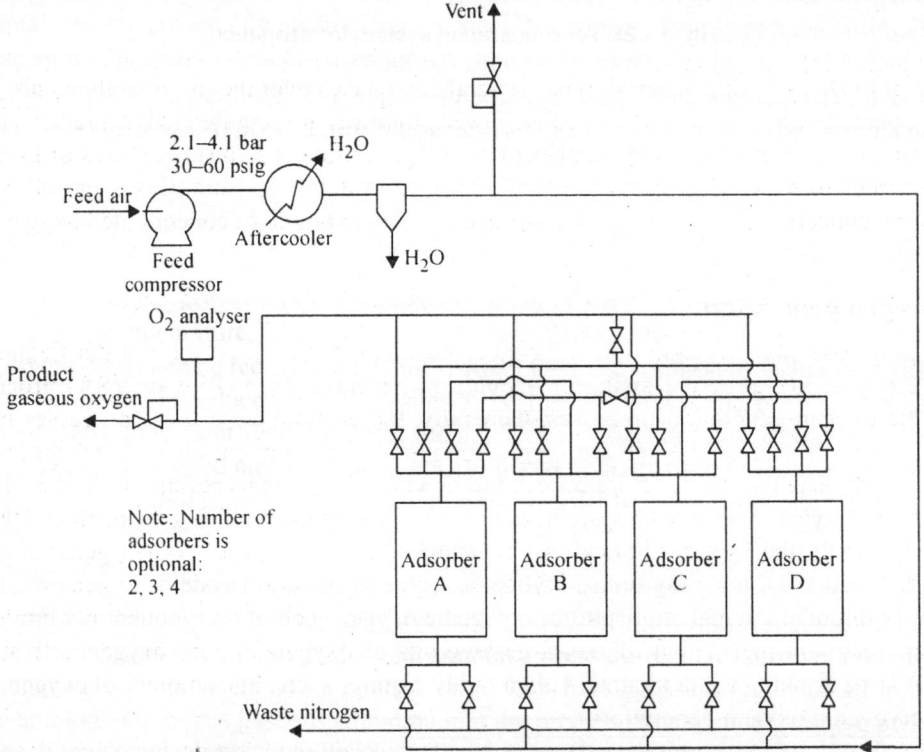

Fig. 13.29. Schematic of pressure swing adsorption system for producing pure oxygen.

Cryogenic oxygen generation system

A cryogenic oxygen generation process includes the following steps:
1. Air preparation—filtration, compression and moderate chilling of the air stream.
2. Air purification—the removal of water vapour, carbon dioxide and hydrocarbons with adsorption on speciality adsorbent material.
3. Air liquefaction—chilling of the air to about $-190°C$ using low-temperature refrigeration resulting in the liquefaction of the purified air, which now contains only oxygen and nitrogen.
4. Distillation—the use of special packed columns to separate the oxygen from the nitrogen. The separated oxygen, which can range from 95 to 99.9 per cent by weight, is mostly in the gas

phase. However, a small portion can be produced as liquid oxygen, which is then stored at the plant and used as a back-up oxygen supply during system downtime.

Regardless of whether air or pure oxygen is used for ozone generation, the efficiency of ozone generators is quite low. When ambient air is used as the feed gas to the generator, typical ozone content in the generator outlet is somewhere between 1 and 2 per cent by weight. With pure oxygen, typical generators produce about 4 to 8 per cent ozone by weight. Recent advances in ozone generation technology, however, have resulted in the manufacture of generators that can produce as much as 14 to 16 per cent ozone by weight with pure oxygen. A schematic of the cryogenic process and a photograph of the cryogenic oxygen generation system at a large water treatment plant are shown in Fig. 13.30.

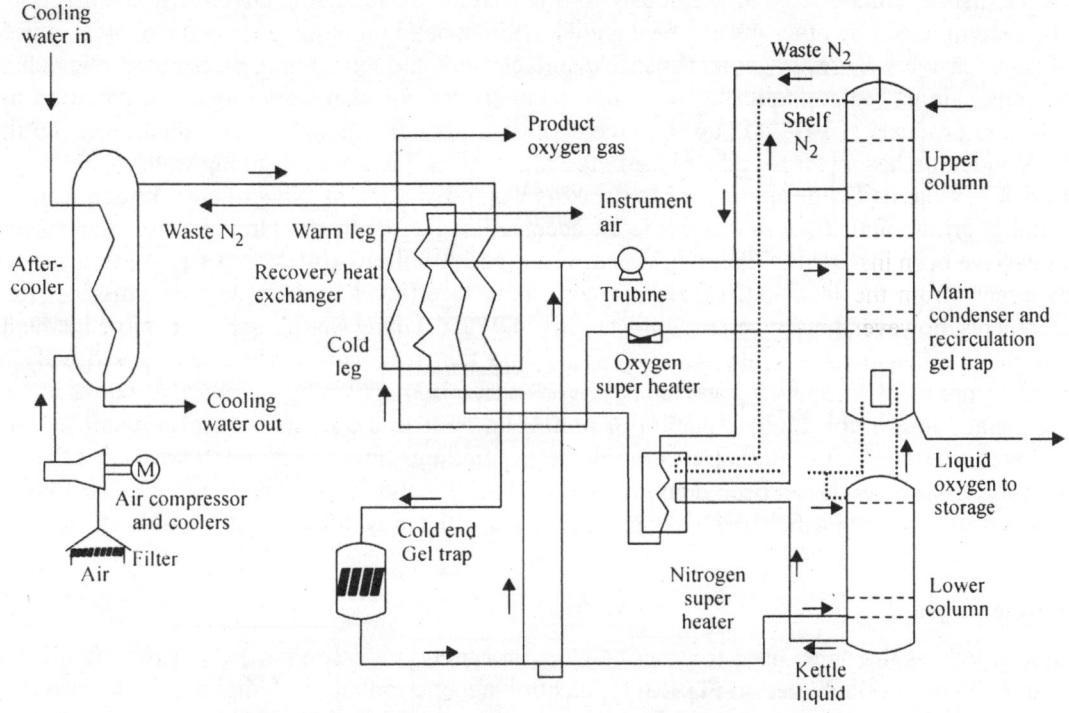

Fig. 13.30. Cryogenic system for producing pure oxygen.

Off-Gas Treatment

Even with the most efficient ozone contactor designs, off-gas ozone concentrations substantially exceed acceptable levels and as a result, off-gas treatment is required. In the United States, the Occupational Safety and Health Administration (OSHA) sets an 8 hours workday ozone exposure limit of 0.1 ppm_v by volume at standard temperature and pressure (STP), which is equivalent to 0.0002 mg/l in air. In general, the concentration in the ozone gas entering the contactor can range anywhere from 5000 to 1,60,000 ppm_v; so ozone contactors would have to achieve removals in excess of 99.998 per cent to meet these standards directly. The efficiencies actually achieved in these reactors range from 90 to 99 per cent, rarely higher. As a result, the ozone concentration in the off-gas from a typical ozone contactor in a water treatment plant exceeds reasonable discharge limits by a substantial margin.

Therefore, the off-gas cannot be vented to the atmosphere before the residual ozone is destroyed.

Ozone in the off-gas stream can be destroyed thermally with or without the use of solid catalysts. When a catalyst is not used, ozone destruction is accomplished by heating the off-gas to a temperature between 300 and 350°C. At this temperature, the required contact time through the destruction unit is less than 5 seconds. Newer destruction units combine the use of speciality metal catalysts with moderate heating to achieve ozone destruction. Depending on the type of catalyst used, the off-gas temperature need only be raised to somewhere between 30 and 70°C.

DISINFECTION WITH ULTRAVIOLET LIGHT

All of the disinfectants discussed previously in this chapter are oxidising chemicals. Disinfection can also be accomplished by other means, heat and electromagnetic radiation among them. Heat is used to disinfect or 'pasteurise', beverages and even to disinfect water through boiling. Electromagnetic radiation, specifically gamma radiation and UV radiation, is also used for disinfection: gamma radiation in the case of food products and UV radiation in the case of air, water and some medical surfaces. Of these, only UV radiation has so far found a place in the routine disinfection of drinking water.

To date UV has had wide application to drinking water disinfection. Nevertheless its application has been rather erratic and it has not become an accepted standard. In the United States, far more UV facilities have been installed to disinfect waste-water. The popularity of UV in waste-water disinfection stems largely from the fact that it does not contribute to effluent toxicity as does chlorine. Recent developments, however, have given the application of UV to drinking water a great deal more momentum. Interest in UV began to increase when it was found that medium-pressure UV, a relatively new, multiwavelength UV technology, was unusually effective in inactivating *C. parvum*. Flaws in earlier testing methodologies for the inactivation of protozoan cysts and oocysts were also identified in that study. Later it became apparent that, contrary to earlier findings, low-pressure UV was also effective in inactivating *C. parvum*. It has been demonstrated that these technologies are effective in inactivating *G. lamblia* as well. Controlling these two pathogens has been particularly difficult using traditional chlorination technology.

Ultraviolet Light

Ultraviolet light is the name used to describe electromagnetic radiation having a wavelength between 100 and 400 nm. As illustrated in Fig. 13.31, electromagnetic radiation of slightly shorter wavelength has been named 'X-rays' and electromagnetic radiation of slightly longer wavelength, visible to the human eye, is referred to as 'visible light'.

Radiation just long enough to be outside the visible range is referred to as infrared radiation. Light in the UV spectrum is often further sub-divided into four segments, vacuum UV, short-wave UV (UV-C), middle-wave UV (UV-B) and long-wave UV (UV-A). In simple terms these classifications can be described as follows:

1. Both UV-A and UV-B activate the melanocytes in the skin to produce melanin ('a tan').
2. UV-B radiation also causes 'sunburn'.
3. UV-C radiation is absorbed by the DNA and is the most likely of the three to cause skin cancer.

If electromagnetic radiation is thought of as photons, then the energy associated with each photon is related to the wavelength of the radiation:

$$E = \frac{hc}{\lambda}$$

... (13.58)

where,

E   =   the energy in each photon, J
h   =   Planck's constant (6.6×10^{-34} J·s)
e   =   speed of light, m/s
λ =   wavelength of radiation, m

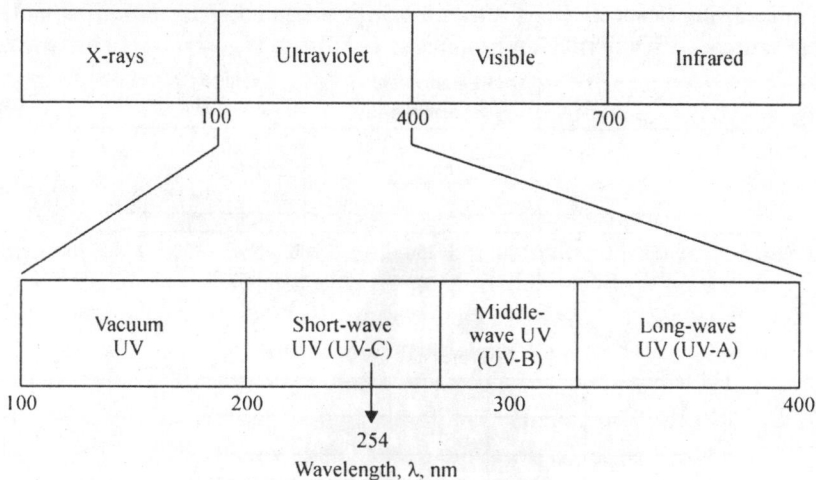

Fig. 13.31. Understanding UV light: (a) the electromagnetic spectrum; (b) output from low-pressure UV lamp; and (c) output from medium-pressure UV lamp.

As a general rule, the more energy associated with each photon in electromagnetic radiation, the more dangerous it is for living organisms. Thus, visible and infrared light have little affect on organisms, whereas both X-rays and gamma rays can be quite dangerous. Beyond these broad considerations, there are other factors that determine the fraction of the UV spectrum that is effective in disinfection. The portion of the UV spectrum that is more effective in disinfection is called the 'germicidal range'. On the lower end, the germicidal range is limited by the absorption of UV radiation by water. As wavelengths decrease, water becomes an increasingly efficient barrier for UV. For practical purposes, vacuum UV, the fraction of UV with a wavelength below 200 nm, cannot penetrate water. So radiation having a wavelength of 200 nm or less is not considered germicidal. It is also well established that UV inactivates micro-organisms by transforming their DNA. This transformation cannot happen unless the UV is at a wavelength at which DNA will absorb it and this absorption does not occur above wavelengths of approximately 300 nm. Therefore the germicidal range for UV. is between approximately 200 and 300 nm (Fig. 13.32a).

Sources of Ultraviolet Light

The UV disinfection units currently used in the water industry employ three different types of lamp technology: (i) low-pressure, low-intensity lamps; (ii) low-pressure, high-intensity lamps (also called low-pressure, high-output lamps); and (iii) medium-pressure, high-intensity lamps.

Low-pressure, low-intensity lamps are most common. The design of these lamps closely approximates that of the common fluorescent light bulb. Low and medium-pressure, high-intensity lamps are new technologies that are able to achieve a higher UV output in an equivalent space. Of the three technologies, medium-pressure UV has the greatest output. The spectrum of the UV light output by both types of low-pressure lamps is essentially the same, a very small amount of the light energy emanating at a wavelength of 188 nm and the vast majority of it emanating at a wavelength of 254 nm. The spectrum of the UV light output by medium-pressure lamps includes a number of wavelengths. These spectra are illustrated and compared with the germicidal range in Figs. 13.32b and 13.32c.

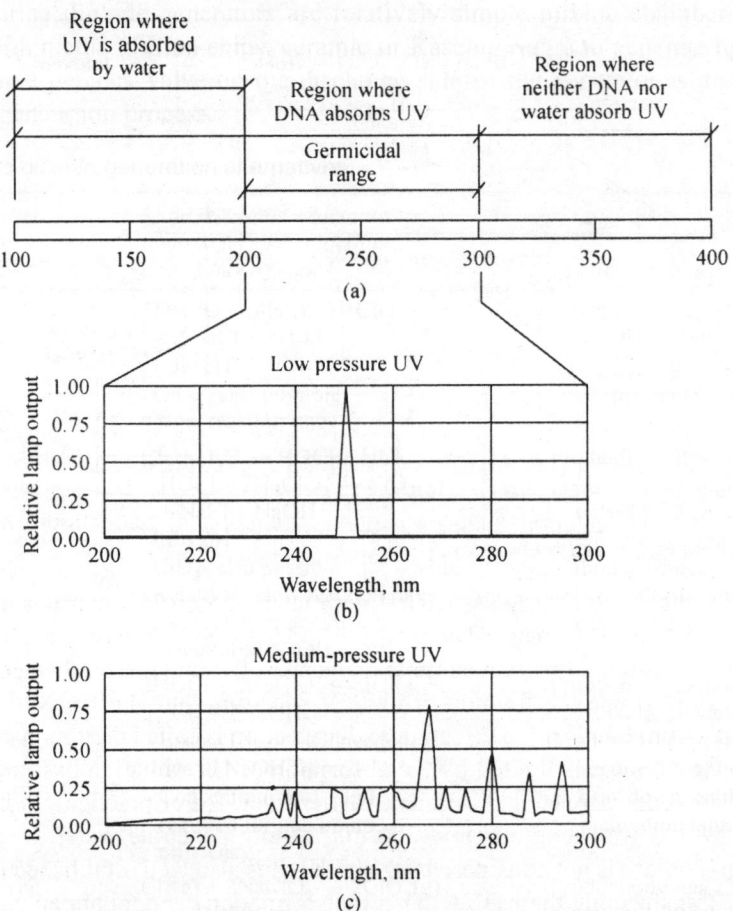

Fig. 13.32. Ultraviolet sources and germicidal range: (a) ultraviolet portion of electromagnetic spectrum; (b) output from low-pressure UV lamp; and (c) output from medium-pressure UV lamp.

Several important characteristics of each of these UV lamps are compared in Table 13.9; however, it must be noted that UV lamp technology is in a state of rapid development. One of the design engineer's more important challenges will be to evaluate those technologies available at the time a design is prepared and to write specifications that will enable new technologies while protecting the owner against innovative but unproven alternatives where the prospect for failure can be significant. Promising technologies emerging at the time of this writing include pulsed UV and narrow-band excimer UV. The pulsed UV lamp produces polychromatic light at very high intensity and the narrow-band excimer lamp produces nearly monochromatic light at wavelengths of 172, 222 and 308 nm.

Table 13.9. Characteristics of three types of UV lamps.

Characteristics	Unit	Type of lamp		
		Low pressure, low intensity	Low pressure, high intensity	Medium pressure
Estimated lamp life	h	8000–10000	8000–10000	3000–5000
Vapour pressure in lamp	Torr	0.007	0.001–0.01	100–10000
Lamp operating temperature	°C	35–45	80–100	400–800
Germicidal output/input	%	30–40	25–35	10–15
Power/lamp	W	40–100	200–500	1000–10000
Sleeve life	yr	4–6	4–6	1–3
Ballast life	yr	10–15	10–15	1–3

Mechanism of Inactivation

More is known about the specific mechanisms of disinfection by UV than for any other disinfectant used in water treatment. The photons in UV light react directly with the nucleic acids in the target organism, damaging them. The genetic code that guides the development of every living organism is made up of nucleic acids. These nucleic acids are either in the form of deoxyribonucleic acid (DNA) or ribonucleic acid (RNA). The DNA serves as the databank of life while the RNA directs the metabolic processes in the cell. Ordinarily DNA is a double-stranded helical structure that includes the nucleotides adenine, guanine, thymine and cytosine. Ordinarily RNA is a single-stranded structure with the nucleotides adenine, guanine, uracil and cytosine. Ultraviolet light damages DNA by dimerising adjacent thymine molecules, inhibiting further transcription of the cell's genetic code (see Fig. 13.33). While not usually fatal to the organism, such dimerisation will prevent its successful reproduction. Ultraviolet light also forms cytosine-cytosine and cytosine-thymine dimers, but these have a lower quantum yield (they occur less frequently).

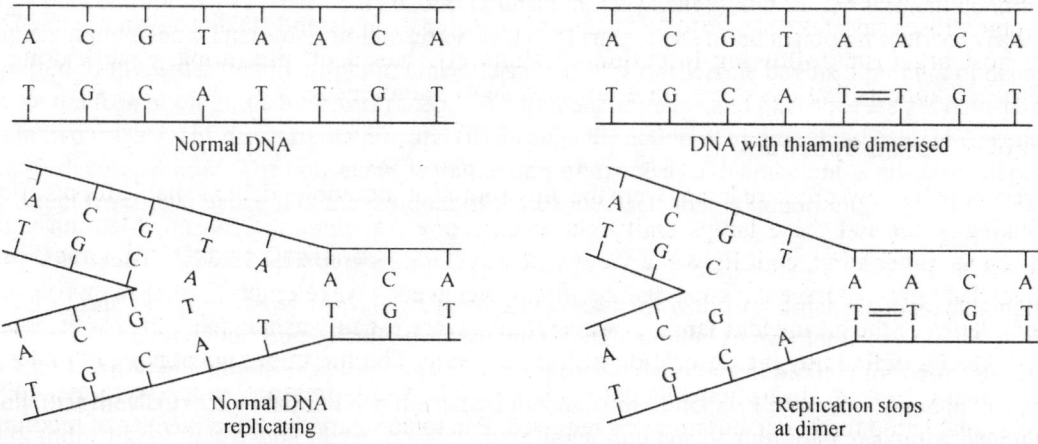

Fig. 13.33. Formation of thiamine dimers by UV light interferes with normal replication of micro-organisms.

As a result, organisms rich in thymine tend to be more sensitive to UV irradiation. For example, *C. parvum* and *G. lamblia* both contain DNA and both are inactivated by UV at relatively low doses (see Table 13.3). Most viruses of significance in drinking water have only RNA (which contains uracil instead of thymine) and thus are less sensitive to UV radiation. Among the most resistant organisms are viruses like rotavirus and adenovirus, which incorporate a special double-stranded RNA. Other factors also influence the rate of inactivation and some are not as well understood.

Ultraviolet radiation can also cause more damage of a more severe kind, breaking chains, cross-linking DNA with itself, cross-linking DNA with other proteins and forming other by-products. These effects have an even lower quantum yield and they are usually observed only at high doses of irradiation.

Re-activation

Re-activation is a more important consideration in UV disinfection than it is with disinfection by other methods. It is important to note that most forms of life evolved with some exposure to the sun and that sunlight includes significant amounts of UV irradiation. As a result, the process of evolution has addressed UV-induced damage by generating mechanisms for repairing the damage it causes. These mechanisms fall into two basic classes: (i) photo-re-activation; and (ii) dark repair. Photore-activation only takes place in the presence of light whereas dark repair has no such requirement. Organisms capable of dark repair generally show much greater UV resistance; however, understanding the importance of photore-activation requires that special tests be conducted, evaluating samples with and without light exposure to understand its effects. Certainly when water is being disinfected for discharge into the environment, only the net inactivation after photore-activation is important. Even in the case of drinking water systems, where light exposure is often more limited, the most conservative approach is to consider photore-activation as well.

Photore-activation was discovered by Kelner (*E. coli*) and Dulbecco (bacteriophage). It is now understood to involve the cleaving of the nucleic acid dimers with the enzyme DNA photolyase. The enzyme first adsorbs to the dimer and then cleaves it with the assistance of photons in the visible wavelength. Eventually, it will probably be possible to determine if an organism is capable of photorepair by using its genetic fingerprint to map its position on the evolutionary tree. It is not safe to assume that any organism is incapable of photorepair; unless through testing it has been demonstrated to be the case. Even some viruses have been shown to be capable of photorepair, apparently taking advantage of enzymes in the host organism following infection. Fortunately, based on preliminary work done with *C. parvum*, it appears that this organism is not capable of photorepair.

Concept of Action Spectrum

Until recent years, low-pressure lamps were the only source of ultraviolet light available for disinfection of drinking water and these lamps emit light at only one wavelength, 254 nm. Medium-pressure lamps, on the other hand, emit light at a variety of wavelengths (see Fig. 13.32c). There is no reason to expect that light will have the same disinfecting power at each wavelength. Earlier in this discussion the boundaries of the germicidal range of wavelengths were broadly established, the lower boundary (200 nm) being defined by the absorption of light by water and the upper boundary (300 nm) being defined by the lack of absorption of light by DNA. When low-pressure lamps were the only UV source available, no further discussion was required. But to compare the effectiveness of medium and low-pressure lamps for disinfection, a better understanding is required. The possible significance of UV irradiation at different wavelengths in disinfection has been of interest for some time. As a result, a number of researchers have looked at this issue. These results are generally expressed in the form of an action spectrum.

To generate the action spectrum, a modification of Eq. 13.5 for UV light of a particular wavelength λ can be used:

$$\left(\frac{dN}{dt}\right)_\lambda = -N\Lambda_\lambda I_\lambda \qquad \qquad ...\,(13.59)$$

where,

$(dN/dt)_\lambda$ = rate of change in number of organisms exposed to light of wavelength λ

N = number of organisms exposed to light, organisms/100 ml

Λ_λ = coefficient of specific lethality for light of wavelength λ, m²/J

I_λ = intensity of light at wavelength λ, W/m²

The action spectrum is a representation of Λ_λ over a range of wavelengths. Often it is displayed as a plot of the ratio $\Lambda_\lambda/\Lambda_{254nm}$ versus wavelength. The action spectrums for *C. parvum* and MS₂ caliphage are compared with the absorption spectrum for DNA in Fig. 13.34. A close correlation between Λ_λ and DNA absorption is observed. The action spectra of a number of organisms have been determined and are similar to the results shown in Fig. 13.34. As a result, many scientists believe that the germicidal efficiency determined for one species of micro-organism to medium-pressure UV may be used to represent the relative response of other micro-organisms as well.

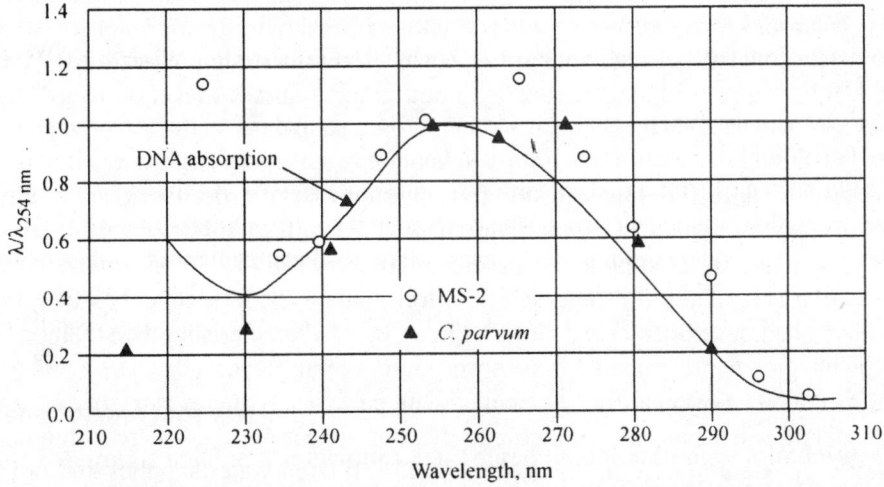

Fig. 13.34. Comparing action spectra for *C. parvum* and MS-2 coliphage with absorption spectrum for DNA.

Influence of Water Quality

The quality of the water being treated can have an important influence on the performance of UV disinfection systems. The two most important impacts stem from the action of dissolved and suspended substances.

Dissolved substances

Pure water absorbs light in the lower UV wavelengths. A number of dissolved substances also have important influence on the absorption of UV irradiation as it passes through the water on its way to the target organism. Among the more significant are iron, nitrate and natural organic matter.

Chlorine, hydrogen peroxide and ozone can also have important effects.

The absorption of light in aqueous solution by dissolved substances is described by the Beer-Lambert law. This relationship, takes the form:

$$\log\left(\frac{I}{I_0}\right) = -\varepsilon(\lambda)Cx \qquad \ldots (13.60)$$

where,

I = light intensity at distance x from light source, mW/cm^2

I_0 = light intensity at light source, mW/cm^2

C = concentration of light-absorbing solute, moles/l

x = light path length, cm

$\varepsilon(\lambda)$ = molar absorptivity of light-absorbing solute at wavelength λ, l/mole·cm

The term on the right-hand side of Eq.13.60 is defined as the absorbance A, which is unitless. As already discussed, the absorbtivity is the absorbance corresponding to a path length of 1 cm, or

$$k(\lambda) = \varepsilon(\lambda)C = \frac{A}{x} \qquad \ldots (13.61)$$

where,

$k(\lambda)$ = absorbtivity, cm^{-1}

The absorbtivity of the water is an important aspect of UV reactor design. Water with higher absorbtivity absorb more UV light and need a higher energy input for an equivalent level of disinfection.

Particulate matter

Particulate matter can also interfere with the transmission of UV light. Particulates are an aspect of water quality that can be particularly important where UV disinfection is concerned. Two mechanisms of particular importance are shading and encasement, as shown in Fig. 13.35. Interference of this kind has been studied at great depth for the case of coliform organisms in secondary waste-water effluents and models have been developed that do an excellent job of characterising the situation. The effect of shading can be integrated into models for the absorption of light. Beyond that, the number of organisms is dominated by the effect of organisms associated with particles. Particles can 'shade' target organisms from UV light via three mechanisms: refraction, reflection and scattering. Where filtration is used, these effects are not very important, but in the treatment of unfiltered water supplies and unfiltered waste-water effluents, these effects can be quite significant.

The effects of particle shading are not particularly significant at low turbidities, as illustrated by the work of Oppenheimer, who examined the inactivation of *G. muris* added to water with turbidities ranging from 0.65 to 7 NTU (see Fig. 13.36). A collimated beam apparatus was used to study the inactivation of *G. muris* with water at three different turbidity levels ranging from 0.65 to 7 NTU. After the UV dose was corrected for apparent absorbance (absorbance including the effects of particle shading), turbidities at these levels seemed to have little significance.

The impact of particulate matter is important with regard to the disinfection of secondary effluent from waste-water treatment facilities, especially when doses above $300 J/m^2$ are employed. At these doses and higher, a tailing phenomenon is observed that can be directly related to the number of particles that have coliform organisms associated with them. Models have been developed that do a good job of characterising this phenomenon.

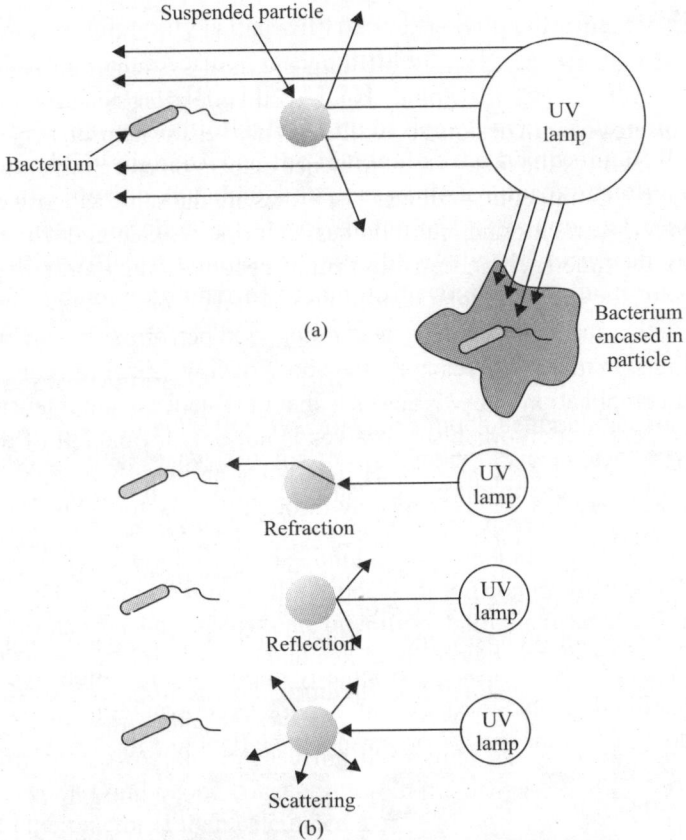

Fig. 13.35. Illustration of mechanisms for interference in disinfection by particles: (a) overview of mechanisms for interference; and (b) mechanisms of 'shading'.

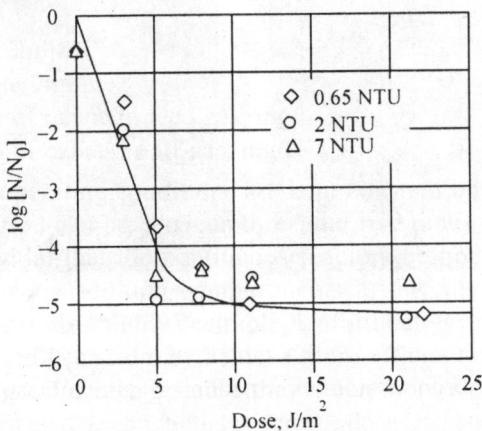

Fig. 13.36. Impact of low levels of turbidity on inactivation of *G. muris* with UV irradiation.

Influence of Dispersion

Ultraviolet disinfection systems, particularly medium-pressure systems, are characterised by overall contact times that are much shorter than other kinds of disinfection systems. In these systems short circuiting and dispersion are difficult design issues. Designing these systems to achieve good performance requires a greater appreciation of the factors that influence dispersion and short circuiting than is required for the design of most other disinfection systems. The issues are the same as those discussed earlier with contactors for disinfection with chlorine, chloramines, chlorine dioxide and ozone; however, with UV disinfection contactors, the time spent in transition zones becomes much more important.

In chlorine contactors, for example, inlet conditions can have a big influence on performance. If the contactor is designed with a sufficiently long aspect ratio, good performance can be achieved in spite of non-ideal inlet conditions. In many UV reactors, the zones of flow transition can dominate most of the contact time. A further complication in UV reactors is that the fluence (light intensity) varies throughout the reactor. So the UV dose that an organism receives is not only a function of the length of time the organism spends in the reactor and the amount of light being emitted by the UV lamps but also of the specific path the organism takes as it makes its way through the reactor. Thus, the issue is not just the contact time the organism receives, but its cumulative exposure to UV. The results of a model developed by Chiu that simulates the path of an organism through a hypothetical UV reactor are illustrated in Fig. 13.37. As shown in Fig. 13.37, the dose the organism gets depends not only on the intensity of the lamps and the time the organism spends in the reactor but also on the specific path the organism takes through the reactor. Such considerations are particularly important in medium-pressure systems, which only use a small number of lamps. In the case of the two traces shown in Fig. 13.37b, the organism on the left was exposed to a dose of 14 J/m^2 and the organism on the right to 138 J/m^2.

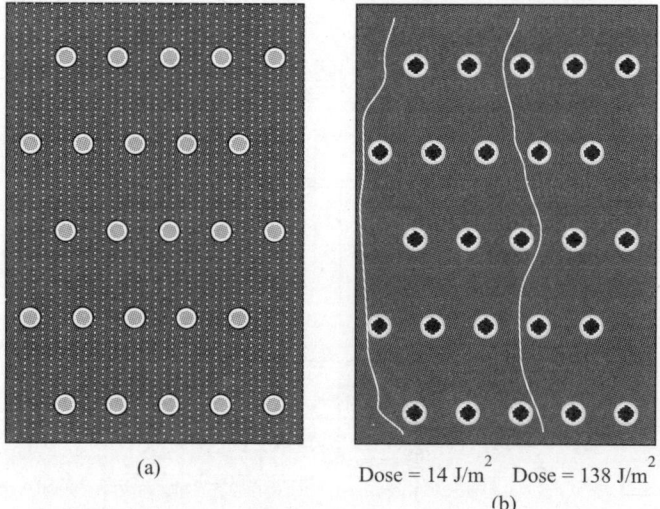

$$(a) \qquad \text{Dose} = 14 \text{ J/m}^2 \quad \text{Dose} = 138 \text{ J/m}^2$$

$$(b)$$

Fig. 13.37. Travelling through UV contactor: (a) flow pattern; and (b) UV intensity and two alternative particle tracks.

Ultraviolet light contactors are also small with respect to their inlet—so small that it is often extremely difficult to conduct a meaningful tracer study. The outcome of the study often depends on precisely where the tracer is introduced. As a result, it is increasingly common for regulators to require full-scale tests of each reactor design to establish, by actual disinfection measurements, how much of a UV dose the design will be credited with delivering. This method of establishing reactor capacity was originally proposed by Qualls and Johnson and is termed biodosimetry. Given biodosimetry to establish the dose credit a UV unit gets, inlet and outlet conditions become a critical aspect of the design of the system that delivers water to the UV unit and transports it away again. The same considerations apply that were discussed in the thought experiment described in Fig. 13.6, but in the case of UV reactors, short circuiting involves not only the time the organism spends in the reactor but also the specific route the organism takes and the cumulative dose it accumulates.

Biodosimetry

The concept of biodosimetry was proposed originally by Qualls and Johnson. The basic technique, illustrated in Fig. 13.38, involves conducting both bench-scale and field-scale tests with a biological test organism. The field-scale test is conducted at design flow and under conditions designed to represent a conservative simulation of full-scale operation. The specifics of the conduct of this test are outlined in the appropriate guidelines.

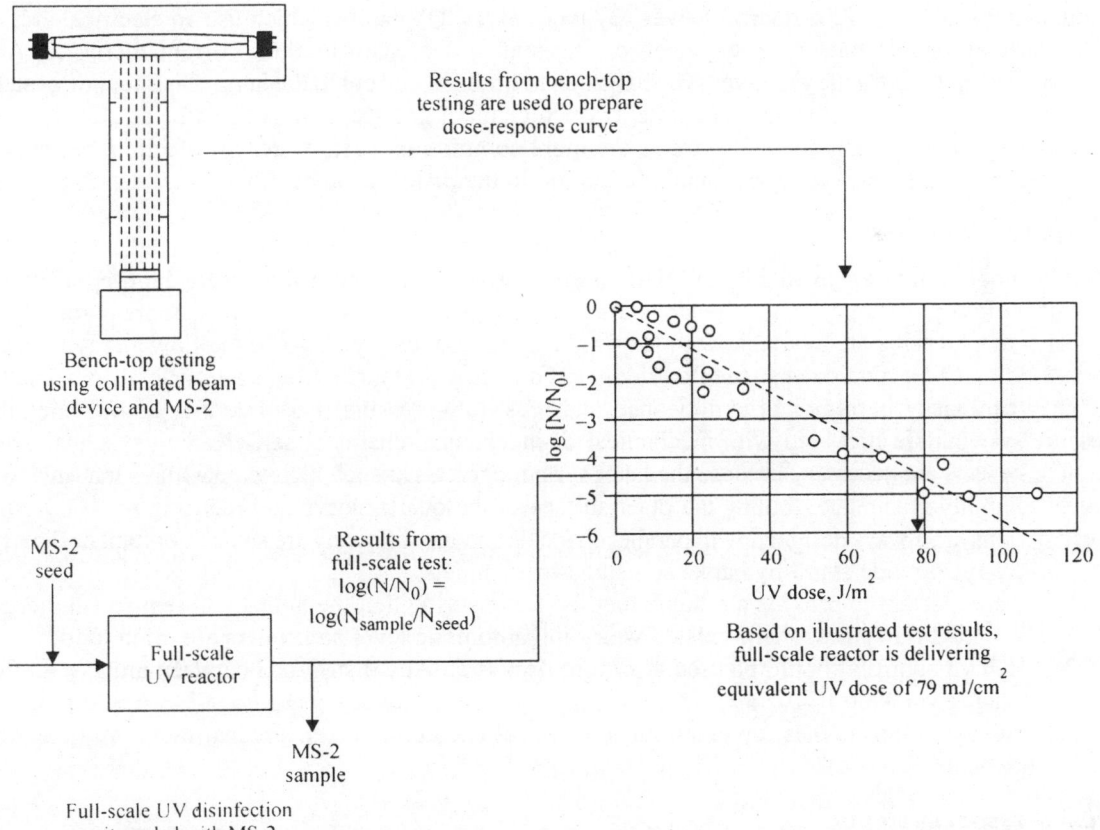

Fig. 13.38. Schematic illustration of the application of biodosimetry as used to determine the performance of a full-scale UV reactor.

The disinfection dose that the unit is credited is determined by the dose that accomplishes the same level of inactivation under carefully controlled laboratory conditions. Because there are so many complications in determining the performance of a given UV reactor design; biodosimetry is used to calibrate the expected performance of full-scale units. Biodosimetry involves the following steps: (i) a laboratory study is conducted to establish the relationship between UV dose and the inactivation of a test organism; and (ii) a field experiment is conducted under design conditions and the same organism is used to establish the performance of the field disinfection unit. The observed survival of the test organism in the full-scale test is compared to the curve established in the laboratory tests to establish the UV dose for which the field unit will be given credit.

Biodosimetry is most effective when it is conducted with an organism that shows approximately the same resistance as the target organism. Establishing the rated fluence of a UV system for an organism with significantly greater resistance increases the risk that the effects of poor dispersion will go unobserved.

Equipment Configurations

A UV disinfection system consists of: (i) the UV lamps; (ii) transparent quartz sleeves that surround the UV lamps, protecting them from the water; (iii) the structure that supports the lamps and sleeves and holds them in place; (iv) the power supply for the system; and, in many cases; (v) the cleaning system to maintain transparency of the quartz sleeves. By themselves, UV lamps, which use an electrical arc, are not electrically stable because their electrical resistance decreases as their current increases. As a consequence, the electrical system must be ballasted to limit the current to the lamp. Disinfection systems are also classified by whether they operate as open-channel gravity flow systems or as closed-vessel pressurised systems. Open-channel systems are more common in waste-water disinfection, but closed-vessel systems are receiving a great deal of attention in the disinfection of drinking water.

Open-channel systems

Open-channel designs are available for low-pressure, low-intensity and low pressure, high-intensity UV systems. In low-pressure systems, the UV lamps are retained in modules or racks that are placed in the flow channel. Designs are available with lamps placed horizontally parallel to the flow and with lamps placed vertically perpendicular to the flow. Conventional low pressure systems are typically designed so that they can easily be removed and cleaned and most low-pressure, high-intensity and all medium-pressure systems are provided with mechanical or mechanical/chemical self-cleaning systems. These cleaning systems are necessary because the latter system operates at such high temperatures that salts with inverse solubility precipitate, fouling the outer surface of the quartz sleeve and reducing net UV output. These systems are always designed with parallel channels and the following are some important design tips:
1. Always provide stand-by banks and stand-by channels.
2. It is important to provide a reliable method for ensuring that the lamps will remain submerged. Weighted flap gates, sharp-crested weirs and automatic level controllers are often used.
3. Positive controls should be used to ensure flow is equally distributed between units (e.g., flow division via weirs).
4. Inlet conditions to these channels must be designed carefully to ensure that short circuiting does not occur.

Closed-vessel systems

Whereas most low-pressure systems are designed with open-channel flow, many low-pressure, high-intensity and medium-pressure systems are designed using closed vessels. These closed-vessel systems have the advantage that they can (and usually do) operate under pressure and this feature makes them particularly attractive in upgrades and retrofits because it is not necessary to 'break head' to use them. These systems, particularly the medium-pressure systems, operate with such a small number of lamps that a great deal more care is required to ensure that short circuiting does not occur. Biodosimetry standards are evolving that will ensure that the individual units perform as specified, but care must be taken to ensure that influent and effluent conditions do not contribute to short circuiting. Some important design tips for closed-vessel system are as follows:

1. All UV reactor designs should have their performance certified via a standard biodosimetry test conducted according to a widely accepted guidelines.
2. Always provide stand-by reactors.
3. Make sure that it is not possible for these units to fire up when dry or become dry while lit.
4. Positive controls should be used to ensure flow is equally distributed between units.
5. Inlet conditions to reactors must be carefully designed to ensure that short circuiting does not occur.

Example 13.1: To obtain 99.70 per cent kill of bacteria, the ozone is to be used in water with a residual of 0.6 mg/l. The reaction constant under these conditions is 3×10^{-2} per second. Calculate the contact time period.

Solution: Now, 99.70 per cent of bacteria are killed i.e., 0.3 per cent of bacteria remains in the water after ozonation.

In the water, the concentration of bacteria is 100 mg/l and after ozonisation, it is 0.30 mg/l. But residual of 0.60 mg/l is given.

$$N_o = \text{Number of organisms initially}$$

$$= 100 \times \frac{0.60}{0.30} = 200 \text{ mg/l}$$

$$N_t = \text{Number of organisms at time t}$$

$$= 0.60 \text{ mg/l}$$

Now,

$$t = \frac{1}{K} \log_{10} \frac{N_o}{N_t}$$

$$= \frac{1}{3 \times 10^{-2}} \log_{10} \frac{200}{0.60} = 84 \text{ seconds.}$$

Example 13.2: A town having population of about 50000 is to be supplied water at the rate of 150 liters per capita per day. The disinfection of water is to be carried out with bleaching powder containing 30 per cent of active chlorine. If the chlorine dose required for infection is 0.3 ppm or 0.3 mg/l, calculate the quantity of bleaching powder per year.

Solution:

$$\text{Total requirement of water} = (150 \times 50000) \text{ liters/day}$$

$$= 7.5 \times 10^6 \text{ liters/day}$$

$$\text{Chlorine dose required for disinfection} = 0.3 \text{ mg/l}$$

Therefore,

$$\text{Quantity of chlorine required} = 0.3 \times \frac{7.5 \times 10^6}{10^6} \text{ kg/day}$$

$$= 2.25 \text{ kg/day.}$$

Now, the bleaching powder contains 30 per cent of active chlorine. It means that 30 kg of chlorine is available from 100 kg of bleaching powder.

Therefore,   Quantity of bleaching powder required $= \dfrac{2.25}{30} \times 100$ kg/day

$$= 7.5 \text{ kg/day.}$$

Requirement of bleaching powder

$$\text{per year} = (365 \times 7.5) \text{ kg}$$
$$= 2737.5 \text{ kg, say 2750 kg.}$$

Example 13.3: The quantity of chlorine use to treat 20000 m^3 of water per day is 8 kg. The residual chlorine after contact period of 10 minutes is found to be 0.20 mg/l. Calculate the dosage in mg/l and the chlorine demand of the water.

Solution:

$$\text{Water treated per day} = 20000 \text{ m}^3$$
$$= 20000 \times 10^3 \text{ liters}$$
$$= 20 \times 10^6 \text{ ml.}$$

Therefore,   Chlorine consumed per day $= 8$ kg $= 8 \times 10^6$ mg

$$\text{chlorine used per liter of water} = \dfrac{8 \times 10^6}{20 \times 10^6} = 0.40 \text{ mg/l.}$$

The given chlorine dosage is therefore 0.40 mg/l.

Now, residual chlorine left = 0.20 mg/l.

Hence the actual chlorine dosage which has reacted in water i.e., the chlorine demand of water is equal to $(0.40 - 0.20) = 0.20$ mg/l.

Chapter 14

Water Transmission and Distribution Systems

INTRODUCTION

Water transmission is the delivery of water from the source to the treatment plant and from the treatment plant to consumers through the distribution system. A distribution system is a network of water lines (pipes) that deliver treated water to consumers. A transmission system consists of the following major components:

1. Pipes (transmission lines) for delivering water.
2. Pumps for pumping and maintaining pressure.
3. Valves to control water flow.
4. Reservoirs and elevated tanks for water storage.
5. Meters to measure the quantity of water supply.
6. Fire hydrants to provide sufficient pressurised water for firefighting.

PIPES

Pipes deliver water to different places. Commonly, pipes are formed of cast iron, reinforced concrete, plastic or copper. Cast iron pipes are ductile or gray cast iron type. A ductile iron pipe differs from a gray cast iron pipe in that it is malleable (can be hammered into a sheet); therefore, it can withstand more stress. Cast iron pipes are corrosion resistant, strong and durable. For further corrosion control, they are lined with cement and tar. They are, mostly, larger than 2 inches in diameter, used for such things as lines to and from pumps. Reinforced concrete pipes are very large. Plastic pipes are light, non-corrosive and available in sizes ¼ to 8 inches. Copper lines, due to their flexibility, are used for service connections and for plumbing. Pipe size depends on population and future development of a community. Mostly it is possible to use 6 to 8 inch pipe to supply a residential area. Domestic service lines are generally ¾ to 1 inch plastic or copper; commercial lines are 2 inches.

PUMPS

A pump is a device for transportation of a liquid from one location to another, for lifting the liquid to a higher elevation or for increasing the liquid pressure. Water is taken from the source to the treatment plant and transported to the customers, under pressure, by the pumps. Good understanding of pumps and pumping is important for an adequate pressurised water supply for domestic use, for commercial use and for firefighting.

Following are the purposes for which pumping is adopted in water supply schemes:
1. To increase or boost pressure at certain points in the distribution system.
2. To lift clear water after treatment to an elevated storage reservoir.
3. To lift raw water from lake, reservoir or river for carrying it to a treatment plant.
4. To lift water available from wells to an elevated storage tank in stages.
5. To make available water at higher pressure during certain processes of treatment.
6. To take out water from basins, sumps, tanks, etc.
7. To throw water directly into the distribution system.

In this chapter, the types of pumps used in water supply schemes will now be briefly described.

Choice of Type of Pumps

Following considerations govern the choice of a particular type of pump in water supply project:
1. Capacity of pumps.
2. Importance of water supply scheme.
3. Initial cost.
4. Location of pump.
5. Maintenance cost.
6. Number of units required.
7. Quality of water to be pumped.
8. Total head of water.
9. Type of power available.
10. Type of supply service—intermittent or continuous.
11. Variations in pumping head and pumping rate.
12. Working or operating conditions such as flexibility in operation, requirement of floor area.

Centrifugal Pumps

Centrifugal pumps are the most commonly used pumps in the water utility; thus, they are discussed in detail and other types only briefly. A centrifugal pump has an impeller that rotates in a casing, sucks water and slings it out by centrifugal force. These pumps are mostly used for lifting water from lower to a higher elevation, boosting pressure in the distribution system and for transportation of water to and from the treatment plant. Figure 14.1 shows the parts of a centrifugal pump: (i) impeller; (ii) shaft; (iii) casing; (iv) bearings and packing; and (v) motor.

Impeller is the rotating disc with vanes of various shapes for different types of flow, such as axial, radial or mixed, depending on the discharge flow type. Rotation of the impeller creates the centrifugal force, which creates a vacuum at the eye of the impeller. Water, then, is sucked from the source due to the atmospheric pressure and forced out to the discharge point (Fig. 14.2). Impellers, with open, semi-enclosed and enclosed vanes, cause a radial flow, which means water is discharged at a right angle to the suction. An impeller with vanes like a propeller causes an axial flow, which sucks water and forces it out in a straight line. Mixed flow or Francis flow, is caused by a conelike impeller with spiral vanes discharging water at a 45° angle from the suction; it is a combination of radial and axial flows.

Shaft is a steel, stainless steel or bronze rod to which an impeller is attached at one end and a motor, the prime driving force, is attached at the other end. The shaft transmits power from the motor to the impeller. It needs to be as straight as possible to reduce the friction losses and vibrations, which cause the higher motor loading; the shaft is kept in line by ball or roller bearings.

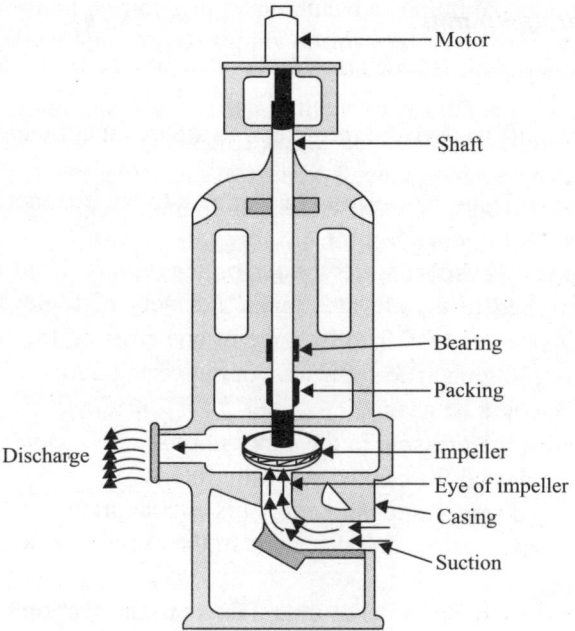

Fig. 14.1. Vertical section of a centrifugal pump.

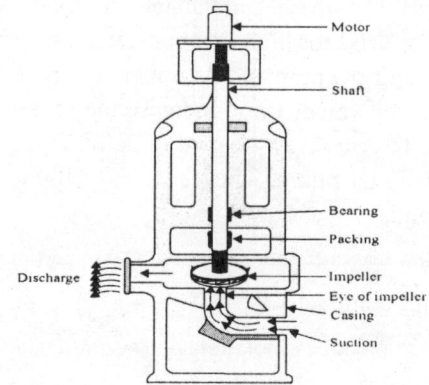

Fig. 14.2. Impeller rotation and discharge.

Casing is the housing for the impeller with the suction and discharge openings. Water enters at the suction opening and leaves at the discharge opening. The impeller housing is sealed off from the atmosphere, with a packing or a seal, to prevent air from getting into the water and excessive water leakage. The packing is formed of braided or woven cotton impregnated with graphite and metallic plastic. Some manufacturers use mechanical seals instead.

The motor is the prime mover of the shaft. It is connected to the shaft with a flexible coupling to reduce excessive wearing of shaft and bearings. Pump capacity is the flow rate or the amount of water delivered per time unit, such as gallon per minute (gpm). A centrifugal pump's capacity is mainly controlled by two factors, revolving speed of the impeller and its diameter.

Classification of centrifugal pumps

Classification of centrifugal pumps is based on the services they perform, the type of flow and the type of casing:

1. Low-service pumps lift water from the source to the treatment plant and from clear well to the filter backwash system.
2. Well pumps lift water from the well and discharge it to the treatment plant. They are multistage centrifugal pumps for high head.
3. High-service pumps transport water under high pressure from the treatment plant to the distribution system and to the elevated tanks. For very high elevations, two or more similar pumps are used in a series with the discharge of one entering the suction of the next. Suppose one pump delivers 200 gpm at 25 ft., head. For delivering 200 gpm to 100 ft. head, four pumps or stages (100/25) would be needed.
4. Booster pumps boost the pressure in the distribution.
5. Backwash pumps are used for backwashing the filters.
6. Sampling pumps send samples to auto analysers and laboratory.
7. Axial flow pumps have suction and discharge in the same direction. They have a propeller-like impeller.
8. Radial flow pumps discharge water at a right angle to the suction.
9. Mixed-flow pumps have discharge at 45° to the suction.
10. Volute pumps have the discharge pipe diameter gradually increasing, which decreases the velocity and increases the pressure. They have spiralshaped guiding vanes on the interior of the casing, called the volute. They are used for high lift, booster or service pumps.
11. Turbine pumps are diffused flow pumps. The impeller is in the center of the circular casing that has fixed vanes to reduce the velocity and increase the pressure.
12. Screw centrifugal pumps have a screw-shaped impeller that combines the characteristics of the screw pump and a centrifugal pump. They are very efficient for high heads and water with particulate matter like sand.

Advantages of centrifugal pumps

1. They are simple, light and compact.
2. They are vertical or horizontal.
3. They are quiet.
4. They produce uniform discharge.
5. They require low maintenance.
6. They don't have chambers with valves.
7. They are not damaged when operated with discharge valve closed.

Disadvantages of centrifugal pumps

1. They have low efficiency due to friction loss and slippage of water.
2. They are not self primed.
3. They have low suction lift, only 15 feet.
4. They have a problem of air leaks into the casing causing cavitation, which is the formation of air pockets due to the fusion of small bubbles. Cavitation causes corrosion.

Displacement Pumps

Displacement pumps have a mechanism such as a piston, plunger, diaphragm, gears, cams or a screw to force the liquid through the pump. They are used, mainly, for feeding chemical solutions and can be divided into two classes:
1. Reciprocating pumps.
2. Rotary pumps.

Reciprocating pumps

Reciprocating pumps have a mechanism to fill a chamber with liquid and to force it out by using a plunger or a piston. They create lift and pressure by displacing liquid. The volume or capacity of the pumped liquid is constant regardless of head. Chemical feed pumps may have two pistons or double-acting plungers that discharge liquid on both the forward and the backward strokes. Plunger pumps are useful for small quantities and high heads. They may have a single or double-acting cylinder with one chamber filling and other discharging with the same stroke of the piston (see Fig. 14.3). Capacity varies by speed (number of strokes/minute) and stroke size (percentage of stroke). These pumps are used for pumping chemical solutions such as liquid chlorine, liquid alum, polymers and potassium permanganate.

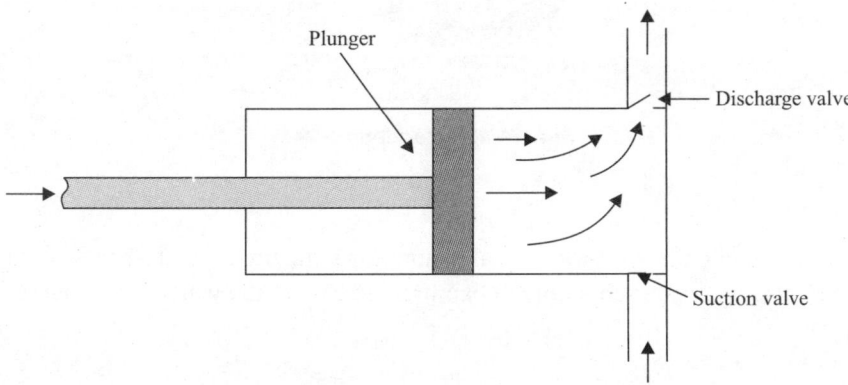

Fig. 14.3. Plunger pump.

Advantages of reciprocating pumps
1. They have high suction lift (up to 22 ft.).
2. They are self-priming.
3. Capacity is not affected by the head.

Disadvantages of reciprocating pumps
1. They are mostly heavy.
2. They are bulky with slow speed (50 to 150 strokes per minute).
3. They produce pulsating flow.
4. Cost is a concern, as they are expensive.
5. A closed discharge valve can cause serious damage to the pump.

Rotary pumps

Rotary pumps are quite simple in design. They have two cams or gears; one is connected to the shaft and is called the driving gear that drives the other gear, the idler gear. A close-fitted casing surrounds the gears and has the suction and discharge ports (see Fig. 14.4). The liquid fills the space around the gears. When the gears rotate, they mesh and squeeze the liquid out to the discharge port. When teeth separate, they create a partial vacuum and suck more liquid. Thus, the liquid is sucked in and squeezed out alternatively by each rotation. Like the reciprocating pumps, these pumps are self-priming, with 22 feet suction lift and a relatively high speed up to 1750 rpms.

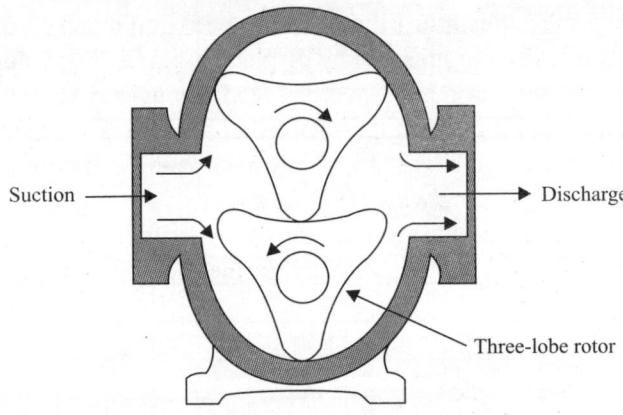

Fig. 14.4. Rotary pump.

Screw pump

Screw pump is a rotary pump that uses a screw-like pumping structure around a shaft instead of gears or cams. The screw rotates in a water containing chamber and carries the water in a progressive cavity to the discharge point.

Air Lift Pumps

In this type of pump, the compressed air is used to lift water. Figure 14.5 shows the section of a typical air lift pump. The air pipe supplies the compressed air from compressor. This compressed air is then released through an air diffuser or foot piece as shown in Fig. 14.5. The air diffuser is located at the bottom of eductor pipe.

The air rises in small bubbles in the eductor pipe and it forms a mixture of air and water. This mixture has low specific gravity than that of water and consequently, the pressure in the eductor pipe becomes less than the pressure in the casing pipe. This difference of pressure causes the water to rise in the eductor pipe and it ultimately results in water coming out through the outlet.

In the final stage, the weight of column of water from A to C equals the weight of column of air and the water from A to B. The length AC of eductor pipe represents the depth of submergence and it is important for the efficient working of the pump. The depth of submergence should be about one-half to two-third of its length.

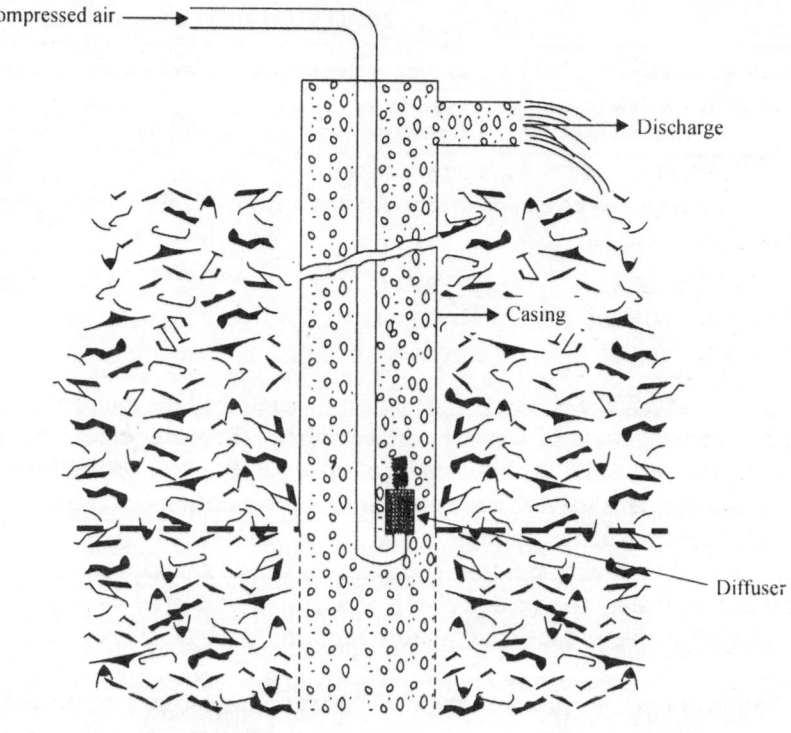

Fig. 14.5. Air lift pump

Advantages

Following are the advantages of the air lift pumps:
1. An air lift pump is found to be cheap, reliable and simple in operation.
2. There are no moving parts of the pump which are in contact with water. Hence this type of pump can be easily used for water containing mud, silt, debris, etc. as well as for highly acidic or alkaline water.
3. The pumping from a number of wells can be done by installing a common compressor unit.
4. The yield of well by this pump is increased by using more quantity of compressed air.

Disadvantages

Following are the disadvantages of the air lift pumps:
1. In order to achieve the length of submergence, the well is sometimes made deeper than required and this factor increases the cost.
2. The efficiency of air lift pumps is relatively low and it varies from 20 per cent to 45 per cent.
3. The nature of flow obtained from an air lift pump is not continuous. But it is intermittent. This is due to the fact that small air bubbles have a tendency to coalesce or to unite and the large bubbles thus formed rise more rapidly than the water.
4. The air lift pumps are less flexible in coping with the variations in demand.

Uses

The air lift pumps are generally adopted for pumping water from deep wells and they can be conveniently used for lifts of about 60 metres or so.

Miscellaneous Pumps

There are various other pumps which can be used to lift water from one level to the other. Following two types of pumps will be considered:

1. Hydraulic ram.
2. Jet pump.

Hydraulic ram

In this type of pump, the advantage is taken of the power generated by the momentum developed due to sudden stoppage of a moving mass of water. In the beginning, the water enters the ram through inlet pipe. At this stage, the delivery valve is closed and the waste valve is kept open. The waste valve is also known as the impetus valve. This valve is maintained in an open position automatically either by gravity or by some other arrangement such as spring.

The water from ram starts flowing out from the waste valve. As this happens, the water attains more and more velocity and ultimately the waste valve is suddenly closed. The momentum or impulse thus created opens the delivery valve and the water is now admitted to the air chamber from which it is lifted up in delivery pipe.

After some duration of time, the pressure in ram decreases. This causes closure of delivery valve and opening of waste valve. The cycle is then repeated. There may be 50 to 200 cycles per minute depending upon the design of ram. The efficiency of ram is about 50 per cent. Figure 14.6 shows a hydraulic ram.

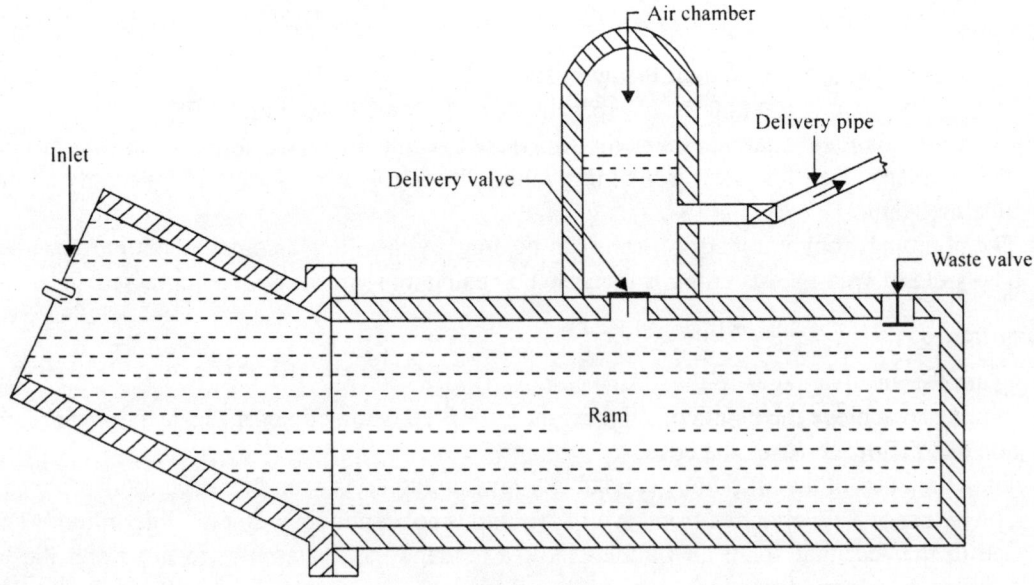

Fig. 14.6. Hydraulic ram.

Advantages

Following are the advantages of the hydraulic ram:
1. The working of ram is simple and once it starts functioning, practically no attention or supervision is required.
2. The ram is durable.
3. The ram is cheap or inexpensive as it does not require fuel for its working.

Disadvantages

Following are the disadvantages of the hydraulic ram:
1. There is considerable wastage of water through waste valve and hence the ram can be installed at places or locations where such loss of water can be tolerated.
2. The ram creates noise when it is working.

Uses

The ram can be used for small water supply projects. It is installed at a lower level than the stream or river and thus natural fall is developed.

The ram can be used to lift water for a height of about 30 meter or so.

Jet pump

In this type of pump, a nozzle is placed at the throat of a venturi tube and it discharges either compressed air or steam or water at high velocity. The jet causes suction and the water is drawn from suction pipe. The high velocity of mixture is converted into pressure head and the water is raised in the discharge pipe. Figure 14.7 shows a jet pump.

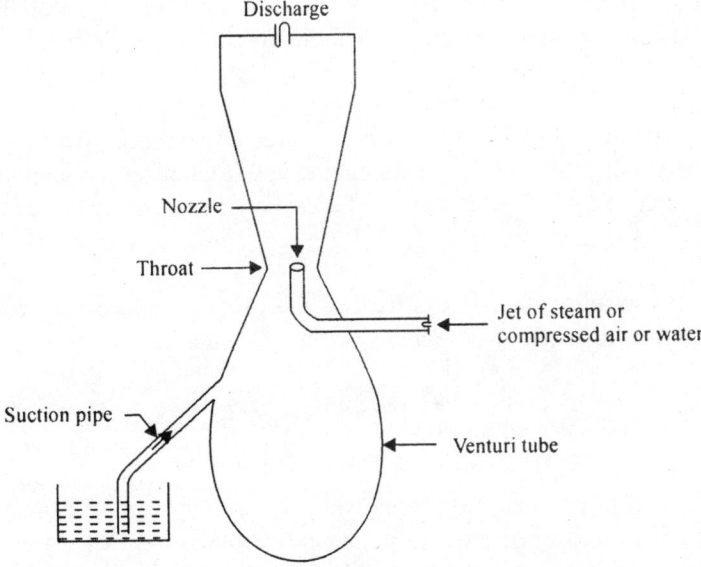

Fig. 14.7. Jet pump.

Advantages

Following are the advantages of the jet pump:
1. The jet pump is portable and it can be carried to any place for installation.
2. The jet pump can be used for water containing sand, silt, etc.

Disadvantages

Following are the disadvantages of the jet pump:
1. It requires compressed air or steam or water at high velocity. It is thus costly.
2. The efficiency of pump is very low i.e., about 15 per cent to 30 per cent.

Uses

The jet pump may be used for pumping water from deep wells with a capacity of about 50 liters per second.

Power for Pumps

Following machines are used to develop power for the working of pumps:
1. Steam engine.
2. Diesel engine.
3. Gasoline engine.
4. Electric motor.

Each of the above machine will now be briefly described.

Steam engine

The steam engine is clumsy and old fashioned. It is practically out of date at present. It consumes more fuel and takes more time to come to the working stage. Also there is considerable loss of energy when it is being stopped for cooling. It is, however, reliable and can be used for large installations where fuel is cheaply available. The efficiency of steam engine is about 60 per cent to 70 per cent.

Diesel engine

The initial cost of this type of engine is high and it produces noise during working. It requires skilled supervision. The engine is not so clumsy as steam engine and it takes up the load at once. It is reliable and consumes less quantity of fuel. The efficiency of diesel engine is about 70 per cent to 80 per cent.

Gasoline engine

In this type of engine, the gasoline or petrol is used as fuel. It is therefore very costly and hence it is rarely adopted. It is suitable for working as stand-by units of pumps.

Electric motor

This is the modern machine to create power for pumps and it is used for small and medium plants. The motor is compact in design and can be switched on immediately. It runs smoothly and it is free from dust and smoke. It is found to be cheap at places where electricity can be developed economically. However, in case of sudden failure of electric plant, the whole system of water supply comes to a standstill and causes great hardship. Hence it is advisable to keep diesel engines or gasoline engines as stand-by units along with electric motors. The efficiency of electric motors is about 90 per cent to 95 per cent.

VALVES

Valves are installed throughout water systems in treatment plants, pumping stations and pipe networks, as well as at storage reservoirs. Their purpose is to control the magnitude or direction of water flow. In order to regulate flow, all valves have a movable part that extends into the pipeline for opening or closing the interior passage. The four basic kinds of valves are slide, rotary, globe and swing; others less common are sphere, diaphragm, sleeve and vertical-lift disk. Valves are also commonly classified by operating purpose (for example, shut-off and altitude) and function (by-pass and flow control) without regard to the kind of device used. Since the primary purpose of this discussion is to describe the use of valves in water systems, the method of presentation is by function with illustrations showing a commonly used kind of valve. Other kinds may often be used for the same purpose.

The means of operating the movable element of a valve are by screw, gears or water pressure. Screw stems are common in gate, globe and needle valves and can be opened or closed by a manually operated handwheel or by a powered operator. In some designs the screw stem rises as the shut-off element closes and in others the element rides up the screw inside the body of the valve as the stem turns; the latter non-rising stem is more common. In large valves, where water pressure prevents use of a screw, a gear train can be employed to allow the shut-off element to be moved slowly by applying a minimum of torque to the stem. The gear-system may be operated manually or by electric, hydraulic or pneumatic power operators. The kinds of valves that may be equipped with a geared operator include butterfly, gate, globe and ball. Water pressure can open or close some kinds of valves by direct pressure on the movable element. The simplest is a hinged swing gate that opens under water pressure and closes under the influence of gravity or back pressure. Another example is the automatic globe valve with a body design that controls movement of the shutoff element by system water pressure.

Shut-off Valves

Valves to stop the flow of water through a pipeline are the most abundant valves in a water system. Pipe networks are sectionalised by installation of shut-off valves so that any area affected by a main break or pipe repair can be isolated with a minimum reduction in service and fire protection. Depending on the district within the city and size of the water mains, valve spacings range from 500 to 1200 ft. (150 to 370 m). Ideally, a minimum of three of the four pipes connected at a junction are valved and arterial mains have a shut-off every 1200 ft. The pipe connecting a fire hydrant to a distribution main contains a valve to facilitate hydrant repair. In treatment plants and pumping stations, shut-off valves are installed in inlet, outlet and by-pass lines so that valves and pumps can be removed for maintenance and repair. Gate valves are the shut-offs; however, rotary butterfly valves may be installed in large diameter pipes.

A gate valve has a solid sliding gate that moves at right angles to the direction of water flow by a screw-operated stem. When installed in a pipeline, the gate is drawn up into the housing of the case by a non-rising stem. The gate is lowered to fit snugly against sides and bottom to block water flow. The modern valve shows the gate encapsulated with rubber and the inside of the valve body coated with epoxy. Guides fit into slots on both sides of the gate to keep it in alignment so the rubber sealing surfaces are compressed when the gate is closed to prevent leakage. Older gate valves, still commonly found in distribution pipe networks, are double-disk, parallel-seat valves constructed of cast iron and subject to incrustation and corrosion. Normally, underground valves in mains are provided with a valve box extending to the street surface. The valve is opened or closed, by using an extension rod to reach down into the enclosure and turning the nut located on top of the valve stem. Larger gate valves may be placed in vaults or manholes to facilitate operation and maintenance. They may be equipped with small by-pass valves to reduce pressure differentials on opening and closing and frequently have spur or beveled gearing to reduce the force required for operation.

A butterfly valve has a movable disk that rotates on a spindle or axle set in the shell. The circular disk rotates in only one direction from full closed to full open and seats against a ring in the casing. The main disadvantages of a butterfly valve result from the disk always being in the flow stream, restricting the use of pipe-cleaning tools. On the other hand, the advantages of this valve are tight shut-off, low head loss, small space requirement and throttling capabilities. The latter is one of the most popular applications of butterfly valves, for example, use in a rate of flow controller to regulate the rate of discharge from gravity filters in a water treatment plant. Recently and more often in larger sizes; rubber-seated butterfly valves are being used in distribution systems. Because pressure differences across a butterfly valve disk tend to close the valve, a mechanical operator must be employed to overcome the torque in opening the valve and to resist this force during closing to prevent slamming. For treatment plant applications, the operator is often a hydraulic cylinder and piston rod assembly used to hold the valve disk in any intermediate position, as well as in opening and closing action. Operators for large valves may also be motor driven.

Check Valves

Check valves are close-fitting platelike structures that are used to control the direction of the flow. They are swing type in horizontal pipes and lift type in vertical pipes. A lift-type check valve is called a foot valve when located at the bottom of a suction lift line of a pump to keep the pump primed (full of water). It is a disc hinged to the wall of a pipe and opens only in the direction of suction.

Small Pressure-Reducing and Pilot Valves

The function of these valves is to reduce high inlet pressure to a pre-determined lower outlet pressure. The applications of small pressure-reducing valves, manufactured in sizes from 0.5 to 2 inch, are to protect house plumbing from excessive main pressures and as pilot valves to control the action of large automatic valves that are operated by water pressure. By control of a variable spring, flow through the valve is regulated causing a head loss resulting in a pressure differential between the inlet and outlet.

Automatic Control Valves

An automatic valve uses water pressure in the system to operate the movable element to close or open the valve. In large valves, an external pilot valve or a spring-loaded diaphragm assembly directs water pressure to the proper side to set the movable element in the valve body. Sensing devices for signalling valve operation may be hydraulic, electric, pneumatic or solenoid. Of the many applications for automatic valves, the common units in a water system are pressure-reducing, altitude, controlled check, surge relief and shut-off valves.

The operating principle of one type of automatic globe valve is shown schematically in Fig. 14.8. The movable element in the upper chamber of the valve is shaped like a piston, with the top surface area greater than the bottom area. When the chamber above the piston is exhausted to the atmosphere, the piston is held open by inlet water pressure in the body of the valve (Fig. 14.8a). Changing the position of the three-way valve on top closes the exhaust and allows inlet water to enter the chamber above the piston. The same magnitude of pressure exerted against both the top and bottom of the piston creates a greater downward force, since the top area of the piston is greater than the bottom area. Thus, when inlet water is allowed to enter the upper chamber, the valve automatically closes (Fig. 14.8b). Air moving in or out of the piston chamber passes through a small opening to slow the movement of the piston and prevent water hammer. The three-way valve on top is usually controlled by a hydraulic or solenoid pilot valve.

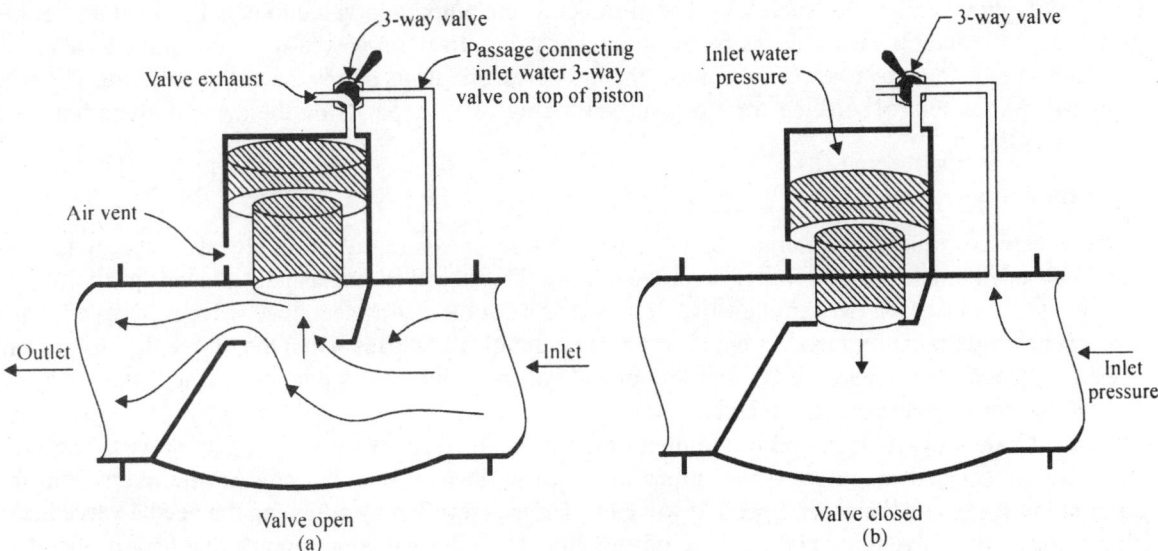

Fig. 14.8. Diagrams illustrating the operation of an automatic globe valve. (a) The valve is held open by inlet water pressure when the chamber above the piston is vented to the atmosphere; and (b) the valve is held closed by inlet pressure exerting a greater force on the top of the piston.

Pressure-Reducing Valves

Pressure-reducing valves are automatic valves operated by pilots to maintain a pre-set outlet pressure against a higher inlet pressure. In water distribution systems, the common application is installation in a main connecting separate pipe networks located on two different elevations. As the water flows through the valve from the higher zone to the lower zone, the water pressure is reduced to prevent excessive pressure in the pipe network at the lower elevation. In pressure-reducing valve, the inlet pressure of the automatic valve is transmitted to the top of the valve piston, which tends to close the valve, while outlet pressure is transmitted to the top of the piston through the pilot valve. The automatic valve's inlet pressure under the diaphragm of the pilot valve presses upward trying to close the pilot. This force is opposed by the spring pushing downward on the diaphragm trying to open the valve. By turning the handwheel, the force of the spring is adjusted, pre-setting the reduced outlet pressure of the automatic valve to the desired value. If the pressure at the inlet of the automatic valve is relatively low, the pressure under the diaphragm of the pilot is lowered, thus opening the pilot valve, which, in turn, relieves the pressure above the valve piston and allows increased flow through the automatic valve. Conversely, a high inlet pressure entering the automatic valve reduces flow through the pilot valve, which increases the pressure on top of the valve piston, reducing flow through the automatic valve. Therefore, the outlet pressure is held constant even though the inlet pressure varies.

Pressure-reducing valves can be equipped with two pilot valves. By this means, the automatic valve can function as both a pressure-reducing and a sustaining valve. It maintains a pre-set outlet pressure, provided the inlet pressure is above a pre-determined value and closes when the inlet pressure drops below a pre-set sustaining pressure to protect water services on the inlet side of the valve. Other control arrangements can be used to operate an automatic valve as a pressure-reducing and relief valve or as a pressure-reducing and check valve. A relief pilot opens the automatic valve to equalise inlet and outlet pressures if the inlet pressure drops below a present value. In the pressure-reducing and check valve, backflow is prevented by closing the valve if the inlet pressure drops below the outlet pressure. The type of inlet pressure control selected for a pressure-reducing valve depends on the desired operation for a given installation.

Altitude Valves

Altitude valves are used to automatically control the flow into and out of an elevated storage tank or standpipe to maintain desired water-level elevations. These valves are usually placed in a valve pit adjacent to the tank riser. An altitude valve designed as a double-acting sequence valve is installed in the pipe connecting the tank to the system. The valve automatically closes when the tank is full to prevent overflow. When the pressure on the distribution side of the valve is less than on the tank side, it opens allowing water to enter (or leave) the tank.

The altitude valve is installed in the inlet-outlet pipe between two isolating gate valves. The pilot valve is connected through two small pipes to the tank riser on one side and to the main from the distribution system on the other. Speed of valve closing is controlled by adjusting the needle valve in the pipe between the valve body and the base of the pilot. The pilot valve on top is a diaphragm-operated, spring-loaded, three-way valve that transmits water pressure to the top of the piston in the main valve. By turning the nut threaded onto the stem, the force of the spring on the diaphragm can be adjusted to pre-set the automatically maintained water level in the storage tank. Water drains out of the chamber above the pilot diaphragm through an open check valve in the small-diameter pipe to the low-pressure system side of the valve. This permits the stem of the pilot to be forced upward when the piston is pushed up by high-pressure water entering the valve from the tank. Therefore, water flows out of the tank through the open valve into the system. During filling of the tank, the direction of flow is reversed, passing from the system into the tank. The pilot does not close the automatic valve during this reverse flow, since the check valve in the pipe to the pilot is closed as a result of the higher pressure from the system side preventing the force of the spring from lowering the diaphragm in the pilot. Thus, in these pilot positions, water is free to flow in and out of the tank and the tank is said to be 'floating' on the system. If water continues to enter the tank, eventually the water level rises to the top and the water pressure plus force from the spring lowers the diaphragm, closing the valve between the chamber under the diaphragm and the top of the valve piston. Water can now flow through a by-pass from the valve body to the top of the piston from the system side. A globe altitude valve is either fully opened or completely closed and therefore, does not act as a throttling valve.

A single-acting altitude valve is installed in the inlet pipe to control only water flow into a storage tank. Outflow is through another pipe with a check valve to prevent system water from entering. Fig. 14.9 illustrates typical arrangements for installing a single-acting sequence valve.

Altitude valves come in a variety of other designs. For example, a differential altitude valve closes when a tank is full and remains closed until the tank level drops to a pre-determined water level before it opens to refill the tank. This operation ensures renewal of the stored water after each filling. An altitude valve may also be a combination valve composed of a swing check built into a butterfly control vane assembly. Outflow is through the check valve. When the water level in the tank drops below a pre-set elevation, the butterfly vane in the connecting pipe to the system opens. Refilling of the tank occurs when the butterfly control vane is open and the system pressure exceeds tank pressure. When the tank is full, the butterfly vane closes; thus, this kind of valve provides a double-acting operating sequence.

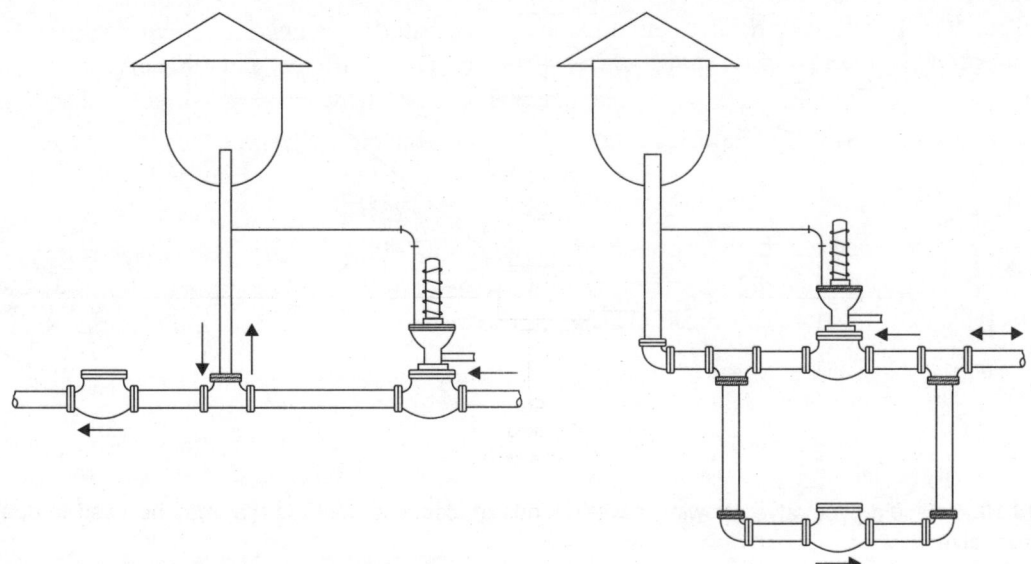

Fig. 14.9. Typical arrangements for installing a single acting altitude valve to automatically control water level in an elevated storage tank.

Solenoid Pilot Valves

The force required to change the position of a three-way valve that controls the operation of an automatic valve is very small and can be produced by a solenoid. A solenoid is a coil of wire wound in a helix so that when electric current flows through the wire a magnetic field is established within the helix. The force created by this magnetic field is sufficient to open or close a pilot valve. Figure 14.10 illustrates several kinds of electrical controls in solenoid valve. When the solenoid pilot is open, the chamber above the valve piston is vented to the atmosphere; simultaneously, a valve in the pipe from the inlet side of the angle valve is closed. This allows the piston in the angle valve to be held open by inlet water pressure. When the solenoid pilot is closed, the vent is sealed and pressure from the inlet water pushing on the larger top area of the piston forces it downward, closing the angle valve.

Air-Release Valves

Air can enter a pipe network from a pump drawing air into the suction pipe, through leaking joints and by entrained or dissolved gases being released from the water. Air pockets increase the resistance to the flow of water by accumulating in the high points of distribution piping, in valve domes and fittings and in discharge lines from pumps. Air-release valves are installed at these locations to discharge the trapped air. The common air-release valve contains a ball that floats at the top of a cylinder, sealing a small opening. When air accumulates in the valve chamber, the ball drops away from the outlet orifice, allowing the air to escape. With return of the water level, the ball reseals the outlet.

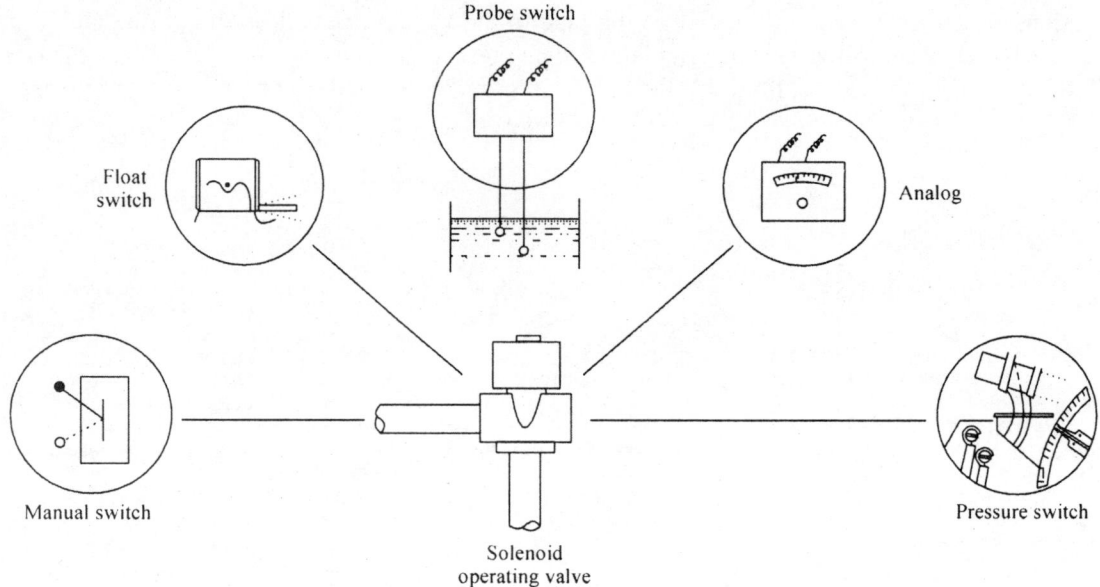

Fig. 14.10. Solenoid pilot valve showing several kinds of electrical controls that may be used to operate a solenoid valve.

BACKFLOW PREVENTERS

The water in a distribution system must be protected against contamination from backflow through customer service lines and other system outlets. A cross-connection refers to an actual or potential connection between a potable water supply and an industrial or residential source of contamination. By practising cross-connection control through enforcement of a plumbing code and inspection of backflow preventers in service connections, a municipal utility or private water purveyor can ensure against distribution contamination under foreseeable circumstances. Of greatest concern is backflow of toxic chemicals and waste-water that may contain pathogens. The risk of back siphonage is reduced by maintaining adequate pressure in the supply mains to prevent reversal of flow. In undersized water distribution networks, low pressures can result from undersized piping, inadequate pumping capacity or excessive peak water consumption. Back siphonage is backflow resulting from negative or reduced pressure in the supply piping. Back siphonage can also result from a break in a pipe, repair of a water main at an elevation lower than the service point and reduced pressure from the suction side of booster pumps. In contrast, back pressure causes reversal of flow when the pressure in a customer's service connection exceeds the pressure in the distribution main supplying the water. An example of backflow resulting from back pressure is chemically treated water from a malfunctioning boiler system being forced back through the feed line into the potable water supply.

The simplest method of preventing backflow is to provide an air space between the free-flowing discharge end of a supply pipe and an unpressurised receiving vessel. To have an acceptable air gap, the end of the discharge pipe has to be at least twice the diameter of the pipe above the highest rim of the receiving vessel, but in no case should this distance be less than 1 inch (Fig. 14.11a). Although air gaps are common in small installations, the loss of water pressure as a result of physical separation in the supply line to a large facility requires re-pressurising the system. Installation of a storage tank and pumps is often more costly than installing a mechanical backflow preventer that transmits supply pressure. In many countries, household water supplies are isolated from the public water-supply system by air gap separation using a storage tank in the attic or on the roof of the dwelling. Flow into the tank is controlled by a float valve in the inlet pipe and an emergency overflow pipe ensures that failure of the inlet valve does not result in flow over the tank rim (Fig. 14.11b). The disadvantage of this arrangement is that a direct pressure connection is necessary for supplying adequate pressure to operate some appliances, like an automatic washing machine and for lawn sprinkling.

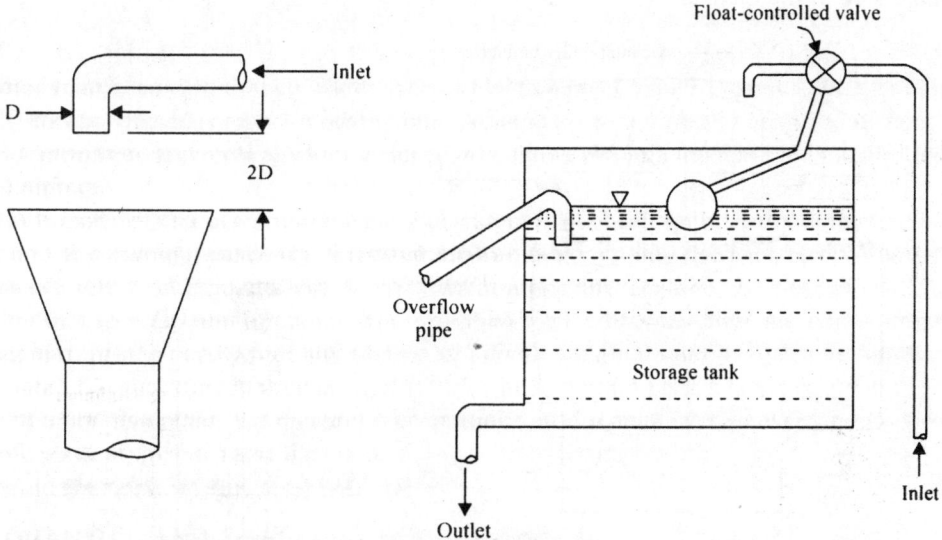

Fig. 14.11. Air-gap separation to prevent backflow into the supply pipe: (a) Minimum recommended air gap is twice the diameter of the supply pipe; and (b) Storage tank with air-gap separation and overflow pipe to prevent overfilling the tank if supply valve fails to close tightly.

The four kinds of mechanical backflow preventers are the atmospheric-vacuum breaker, pressure-vacuum breaker, double check valve and reduced-pressure-principle device. The one selected depends on the type of installation and the hazard involved if backflow occurs. For direct water connections subject to back pressure, only the reduced-pressure-principle backflow preventer is considered adequate as an alternate to an air gap separation. Still, air gaps are recommended in industrial applications where toxic chemicals are being mixed with the water, for example, vats containing acids and solutions for metal plating. For applications not subject to back pressure, vacuum breakers can be installed. The atmospheric-vacuum breaker is used in flush valve toilets and either an atmospheric or a pressure-vacuum breaker may be connected to a lawn sprinkler system.

Vacuum and Pressure-Vacuum Breakers

An atmospheric-vacuum breaker has an inside moving element that prevents water from discharging through the top of the breaker during flow and drops down to provide a vent opening when flow stops. As diagramed in Fig. 14.12a, the check-float element is open, raised by the pressure of water flowing up through the valve. When flow stops, the element drops onto the valve seat by force of gravity. This prevents back siphonage by allowing air to enter and break the siphon in the elevated pipe loop. A typical installation for lawn sprinklers is shown in Fig. 14.12b. Another application of an atmospheric-vacuum breaker is installation on the overhead pipe of a water filling station where water can be added to mobile chemical tanks to mix fertiliser, herbicide or other chemical solutions. Since the hose attached to the end of the fill pipe can be submerged, back siphonage of the solution is possible unless the system is designed to prevent backflow. A vacuum breaker on the top of the riser pipe will break the siphon in the overhead pipe when the water supply is shut off, thus, preventing backflow. An atmospheric breaker should not remain under pressure for long periods of time and cannot have a shutoff valve installed on the discharge side of the breaker.

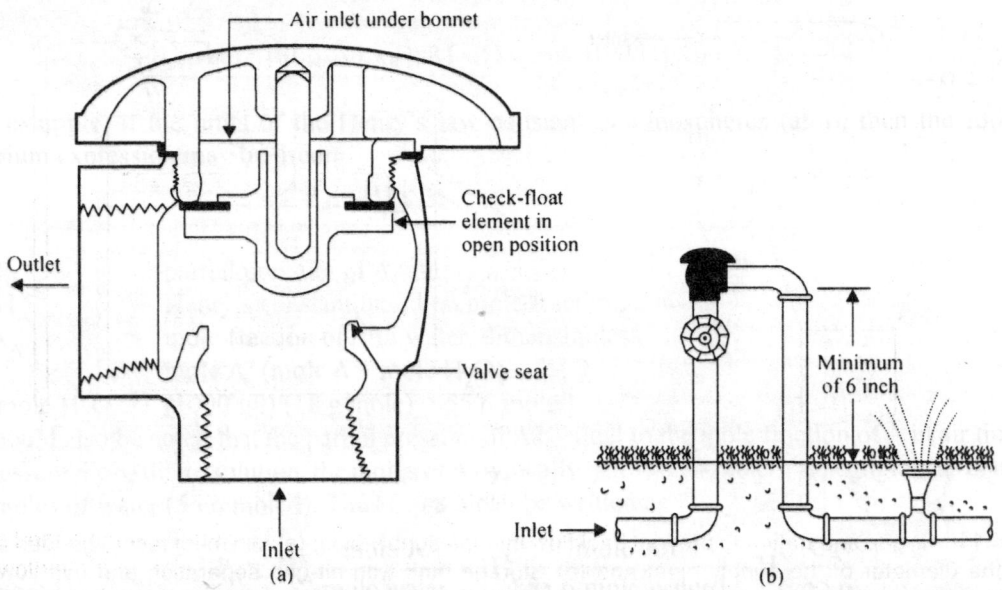

Fig. 14.12. Atmospheric vacuum breaker. (a) The check-float element seals the air inlet with water flow through the unit. When flow stops, the element drops to cover the inlet and allows air to enter to break the siphon; and (b) typical installation on a sprinkler system.

A pressure-vacuum breaker contains an assembly consisting of a spring-loaded check valve and a spring-loaded air valve. The check valve prevents backflow and the air valve opens to admit air when the pressure within the body of the breaker approaches atmospheric pressure. If the check valve does not close tight because of interference from foreign matter, air is drawn in through the automatically operating vent valve to preclude backflow. The breaker body is fitted with two test cocks for checking valve tightness against reverse flow and a shut-off valve on each side of the breaker. Pressure-vacuum breakers are installed where low inlet water connections are not subject to back pressure; however, they may be attached to direct water connections subject to backflow provided the cross-connection is with water containing only non-toxic substances.

Double Check Valve Assembly

A check valve backflow preventer is composed of two single, independently acting check valves. An installation also has two tightly closing shut-off valves located on each side of the four test cocks for checking tightness of the valves against backflow. As with pressure-vacuum breakers, double check valves are used to protect against backflow of non-toxic substances in direct water connections subject to backflow and to protect against backflow of non-toxic and toxic substances in water connections not subject to back pressure.

Reduced-Pressure-Principle Backflow Preventer

This backflow device consists of two independent, springloaded check valves with an automatically operating, pressure differential relief valve located between the two checks. The two isolating gate valves are for testing the operation of the three interior valves and to facilitate removal of the unit for maintenance.

Reduced-pressure-principle backflow preventers are comparable in safety to air gap separation. In addition, they have the advantage of transmitting water pressure to the user's piping system. This allows operation of water fixtures with main pressure, including an automatic sprinkler system for fire protection. Common preventer installations are in service connections to industries, chemical plants, hospitals, mortuaries and irrigation systems.

All mechanical backflow preventers must be inspected periodically to ensure proper operation of the interior valves. The procedure involves attaching a differential pressure gauge to selected test cocks and closing one of the isolating shut-off valves to apply backflow water pressure to the discharge side of an internal valve. Then, by release of the water pressure from the inlet side of the valve, tightness of the closure is checked by the ability of the valve to hold backflow pressure without leaking.

RESERVOIRS AND ELEVATED TANKS

Reservoirs and elevated tanks are used for storage of water and adequate pressure. Every water system keeps a sufficient amount of water for emergencies, such as breakdown of treatment or firefighting.

Reservoirs

Reservoirs provide storage for ground-level water. The reservoir at the treatment plant, which receives the filter effluent, is called a clear well. Other reservoirs are located at different points in the distribution system for an adequate supply of water. They are reinforced concrete or steel structures. Concrete reservoirs are built partially underground. The reservoir capacity is generally sufficient to meet the water supply for 4 to 6 hours. Advantages of reservoirs are that they provide a reserve supply of water, uniform pumping, proper detention time for disinfectant and mixing of water from different sources. For proper protection, a reservoir should be partly above ground level, away from any sewer line (within 50 feet, the sewer line should be extra heavy cast iron), crack free, covered and equipped with overflow capability, vent, drains and manholes. It must be painted inside and outside to prevent corrosion. Vents and overflows are screened with fine mesh to prevent the entry of birds and insects. The drain of a reservoir should never be directly connected to the sewer lines.

Elevated Tanks

Elevated reservoirs provide pressure due to height. The main advantage of an elevated tank is the availability of reserve water under pressure without pumping. Usually, water at a rate of 60 gallons/person/day must be elevated for fire protection. Minimum required capacity of an elevated tank is 50000 gallons. Generally, an elevated tank is located close to high-consumption and high-value areas for better fire protection. An elevated tank should be properly maintained by painting inside and outside. It should also have cathodic protection and be covered.

They are connected to the distribution system and filled when water demand is low and emptied when demand is high. The emptying tank is called a floating tank.

METERS

Meters are the devices to measure the quantity of water used by the customer. Meters may be displacement, current, proportional or compound in type.

Displacement Meters

Displacement meters are used for domestic customers. A displacement meter has a measuring chamber of a definite capacity. As the chamber fills and empties while the water flows through it, the meter records the quantity of that water measured by a nutating disc, oscillating piston or a rotating gear. They are good for low flows.

Current Meters

In current meters, the velocity of the passing water causes a bladed wheel to rotate and to record the volume. The higher the velocity, the more is the volume and vice versa. They are the best for the large and sustained flows.

Proportional Meter

A proportional meter has a certain proportion of the flow passing through it and then through a displacement-type measuring device, which records the total volume. Proportional meters are good for large flows; however, they are expensive and hard to repair.

Compound Meter

It is a combination of a displacement and a current meter. A small flow is measured by the displacement mechanism and a large flow by the current mechanism. They are used for variable flows.

FIRE HYDRANTS

Hydrants provide access to underground water mains for the purposes of extinguishing fires, washing down streets and flushing out water mains. Principal parts of a hydrant; the cast-iron barrel is fitted with outlets on top and a shutoff valve at the base is operated by a long valve stem that terminates above the barrel. A typical unit has two 2½ inch-diameter hose nozzles and one 4½ inch pumper outlet for a suction line. The barrel and valve stem are designed such that accidental breaking of the barrel at ground level will not unseat the valve, thus preventing water loss. Hydrants are installed along streets behind the curb line a sufficient distance, usually 2 ft., to avoid damage from overhanging vehicles. The pipe connecting a hydrant to a distribution main is normally not less than 6 inch (150 mm) in diameter and includes a gate valve allowing isolation of the hydrant for maintenance purposes. A firm gravel or broken-rock footing is necessary to prevent settling and to permit drainage of water from the barrel after hydrant use. In cold climates the water remaining in a hydrant can freeze and break the barrel. Where groundwater stands at levels above the hydrant drain, generally the drain is plugged at the time of installation and for service in cold climates, is pumped out after use. A fire hydrant has at least two 2.5-inch outlets and a gate valve between itself and the water main. It should be able to deliver 600 gpm with a head loss not over 2.5 pounds in the hydrant and 5 pounds from the main to the outlet. There should be a good drainage of water around it after the fire hydrant valve is closed.

WATER DISTRIBUTION SYSTEMS

The objectives of a municipal water system are to provide safe, potable water for domestic use; adequate quantity of water at sufficient pressure for fire protection and industrial water for manufacturing. A typical waterworks consists of a source-treatment-pumping and distribution system. Sources for municipal supplies are deep wells, shallow wells, rivers, lakes and reservoirs. About two-thirds of the water for public supplies comes from surface-water sources. Large cities generally use major rivers or lakes to meet their high demand, whereas the majority of towns use well water if available. Often groundwater is of adequate quality to preclude treatment other than chlorination and fluoridation. Wells can then be located at several points within the municipality and water can be pumped directly into the distribution system. However, where extensive processing is needed, the well pumps or low-lift pumps from the surface water intake, convey the raw water to the treatment plant site. A large reservoir of treated water (clear-well storage) provides reserves for the high demand periods and the equalising of pumping rates. The high-lift pumps deliver treated water under high pressure through transmission mains to distribution piping and storage.

The distribution consists of a gridiron pattern of water mains to deliver water for domestic, commercial, industrial and fire fighting purposes. Elevated storage tanks or underground reservoirs with booster pumps, reserve water for peak periods of consumption and fire demand. A short lateral line connects each fire hydrant to a distribution main. Shut-off valves are located at strategic points throughout the piping system to provide control of any section or service outlet, including hydrants. These valves are used to isolate units requiring maintenance and to insure that main breaks affect only a small section. A service connection to a residence includes a corporation stop tapped into the water main, a service line to a shut-off valve at the curb and the owner's line into the dwelling, which incorporates a water meter and a pressure regulator or relief valve if necessary.

WATER QUANTITY AND PRESSURE REQUIREMENTS

The amount of water required by a municipality depends on industrial use, climate and economic considerations. Although industries in the rural countryside frequently maintain private water systems, major plants in urban areas rely on the municipal waterworks.

Water flows used in waterworks design depend on the magnitude and variations in municipal water consumption and the reserve needed for fire fighting. Quantities of water required for fire demand, are of significant magnitude and frequently govern design of distribution piping, pumping and storage facilities. Water intakes, wells, treatment plant, pumping and transmission lines are sized for peak demand, normally maximum daily use where hourly variations are handled by storage. Stand-by units in the source-treatment -pumping system may be installed for emergency use, for convenience of maintenance or to serve as capacity for future expansion. The required design flow of maximum daily consumption plus fire flow frequently determines the size of distribution mains and results in additional pumping capacity and a need for storage reserves, in addition to that required to equalise pumping rates. If the maximum hourly consumption exceeds the maximum daily plus fire fighting demand, it may be the controlling criterion in sizing some units.

The recommended water pressure in a distribution system is 65 to 75 psi (450 to 520 kPa), which is considered adequate to compensate for local fluctuations in consumption. This level of pressure can provide for ordinary consumption in buildings up to ten stories in height, as well as sufficient supply for automatic sprinkler systems for fire protection in buildings of four or five stories. For a residential service connection, the minimum pressure in the water distribution main should be 40 psi (280 kPa).

Pressures in excess of 100 psi (690 kPa) are undesirable and the maximum allowable pressure is 150 psi (1030 kPa). At excessive levels, leaks occur in domestic plumbing requiring pressure reducers in service connections and undue stress is placed on mains in the ground. Pipe and fittings used in ordinary water distribution systems are designed for a maximum working pressure of 150 psi.

Pressure

The pressure in a distribution system must be high enough to permit pumpers of the fire department to obtain adequate flows from hydrants. A positive water pressure is needed during withdrawal of fire flow from hydrants to overcome friction losses in the hydrant and section hose. The ISO specifies a minimum residual pressure of 20 psi (140 kPa) during fire flow in analysing the adequacy of a water system.

Water Supply Capacity

In evaluating a system, pumps should be credited at their effective capacities when discharging at normal operating pressures. The pumping capacity, in conjunction with storage, should be sufficient to maintain the maximum daily use rate plus maximum required fire flow with the single most important pump out of service. The fire flow must be sustainable at the required pressure of 20 psi (140 kPa) for the required duration. Storage is frequently used to equalise pumping rates into the distribution system as well as provide water for fire fighting. Since the volume of stored water fluctuates, only the normal minimum daily amount maintained is considered available for fire fighting. In determining the fire flow from storage, it is necessary to calculate the rate of delivery during a specified period. Even though the amount available in storage may be great, the flow to a hydrant cannot exceed the carrying capacity of the mains and the residual pressure at the point of use cannot be less than 20 psi (140 kPa).

Although a gravity system, that is, delivering water without the use of pumps, is desirable from a fire protection standpoint because of reliability, well-designed and properly safeguarded pumping systems can be developed to such a high degree that no distinction is made between the reliability of gravity-fed and pump-fed systems. Where electrical power is used, the supply should be so arranged that a failure in any power line or repair of a transformer or other power device, does not prevent delivery of required fire flow. Electric power should be provided to all pumping stations and treatment facilities by two separate lines from different directions.

Reliability of water system operations is essential to meet fire demand. Pumping stations, treatment plants and operations control centers are constructed for protection from fire, flooding and other disasters or accidents. Audible alarms and other notification systems by radio or automatic dialing telephones are installed to warn water personnel of problems. To take remedial actions, suitable equipment and transportation for emergency crews should be on call to assist the fire department to ensure water service in case of a system malfunction or failure, such as a broken main.

Distribution System

Proper layout of supply mains, arteries and secondary distribution feeders is essential for delivering required fire flows in all built-up parts of the municipality with usage at the maximum daily rate. Consideration must be given to the greatest effect that a break, joint separation or other main failure could have on the supply of water to a system. With the most serious failure, no defficiency is considered if the remaining mains from the source of supply and storage can provide the fire flow for the specified duration, during a period of three days with usage at the maximum daily rate.

Supply mains, arteries and secondary feeders should extend throughout the system properly spaced about every 3000 ft (910 m)—and looped for mutual support and reliability of service. The gridiron pattern of small distribution mains supplying residential districts should consist of mains at least 6 inch (150 mm) in diameter. Where long lengths are necessary, exceeding about 600 ft (180 m) 8 inch (200 mm) or larger intersecting mains should be used. In new construction, 8 inch or larger pipes are used where dead ends and poor gridiron are likely to exist for a considerable time during development or because of layout of streets and topography. Hydrants for fire protection should never be located on the dead end of 6 inch or smaller mains. In commercial districts, the minimum size main should be 8 inch with intersecting lines in each street with 12 inch (300 mm) or larger mains used on principal streets and for all long lines that are not connected to other mains at intervals close enough for mutual support.

A distribution system is equipped with a sufficient number of valves located so that a pipeline break does not affect more than ¼ mile of arterial mains, 500 ft (150 m) of mains in high-value districts or 800 ft (240 m) of mains in other districts.

DESIGN LAYOUTS OF DISTRIBUTION SYSTEMS

The arrangement of a water system is dictated by the source of water supply, topography of the distribution area and variations in water consumption. If water is supplied at only one point in a pipe network, elevated storage or ground-level storage with booster pumping is required in remote areas to maintain residual pressures. Also, the arterial mains between supply and storage must be large enough to transmit water without excessive head loss. On the other hand, if water enters a distribution network at several points from individual wells, the storage capacity can be reduced and the pipe sizes can be smaller because the water supply does not have to travel as far before consumption. Topography can be the major factor in system design. Consider the two extremes, one where the water is supplied at a high elevation and flows by gravity through the pipe network and the other where the supply enters at a low elevation and must be pumped up into the pipe network and storage tanks. Many distribution systems have independent pipe networks at different elevations for pressure control; these may be completely separated or connected by pipes with pressure-reducing valves or through a common reservoir that acts as storage on the lower network and pumping suction for the upper system. The pattern for water consumption is directly a function of industrial, commercial and residential demands. For municipalities, land-use planning and zoning are commonly employed to control variations of water consumption in the distribution network. Climate has a definite effect, since lawn watering creates a major demand in residential areas in semi-arid regions. Each water distribution system is unique and is influenced by local conditions. Since pumping capacity, sizes of network pipes and volume of storage are all inter-related, increase in one of these can compensate for deficiencies in the others. Although founded on economic considerations, the options for expansion or modification or a system are governed to a considerable extent by the arrangement of existing facilities. Furthermore, the costs of construction and operation vary with time. For example, as a result of increasing energy costs, larger pipes to reduce pumping pressures and bigger storage reservoirs to equalise pumping rates are desirable options. In all cases, however, the primary engineering objectives are to provide a stable hydraulic gradient pattern for maintaining adequate pressure throughout the service area and sufficient pumping and storage capacities to meet fire demands and emergencies, such as main breaks and power failures.

The following simplified distribution systems illustrate the basic principles of design. In Fig. 14.13a, an arterial pipe network for a small system is shown with a high-service pumping station and an elevated storage tank. The ground-level storage at the pumping station receives water after treatment or directly from wells. By providing larger mains between the pumping station and the elevated tank, an adequate quantity of stored water can be maintained to supply peak demands in the area of the system around the elevated tank. The hydraulic gradient over the system, produced by the high-service pumps, is supported by the water level in the elevated storage tank. In Fig. 14.13b, wells distributed throughout the pipe network pump water directly into the system at several locations, allowing the installation of smaller diameter pipes. The primary functions of the storage tank are to stabilise pressures and equalise pumping rates. Wells are operated as needed to provide water and stand by power is installed at a sufficient number of locations to meet demand in emergencies. Some wells are used only during part of the year, for example, during a dry summer season when consumption reaches a maximum. The hydraulic gradient is supported by the operating wells and the level in the elevated storage tank.

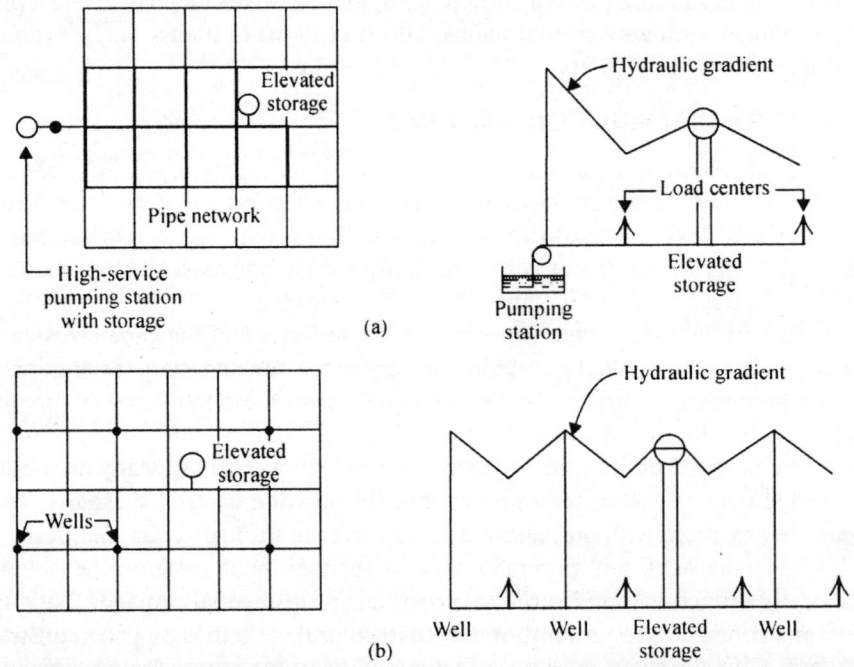

Fig. 14.13. Simplified layouts of distribution systems. (a) The high service pumping and elevated storage tank support the hydraulic gradient over the distribution system; and (b) wells located throughout the pipe network support the hydraulic gradient.

EVALUATION OF DISTRIBUTION SYSTEMS

Quantity

The supply source plus storage facilities should be capable of yielding enough water to meet both the current daily demands and the anticipated consumption ten years hence. To quantitatively evaluate water usage, a record of average daily peak daily and peak hourly rates of consumption of the past ten years should be recorded. Projected future needs can be estimated from these values and other factors relating to community growth. The minimum amount of water available from a supply source should always be sufficient to insure uninterrupted service.

Consideration must be given to the probability of a succession of drought years equal to the previous worst drought experience and the possibility of lowering groundwater levels. For surface supplies, the tributary watershed should be able to yield the estimated maximum daily demand ten years in the future. As a general rule, the storage capacity of an impounding reservoir should be equal to at least 30 days maximum daily demand five years into the future. Ideally, for well supplies there should be no mining of water; that is, neither the static groundwater level nor the specific capacity of the wells (gpm/ft of drawdown) should decrease appreciably as demand increases. Preferably these values should be constant over a period of five years except for minor variations that correct themselves within one week.

Intake Capacity

A surface-water intake must be large enough to deliver sufficient water to meet municipal use and treatment plant needs (for example, filter backwashing) during any day of peak demand. With respect to fire flow requirements, if no storage is available the intake capacity must be large enough to meet fire demand, maximum hourly flow and implant process needs simultaneously. On the other hand, where the quantity of distribution storage is sufficient to meet all fire flow requirements, the intake capacity involves comparing total storage of the system to the maximum amount of water needed, for both present and future demands. Intake facilities should be sized to meet maximum needs projected for at least five years into the future. A water intake system must be reliable; it must be located, protected or duplicated such that no interruption of service to customers or to fire protection occurs by reason of floods, ice or other weather conditions or for reasons of breakdown, equipment repair or power failure. In other words, intake facilities should be so reliable that no conceivable interruption in any part of the facility could cause curtailment of water supply service to a community.

Pumping Capacity

In a typical surface-water supply system, low-lift pumps draw water from the source and transport it to the treatment plant. After processing, high-lift pumps deliver the water from clear-well storage to the distribution system. In the case of a groundwater supply, the well pumps deliver the raw water to treatment. However, if processing is not needed, the wells may discharge water directly to the transmission mains. In large communities or in areas with widely varying elevations in topography, booster pumping stations may be needed to increase pressure in the distribution network and to extend the system for greater distances from the main pumping station.

With due consideration for the amount of water storage available, a pumping system must have sufficient capacity to provide the amount of water at pressures and flow rates needed to meet both daily and hourly peak demand with required fire flow. Pumping facilities must also be able to meet demands, taking into account common system failures and maintenance requirements. For example, the specified maximum pumping capacity should be obtainable with the largest pump out of service. In many instances system storage, either elevated or ground level with booster pumping, is a component part of the pumping system to meet peak hourly demands.

Pumping facilities should be sufficiently reliable through duplication of units, stand-by equipment and alternate sources of power such that no interruptions of service occur for any reason. In the event of power outage, stand-by power sources must be capable of sufficiently quick response to prevent exhaustion of water available in distribution storage during peak hourly demands including required fire flow. Where unattended automatic booster stations exist, the control system should report back to a central station both the condition of operation and any departure from normal.

Piping Network

Arterial and secondary feeder mains should be designed to supply water service for 40 or more years after installation. Actual useful service life of mains under normal conditions is 50 to 100 years. Sub-mains should be at least 6 inch in diameter in residential districts and the minimum size in important districts should be 8 inch in diameter with intersecting mains 12 inch. Distribution lines are laid out in a gridiron pattern, avoiding dead ends by proper looping. An adequate number of valves should be inserted to permit shut-off in case of break so that no more than one block is out of service. The distribution of hydrants is based on Insurance Services Office standards.

WATER QUALITY IN THE DISTRIBUTION SYSTEM

It is important to maintain the proper water quality in the distribution system until the last customer is served. Water quality deteriorates due to stagnancy, dead ends, cross-connections and corrosion.

Stagnancy

Stagnancy results when water stays too long in the lines. The water loses some of the residual disinfectant and dissolved oxygen. When water is stagnant, microbes such as iron, sulphur, nitrogen or even coliform bacteria begin to multiply and produce objectionable by-products.

Bacteria occur in a thin layer of organic and inorganic deposition, called biofilm, on the inner surface of water lines, where they multiply. Part of this film can be eroded by certain hydraulic conditions, such as high velocity and water hammer to release these bacteria, resulting in high heterotrophic plate counts and even a positive coliform test. The best remedy, to stop the re-growth of microbes, is the adequate unidirectional (in the same direction as the normal flow) flushing of the distribution system and continuously maintaining an appropriate amount of residual disinfectant. Unidirectional flushing is important to maintain proper water quality by removing biofilm, maintaining disinfectant residual and reducing erosion,

If the water pH is low and chloramines are the residual disinfectant, then nitrification bacteria convert ammonia to nitrites and nitrites to nitrates (nitrification).

$$\text{Ammonia} \longrightarrow \text{Nitrites } (NO_2^-) \longrightarrow \text{Nitrates } (NO_3^-)$$
$$\underset{\text{Nitrification}}{}$$

To stop nitrification, the pH of water needs to be raised above 9.0. Water should be kept moving and lines should be kept flushed regularly.

Dead Ends

Dead ends are the endings of the pipes without any movement of the water. There should be a regular flushing of the dead ends and cleaning as required to keep the water fresh.

Cross-Connection

Cross-connection is any physical connection between the treated and non-treated water. Examples of cross-connections are a direct pipe connection from a private water supply (well water) to the municipal water supply; discharge of the potable water supply below the water level in a swimming pool and submerged inlets in lavatories, bath tubs, fountains and spray tanks. When there is a high demand of water and pressure is low in the water lines, the non-treated water is sucked into the public water supply system, sometimes resulting in serious health problems.

Cross-connection contaminations of drinking water with pesticides are reported. Sprinkler systems have also caused cross-connections. There should always be backflow prevention, such as an air gap or a check valve, to protect the public water supply.

Corrosion

Corrosion of water lines and plumbing fixtures can cause the leaching of harmful metals such as lead and copper into water, growth of objectionable microbes and biofilm formation. Water should be non-corrosive.

It does not matter how properly the water is treated unless it is delivered as good as it is treated. For proper delivery, proper maintenance of system and pressure are important. Transmission system problems and their solution are given in Table 14.1.

Table 14.1. Transmission system problems and their solutions.

Problems	Possible causes	Possible solutions
Water smells like a pesticide	There is a cross-connection. If pesticide is being sprayed in an area, look for the water line in the tank. A hose under the water in the tank is causing the problem	Remove the hose and flush the area. Check the reservoir for any contamination. If contaminated, flush it and clean it thoroughly before putting it back in service
High heterotrophic bacterial count in the distribution system when chloramines are used for post-disinfection	Nitrification, which is the conversion of ammonia into nitrites and then into nitrates by bacteria. It is indicated by pH around 8, stagnant smell of water, high nitrites and nitrates and red water complaints	Increase the pH to an appropriate level (about 9) to solve the nitrification problem
	Dead ends. This is indicated by red and stagnant water complaints and high HPC count. It occurs when chlorine gets low and dissolved oxygen is depleted in the stagnant water. Iron and sulphur bacteria become active and corrosion of pipes starts	Unidirectional (in one direction throughout the system to prevent the erosion of deposits in the pipe) flushing of the area is the solution to this problem
Coliform bacteria are in the distribution system.	Contaminated sample	Make sure that all equipment is properly sterilised; there is no contamination of the sample or the equipment while collecting or testing the sample. Carefully, re-sample the site, test by two different methods (colilert and membrane filter) to re-confirm the results
	Breakdown of treatment	Check all bacteriological test results of the treatment train; be sure the water is properly disinfected. Increase the disinfectant residual in the distribution system
	Change in the direction of water flow in the lines	It will cause the erosion of the biofilm. Correct all flows to unidirectional

(Contd...)

Problems	Possible causes	Possible solutions
	Sewage cross-connection	Check for any water or sewer line breaks in the area. Correct the situation and flush the area
	Re-growth of coliform in biofilm	Biofilm has certain coliform bacteria which start growing as soon as conditions are favourable, such as low disinfectant residual and low dissolved oxygen. Flush the area thoroughly, increase disinfectant residual and check for coliforms. It may need the pigging of lines to remove biofilm. Consistently, maintain the clean system and proper disinfectant residual
Pinkish deposits are on faucets	High calcium carbonate in the water	The colour is due to a fungus growing on calcium carbonate deposits. Adjust the alkalinity and pH to reduce the calcium carbonate deposition potential
There is a high chloride content	Inadequate ferrous treatment. Ferrous should be added before alum or lime	Determine the needed dose of ferrous to remove the desired amount of chlorite. Apply three parts of ferrous for each part of chlorite
There is a high chlorate content	An old sodium hypochlorite solution used for chlorination can have high chlorates	Always use a fresh sodium hypochlorite solution
	Concentrated sodium hypochlorite has a high amount of chlorates	Normally sodium hypochlorite is 15%, dilute it to 7.5%

Chapter 15

Advanced Waste-Water Treatment

INTRODUCTION

Advanced waste-water treatment refers to methods and processes that remove more contaminants from waste-water than are taken out by conventional biological treatment. The term may be applied to any system that follows secondary treatment or that includes phosphorus removal or nitrification in conventional secondary treatment. The expression tertiary treatment is often used as a synonym, but the two terms do not have precisely the same meaning. Tertiary suggests a third step that is applied after primary and secondary processing. Waste-water reclamation consists of a combination of conventional and advanced treatment processes employed to return a waste-water to nearly original quality, reclaiming the water. Common advanced waste-water treatment processes remove phosphorus, oxidise ammonia to nitrate and inactivate pathogenic bacteria and viruses. In water reclamation, the objectives may be expanded to include removal of heavy metals, organic chemicals, inorganic salts and elimination of all pathogens.

LIMITATIONS OF CONVENTIONAL TREATMENT

Contamination of municipal waste-water results from human excreta, food preparation wastes and a variety of organic and inorganic industrial wastes. Conventional waste-water treatment is a combination of physical and biological processes to remove organic matter. For acceptable processing, at least 85 per cent of the BOD and suspended solids are removed to produce an effluent of 30 mg/l of BOD and 30 mg/l of suspended solids or less. Some treatment plants use chlorination to reduce the numbers of viable bacteria and viruses in the effluent waste-water. Conventional treatment typically results in negligible reduction in ammonia and phosphorus, incomplete disinfection, removal of toxins present in the raw waste-water to varying degrees and no removal of soluble non-biodegradable chemicals.

The primary pollutional effect of phosphorus in surface water is eutrophication. Since phosphorus is the growth-limiting plant nutrient in natural water, discharge of waste-water high in soluble phosphates leads to accelerated fertilisation. The attendant results in lakes and reservoirs are excessive growth of algae causing reduced water transparency, depletion of dissolved oxygen, release of foul odours, loss of finer fish species and dense growths of aquatic weeds in shallow bays.

Flowing water with turbidity that blocks the sunlight necessary for photosynthesis are not subject to eutrophication; however, estuaries and slow-moving rivers with a clear water can be adversely affected in the same way as impounded water.

Unionised ammonia is toxic to fish and other aquatic animals. The amount of unionised ammonia is based on the pH of the water, since ammonia is converted to the non-toxic ammonium ion with decreasing pH. The criterion for salomonid fish is 0.02 mg/l of unionised ammonia and for tolerant fish species it is 0.08 mg/l. These values are equivalent to total ammonia-nitrogen concentrations of approximately 0.5 mg/l and 5.0 mg/l at pH 8, respectively. As a result of nitrification, ammonia can also result in dissolved oxygen uptake in flowing and impounded water. Theoretically, 1.0 mg of ammonia nitrogen can exert an oxygen demand of 4.6 mg when converted to nitrate nitrogen. Nitrification of ammonia from a waste-water discharge rarely occurs to this extent in natural water because of competing reactions, such as algal photosynthesis and environmental conditions adverse to nitrifying bacteria. In oligotrophic water, the contribution of nitrogen in waste-water can increase the rate of eutrophication.

Protection of public health is the primary concern regarding pathogens in waste-water discharged to surface water used for body-contact recreation and drinking water supplies. Conventional biological treatment removes 99 to 99.9 per cent of pathogenic micro-organisms in raw waste-water. In addition, effluent chlorination can kill an additional 99.99 per cent or more of bacteria and an unknown number of viruses and protozoa. Protozoal cysts and helminth eggs are resistant to the combined chlorine residual formed in waste-water. Although a treated waste-water may be considered satisfactory for discharge if the fecal coliform count is reduced to 200 per 100 ml, effluent chlorination should not be considered or referred to as disinfection. In addition to resistant pathogens, viruses and bacteria can be harboured and protected from the effects of chlorine in suspended organic matter. Since chlorine residual is toxic to aquatic life, dechlorination can be performed by the addition of sulphur dioxide.

Soluble organic and inorganic chemicals are difficult to remove by biological treatment. Some cause aesthetic problems like foam and colour, while others are injurious to aquatic life and in sufficiently high concentrations prevent safe reuse of receiving water. The chemicals are best removed at their point of origin by instituting an industrial pre-treatment programme. The traditional approach to detection used numerous chemical analyses to identify the presence of specific compounds. As the list of potential toxins increased, testing for specific compounds became increasingly difficult. Biomonitoring was developed to reduce the cost of broad spectrum effluent toxicity monitoring.

SUSPENDED SOLIDS REMOVAL

The inability of gravity sedimentation in final clarifiers to remove small particles is the major limitation of suspended solids, BOD and organic phosphorus removal by conventional waste-water treatment. Where needed, tertiary granular-media filtration can be employed to upgrade effluent quality. The filters are similar to those used in water treatment. Design must take into account, however, the higher suspended solids content and fluctuating rates of waste-water flow not common to water processing. The preferred filters are coal-sand dual media or mixed media containing anthracite coal, garnet and sand. A sand medium alone cannot accommodate the quantity of coarse solids in waste-water because of surface plugging. Multimedia beds allow in-depth filtration and greater solids holding capacity, resulting in longer filter runs. Efficient backwashing requires auxiliary scour since hydraulic suspension of the media, by upward flow of water through the bed, does not provide adequate cleaning. Either air scrubbing or a rotating agitator improves scouring action. The filters may be either gravity or pressure units depending on size of the treatment plant. Chlorination prior to filtration prevents growth within the filter but does not contribute to disinfection.

Flow schemes for tertiary filtration shown in Fig. 15.1 range from plain filtration to traditional filtration with chemical coagulation, flocculation and sedimentation. In plain filtration, the secondary effluent from biological processing and final settling is applied directly to granular-media filters. Normally, at least two and usually four filter cells are provided for operational flexibility to accommodate diurnal flow variations. The total filter area should be sufficient to allow peak design flow with one unit out of service for backwashing or repair. When a filter is cleaned, the surge of dirty wash water is collected in an equalising tank and returned to the plant influent at a constant rate for treatment. Plain filtration can reduce suspended solids to 10 mg/l or less; however, small particles including micro-organisms pass through the granular media.

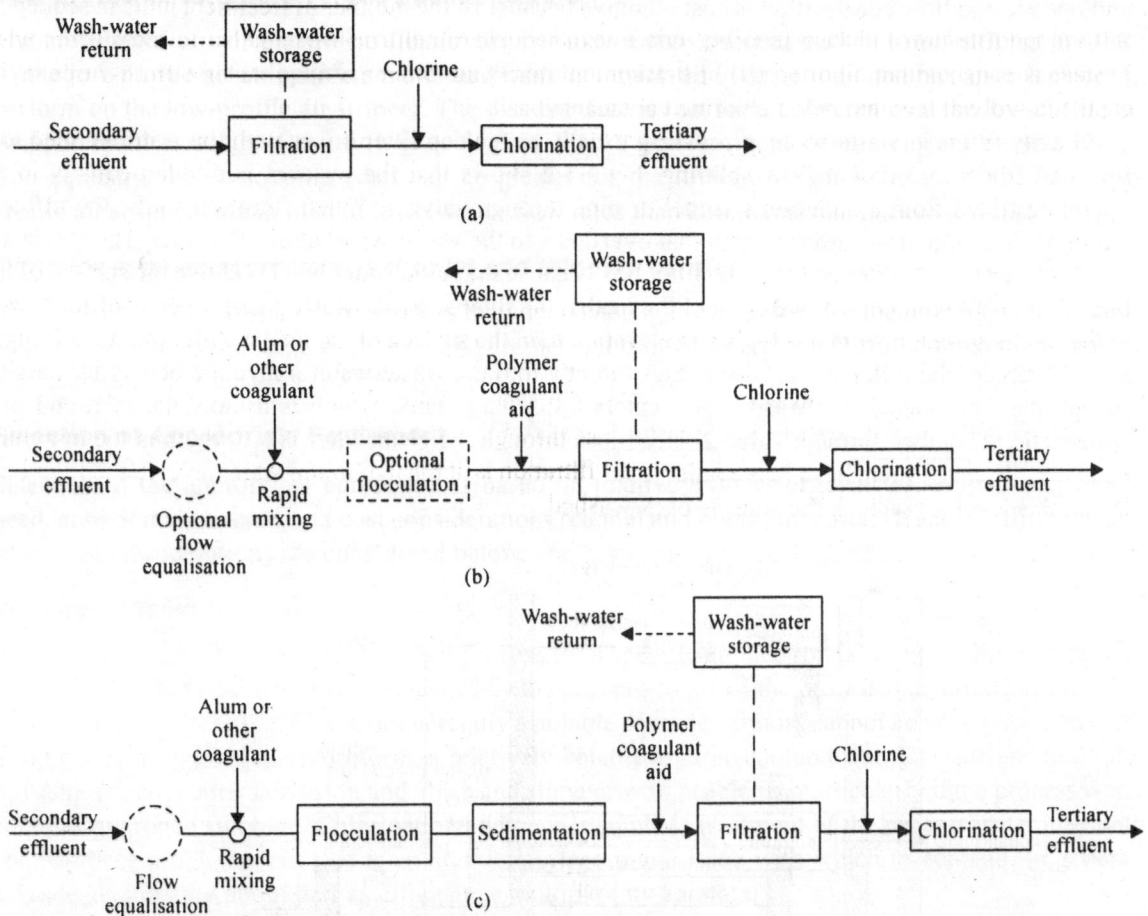

Fig. 15.1. Flow diagrams of tertiary filtration of biologically treated waste-water: (a) Plain filtration to reduce suspended solids without chemical pre-treatment followed by chlorination; (b) direct filtration with chemical coagulation and optional flocculation followed by chlorination; and (c) traditional filtration with chemical coagulation, flocculation, sedimentation, filtration and chlorination preceded by flow equalisation.

Direct filtration with chemical coagulation and optional flocculation is the most common tertiary filtration process (Fig. 15.1b). The impurities, including tiny particles, removed from the water are collected and stored in the filter media and removed by backwashing. Although rapid mixing of alum or other coagulant is necessary, subsequent flocculation is reduced to less than 30 minutes or eliminated. Contact flocculation of the chemically coagulated particles in the waste-water takes place in the granular media of the filter. Polymer as a coagulant aid is usually added just ahead of filtration or in the flocculation tank. This tertiary system is either sized to process the diurnal flow variation or a flow equalisation basin is installed to eliminate or attenuate the diurnal flow pattern. Maintaining a relatively uniform rate of flow allows improved chemical feed control and process reliability. The traditional filtration system shown in Fig. 15.1c with rapid mixing, flocculation, sedimentation and filtration can process a waste-water with higher and more variable concentrations of suspended solids. It is more costly to construct and operate and flow equalisation is cost effective because of the number of treatment units in sequence. Both direct filtration and the traditional system can provide reliable disinfection by physically removing protozoal cysts and helminth eggs by filtration and inactivate the remaining bacteria and viruses in the clear filtered water by extended chlorine contact.

Gravity filters in waste-water processing usually have deep filter boxes with the water applied to a series of filters by influent flow splitting. Fig. 15.2 shows that the influent is divided equally to all operating filters from a common channel or pipe through valve 1. When filtration starts, the influent weir box discharge into a wash trough that overflows to the waste-water above the filter. The media are not disturbed by the water entering the filter box because the static water level is above the surface of the bed. To prevent accidental dewatering of the media, the filtered waste-water passes over an effluent weir in the discharge chamber that is higher in elevation than the surface of the media. After the wash troughs are submerged, the influent weir discharges directly into the waste-water above the bed. After passing though the filter media, the waste-water enters a discharge tank, which is usually the inlet end of a chlorination chamber, through valve 2. Head loss through the clean filter is h_c, which is the minimum head loss. The maximum head loss available for filtration is at the maximum water level in the filter box. When this level is reached, the filter is back washed.

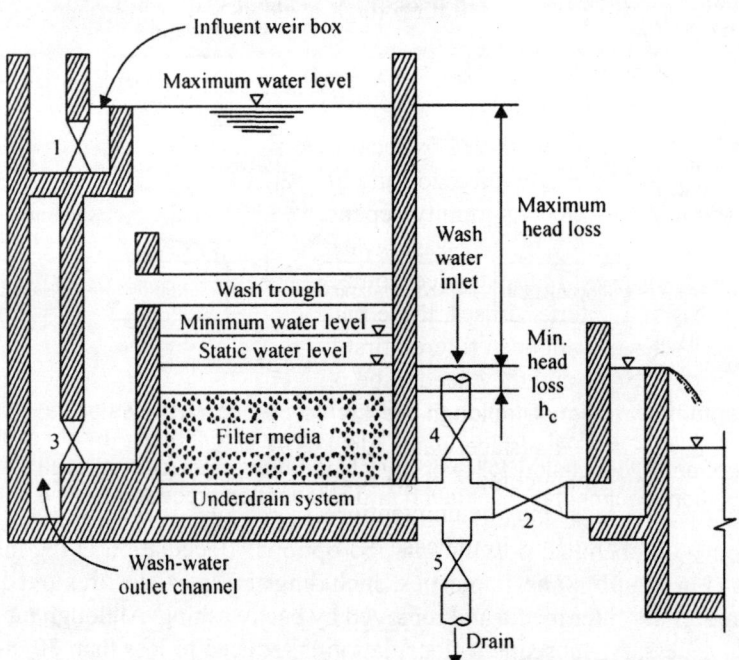

Fig. 15.2. Cross-sectional view of a gravity filter with a deep filter box.

Cleaning of a filter is started by closing the influent valve 1 and opening the drain valve 3 to the wash-water outlet channel. This drains out the water overlying the filter and lowers the level to the bottom of the wash troughs. After valve 2 is closed to the discharge tank, the wash-water valve 4 is opened allowing filtered waste-water to flush impurities out of the fluidised bed. Mechanical agitation or air scouring for dislodging the impurities from the media precedes hydraulic backwashing and may continue as the backwash water rises in the bed. The clean media are allowed to re-stratify in quiescent water before re-starting filtration.

A deep filter box, rather than the traditional pressure suction water filtration system, prevents the possibility of causing air binding in the bed as a result of creating a negative pressure in the filter. Another advantage of a deep filter box is eliminating the installation of costly rate of flow controllers and effluent piping. In flow splitting, effluent flow meters are not necessary since the influent to each filter is the total plant influent divided by the number of units in operation. Also, head loss in a filter can be easily observed by the water level in the filter box.

Hydraulic loading and time of filtration before a bed requires backwashing are interrelated. Design rates in waste-water filtration are generally in the range of 3 to 4 gpm/sq ft (2.0 to 2.7 l/m^2·s). Proper design of a filtration system depends on providing sufficient bed area. Although removal efficiency is relatively independent of the hydraulic loading rate, too small a surface area causes short filter runs. Excess backwash water cycled to the head of the plant and filter downtime result in inefficient operation. A minimum filtration time of 24 hours between backwashes is desirable under normal conditions of waste-water flow and suspended solids concentration. Effective suspended solids removal relies on the design of the installation and the characteristics of the applied waste-water. In general, the best effluent quality achievable by plain filtration is 5 to 10 mg/l of suspended solids. If further reduction is desired, chemical coagulation must precede filtration to flocculate the colloidal solids and reduce the effluent suspended solids to less than 5 mg/l.

PATHOGEN REMOVAL

Many infectious diseases are transmitted by fecal wastes and the pathogens encompass all categories of micro-organisms: viruses, bacteria, protozoa and helminths. The kinds and concentrations of pathogens in domestic waste-water from a community depend on the health of the population.

Disinfection of biologically treated waste-water requires coagulation and granular-media filtration followed by chlorination with an extended contact time. Although chlorination is effective in killing bacteria and inactivating enteric viruses, these pathogens can be protected in suspended and colloidal solids if the waste-water has not been filtered first for turbidity removal. Cysts of protozoa and helminth eggs are resistant to chlorine and they need to be physically removed by effective chemical coagulation and granular-media filtration. Even though the disinfection of waste-water by tertiary treatment is not as well defined as disinfection of surface water for potable water supplies, the principles for removal of pathogens are the same. Human contact in the reuse of waste-water effluents containing enteric viruses is of major concern. The reuse may be unintentional, for instance, discharge to a dry stream bed where children can play or intentional as in the case of landscape watering. The tertiary treatment scheme for virus removal is shown in Fig. 15.3.

TOXIN REMOVAL

Numerous organic chemicals and several inorganic ions, mostly heavy metals, are classified as toxic water pollutants. To qualify as a toxin, a substance must be hazardous to aquatic life or human health and be known to be present in polluted water. Toxicity to aquatic life may be related to either acute or chronic effects on the organisms themselves or to human by biological accumulation in seafood. Persistence in the environment and treatability are also important factors. Toxicity to human may be related to either carcinogenicity or chronic disease through long-term consumption of contaminants. Currently, the number of priority toxic water pollutants consists of over 100 substances. While some are well-documented toxins, others have limited data supporting their hazard in the aquatic environment and human health.

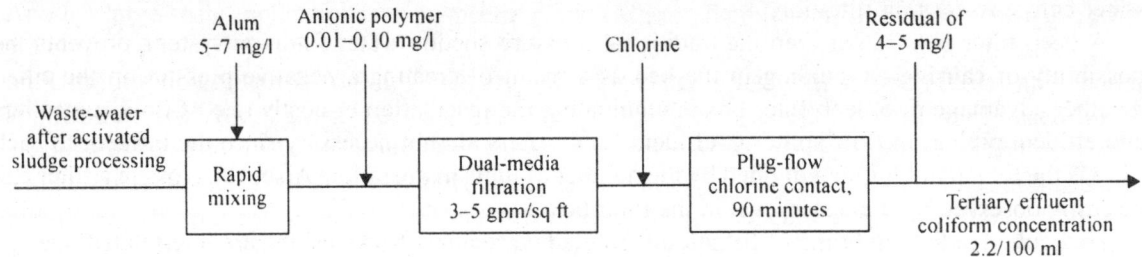

Fig. 15.3. Tertiary treatment scheme for virus removal.

Gross toxicity of a raw waste-water is evidenced by upset of the biological processes in waste-water treatment. This is an indication of industrial toxins being discharged to the sewer collection system. Heavy metals are partially removed by entrapment and adsorption onto settled solids or biological floc; nevertheless, a portion appears in the effluent. For cadmium, chromium, copper, lead, nickel and zinc, the removal percentages are about 70 per cent. Very few data are available on the removal of toxic organic compounds. Some may be biologically decomposed and others entrapped in settled solids or volatilised, but many are likely to be only slightly affected and be present in the treated waste-water.

The general goal of municipal waste-water treatment with pre-treatment of industrial waste-water is to reduce the discharge of toxic pollutants to an insignificant level. Toxicity reduction evaluation of a treatment plant is the first step. The objectives are to: evaluate plant operation and performance to identify and correct treatment deficiencies causing effluent toxicity, identify the toxic substances in the effluent, trace the toxins to their sources in the waste-water collection system and implement appropriate remedial measures to reduce effluent toxicity.

The conventional approach to evaluating toxicity is to test raw and treated waste-water samples for specific substances. This can be very costly unless the toxins likely to be present can be reduced to a reasonable number. Initial tests should be for those substances related to wastes from industries served by the sewer system. After the toxic pollutants have been quantitatively identified, remedial measures must be taken by pre-treatment of the industrial waste-water at the factory and improved waste-water processing at the treatment plant. Potential deficiencies of this analytical approach through chemical testing are the possible presence of undetected toxins and combinations of synergistic toxins.

The second approach to evaluating toxicity is by biomonitoring of effluent. Bioassays to determine toxicity are performed by exposing selected aquatic organisms to waste-water effluent controlled laboratory environment. In the static test for acute toxicity, effluent is placed in laboratory containers, the test organisms added and the containers held in a controlled environment with termination after 24 hours. A control test in prepared water is conducted in parallel for quality assurance of the test organisms and test procedures. An enhanced bioassay method is the flow-through test where fresh effluent is continuously supplied to the laboratory containers with continuous outflow. The effluent is usually a 24 hour composite sample that has been filtered, aerated and stabilised at a temperature of 20°C for testing warm-water species (12°C for coldwater species). From the several kinds of vertebrates and invertebrates, the common warm-water test organisms are fathead minnow and water flea (*Daphnia*). The fathead minnow, which is a popular bait fish, feeds primarily on algae and grows to an average length of 50 mm with a normal life span less than 3 years. *Daphnia* are invertebrates that feed on algae, grow to a maximum length of 4 to 6 mm and have a life span of 40 to 60 days. 10 to 20 of each species are tested in several containers under controlled temperature, light intensity, dissolved oxygen, pH and food supply. The results of a screening test are expressed as percentages of surviving minnows and water fleas.

Definitive bioassays are conducted for longer test periods at effluent dilutions in a geometric series. For chronic toxicity, the recommended test durations are extended up to 7 days by renewing the diluted effluent water every 24 hours in static tests or continuously in the flow-through tests. The dilution water is either a synthetic water or filtered water from the receiving water away from the point of waste-water discharge. Based on the results, the no-observed-effect-concentration (NOEC) is determined for effluent in the diluted test samples. This is compared against the waste-water concentration after initial mixing of the waste-water in the receiving water body. The initial waste-water concentration should be less than the NOEC with a margin of safety.

Pre-treatment of industrial waste-water is essential to control toxins in waste-water. Since many hazardous substances are biocides, reductions in their concentrations in raw waste-water are essential to effective biological processing. If the evidence of toxicity is less dramatic, presence of toxins in the raw waste-water and the treatment plant effluent may pass unnoticed. A typical list of toxic pollutants includes 13 metals, cyanide, 15 polynuclear aromatic hydrocarbons, 13 aromatics, 10 phenols, 21 aliphatics, 6 phthalates, 2 nitrogen compounds, 2 oxygenated compounds and pesticides.

PHOSPHORUS IN WASTE-WATER

The common forms of phosphorus in waste-water are orthophosphate ($PO_4^=$), polyphosphates (polymers of phosphoric acid) and organically bound phosphates. Polyphosphates, such as hexametaphosphate, gradually hydrolyse in water to the soluble ortho form and bacterial decomposition of organic compounds also releases orthophosphate. With the majority of compounds in waste-water soluble, phosphorus is removed only sparingly by plain sedimentation. Secondary biological treatment removes phosphorus by biological uptake; however, relative to the quantities of nitrogen and carbon, the amount of phosphorus is greater than necessary for biological synthesis. Consequently, conventional treatment removes only about 20 to 40 per cent of the influent phosphorus.

Remedial action for phosphorus pollution is the treatment of waste-water that discharge directly into lakes and rivers or streams that flow into lakes. Several states have adopted effluent standards for phosphorus. Effluent limits range from 0.1 to 2.0 mg/l as P, with many established at 1.0 mg/l. To protect the lakes and surface water of the Great Lakes and Chesapeake Bay drainage basins, phosphorus removal has been implemented at many waste-water treatment facilities.

Phosphorus Removal in Conventional Treatment

The nutrient composition of an average sanitary waste-water based on 120 gpcd (450 l/person·d) is listed in Table 15.1. Phosphorus enters the sewer in the form of soluble and organically bound phosphates. Biological activity in the sewer releases organically bound phosphates into solution that are not removed by plain sedimentation of raw waste-water. The amount of organically bound phosphates released into a soluble form varies with the sewer length, waste-water temperature and biological conditions.

Table 15.1. Approximate nutrient composition of average sanitary waste-water based on 120 gpcd (450 l/person·d).

Parameter	Raw	After settling	Biologically treated
Organic content (mg/l)			
Suspended solids	240	120	30
Biochemical oxygen demand	200	130	30
Nitrogen content (mg/l as N)			
Inorganic nitrogen	22	22	24
Organic nitrogen	13	8	2
Total nitrogen	35	30	26
Phosphorus content (mg/l as P)			
Inorganic phosphorus	4	4	4
Organic phosphorus	3	2	1
Total phosphorus	7	6	5

Based on the tabulated values, the total phosphorus is reduced from 7 mg/l to 6 mg/l by sedimentation. Secondary biological treatment removes phosphorus by biological uptake; however, the amount of phosphorus is surplus relative to the quantity of nitrogen and carbon necessary for synthesis. In general, the amount of phosphorus in the excess biological floc produced in activated-sludge treatment of a waste-water is equal to about 1 per cent of the BOD applied. Based on this, the total phosphorus is further reduced from 6 mg/l to approximately 5 mg/l. As a result, in Table 15.1, the total phosphorus of 7 mg/l in the raw waste-water is reduced to 5 mg/l in the biologically treated effluent.

Figure 15.4 is a diagram tracing phosphorus through a hypothetical treatment plant. The phosphorus concentration in the influent waste-water is assigned a value of 100 per cent. Phosphorus removal with the settled primary sludge is taken as 5 per cent and phosphorus synthesised in secondary biological aeration is assumed to be an additional 20 per cent. Thus, the quantity in the raw sludge is 35 per cent of the influent phosphorus. Yet the only portion of phosphorus extracted is that portion ultimately disposed of with the sludge solids. Stabilisation and dewatering of the raw sludge returns some of the phosphorus back to waste-water processing. Mechanical dewatering of raw waste sludge followed by land burial or incineration results in little or no recycle of phosphorus. Anaerobic or aerobic digestion, on the other hand, returns supernatant containing soluble phosphorus to the influent of the treatment plant. Figure 15.4 assumes that anaerobic digestion and sludge dewatering returns 5 per cent of the influent phosphorus, which then passes through the treatment system increasing the effluent phosphorus content to 70 per cent of the influent. Of course, variations in characteristics of the waste-water and different methods of treatment can result in higher or lower removal efficiencies. Still, phosphorus removal in conventional biological treatment systems is generally within the range of 20 to 40 per cent.

CHEMICAL-BIOLOGICAL PHOSPHORUS REMOVAL

Chemical precipitation using aluminium or iron coagulants is effective in phosphate removal. Although coagulation reactions are complex and only partially understood, the primary action appears to be the combining of orthophosphate with the metal cation. Polyphosphates and organic phosphorus compounds are probably removed by being entrapped or adsorbed, in the floc particles. Aluminium ions combine with phosphate ions as follows:

$$Al_2(SO_4)_3 \cdot 14.3H_2O + 2PO_4^{\equiv} = 2AlPO_4 \downarrow + 3SO_4^{=} + 14.3H_2O \qquad \dots (15.1)$$

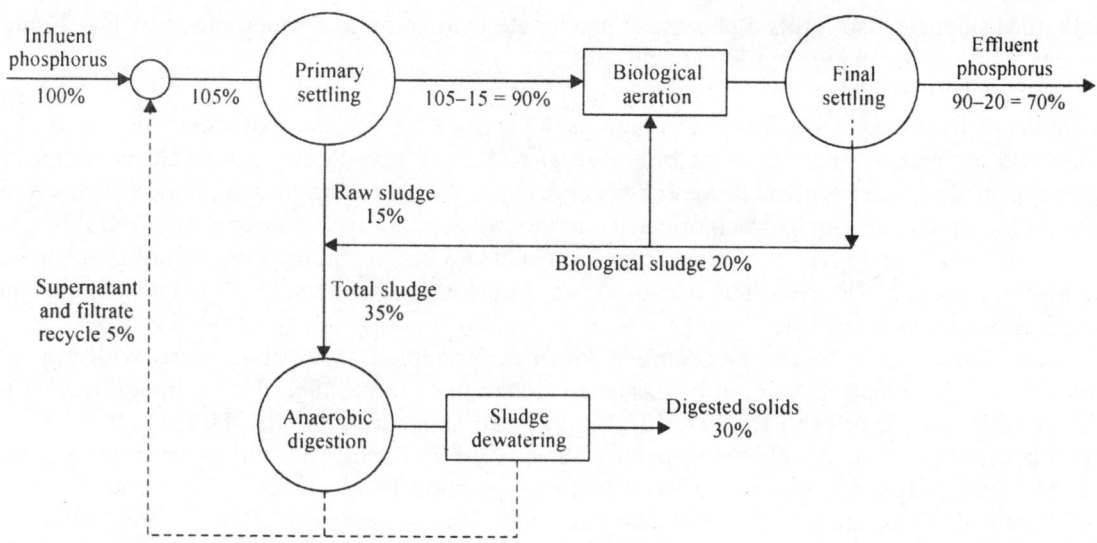

Fig. 15.4. Diagram tracing waste-water phosphorus through a hypothetical treatment plant. Of 100 per cent influent phosphorus, 70 per cent is in the treated waste-water and 30 per cent is in the digested sludge solids.

The molar ratio for Al to P is 1 to 1 and the weight ratio of commercial alum to phosphorus is 9.7 to 1.0. Coagulation studies have shown that greater than this alum dosage is necessary to precipitate phosphorus from waste-water. One of the competing reactions, which accounts in part for the excess alum requirement, is with natural alkalinity as follows:

$$Al_2(SO_4)_3 \cdot 14.3H_2O + 6HCO_3^- = 2Al(OH)_3\downarrow + 3SO_4^= + 6CO_2 + 14.3H_2O \qquad \ldots (15.2)$$

As a result, phosphorus reductions of 75 per cent, 85 per cent and 95 per cent require alum to phosphorus weight ratios of about 13 to 1, 16 to 1 and 22 to 1, respectively. For example, to achieve 85 per cent phosphorus removal from a waste-water containing 10 mg/l of P, the alum dosage needed is approximately $16 \times 10 = 160$ mg/l, which is substantially greater than the $9.7 \times 10 = 97$ mg/l stoichiometric quantity of alum based on Eq. 15.1.

Iron coagulants precipitate orthophosphate by combining with the ferric ion as shown in Eq. 15.3 at a molar ratio of 1 to 1.

$$FeCl_3 \cdot 6H_2O + PO_4^{\equiv} = FePO_4\downarrow + 3Cl^- + 6H_2O \qquad \ldots (15.3)$$

Just as with aluminium, a greater amount of iron is required in actual coagulation than this chemical reaction predicts. One of the competing reactions with natural alkalinity is

$$FeCl_3 \cdot 6H_2O + 3HCO_3^- = Fe(OH)_3\downarrow + 3CO_2 + 3Cl^- + 6H_2O \qquad \ldots (15.4)$$

Provided that the waste-water has sufficient natural alkalinity, ferric salts applied without coagulant aids results in phosphorus removal at Fe to P dosages of 1.8 to 1.0 or greater. This is equivalent to an application of approximately 150 mg/l of commercial ferric chloride for treatment of a waste-water containing 10 mg/l of P. Since the reaction of ferric chloride with natural alkalinity is relatively slow, lime or some other alkali may be applied to raise the pH and supply the hydroxyl ion for coagulation as follows:

$$2FeCl_3 \cdot 6H_2O + 3Ca(OH)_2 = 2Fe(OH)_3\downarrow + 3CaCl_2 + 6H_2O \qquad \ldots (15.5)$$

Ferrous sulphate also forms a phosphate precipitate with an Fe to P molar ration of 1 to 1 and the dosages for coagulation are similar to ferric salts.

Commercially available iron salts are ferric chloride, ferric sulphate, ferrous sulphate and waste pickle liquor from the steel industry. The latter is the least expensive and most common source of iron coagulants for waste-water treatment in industrial regions. Pickle liquor is variable in composition depending on the metal treatment process. Ferrous sulphate from pickling with sulphuric acid and ferrous chloride from pickling with hydrochloric acid are the two common waste liquors from metal finishing. Waste liquors have an iron content from 5 to 10 per cent and free acid ranges from a low of 0.5 per cent to a high of 15 per cent. Preparation prior to use includes neutralisation and pH adjustment of the liquors with lime or sodium hydroxide.

Chemical-biological treatment combines chemical precipitation of phosphorus with biological removal of organic matter. Alum or iron salts are added prior to primary clarification, directly to the biological process or prior to final clarification. For all application points, the amount of chemical added is about the same to achieve a specific phosphorus removal. Addition to the primary clarifier enhances both suspended solids and BOD removal resulting in 75 per cent solids and 50 per cent BOD removal. Thus, subsequent treatment capacities increase and tend to be hydraulically rather than BOD limited.

In activated-sludge aeration, the coagulant can be added to the aerating mixed liquor. Although the resulting chemical-biological floc has fewer protozoa, BOD removal efficiency is not adversely influenced. In trickling filtration or rotating biological contactors, the coagulant is usually mixed with the process effluent just prior to final sedimentation. Depending on coagulant dosage, the production of sludge solids in chemical-biological treatment is generally in the range of 20 to 60 per cent more than the biological solids produced without chemical addition. In part, this increase in solids production is the result of improved effluent clarification because of chemical coagulation. The volume of sludge, on the other hand, is usually a smaller percentage increase because of the higher density of the chemical-biological sludge.

NITROGEN IN WASTE-WATER

The common forms of nitrogen are organic, ammonia, nitrate, nitrite and gaseous nitrogen. Decomposition of nitrogenous organic matter releases ammonia to solution (Eq. 15.6). Under aerobic conditions, nitrifying bacteria perform Eq. 15.7 to oxidise ammonia to nitrite and subsequently to nitrate. Bacterial denitrification, Eq. 15.8, occurs under anaerobic or anoxic conditions when organic matter (AH_2) is oxidised and nitrate is used as a hydrogen acceptor releasing nitrogen gas.

$$\text{Organic nitrogen compounds} \xrightarrow[\text{decomposition}]{\text{Bacterial}} NH_3 \text{ (ammonia)} \qquad \text{... (15.6)}$$

$$NH_3 + O_2 \xrightarrow[\text{nitrification}]{\text{Aerobic}} NH_3 \text{ (nitrate)} \qquad \text{... (15.7)}$$

$$NO_3 + AH_2 \xrightarrow[\text{denitrification}]{\text{Anaerobic}} A + H_2O + N_2\uparrow \qquad \text{... (15.8)}$$

Nitrogen in municipal waste-water results from human excreta, ground garbage and industrial wastes, particularly from food processing. Approximately 40 per cent is in the form of ammonia and 60 per cent is bound in organic matter with negligible nitrate. The total nitrogen contribution is in the range of 8 to 12 lb per capita per year (4 to 6 kg/person·year) and the average concentration in domestic waste-water is 35 mg/l.

Nitrogen Removal in Conventional Treatment

The nitrogen compounds in an average sanitary waste-water are listed in Table 15.1. With most of the nitrogen in soluble and colloidal organic forms, the amount removed by primary sedimentation is limited to about 15 per cent. Based on the tabulated values, the uptake in subsequent biological treatment is only another 10 per cent. In general, the amount of nitrogen in excess biological floc produced in activated-sludge treatment of a waste-water is equal to about 4 per cent of the BOD applied. With a total reduction of only 25 per cent, the effluent contains 26 mg/l of the original 35 mg/l. Approximately 2 mg/l is organic nitrogen bound in the effluent suspended solids. The remaining 24 mg/l is in the form of ammonia, except when nitrification occurs during aeration. Oxidation of a portion of the nitrogen to nitrate is most likely to occur in an activated-sludge process treating a warm waste-water at a low BOD loading.

Figure 15.5 is a diagram tracing nitrogen through a hypothetical treatment plant. The nitrogen concentration in the influent waste-water is assigned a value of 100 per cent. Primary removal by sedimentation of raw organic matter is 15 per cent and removal by biological synthesis in secondary aeration is assumed to be an additional 10 per cent. Processing of the raw sludge can release some of the nitrogen extracted from the waste-water. In this treatment scheme, the sludge is stabilised by anaerobic digestion and the ammonia released from decomposition of the sludge solids is returned to the influent of the treatment plant in supernatant from the digester. Assuming 40 per cent of the organic nitrogen in the sludge is converted to ammonia, 10 per cent of the original 25 per cent is recycled to the treatment plant and appears in the effluent, which then contains 85 per cent of the influent nitrogen. Depending on variations in nitrogen content of the waste-water and methods of waste-water and sludge processing, nitrogen removal in conventional biological treatment systems ranges from nearly zero up to 40 per cent.

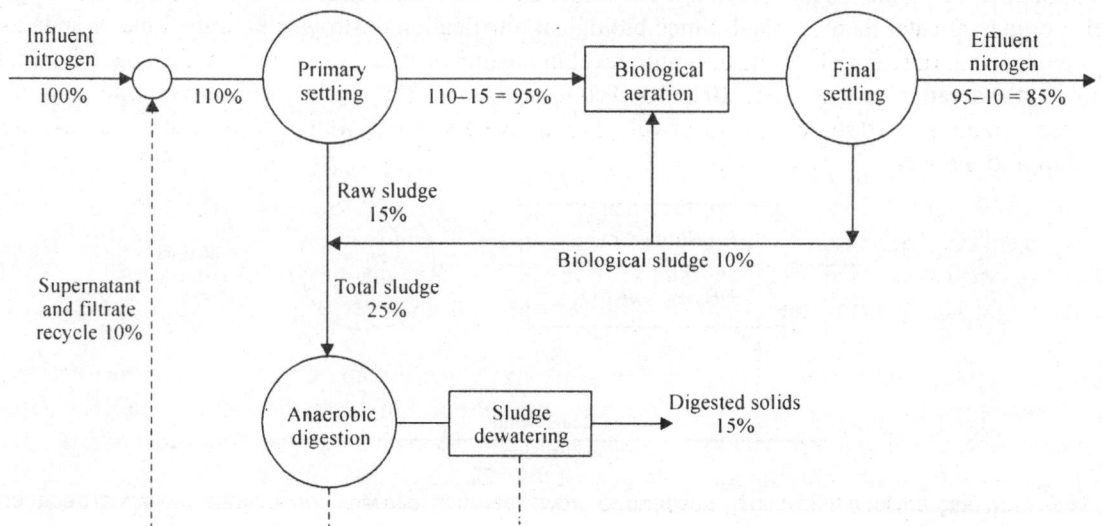

Fig. 15.5. Diagram tracing waste-water nitrogen through a hypothetical treatment plant. Of the 100 per cent influent nitrogen, 85 per cent is in the treated waste-water and 15 per cent is in the digested sludge solids.

BIOLOGICAL NITRIFICATION AND DENITRIFICATION

Nitrification of a waste-water is practiced where the ammonia content of the effluent causes pollution of the receiving watercourse. The process does not remove the nitrogen but converts it to the nitrate form

(Eq. 15.7). Nitrification-denitrification, which reduces the total nitrogen content, includes conversion of the nitrate to gaseous nitrogen (Eq. 15.8)

Nitrification

Nitrification is usually a separate process following conventional biological treatment. Although nitrification may be possible to perform along with organic matter removal in a single-stage extended aeration unit in a warm climate, two-step treatment is necessary for reliable operation at a reduced waste-water temperature. The conventional biological treatment removes BOD, without oxidation of ammonia nitrogen, to produce a suitable effluent for nitrification. The high ammonia content and low BOD provide greater growth potential for the nitrifiers relative to the heterotrophs. This allows operation of the nitrification process at an increased sludge age to compensate for lower operating temperature and to ensure that the growth rate of nitrifying bacteria is rapid enough to replace those lost through washout in the plant effluent. Although synthetic-media filtration in biological towers can perform nitrification, the most reliable system is suspended-growth aeration.

Figure 15.6 is the common flow scheme for nitrification following biological treatment of a waste-water for organic matter reduction. After the nitrifying bacteria oxidise the ammonia in the aeration tank, the activated sludge containing high population of nitrifiers is settled in the final clarifier for return to the aeration tank. Sludge can be wasted, if necessary, to remove excess bacterial growth from the system. Because the rate of oxidation of ammonia is nearly linear, the tank configuration is plug flow to minimise short-circuiting. The important parameters in bacterial nitrification kinetics are temperature, pH and dissolved oxygen concentration. Reaction rate is decreased markedly at reduced temperatures with about 8°C being the reasonable minimum value. Optimum pH is near 8.4 and the dissolved oxygen level should be greater than 1.0 mg/l. Since biological nitrification destroys alkalinity, lime or soda ash may be needed to raise the pH to the optimum level in the nitrification tank. Ammonia nitrogen loadings applied to the aeration tank are 10 to 20 lb/1000 cu ft/day (160 to 320 g/m^3·d) with corresponding waste-water temperatures of 10 to 20°C, respectively. For an average waste-water effluent, this is an aeration period of 4 to 6 hours.

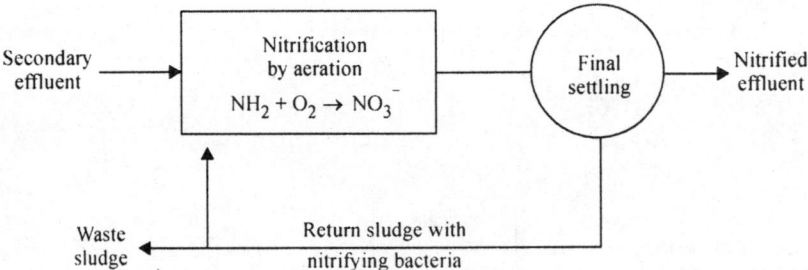

Fig. 15.6. Flow diagram for nitrification by suspended-growth aeration following conventional biological treatment.

Denitrification

Nitrate can be reduced to nitrogen gas by facultative heterotrophic bacteria in an anoxic environment. An organic carbon source, AH$_2$ in Eq. 15.8, is needed to act as a hydrogen donor and to supply carbon for biological synthesis. Although any biodegradable organic substance can serve as a carbon source, methanol is common because of its availability, ease of application and ability to be applied without leaving a residual BOD in the process effluent. As with all other chemical carbon sources, methanol is expensive. In fact, the cost of methanol makes the widespread application of this denitrification process unrealistic in waste-water treatment.

The denitrification reaction between methanol and nitrate is as follows:

$$5CH_3OH + 6NO^-_3 = 3N_2\uparrow + 5CO_2\uparrow + 7H_2O + 6OH^-$$ ... (15.9)

Since the process effluent from nitrification also contains dissolved oxygen and nitrite, total methanol required as a hydrogen donor in denitrification is given in Eq. 15.10. In addition, methanol is used as a carbon source in bacterial synthesis. Based on approximately 30 per cent excess methanol needed for synthesis, the total methanol demand is calculated using Eq. 15.11.

$$CH_3OH = 0.7DO + 1.1 NO_2\text{-}N + 2.0NO_3\text{-}N$$ ... (15.10)

$$CH_3OH = 0.9DO + 1.5NO_2\text{-}N + 2.5NO_3\text{-}N$$ ... (15.11)

where,

CH_3OH = methanol, milligrams per liter
DO = dissolved oxygen, milligrams per liter
NO_2-N = nitrite nitrogen, milligrams per liter
NO_3-N = nitrate nitrogen, milligrams per liter

The recommended denitrification system consists of a plug-flow tank with underwater mixers followed by a clarifier for sludge separation and return, (Fig. 15.7). The level of agitation in the denitrification chambers must keep the microbial floc in suspension, but controlled to prevent undue aeration. Since nitrogen gas released from solution can float the biological floc, the last chamber must strip the nitrogen gas from solution for efficient final clarification. This may be done using an aeration chamber, which also has the advantage of oxygenating the plant effluent or by a degasifier. The detention time required for denitrification of a domestic waste-water is usually in the range of 2 to 4 hours depending on nitrate loading and temperature. Since methanol is expensive, denitrification following nitrification is generally performed only where the receiving watercourse is used as a source for public water supply and an effluent nitrogen concentration of less than 10 mg/l requires strict control.

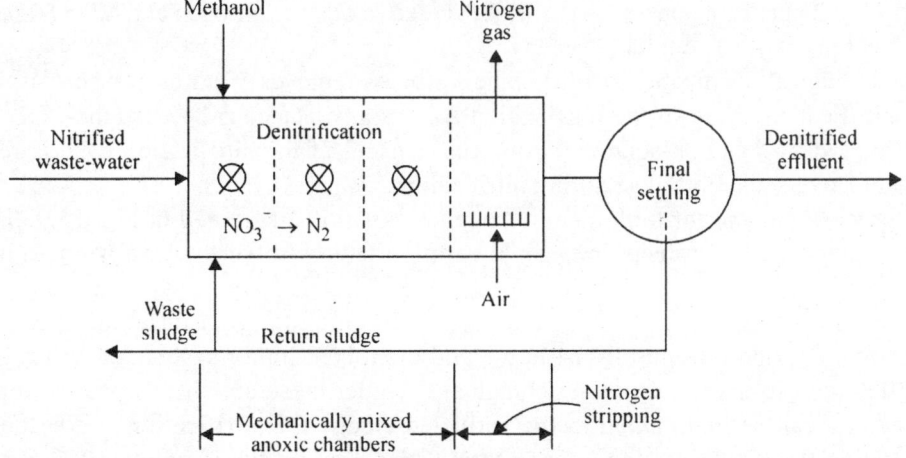

Fig. 15.7. Flow diagram for denitrification by anoxic (anaerobic) suspended growth.

Biological Nitrification-Denitrification

Unoxidised organic matter can be used as an oxygen acceptor (hydrogen donor) for conversion of nitrate to nitrogen gas. This reaction satisfies a portion of the BOD in a waste-water.

$$\begin{matrix} \text{Unoxidised} & & \text{Oxidised} & \\ \text{organic matter} & + \ NO_3^- \longrightarrow & \text{organic matter} & + \ N_2\uparrow \\ \text{(BOD)} & & \text{(reduced BOD)} & \end{matrix} \qquad \text{... (15.12)}$$

The flow scheme of a biological nitrification-denitrification process requires mixing of raw organic matter with nitrified waste-water; hence, an aerobic zone is needed for nitrification and an anoxic zone for denitrification. (The term anoxic means low or lacking in oxygen).

The plug-flow activated-sludge system, diagramed in Fig. 15.8, blends nitrified recirculation flow with raw settled waste-water (primary effluent) in an anoxic zone. In this mechanically mixed chamber, or segregated zone of a long narrow tank, the biological floc in the return activated sludge uses recirculated nitrate as an oxygen source, releasing nitrogen gas.

In the subsequent chambers mixed by diffused or mechanical aeration, dissolved oxygen is taken up by the biological floc for nitrification of the ammonia in the waste-water. Both the anoxic and aerobic chambers reduce the waste-water BOD—the anoxic zone by denitrification and the aerobic zone by uptake of dissolved oxygen. The final clarifier contributes to denitrification as the bacterial floc extract the oxygen from nitrate in the metabolism of the organic solids during sedimentation. The majority of nitrogen removal, however, occurs in the anoxic chamber and the degree of nitrogen removal is controlled by the rate of recirculated flow.

The amount of return nitrate that can be reduced depends on the maximum rate of denitrification possible in the anoxic zone. Too great a recirculation carries nitrate through the anoxic zone into the aerobic chambers and similarly, too short an aeration period can reduce the degree of nitrification. Other factors, such as the relative concentration of nitrogen to BOD in the waste-water and temperature, also influence the method of process operation. Figure 15.8 illustrates typical treatment of a settled domestic waste-water at a moderate temperature with a total retention time of approximately 8 hours in the biological chambers. Influent nitrogen is assigned a value of 100 per cent. Of this, 30 to 35 per cent is converted to nitrogen gas, 20 to 25 per cent appears in the waste sludge as organic nitrogen and 40 to 50 per cent is in the plant effluent primarily as nitrate nitrogen.

Thus, this biological nitrification-denitrification process removes 50 to 60 per cent of the influent nitrogen. Denitrification ahead of the nitrification zone is advantageous because the BOD in the raw waste-water is used as a carbon source, the oxygen demand of the nitrification zone is reduced and denitrification recovers alkalinity lost during nitrification.

Another process scheme for biological nitrogen removal is diagramed in Fig. 15.9. The plant is a Carrousel system that has a deep, long, oval aeration tank with vertical walls and a mechanical surface aerator to propel the waste-water around in the channel. A similar system is a deep, vertical wall, oxidation ditch with horizontal rotor aerators. When treating an unsettled domestic waste-water, the aeration period is generally 24 hours and the mixed liquor suspended solids maintained at 3000 to 5000 mg/l to operate at a long sludge age. Under these conditions; viable population of nitrifying bacteria can be maintained in the activated sludge for nitrification. Nevertheless, cold-weather operation that decreases the temperature of the aerating liquid below 10°C can adversely affect nitrification.

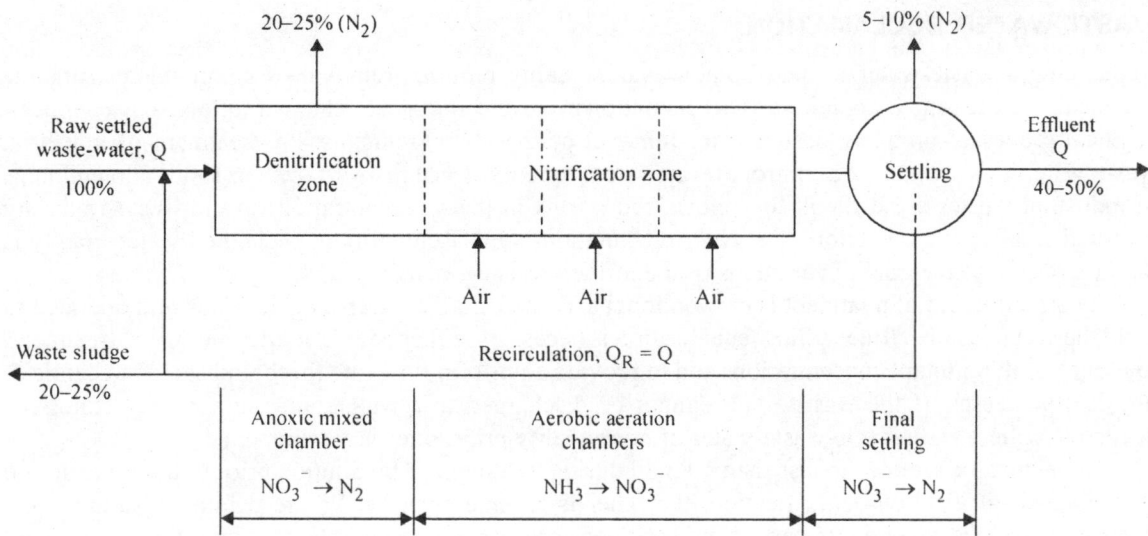

Fig. 15.8. Diagram tracing waste-water nitrogen through a hypothetical biological nitrification-denitrification system. Of 100 per cent influent nitrogen, 30 to 35 per cent is released as nitrogen gas, 20 to 25 per cent is organic nitrogen in the waste sludge and 40 to 50 per cent appears in the effluent primarily as nitrate nitrogen.

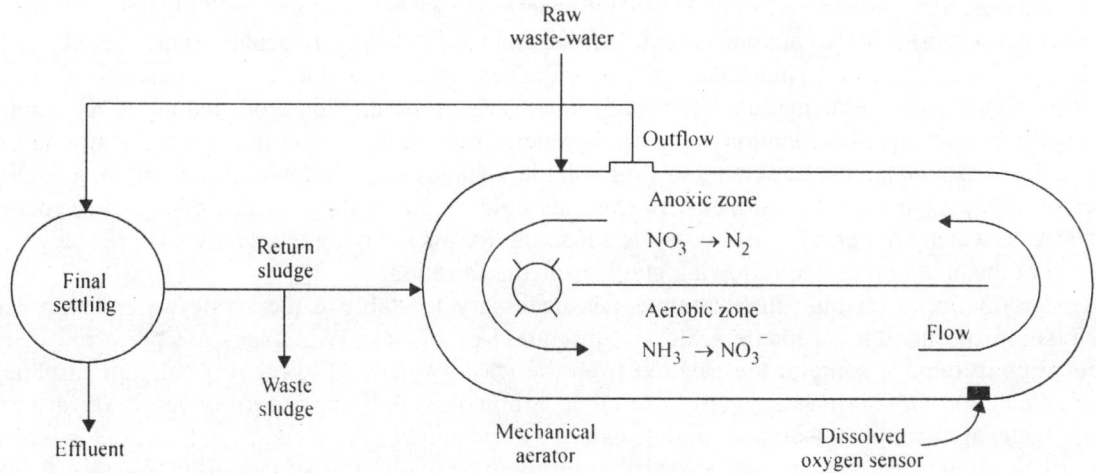

Fig. 15.9. Schematic diagram of a Carrousel system (deep, vertical wall, oxidation ditch) operating as an extended aeration process with concurrent nitrification and denitrification.

Denitrification in a Carrousel system is possible with proper operational control. By reducing the degree of aeration, the aerobic zone along the channel can be shortened to create an anoxic zone before the mixed liquor recirculates back to the aerator. A dissolved oxygen sensor located midway along the channel can be used to control the aerator. Consequently, the first half of the channel can be aerobic enough for nitrification and the second half deficient so that oxygen is biologically removed from nitrate to satisfy waste-water BOD. Total nitrogen removal in the range of 50 per cent to over 70 per cent is possible in this combined nitrification-denitrification process.

WASTE-WATER RECLAMATION

Reclaiming a waste-water to near potable water quality requires removal of heavy metals, organic chemicals, viruses and inorganic salts. Reclamation involves biological treatment followed by chemical-physical processes uniquely designed for removal of these contaminants. Pre-treatment of industrial waste-water is essential in the control of waste-water quality. Since many of the hazardous contaminants in industrial wastes are difficult to remove, reductions in their concentrations in the waste-water are essential to effective operation of a water reclamation plant. Reliability of reclaimed water quality is directly related to the quality variations of the influent waste-water.

The preferred initial treatment is conventional activated sludge processing designed and operated to yield the best possible effluent. Flow equalisation is necessary, either before or after biological treatment, to reduce peak pollutant concentrations and to provide a uniform flow rate for the subsequent chemical-physical processes. If the waste-water cannot be discharged to a watercourse, emergency storage is needed to collect and recycle waste-water of poor quality prior to tertiary treatment.

Nitrate must be reduced to less than 10 mg/l during treatment or by dilution prior to use. One option is biological nitrification-denitrification; the other is reverse osmosis. Where demineralisation using reverse osmosis is required as part of the treatment process, excess nitrate may be removed to avoid biological denitrification.

Lime precipitation of the biological effluent is the first step in chemical-physical processing. High-lime treatment is effective in precipitation of heavy metals, suspended organics, some dissolved organics and phosphates. An alkaline pH of about 11 also kills bacteria and inactivates viruses. Although flocculator-clarifiers have been used, the common scheme is to separate rapid mixing, flocculation and sedimentation in a series of tanks. The lime sludge can be gravity thickened and mechanically dewatered for disposal or for purification and recalcination for reuse of lime. The carbon dioxide produced in the recalcining process can be used in re-carbonation of the waste-water. After sedimentation and before re-carbonation, the alkaline waste-water can be aerated to strip volatile organics and ammonia, commonly in a cooling tower. Air stripping, however, cannot be performed at a cold air temperature because of excessive cooling of the waste-water. One process as a possible substitute for air stripping is selective ion exchange for removal of the ammonium ion, following granular-media filtration.

Re-carbonation after high-lime treatment is necessary to stabilise the waste-water to prevent precipitation of calcium carbonate scale in subsequent processes. Two stages are preferred in re-carbonation to remove some of the calcium from the waste-water and to recover calcium carbonate for re-calcination. In the first stage, only sufficient carbon dioxide is applied to convert the hydroxide to carbonate, at about pH 8.4 and precipitate calcium carbonate. In the second stage, the pH is further reduced to about 7.5 in order to convert the carbonate remaining in the settled waste-water to bicarbonate. Granular-media filtration is used to remove non-settleable solids prior to subsequent treatment.

Granular activated-carbon adsorption following filtration removes trace concentrations of a variety of non-polar organic compounds. Disinfection by chlorination or ozonation is also integrated with carbon adsorption; for example, the sequence may be carbon adsorption-ozonation-carbon adsorption.

Depending on the waste-water concentration of total dissolved solids and usage limitations (1000 mg/l for drinking water), reverse osmosis may be required. Besides reducing the content of the undesirable ions of sodium, chloride, nitrate and trace metals, passage of water through a membrane removes non-specific dissolved organic compounds. Typical flow diagram of water reclamation plant is shown in Fig. 15.10.

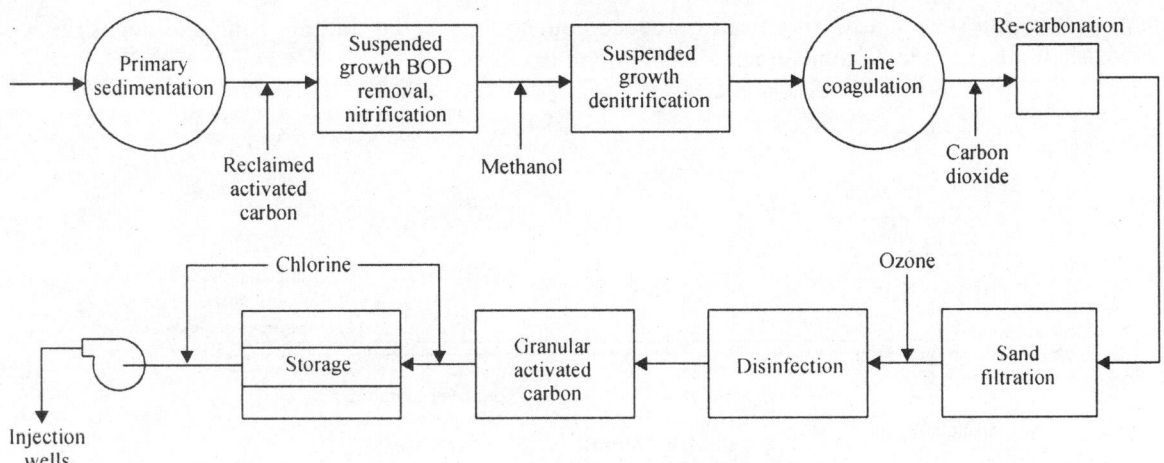

Fig. 15.10. Flow diagram of water reclamation plant.

OCEAN DISCHARGE

The major beneficial uses of ocean water to be protected are water contact and non-contact recreation, commercial and sport fishing, marine habitat, shellfish harvesting (mussels, clams and oysters) and industrial water supply.

Discharge of waste-water to ocean water requires construction of an ocean outfall, as illustrated in Fig. 15.11, composed of a pipeline to relatively deep water with a diffuser at the end with a series of ports spaced to provide initial dilution. By this method of discharge; sufficient initial dilution can be provided to minimise the concentration of substances not removed in biological treatment and effluent chlorination. The location of a discharge is determined by an assessment of the oceanographic characteristics and current patterns to assure pathogens are not present in shellfish harvesting areas or water-contact recreational areas. Also, the location should provide maximum protection of the marine environment.

Water Quality in the Zone of Initial Dilution

Compliance with bacterial, physical and chemical water quality standards is determined from samples collected at boat stations representative of the zone of initial dilution (Fig. 15.11). For water-contact standards, coliform sampling is generally conducted from the shoreline into the ocean for a distance of 1000 ft or the 30 ft depth contour, whichever is further.

A typical standard for coliforms is an average of less than 1000 total coliforms per 100 ml in any 30 day period and no single sample exceeding 10000 per 100 ml and for fecal coliforms is not to exceed to geometric mean of 200 per 100 ml with no more than 10 per cent exceeding 400 per 100 ml in any 60 day period. At all areas where shellfish are being harvested for human consumption, the median total coliform count is not to exceed 70 per 100 ml with no more than 10 per cent of the tests exceeding 230 per 100 ml. Waste-water discharges must be essentially free of floatable materials, settled materials, toxins, turbidity and colour that can interfere with the indigenous marine life and a healthy and diverse marine community.

The physical characteristics in the zone of initial dilution for compliance are no visible floating particulates, grease or oil and no aesthetically undesirable discolouration of the ocean surface. Natural light penetration shall not be significantly reduced outside the initial dilution zone and deposition of solids shall not degrade biological benthic communities.

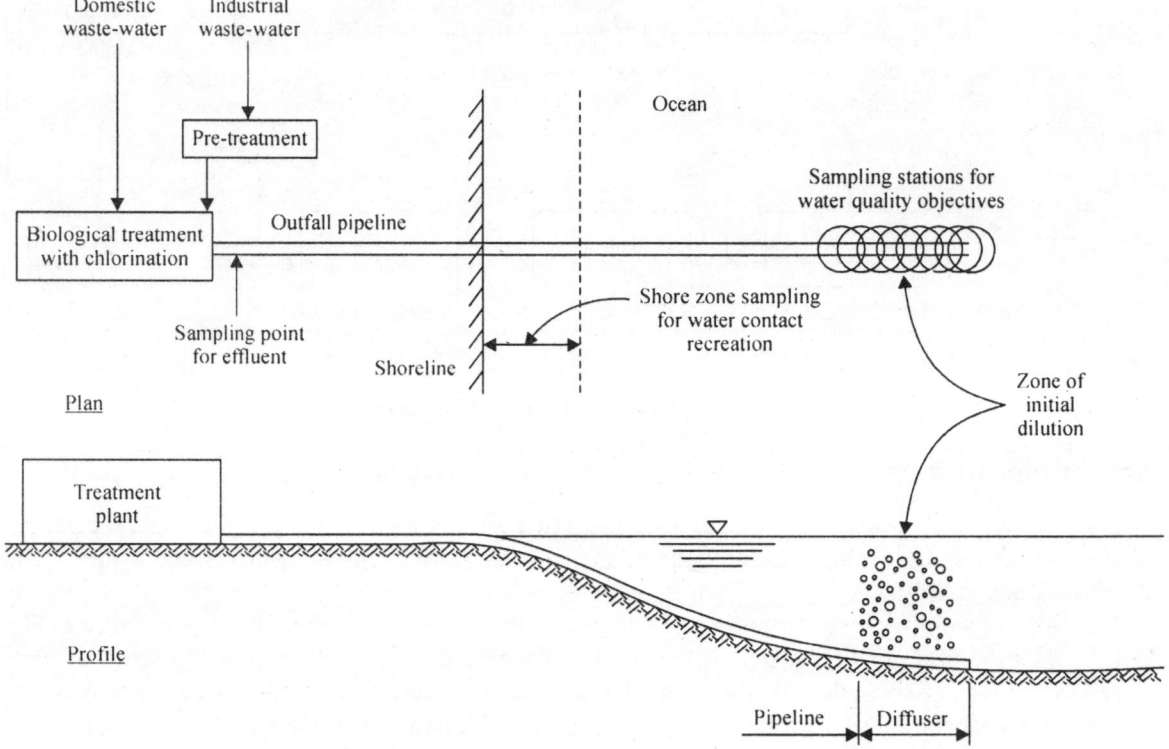

Fig. 15.11. Schematic plan and profile diagrams of marine discharge.

The chemical changes limit the decrease of dissolved oxygen to no more than 10 per cent, lower the pH more than 0.2 units, increase the dissolved sulphide concentration, increase the concentration of degrading substances in sediments or increase nutrients causing objectionable aquatic plant growths. Change in the biological characteristics of the water should neither degrade vertebrate, invertebrate and plant species; nor alter the natural taste, odour and colour of fish or shellfish for human consumption; nor increase the bioaccumulation of toxins in fish or shellfish harmful to human health. Radioactivity cannot degrade marine life.

Effluent Quality Limits for Waste-Water Discharge

Quality requirements for waste-water discharges are specified as effluent limitations for major waste-water constituents and toxins for protection of marine aquatic life and non-carcinogens and carcinogens for protection of human health. These quality requirements apply to community and industrial waste-water and other discharges, for example, cooling water from power plants. The limitations of major waste-water constituents specified in Table 15.2 are grease and oil, suspended solids, settled solids, turbidity, pH and acute toxicity.

Table 15.2. Effluent quality limits of major waste-water constituents for ocean discharge to protect marine aquatic life.

Parameter	Monthly (30 day average)	Weekly (7 day average)	Maximum (at any time)
Grease and oil, mg/l	25	40	75
Suspended solids, mg/l	60 with a minimum removal of 75 per cent		
Settleable solids, ml/l	1.0	1.5	3.0
Turbidity, NTU	75	100	225
pH	Within limits of 6.0–9.0 at all times		
Acute toxicity, TUa[a]	1.5	2.0	2.5

$$^a TUa = \frac{100}{96 \text{ hours LC } 50}$$

where,

TUa = toxicity units acute

LC = lethal concentration 50 per cent

The limitations of specific toxins, mostly heavy metals and pesticides, are listed separately and BOD is included for domestic waste-water. Fish bioassays for acute toxicity are conducted with the threespine stickle-back, *Gasterosteus aculeatus*. The test species for chronic toxicity bioassays preferably include a fish (silversides), an invertebrate (shrimp, oyster) and an aquatic plant (red algae, giant kelp). Depending on the species, the duration of the toxicity tests are from 48 hours to 7 days.

The state of California lists effluent quality limitations for chemicals that are non-carcinogens and carcinogens for protection of human health. The allowable limits in the effluent from a particular plant can differ from the listed concentrations based on calculated initial dilution. The dilution factor is determined using an approved mathematical model.

The characteristics of the outfall for inputs to the model include length of diffuser, number and spacing of ports, port diameter and angle from the horizontal average depth of ports under mean sea level and rate of waste-water discharge. If the dilution factor is sufficiently high, the limitations of selected chemicals in the waste-water discharge can be increased to utilise the greater dilution by initial mixing. The effluent limitations for several of the chemicals listed for protection of human health have been established at the minimum detection level for a specified testing technique. For this reason, the allowable limits of non-carcinogens may be increased to a practical quantification level equal to 10 times the detection level and carcinogens increased to 5 times the detection level.

Chapter 16

Biological Waste Treatment

INTRODUCTION

It is recognised by most professionals in the field of waste-water treatment that the most efficient method for eliminating or removing organic material from waste-water is by utilising biological treatment systems. There are three basic systems most often utilised for this removal: activated sludge, trickling filters and aerobic oxidation lagoons. While these three systems may differ somewhat regarding the detention times, oxygen requirements and mode of utilisation of biological slimes, the essential biochemistry that occurs within each system is identical.

PRINCIPLES OF BIOLOGICAL WASTE TREATMENT PROCESSES

In general, the purpose of the biological treatment unit is to remove organic material either by oxidation to carbon dioxide, water and other derivatives or by conversion of the organic material into a settleable form which can be removed by gravity sedimentation. The production of carbon dioxide, water and ammonia is referred to as the respiration stage, while the conversion of the organic material into new bacterial cells which can be settled out is referred to as synthesis.

Since these are aerobic biological treatment processes, the final hydrogen acceptor for the oxidation of organic material is oxygen. During this hydrogen transfer the liberation of energy from the organic molecule is utilised for synthesis and for energy requirements of cellular substances.

The quantity of oxygen that is necessary to oxidise the organic material depends mainly upon the biological oxygen demand (BOD) satisfied during the biological treatment process.

Various treatment methods have been estimated as satisfying the following percentages of applied BOD; conventional activated sludge, 90–95 per cent; high rate trickling filters, 65–85 per cent and low rate trickling filters, 80–90 per cent. The removal of organic material is accompanied by oxidation and synthesis of cells. Gellman and Heukelekian have shown that the amount of new cell material produced per kg of BOD added varies with the chemical composition of the substrate. Hoover and Porges found 52 per cent by weight of cell yield from the oxidation of skimmed milk. Gellman and Heukelekiarn reported a yield of 0.5 kg of volatile suspended solids per pound of BOD fed to the system for several industrial waste. Helmers reported that solids production varied with BOD removal. Okun found that the total quantity of new cell material formed was a function of the loading applied to the system. At high loading ratios high growth rates were obtained, while at low loadings sludge destruction was evident.

Table 16.1 shows the effect of various organic substrates on the relative proportions of oxidation and synthesis in 24 hours, batch-fed activated sludge systems.

Table 16.1. Division of substrate between oxidation and synthesis[a].

Class of compound	Per cent		
	Range present	Oxidised mean	Converted to sludge
Carbohydrates	5–25	13	65–85
Alcohols	25–38	30	52–66
Amino acids	22–58	42	32–68
Organic acids	30–80	50	10–60

[a] 24 hrs., batch-fed activated sludge system.

The variabilities among the yields reported can be explained by visualising that if the organic loading to the unit is quickly assimilated by the micro-organisms then the organisms metabolise themselves, thereby oxidising cellular material normally contributed to sludge yield. The energy yields of different compounds are not the same; consequently, more or less of a particular substrate may be used to satisfy the energy requirements of the system. Thus, the removal of organic material from liquid waste is achieved by complete destruction (oxidation), which yields energy and by synthesis, which uses the energy produced during the oxidation of organic matter. The object of this chapter is to briefly review the three basic biological waste treatment systems and to deal specifically with their methods of operation and the results which we have so far been able to achieve.

Primary Treatment

Before undergoing secondary biological treatment a waste-water is usually subjected to a preliminary form of treatment, referred to as 'primary,' in order to remove suspended and other insoluble matter. These systems usually include rough screens or racks, constant velocity grit removal tanks and primary settling tanks.

Coarse screens or racks

Racks are usually made of long, parallel-shaped bars placed on a slope of approximately 45° from the horizontal and 2–4 inch apart. Their purpose is to remove larger particles of floating or suspended matter from the waste stream. Quantities of screenings removed from domestic waste-water vary from 1 to 15 ft^3 of screenings per million gallons, depending upon the size of openings. There is no measurable reduction in BOD or suspended solids after screening, but the procedure prevents the clogging of treatment equipment, overcomes accumulations of unsightly deposits and intercepts esthetically undesirable floating matter.

Grit chambers

A grit chamber can be defined as a modified settling basin in which the horizontal velocity of flow is so controlled that only heavier solids such as grit and sand are removed while the lighter organic solids are carried forward in suspension. Grit removal is extremely important in protecting working equipment such as pumps and in preventing accumulations of undigestible solids in sludge disposal systems. From 1 to 12 ft^3/l million gallon of grit can be collected from systems, depending upon the type of sewers, weather conditions and the age of the systems.

Primary settling tanks

Primary tanks are utilised in all trickling filter plants and in some activated sludge plants. Their purpose is to reduce the content of settleable solids so as to reduce the load on subsequent treatment systems and to prevent the formation of sludge deposits. They may be either manually or mechanically cleaned, have detention times of 2 hours or less and can be responsible for as much as 35 per cent of the BOD removal and 60 per cent of the suspended solids removal.

The necessity for primary treatment prior to activated sludge systems is still heatedly discussed. On the one hand, the ease of removal of the primary sludge removed approximately 400 ft³/l million gallon at 4–6 per cent and its subsequent effect on secondary treatment make it an attractive addition to a large system. On the other hand, the cost of disposing of the primary sludge removed (approximately 400 ft³/l million gal at 4–6 per cent solids) by anaerobic systems has caused some investigators to advocate its discontinuance.

There is no doubt that it is necessary to use primary treatment prior to trickling filtration systems to avoid clogging of the filter media.

Activated Sludge

The activated sludge process of waste-water purification is one of the most common processes for the secondary treatment of wastes. The activated sludge consists of a gelatinous matrix in which filamentous and unicellular bacteria are imbedded and on which protozoa crawl and feed. The bacterial genera which predominate depend on the characteristics of organic matter in the waste-water, e.g., *Pseudomonas* for hydrocarbon and carbohydrate waste and *Alcaligenes*, *Bacillus* and *Flavobacterium* for proteinaceous wastes. The process consists of mixing activated sludge, recirculated from a final settling tank, with incoming raw or primary sewage to form a mixed liquor, which is subsequently aerated and from which activated sludge is later settled.

When a plant is first started it can be seeded with an activated sludge from a currently operating plant. If however, no seed sludge is available, then one can be built up over a short period of time (4–6 weeks) by simply continually aerating, settling and returning the residue of the sewage.

Description of conventional process

A flow diagram of a conventional activated sludge plant is shown in Fig. 16.1.

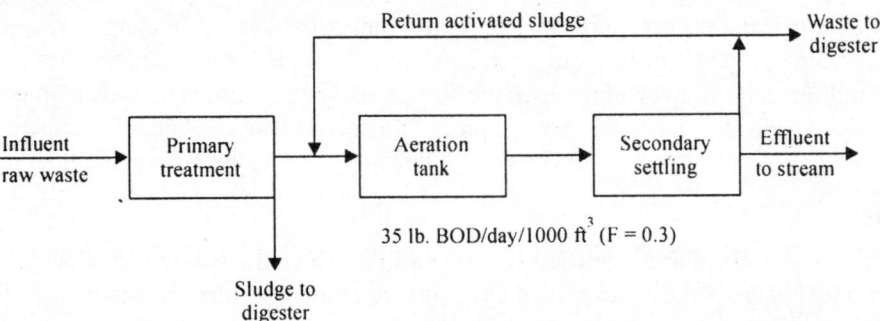

Fig. 16.1. Flow diagram of a conventional activated sludge plant. F refers to sludge loading ratio.

In the process, primary treated sewage is mixed with a portion of returned activated sludge and aerated for 4–6 hours.

The process can be said to consist of the following steps:
1. Mixing of the activated sludge with the sewage to be treated.
2. Aeration and agitation of this mixed liquor for the required length of time.
3. Separation of the activated sludge from the effluent and the subsequent return of a portion of the settled sludge to be mixed with the incoming sewage.

Theory of activated sludge operation

The mechanism of removal of organic material from sewage by activated sludge can be generalised in the following way:

$$\text{Organic material + bacteria} \xrightarrow[\text{O}_2]{\text{Sufficient}} CO_2 + H_2O + NH_3 + \text{energy} \qquad \text{... (16.1)}$$

The release of energy results in the formation of new cellular material; this formation is commonly referred to as synthesis, while the production of carbon dioxide, water and ammonia is referred to as respiration. Thus, we can then say that the oxidation of organic material results in respiration and synthesis.

The bacteria necessary to utilise the organic material are present in the activated sludge flocs. These flocs are gelatinous matrices in which unicellular and filamentous bacteria are present. Protozoa and metazoa are usually present on the surface of the flocs.

When sewage is first mixed with activated sludge a portion of material is stored away (adsorbed) until the bacteria find it necessary to use it as food. The remaining portion of organic material is then oxidised and results in synthesis and respiration. Thus, the total removal of organic material is accomplished in two major parts, adsorption and oxidation. Adsorption occurs within the first section of the aeration tank during the first hour of aeration. The total removal of organic material by the activated sludge process is from 90 to 95 per cent.

Organic loading parameters

The main treatment units involved in the activated sludge process are shown in Fig. 16.1. They include the aeration tank and the sludge separation or secondary settling tank. Primary effluent or the untreated waste-water containing colloidal and soluble organic material, enters the aeration tank where it is attacked and stabilised by the mixed flora and fauna known as activated sludge. This activated sludge, when combined with the influent waste-water, is known collectively as 'mixed liquor,' and the sludge solids are designated as either the 'mixed liquor suspended solids' (MLSS) or 'mixed liquor volatile suspended solids' (MLVSS).

The MLVSS is the volatile portion of the MLSS and is usually related by: MLVSS = 0.80 (MLSS). Most investigators feel that the use of MLVSS more closely represents the total active (biological) mass in the system. The overflow from the secondary settling tank leaves the process as effluent. Since there is usually a net production of biological cellular material by the aerobic treatment process, some sludge must be removed from the system. It is either removed intentionally by separation of sludge or occurs unintentionally by the loss of solids in the process effluent.

The nutrient substrate level or food value of a waste-water, is measured in terms of the biochemical oxygen demand (BOD) of the constituents present in the waste. The relationship existing between the quantity of substrate and the BOD removal efficiency of the process has led to the development of many concepts proposed for use as loading parameters. Loadings have been expressed in terms of pounds of BOD/1000 ft^3 of aeration tank volume. However, constant BOD removal efficiencies (87–90 per cent) have been reported at loadings of 30–120 lb/1000 ft^3, thus indicating the ineffectiveness of this ratio as a general loading parameter. Parameters based on tank volumes do not take into account the amount of solids under aeration or specify the length of the aeration period.

Gould, suggested the term 'sludge age' as a measure of the quantity of organic material entering the process. This term was defined as the average length of time a particle of suspended solids remains under aeration. It was expressed as the ratio of the pounds of mixed liquor solids under aeration to the pounds of suspended solids entering the system per day and thus had a unit of days.

When dealing with sewage, sludge age is adequate since the suspended solids concentration in milligrams per liter usually is about equal to the BOD concentration in milligrams per liter. The parameter is modified for relatively soluble organic wastes by substitution of the pounds of BOD entering the system per day in place of the pounds of suspended solids. This substitution makes the sludge age equivalent to the ratio of the pounds of mixed liquor solids under aeration per pound of BOD entering the process per day.

Harris and others were actually the first to propose, if not actually formulate, a loading parameter combining the three major factors applicable to the activated sludge process. The factors evaluated were the BOD of the applied waste, the quantity of sludge present in the aeration tank and the period of aeration. Since then, ample evidence has been presented to demonstrate that the effectiveness of the activated sludge process and the amount of aid required for its operation are primarily dependent upon the daily BOD input/mixed liquor volatile solids ratio.

In the activated sludge process, it is difficult to obtain an exact measure of the amount of active cell material held in a system. However, because of the simplicity of the test procedures involved, it has become standard practice to measure the weight of suspended 'solids' (MLSS) contained in the mixed liquor and assume that there is a relationship between the amount of solids in the mixed liquor and the number of active organisms present.

The Water Pollution Control Federation has recommended the 'sludge loading ratio' (SLR) as the loading parameter to be used for activated sludge. The SLR is expressed as pounds of BOD applied per day per pound of MLVSS and thus amounts to a food-to-organisms ratio.

The concentration of suspended solids in the mixed liquor depends upon the waste being treated and the aeration capacity of the plant. Normally, the solids concentration is between 1000 and 2000 mg/liter. Calculations of the required solids concentration necessary to obtain efficient treatment under specific conditions can be made by utilising the SLR.

The equation for the calculation of the SLR is:

$$SLR = \frac{24La}{Sat(1+R)} \qquad \qquad ... (16.2)$$

where 'La' is the BOD in milligrams per liter, 't' is the detention time in hours, 'Sa' is the concentration of MLVSS in milligrams per liter and 'R' is the recycle ratio.

It is usually expected that the sludge loading ratio should never exceed a value of 0.3/day in normal sewage plant operation utilising conventional activated sludge.

If we recognise that the area of contact surface and the opportunity for contact are two of the most important factors in the activated sludge process, the sludge loading ratio can be readily accepted as a general parameter of loading intensity. It can be considered as: (i) the weight of removable substrate applied in a unit of time to; (ii) a unit of contact surface for; and (iii) a unit of contact time. The first factor: (i) is readily determined by analytical procedures (BOD) and measurements of sewage flow. The second factor; (ii) can be evaluated only indirectly by the volatile suspended solids concentration in the mixed liquor, while the third factor; and (iii) is usually represented as a statistical average for activated sludge plants.

Even though much research has gone into the development of this loading parameter, situations are encountered in plant practice that indicate the ineffectiveness of this ratio.

The quality of the material making up the BOD is extremely important in plant performance and also for design purposes. BOD is not a simple entity, but must be evaluated in light of the composition of the substrate contributing the BOD. Some domestic sewage treatment plants experience process trouble at one SLR but other plants at the same SLR operate extremely well.

Sludge volume index

Knowledge of the volume of sludge present or of the solids (on a dry basis) in the aeration tank is not sufficient for good process control. A combination of these two is, however, an essential feature of good process control. This combination, called the sludge volume index (SVI), is the volume in milliliters occupied by 1 g (dry weight) of sludge after 30 minutes of settling and is sometimes referred to as the Mohlman index. The test is performed by settling a 1 liter sample of mixed liquor for 30 minutes in a 1000 ml graduated cylinder. The volume occupied by the sludge is reported as per cent or in milliliters; the suspended solids are determined and reported in per cent by weight or in milligrams per liter. The SVI can be expressed as:

$$SVI = \frac{\text{Per cent volume of sludge settled in 30 minutes}}{\text{Per cent suspended solids}} \qquad ... (16.3)$$

Although the 30 minutes period has been adopted as a standard, variations in sludge settling rates taking place in this time, due to the effect of different mixed liquor solids concentrations, have been reported. A modification of this test has been developed by Isenberg which somewhat eliminates these variations and is usually used for all the index calculations in experimental work. This modification calls for an adjustment of the mixed liquor solids concentration to between 1000 and 1500 mg/liter before the SVI determination is made.

A well-settling sludge may have a Mohlman index between 50 and 100, but an index of 200 is indicative of a sludge with poor settling characteristics.

Knowledge of the sludge volume index and the SLR is necessary for good process control since there is a critical value of sludge volume index below which the volume of settled sludge in the final tanks will exceed the return sludge rate. If the SVI rises to 200 then the return sludge rate must be increased to maintain a constant solids concentration under aeration. If the concentration of mixed solids decreases, the sludge loading ratio will increase, thereby increasing the bulking tendency of the sludge and compounding the problem.

Sludge density index

The Water Pollution Control Federation has recommended the use of the sludge density index (SDI) rather than the sludge volume index (SVI). This can be defined as the weight of a specific volume of sludge after it has settled for 30 minutes. It is calculated as follows:

$$SDI = \frac{100}{SVI} \qquad ... (16.4)$$

SDI values are more amenable to graphical expression than are SVI values. Another reason for their use is to satisfy the recommendations of Water Pollution Control Federation and to encourage standardisation of this nomenclature for usage in the field. This means that a sludge with good settling characteristics has an SDI of between 2.0 and 1.0, while an SDI of 0.5 indicates a 'bulky' or 'non-settleable' sludge.

Activated sludge bulking

During normal operation, mixed liquor flows into the final settling tanks from the aeration tanks. The activated sludge forms flocs, settles and the effluent flows over the weirs of the final tank. At times the activated sludge does not settle well; its volume becomes greater in comparison to its density and the return sludge pumps cannot keep up with the large volumes of light sludge settling in the final tanks. If this condition persists, the sludge in the final tanks will spill over the weirs and the BOD of the final effluent will increase. This phenomenon, known as 'sludge bulking', is a major problem of the activated sludge process. Quite frequently, bulking occurs unexpectedly when the plant seems to be operating at its peak efficiency and producing an excellent effluent.

There have been many theories advanced as to the cause of bulking, none of them completely satisfactory. It was observed by many early workers that the organism Sphaerotilus natans was often present in large numbers when bulking occurred. *S. natans* is a sheathed, filamentous bacterium and it was reasoned that its growth in excessive numbers caused the sludge to be less dense and hence caused it to bulk. Heukelekian stated that bulking 'produced by carbohydrates is a direct response of *Sphaerotilus* to a relatively long contact with an available energy food.' Babbitt and Baumann, on the other hand, have stated that 'these organisms are a result, not a cause of bulking.' A causal relationship between the presence of *Sphaerotilus* and bulking has yet to be established, although Finstein has shown a correlation between the number of measurable filaments in a floc and high SVI values. This suggests that the filaments exert control over the sludge volume index.

Bulking has been associated with such characteristics of the raw waste as septicity, heavy organic load, trade wastes, mineral oil and excessive carbonaceous content. In the treatment plant, overaeration, underaeration, poor mixing, short circuiting and too high or too low mixed liquor solids have been listed as causes of bulking. Other causes within the plant have been listed as septic return sludge and excessive detention periods.

Although the characteristics of the raw waste are, without question, associated with the causes of bulking, the investigations indicates that in-plant causes of bulking are more significant than those attributed to raw waste characteristics.

The many different measures used to control bulking reflect the incomplete knowledge of its causes. Reducing the suspended solids in the aeration tank, increasing the quantity of air, increasing the solids in the aeration tanks, use of inert materials, use of iron compounds, chlorination of the return sludge, re-aeration of return sludge and addition of lime to the mixed liquor are some of the methods which have been recommended at various times.

Since most treatment operations are concerned with maintaining a constant SLR or sludge age, it becomes apparent that perhaps there are inconsistencies contained within the formulation of the parameter.

Recent work has shown that differences inherent within the SLR factor are related primarily to solids concentration values alone. The MLVSS value which is maintained is the single most important operational parameter in activated sludge operation.

Active mass approximation and loading parameters

The food-to-organisms ratio designated as the sludge loading ratio (SLR), mentioned earlier, is the parameter currently used in waste-water treatment practice. The SLR corresponds to the symbol F which is sometimes used in mathematical formulations of biological treatment processes. This symbolic representation of a biological system depends on an understanding of the fundamental biological concepts, within the activated sludge process. Without these concepts, it is extremely easy to represent an analysis that is mathematically correct but biologically incorrect.

The constant assumption made in most mathematical formulations is that the suspended solids or volatile suspended solids contained in the mixed liquor represent the total number of bacteria present and are therefore related to the total enzymatic activity of the system. This assumption is not entirely correct. For any given system it may be possible to use some solids parameter as an index of active mass, but it must be used with extreme caution. In an unchanging ecological system the solids concentration may be adequate, but as a general estimation of active mass it is sorely lacking.

Hartmann reported that the weight of nitrogen in the sludge was a better estimation of the quantity of bacteria and was hence the optimum loading parameter for activated sludge systems. Hartmann designated his loadings as BOD per milligram sludge nitrogen.

More recently, interest has been expressed in the possibility of utilising pounds of BOD applied per day per unit weight or deoxyribonucleic acid (DNA) as a loading parameter. This parameter would appear to provide a more accurate measure of the food-to-organisms ratio since the amount of DNA present gives a fairly accurate measure of the amount of cell material present in the system.

Recently, it was shown that a parameter based upon DNA is more responsive than any others currently employed in predicting certain types of operational upsets.

Anaerobic Digestion

When primary treatment is employed prior to secondary systems, a sludge is produced which is usually disposed of by anaerobic digestion. Excess activated sludge may also be treated in such fashion. The main purpose of sludge digestion is to produce an innocuous residue which can be easily disposed of and also to reduce the volume of sludge which must ultimately be removed.

In general, the decomposition of complex organic matter is accompanied by production of intermediate and end products. Such compounds as methane, hydrogen, organic acids and alcohol are the main products of decomposition of carbonaceous organic materials under anaerobic conditions. Similarly, degradation of proteins will result in compounds such as ammonia, amino acids, amides, peptones, hydrogen sulphide, indole, skatole and mercaptans.

Sewage solids subjected to anaerobic decomposition pass through three stages: (i) a period of intensive acid production (acidification); (ii) a period of acid digestion (liquefaction); and (iii) a period of intensive digestion and stabilisation (gasification). Each step is represented by the production of typical intermediate and end products. Under normal operating conditions all three stages occur simultaneously.

The term 'liquefaction' as applied to digestion connotes the transformation of large solid particles into either a soluble or finely dissolved form. This process is brought about by hydrolysis utilising extracellular enzymes. It is during this period that intermediate products of fermentation accumulate and gasification is at a minimum. When sludge is digested without a seed source (a sludge that has been digested under similar environmental conditions) this condition is greatly exaggerated. With seeded sludge, liquefaction is in balance with gasification and there is usually no undue accumulation of intermediate products. Although the terms 'liquifaction' and 'hydrolysis' are used interchangeably they are not strictly synonymous. Hydrolysis is a well-defined chemical term designating the addition of water to the molecule to break-down complex substances into simpler ones. Liquefaction does not have such an exact connotation and refers merely to the transfer of substances from a solid sludge stage to a liquid phase.

The primary gases produced during the gasification stage are methane and carbon dioxide. These two gases normally form more than 95 per cent of the gas evolved. The average heat value of the gas is approximately 700 BTU/ft^3. A good indicator of the degree of digestion is the percentage of methane gas contained in the total digester gas. Usually, when the methane production is low (under 65 per cent) the digestion is poor. The maximum volume of gas that is generated from a heated anaerobic digester is approximately 10 ft^3/pound of volatile solids added to the tank or approximately 0.7 ft^3/capita per day.

Methane production results from the breakdown of many compounds by numerous interdependent and interaction reactions which take place in an orderly and integrated fashion.

Methane organisms which produce methane do not utilise such substances as cellulose, glucose, proteins, amino acids or fats but they do utilise a restricted group of simple compounds consisting of lower fatty acids (formic, acetic, propionic, n-butyric, etc.). The transformation of complex organic materials contained in sludge to methane and carbon dioxide is brought about in two stages by two different groups of bacteria. The complex organics are converted by a variety of common bacteria to volatile acids and alcohols without the production of methane (acid production). These products are then converted to methane (methane fermentation stage) by a restricted and specialised group of bacteria among which are: *Methanobacterium omelianskii*, which utilise primary and secondary alcohols; *Mbact. suboxydans*, which partially utilises butyrate, valeric acid and other four and six-carbon fatty acids to produce acetic and propionic acid and *Mbact. propionicum, Mbact. mayei* and *Mbact. barkerii,* which utilise the simpler organic acids and alcohols and produce methane and carbon dioxide.

The organisms responsible for active and thorough digestion of waste solids require an environment in which the pH is about 7. The optimum pH value for digestion varies slightly with the characteristics of water supply and the types of waste present. Insufficient seed or low temperature results in retardation of gasification and accumulation of acidic intermediate decomposition products.

In the normal digestion of domestic sewage sludge small amounts of carbon monoxide, as well as hydrogen sulphide, may be present. Oxygen should not be present in the digester gas. Its presence is indicative of an air leak in the digester or an error in the gas sampling procedures. Although some claims of improved digestion with aeration have been made these can usually be attributed to increased mixing of the tank contents with the air. The introduction of large quantities of air on a regular basis will adversely affect methane fermentation.

Some of the most important considerations affecting sludge digestion are: food supply, which is influenced by the type of primary sludge generated in the primary treatment system; time of digestion; utilisation of seed sludge; temperature; mixing; pH; the volatile acids to alkalinity ratio and the quantity and amount of chemicals added to the digestion system. The primary indices of digestive action are:

Gas production

A good rule of thumb is that 10 ft^3 of gas should be produced per pound of volatile solids added to the tank. The gas produced should be approximately 70 per cent methane and 30 per cent carbon dioxide with only traces of miscellaneous gases.

Volatile solids

The sludge produced should have a solids content of approximately 5 per cent with a volatile content of roughly 50 per cent.

Volatile acids

The volatile acid content of the digested sludge should be approximately 500 mg/liter or less. Levels of higher acidity can be corrected by utilising chemicals such as lime.

Care should be taken, however, not to utilise great amounts of chemicals but rather to change environmental conditions to secure better digestion.

Sludge characteristics

The digested sludge should be black in colour, easily dewatered and have a 'tarry' odour that is not repellent. In high rate digestion, rates of digestion can be substantially increased by: (i) thorough mixing of the tank contents either mechanically or by gas recirculation; and (ii) optimum loading of the digesters by regulation of the solids in the feed sludge so that as dense a sludge as possible is fed to the digesters. Detention times can be decreased to as little as five days when adequate heating is provided.

The rate of activities of the organisms responsible for digestion is greatly influenced by temperature. The time required for digestion is indicated by the quantity of gas produced and the increased amount of volatile matter destroyed. The total amount of gas is not appreciably different at the end of digestion but is produced in a shorter time at higher temperatures. Increased gas production follows an increase in the amount of volatile matter destroyed. The best temperature for mesophillic digestion is about 85°C. Temperatures up to 95°F increase the rate of digestion slightly but may be more difficult to maintain throughout the year and may result in problems. The volume of gas may be greater in the presence of some organic industrial waste, while other wastes may reduce the amount of gas. The carbon dioxide in the gas ranges from 15 to 35 per cent and is affected by the degree of digestion and the types of trade waste present.

ACTIVATED SLUDGE PROCESS MODIFICATIONS

Tapered Aeration

In an aeration tank having a definite pattern of longitudinal flow, the impact of the high BOD of the influent entering the head end of the tank will create a relatively high oxygen demand in this point in the mixed liquor. As the oxygen demand of the waste is gradually decreased, the demand for oxygen becomes less and less. If one can envision a plug flow system passing down the length of the tank, then the tapered aeration process (Fig. 16.2) can be seen to take the gradually decreasing oxygen demand of the mixed liquor into account by making more oxygen available at the head end of the tank by the use of more diffusers.

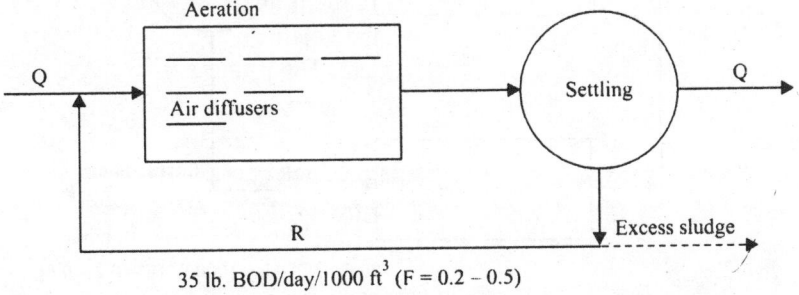

Fig. 16.2. Tapered aeration. Q, R and F refer to flow through plant, recirculated effluent and sludge loading ratio, respectively.

As with the conventional activated sludge process, tapered aeration has a volumetric loading of about 35 lb BOD/day per 1000 ft³ of aeration tank capacity and an F value of 0.2–0.5, which is in the same order of magnitude as conventional activated sludge.

Step Aeration

Another activated sludge modification which is capable of handling shock loadings as well as evening out the oxygen demand in the mixed liquor entails the introduction of the waste flow at intervals throughout the length of the tank. This process is termed step aeration. This system (Fig. 16.3) has a volumetric loading of greater than 50 lb/day per 1000 ft³ of aeration tank capacity. The F ratio, however, is the same magnitude as that used in conventional activated sludge plant operation.

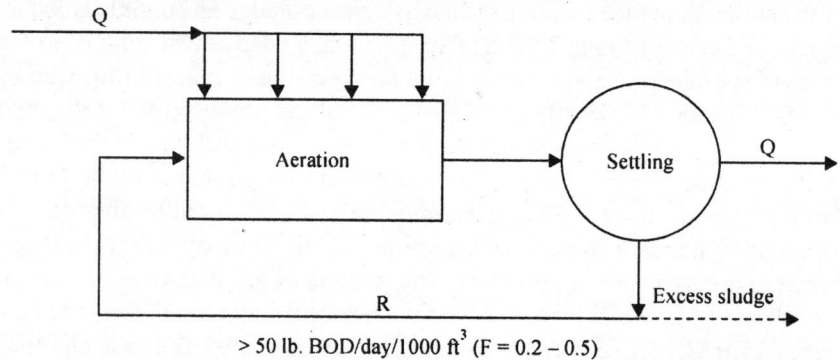

> 50 lb. BOD/day/1000 ft³ (F = 0.2 – 0.5)

Fig. 16.3. Step aeration (step loading). Q. R and F refer to flow through plant, recirculated effluent and sludge loading ratio, respectively.

Contact Stabilisation or Sludge Re-aeration

Contact stabilisation is yet another modified process that permits up to twice the volumetric loading of the conventional process. In contact stabilisation, the volumetric loading for the system is in the order of 70 lb BOD/day per 1000 ft³ of aeration tank capacity. This system is shown in Fig. 16.4.

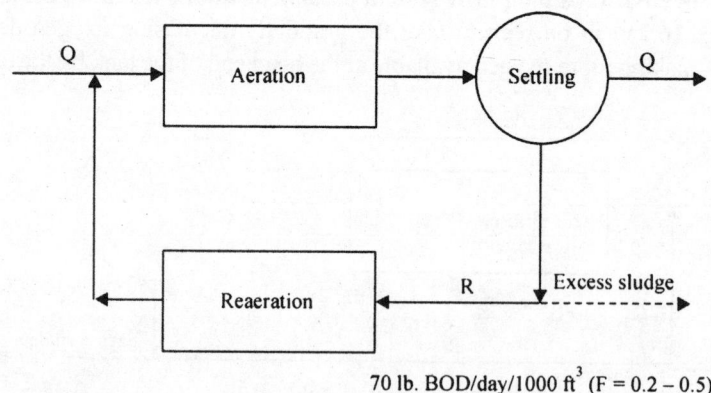

70 lb. BOD/day/1000 ft³ (F = 0.2 – 0.5)

Fig. 16.4. Contact stabilisation, Q, R and F refer to flow through plant, recirculated effluent and sludge loading ratio, respectively.

The mixed liquor, displaced to the settling unit, is settled out and pumped to the re-aeration unit where it is aerated without further waste-water addition. The organisms thus enter a declining phase and when finally discharged into the aeration tank, they have the capability of removing large amounts of substrate BOD by the anabolic processes involved in the assimilation and storage of substrate previously discussed.

The actual contact time between waste-water and mixed liquor is maintained between 30 minute and 1 hour by the design of the system. Unlike the step aeration process, in which this contact time can be varied by the operator regulating the point(s) of waste entry, the contact stabilisation process does not give extremely great flexibility in load assimilation. It is primarily used in the design of 'package' treatment systems as well as in industrial waste applications.

Hatfield Process

The Hatfield process, shown in Fig. 16.5, differs from contact stabilisation in that anaerobic digester supernatant or, in some cases, digested sludge is fed to the re-aeration tank. Proponents of this process feel that in a waste flow containing large amounts of highly carbonaceous industrial material, the supplying of anaerobic digester effluent to the re-aeration unit fortifies the active sludge solids with amino acids and other nitrogenous substances. In addition, there is some thought that the addition of a heavier type of solid to the aeration tank would act to prevent the non-settleability or bulking of the activated sludge. As with the contact stabilisation process, the Hatfield process has the advantage of being able to maintain a large weight of organisms under aeration in a relatively small aeration system.

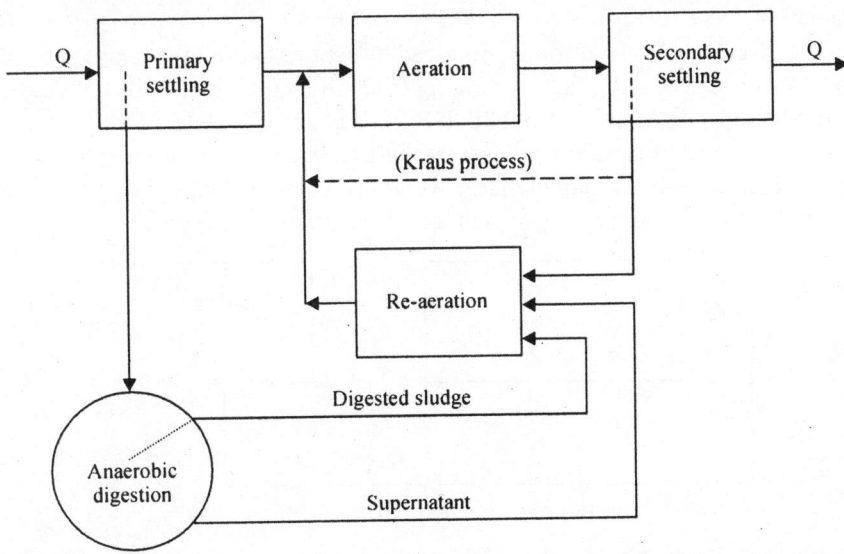

Fig. 16.5. Hatfield process. Q and R refer to flow through plant and recirculated effluent, respectively.

Kraus Process

Also shown in Fig. 16.5 is the Kraus process modification. In this process some of the return sludge by-passes the re-aeration unit and is delivered directly to the mixed liquor aeration tank. Therefore, in certain aspects, the Kraus process is a hybrid between the conventional activated sludge process and the Hatfield system.

Short-Term or Modified Aeration

The short-term aeration processes have extremely high loading factors, varying from about 0.5 up to 5 lb BOD/day per lb MLVSS. Modified aeration has a loading factor range of about 2–5 lb BOD/day per lb MLVSS. The volumetric loading for modified aeration is about 100 lb BOD/day per 1000 ft³ of aeration tank capacity.

The short-term aeration processes offer considerable economy of construction due to the very small aeration tank capacities that are required. However, it will be noted that aeration systems are able to contain only relatively low organism weights and that the effluent quality will suffer accordingly, since the lower the organism weight in a system, the greater will be the amount of unused BOD in the process effluent. However, this effect is not directly proportional to the weight of organisms contained in the system and effluent quality reduction becomes serious only when the weight of organisms is extremely small, as is the case for the modified aeration and supraactivation systems.

Short-term aeration systems produce a relatively large amount of net growth of MLVSS. Sludge disposal becomes a major problem if the sludge is intentionally removed from the system. If sludge is allowed to remain within the system the BOD removal efficiency would range between 50 and 75 per cent since excess sludge would pass out in the process effluent. Removing the sludge intentionally, although creating a sludge handling problem, would increase the BOD removal efficiency to as high as 90 per cent.

Extended Aeration

The extended aeration process (Fig. 16.6) is characterised by a low loading spectrum. The objective of this process is to oxidise the biological solids produced by synthesis from the removal of BOD. Extended aeration typically operates at loading factors ranging from about 0.05 to 0.2 lb BOD/day per lb MLVSS and the volumetric loading is generally about 20 lb BOD/day per 1000 ft³ of aeration tank capacity. One unique feature of the extended aeration process is the system's ability, because of its relatively large aeration tank volume, to contain a relatively large weight of volatile sludge. The low volumetric loading rate, together with the large weight of organisms, combine to give the typically low loading factors.

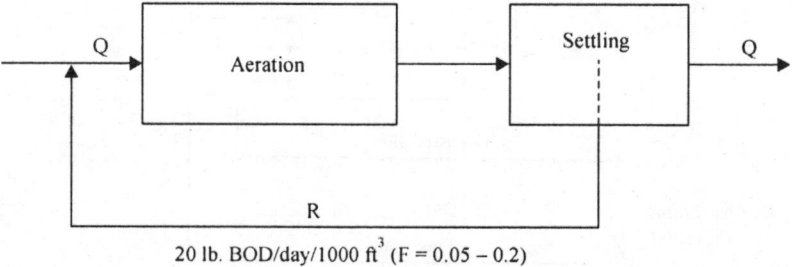

20 lb. BOD/day/1000 ft³ (F = 0.05 – 0.2)

Fig. 16.6. Extended aeration. Q, R and F refer to flow through plant, recirculated effluent and sludge loading ratio, respectively.

As usually operated, the extended aeration process has no intentional wasting of sludge from the system because, theoretically, no excess activated sludge is produced. Practically, however, net growth is produced and a large amount of it is wasted from the system, unintentionally, in the effluent.

This escape of solids results in lowered efficiencies in the range of 75–85 per cent. If sludge is removed intentionally there are indications that removal efficiencies can be improved to those of conventional activated sludge.

Because of the relatively large concentration of micro-organisms carried within the aeration tank and the extended length of aeration time, endogenous respiration plays a major role in sludge quality. The volatile portion of the sludge remaining is not degraded at the same rate as normal activated sludge and thereby results in a lower BOD exertion per unit weight of solids. Effluents from this process, therefore, contain higher suspended solids contents but relatively less BOD than conventional effluents containing equivalent solids concentrations. For this reason effluents from extended aeration systems often meet regulatory agency requirements for BOD levels but contain unsatisfactory levels of suspended solids.

TRICKLING FILTERS

The second major biological treatment system is called the trickling filter. In essence, the name is a misnomer since the biological unit neither filters nor does it trickle. The major difference between the trickling filter and the activated sludge system, as far as its ecological patterns are concerned, is that the trickling filter utilises a succession of biological communities established at different levels within the trickling filter and associated with correspondingly different degrees of purification. The activated sludge system, however, has the same biological community within the floc at any one time and this community is associated both with the raw untreated waste entering the basin and with the purified effluent.

Within the trickling filter the interrelationships and activities of the different members of the biological community are similar to those outlined for the activated sludge system, although they are limited to the extent that stratification occurs. In general, there seems to be an underlying consensus of opinion in the field of pollution control stating that trickling filters are more easily adapted or utilised to treat shock loadings of waste sources.

Description

A trickling filter is a bed of coarse, rough, hard, impervious material over which the sewage is sprayed or otherwise distributed (by a distributor) through the air. A biological slime which grows on the filter packing is responsible for the biological reactions. The sewage then flows downward through the filter in contact with the air. The filter is usually 3–12 ft deep and is provided with an underdrainage system to remove the filter effluent and provide ventilation.

The main function of a trickling filter is to remove unstable, organic, pollutional materials in the form of dissolved and finely divided organic solids and to oxidise these solids biologically to form more stable materials. There are many variations in flow pattern for trickling filtration plants. The most common patterns are given in Figs. 16.7a–d. No universally applicable flow system exists and it is not unusual for one plant to operate on many different patterns during the course of a year.

Theory of Filter Operation

The trickling filter process depends upon the biochemical oxidation of complex organic material in sewage. Soon after a filter is placed in operation, the surface of the filter medium becomes covered with zoogleal slime, a viscous, jelly-like substance containing bacteria and other biota. Under favourable environmental conditions, the slime adsorbs and utilises suspended, colloidal and dissolved organic matter from the sewage, which passes in a relatively thin film over the slime's surface. Eventually a population equilibrium is reached. As biota die, they are discharged from the filter together with the more or less partly decomposed organic matter.

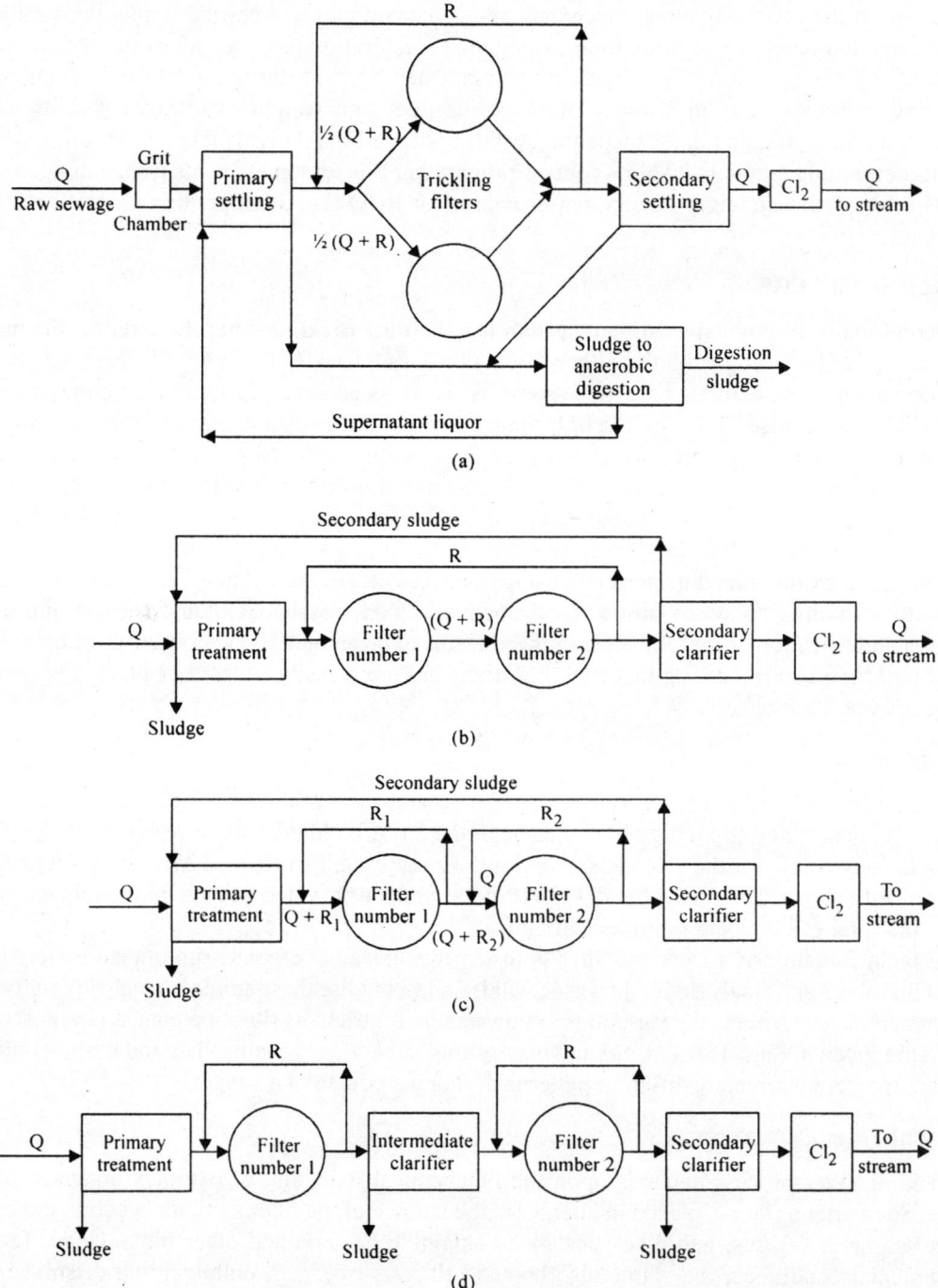

Fig. 16.7. (a) Trickling filters in parallel with direct recirculation; (b) two-stage plant with direct recirculation to primary stage; (c) two-stage plant with direct recirculation within each stage; and (d) two-stage plant with recirculation and clarification between stages. Q and R refer to flow through plant and recirculated effluent, respectively.

This 'sloughing off' of material may occur periodically or continuously depending upon the type of filter. A low or standard rate filter sloughs periodically, while a high rate filter sloughs continuously. Generally, secondary settling is provided to retain the settleable solids sloughed from the filter.

The essential features necessary to the process are: (i) surface area must be provided for biological growth and the surface area/volume ratio must be large; (ii) free oxygen must be available so that aerobic conditions exist; and (iii) waste must be amenable to biological treatment; sewage is no problem, but industrial wastes may be.

BOD and Suspended Solids Removal

Trickling filters are capable of providing adequate treatment of wastes susceptible to aerobic biologic processes where the production of a plant effluent of 20–30 mg/liter BOD is acceptable. Table 16.2 shows some generally expected BOD and suspended solids removals.

Table 16.2. Comparison of BOD and suspended solids (SS) removal for various processes.

Type of plant	Expected BOD removal, %	Expected SS removal, %
Primary treatment	15–35	45–65
Primary and trickling filtration treatment	80–90	80–95

BOD removals are affected by:
1. Quantity and quality of waste: A definite ratio exists between the quantity of a waste applied to a trickling filter and the BOD removal efficiency. In general, the greater the quantity of waste applied, the lower the BOD removal efficiency. In order to obtain a reasonable degree of treatment the waste should be free from constituents that are toxic to the filter organisms, such as cyanide, copper, chromium and other heavy metals.
2. Temperature: Greater BOD removals should be expected during the summer months due to the increased activity of the micro-organisms.

Construction Features

The shape of a filter is related to the type of distributor used. Most of the old plants have rotary distributors, the filters are usually circular. Plants utilising fixed nozzle distributors usually have rectangular filters.

Some filters have been built without retaining walls for the media. However, such construction is seldom economical. The majority of filters have circumferentially reinforced concrete walls, usually 8–12 inch thick.

The choice of the filter medium is often governed by the material locally available and the cost of transporting it. Field stone, gravel, blast furnace slag, redwood blocks and synthetic inert materials have all been used. The medium should be hard, clean, free of dust, insoluble in sewage and approximately cubical in shape to obtain a large surface area/volume ratio. Ninety-five per cent or more of the medium should pass a $4\frac{1}{2}$ inch square screen but be retained on a 2 inch square screen. According to new regulations, the depth of filtering medium at any point in the filter should not be less than 6 ft or greater than 9 ft for standard rate filters. The minimum depth of medium for high rate filters is set at 5 ft.

The medium must serve two primary purposes: (i) it must provide surface area for slime growth; and (ii) it must leave sufficient voids for free circulation of air.

The underdrainage system in the filter serves two primary purposes: (i) to carry sewage effluent passing through the filter away for further treatment; and (ii) to provide ventilation and maintenance of aerobic conditions. The entire underdrainage system should be designed to permit free passage of air and provisions should be made for flushing out of the lateral channels. The inlet openings should have an unsubmerged gross combined area equal to at least 5 per cent of the surface area of the filter.

The floor of the filter must be strong enough to support the underdrainage system, the filter medium and the water load if the filter is to be flooded.

Distribution Systems

Sewage is applied to the filters by either fixed nozzles or rotary arm distributors. Fixed nozzle filters are usually rectangular in shape and utilise deflector plates along with the orifices, as shown in Fig. 16.8.

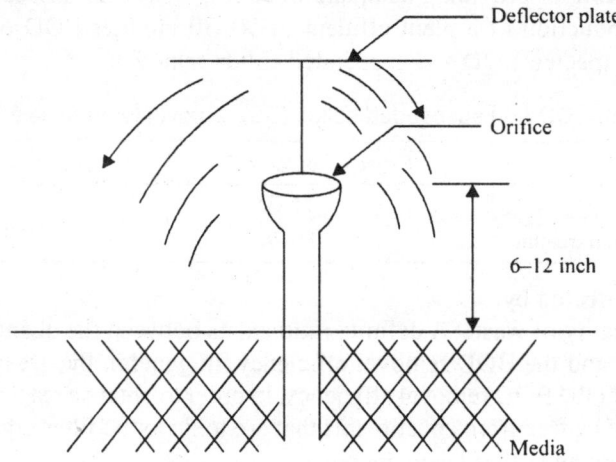

Fig. 16.8. Fixed nozzle distributor.

Rotary distributors are the most commonly employed. The arms are rotated by the reaction of the sewage discharge or, in some cases, are motor-driven. The arms are usually set 6 inch above the filter bed.

Loadings

Trickling filters are classified according to the applied hydraulic and organic loadings. The hydraulic load is the total volume of sewage (including recirculation) applied to the filter per day per square feet of surface area [gal/day (gpd) per ft^2 or million gpd/acre]. The organic load is the pounds of 5 day BOD applied to the filter per day per cubic foot of filter medium (lb BOD/day per ft^3 medium) or (lb/day per 1000 ft^3), Table 16.3 shows a differentiation between standard rate and high rate filters.

Table 16.3. Classification of trickling filters.

Type of filter	Hydraulic load. gpd/ft²	Organic load, lb BOD/day per 1000 ft³	Application	Sloughing	BOD removal through filter, %
Standard rate	25–100	5–25	Intermittent	Largely periodic	80–85
High rate	200–1000	25–300	Continuous	Continuous	65–80
Roughing filters	500	300	Continuous	Continuous	25–65

Some filters are called 'roughing filters'. These are usually high rate filters receiving high organic loadings. Although they may give a high pound per unit volume of organic load removal, their settled effluent still contains substantial BOD. They are used to provide intermediate treatment or as the first of a multistage biological treatment. Most high rate filters utilise recirculation, which is the recycling of filter effluent through the filter. In such cases the ratio of recycled flow to sewage flow is known as the recirculation ratio. Among the useful purposes of recirculation are: (i) reducing 'out of service' periods to a minimum by adjusting recirculation to influent flow; (ii) keeping self-propelled distributors turning by adjusting recirculation to influent flow; (iii) lowering film thickness and fly breeding by film sloughing; and (iv) improving the quality of the effluent, at constant efficiency of treatment, by reducing the concentration of applied sewage. Some high rate filters are given special names which signify specific patented recirculation or distribution schemes.

Biofilter

The biofilter (Fig. 16.9) employs recirculation and a high rate of application to a shallow trickling filter. The recirculation in this case involves bringing the effluent of the filter or the secondary settling tank back to the primary sedimentation basin. This requires designing the primary settling tank to accommodate not only the average daily sewage flow but also the recirculated flow, which may be as much as 10 times the average flow.

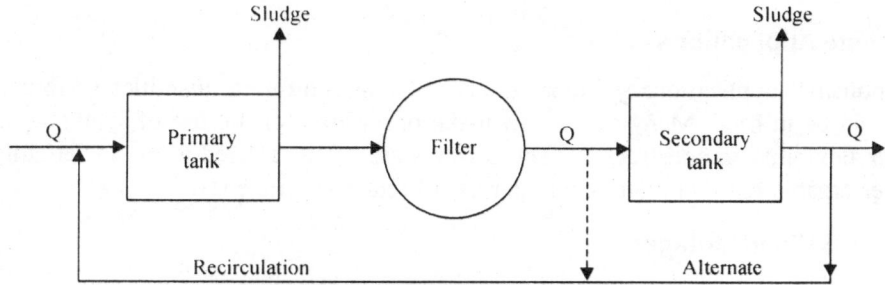

Fig. 16.9. Biofilter. Q refers to flow through plant.

Accelofilter

This filter (Fig. 16.10) employs recirculation of the filter effluent directly back to the filter.

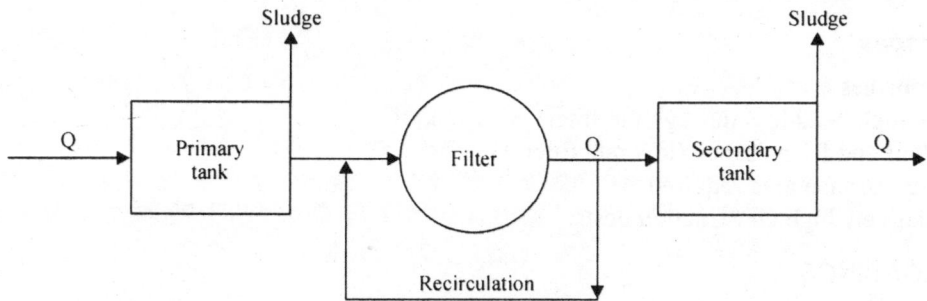

Fig. 16.10. Accelofilter. Q refers to flow through plant.

Aerofilter

The aerofilter (Fig. 16.11) is a filter over which the sewage is distributed by maintaining a continuous rain-like application of the sewage over the filter bed. A disk distributor revolving at speeds as high as 380 rpm and set 20 inch above the surface of the filter is sometimes used on small beds. For large beds 10 or more revolving distributor arms are used to obtain the uniform rain-like application. Aerofilters operate at hydraulic loadings greater than 300 gpd/ft^2 of surface area.

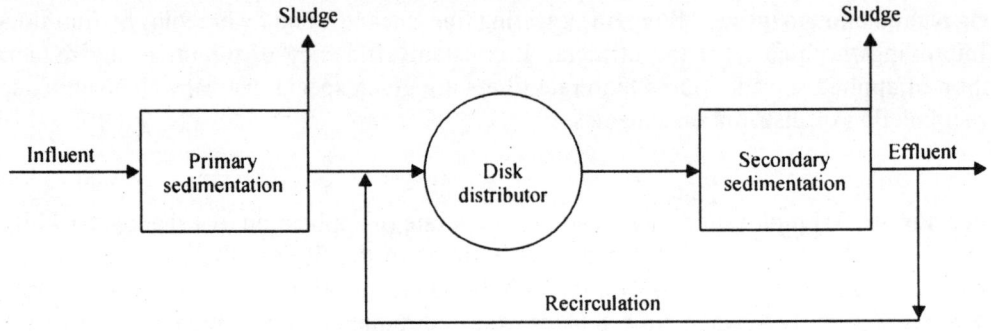

Fig. 16.11. Aerofilter.

Industrial Waste Applications

For certain industrial applications sectional filters utilising controlled quantities of forced air within each section can be utilised. Many states now make provisions for the use of synthetic media which allows greater flexibility in construction. These filters are often referred to as 'tower' filters and can receive greater organic loads at relatively high rates of removal efficiency.

Advantages and Disadvantages

Advantages

Among the advantages of a trickling filter are:
1. Relatively high nitrifying effect, i.e., biological oxidation of amino nitrogen to nitrates.
2. Dependability to give a good effluent under wide variations in influent quality.
3. Relatively low operating and maintenance cost.
4. Ability to function under extreme weather conditions.

Disadvantages

The disadvantages are:
1. The high head loss through the filter.
2. Odour and fly nuisance (low rate filters).
3. Large surface area required.
4. Relatively high construction costs.

SHOCK LOADINGS

An important consideration in the design of a biological sewage treatment plant should be the ability of the system to cope with immediate or rapidly occurring changes in the chemical or physical environment of the biological population. Such environmental changes may be broadly categorised as shock loads, the effects of which may vary from slight malfunctioning of the system to a complete cessation of metabolic activity resulting in the shutdown of the system.

Shock loading of systems may take any of several forms. The most common type is the quantitative shock, which involves a change in the concentration or the waste or the organic loading of the system. A rapid decrease in organic loading, termed a hydraulic shock, might result from an influx of storm water which dilute the influent waste. In this same category is the rapid increase in loading which might result from the removal of a blockage in the sewer system. It is important to bear in mind that the chemical nature and structure of the waste are unchanged from those normally handled by the system and thus the existing population requires no acclimation period for this type of shock.

A toxic shock is the result of materials which damage or inhibit the existing metabolic pathways or disrupt the physiological condition of the microbial population. The occurrence of heavy metals, cyanide compounds or materials producing rapid pH changes would be considered toxic shocks. The treatability of such a waste depends on a prolonged acclimation period and even then adaptation of the micro-organisms to the new substrate is not guaranteed. Qualitative shock loads are associated with a change in the structural configuration of the carbon source and do not of themselves entail any change in the organic loading of the system. This type of shock could occur when a new waste is introduced into the system, such as cannery or meat-packing wastes.

Trickling filters and the activated sludge process are the most commonly encountered biological treatment systems. Since quantitative shock loads involve an increase in the hydraulic or organic loading of the system, the apparent method of treatment would be to increase the capacity of the plant by having additional sedimentation tanks, filters or aeration tanks on a standby basis. It is immediately apparent that this is an extremely uneconomical approach to the problem.

Inherently, a high rate trickling filter has the potential to cope with quantitative shocks since it employs recirculation to maintain the hydraulic loading on the system. Recirculation may be increased to dilute high organic loads or decreased to accommodate hydraulic shocks. By these adjustments the trickling filter effluent can be maintained at a fairly stable level irrespective of large variations in the influent.

The standard activated sludge process does not provide sufficient latitude in operation to handle large quantitative shocks. Sludge bulking and inadequate treatment times caused by the shock result in the release of an unsatisfactory effluent. This inability of the activated sludge process to cope with shock loads has led to numerous modifications of the system in an attempt to alleviate this problem.

The step aeration system (Fig. 16.3) provides for the introduction of the primary effluent at several points along the course of the mixed liquor. All of the return sludge is introduced at the head of the plant. By varying the feed to the various passes, the BOD/suspended solids ratio may be completely controlled, thus evening out the oxygen demand in the mixed liquor. Step aeration provides great flexibility in the handling of quantitative shocks, but BOD removal is sacrificed.

Other variation of the activated sludge (AS) process is commonly known by the names sludge re-aeration, contact stabilisation or biosorption (Fig. 16.12). By separating the adsorption and oxidation phases, double the volumetric loading of step aeration is facilitated. This ability to handle larger loadings than the standard process is the property which provides for adequate treatment of shock loads, but again, it is at the expense of BOD removal. It should also be noted that this system is not as flexible as step aeration.

A somewhat different approach to the problem of quantitative shock loading has been the use of multistage biological treatment. To this end, systems have been designed with high rate biological filters in parallel with the AS process or trickling filters in series with the AS process. Systems such as these take advantage of the trickling filter's ability to handle shock loads and the ability of AS to produce an effluent of high quality and in this fashion eliminate some of the shortcomings of each.

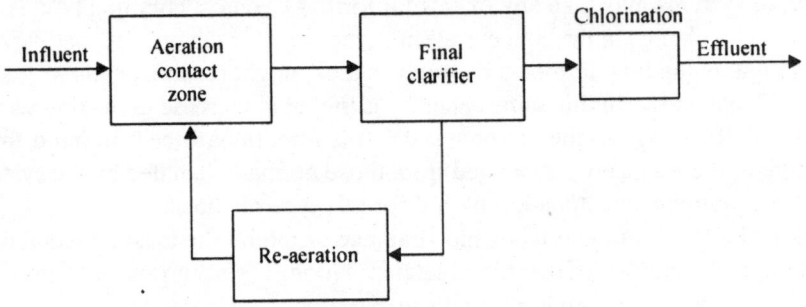

Fig. 16.12. Sludge re-aeration or biosorption.

Along these lines, AS systems have been used sequentially. Such serial operation of AS systems permits plant operation to be easily adjusted to shock overloads.

The activated sludge aeration system is a combined approach to shock loading of a quantitative type. It employs a step aeration system run in parallel with a conventional AS process, the step aeration system serving as a source of return sludge for both legs of the circuit. This combination provides the flexibility of step aeration and the high quality effluent of the AS process.

Quantitative shock loads may be successfully handled, in general, by processes designed for increased volumetric and/or organic loading without provision for duplication of system components. This ability has been provided by the use of recirculation, separation of the adsorption and oxidation phases of treatment or by multistage biological treatment, in general, at the expense of BOD removal.

The treatment of qualitative or toxic shock loads poses very different problems since an adaptation phase is required for the biological population to adjust to the new substrate.

The utility of trickling filters is open to question when non-quantitative shocks are considered. One opinion is that they are more resistant to shocks of toxic and organic substances than AS, because of their relatively short contact time and that they recover more quickly from the effects of deleterious substances or unfavourable conditions. Rhodes maintains that substances which are detrimental to biological life (toxic shocks) have the same effect on micro-organisms whether they live on filter media or in an AS floc. The utility of trickling filters is also questioned on the grounds that it is more time-consuming and costly to clean filter material and renew biological activity on a trickling filter than it is to renew the activity of the biological mass in the activated sludge process. Since the successful response of a system to a qualitative shock must be initiated and completed during the time of substrate-biological solids contact, activated sludge is better than a trickling filter from a biochemical standpoint. The treatability of a toxic shock depends on a prolonged acclimation period which is very much greater than the delay time in the system and hence neither method can adequately treat such a shock.

It has been shown, in practice, that the sludge re-aeration or biosorption system (see Fig. 16.12) recovers from wide variations in pH of short duration in a short period of time and that long-term shocks (36 hours) require recovery times of 5 days. This is a measure of the system's ability to resist a toxic shock and has nothing to do with its ability to treat it. Similarly, it has been found that biological treatment systems can be acclimated to specific toxic shocks not exceeding some maximum concentration, but this is in the realm of industrial waste treatment, including such patented systems as the 'integrated waste treatment system' for the removal of cyanide and chromium from an acid medium. It seems that the best way to handle a toxic shock is prevention of it at the source, i.e., before the material enters the biological treatment system.

Biological treatment systems can and do satisfactorily treat qualitative shocks. Stage operation of activated sludge systems can be used since there is a highly active sludge available in the second stage which is not endangered, since the new substrate is caught up in the first stage.

Especially suited to the treatment of qualitative shocks are the Hatfield process (Fig. 16.5) and its modification, the Kraus process. In the Hatfield process the supernatant from an anaerobic digester or digested sludge is fed to the re-aeration tank and subsequently to the aeration tank. This provides an extracellular source of nitrogen, which is essential to the successful response of the system to a shock high in carbonaceous materials. The Kraus process differs only in that some of the return sludge by-passes the re-aeration tank and goes directly to the mixed liquor aeration tank. This is shown by the dotted line in Fig. 16.5. That biological treatment can treat quantitative and qualitative shock loads is demonstrated by the number of plants using the methods already discussed. One of these plants is of special interest in that it was designed to optimise treatment of shock loads. This plant utilises a Kraus nitrified sludge interchange process which is provided with a balanced nutrient through the continuous introduction of a mixture of activated sludge and waste digester liquor into the aerated sewage. The nutrient sludge is produced by a 24 hours aeration period in nitrification tanks. The aeration tanks are arranged so that they may be used as either two or four-pass units. Air is introduced, at different depths, from each side of the tank to maintain turbulent mixing.

The processes employed at this plant allow great flexibility in the handling of shock loads, as they provide a step aeration for quantitative shocks and an extracellular source of nitrogen for qualitative shocks. It is a prime example of the ability of biological systems to treat shock loads.

Chapter 17

Residual Management

INTRODUCTION

In most water treatment processes the objective is to remove certain materials from the water, to purify it. These materials are referred to as residuals and consist of the liquid-solid and gaseous-phase by-products removed during the water treatment process along with any transport water that is removed with them. These residuals include the turbidity-causing materials in raw water, organic and inorganic solids, algae, bacteria, viruses, colloids, precipitated from the raw water and those added in treatment and dissolved salts. Sludge is the term used to refer to the solid or liquid–solid, portion of some types of water plant residuals such as the underflow from sedimentation basins.

Residuals management is a term used to describe the planning, design and operation of facilities to reuse or dispose of water treatment residuals. From a technical standpoint the objective in residuals management is usually to minimise the amount of material that must ultimately be disposed of by recovering recyclable materials and reducing the water content of the residuals. In most cases, the cost of transporting and ultimately disposing of the residuals makes up the major fraction of residuals management costs and the most economical solution is to reduce the quantity of material for ultimate disposal. Other considerations include minimising environmental impacts and meeting discharge requirements established by regulatory agencies.

Residuals management can have an important impact on the design and operation of many water treatment plants. For existing plants, residuals management systems may limit overall plant capacity if not designed and operated properly. Frequently, residuals are stored temporarily in the process train before removal for treatment, recycle and/or disposal. Residual removal must be optimised for the process train and co-ordinated with the residuals management systems to maintain water quality.

The anticipated type and quantity of residuals may influence the selection of water treatment processes for new plants. For instance, concentrate is the main residual from a reverse osmosis membrane water treatment plant. The characteristics of concentrate (e.g., high salt and total dissolved solids concentrations) make it difficult to dispose of and may influence the selection of reverse osmosis as a treatment option.

In light of the many issues associated with and the importance of residuals management, the purpose of this chapter is to: (i) define the nature of the problem, including the sources of residuals; (ii) review the physical and chemical properties used to characterise water treatment plant residuals; (iii) consider the residuals and their properties, produced by the principal treatment processes; (iv) review options available for the management of residual liquid streams; (v) review the options available for the management of residual concentrates and brines; (vi) review the options available for the management of residual sludges; and (vii) review options available for the ultimate reuse and/or disposal of residuals after processing.

DEFINING THE PROBLEM

The problem of residuals management can be quantified with respect to: (i) the quantities and costs of handling residuals; (ii) the constituents of concern and (iii) the environmental and regulatory constraints any engineered solution must meet. Before considering these topics, it is appropriate to consider the sources of residuals in water treatment processes.

Sources of Residuals

The residuals generated from the treatment of water can be classified as: (i) sludges from water treatment processes; (ii) liquid wastes from water treatment processes; (iii) liquid wastes resulting from processes used to thicken process sludges and, to treat liquid wastes; and (iv) gaseous wastes from specialised water treatment processes. The sources of these residuals and a brief description are, presented in Table 17.1. The specific types of sludges and liquid waste streams will depend on the type of treatment train as illustrated on Fig. 17.1.

Table 17.1. Sources of solid, liquid and gaseous residuals from the treatment of water.

Source of residual	Description
	Treatment process residuals
Coagulant/polymer (e.g., alum and iron) pecipitation	Sludges resulting from the chemical precipitation of surface water that may contain clay, slit, colloidal material and micro-organisms with chemicals and polymers
Water softening water	Lime sludge resulting from the removal of calcium and magnesium from hard during precipative softening
Coarse screens	Coarse screens prevent the entry of debris and fish into the intake structure. The coarse solids retained on the screens include rags, stringy material and large wood pieces
Travelling screens	Travelling screens are used to prevent grit, sand and small rocks that have come through the intake from continuing into the treatment facility. Screening include grit, sand and small rocks
Presedimentation	Sludge resulting from presettling to remove gross amounts of sediment prior to conventional treatment
Slow sand filter scrapings	Semisolid material resulting from the scraping of the surface of slow sand filters
	Treatment process liquid wastes
Filter waste washwater	Waste washwater from backwashing filters to remove residual solids. Waste washwater is high in turbidity and may contain pathogenic organisms such as Giardia and Cryptosoridium
Filter-to-waste-water	Water used to condition filters after backwashing that has particles and turbidity above regulatory action levels
Membrane (nanofiltration or reverse-osmosis) concentrate	High-salinity solution produced by the concentration of salts from brackish or saline waters

(Contd...)

Source of residual	Description
Ion exchange brines	Regenerant wastes from brine and rinse water typically containing sodium, chloride, and hardness ions. Usually high in TDS but low in suspended solids
Slow sand filter washwater	Washwater high in turbidity that may contain pathogenic organisms such as Giardia and Crytosoridium resulting from the cleaning of slow sand filter scrapings. (Note in many facilities the scraped sand is not washed for re-use on the filter beds but is used for other purposes)
	Thickening/dewatering process liquid wastes
Centrate	Liquid resulting from centrifugal thickening of sludge
Filtrate	Liquid resulting from plate and frame thickening of sludge
Pressate	Liquid resulting from belt press thickening of sludge
Supernatant flow	Clear water decanted off residual solids resulting from the gravity and flotation thickening of sludge
Leachate	Liquid underflow (percolate) from sand and other types of drying beds
	Treatment process gaseous wastes
Stripping towers	Off-gas from stripping operations contains contaminants that may need to be removed before gas can be discharged

Magnitude of the Problem

As much as 3 to 5 per cent of the volume of the raw water entering a conventional water treatment plant may end up as solid and liquid residuals. The bulk of that volume will be the filter waste washwater, which typically contains less than 10 per cent of the removed solids in a conventional treatment plant. Underflow from sedimentation basins typically contain on the order of 0.1 to 0.3 per cent of the plant flow but contain most of the solids removed. In a direct or in-line filtration plant, however, all solids removal is accomplished in the filters. Typical values for the quantities of residuals produced by various treatment processes are summarised in Table 17.2. The costs of handling these residuals are dependent on the type of handling provided and the nature of the residuals. In general, the major portion of the cost with residuals management is associated with transport and ultimate disposal.

Table 17.2. Typical production of residuals in water treatment facilities as per cent of plant flow.

Type of residual	Per cent of plant flow	
	Range	Typical
Alum sludge	0.08–0.3	0.1
Iron sludge	0.08–0.3	0.1
Filter backwash water	2–5	4
Microfiltration backwash water	2–8	6
Reverse-osmosis or nanofiltration concentrate	10–50	10–50
Lime-softening sludge	0.6–6	4
Ion exchange brine	1.5–10	5–8

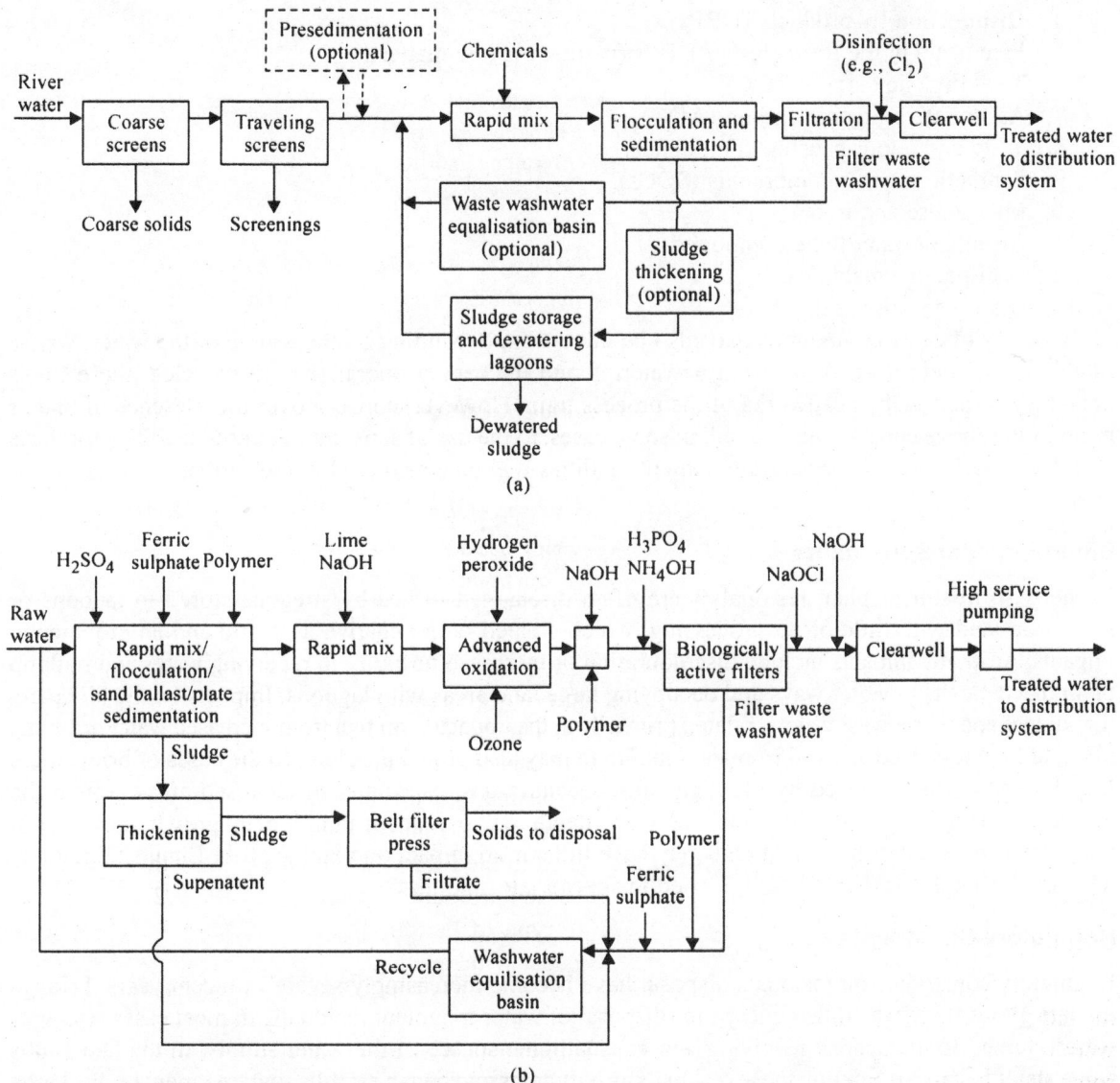

Fig. 17.1. Typical water treatment process flow diagrams: (a) small plant with sludge storage lagoons. Future options include the addition of a waste washwater recovery basin and sludge thickening before discharge to sludge lagoons; and (b) large plant with mechanically intensive sludge-processing facilities.

Constituents of Concern

Residual constituents of concern contained in the sludges and liquid wastes from treatment processes and thickening operations may include the following:

1. Pathogenic micro-organisms.
2. Giardia cysts and Cryptosporidium oocysts.
3. Turbidity/particles.

4. Disinfection by-products (DBPs).
5. Precursers in the formation of DBPs (natural organic matter).
6. Total organic carbon (TOC).
7. Assimilative organic carbon (AOC).
8. Taste-and odour-causing compounds.
9. Synthetic organic compounds (SOCs).
10. Manganese and iron.
11. Arsenic or other toxic compounds.
12. Radio-active materials.
13. Dissolved solids/salt.

A variety of other compounds may also be of concern depending on the source of the water. Where liquid wastes and return flows from dewatering and thickening operations are recycled, these flows must be returned to the headworks of the process train. However, concern over the presence of one or more of the above constituents has led, in some cases, to the use of separate treatment facilities for these liquid wastes. The use of separate treatment facilities will be considered in the further sections of this chapter.

Environmental Constraints

In the past, treatment plant residuals were often discharged to nearby streams, stored in lagoons or spread on land with little or no processing, which created both negative aesthetic and environmental impacts. Aesthetic impacts include discolouration or increased turbidity in receiving water and buildup of sludge deposits in water ways and occupying large land areas with lagoons. Impacts on the biota are, for sludges and waste washwaters, related primarily to the impact(s) on fish from increased water turbidity, pH; and hardness. Redissolved iron and aluminum may also pose a problem. In the cases of brine, there may be toxic effects caused by the high salt concentrations, especially in localised areas around the discharge. Most sludges, if spread on land to any depth, will prevent or inhibit plant growth; however, if adequately mixed into the soil, sludge may have little or no impact on plant growth. Lime sludges may have beneficial impacts on the soil, if used in appropriate amounts.

Regulatory Constraints

Regulatory constraints on residuals disposal have become increasingly severe in recent years. Prior to the late 1960s there was little concern for disposal of water treatment residuals. In most cases residuals were returned to the nearest receiving water, usually the source of the water supply. In the late 1960s some states began considering these residuals as pollutants and began establishing treatment or discharge standards for them.

The Water Pollution Control Act classified water treatment plant residuals as pollutants and categorised them as industrial waste. As such, they are now required to meet standards for best practicable control technology (BPT) currently available and best available technology (BAT) economically achievable. There has also been legislation, to control toxic and hazardous substances. Such regulations, while protecting public and environmental health, can severely limit the available residuals management options and add to the cost of disposal.

An understanding of the physical, chemical and biological properties of the residuals produced by treatment processes is fundamental to determining appropriate management techniques and to design facilities to implement those techniques. In this section, the physical, chemical and biological properties used to characterise water treatment plant residuals are reviewed. Additional information on the individual residuals is presented in the following sections dealing with coagulation sludges, lime sludges, filter waste washwater and softening and demineralisation concentrates.

Physical Properties

In general, the physical properties of water treatment plant residuals are the most important for sizing and design of residuals management facilities. The physical properties used most commonly to characterise residuals are summarised in Table 17.3. Total solids is one of the most important physical parameters. Sludge density is dependent on the moisture content, varying from the density of water (1000 kg/m^3, 62.4 lb/ft^3) for sludges below about 1 per cent to 1200 to 1520 kg/m^3 (75 to 95 lb/ft^3) for sludges greater than 2 to 3 per cent. A reasonable estimate of the density of wet inorganic sludges, typical of alum or iron salts, can be made by assuming the density of the dry solids is about 2300 kg/m^3 (145lb/ft^3).

Specific gravity of sludge

The volume of sludge depends mainly on its water content and only slightly on the character of the solid matter. For example, a 5 per cent sludge contains 95 per cent water by weight. If the solid matter is composed of fixed (mineral) solids and volatile (organic) solids, the specific gravity of all the solid matter can be computed as:

$$\frac{W_s}{S_s \rho_w} = \frac{W_f}{S_f \rho_w} + \frac{W_v}{S_v \rho_w} \qquad \qquad ...(17.1)$$

where,

$\quad W_s \quad = \quad$ weight of total dry solids, kg
$\quad S_s \quad = \quad$ specific gravity of total solids
$\quad \rho_w \quad = \quad$ density of water, kg/m^3
$\quad W_f \quad = \quad$ weight of fixed solids (mineral matter), kg
$\quad S_f \quad = \quad$ specific gravity of fixed solids
$\quad W_v \quad = \quad$ weight of volatile solids, kg
$\quad S_v \quad = \quad$ specific gravity of volatile solids

Thus, if 90 per cent by weight of the solid matter in a sludge containing 95 per cent water is composed of fixed mineral solids with a specific gravity of 2.5 and 10 per cent is composed of volatile solids with a specific gravity of 1.0, then the specific gravity of all solids, S$_s$, would be equal to 2.17, computed using Eq. 17.1:

$$\frac{1}{S_s} = \frac{0.90}{2.5} + \frac{0.10}{1.0} = 0.46$$

$$S_s = \frac{1.0}{0.46} = 2.17$$

If the specific gravity of the water is taken to be 1.00, the specific gravity of the sludge, S$_{sl}$, is 1.03, as follows:

$$\frac{1}{S_{sl}} = \frac{0.05}{2.17} + \frac{0.95}{1.00} = 0.97$$

$$S_{sl} = \frac{1.0}{0.97} = 1.03$$

Volume of sludge

The volume of a sludge may be computed with the following expression:

$$V = \frac{W_s}{\rho_w S_{sl} P_s} \qquad \qquad ...(17.2)$$

where,

W_s = weight of total dry solids, kg
ρ_w = density of water, kg/m^3
S_{sl} = specific gravity of sludge
P_s = per cent solids expressed as a decimal

For approximate sludge volume calculations for a given solids content, it is simple to remember that the volume varies inversely with the per cent of solid matter contained in the sludge, as given by

$$\frac{V_1}{V_2} = \frac{P_2}{P_1} \quad \text{(approximate)} \qquad \qquad ...(17.3)$$

where,

V_1, V_2 = sludge volumes
P_1, P_2 = per cent solid matter

Specific resistance

Specific resistance, dynamic viscosity, initial settling velocity and other physical properties are dependent on solids concentrations and the relative proportions of coagulant and other materials in the sludge. Specific resistance is a measure of the rate at which a sludge can be dewatered. Although developed for the vacuum filtration process, specific resistance has been found to be a useful parameter for assessing the dewaterability of sludges by gravity settling, centrifugation, belt filtration, plate and frame pressure filtration and sand beds. The specific resistance of a sludge can be determined from laboratory data on the time for a given volume of water to be filtered, obtained using a Buchner funnel (see Fig. 17.2) or other filters specifically designed for the purpose. Sludges with values greater than 100×10^{10} m/kg are considered to be difficult to dewater.

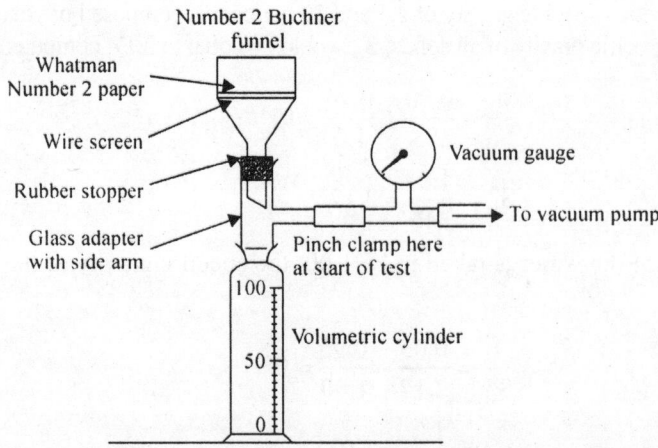

Fig. 17.2. Buchner funnel test apparatus used for determination of specific resistance of sludge.

Other physical properties

The variation in physical properties among sludges of various compositions is due to the physical structure of the sludge. A coagulation sludge is made up of the suspended material in the raw water, metal hydroxides added during coagulation treatment and a large amount of bound and entrapped water in a loose structure. The suspended materials are clay and sediment particles, colour-causing colloids, algae and other similar materials. Clays and sediments are structurally solid and have a specific gravity of around 2.6; the other materials are agglomerations of individual metal hydroxide molecules with various ions and water molecules all loosely held together by electrostatic bonds. Metal hydroxides become attached to the suspended materials by electrostatic bonds and also physically entrap suspended materials as well as water molecules. In the coagulation–flocculation process the suspended particles and metal hydroxides are brought together to form the flocs that then settle and make sludges. When the individual flocs come in contact with each other in the sludge, they become loosely bound by the same electrostatic forces that hold the individual flocs together.

The extent of the bonding depends on the extent of the contact between the flocs, which is limited by the entrapped water that separates the flocs. As more and more water is removed by draining, pressure or other means, the particles contact more and the sludge becomes increasingly solid. Because the metal hydroxides have water molecules in their structure, direct particle contact is more difficult than for other suspended materials that do not have water in their structure. Therefore, sludges with high proportions of metal hydroxides are not as easily dewatered as are sludges that have higher proportions of other materials unless the sludge is conditioned with large amounts of lime or an appropriate polymer dose. The polymer molecules form bridges between floc particles and improve the bond between particles. If polymers are added to coagulation sludges, either as sludge conditioners or as a part of the coagulation process, the sludge produced will have a more solid structure.

Chemical Properties

The chemical properties of residual sludges are related directly to the chemical content of the raw water and the coagulant chemicals. Important chemical characteristics are summarised in Table 17.3. The BOD, COD, TOC and related organic content are representative of the dissolved and suspended organic materials and algae removed from the water. The inorganic solids are derived from the coagulant chemicals and the clay and sediments removed from the raw water. The pH and dissolved solids in the liquid portion of the sludge are about the same as those in the water being treated. In a complete chemical analysis of an alum sludge, a total of 72 elements were detected. The major elements found were (in order of decreasing predominance by weight) silicon, aluminum, iron, titanium, calcium, potassium, magnesium and manganese.

Table 17.3. Physical, chemical and biological properties used to characterise water treatment plant residuals.

Parameter	*Unit of expression*	*Description*
Physical		
Total solids	%	Measure of total mass of material that must be handled on dry basis as percent of combined mass of solute and material
Dry density	kg/m^3	Measure of mass per unit volume on dry basis
Wet density	kg/m^3	Measure of mass per unit volume on wet basis

(Contd...)

Parameter	Unit of expression	Description
Specific gravity of dry solids	Unitless	Mass relative to mass of water
Specific resistance	m/kg	Measure of rate at which sludge can be dewatered (see Eq. 17.6)
Dynamic viscosity	$N \cdot s/m^2$	Measure of resistance to tangential or shear stress
Initial settling velocity	mm/s	Initial settling rate of water-solids suspension
Chemical		
BOD	mg/l	Estimate of readily biodegradable organic content
COD	mg/l	Measure of oxygen equivalent of organic matter determined by chemical oxidation
pH	Unitless	Measure of effective acidity or alkalinity of solution
Alum content	% or mg/l	Derived from addition of coagulating chemical
Calcium, magnesium content	% or mg/l	Derived from addition of lime for water softening
Iron content	% or mg/l	Derived from addition of coagulating chemical
Silica and inert material	% or mg/l	Material present in surface water supplies
Trace constituents	µg/l or mg/l	Detection of specific constituents of concern
Biological		
Bacterial	no./100 ml	Variable depending on source of water and season
Protozoan cysts and oocysts	no./100 ml	Variable depending on source of water and season
Helminths	no./100 ml	Variable depending on source of water and season
Viruses	no./100 ml	Variable depending on source of water and season

Biological Properties

Water treatment plant residuals may contain a variety of micro-organisms, depending on the source, the quality of the raw water, the treatment process employed (e.g., prechlorination) and the time of year. Coagulation sludges, filter waste washwater and membrane concentrates will contain bacteria, protozoan cysts and oocysts and viruses removed during treatment. It is not possible to generalise on the number of micro-organisms that may be present per unit mass or volume for the reasons cited above.

COAGULATION SLUDGES

Coagulation sludges are produced by the coagulation and settling of natural turbidity by added coagulant chemicals. In a treatment plant coagulation sludges are collected in the sedimentation basins and on the filters. The amount and properties of the sludge collected in the basins and the filters depends upon the water quality, type and dose of coagulant used, efficiency of operation, plant design and other factors. For typical plants using alum as the coagulant, between 60 and 90 per cent of the total residuals will be collected in the sedimentation basins with the remainder in the filters. The residuals collected on the filters are removed from the filters during backwashing and, if the waste washwater is recovered, are removed from the waste washwater by settling.

Sludge from the sedimentation basins may be removed continuously or, more commonly, on an intermittent basis. Sludge removal may be accomplished using various mechanical devices or by draining and manually washing down the basin. If basins are manually cleaned, the frequency may be once every 3 months or more. Mechanical cleaning equipment is usually designed to operate between once a week and once every few hours or continuously. Sludge removal frequency is decided based on balancing the competing interests of maintaining water quality in the process train, available sludge storage in the process train, the ability of the residuals management system to accept additional inflows and disposal costs.

Filters are typically backwashed every 24 to 72 hours, resulting in a relatively large volume of waste washwater produced in a short time. However, some proprietary filter types (such as automatic backwash filters) may backwash as frequently as every 2 to 6 hours. Waste washwater recovery facilities must be designed to accept high, intermittent flows.

Coagulation sludges are grouped according to the type of primary coagulant employed. The principal types of coagulant employed are (i) hydrolysing metal salts of alum and iron; (ii) prehydrolysed metal salts such as polyaluminum chloride (PACl) and polyiron chloride (PICl) and (iii) synthetic organic polymers.

Estimating Quantities of Coagulant Sludges

Typical overall values for the quantities of coagulant sludge produced were summarised previously in Table 17.2. For design purposes, the amount of sludge anticipated at a plant can be estimated based on the quality of the raw water and the type of chemical treatment. The suspended solids fraction of the sludge may be safely assumed to be equal to the suspended solids of the raw water or, if suspended solids data are not available, it may be estimated from turbidity data. For turbidities less than 100 nephelometric turbidity units (NTU) the suspended solids in milligrams per liter have been shown to be approximately equal to the turbidity in NTU. It should be noted that the relationship between turbidity and suspended solids can be quite variable, depending on the organic content of the source water.

Typical precipitation reactions for alum and iron, when used as coagulants, are as follows:

$$Al_2(SO_4)_3 \cdot 14H_2O + 3Ca(HCO_3)_2 \longrightarrow 2Al(OH)_3 + 3CaSO_4 + 6CO_2 + (14)H_2O \qquad ... (17.4)$$

$$Fe_2(SO_4)_3 + 3Ca(HCO_3)_2 \longrightarrow 2Fe(OH)_3 + 3CaSO_4 + 6CO_2 \qquad ... (17.5)$$

$$2FeCl_3 + 3Ca(HCO_3)_2 \longrightarrow 2Fe(OH)_3 + 3CaCl_2 + 6CO_2 \qquad ... (17.6)$$

For alum or iron sludges, the precipitated coagulants are largely aluminum and iron hydroxide, respectively. Both of these will include some bound water that will not be released in a standard total suspended solids test. Bound water must be considered in estimating solids loadings on process facilities. In the case of alum sludge, the amount of bound water will vary from 1.25 to 3.0 times the amount of aluminum hydroxide. Using Eq. 17.7, it can be calculated that a total of 0.33 kg of sludge on a dry-solids basis will be produced for each kilogram of alum $[Al_2 (SO_4)_3 \cdot 14H_2O]$ added. If a similar proportion of bound water is assumed for iron hydroxide sludges, the value for ferric sulphate $[Fe_2 (SO_4)_3]$ would be about 0.59 kg sludge/kg ferric sulphate $[Fe_2(SO_4)_3]$ added and 0.48 kg sludge/kg ferric chloride ($FeCl_3$) added. Typical values that can be used to estimate the quantity of alum and iron sludges are given in Table 17.4.

Prehydrolised PACl $[(Al)_a(OH)_b(Cl)_c(SO_4)_d]$ is supplied in solution form containing varying amounts of aluminum. The amount of sludge produced can be estimated using the relationship:

$$mgAl(OH)_3/mg \text{ PACl added} = (mg \text{ PACl}/l) \text{ (\% Al in PACl}/100) \text{ (mg Al(OH)}_3/mg \text{ Al)} \quad ... (17.7)$$

Typical values that can be used to estimate the quantity of PACl sludge are given in Table 17.4.

For polymer sludges or sludges with polymer used as coagulant aid, the amount of polymer added should also be included in the calculation of the total amount of sludge produced. Other coagulant aids, such as bentonite or actived silica, should also be considered in the calculation as well as any other chemicals or materials, such as activated carbon, that may be collected in the basins or filters.

Table 17.4. Typical values that can be used to estimate quantities of sludge resulting from addition of coagulating chemicals and polymers, turbidity removal and softening in water treatment processes.

Process	Unit	Range	Typical value
Coagulation			
Alum, $Al_2(SO_4)_3 \cdot 14H_2O$	kg dry sludge/kg coagulant	0.33–0.44	0.33
Ferric sulphate, $Fe_2(SO_4)_3$	kg dry sludge/kg coagulant	0.59–0.8	0.59
Ferric chloride, $FeCl_3$	kg dry sludge/kg coagulant	0.48–1.0	0.48
PACl	kg dry sludge/kg PACl	$(0.0372–0.0489) \times Al$ (%)	$(0.0489) \times (Al,$ %)
Polymer addition	kg dry sludge/kg coagulant	1.0	1.0
Turbidity removal	mg TSS/NTU removed	0.9–1.5	1.25
Softening			
Ca^{2+a}	kg dry sludge/kg Ca^{2+} removed	2.0	2.0
Mg^{2+b}	kg dry sludge/kg M^{2+} removed	2.6	2.6

[a] Sludge is expressed as $CaCO_3$.
[b] sludge is expressed as $Mg(OH)_2$.

The total sludge mass and volume produced can be calculated as

$$\text{Total sludge} = \text{sludge from turbidity} + \text{sludge from alum}$$
$$+ \text{sludge from other chemicals or materials} \qquad \dots (17.8)$$

Physical Properties of Coagulant Sludges

The physical properties of alum and iron sludges are summarised in Table 17.5. The solids concentrations and physical properties are the most important properties for sizing and design of residuals management facilities. Solids concentrations depend on the design and operation of the sedimentation basins in addition to the type of sludge and its composition. For example, alum sludge from an upflow clarifier would typically be drawn off at a concentration of 0.1 to 0.3 per cent solids, compared to sludge from a horizontal-flow basin at 0.2 to 1.0 per cent or more. Sludge may thicken to 4 to 6 per cent solids if it is allowed to accumulate for a month or longer in a horizontal-flow sedimentation basin. Sludges that have relatively high proportions of alum or iron coagulant, as would result from treating low-turbidity water, will have lower solids concentrations than will those with relatively higher proportions of turbidity or silt. Coagulation of water having substantial algae concentrations will also result in light, low-solids-concentration sludges. The addition of polymers generally tends to produce higher solids concentrations.

As coagulation sludges are dewatered and dried, there is a gradual transformation from a liquid to a solid. For the purposes of designing residuals management facilities, it is important to know when that transformation occurs as it will determine the type of equipment required to handle the sludge. As a liquid, the sludge can be pumped, piped and transported in tank trucks, while as a solid it must be shoveled and transported on a conveyor or in open trucks.

Unfortunately, the transition is not sharply defined and the transition point is not the same for all sludges. Coagulation sludges that have high proportions of gelatinous aluminum or iron hydroxides will act as liquids at higher solids concentrations than will those that contain more clay and sediments. These sludges are also thixotropic; that is, on standing they will seem to solidify, but when disturbed they go back to a liquid. The minimum concentration at which sludge can be considered a solid is about 16 per cent solids; however, for design purposes a value in the range of 40 to 50 per cent is recommended.

Table 17.5. Typical physical properties and chemical constituents of alum and iron sludges from chemical precipitation.

| Item | Unit | Type of sludge | |
		Alum	Iron
Physical properties			
Volume	% water treated	0.1–0.3	0.1–0.3
Total solids	%	0.1–4	0.25–3.5
Dry density	kg/m^3	1200–1500	1200–1800
Wet density	kg/m^3	1025–1100	1050–1200
Specific resistance[a]	m/kg	10–50 × 1011	40–150 ×1011
Viscosity at 20°C	N · s/m^2	2–4 × 10^{-3}	2–4 × 10^{-3}
Initial settling velocity	m/h	2.2–5.5	1–5
Chemical constituents			
BOD	mg/l	30–300	30–300
COD	mg/l	30–5000	30–5000
pH	Unitless	6–8	6–8
Solids			
$Al_2O_3 \cdot 5.5\ H_2O$	%	15–40	
Fe	%		4–21
Silicates and inert materials	%	35–70	35–70
Organics	%	5–15	5–15

[a] Values of specific resistance reported in literature in units of s^2/g must be multiplied by 9.81 × 10^3 [(s^2/g)(9.81 m/s^2) (10^3 g/kg)] = m/kgl to obtain units of m/kg.

Chemical Properties of Coagulant Sludges

The chemical characteristics of coagulant sludges are directly related to the chemical content of the raw water and the coagulant chemicals. Typical data on the chemical characteristics of coagulant sludges are given in Table 17.5.

LIME SLUDGES

Lime sludges are produced from the precipitation of calcium carbonate and magnesium hydroxide in the lime-soda softening process. Lime sludges may be essentially pure chemical sludges or they may include suspended materials from the raw water if turbidity removal is combined with softening. Similar sludges are produced in the magnesium carbonate softening process.

Estimating Quantities of Lime Sludges

Typical quantities of lime sludge produced from water softening are reported in Table 17.2. For design purposes, the amount of sludge anticipated at a plant can be estimated based on the chemical treatment and raw-water quality. The sludge is essentially composed of precipitated calcium carbonate and magnesium hydroxide, any turbidity or suspended solids that are removed during softening and any insoluble impurities present in the treatment chemicals, such as lime grit. The suspended contribution can be estimated from turbidity data, as discussed above for coagulation sludges.

The amounts of precipitated calcium carbonate and magnesium hydroxide can be estimated directly from the anticipated calcium and magnesium removals. It is not necessary to include any water of hydration in the formula of the precipitates as in the case of coagulation sludges, so removal of 1.0 mg of calcium (expressed as Ca) results in 2.5 mg of $CaCO_3$ in the sludge. Similarly, removal of 1.0 mg of magnesium (expressed as Mg) results in 2.4 mg of $Mg(OH)_2$ in the sludge. A graphical means of estimating sludge production on the total hardness removed is given on Fig. 17.3.

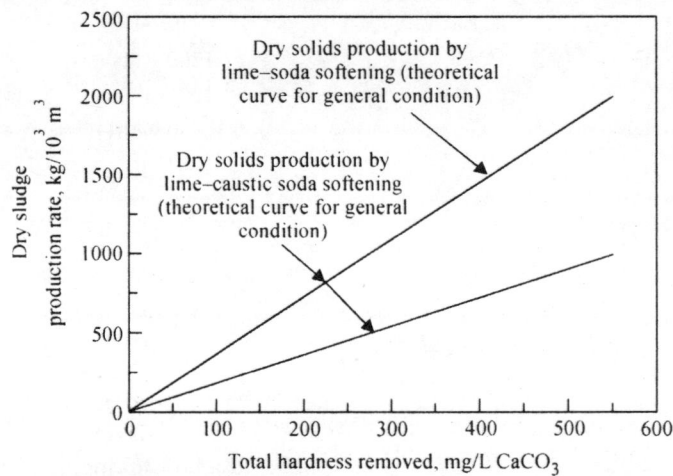

Fig. 17.3. Dry sludge production rate versus total hardness removed.

Physical Properties of Lime Sludges

The physical properties of lime sludges, as reported in Table 17.6, are the most important factors in sizing treatment facilities. Although not very common, in larger plants, it is sometimes economical to recover lime from the sludge, in which case the chemical content of the sludge also becomes important. Solids concentrations are dependent on the treatment process and on the proportions of the various chemical precipitates in the sludge. Sludges that are high in $CaCO_3$ typically have higher solids concentrations than sludges with more $Mg(OH)_2$ because $CaCO_3$ is a fine-grained, dense precipitate while $Mg(OH)_2$ is a more gelatinous material. The typical treatment process utilising either upflow or horizontal-flow sedimentation basins following chemical addition and reaction produces fine-grained precipitates similar in nature to mud or silt deposits.

Table 17.6. Typical physical properties and chemical constituents of lime-softening sludge.

Item	Unit	Range of values
Physical properties		
Volume	% water treated	0.3–6
Total solids	%	2–15
Dry density	kg/m³	1100
Wet density	kg/m³	1920
Specific resistance	m/kg	12×10^{10}

(Contd...)

Item	Unit	Range of values
Viscosity	$N \cdot s/m^2$	$5–7 \times 10^{-3}$
Initial settling velocity	m/h	0.4–36
Chemical constituents		
BOD	mg/l	0–low
COD	mg/l	0–low
pH	Unitless	10.5–11.5
Total dissolved solids	%	2–15
Solids		
CaO_3	%	40–80
Silicates and inert materials	%	8–12
Organics	%	5–8

Chemical Properties of Lime Sludges

Lime sludges typically have a high pH (10.5 to 11.5) and are white unless coloured by turbidity, iron or manganese. Generally lime sludges are odourless, with little or no organic matter. Because of the high pH, lime sludges do not contain significant numbers of viable micro-organisms. Typical chemical characteristics are summarised in Table 17.6. The specific chemical content of sludge from any given plant can be determined from the raw-water quality and the chemical treatment used.

DIATOMACEOUS EARTH SLUDGES

Sludges are produced during backwash of diatomaceous earth filters. Generally the diatomaceous earth (DE) filter process is operated until the filter cake contains about two parts of DE for each part of impurities removed from the water. As a result, the sludge characteristics are predominantly those of the DE. The volume of backwash water will vary from 2 to 5 per cent of the plant flow. The corresponding solids concentration will typically vary from 6000 to 8000 mg/l, comprised primarily of DE. The total dry sludge is equivalent to the DE added in the process plus the suspended materials removed in the filter. The dry density of DE is about 160 kg/m^3.

GRANULAR MEDIA FILTER AND MEMBRANE FILTER WASTE WASHWATER

Waste washwater from the cleaning of granular or membrane filters is the most common type of liquid waste produced at water treatment plants.

Estimating Quantities of Filter Waste Washwater

It is difficult to estimate the quantity of waste washwater because it will depend on the raw-water quality, the degree and effectiveness of the treatment processes preceding the filtration step, the duration of the filter run and the duration and type of backwash cycle employed. Based on the operating experience from a variety of water treatment plants, the quantity of waste washwater for both granular and membrane filters will typically comprise from 2 to 5 per cent of the total amount of water processed. Some designers use 5 per cent as a design value for the quantity of waste washwater. The volume of washwater to be handled depends on the frequency and duration of the backwash cycle. Because the backwash cycle from membrane filters is shorter and more frequent than that from granular filters, the volume of water to be handled from each backwash will be smaller.

Estimates of filter waste washwater quantities and frequencies may be obtained from pilot studies and filter design criteria. Pilot studies can be used to obtain critical information on backwashing rates and frequencies, which may be scaled up to address backwash duration at full scale. Using the information from pilot plant studies, the frequency, volume and flow rate of waste washwater may be estimated.

Filter design criteria that are relevant to determining waste washwater frequency in granular filters are the unit filter run volume (UFRV) and the unit backwash volume (UBWV). These concepts are used to determine the effective filtration rate (v_{EFF}) and the recovery or production efficiency (Rec= $v_{EFF}v$) for a filter. The design criterion for production efficiency is typically 95 per cent or greater. Typically, waste washwater quantities are 8 m^3/m^2 (200 gal/ft^2). To achieve a filter production efficiency of 95 per cent, the UFRV would have to be at least 200 m^3/m^2 (5000 gal/ft^2) a run. At filtration rate 0.2 m^3/m^2 (5 gpm/ ft^2), a filter run would have to last at least 1000 minute between backwash cycles.

The calculation for recovery in membrane filters is the same as for granular filters but the terminology is different. Recovery (r) is the ratio of net-to-gross water production over a filter run, which is the volume of water fed to the membrane over a filter run, (V_F) less the volume of water used during a backwash (V_{BW}) quantity divided by V_F. Recovery in membrane filtration is typically 95 to 98 per cent, which is comparable to granular filters.

Another liquid waste stream from granular filters occurs when a filter is initially brought online after backwashing and the initial filter effluent is wasted, called filter-to-waste. During this initial period of operation, the filter is clean and does not have the same ability to remove particles as it does when fully ripened. The initial flow from a clean filter is typically diverted from the filter effluent to reduce the chance of undesirable constituents passing through the filter outlet and on through the process train. Filter-to-waste flow typically occurs for 15 minute to 1 hour after a filter is backwashed, but the specific time a filter operates in a filter-to-waste mode is based on the filter effluent quality.

The filter-to-waste flow may be captured and recycled through the treatment plant headworks or, in some cases, directly upstream of the filters. Filter-to-waste water quality is different than both filter waste washwater and supernatant from dewatering processes so it may need to be separated from these other waste streams.

Physico-chemical Properties of Waste Washwater

The physico-chemical properties of waste washwater are reported in Table 17.7. As reported in Table 17.7, the average total suspended solids concentration is typically on the order 100 to 1000 mg/l. Thus, the physical properties of waste washwater are similar to those for water. Because of the low concentration of solids, waste washwater has historically been: (i) returned directly to the headworks of the treatment plant when comprising less than 10 per cent of the plant flow; (ii) discharged to a flow equalisation basin and then returned to the headworks of the treatment plant; (iii) discharged to waste washwater recovery ponds, basins or lagoons where it is allowed to settle for 24 hour or more before being decanted and returned to the headworks of the treatment plant; and (iv) discharged to surface water with the appropriate National Pollutant Discharge Elimination System (NPDES) permit in place. The current trend for handling waste washwater is to have a separate treatment facility because of concern over the presence and recycling of micro-organisms, potential increases in the concentration of disinfection by-products, as well as other concerns such as taste and odour.

REVERSE OSMOSIS AND NANOFILTRATION CONCENTRATE

Increasingly, greater use is being made of membrane processes for the treatment of water. The principal types of membrane systems are ultrafiltration, nanofiltration and reverse osmosis. Nanofiltration is typically used for water softening, whereas reverse osmosis is used most commonly for demineralisation of brackish water and seawater. In general, these processes produce wastes that are high in dissolved solids but low in suspended solids.

Table 17.7. Typical physical properties and chemical constituents of granular filter and membrane filter waste washwater.

		Range of values	
Item	Unit	Granular filter	Microfiltration
Physical properties			
Volume	% water treated	1–5	2–8
Total solids	mg/l	100–1000	100–1000
Specific gravity	sg	1.00–1.025	1.00–1.025
Specific resistance	m/kg	$11–120 \times .10^{10}$	$11–120 \times 10^{10}$
Viscosity at 20°C	$N \cdot s/m^2$	$1–1.2 \times .10^{-3}$	$1–1.2 \times .10^{-3}$
Initial settling velocity	m/h	0.06–0.15	0.06–0.15
Chemical constituents			
BOD	mg/l	2–10	2–10
COD	mg/l	20–200	20–200
pH	Unitless	7.2–7.8	7.2–7.8
Solids			
Al_2O_3 or Fe	%	20–50	20–50
Silicates and inert materials	%	30–40	30–40
Organics	%	15–22	15–22

Estimating Quantitites of Membrane Concentrate

Typical quantities of concentrate produced in reverse osmosis were shown in Table 17.2. The quantity of concentrate produced depends on the operating characteristics of the membrane and the water and solute mass transfer coefficients.

Physico-chemical Properties of Membrane Concentrate

Concentrate may be clear or coloured, with the specific gravity being dependent on the salt concentration but typically in the range of 1.02 to 1.035. Any residual suspended material present in the water to be treated would also be included in the waste concentrate. The detailed chemical content of a waste concentrate can be determined from a mass balance on the process and, as discussed above, depends on the quality of the water to be treated, the water and solute mass transfer coefficients (specific for the membrane) and the detailed design and operation of the system.

ION EXCHANGE BRINE

Ion exchange brines resulting from the softening of water are considered in this section. Specific quantities and characteristics of the brine from a particular softening plant will depend on the resin selected, the raw-water characteristics and the operation of the regeneration process.

Estimating Quantities of Ion Exchange Brine

Typical quantities of brine produced in ion exchange were given previously in Table 17.2 and will vary from 1.5 to 10 per cent of the plant flow. Quantities of brine required for regeneration are determined during design of an ion exchange facility. The brine to be disposed of is equal in volume to that used for, regeneration, but instead of being entirely composed of water and regeneration salts, it will be a mixture of the excess sodium required to drive the regeneration process and the ions removed by the ion exchange process. In addition, there will be rinse water used to flush the brine out of the resin between the regeneration cycle and operation and there may be waste washwater from an initial backwash cycle to remove any suspended materials collected by the bed. Depending on the design of the facilities, the various waste streams may be separated or combined. Ideally the brine and the freshwater streams are segregated, as are less contaminated portions of the regenerant brine, so as to allow some brine recovery.

Chemical Properties of Ion Exchange Brine

The chemical properties of typical ion exchange brines are summarised in Table 17.8. Physically these waste brines are clear, with the specific gravity being dependent on the salt concentration but typically in the range of 1.02 to 1.035. If waste washwater is included, any suspended material present in the raw water would also be included in the waste. The detailed chemical content of a waste brine can be determined from a mass balance on the process and, as discussed above, depends on the raw-water quality and the detailed design and operation of the system.

Table 17.8. Typical chemical properties of ion exchange brine from softening[a].

Item	Unit	Range of values
BOD	mg/l	30–300
COD	mg/l	30–5000
pH	Unitless	6–8
Total dissolved solids	mg/l	15000–30000
Solids		
Ca^{2+}	mg/l	3000–6000
Mg^{2+}	mg/l	1000–2000
Na^+	mg/l	2000–5000
Cl^-	mg/l	9000–22000
SO_4^{2-}	%	5–15

[a] The volume of brine as a function of the water treated will vary from 1.5 to 10 per cent.

MANAGEMENT OF RESIDUAL LIQUID STREAMS

In addition to sludges, a number of residual liquid streams result from the treatment of water. As identified in Table 17.1, the principal liquid waste streams, excluding membrane concentrates and ion exchange brines, are filter waste washwater and filter-to-waste water for water treatment plants with granular filters and filter waste washwater for water treatment plants that use membrane filtration. Other waste streams are comprised of recycle flows from sludge-processing operations, including centrate, filtrate, pressate, supernatant flow and leachate. The combined volume of these waste streams may approach 4 to 5 per cent of the total water treated, depending on the processes employed. In the past, these streams were returned to the headworks, discharged to nearby water bodies, land applied or discharged to waste-water collection systems. Because of new regulations, many of these past practices are no longer acceptable. As a result, the management of these liquid waste streams is a major issue in the design and operation of most water treatment plants.

Concerns with Recycle Waste Streams

As noted in the introduction to this chapter, the concerns with recycle flows are related to the constituents contained in them. The principal constituents of concern are already discussed and the options for dealing with these constituents are considered below.

Flow Equalisation Lagoons or Basins

Flow equalization is used to reduce the impact of the intermittent high volume flows from backwashing operations. By returning the waste washwater at a more constant rate, the impact on treatment process performance is minimised.

When the equalisation basin also functions as a settling basin, the impact of the return flow is further mitigated. When the suspended material in the raw water has been effectively coagulated and flocculated prior to filtration, the solids in the waste washwater generally settle rapidly. To achieve a supernatant turbidity of about 5 NTU with this type of waste washwater, the equalisation basin should provide a minimum detention time of 1 to 2 hour. Coagulants and coagulant aids such as alum and cationic polymer may be added to improve the settling characteristics of the solids in the waste washwater.

Treatment of Recycle Waste Streams

Because of concern over the constituents in the return flows, separate treatment facilities are now used at some water treatment plants to process recycle flows. Treatment options include:
1. Flow equalisation without or with chemical addition.
2. Lagoons without or with chemical addition.
3. Batch sedimentation without or with chemical addition.
4. High-rate sedimentation without or with chemical addition and preflocculation.
5. Dissolved air flotation.
6. Granular filtration.
7. Membrane filtration.
8. Disinfection.
9. UV oxidation.

Because of the larger area required for waste washwater storage basins, the use of high-rate sedimentation (see Fig. 17.4) has become common in larger water treatment plants. The sludge resulting from the high-rate sedimentation process as well as from the treatment options identified above is typically combined with other plant sludges for further treatment.

Disposal of Liquid Streams

In some cases, residual liquid waste streams have been discharged to surface water and/or to waste-water collection systems. The ability to use either of these options is site specific.

Disposal to surface waters

Surface water discharges are regulated under the Clean Water Act through the NPDES. These laws consider water treatment and supply to be an industry and, therefore, consider water treatment residuals, such as concentrate, an industrial waste. The NPDES permits can specify a variety water quality requirements, depending on classification of the receiving water body (e.g., potable water source, trout stream). State and local governments may impose additional restrictions on surface water discharges.

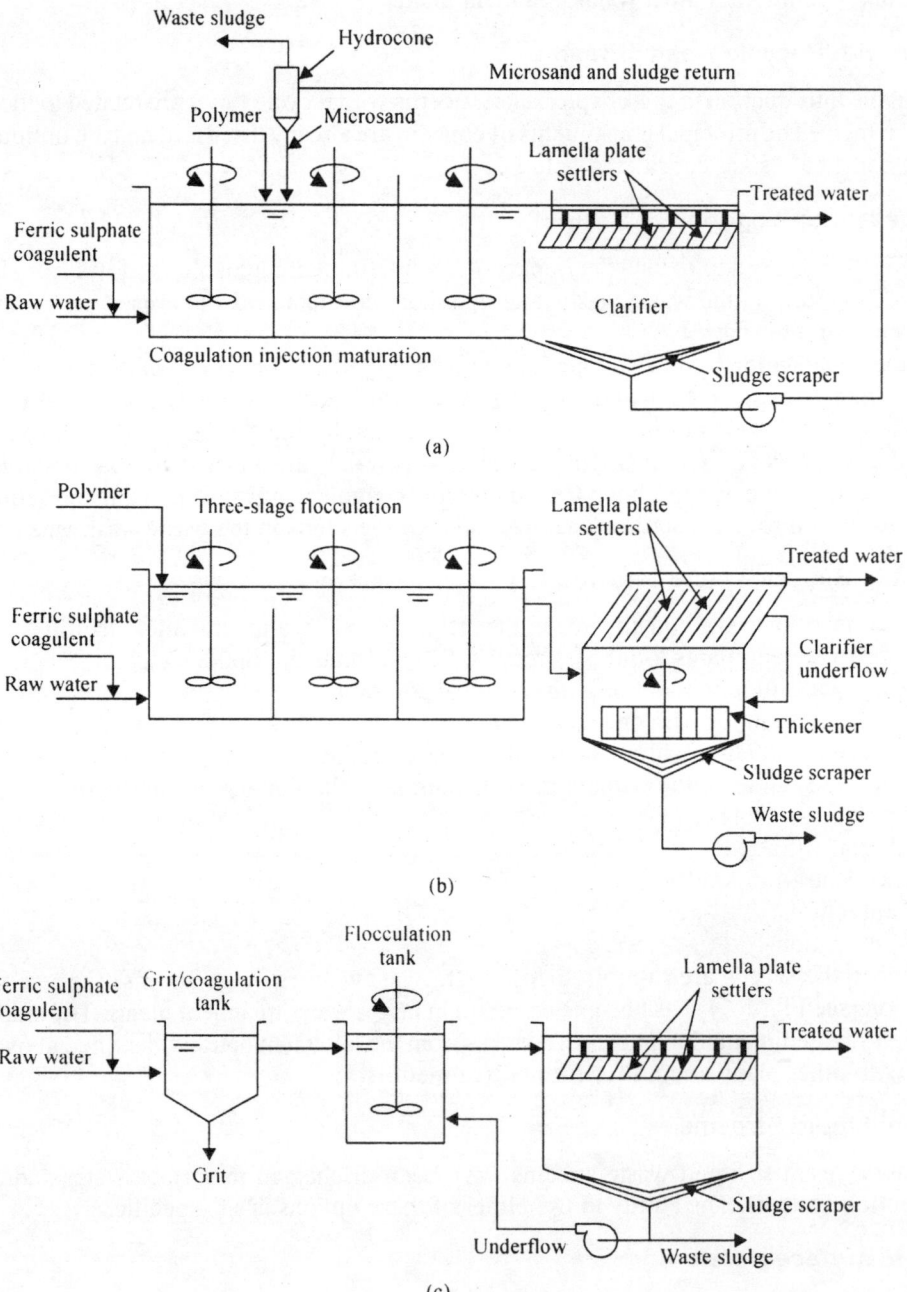

Fig. 17.4. High-rate clarification processes: (a) ballasted flocculation; (b) lamella plate clarification; (c) dense sludge.

Discharge to waste-water collection system

The same laws that govern surface water discharge apply to waste-water collection system disposal. Pre-treatment of the residual prior to discharge to the waste-water plant may be required because of state regulations or conditions imposed by the waste-water plant. In general, local pre-treatment guidelines will cover the discharge from a water treatment plant to the waste-water collection system.

The capacity of the collection system or the waste-water treatment plant and the types of processes and operations at the waste-water facility may limit the amounts and type of liquids and/or solids that may be added to the waste-water system. The viability of sewer discharge is affected by the chemical characteristics of the residual stream, particularly with respect to whether high TDS, low dissolved oxygen or high metals content may be toxic to the biological process at the waste-water plant.

Direct discharge to a waste-water collection system has a low capital cost and may also have a low operation and maintenance cost depending on monitoring requirements and sewer use fees. An advantage of this method is simpler permitting requirements. Discharge to a collection system is the easiest disposal method if a local waste-water treatment plant is willing to accept the waste, an issue that is often facilitated when a municipality operates both the water and waste-water systems.

A condition of discharge may be continuous monitoring of the organic strength and solids content of the residual flow. An attempt should be made to assess the impact of the residuals on the waste-water treatment facility prior to the selection of this alternative. It is possible that disposal of alum sludge may enhance phosphate removal at the waste-water treatment plant if any of the alum activity remains. Residual coagulant activity may also enhance primary sedimentation.

MANAGEMENT OF MEMBRANE CONCENTRATE

Concentrate is produced when nanofiltration (NF) and reverse-osmosis (RO) membranes are utilised in the treatment process. As noted previously, these processes produce a concentrate that is high in TDS but low in suspended solids. Concerns with membrane concentrates, concentrate disposal methods and concentration methods are considered in the following discussion. Ion exchange brines are also considered in the following section.

Concerns with Membrane Concentrates

Concerns that must be addressed in the management of membrane concentrates include: (i) the volume of concentrate; and (ii) environmental classification and regulations. The management of cleaning solutions is also an important consideration, as discussed below.

Volume of concentrate

The management of membrane concentrate is often problematic because the volume of the concentrate stream is considerably larger than the waste stream from virtually all other water treatment processes. Recovery typically ranges from 50 to 85 per cent. A 10000 m^3/d NF plant operating at 85 per cent recovery has a concentrate waste stream of 1800 m^3/d; a seawater RO plant operating at 50 per cent capacity has a concentrate stream equal in size to the product water stream. With waste streams of this magnitude, concentrate disposal problems can be formidable.

Concentrate classification

A second problem is the classification of concentrate as an industrial waste. Concentrate disposal is regulated under several different environmental laws and the interaction between these regulatory requirements can be complex. Regulatory considerations are often as important as cost and technical considerations for determining viable concentrate disposal options.

Cleaning solutions

Although the concentrate is by far the most voluminous waste stream, RO plants must also dispose of spent cleaning solutions. Frequently, the cleaning solutions are acidic or basic solutions with added detergents or surfactants. In many cases, the cleaning solution volume is so small compared to the concentrate stream that the cleaning solution is diluted into and disposed with the concentrate. In some cases, treatment of the cleaning solution may be required prior to disposal, but treatment may consist only of pH neutralisation. Detergents and surfactants should be selected with disposal issues in mind.

Methods of Thickening Concentrates

Because the volume of the concentrate stream from membrane processes that must be disposed of is larger than the waste stream from virtually all other water treatment processes, a number of alternative processes have been developed to further thicken (concentrate) the concentrate. Included among the thickening methods that have been developed are: (i) membrane concentration; (ii) evaporation/distillation; (iii) crystallisation; (iv) solar evaporation; and (v) crystallisation.

Membrane concentration

Two-and three-stage RO membrane concentration steps (see Fig. 17.5) have been used to increase the concentration of the brine to TDS values greater than 35000 mg/l. The concentrated brine can then be processed further by crystallisation and solar evaporation.

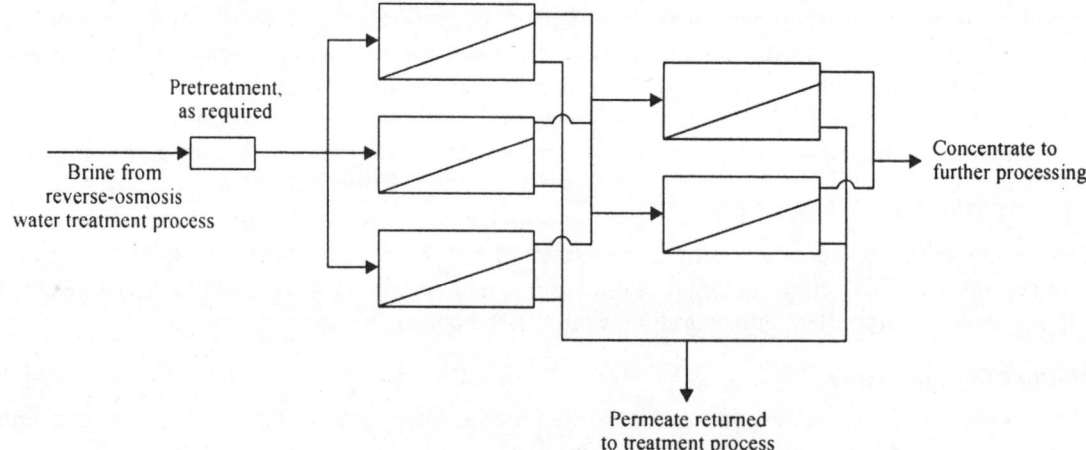

Fig. 17.5. Two-stage reverse osmosis process.

Evaporation/distillation

Residual concentrates and ion exchange brines (discussed in the following section) can be concentrated further by evaporation/distillation. Evaporation/distillation technologies that potentially could be used to concentrate residual brines include: (i) boiling with submerged tube heating surface; (ii) boiling with long-tube vertical evaporator; (iii) flash evaporation; (iv) forced circulation with vapour compression;

(v) solar evaporation; (vi) rotating-surface evaporation; (vii) wiped-surface evaporation; (viii) vapour reheating process, (ix) direct heat transfer using an immiscible liquid; and (x) condensing-vapour heat transfer by vapour other than steam. Of these types of evaporation/distillation processes, multistage flash evaporation, multiple-effect evaporation and vapour compression distillation appear most feasible for the processing of residual concentrates. A vertical-tube falling-film evaporator is illustrated on Fig. 17.6.

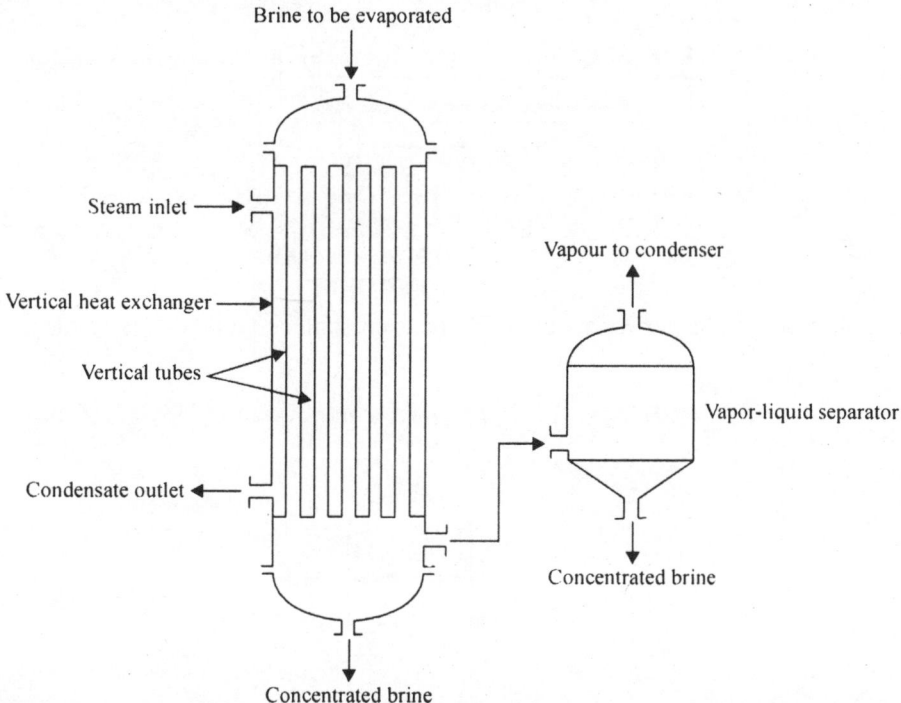

Fig. 17.6. Schematic diagram of vertical-tube falling-film evaporator that can be used as a brine concentrator. Brine to be concentrated flows downward in a thin film along the inner walls of the vertical tubes and is partially evaporated. The evaporated vapour and concentrated brine flow into a vapour-liquid separator where the vapour is separated from the concentrated brine.

Solar evaporation

Where climatic conditions are favourable, the use of evaporation ponds may be feasible. Important factors that affect the performance of evaporation ponds include relative humidity, wind velocity, barometric pressure, water temperature and the salt content of the brine. In some locations, glass covered solar ponds similar to those used for desalination in many of the dry Mediterranean countries are used to further concentrate brines by evaporation (see Fig. 17.7).

Crystallisation

The crystallisation process involves the conversion of thickened concentrate and brine into crystals that can be dewatered with a centrifuge or belt press. A typical brine crystalliser is illustrated on Fig. 17.8. The disposal of brine crystals is by landfilling.

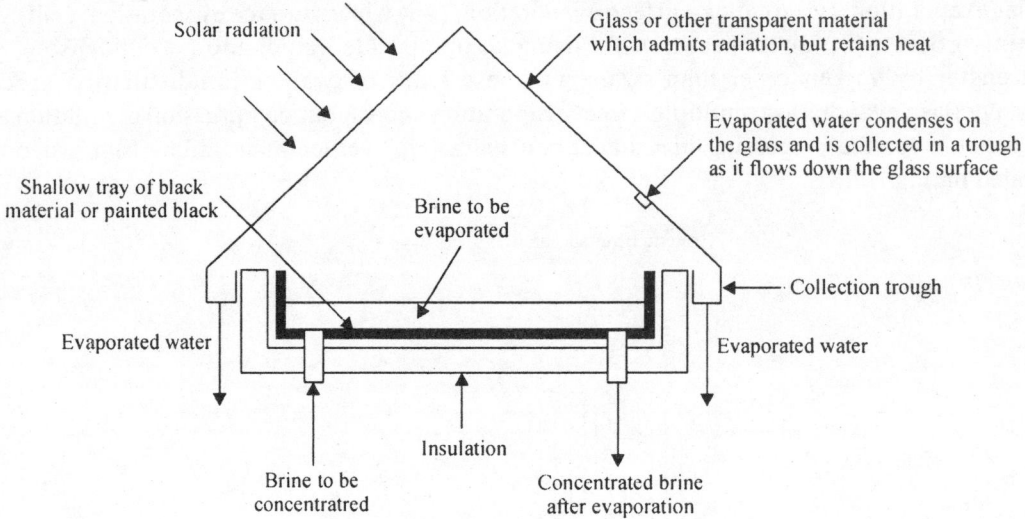

Fig. 17.7. Schematic of brine solar evaporator. Solar radiation absorbed on dark tray heats the brine to be evaporated.

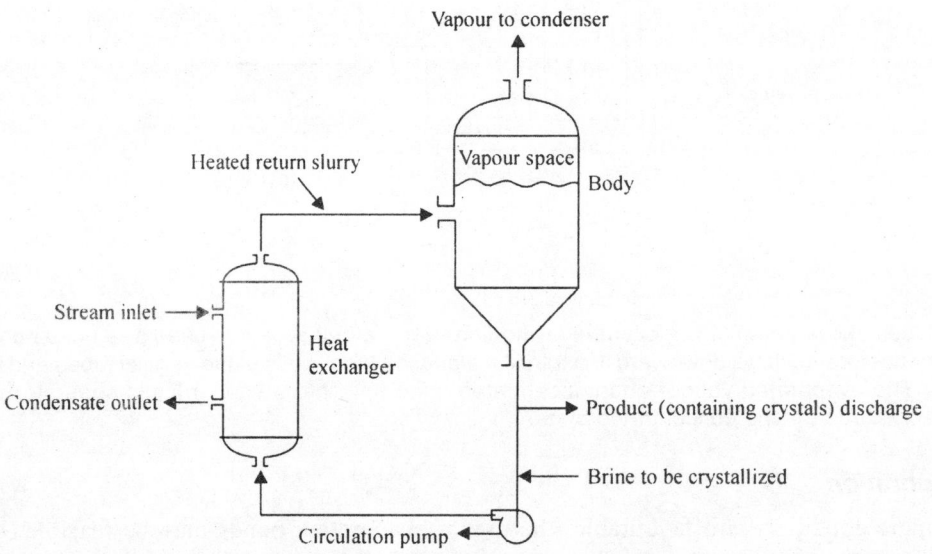

Fig. 17.8. Schematic of forced circulation brine crystalliser. The heated return slurry causes the liquid in the body to boil. As the liquid boils and vapour is released, the concentration of the constituents in the liquid increases until supersaturation occurs and crystals form.

Ultimate Disposal Methods for Membrane Concentrates

Conventional methods used for the disposal of membrane concentrates include:

1. Disposal to surface water.

2. Discharge to waste-water collection systems.
3. Deep-well injection.

Disposal to surface water

Despite complex regulatory requirements discussed previously in connection with the management of liquid waste streams, surface water discharge is often the most cost-effective disposal option for RO plants, especially for those plants located in coastal areas, where brackish or saline receiving water are a viable discharge option. The advantage of a surface water disposal system is the relatively low capital and operation and maintenance costs. Disadvantages include the uncertainty of continued allowance of this practice in the future and the potential for creating a water pollution problem. Under the NPDES, extensive monitoring of the concentrate and the discharge water body is required.

Water quality issues

Surface water discharge is dependent on the quality of the concentrate. Nominally, the waste stream is comprised of inorganic solutes from the source water that have been concentrated.

Thus, a coastal plant drawing water from a brackish aquifer and discharging concentrate to the ocean would appear to have little environmental impact. In addition to TDS, however, it is important to consider the toxicity of individual heavy metals, whose concentration is increased by the same factor. Additionally, many concentrate streams are anaerobic, which can be toxic to fish in the receiving water without sufficient dilution. Toxicity can initially be assessed by comparing predicted heavy-metal concentrations to regulated limits, but bioassays are often required before permits are issued. In many cases, economic considerations favour discharge to a brackish river or bay near the plant over an ocean outfall. However, the difference in salinity between the concentrate and the receiving water is more important in these locations.

Design considerations

Considerations for the design of a surface water disposal system include quality of the concentrate, pumping requirements, flow equalisation and outfall location and design. Outfall location is also an extremely important concern. The outfall should be located such that it discharges to a point of maximum dispersion. Similarly, the outfall should be designed to disperse the concentrate across the well-mixed zone of the water body. The location and design of the outfall significantly impact the pumping requirements. Consideration should be given to equalising the residual flow to minimise pump and motor sizing and limit discharge of large residual slugs to the receiving water.

Discharge to waste-water collection system

As with the disposal of liquid streams, local pre-treatment guidelines will cover the discharge of membrane concentrates to the waste-water collection system. In the case of RO plant residuals, which are primarily concentrated inorganic solutes, the biological process provides little treatment for the concentrate stream. In some situations it may be advantageous to discharge the concentrate to the waste-water plant effluent rather than the influent. Blending concentrate with waste-water effluent avoids toxicity issues in the waste-water process while still using the effluent to dilute the concentrate prior to discharge to a surface water. Dilution before discharge is a legitimate treatment strategy for a waste consisting primarily of concentrated inorganic solutes from a natural source water. Design of a waste-water disposal system must provide for controlled discharges to eliminate the possibility of large slugs of residuals upsetting the waste-water treatment facility. Discharge should be co-ordinated with the waste-water treatment plant operators so that they may optimise the performance of their process units.

Deep-well injection of membrane concentrate

Discharge of clear membrane concentrate and ion exchange brine by means of deep-well injection into a brackish or saline aquifer is regulated by EPA environmental regulations and is dependent on the geology and groundwater hydrology of the area. Deep-well injection involves pumping the concentrate or brine stream into an injection well, typically thousands of meters deep. A typical injection well is shown on Fig. 17.9. The injection zone is typically a brackish or saline aquifer with no potential for use as a potable water supply, which is overlain by thick layers of impermeable rock that prevent contamination of shallower freshwater aquifers. Deep-well injection is used by about 10 per cent of RO plants, although its use is becoming more common. Preference for deep well disposal has arisen because of the existence of a reliable injection zone and public and regulatory resistance to surface water discharge.

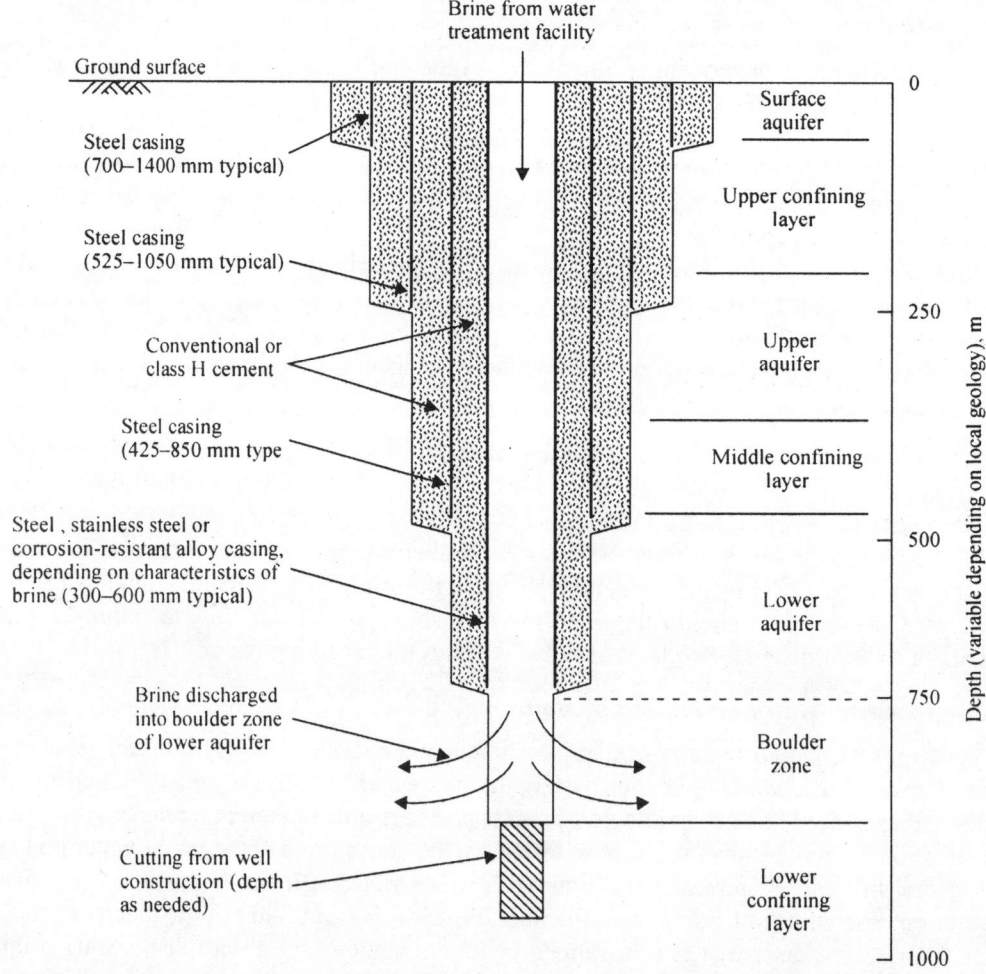

Fig. 17.9. Schematic of typical well used to inject brine into sub-surface aquifers.

Well construction is governed by regulations for deep-well injection of industrial wastes. Wells are constructed of three to four casings, with the space between each casing filled with cement grout. Each casing typically ends at a different depth. Depending on the local groundwater hydrology, there may be a significant potential for groundwater contamination and multiple casings are designed to prevent any leakage from one aquifer to the next. Deep-well injection systems tend to be fairly expensive due to well-drilling cost and maintenance costs. The high pressure at the bottom of the injection well and the saline solution tend to enhance the corrosion potential of the well screen and casing. Selection of materials resistant to corrosion under those conditions may prolong the operating life of an injection facility.

MANAGEMENT OF ION EXCHANGE BRINES

The principal source of ion exchange brines is from the softening of hard water. In general, large ion exchange water-softening plants have been located on or near coastal areas so that the resulting brines can be discharged to the ocean.

Processing of Brines

The management of brines will often involve some form of thickening before disposal. The thickening methods discussed previously in connection with management of membrane concentrates are also used for brines.

Ultimate Disposal of Brines

The principal methods used to for the disposal of brines involves discharge to brackish or saline receiving water, deep-well injection and in the case of small facilities to waste-water collection systems. The same considerations discussed previously is for the ultimate disposal of membrane concentrate also apply to ion exchange brines.

MANAGEMENT OF RESIDUAL SLUDGES

Based on an understanding of the characteristics of the residuals, development of a complete residuals process treatment train requires an understanding of the technologies that are available for processing the residuals. The major unit operations and processes that are employed for the management of residuals will be reviewed in this section. A generalised process diagram showing the various unit processes that may be used in residuals management and the sequence in which they may be assembled to form complete treatment systems are shown in Fig. 17.10. Some of the processes shown in Fig. 17.10 are omitted in the following discussion because either they are seldom used or insufficient data are available on their application.

A complete residuals management system is made up of one unit process from one or more of the process steps shown (e.g., thickening/dewatering, conditioning) and must include one of the unit processes from the final reuse and/or disposal step. Some typical residuals management processes are as follows:

1. For alum sludge, gravity thickening, chemical conditioning, centrifugation and final disposal to sanitary or monofill landfill.
2. For alum sludge, sludge lagoons, decant recovery and recycle and final disposal to a sanitary or monofill landfill or waste-water collection system.
3. For lime sludge, gravity thickening, filter press dewatering, heat drying, lime calcining and reuse.
4. For lime sludge, sludge lagoons, drying beds, cropland application or monofill landfill.
5. For membrane concentrate, final disposal directly to brackish surface water, the ocean, deep-well injection or waste-water collection system.
6. For ion exchange brines, membrane concentration, thermal brine concentration and evaporation ponds.

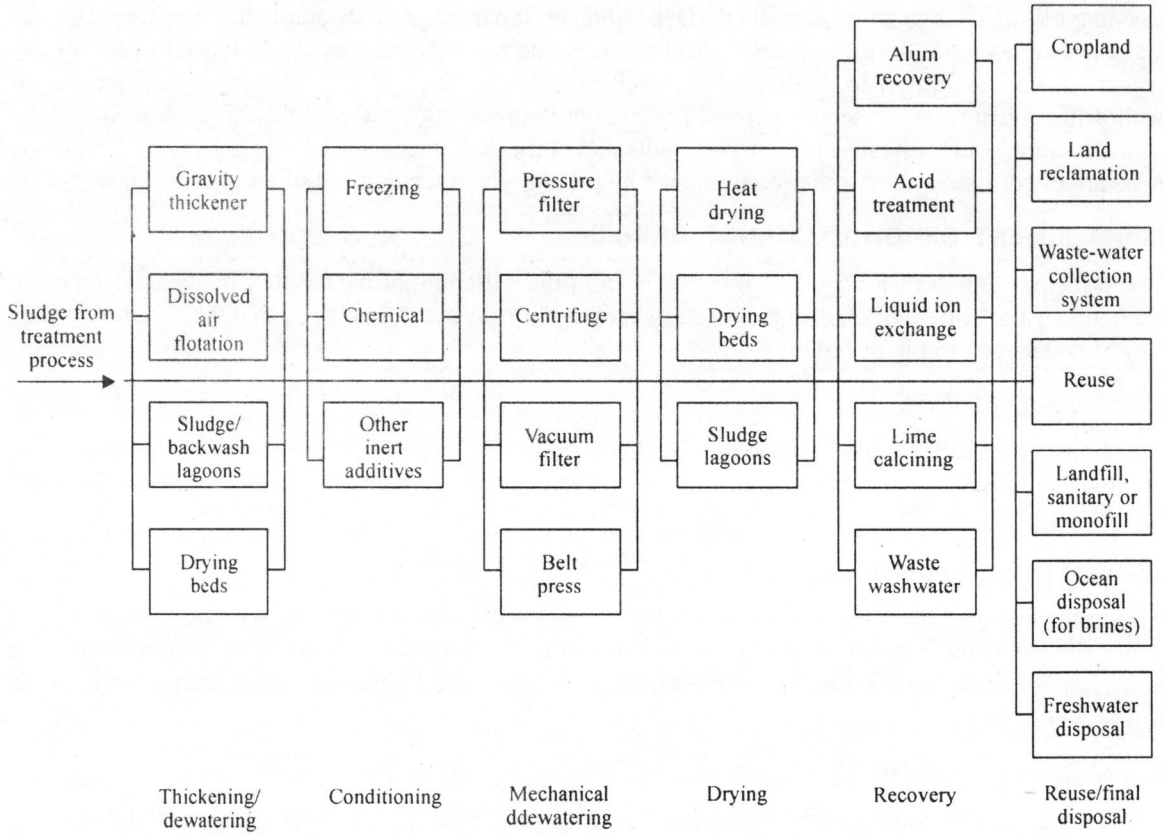

Fig. 17.10. Unit operations and processes for management of water treatment plant sludges.

Unit operations that have proven to be the most successful and to have significant capabilities for dewatering sludges from water treatment plants are drying beds, vacuum filtration, pressure filter press, belt filter press, centrifugation, alum and lime recovery and pellet flocculation. The above technologies are considered in the following discussion.

Thickening/Dewatering

Thickening to increase the solids content of sludge involves the removal of excess water by decanting and the concentration of the solids by settling. The decanted water is usually recovered unless the water contains objectionable tastes or odours or large numbers of algae or other micro-organisms while the solids are processed further or disposed of. Gravity thickening is used most commonly as the first step in the residuals management process. The most common methods of gravity thickening for sludges are lagoon settling with a decantation operation and conventional gravity thickening in specifically designed reactors. For coagulation or softening sludges, the primary process involved is compaction thickening of the sludge, while for filter waste washwater the processes of settling and hindered settling are most important.

Mechanical gravity thickening

Gravity thickening is typically accomplished in a circular tank designed and operated similarly to a solids-contact clarifier or sedimentation tank (see Fig. 17.11). Sludge is introduced into the tank and allowed to settle and compact. Gentle agitation of the sludge prior to settling creates channels in the sludge matrix for water to escape and promote densification of the solids. The thickened sludge is collected and withdrawn at the bottom of the tank. A properly designed and operated gravity thickening system can produce softening sludges in excess of the 2 to 6 per cent associated with alum sludge, depending upon the calcium carbonate and magnesium hydroxide content in the sludge. Lime sludges that are predominantly calcium carbonate can be thickened to 30 per cent solids and higher. Typical performance and design data on gravity thickening are presented in Table 17.9.

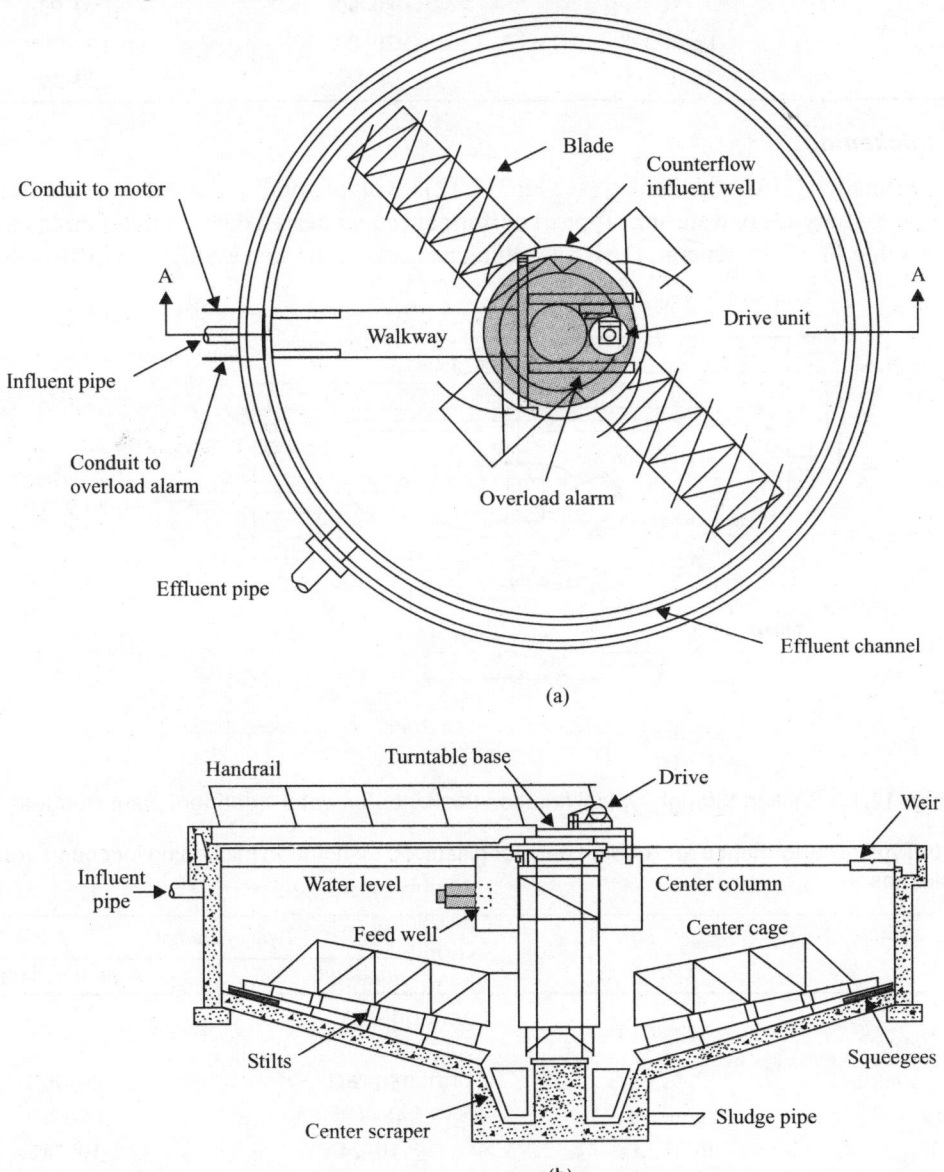

(a)

(b)

Fig. 17.11. Typical mechanical gravity thickener for water treatment plant sludges: (a) plan view; (b) section through thickener.

Table 17.9. Typical performance and design data for gravity mechanical thickening of coagulant and lime sludges.

Parameter	Unit	Type of sludge	
		Coagulant	Lime softening
Feed solids	%	0.2–1	1–4
Thickened solids	%	2–3	>5
Solids recovery	%	80–90	80–90
Solids loading	kg/m² · d	20–80	100–200
	lb/ft² · d	4–16	20–40

Flotation thickening

Dissolved air flotation (DAF) thickening (see Fig. 17.12) involving both sedimentation and flotation has been used successfully for dewatering. Typical performance and design data on flotation thickening are presented in Table 17.10. In general, DAF thickening has been most successful with hydroxide sludges.

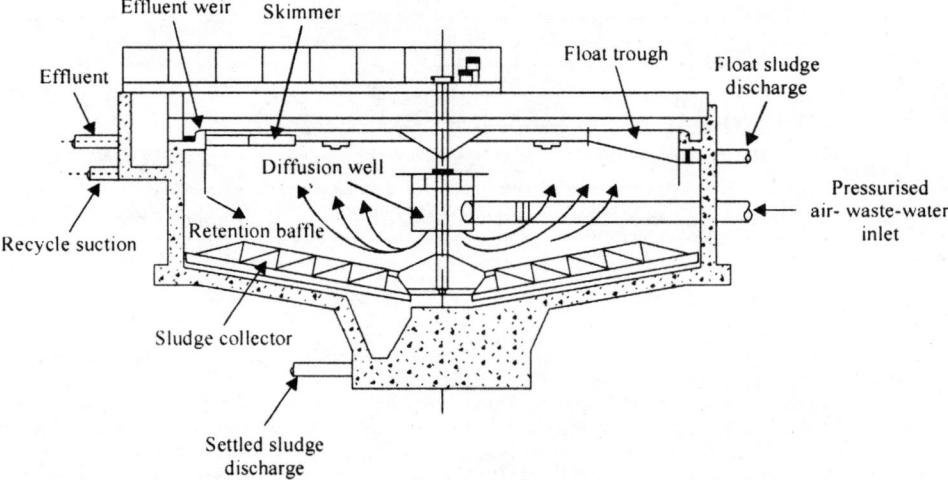

Fig. 17.12. Section through typical flotation thickener for water treatment plant sludges.

Table 17.10. Typical performance and design data for dissolved air flotation thickening for coagulant and lime-softening sludges.

Parameter	Unit	Type of sludge	
		Coagulant	Lime softening
Feed solids	%	0.5–1	0.5–1
Thickened solids	%	3–5	3–5
Solids recovery	%	80–90	80–90
Solids loading	kg/m² · d	48–120	48–120
	lb/ft² · d	10–24	10–24
Volumetric loading	m³/m² · d	110–150	110–150
	gal/ft² . d	2800–3600	2800–3600

Sludge lagoons

A non-mechanical means of handling water treatment plant sludges consists of dewatering in sludge lagoons or drying beds. If land is readily available and inexpensive, the use of sludge lagoons is a cost-effective means of storing and thickening residuals. Lagoons are commonly lined earthen basins equipped with inlet control devices and overflow structures (see Fig. 17.13). Wastes with settleable solids are discharged into the lagoons from which the solids are separated by gravity sedimentation. Sludge lagoons can be classified by their mode of operation: permanent lagoons and dewatering lagoons. Permanent lagoons act as a final disposal site for settled water solids whereas dewatering lagoons are cleaned periodically. Lime sludges have been dewatered to 50 per cent solids concentration by the sludge lagoon handling method.

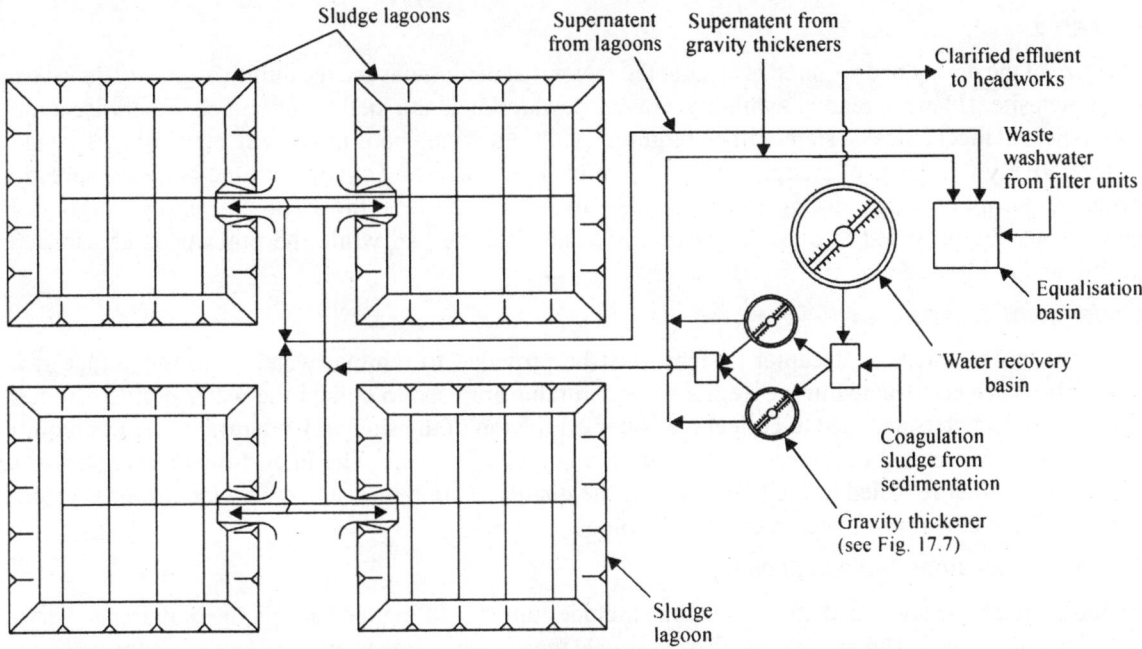

Fig. 17.13. Schematic diagram of typical sludge storage lagoons.

Lime-softening sludges dewater more readily than alum sludges. In addition, most softening plant sludges will air dry well in lagoons; therefore it is important to design the lagoon so that the sludge does not remain submerged after initial filling, as lime sludge does not compact well when under water. Typically, sludge lagoons for lime wastes should be such that they can be filled and allowed to dry before being refilled. Typical values to which lime sludges can be dewatered (in percent solids concentration) vary from 30 to more than 50 per cent. It should be noted that the final solids concentrations obtained by the various dewatering methods will vary depending on the type of sludge and sludge conditioning employed.

A common approach used at many water treatment plants is to use lagoons not only as thickeners with continuous decanting but also as drying beds after a predetermined filling period. Three months of filling and an average drying cycle of 3 months are the most common design parametes used. The required lagoon area can be determined using a sludge loading rate of 40 and 80 kg dry solids/m² of lagoon area (8.2 to 16.4 lb / ft²) for wet and dry regions, respectively.

The actual area required for a lagoon would be at least 1.5 times the area computed because of the additional area required for berms and access roads.

Gravity dewatering on drying beds

Gravity dewatering involves placement of the sludge to be dewatered on a sand or wedge wire filter surface and the subsequent drainage of water from the sludge through the filter material. The process will produce a relatively dry, solid sludge for further treatment or disposal. Gravity dewatering may be combined with other drying and dewatering operations to produce a sludge of any desired dryness. Gravity dewatering is applicable to dewatering of sludge discharged directly from sedimentation basins or following thickening.

Bed area

The size of the drying bed required is usually the factor that determines the feasibility of gravity dewatering at a given site. If land is readily available, gravity dewatering is the method of choice; otherwise a more sophisticated mechanical system will be required. Multiple drying beds must be sized to allow spreading of a relatively thin layer of sludge 0.15 to 0.3 m (6 to 12 inch) and sufficient time between spreading cycles to permit drainage, drying and removal of the sludge. At least three and preferably four or more beds should be provided to allow discharge of sludge to one bed while the other beds are draining, drying and being cleaned.

Underdrains and decanters

An underdrain system or decanter system must be provided to remove water from the sludge if the drying beds are constructed in wet regions. Underdrains are used to collect the water drained from the sludge and decanters are used to collect the water off the top of the sludge. Underdrains are not required in most dry regions, but decanters are helpful in any type of climate. The underflow or decanted water can then be either recycled to the plant inlet, if the quality is good or discharged. Underdrains typically consist of gravel and perforated clay or PVC pipes.

Cycle time, weather and conditioning

As indicated above under bed area, cycle time includes time for filling the bed, sludge draining and drying and cleaning the bed. The major portion of the cycle time is the drainage and drying time. Ideally the time required for draining and drying should be determined by bench or pilot testing, with due consideration given to variations in climate and sludge characteristics. In the absence of actual testing, the engineer must estimate the extent to which the sludge will drain. The use of polymers to condition alum sludge can reduce draining time but will probably not substantially increase the drained solids concentration.

After draining is completed, further dewatering occurs by evaporation. The time required to reach the desired dryness can be calculated from evaporation/precipitation data. Once the sludge is drained, rainfall will drain through or be decanted from the surface of the bed, rather than rewetting the sludge. For conservative design the net evaporation rate (evaporation minus precipitation) should be used as a reference for sizing drying beds.

Conditioning

As in thickening, successful dewatering often depends on proper conditioning of the sludge in advance. The objectives of conditioning are to improve the physical properties of the sludge so that water will be released easily from the sludge, improve the structural properties of the sludge to allow free draining of the released water, improve the solids recovery of the process (i.e., to reduce the fraction of solids lost in the removed water) and minimise dewatering process cycle times.

Chemical addition

Polymers are the most commonly used conditioners for dewatering water treatment sludges. Based on full-scale operating experience, it has been found that most types of polymers will improve the dewatering characteristics of sludges. The selection of a polymer for a given application should be based on bench tests or, preferably, pilot-or full-scale tests. Also, in general, higher molecular weight polymers are more effective, except that the viscosity of very high molecular weight polymers may cause handling problems.

Successful use of polymers is dependent on good dispersion of the polymer into the sludge to be conditioned. A typical blending device for polymers is shown on Fig. 17.14. As with any chemical addition, provision must be made for good initial mixing. Sizing of polymer feeding equipment should be based on bench-scale determination of dosage requirements. If that is not practical (such as in the design of a new plant), facilities can be sized based on estimated sludge solids concentrations and quantities. Polymer doses required are typically in the range of 10×10^{-4} to 100×10^{-4} kg polymer/kg sludge solids for metal hydroxide sludges.

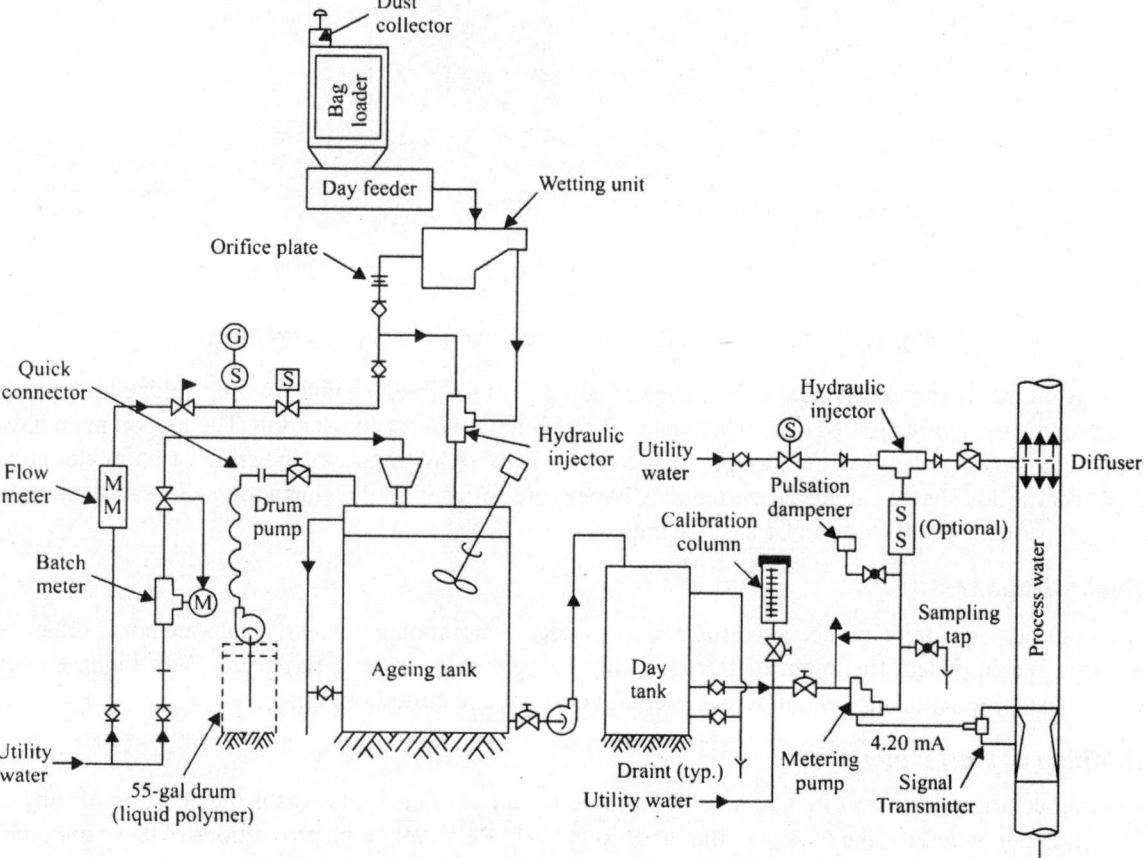

Fig. 17.14. Blending diagram for organic polymers used to condition sludge.

Freezing

Freezing is very effective for metal hydroxide sludges such as alum and iron sludges (see Fig. 17.15). The effect is to destroy completely the gelatinous structure, leaving the sludge (after thawing) in the form of a fairly coarse granular material like sand or coffee grounds. The process is irreversible.

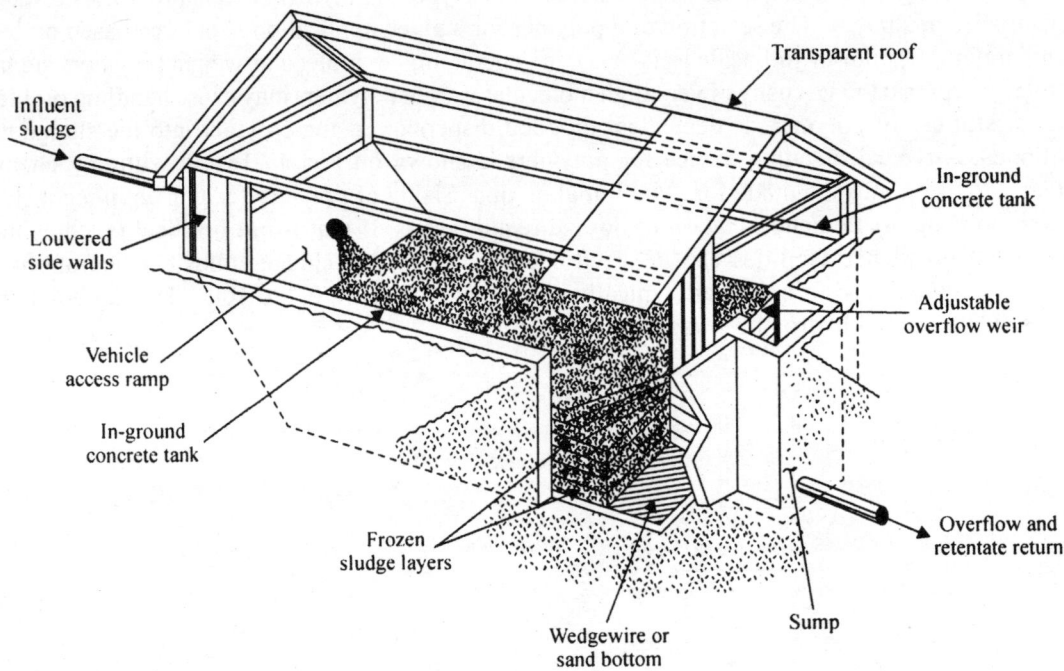

Fig. 17.15. Typical installation for conditioning of sludge by freezing.

Unfortunately the mechanical efficiencies of equipment for freezing and thawing sludge are low, so this process is usually applied only where natural freezing will occur in a lagoon. The lagoon must have sufficient capacity to allow the sludge to sit over the winter. With natural freezing of alum sludge, a 2 per cent solids sludge can be converted to a 20 per cent solids granular slurry that will readily drain to over 30 per cent solids and can be easily handled.

Heat treatment

Although heat treatment has been investigated as a sludge-conditioning process, results are not as dramatic as with freezing. Heat treatment of storage is not being employed on a full scale. With rising energy costs, heat treatment is not an attractive alternative for sludge conditioning.

Addition of inert materials

Another conditioning step often applied in pressure filtration of alum sludges is the addition of lime or inert granular materials like fly-ash or diatomaceous earth. Relatively high proportions of these materials are required.

Mechanical Dewatering

Dewatering includes all those processes intended to remove free water from sludges beyond that which can be removed by decanting from a thickener.

The objective is to reduce the sludge volume and produce a sludge that can be easily handled for further processing. As the use of open storage lagoons and drying beds becomes less feasible for dewatering, some form of mechanical dewatering is now used at most large treatment plants. The principal types of mechanical dewatering devices now used are: (i) vacuum filtration; (ii) plate and frame filter presses; (iii) belt filter presses; and (iv) centrifuges.

Vacuum filtration

The type of vacuum filter employed almost exclusively is the rotary drum vacuum filter. There are two basic types of rotary drum filters: (i) travelling media; and (ii) precoat media filters. The precoat filter is used mainly for dewatering coagulated sludges such as alum sludge (see Fig. 17.16). For alum and ferric hydroxide sludges, successful operation of vacuum filters requires the use of polymer or lime as sludge-conditioning chemicals. Lime sludges, however, generally do not require conditioning prior to vacuum filtration.

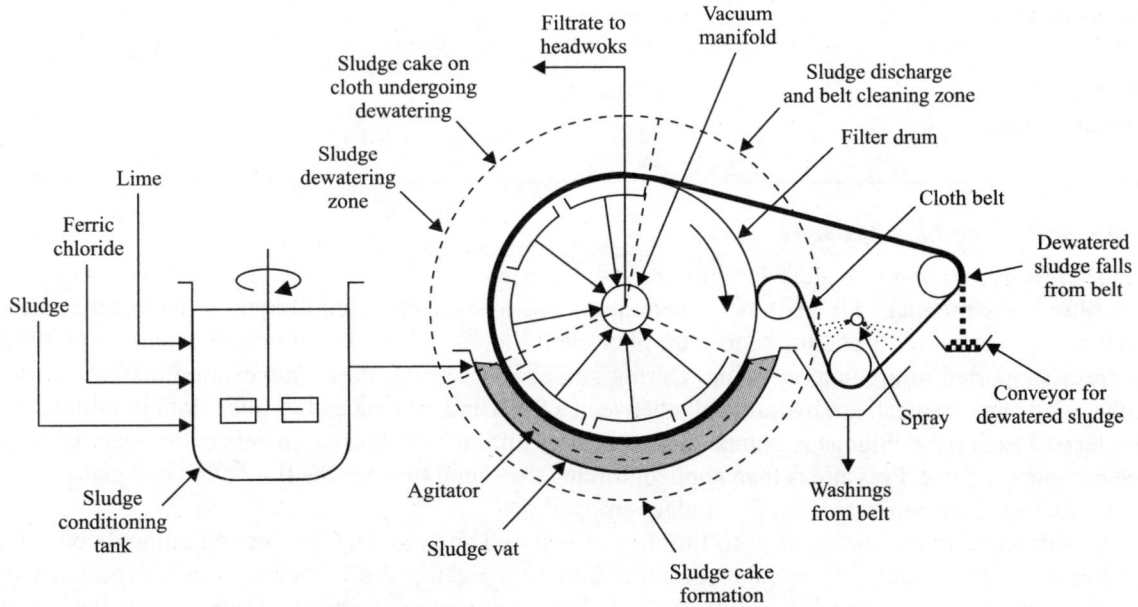

Fig. 17.16. Schematic diagram of typical vacuum filtration installation.

The variables to be considered in designing a vacuum filter system are the size and type of filter, the cake discharge mechanism, the filter media, vacuum level, cycle time and sludge conditioning. For a specific sludge, filter leaf tests are used to help determine the appropriate conditioner and dose, filter media and cycle time. Typical performance and design data for sludge dewatering with vacuum filtration are reported in Table 17.11.

Table 17.11. Typical performance and design data for precoat rotary vacuum filter for dewatering water treatment plant sludges.

Parameter	Unit	Range of values
Feed solids	%	2–6
Feed rate	l/m²h	0.7–2.1
	gal/ft² · h	2–6
Solids recovery	%	96–99+
Dry-solids yield	kg/m² · h	0.2–0.3
	lb/m² · h	1.0–1.5
Thickened solids		
Alum sludges	%	15–25
Lime sludges	%	20–40
Filtrate suspended solids	mg/l	10–20
Precoat recovery	%	30–35
Precoat rate	kg/m² · h	0.02–0.04
	lb/ft² · h	0.1–0.2
Precoat thickness	mm	38.1–63.5
	inch	1.5–2.5
Drum speed	rev/minute	0.2–0.3
Operating vacuum	mm Hg	127–508
	inch Hg	5–20

Plate and frame filter presses

Filter press dewatering is achieved by forcing the water from the sludge under high pressure. Although the filter press produces high solids concentration and low chemical consumption, its disadvantages include high labour costs and limitations on filter cloth life. A filter press consists of a number of plates or trays supported in a common frame. During sludge dewatering, these frames are pressed together either electromechanically or hydraulically between a fixed and moving end. A filter cloth is mounted on the face of each plate. Sludge is pumped into the press until the cavities or chambers between the trays are completely filled. Pressure is then applied, forcing the liquid through the filter cloth and plate outlet. The plates are then separated and the sludge removed.

Conditioning of the sludge prior to filtration is required and the degree of conditioning dictates the performance. In general, a filter cake of about 30 to 40 per cent solids concentration is expected after pressure filtration with lime and polymers as sludge-conditioning chemicals. Lime sludges have been reported to readily dewater to above 50 per cent solids without sludge conditioning. The filtrate may contain less than 10 mg/l of suspended solids if the sludge is conditioned properly. Both capital and operation and maintenance (O&M) costs for this process are high.

Gravity and pressure belt filters

The application of belt filters for water sludge thickening and dewatering is a relatively recent application. Thickening with a gravity belt filter involves two operational steps: (i) chemical conditioning of the sludge; and (ii) gravity drainage using a single belt as illustrated on Fig. 17.17a. In some designs a vacuum is applied to the under side of the belt to enhance dewatering. Sludge dewatering with a belt filter press involves three operational steps: (i) chemical conditioning of the sludge; (ii) gravity drainage; and (iii) mechanical application of pressure. To accomplish the application of pressure two or more belts are used, depending on the manufacturer. A belt filter press employing two belts for dewatering is illustrated schematically on Fig. 17.17b. For both types of belt filters, the key to successful performance is the sludge chemical conditioning step.

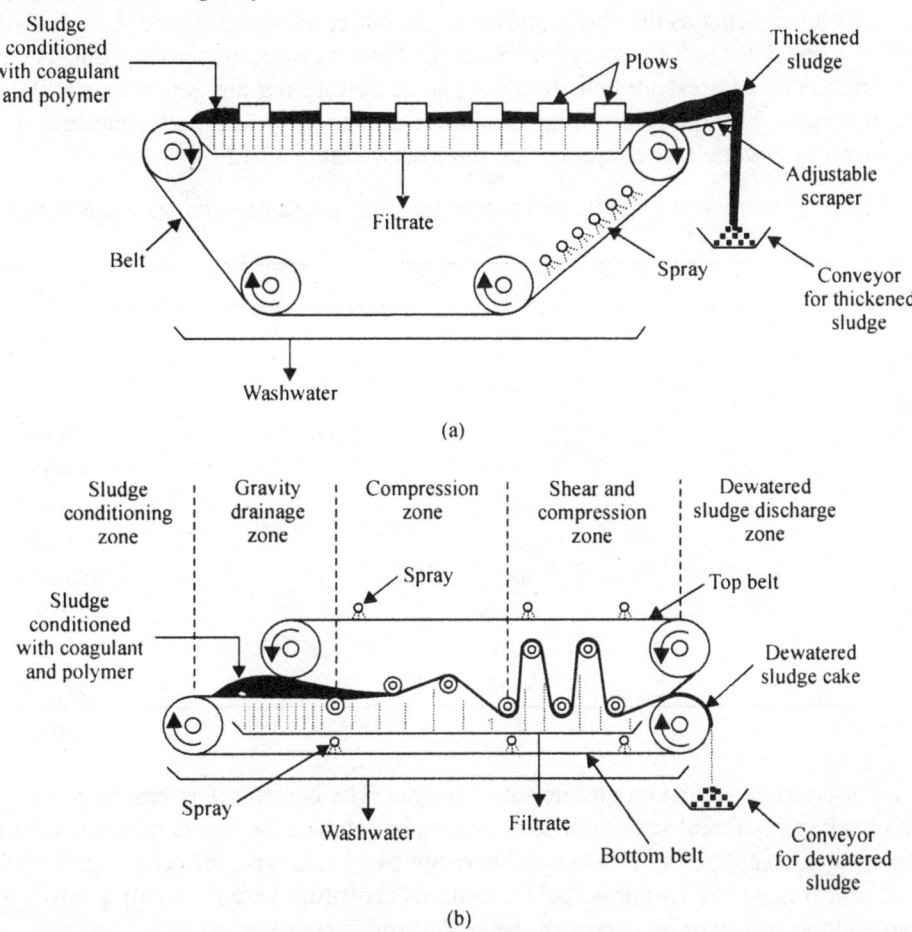

Fig. 17.17. Belt filters for sludge thickening and dewatering; (a) schematic gravity belt thickener, and (b) schemmatic belt filter press for dewatering sludge.

In thickening dilute sludges, both coagulant and polymer addition is employed. Coagulant addition is used to concentrate the solids. Polymer addition is used to coagulate and flocculate the sludge before it is applied to the gravity belt thickener. Once applied to the belt thickener, the sludge is distributed uniformly across the width of the belt and moves with the belt. Fixed guide veins or plows located just above the surface of the moving belt create clear zones for free water released from the sludge to drain through the belt. Typically, from 70 to 80 per cent of the free water is drained within the first meter. Thickened solids, scraped from the belt, are collected in a hopper for further processing, transport or disposal. Thickened sludge cake with up 20 per cent solids is possible with proper conditioning.

Sludge dewatering as illustrated in Fig. 17.17b involves subjecting chemically conditioned sludge to gravity drainage, mechanical pressing and shearing. Often shear and compression dewatering are shown as occurring together. Chemical conditioning typically involves the use of organic polymers. Complete and thorough mixing of the polymer and the sludge is the key to successful dewatering with belt presses. Once the coagulated solids are applied to the filter; gravity drainage occurs. The partially dewatered sludge then moves into the compression zone where it is squeezed between two moving belts. Additional dewatering occurs by shearing as the sludge moves to the outlet. Dewatered sludge, which falls off belt, is transported by a conveyor belt to storage facilities for further processing or disposal. The capital cost and space requirements for pressure belt filtration sludge dewatering are generally significantly lower than for plate and frame pressure filter sludge dewatering systems. Typical performance and design data for sludge dewatering with belt filter presses are reported in Table 17.12.

Table 17.12. Typical performance and design data for belt filter press dewatering of water treatment plant sludges.

Parameter	Unit	Range of values
Feed solids	%	4–30
Thickened sludge		
Alum sludges	%	15–30
Lime sludge	%	25–60
Solids recovery	%	95–99+
Cake yield	$kg/m^2 \cdot h$	0.8–4.0
	$lb/ft^2 \cdot h$	4–20
Filtrate solids	mg/l	950–1500
Filter speed	rev/min	0.2–0.5
Operating pressure	kPa	550–830
	lb/in^2	80–120

Centrifuges

Centrifuges are used to both thicken and dewater sludges. The centrifuge is basically a sedimentation device in which the solids/liquid separation is improved by rotating the liquid at high speeds to increase the gravitational forces applied on the sludge. There are two basic types of centrifuges: (i) solid-bowl; and (ii) basket centrifuges. The two principal elements of centrifuges are the rotating bowl, which is the settling vessel and the conveyor discharge of the settled solids (see Fig. 17.18).

Application of centrifuges in the water treatment field has normally been in dewatering lime-softening sludges. Solids concentrations of 35 to 50 per cent are typical with lime sludges, although higher values have been reported. In centrifugation of predominantly alum sludges, solids concentrations of 12 to 15 per cent have been obtained. Effective dewatering of alum sludge by centrifugation requires conditioning of the sludge with polymers and lime. 20 to 25 per cent solids may be obtained from 3 to 4 per cent solids alum sludge. Polymer doses of approximately 1 to 2 g/kg (2 to 4 lb/tonne) of feed solids are typical. Feed solids concentration for alum sludge centrifugation is in the range of 1 to 6 per cent and 10 to 25 per cent for lime sludge. Typical performance and design data for centrifuge thickening of coagulant and lime-softening sludges are reported in Table 17.13. Both capital and operation and maintenance costs for this process are relatively high.

Table 17.13. Typical performance and design data for centrifuge thickening of coagulant and lime-softening sludges

Parameter	Unit	Coagulant	Lime softening
		Type of sludge	
Feed solids	%	1–6	10–25
Thickened solids	%	12–15[a]	35–50
Solids recovery	%	90–96	90–97
Polymer dosage	g/kg	1–2	–
	lb/tonne	2–4	–

[a] Up to 25 per cent solids has been achieved with the use of conditioning chemicals.

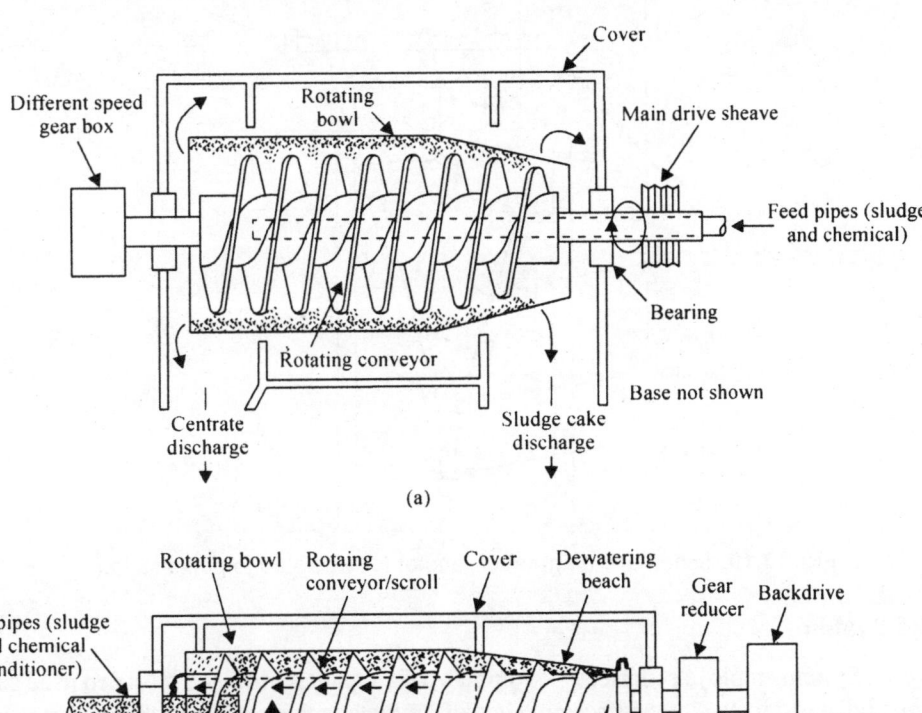

(a)

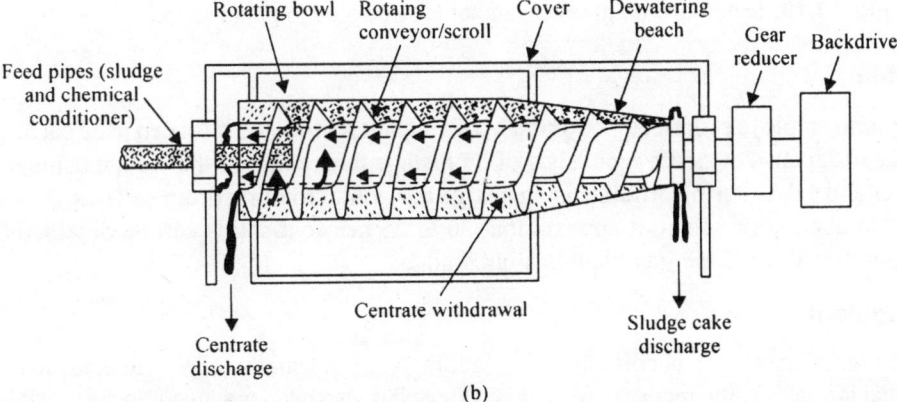

(b)

Fig. 17.18. Typical centrifuges used for dewatering of water treatment plant sludges: (a) continuous countercurrent solid bowl; and (b) continuous concurrent solid bowl.

Lime Sludge Pelletisation

Taking advantage of the observation that sludge pelletisation occured during the suspended-bed cold-softening water treatment process led to the development of the lime sludge pelletiser shown in Fig. 17.19. Operationally, the conical vessel is charged with a silica catalyst. Lime sludge injected into the reactor reacts with calcium bicarbonate (hardness) in the carrier water to form calcium carbonate precipitates on the silica catalyst. The softened water is removed from the top of the reactor. The calcium pellets removed from the reactor are discharged to a storage and drainage facility. The solids content of the calcium carbonate pellets is about 60 per cent.

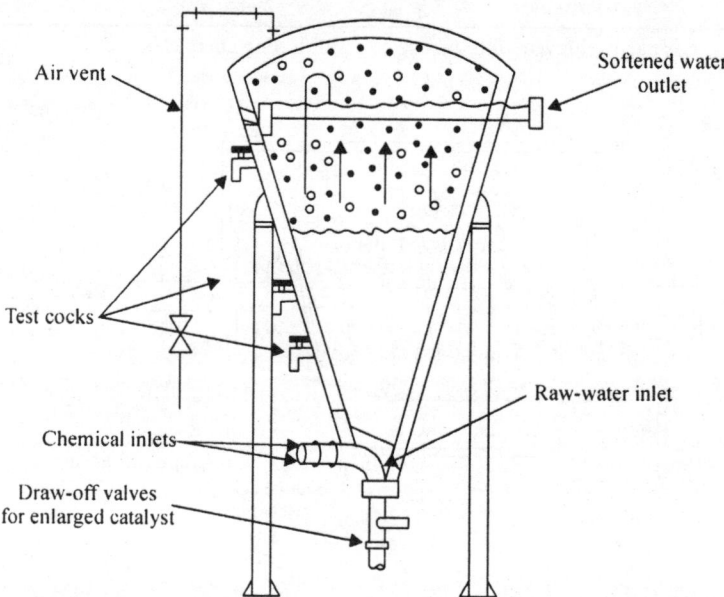

Fig. 17.19. Schematic diagram of reactor for lime sludge pelletisation.

Aqua Pellet System

The Aqua Pellet System employs proprietary equipment of a Japanese firm and has been used successfully to treat alum sludge in Japan. The process consists of multistage thickening of the sludge using sodium silicate and polymer and a dewatering process using a large horizontal rotating drum called a Dehydrum. Based on a full-scale operating system, it appears that 20 to 25 per cent solids can be obtained by the Dehydrum starting with a 0.5 to 2 per cent solids alum sludge.

Recovery of Coagulant

Aluminum and iron recovery can be accomplished by adding acid (normally sulphuric acid) to sludges to solubilise the metal ion salts. Lime recovery from lime sludge has also been practiced by the recalcination process.

Alum and iron recovery

Alum and iron recovery can be accomplished by acidification with sulphuric acid.

In simplified form, the reactions involved are:

$$2Al(OH)_3 + 3H_2SO_4 \longrightarrow Al_2(SO_4)_3 + 6H_2O \qquad \text{... (17.9)}$$

$$2Fe(OH)_3 + 3H_2SO_4 \longrightarrow Fe_2(SO_4)_3 + 6H_2O \qquad \text{... (17.10)}$$

Normally, over 80 per cent of alum and iron recovery is achieved at a pH of about 2.5. Unfortunately, heavy metals, manganese and other organic compounds are often found in the recovered alum and iron. The presence of potential contaminants, as well as rising costs, has limited the recovery of alum and iron as a viable processing alternative.

Lime recovery

Lime recovery by recalcination has been practiced in a number of locations in the United States and European Countries. Quicklime (CaO) can be recovered from softening sludges after purification and dewatering. To recover CaO_3, water-softening sludges are burned at a temperature of 1010°C (1850°F). The pertinent reactions are:

$$CaCO_3 \longrightarrow CaO + CO_2 \qquad \text{... (17.11)}$$

$$Mg(OH)_2 \longrightarrow MgO + H_2O \qquad \text{... (17.12)}$$

The types of furnaces that have been used for recalcining include the rotary kiln, flash calciner, fluidised-bed calciner and multiple-hearth calciner.

The primary consideration in selecting the recalcination process is the overall economics of the process because the process is energy intensive (10×10^9 to 15×10^9 J/kg of sludge).

Magnesium bicarbonate recovery

The magnesium bicarbonate recovery process was developed as a part of an overall treatment process that would use magnesium bicarbonate (precipitating magnesium hydroxide and calcium carbonate) as the primary coagulant. The pertinent reactions are:

$$Mg(HCO_3)_2 + 2Ca(OH)_2 \longrightarrow Mg(OH)_2 + 2CaCO_3$$

$$\text{(precipitation)} \qquad \text{... (17.13)}$$

$$Mg(OH)_2 + 2CO_2 \longrightarrow Mg(HCO_3)_2$$

$$\text{(magnesium bicarbonate recovery)} \qquad \text{... (17.14)}$$

During the late 1960s and the early 1970s, recycling magnesium carbonate was a developing and promising technology. Although the magnesium bicarbonate recovery process is technically feasible, based on pilot plant studies, the economics could not justify scale-up to a full-size facility. Presently, there are no known existing full-scale water treatment facilities that employ the magnesium bicarbonate recovery process; however, recovery of magnesium bicarbonate is being practiced in the paper and pulp industry.

ULTIMATE REUSE AND DISPOSAL OF SEMISOLID RESIDUALS

Several alternatives are available for the disposal or reuse of water treatment plant residuals. In practice, the options available for ultimate disposal or reuse of water treatment plant residuals frequently dictate the type of inplant handling system necessary. Selection of an alternative should be based on economic as well as regulatory considerations. The type of sludge and sludge characteristics are also important criteria to be used in developing disposal or reuse alternatives. It is critical that the ultimate solids disposal or reuse programme be a reliable, environmentally sound practice to ensure that it does not affect the primary goal of the treatment plant—the production of potable water. Alternatives available for disposal or reuse of water treatment plant residuals include:

1. Landfilling.
2. Disposal on land (reuse as a soil amendment).
3. Discharge to a waste-water collection system.
4. Codisposal with waste-water biosolids.
5. Reuse in building or fill materials.

Landfilling, land spreading and lagoon storage followed by landfilling or spreading are typical land disposal options. Residuals disposed of in a waste-water collection system end up in the waste-water treatment plant, where they are removed and disposed of with waste-water sludge. Codisposal involves the mixing of water treatment plant residuals with waste-water treatment plant sludges followed by disposal or reuse. Reuse as building or fill material is site specific. However, before discussing the various disposal methods, it is appropriate to consider the impacts of arsenic in residuals.

Arsenic in Residuals

As the regulations for arsenic in drinking water have become more stringent, water treatment processes have shifted to include greater arsenic removal from raw water, increasing arsenic concentrations in residuals. Arsenic in water treatment plant residuals comes from two sources: the raw water and the chemicals used for treating the raw water. The amount of arsenic contributed from treatment chemicals may be considered minor but earlier research found up to 40 per cent of the arsenic in water treatment plant residuals was contributed by iron coagulants.

The effectiveness of arsenic removal from water is dependent upon both the pH of the water and the oxidation state of the arsenic. Once arsenic is in the treatment plant residuals, changes in pH or changes that result in a reducing environment may cause the arsenic to resolubilise. Processes commonly used for water treatment plant residuals that may cause pH or oxidation state changes include dewatering, lagooning and landfilling.

Landfilling

The most common disposal method for water treatment plant sludge in the United States and other European Countries is landfilling in a commercial non-hazardous landfill (or monofill that receives only drinking water treatment plant residuals) or a hazardous waste landfill. Water treatment plant sludge is tested to determine if it is a hazardous (Resource Conservation and Recovery Act, RCRA, subtitle C) or non-hazardous (RCRA subtitle D) waste to determine which type of landfill is appropriate for final disposal.

Water treatment plant sludge testing is performed to meet the EPA requirement for solid waste characterisation by the toxicity characteristic leaching procedure (TCLP). The TCLP test exposes a waste to a mildly acidic solution similar to what might be found in a municipal landfill. If the waste leachate contains any of the regulated compounds at or above the minimum concentration in leachate for toxicity characteristics, it is considered to be toxic and therefore a hazardous waste.

The Waste Extraction Test (WET) uses a slightly more aggressive leaching procedure than is used by the TCLP test, as shown in Table 17.14. Both the TCLP test and the WET are designed to simulate landfill leaching. If the leachate contains any of the regulated compounds on the List of Inorganic Persistent and Bioaccumulative Toxic Substances and Their Soluble Threshold Limit Concentration and the concentration of the compound is equal to or exceeds its listed soluble threshold limit concentration

(STLC) or total threshold limit concentration (TTLC), the waste is considered toxic and therefore a hazardous waste.

There is concern that the current testing procedure, the TCLP, is not adequate for determining the long-term leachability of a material in its final disposal site. A variety of conditions can exist (wet, dry, acidic, basic) at a disposal site, making it unlikely that one test can assess all possible conditions. Research by the Department of Energy's Mixed Waste Focus Area and the U.S. EPA Office of Solid Waste evaluated six different testing protocols using a tiered approach. The conclusion reached in this study was that a better picture of how waste would behave at its final disposal site was obtainable but was expensive to achieve.

Land Application

Land application of water treatment plant residuals is a disposal method that is regulated in the United States by the federal government under the Resource Conservation and Recovery Act (RCRA) as well as state and local governments. Sludges to be spread on land must be tested to determine if they are a hazardous (RCRA subtitle C) or non-hazardous (RCRA subtitle D) waste by either the TCLP or WET, which are compared in Table 17.14.

Table 17.14. Comparison of toxicity characteristic leaching procedure (TCLP) and the waste extraction test (WET).

Parameter	Test procedure	
	TCLP[a]	WET[b]
Extraction fluid	Acetic acid	Citric acid
Extraction fluid pH	4.93	5.00
Extraction duration, h	18	48
Dilution of waste to extraction fluid of solid portion of waste	20 fold	10 fold
Anaerobic conditions	No	Yes, by purging with N_2 gas prior to agitation
Inorganic constituents measured	8	19
Organic constituents measured	23	18
Aggressiveness for inorganic constituents	Less	More

Residuals that have been land applied include coagulant sludges, lime-softening sludges, nanofiltration concentrate and slow sand filter washings. Benefits from land application of coagulant sludges have not been clearly demonstrated, while concerns have been reported. Specific concerns raised include aluminum having a negative impact on barley growth in soils where the pH is below 5.5; high levels of aluminum reducing the availability of phosphorus and increasing soil compaction; and iron becoming concentrated in grazing land resulting in a negative effect on copper metabolism, especially in sheep.

Depending on local soil conditions, the spreading of lime sludges may have beneficial impacts on the soil and crop yields when used in appropriate amounts. Nitrogen fertilisers typically lower soil pH, resulting in a decrease in calcium availability and reduced crop production. The addition of lime sludge raises the soil pH comparable to commercially available agricultural limestone materials. The effectiveness of lime sludge, measured in terms of the total neutralising power (TNP), is typically around 100, which is comparable to the TNP of commercial agricultural lime products. A small number of facilities use land application or irrigation for concentrate disposal.

Land application, is often limited by salinity, which can accumulate in soil and prevent plants from growing or leach into underlying freshwater aquifers. Land application disposal methods are more appropriate for low-pressure systems that primarily remove hardness or NOM; the concentrate from these systems have lower salinity.

Lagooning Prior to Disposal

As noted above, lagooning of water treatment plant residuals is typically an intermediate step prior to final disposal in a landfill. Lagooned residuals separate into sludge and supernatant. The sludge settles to the bottom of the lagoon and becomes more compressed and less porous, anoxic and anaerobic conditions may develop leading to a reducing environment in the settled sludge layer. Under reducing conditions, arsenic may gain electrons, going from As (III) to As (IV) and resolublise into the lagoon supernatant.

Based on recent research, it has been found that the release of arsenic from settled sludge in lagoons into the supernatant is associated with the release of iron and generally follows a lowering of the redox potential. Arsenic release being associated with the release of iron was correlated with the change in iron concentration, not the total iron concentration and was found to occur for both ferric and alum sludges. Results from lagoon simulations were used to conclude that reduced pH and biodegradable organic matter cause an increase in arsenic release from sludge.

Seven different sludges in a simulated lagoon study were tested at periodic intervals for toxicity using both the TCLP and the WET and the test results were compared to the arsenic levels measured in the lagoon supernatant. In all cases, the WET results were much higher, up to 700 times higher, than the measured arsenic levels in the supernatant. The TCLP results tracked more closely to the actual measured concentrations, exceeding the measured concentrations four times and understating the measured concentrations three times.

The conclusion drawn from TCLP testing on sludges from settling ponds and lagoons in arsenic removal studies is that these sludges did not qualify as hazardous waste as the arsenic levels were under 5.0 mg/l. However, some of the same sludges were also tested using the WET and were found to be hazardous because arsenic levels above 5.0 mg/l were measured.

PROCESS SELECTION

The selection of process steps and unit processes for a specific installation depends on factors such as availability and cost of raw water, space available at the site, local weather conditions, type of residual, cost of reuse versus disposal, sophistication of the plant operations personnel, distance to an ultimate disposal site and types of ultimate disposal available.

In most cases it will be possible to rule out most of the options by inspection (e.g., a small site surrounded by residential development would not have space available for sludge lagoons or drying beds, an inland site could not consider ocean disposal), leaving a limited number of options that can be evaluated in more depth to determine the least cost alternative. With most water treatment plant residuals, the general treatment objective is to recover as much of the usable liquid as possible and reduce the volume as much as possible to reduce the costs of subsequent recovery or disposal steps. Because most of the residuals from the treatment process are fairly dilute, the simplest and most economical first step in liquid recovery and reducing volume is gravity thickening.

Operational Factors

Operational factors that must be evaluated are the plant location, size and reliability. An extremely complicated system would probably not be either economical or effective for a small, simple plant. Similarly, the plant reliability history must be evaluated to determine if a complex system would be well operated and maintained at the particular facility.

Economic Factors

Economic factors that must be considered are capital, operation and maintenance and solids disposal costs. Capital costs should include such items as construction costs, trucks and special equipment needed for the process. Operation and maintenance costs should include power, chemicals, labour, parts replacement and equipment repair costs. Disposal costs are typically such items as fees at the landfill or waste-water collection system discharge fees.

Chapter 18

Sewage

INTRODUCTION

Sewage is waste-water generated from domestic activities including kitchen, bathroom, toilet and floor washing. It is basically an urban issue. In rural areas sewage is insignificant. Fast urbanisation and increasing standards of living resulting in steep increase in generation of sewage in the country. Due to paucity of resources, the sewage is not properly collected and treated in most of the urban centers of India. Nearly 2/3rd of the pollution problem is due to discharge of untreated or partially treated sewage in our country. Conventionally sewage is collected through a vast network of sewerage system and transported to a centralised treatment plant, which is resource intensive. Instead of transporting it to long distance for centralised treatment. It is advisable in promoting decentralised treatment at local level using technology based on natural processes.

Sewage, after proper treatment can be used in pisciculture, irrigation, forestry and horticulture. Its conventional treatment generates sludge, which acts as manure. The sludge can also be used for energy recovery. Some sewage treatment plants in the country are recovering this energy and utilising it.

The sewage is generated by all of us, therefore, with little care we can somewhat reduce its amount. 'We must save water because water is scarce and limited', all of us know it but saving water also will reduce the quantity of sewage.

CHARACTERISTICS OF DOMESTIC SEWAGE

The design of a sewage treatment works will be dependent on the quality and quantity of the waste to be treated. The following are some of the important characteristics of domestic sewage:

Organic Matter

Organic matter is the most important polluting constituent of sewage in respect of its effects on receiving water bodies. It is mainly composed of proteins, carbohydrates and fats. Organic matter is commonly measured in terms of BOD and COD. If untreated sewage is discharged into natural water bodies, biological stabilisation of organic matter leads to depletion of oxygen in water bodies.

Nitrogen and Phosphorus

Nitrogen and phosphorus are also very important polluting constituents of sewage because of their role in algal growth and eutrophication of water bodies. Nitrogen is present in fresh domestic sewage in the form of proteinaceous matter urea (i.e., organic nitrogen). Its decomposition by bacteria readily changes it into ammonia. In aerobic environment ammonia nitrogen is oxidised into nitrites and nitrates. Nitrates can be used by algae to form plant proteins.

Nitrogen is commonly measured as TKN (organic + ammonical) as sewage characteristics. Nitrate and nitrite forms of nitrogen are also measured when quality of receiving/affected water (streams, underground water) is monitored.

Phosphorus is usually present in orthophosphate, polyphosphate and organic phosphate forms. Organically bound phosphorus is of little importance in domestic sewage whereas polyphosphate forms undergo hydrolysis to revert into the orthophosphate forms, although this conversion is quite slow.

Suspended Solids

Suspended solids represent that fraction of total solids in any waste-water that can be settled gravitationally. Suspended solids can further be classified into organic (volatile) and inorganic (fixed) fractions. Organic matter is present in the form of either settleable form or non-settleable (dissolved or colloidal) form. If the organic fraction of suspended solids present in sewage is discharged untreated into streams, it leads to sludge deposits and subsequently to anaerobic conditions.

Dissolved Oxygen

Dissolved oxygen, as such, does not have any significance as a sewage characteristics. However, it is the most important pollution assessment parameter of the receiving water bodies. Stabilisation of organic matter, when discharged untreated or partially treated in receiving water, leads to depletion of their dissolved oxygen. Nutrients (nitrogen and phosphorus) addition due to discharge of untreated or treated sewage may lead to algal growth in streams. During day time, algae undergo photosynthesis process and the oxygen released by this process is much more than their respiration requirements resulting in a net addition of dissolved oxygen to water. However, during night time photosynthesis process is stopped whereas respiration requirement continues. This leads to depletion of dissolved oxygen in water. Thus, it is observed that all the polluting constituents of sewage explained above have their direct or indirect effect on dissolved oxygen of receiving water.

Bacterial Parameter (Fecal Coliform)

Although organic matter, in dissolved as well as suspended form, is the most important parameter of sewage as far as ecology of receiving water bodies is concerned. Bacterial parameters, such as Fecal Coliform (FC), which serve as indicators of fecal pollution are also very important when human health is the prime concern.

Sewage from large and small towns is discharged either into a water body, which is used for various purposes such as source of drinking water supply and bathing or discharged on land for irrigation, where human beings come in contact with it. Population consuming water from such sources which receive sewage discharges and persons involved in agricultural activities where sewage is applied become vulnerable to infection from pathogenic organisms (mainly bacteria and viruses) which are discharged by human beings who are infected with disease or who are carriers of a particular disease. Thus, to check quality of receiving water for various uses and to assess acceptability of degree of treatment given to sewage, assessment of bacterial quality also becomes important. Because specific identification of pathogenic bacteria is extremely difficult, the coliform group of organisms is used as an indicator of the presence in waste-water of pathogenic organisms. Coliform bacteria are found in intestinal tract of human beings. Each person discharges about 100 to 400 billion coliform bacteria per day. Presence of coliform organisms is taken as an indication of presence of pathogenic organisms and absence of coliform organism is taken as an indication that water is free from disease producing organisms.

Coliform group of bacteria include genera *Escherichia* and *Aerobacter*. *Aerobacter* and certain *Escherichia* can also grow in soil and, therefore, use of coliform group of bacteria as indicator of human waste becomes complicated. Difficulty in determining *E. coli* to the exclusion of the soil coliform led to use of entire group of coliform as indicator of fecal pollution. Separate determination of total coliform (TC), fecal coliform (FC) and fecal streptococci (FS) is now possible. Presence of FC and pathogenic organism together is well established and FC is the widely used bacterial parameter as indicator of fecal pollution. Determination of FS in water and waste-water is also in practice because FC/FS ratio further helps in identification of source. FC/FS ratio for human beings is more than 4, whereas FC/FS ratio for domestic animals is less than 1. Thus FC/FS ratio can be used to find whether suspected contamination of water is derived from human or animal waste. When FC/FS ratio is obtained between 1 to 2 interpretations become difficult. Incidentally, the rate of removal or death of coliform bacteria in water and waste-water is parallel to the respective rates for pathogenic intestinal bacteria which makes the use of coliform organisms as indicator of fecal pollution very important. FC is therefore a very important parameter in determing bacterial quality of water and waste-water.

EFFECT OF SEWAGE POLLUTION ON SURFACE WATER BODIES

Organic Pollution

All organic materials or wastes can be broken down or decomposed by microbial and other biological activity (biodegradation). Although some inorganic substances are included in this category, most are organic compounds that can exhibit a biochemical oxygen demand (BOD) because oxygen is used in the degradation process. Oxygen is a basic requirement of almost all aquatic life except anaerobic microbes. If sufficient oxygen is not available to the aquatic life, the ecosystem will be adversely affected. Typical sources of organic pollution include sewage from domestic and animal sources; industrial wastes from food processing, paper mills, tanneries, distilleries, sugar and other agro-based industries.

This category of pollution becomes a problem when the oxygen required for biodegradation due to organic pollution is greater than the available oxygen in the water body. Natural systems do have a limited capacity to accommodate self-purification through biodegradation by employing re-oxygenation processes. However, in many situations the anthropogenic pollution overwhelms the given system.

Effect of Nutrients

The nutrients are always present in water and thus it supports aquatic life. Here the primary focus is on fertilising chemicals such as nitrates and phosphates. While important for plant growth, too much of nutrients encourage the overabundance of plant life and can result in environmental damage called 'eutrophication'. This can occur at both microscopic level in form of algae or macroscopic level in form of larger aquatic weeds. The diurnal change in dissolved oxygen is of serious concern.

During day time oxygen remain supersaturated due to photosynthetic contribution of oxygen. But during night the oxygen is depleted as the algal mass consumes significant amount of oxygen. Nitrates and phosphates contributed through anthropogenic sources such as sewage, agricultural run-off and run-off from un-sewered residential areas.

Effect of High Dissolved Solids (TDS)

As water is best solvent known on the earth, it can dissolve variety of substances to which it come in contact during hydrological cycle. In natural waters, the dissolved solids mainly consist of bicarbonates, carbonates, sulphates, chlorides, nitrates and phosphates of calcium, magnesium, sodium, potassium with traces of iron, manganese and other minerals. The amount of dissolved solid is important consideration in determining its suitability for irrigation, drinking and industrial uses. In general, water with a total dissolved solids <500 mg/l are most suitable for drinking. Higher dissolved solids may leads to impairment in physiological processes in the human body. For irrigation water dissolved solid are very important criteria due their gradual accumulation resulting in salinisation of soil, thus, rendering the agriculture land non-productive.

Dissolved solids are undesirable in industrial water due to many reasons. They form scales, cause foaming in boilers, accelerate corrosion and interfere with the colour and tastes of many finished products.

Effect of Toxic Pollutants on Water Quality

The toxic Pollutants are mainly heavy metals, pesticides and other industrial xenobiotic pollutants. The ability of a water body to support aquatic life, as well as its suitability for other uses depends on many trace elements. Some metals e.g., Mn, Zn and Cu present in trace quantity are important for the life as it helps and regulates many physiological functions of the body. The same metals, however, causes severe toxicological effects on human health and the aquatic ecosystem. Water pollution by heavy metals resulting from anthropogenic impact is causing serious ecological problems in many parts of the world. This situation is aggravated by the lack of natural elimination processes for metals. Thus, metals shift from one compartment of environment to another, including the biota, often with detrimental effects. Where sufficient accumulation of the metals in biota occurs through food chain transfer, there is also an increasing toxicological risk for man. As a result of absorption and accumulation, the concentration of metals in bottom sediments is much higher than in the water above, which may cause secondary pollution problem. The toxicity of metals in water depends on the degree of oxidation of a given metal ion together with the forms in which it occurs. As a rule, the ionic form of a metal is the most toxic form. However the toxicity is reduced if the ions are bound into complexes with, for example, natural organic matter. Under certain conditions, metallo-organic, low-molecular compounds formed in natural water exhibit toxicities greater than the uncombined forms. An example is the highly toxic alkyl-derivatives of mercury (methylmercury) from inorganic mercury by aquatic micro-organisms. A famous episode of Minamata disease occurred in Japan in fifties due to consumption of fish contaminated by methyl mercury. Metals in natural water can exist in truly dissolved, colloidal and suspended forms. The proportion of these forms varies for different metals and for different water bodies.

Many thousands of organic compounds enter water bodies as a result of human activities. Monitoring every individual compound is not feasible. However, it is possible to select priority organic pollutants based on their prevalence, toxicity and other properties. Mineral oils, petroleum products, phenols, pesticides, polychlorinated biphenyls (PCBs) and surfactants are examples of such compounds. However, these compounds are not universally monitored because their determination requires sophisticated instrumentation and highly trained personnel.

Therefore, they are evaluated in terms of toxicity as a summary parameter. Many of these compounds are highly toxic and sometimes are carcenogenic and mutagenic in nature. Some selected compounds are measured by gas chromatography method.

Ecological Health

A large number of areas in our aquatic environment support rare species and ecologically very sensitive. They need special protection. Since, the Water Act, 1974 provides for maintenance and restoration of 'wholesomeness' of aquatic resources, which is directly related to ecological health of the water bodies, it is important that ecological health of the water bodies is given first priority in the water quality goal.

HEALTH DIMENSION OF SEWAGE (POLLUTED WATER)

Water-related diseases are a human tragedy, killing millions of people each year, preventing millions more from leading healthy lives and undermining development efforts. About 2.3 billion people in the world suffer from diseases that are linked to water.

Providing clean supplies of water and ensuring proper sanitation facilities would save millions of lives by reducing the prevalence of water-related diseases. Thus, finding solutions to these problems should become a high priority for developing countries and assistance agencies.

While water-related diseases vary substantially in their nature, transmission, effects and management, adverse health effects related to water can be organised into three categories: water-borne diseases, including those caused by both fecal-oral organisms and those caused by toxic substances; water-based diseases; and water-related vector diseases. Another category—water-scarce (also called water-washed)—diseases consist of diseases that develop where clean freshwater is scarce.

Water-Borne Diseases

Water-borne diseases are 'dirty-water' diseases-those caused by water that has been contaminated by human, animal or chemical wastes. Worldwide, the lack of sanitary waste disposal and of clean water for drinking, cooking and washing is to blame for over 12 million deaths a year.

Water-borne diseases include cholera, typhoid, shigella, polio, meningitis and hepatitis A and E. Human beings and animals can act as hosts to the bacterial, viral or protozoal organisms that cause these diseases. Millions of people have little access to sanitary waste disposal or to clean water for personal hygiene. An estimated 3 billion people lack a sanitary toilet, for example, over 1.2 billion people are at risk because they lack, access to safe freshwater.

Where proper sanitation facilities are lacking, water-borne diseases can spread rapidly. Untreated excreta carrying disease organisms wash or leach into freshwater sources, contaminating drinking water and food. The extent to which disease organisms occur in specific freshwater sources depends on the amount of human and animal excreta that they contain.

Diarrheal disease, the major water-borne disease, is prevalent in many countries where sewage treatment is inadequate. Instead, human wastes are disposed of in open latrines, ditches, canals and water courses, or they are spread on cropland. An estimated 4 billion cases of diarrheal disease occur every year, causing 3 million to 4 million deaths, mostly among children.

Using contaminated sewage for fertiliser can result in epidemics of such diseases as cholera. These diseases can even become chronic where clean water supplies are lacking. In the early 1990s, for example, raw sewage water that was used to fertilise vegetable fields caused outbreaks of cholera in Chile and Peru. In Buenos Aires, Argentina, a slum neighbourhood faced continual outbreaks of cholera, hepatitis, and meningitis because only 4 per cent of homes had either water mains or proper toilets, while poor diets and little access to medical services aggravated the health problems.

Toxic substances that find their way into freshwater are another cause of water-borne diseases. Increasingly, agricultural chemicals, fertilisers, pesticides and industrial wastes are being found in freshwater supplies. Such chemicals, even in low concentrations, can build up over time and eventually, can cause chronic diseases such as cancers among people who use the water.

Health problems from nitrates in water sources are becoming a serious problem almost everywhere. In over 150 countries nitrates from fertilisers have seeped into water wells, fouling the drinking water. Excessive concentrations of nitrates cause blood disorders. Also, high levels of nitrates and phosphates in water encourage growth of blue-green algae, leading to deoxygenation (eutrophication). Oxygen is required for metabolism by the organisms that serve as purifiers, breaking down organic matter, such as human wastes, that pollute the water. Therefore the amount of oxygen contained in water is a key indicator of water quality.

Pesticides such as DDT and heptachlor, which are used in agriculture, often wash off in irrigation water. Their presence in water and food products has alarming implications for human health because they are known to cause cancer and also may cause low sperm counts and neurological disease.

The seepage of toxic pollutants into ground and surface water reservoirs used for drinking and household use causes health problems in industrialised countries as well.

Improving public sanitation and providing a clean water supply are the two steps needed to prevent most water-borne diseases and deaths. In particular, constructing sanitary latrines and treating waste-water to allow for biodegradation of human wastes will help curb diseases caused by pollution. At the least, solids should be settled out of waste-water so that it is less contaminated. It is important that a clean water supply and the construction of proper sanitary facilities be provided together because they reinforce each other to limit the spread of infection.

While the cost of building freshwater supply systems and sanitation facilities is high, the costs of not doing so can become staggering.

Water-Based Diseases

Aquatic organisms that spend part of their life cycle in the water and another part as parasites of animals cause water-based diseases. These organisms can thrive in either polluted or unpolluted water. As parasites, they usually take the form of worms, using intermediate animal vectors such as snails to thrive, and then directly infecting human either by boring through the skin or by being swallowed.

Water-based diseases include guinea worm (dracunculiasis), paragonimiasis, clonorchiasis and schistosomiasis (bilharzia). These diseases are caused by a variety of flukes, tapeworms, roundworms and tissue nematodes, often collectively referred to as helminths, that infect human. Although these diseases usually are not fatal, they can be extremely painful, preventing people from working and sometimes even making movement impossible. The prevalence of water-based diseases often increases when dams are constructed, because the stagnant water behind dams is ideal for snails, the intermediary host for many types of worms.

Individuals can prevent infection from water-based diseases by washing vegetables in clean water and thoroughly cooking food. They can refrain from entering infected rivers, because many parasites bore through the feet and legs. In areas where guinea worm is endemic, people can use a piece of cloth or nylon gauze to filter out guinea worm larvae, if clean water is unavailable. As with water-washed diseases, providing hygienic disposal of human wastes helps control water-based diseases. Also, for irrigation channels and other constructed waterways building fast-flowing streams makes it more difficult for snails to survive, thus eliminating the intermediary host.

Water-Related Vector Diseases

Millions of people suffer from infections that are transmitted by vectors-insects or other animals capable of transmitting an infection, such as mosquitoes and tsetse flies-that breed and live in or near both polluted and unpolluted water. Such vectors infect human with malaria, yellow fever, dengue fever, sleeping sickness and filariasis.

The incidence of water-related vector diseases appears to be increasing. There are many reasons: people are developing resistance to antimalarial drugs; mosquitoes are developing resistance to DDT, the major insecticide used; environmental changes are creating new breeding sites; migration, climate change and creation of new habitats mean that fewer people build up natural immunity to the disease; and many malaria control programmes have slowed or been abandoned.

Lack of appropriate water management, along with failure to take preventive measures, contributes to the rising incidence of malaria, filariasis and onchocerciasis.

The solution to water-related vector diseases would appear to be clear—eliminate the insects that transmit the diseases. This is easier said than done, however, as pesticides themselves may be harmful to health if they get into drinking water or irrigation water. Also, many insects develop resistance to pesticides and diseases can emerge again in new forms. Alternative techniques to control these diseases include the use of bednets and introducing natural predators and sterile insects. An inexpensive approach to controlling insect vectors involves the use of polystyrene spheres floating on the top of bodies of static water. Because the spheres cover the surface of the water, the mosquito larvae die from lack of air. Water related diseases and causative factors are shown in Table 18.1.

Table 18.1. Water related diseases and Causative factors.

Name of the disease	Causative organism
Waterborne diseases	
Bacterial	
Typhoid	*Salmonella typhi*
Cholera	Vibrio cholerae
Paratyphoid	*Slmonella parayphi*
Gastroenteritis	Enterotoxigenic *Escherichia coli*
Bacterial dysentery	Variety of *Escherichia coli*
Viral	
Infectious hepatitis	Hepatitis-A virus
Poliomycetis	Polio-virus
Diarrhoeal diseases	Rota-virus, Norwalk agent, other virus
Other symptoms of enteric diseases	Echono-virus, Coxsackie-virus
Protozoan	
Amoebic dysentery	*Entamoeba hystolitica*
Water-washed diseases	
Scabies	Various skin fungus species
Trachoma	Trachoma infecting eyes

(Contd...)

Name of the disease	Causative organism
Bacillary dysentery	E. coli
Water-based diseases	
Schistosomiasis	Schistosoma sp.
Guinea worm	Guinea worm
Infection through water related insect vectors	
Sleeping sickness	Trapanosoma through tsetse fly
Malaria	Plasmodium through Anaphelis
Infections primarily due to defective sanitation	
Hookworm	Hook worm, Ascaris

Another way to control the vectors is species sanitation—using biological methods and habitat management to reduce or eliminate the natural breeding grounds of the disease vectors. Such methods can include: filling and draining unneeded bodies of stagnant water; covering water storage containers; eliminating mosquito breeding sites by periodically clearing canals, reservoirs, and fish ponds of weeds; installing sprinkler and trickle irrigation instead of canals; and lining canals to prevent silt deposits from forming and impeding the flow of water. Also, integrating education about disease prevention into health services and encouraging community discussion of prevention would help people to control vectors and to identify and eliminate inconspicuous breeding sites.

Water-Scarce Diseases

Many other diseases-including trachoma, leprosy, tuberculosis, whooping cough, tetanus and diphtheria-are considered water-scarce (also known as water-washed) in that they thrive in conditions where freshwater is scarce and sanitation is poor. Infections are transmitted when too little fresh water is available for washing hands. These diseases, which are rampant throughout most of the world, can be effectively controlled with better hygiene, for which adequate freshwater is necessary.

Some parasitic diseases not usually considered water-related and previously limited in their reach have been rapidly expanding as population grow and water supplies become more polluted. For example, cysticercosis, a disease usually produced by tapeworms found in undercooked pork and limited to rural areas, expanded rapidly in Mexico City in the early 1980s.

TECHNOLOGICAL OPTIONS FOR TREATMENT OF MUNICIPAL WASTE-WATER

There are a large variety of treatment techniques designed to remove pollutants from waste-water. The objective of waste-water treatment is to separate wastes from water. In one sense, all waste-water treatment processes can be considered separation processes. There are physical, chemical and biological separation processes. Sedimentation and screening are examples of physical processes.

Coagulation, ion exchange and pH adjustment are typical chemical processes, while various forms of biological digestion belong to the category of biological processes. In the biological processes living organisms, while in the physical and chemical processes physical and chemical properties are utilised for waste separation metabolises organic wastes. Table 18.2 shows the major elements of waste-water management systems and associated tasks.

Treatment of sewage is accomplished by adopting various treatment schemes, each incorporating one or several different treatment units such as screens, grit chambers, plain sedimentation, chemical precipitation, trickling filter, activated sludge, anaerobic digestion, up flow anaerobic sludge blanket (UASB) reactor, waste stabilisation pond and maturation pond.

The findings revealed that a majority of the treatment plants are based on Primary Settling followed by Activated Sludge Process (PS + ASP) technology (with anaerobic digesters for sludge), Oxidation Pond or Waste Stabilisation Pond (OP or WSP) technology and UASB followed by Polishing Pond (UASB + PP) technology. Findings have also revealed that most of the STPs are not being utilised to the full capacity due to various reasons.

Table 18.2. Major elements of waste-water management systems and associated tasks.

Elements of waste-water management	Associated tasks
Source of generation	Quantification of waste-water, evaluation of techniques of waste-water reduction and determination of waste-water characteristics
Source control	Design of onsite systems to provide partial treatment of the waste-water
Collection	Design of sewers used to remove waste-water from the various sources of generation
Transmission and pumping	Design of large sewers used to transport waste-water to treatment facilities
Treatment	Selection, analysis and design of treatment operations and processes to meet specified treatment objectives related to the removal of waste-water contaminants of concern
Disposal and reuse	Design of facilities used for the disposal and reuse of treated effluent in the aquatic and land environment and the disposal and reuse of sludge on land

It has been found that low capital and low operational cost sewage treatment method such as Waste Stabilisation Ponds (OP or WSP) technology and low operational cost sewage treatment method such as (UASB + PP) technology are quite effective in BOD removal as well as Fecal Coliform (FC) removal. Overall efficiency of STPs based on these low cost technologies in terms of BOD and FC removal can be further improved if effluent suspended solids (SS) are controlled by improvement in final outlet structures. These technologies are best suited for towns and small cities.

In such situations where sewage of a large city is discharged into a receiving water body having insufficient dilution and/or requires to be maintained at high bacteriological quality, the conventional sewage treatment schemes based on (PS + ASP) technology need augmentation with tertiary treatment units for further removal of BOD and FC. Low cost tertiary treatment method such as series of polishing ponds is the best option for tertiary treatment. However, if land availability is a constraint then other tertiary treatment options such as coagulant aided flocculation + tertiary sedimentation (TS), TS + filtration, TS + chlorination may be adopted.

Conventional Waste-water Treatment

Conventional waste-water treatment consists of pretreatment, primary sedimentation, secondary biological treatment, secondary sedimentation and chlorination before being discharge. Historically, biological techniques have been widely utilised since they are generally economical to build and operate as composed to physico-chemical techniques. Moreover, they are more efficient as natural means of treatment are utilised in optimised conditions.

Treatment systems could be classified according to the degree of pollutant removal into pretreatment, primary, secondary, tertiary and ultimate treatment. They could be classified according to the means of pollutant removal into biological or physico-chemical treatment. Essentially, pretreatment and primary treatment involves screening and grit removal, equalisation and the removal of high concentration of solids that might decrease the efficiency of subsequent treatment processes. The term secondary treatment is commonly used to describe any of the following biological processes: activated sludge, extended aeration, trickling filters, aerobic and anaerobic lagoons and anaerobic and facultative (mixed) ponds. In the typical aerobic process the removal of oxygen-demanding dissolved organics through micro-organisms takes place.

$$\text{Organic material} + \text{bacteria} + \text{oxygen} \nearrow \text{More bacteria}$$
$$\searrow CO_2 + H_2O + \text{stabilised residual}$$

In an activated sludge process, the incoming waste effluent is continuously fed into biological reactor (aeration tank) in which bacterial mass, in a desired concentration, is maintained in suspension. Organic matter in the incoming effluent is partially oxidised by the bacterial mass and partially converted to excess sludge. The sludge in the out-flow of aeration tank is then separated in a clarifier. This sludge is continuously recycled back to the aeration tanks, however, a portion of sludge (excess sludge) is sent to the sludge beds for drying and in this way a desired concentration is maintained. The conventional type activated sludge process could remove as much as 85 per cent of the BOD load.

The extended aeration is essentially similar to the activated sludge process, but yields less sludge for disposal. Through sufficient retention time, biological solids are oxidised, thus minimising resultant sludge.

In aerobic lagoons, oxygen is usually supplied through surface aerators that keep solids in suspension, allowing for about 50 to 60 per cent BOD removal.

Trickling filters are packed with rocks, on the surface of which bacteria are allowed to grow, while waste-water is trickled over through nozzles, allowing for consumption of dissolved organics by bacteria. The relative effectiveness in BOD removal of trickling filters is relatively low compared to other secondary treatment systems.

Tertiary treatment aims at further removal of BOD, suspended solids etc., as well as colour, nitrates, phosphates and other pollutants not adequately removed by secondary treatment processes. Tertiary treatment could involve carbon adsorption, coagulation and sedimentation, ion exchange, membrane filtration and other processes. Treatment processes and purpose of each process in a treatment system is shown in Table 18.3.

Table 18.3. Treatment processes and purpose of each process in a treatment system.

Principal purposes of unit processes	Unit processes
Grit Removal	Grit chambers
Removal or grinding of coarse solids	Bar screens
Odour control	Perchlorination, ozonation
Gross solids-liquid suspension, BOD reduction	Plain primary settling
Gross removal of soluble BOD and COD from raw waste-water	Biological treatment
Removal of oxidised particulates and biological solids	Plain secondary settling
Decomposition or stabilisation of organic solids, conditioning of sludge for dewatering	Anaerobic sludge digestion

(Contd...)

Principal purposes of unit processes	Unit processes
Ultimate sludge disposal	Sludge drying beds, land disposal, land reclamation
Removal of colloidal solids and turbidity from waste-water	Chemical treatment, sedimentation, mixed-media filtration
Phosphates removal	Chemical coagulation, flocculation and settling
Nitrate removal	Ammonia stripping
Removal of suspended and colloidal materials	Mixed-media filtration
Disinfections	Chlorination, UV treatment

OPTIONS FOR ADOPTION OF NEW TREATMENT METHODS

Upflow Anaerobic Sludge Blanket (UASB)

Among the high rate reactors for waste-water treatment, the UASB process has gained popularity in recent years all over the world. Under the Ganga Action Plan, this system is installed at Kanpur and Mirjapur. Several distilleries in the country have also adopted UASB treatment system because of its advantage over the conventional treatment. In the last 20 years, over 150 UASB units have been built in the world for treating high BOD industrial wastes (distilleries sugar, milk etc.). Since 1982, their use has been extended to include typical municipal sewage which has a relatively low BOD of only 200–300 mg/l. The world's first full-scale demonstration plant for municipal sewage was built in Kanpur, India in 1989 (5 mld capacity) under the Indo-Dutch project.

Advantages

The hydraulic retention time is only 8–10 hours, no prior sedimentation is required. The anaerobic unit does not need to be filled with any stones or other media. The upflowing sewage itself forms millions of small 'granules' which are held in suspension and thus provide large surface area. No mixer or aerators are required, thus conserving energy and operation cost, the gas produced can be collected and used. Daily operation of UASB requires minimum attention. No special instrumentation is necessary for control and surplus sludge is easy to dry.

Constraints

The most difficult problem with the UASB system is corrosion. Hence, all construction materials used to be carefully chosen.

Two-Stage, Aerobic Unitank System (TSU-System)

The two stage, aerobic Unitank system and the tri-stage, anaerobic-aerobic Unitank system with biological nitrogen removal (3SU - N System) has been developed in Europe. This is a cost-effective alternative to conventional activated sludge systems. The main advantages are reduction in capital and operational costs, flexible and reliable operation and high process performance. After the preliminary treatment (screening, grit removal equalisation, no primary settling) the waste-water is first treated in a high loaded combined aeration-sedimentation stage. The BOD reduction is about 80–85 per cent. The partially purified water then flows by gravity to a low loaded combined aeration-sedimentation stage where the residual BOD is removed to obtain a high quality effluent resulting in more than 98 per cent removal of BOD.

Advantages

1. Less capital costs, no primary settling, less total aeration volume, no separate sedimentation tanks, no sludge scraping, no sludge recycling facilities, rectangular tanks, compact construction possible, full use of available land, cheaper and easier to construct as compared to circular tanks, economical lengths of connecting pipes and channels, compact system: smaller land area required.
2. Less operational costs, less energy for aeration, no energy for sludge recycle, less maintenance costs (less moving parts).
3. Better process performance, high treatment efficiency, control of sludge bulking, simple and reliable process, reduced need for supervision.
4. Easily controlled by microprocessor.
5. Flexible operation, flexibility of temporary operation with half capacity, restoration of full capacity without long time lag, possible applications, brewing and malting waste-water, treatment, municipal waste-water treatment, food processing waste-water treatment, industrial waste-water treatment, aerobic post treatment of anaerobic effluents from distilleries.

Root Zone Treatment

The process is a natural way of treating industrial or domestic wastes. The method developed in Sixties in Germany, is now commercialised for treatment of domestic and industrial waste-water, economically and efficiently. It has got three integrated components; reeds, reed bed and microbial organisms. In this system, contaminated water is allowed to flow underground through the root zones of especially designed reed beds. The reeds and the reed bed on the soil surface provide an efficient treatment system. The reed bed serves as the host for more than 2000 species of bacteria and thousands of fungal species. These microbial organisms oxidise the organic matter both aerobically as well as anaerobically. Phosphate, sulphur and carbon compounds, nitrogenous materials reduce to their elemental forms. Heavy metals precipitate from solution and are bound into the soil matrix. Due to the high biodiversity of microbes, the root-zone system is capable of shock loads.

The root zone system is suitable for concentrations from a few mg/l upto 20000 mg/l of COD and 4000 mg/l of nitrogen. It can be built for effluent throughout from about $1m^3$/day to more than $10000 \, m^3$/day. For domestic sewage the land requirement is around $0.2 \, m^2$/person. But for the larger area requirement as compared to conventional methods, the root zone treatment system offers an ideal option for biological effluents because of its simplicity and ruggedness. Even in areas where land is a constraint, the system could be adopted with innovations like vertical treatment facility.

Land Treatment for Waste Management

While indiscriminate discharge of wastes on land is an issue of serious environmental concern, it needs to be recognised that land is the best available sink for ultimate disposal of wastes. This becomes particularly relevant in the context of a developing country where it is unlikely that all the wastes would be provided fullest treatment at source before their disposal.

Controlled application of wastes on land can help in achieving a desired degree of treatment through the physical, chemical and biological processes within the plant-soil water matrix. Partially treated waste water can be further treated through land application and land can serve as a 'living filter' comprising interaction of soil, vegetation cover and soil micro-organisms.

The various purposes for which land application could be resorted are:
1. Extraction of useful constituents in the wastes to provide plant nutrients or soil amendments.

2. Revegetation and reclamation of degraded lands.
3. Dedicated disposal of recalcitrant wastes.

Depending on the methods of application and percolation, the land treatment of wastes may be of three types viz: slow rate system, rapid infiltration system and overland flow system.

To ensure safety and precautionary measures in land treatment of wastes, it is essential to ascertain the background concentration of pollutants, possible fate of the pollutants added to the land and the risks involved in terms of assimilative capacities and acceptable limits. Decisions in this regard are to be necessarily guided by a clear understanding of the reaction processes and transport phenomena within and among various sinks namely living systems, soil, water and air. Pilot projects undertaken in selected areas have shown encouraging results based on which it is possible to establish cost effective approaches for waste management through land treatment.

SEWAGE UTILISATION

Land Application of Waste-water

Broadly, land application can be defined as a technique which utilises the interaction between natural soil, vegetation and waste-water to upgrade the quality of waste-water. The traditional sewage farming with innovations to suit location specific conditions could be a cost effective method for treatment and utilisation of waste-water. The value of waste-water as a substitute of organic manure in agriculture (also of water in arid regions) has been recognised for over a century but its use has been restricted by the constraints of social acceptability and the high incidence of diseases in human beings. The municipal sewage has very high economic value. In our country, nearly half of the sewage generated is used for irrigation. The major constraints in sewage farming practices are as follows:

1. Application of raw (untreated) sewage on land causes serious problems of stinking odour, water logging and mosquito breeding.
2. Long term application of sewage effluents and/or sludge results in accumulation of chlorides, sulphates and toxic elements like cadmium, nickel, copper, chromium, manganese, arsenic and mercury in the soil and consequently reduce crop growth. Irrigation generally results in gradual building up of salinity and this is accelerated by the use of municipal waste-water. Changes in soil texture and consequent water logging also may occur in certain areas.
3. Depending upon the soil texture and the flow velocity of water through the soil layers, the nutrients (especially nitrates), organic toxic substances and also the pathogens (bacteria and viruses) move to the groundwater.

Human population engaged in agriculture and fish farms supplied with municipal waste-water and sludge are directly exposed to the pathogens which cause different diseases. Several reports show that upto 70 per cent of the farm workers suffer from helminth infection. Further risks to human health arise from the consumption of food contaminated with pathogens and toxic substances directly or through the food chain.

Studies show that the municipal sewage can be used profitably provided that the treatment procedures ensure that the sludge and municipal waste-water do not contain significant amounts of pesticides, detergents, heavy metals and pathogenic organisms. Conventionally, treated sewage appears best suitable for raising tree plantations, horticultural use (watering public gardens and roadside trees) and growing such plants which are tolerant to various pollutants and are not consumed directly by human and cattle.

The effects of using municipal waste-water in forestry are not well known and therefore, long term studies are required on the impacts of sewage application on the tree growth, other biota and soil characteristics. The utilisation of sewage is also limited by the climate and soil types. Whereas sewage irrigation can be readily recommended in areas with limited water resources seasonally or throughout the year, it is not possible to utilise the effluents in high rainfall regions and during the rainy season elsewhere. Soils prone to salinity and water logging are not suitable whereas many wastelands can possibly be reclaimed with the sludge and sewage effluents.

Another major problem in sewage utilisation is that of the long distances to which the sewage or the treated effluents have to be transported as the areas under agriculture and forestry are far off from the urban centers. Decisions about the location of treatment plants have to take into account a number of factors like the location of the urban centers and their physiographic features. The periodic failure of the treatment plants as well as their overflow during the rainy season also create problems in the utilisation and hence, better management of the treatment facilities is essential.

Use of Sewage in Pisciculture

Sewage contains all the essential major and minor fertilising elements normally used in fish culture. Being in a digested and hence available form, its nutrients promote rapid growth of fish food organisms, which in turn results in greater production of fish per unit area. Fish spawn immediately after stocking needs plentiful supply of natural food in the form of planktons of restricted size, preferably rotifers and cladocerans. Unfortunately, fish culture has not yet been regarded as a means for recycling sewage effluents. It is extensively used in certain parts of the country as a convenient and cheap means of fertilising the ponds and as such little money is spent on proper treatment. Sewage fed fish culturing is still to gain popularity on a wider scale although these are in the experimental stage at various research centers.

A number of field and experimental studies have demonstrated that the utilisation of the nutrients in the domestic sewage by aquaculture is profitable and that using a favourable fish species, with judicious management and correct harvesting techniques, very high yields of fish can be obtained.

Sewage Utilisation in Forestry

Though considerable effort has been made towards the utilisation of municipal waste-water and sludge in natural forests as well as plantations in North America, it has received hardly any attention in India. Often suggestions have been made for applying sludge and irrigation with sewage effluents in tree plantations, orchards, gardens, lawns, golf courses and similar areas, there is no information on the suitable species, their responses at different growth stages and adverse impacts, if any. A study showed that sewage with a high concentration of heavy metals can be better used in forestry as the woody species normally grown are sturdy and the problems of toxicity, heavy metals and salinity stress are relatively negligible. As these are not consumed directly by human or animals, no major hazards to life should be expected. Eucalyptus, Leuacnea and Poplar species have been recommended for plantation under sewage irrigation through ridges and trenches where water is not allowed to stagnate.

Use of Vermiculture for Waste Management

Recently, verticulture technology (use of earth worms for bioconversion of wastes) has been used for the management of garbage, kitchen wastes, organic wastes from food industries etc. The effect of organic matter on earthworm population and the ability of earthworms to promote the decomposition of organic matter have been described for decades. A combination of recycling and resource recovery through biogas and vermiculture could yield fuel methane fertiliser (biogas plant effluents and nutrient rich vermucasts and feed worm biomass). A project on development of design criteria for a small community sewage treatment plant based on vermiculture technology. According to the findings sewage with less than 700 g/m^3 COD, a vemiflter can be designed with a hydraulic loading of only 0.5 m^3.

For dilute waste-water, hydraulic loading is the controlling factor sewage with less than 700 g/m^3 COD, a vermifilter can be designed with a hydraulic loading of only 0.5 m^3. For dilute waste water, hydraulic loading is the controlling factor governing the requirement of vermifilter area. On the other hand, for strong waste-water containing more than 500 g/m^3 organics (equivalent to 700 g/m^3 COD), the area requirement is to be governed by the organic loading. This needs further research to optimise the technical and economical aspects.

DECENTRALISED SMALL SCALE TREATMENT SYSTEMS

Promoting the development of decentralised waste-water treatment and recovery technologies that are linked with urban agriculture systems, at the neighbourhood level, appear to be a national approach to solving the human and environmental health dilemmas that result from under managed waste-water. Decentralised small scale systems must be considered in planning and upgrading urban environment. Gravity flow, small bore sewage, and water borne conveyance systems offer the potential to decentralise urban environments into catchment systems, each with their own integrated treatment plants and at low costs. These systems would be based on the topography of the local water shed and would result in small-scale facilities equally dispersed through environment. Pathogenic reduction and nutrient recovery would occur through the use of integrated biological processes, which are also low cost. This approach would allow for independent, self maintained and self sustained facilities that are capable of recovering waste-water resources and immediately reusing them in decentralised urban farms. Table 18.4 shows the merits and demerits of different low-cost waste-water treatment systems.

Table 18.4. Merits and demerits of different low-cost waste-water treatment systems.

Type	Kind of treatment	Use for type of waste-water	Advantages	Disadvantages
Septic tank	Sedimentation, sludge stabilisation	Waste-water of settleable solids, especially domestic	Simple, durable, little space because of being underground	Low treatment efficiency, effluent not odourless
Imhoff tank	Sedimentation, sludge stabilisation	Waste-water of settleable solids, especially domestic	Durable, little space because of being underground, odourless effluent	Less simple than septic tank, needs very regular desludging
Anaerobic filter	Anaerobic degradation of suspended and dissolved solids	Pre-settled domestic waste-water of narrow COD/BOD ratio	Simple and fairly durable if well constructed and waste-water has been properly pre-treated, high treatment efficiency, little permanent space required because of being underground	Costly in construction because of special filter material, blockage of filter possible, effluent smells slightly despite high treatment efficiency
Baffled septic tank	Anaerobic degradation of	Pre-settled domestic waste-water of	Simple and durable, high treatment	Less efficient with weak waste-water, longer start-

(Contd...)

Type	Kind of treatment	Use for type of waste-water	Advantages	Disadvantages
	suspended and dissolved solids	narrow COD/BOD ratio	efficiency, less space required because of being underground, hardly any blockage, relatively cheap compared to anaerobic filter	up phase than anaerobic filter
Root zone treatment system	Aerobic facultative anaerobic degradation of dissolved and fine suspended solids, pathogen removal	Suitable for domestic waste-water where settleable solids and most suspended solids already removed by pretreatment	High treatment efficiency when properly constructed, pleasant landscaping possible, no waste-water above ground, no nuisance of odour	High space requirement, great knowledge and care required during construction, intensive maintenance and supervision during first 1–2 years
Anaerobic pond	Sedimentation, anaerobic degradation and sludge stabilisation	Domestic and strong and medium waste-water	Simple in construction, flexible in respect to degree of treatment, little maintenance	Waste-water pond occupies open land, there is always some odour, can even be stinky, mosquitoes are difficult to control
Aerobic pond	Aerobic degradation, pathogen removal	Pre-treated domestic waste-water	Simple in construction, reliable in performance if proper dimensioned, high pathogen removal rate, can be used to create an almost natural environment, fish farming possible when large in size and low loaded	Large space requirement, mosquitoes and odour can become a nuisance if undersised, algae can raise effluent BOD
Duck-weed pond	Anaerobic except aerobic at top, degradation of suspended and dissolved solids, nutrient removal	Sullage or Pretreated sewage	Simple in construction, revenue generation through pisciculture, suitable for rural and semi-rural area	High space requirement, possibility of odour can not be ruled out, proper harvesting of duckweed is must

COLLECTION, CONVEYANCE AND DESIGN OF SEWAGE

Wastes or refuse are of two types: solid waste and liquid waste. Night soil consists of human or animal excreta and ureas, originating at privies, water-closets, urinals and stables. Garbage indicates dry refuse from a town and includes sweepings from houses, streets, markets and such other public places, waste paper, leaves, grass, parings from vegetables, decaying fruit etc., constitute garbage. Sullage indicates waste-water from bath rooms, kitchens, washing places and wash basins etc. Sewage indicates the liquid waste from the community. It includes sullage, discharge from latrines, urinals, stables and the industrial waste. It is extremely putrescible; its decomposition produces large quantities of malodorous gases and it may contain numerous pathogenic or disease producing bacteria. The refuse, consisting of all above item, formed in any sanitary system should be rapidly and safely carried to its disposal site so as to maintain a clean environment. Excreta and sewage is satisfactorily disposed of if the following conditions are satisfied.

1. The waste does not pollute the ground surface, nor is it exposed to atmosphere when inadequately treated.
2. It should also not be accessible to children or household pets.
3. It does not pollute or contaminate drinking water supply.
4. It does not give rise to odour nuisance.
5. It does not give unsightly appearance.
6. It does not give rise to mosquito nuisance. It should also not be accessible to insects and rodents.
7. It does not pollute or contaminate the water of bathing beach or streams used for domestic water supply.

SYSTEMS FOR COLLECTION, CONVEYANCE AND DISPOSAL

Depending upon the type of waste, two systems may be employed for its collection, conveyance and disposal:

1. Conservancy system.
2. Water carriage system.

Presence of Particles

The water carried by water mains is practically free from particles of any solid matter—organic and inorganic. The sewage, on the other hand, contains such particles in suspension and the heavy particles settle down at the bottom of sewers which may ultimately result in the clogging of sewers. The sewers are therefore to be laid down at gradient and they should be capable of resisting the wear and tear due to abrasion of these particles.

Conservancy System

In this system, the different types of refuse are collected separately and then each type is carried and suitably disposed of. This system is sometimes referred to as the dry system.

The garbage or dry refuse is collected from roads and streets in pans or baskets. It is then conveyed by carts, trucks, etc., to some suitable place. The garbage is separated into two categories, namely, flammable and inflammable matters. The former is burnt into incinerators and the latter is buried into low lying areas for the reclamation of soil. The night soil is collected in pans from lavatories and the sewage is carried by labour in carts, trucks, etc. It is then buried into the ground and is thus converted into manure.

The storm water and sullage are collected and conveyed separately by closed or open channels. They are discharged in natural rivers or streams. The conservancy system is out of date at present for modern cities. It is however adopted for small towns, villages, undeveloped areas of big cities, etc. where there is scarcity of water for the adoption of water carriage system.

Presence of Solid Matter

Water flowing through the water mains is practically free from solid matter, while the sewage flowing through sewers contain particles of solid matter (both organic as well as inorganic). These solid particles settle at the bottom and have to be dragged during the sewage transport. In order that the sewers are not clogged, they are to be laid at such a gradient that self cleansing velocity is achieved at all value of discharges. Also the inner surface of the sewers must be resistant to the abrasive action of these solid particles.

Pressure

Water in the water mains flow under pressure. Hence the water mains can be carried, within certain limits, up and down the hill or gradient. The hydraulic gradient line lies very much above the pipe surface. On the other hand in most cases, sewers may be considered as open channels, wherein the sewage runs under gravity. The sewers seldom run full and the H.G. line falls within the sewer. Hence the sewers must be laid at continous downward gradient. Sewers run under pressure only when they are designed as force mains and as inverted siphons.

Disadvantages of Conservancy System

The conservancy system has the following disadvantages:

Hygiene and sanitary aspect

The conservancy system is highly unhygienic and cause insanitary conditions since the excreta starts decomposing within few hours of its production. Even if it is assumed that cleaning will take place twice in a day, the excreta remaining in the previes will emit bad smell and will give rise to fly nuisance.

Transportation aspect

Transportation of night soil takes place in open carts through streets and other crowded localities. This is highly undesirable.

Labour aspect

The working of the system depends entirely on the mercy of labour (sweepers). If they go on strike even for one day for any reason whatsoever, the previes cannot be used because of foul smell. The whole locality will smell very badly.

Building design aspect

The lavatories or previes are to be located outside the house and slightly away from the main building. The compact design is therefore not possible.

Condition of drains

Insanitation may be there due to carriage of sullage through open drains laid in the streets.

Human aspect

In the present day world, when man has progressed much, it is highly humiliating to ask human beings to transport night soil in pails on their heads.

Risk of epidemic

Due to improper or careless disposal of night soil, there are more chances of outbreak of epidemic.

Pollution problems

The liquid wastes from lavatories etc., during their washing, may soak in the ground, thus contaminating the soil. If the ground water is at a shallow depth, it may also be polluted due to percolation of waste-water.

Cost consideration

Though the system is quite cheap in the beginning, its maintenance and establishment costs (i.e. recurring expenditure) are very high.

Disposal land requirement

The system requires considerable land for the disposal of sewage.

Minimum and Maximum Velocities

The silting or deposition of particles of solid matter is undesirable in sewers and hence the sewers should be laid at such a gradient that a minimum velocity which will prevent the silting of particles in sewers is developed over a wide variation in discharge of sewage. Such a minimum velocity is known as the self-cleansing velocity and for keeping the sewers free from any trouble, this velocity should be developed at least once in a day, preferably twice in a day. The self-cleansing velocity depends on the nature of suspended matter in sewage and the size of sewer. Table 18.5 shows the self-cleansing velocities for different materials in suspension as recommended by Beardmore and Table 18.6 shows the self-cleansing velocities for sewers of different sizes as recommended by Badwin Latham. Usually a self-cleansing velocity of about 800 mm to 900 mm per second is adopted for normal sewage.

Table 18.5. Self-cleansing velocities.

Name of material	Self-cleansing velocity in mm per second
Angular stones	1000
Round pebbles	600
Fine gravel	300
Coarse sand	200
Fine sand and clay	150

Table 18.6. Self-cleansing velocities.

Diameter of sewer in mm	Self-cleansing velocity in mm per second
150–250	1000
300–600	750
Above 600	600

The maximum velocity of flow is also to be taken into consideration. If the velocity of flow exceeds a certain limit, the particles of solid matter start to damage the inside smooth surface of sewers or in other words, a scouring action takes place. The maximum permissible velocity at which no such scouring action will occur is known as the non-scouring velocity and it will mainly depend on the material used in the construction of sewers. Table 18.7 shows the non-scouring velocities for common sewer materials.

Water Carriage System

In this system, the water is used as medium to convey the sewage to the point of its treatment or final disposal. The quantity of water to be mixed with solid matter is quite sufficient and the dilution ratio of solid matter with water is so great that the mixture behaves more or less like water. The sewage is conveyed in suitably designed and maintained sewers.

Table 18.7. Non-scouring velocities.

Material of sewer	Non-scouring velocity in mm per second
Earthen channels	600–1200
Brick-lined sewers	1500–2400
Cement concrete sewers	2400–3000
Stoneware sewers	3000–4500
Cast-iron sewers	3500–4500
Vitrified tiles and glazed bricks	4500–5500

In this system, the garbage is collected and conveyed as in case of conservancy system. The storm water may be carried separately or may be allowed to flow with the sewage.

In this system, specially designed latrines, called water closets (W.C.) are used which are flushed with 5 to 10 litres of water after its use by every person. The human excreta is thus flushed away and led to suitable designed and maintained sewers. The wastes from kitchens, baths, wash basins etc. are also led to the sewers. The sewers are the underground closed pipes which are laid on suitable longitudinal gradient so that flow takes under gravity and proper flow velocity is maintained to keep the sewer clean. The sewers lead the sewarage so collected, to a suitable site where it is treated suitably and then is disposed of by irrigation or by dilution. It should be noted that the garbage is collected separately and conveyed in the same manner as is done in the case of conservancy system. If garbage is permitted in the sewers, they may be clogged.

Self cleansing and non-scouring velocities

Following points should be noted in connection with the self-cleansing and non-scouring velocities:

1. In the design of sewers, the discharge is known and hence the velocity of flow and gradient of sewers are to be properly correlated to achieve the desired results.
2. In a flat country, the design of sewers should be such that the self-cleansing velocity is developed at the time of maximum discharge. But provision is made in the design that a velocity of about 400 mm per second develops at the time of minimum discharge. If it is not possible to achieve a minimum velocity of about 400 mm per second, the flushing tanks will have to be provided to keep the sewers clean.
3. In a rough country, the sewers are designed to achieve non-scouring velocity at the time of maximum discharge and self-cleansing velocity at the time of minimum discharge. If fall is too great, the drop manholes are to be constructed to bring down the velocities of flow within these limits.
4. For combined sewers, it becomes difficult to achieve self-cleansing velocity at the time of minimum discharge. For this purpose, special forms of sewers should be adopted.

The initial cost of the installation of water carriage system is very high and it becomes difficult to adopt it when the financial condition of the area is very poor. However the water carriage system is the modern method of conveyance of sewage and it is to be recommended wherever possible. It can even be adopted in stages as the town develops.

Hydraulic Formula

The design of sewers is done on the basis of the following empirical hydraulic formulae:

1. Chezy's formula
2. Kutter's formula
3. Bazin's formula
4. Manning's formula
5. Crimp and Bruge's formula
6. Hazen and William's formula.

Apart from these formulae, there are various charts, diagrams, graphs and tables with the help of which the hydraulic design of sewers can be done. The factors that influence the flow of sewage in the sewers are:

1. Slope of sewer.
2. Geometry of sewer.
3. Roughness of interiors surface of sewer.
4. Bends, transitions, obstructions etc.
5. Flow conditions.
6. Characteristics of sewage.

Chezy's Formula

Chezy's gave the following formula:

$$V = C\sqrt{R.S} \qquad \qquad ...(18.1)$$

where,

V = velocity of flow (m/sec.)

S = hydraulic gradient or slope of the sewer

R = hydraulic mean radius (m) = A/P

A = area of cross-section (m²)

P = wetted perimeter.

C = Chezy's constant

The Chezy's constant C is very complex in nature and it depends upon several factors, such as roughness of inner surface of sewer, hydraulic mean radius, size and shape of sewer, slope etc. Generally, the value of Chezey's constant C is found either by Kutter's formula or by Bazin's formula. Knowing the velocity of flow V from Eq. 18.1, the channel section is designed by the general formula:

$$Q = A \times V$$

where,

Q = discharge in m³/sec.

Advantages of water carriage system

The water carriage system is the most modern system of drainage and has the following advantages:

Hygienes and sanitary aspect

The system is very hygienic since the night soil and other waste-water is conveyed through closed conduits which are not directly exposed to the atmosphere. There is no bad smell because of continuous flow.

Epidemic aspect

There are no chances of outbreak of epidemic because flies and other insects do not have direct access to the sewage.

Pollution aspect

The liquid wastes etc., are directly conveyed through the sewers and therefore there are no changes of the waste-water being soaked in the ground thus contaminating the soil. The waste-water does not percolate down to join the ground water.

There are no chances of pollution of water of wells in individual houses if any.

Compactness in design

Since the latrines are flushed after every use, excreta does not remain and there are no foul smells. The latrines can therefore be attached to the living and bed rooms. This permits a compact design. The lavatories can be accommodated in any part of the house.

Labour aspect

The labour required for the operation and maintenance is extremely small. In fact, the functioning of the system is practically automatic, except for the operation of certain pumps etc. Therefore, there is no labour problem. In the individual houses, the latrines/lavatories can be coveniently cleaned by occupants themselves.

Treatment aspect

The system permits the use of modern methods of treatment of the sewerage collected through the sewers. The treated waste-water and sewage can be safely disposed off without any risk.

Land disposal requirements

Because of treatment facilities, the land required for the disposal of the treated waste-water is very much smaller than that required for the conservancy system.

Cost consideration

Though the initial cost of installation of the system are very high, the running costs are very small since manual labour is very much reduced.

Bazin's Formula

According to Bazin, the value of constant C in Chezy's formula can be obtained by the following equation:

$$C = \frac{157.6}{1.81 + \dfrac{K}{\sqrt{m}}}$$

where,

       m = hydraulic mean depth in meters

       K = Bazin's constant.

This formula was developed by Bazin in 1897 and some of the values of K, as recommended by Bazin, are given in Table 18.8.

Table 18.8. Bazin's constant.

Nature of inside surface of sewer	Value of K
Very smooth surface	0.109
Smooth brick and concrete surfaces	0.290
Smooth rubble masonry surface	0.833
Good earthen channels	1.540
Rough brick and concrete surfaces	0.500
Rough earthen channels	3.170

Manning's Formula

Manning gave the following expression for velocity of flow. The formula is widely used in U.S.A as well as in India:

$$V = \frac{1}{N} R^{2/3} S^{1/2} \qquad \qquad ...(18.2)$$

where,

V, N, R, S have the same meanings, as above. The value of rugosity coefficient N is the same as suggested by Kutter and may be taken from Table 18.9. Gives the comparative idea of conservancy or dry system and water carriage or wet system.

Table 18.9. Comparison between conservance and water carriage system.

Conservancy system	Water carriage system
It does not permit compact design of structures.	It permits compact design of structures
It is laid above ground. Hence it is visible, but non-hygienic	It is necessarily laid below ground. Hence it is not visible, but hygienic
It requires small quantity of water to the extent of about 30 to 40 liters per capita per day	It requires large quantity of water to the extent of about 100 to 120 liters per capita per day
There exists putrefaction	There are no chances for putrefaction
It has been normally considered as system for rural conditions	It has come up basically as an urban system
The labour force required is much more	Only few labourers are required
There is presence of segregation	There is absence of segregation
It is cheap in initial cost, but expensive in maintenance works	It is expensive in initial cost, but maintenance costs are low
There are chances for the outbreak of epidemic	The risk of outbreak of epidemic is greatly reduced
It does not require the help of skilled or technical personnel	It requires the help of skilled or technical personnel for laying, maintenance and running of treatment units
The city remains dirty and foul smelling	The city appears neat and clean
It is likely that underground sources of water may be polluted due to soaking of liquid wastes from the latrines	There is practically no risk of pollution of underground sources of water as sewage is carried in closed sewers and above the water pipes

Kutter's Formula

According to Kutter, the value of constant C in Chezy's formula can be obtained by the following equation:

$$C = \frac{23 + \dfrac{0.00155}{i} + \dfrac{1}{N}}{1 + \left(23 + \dfrac{0.00155}{i}\right)\dfrac{N}{\sqrt{m}}}$$

where, m and i are as above and

N = roughness coefficient or rugosity factor.

This formula was developed by Swiss engineers named Ganguillet and Kutter in 1869 and some of the values of N, as recommended by them, are given in Table 18.10.

Table 18.10. Rugosity factor.

Nature of inside surface of channel	Value of N
Smooth cement plastered surface	0.010
Iron or steel surface or wood-stave pipe	0.011
Surface of concrete or unplaned timber	0.012
Cast-iron pipe	0.013
Stoneware pipe well laid	0.013
Stoneware pipe roughly laid	0.015
Ordinary brickwork	0.015
Rough brickwork or stone masonry	0.017
Smooth earthen channels	0.020
Corrugated iron pipe	0.021
Earthen channels in average condition	0.025
Earthen channels in bad condition	0.030

Classification of Water Carriage System

The water carriage system can be divided into the following types:

1. Separate system
2. Combined system
3. Partially separate system

The above three types are commonly referred to as the three systems of sewerage.

Separate system

The separate system provides two separate systems of sewers—the one intended for the conveyance of foul sewage only, such as faecal matter, domestic waste-water, the washings and drainings of places such as slaughter houses, laundries, stables and the waste-water derived from the manufacturing processes; and the other for the rain water, including the surface washing from certain streets, overflow from public baths and foundations etc. The sewage from the first system of sewers can be led to the treatment works, while the flow from the second system of sewers can be discharged directly to natural streams etc., without any treatment.

Crimp and Bruge's formula

This formula is commonly used in England

$$V = 83.47 \, R^{2/3} \, S^{1/2} \qquad \qquad \dots (18.3)$$

where,

V, R and S have the same meanings, as defined earlier. Comparing this with Manning's formula, we have:

$$V = 83.47 \, R^{2/3} \, S^{1/2} = \frac{R^{2/3} \, S^{1/2}}{N} \qquad \qquad \dots (18.4)$$

which gives $N = 1/83.47 = 0.012$. Hence Manning's formula becomes Crimp and Bruge's formula when $N = 0.012$.

For a circular pipe,
$$R = \frac{A}{P} = \frac{\frac{\pi}{4} D^2}{\pi D} = \frac{D}{4}$$

$$\therefore \qquad V = 83.47 \left(\frac{D}{4}\right)^{2/3} S^{1/2}$$

Now,
$$Q = A \times V = \frac{\pi}{4} D^2 \times 83.47 \left(\frac{D}{4}\right)^{2/3} S^{1/2}$$

or
$$Q = 26.02 D^{8/3} S^{1/2} \qquad \qquad \dots (18.5)$$

Advantages

Following are the advantages of this system:

1. The load on treatment units becomes less.
2. The natural water is not unnecessarily polluted.
3. The sewers are small in size.
4. The storm water can be discharged into natural streams without any treatment.
5. The system proves to be economical when pumping is required for the lifting of sewage.

Hazen and Williams' formula

The formula is as follows:

$$v = 0.85 \, C \, m^{0.63} \, i^{0.54}$$

where, v, m and i are as above and C is a coefficient.

This formula was developed by Hazen and Williams and it is mainly used for flow under pressure. It may however be adopted in sewer design and the values of coefficient C, as recommended by Hazen and Williams, are mentioned in Table 18.11.

Disadvantages

1. Since the sewers are of small size, it is difficult to clean them.
2. They are likely to get chocked.
3. Two sets of sewers may ultimately prove to be costly.
4. There is a likelihood of connections being wrongly made through a confusion of the systems.

5. Storm water sewers or drains comes in use only during the rainy season. During other part of the year, these may serve as dumping place for garbage and may get chocked.

6. Because of lesser air contact in small size sewers, foul smell may be there due to the sewage gas formed.

Table 18.11. Coefficient of Hazen and Williams' formula.

Type of material	Value of C
Old cast-iron pipes; Brick sewers in good condition	100
Stoneware pipes in good condition; Cement-lined pipes; new riveted steel pipes	110
Wood-stave pipes	120
New cast-iron pipes	130
Pipes with very smooth inside surface	140

Table 18.12. Hazen and William's coefficient C.

Type of material	C
Steel pipe under future conditions	95
Old C.I. pipes; brick sewers in good condition	100
Stoneware pipes in good condition	110
Cement lined pipes	110
New riveted steel pipe	110
Wood stave pipe	120
New C. I. pipes	130
Pipes with very smooth inside surface	140
Asbestos cement pipes	140

Combined system

In this system, only one set of sewers is laid and it carries both, namely, sewage and storm water. The sewage and storm water are carried to the sewage treatment plant.

Sizes of Sewers

The minimum size of a sewer depends upon the practice followed in the locality. Usually the sewers of 100 mm diameter are allowed upto a maximum length of 6 meters or so. But when the length of sewer line exceeds about 6 meters, a sewer of minimum diameter 150 mm is allowed. The smaller the diameter of sewer, the greater will be the slope and hence, in order to take advantage of available fall, the sewers of larger diameter are sometimes used.

The design of sewers should be made in such a way that it ends in sections of sewers which are commercially available. The non-commercial sizes are difficult to obtain and they prove to be costly. For sewers to be constructed on site of work, this problem does not arise.

There is no upper limit for the size of a sewer. It is, however, submitted that it is desirable to lay duplicate sewer line when sewer diameter exceeds about 3 metres or so.

Time of Concentration

Meaning of the term

This term is used in connection with the design of storm water drains. As the rain falls on the ground, all the area to be served by the sewer does not start to contribute immediately to the flow of sewer. But the flow is built-up gradually as follows:

1. The area just near the sewer line will start contributing first and it will go on increasing as more and more area starts to contribute.
2. When the whole area is contributing to the flow of sewer, the maximum limit of flow will be reached and it will be equal to the rate of precipitation of rain water.
3. The maximum flow continues until the storm stops. The flow then gradually falls down as the area near the sewer line stops contributing firstly, while flow continues to come for considerable time from the distant areas.

The total time required by the flow to reach to the maximum limit is known as the time of concentration and it consists of two parts:

1. Time of entry.
2. Time of flow.

Time of entry

The time required by storm water to reach to the uppermost inlet of sewer line is known as the time of entry. The value of time of entry depends on many circumstances, the chief being size, shape and slope of area. The various authors have given different values for time of entry. Lloyd—Davies gave time of entry for paved surfaces as 1 to 3 minutes. According to Horner, time of entry is about 2 to 5 minutes and according to Metcalf and Eddy, it is about 3 to 20 minutes. Appleby has recommended time of entry from 7 to 15 minutes. Thus extreme care is to be taken to assume the proper value of time of entry. Generally the time of entry for paved surfaces is taken as about one minute for every 6 metres length.

Time of flow

The time taken by storm water to flow down the sewer upto a particular point is known as the time of flow and it depends on length, size, slope and smoothness of sewer. The time of flow can be easily determined by applying the principles of hydraulics. If velocity of flow and distance to be traversed are known, the time of flow can be determined by the relation:

$$\text{Time of flow} = \frac{\text{distance to be traversed}}{\text{velocity of flow}}$$

The addition of time of entry and time of flow gives the time of concentration.

Importance

The importance of time of concentration in the design of storm water sewers lies in the fact that out of all the storms of equal frequency of occurrence, that storm which has duration equal to the time of concentration, produces the maximum flow in sewer.

Advantages

1. The system requires only one set of sewers. Hence the maintenance costs are reduced.
2. The sewers are of larger size and therefore the chances of their choking are rare. Also, it is easy to clean them.

3. The strength of the sewage is reduced by dilution.
4. There is more air in the larger sewers than in smaller ones of the separate system. Hence the sewer gas that may be formed gets diluted. Thus the chances of foul smell are reduced.

Disadvantages

Following are the disadvantages of this system:
1. During extraordinary heavy storms, the combined sewer may overflow and it may thus put public health in danger.
2. The combined sewer, if not properly designed, gets easily silted and it may even become foul in dry weather.
3. The load on treatment plant increases.
4. The sewers are large in diameter.
5. The storm water is unnecessarily polluted.
6. The system proves to be uneconomical when pumping is required for the lifting of sewage.

Partially Combined System

In this system, only one set of underground sewers is laid. These sewers admit the foul sewage as well as the early washings by rains. As soon as the quantity of storm water exceeds a certain limit, the storm water overflows and is thus collected and conveyed in open drains to the natural streams. The foul sewage, however, continues to flow in the sewers.

Design procedure

In the design of sewers, the following procedure is generally adopted:
1. Formation of zones: The area to be served by the drainage system is divided into different zones. The general layout of roads is to be properly studied for the location of sewers. The zones are marked on the map and the contours are also drawn on the map.
2. Arrangement of sewers: The proposed arrangement for sewers for different zones is then worked out. The low lying areas are isolated and pumping stations are installed for them. The flow of sewage starts from high level zones. The various sewers such as branch sewers, main sewers, trunk sewers, outfall sewers, etc., are marked on the map.

Advantages

Following are the advantages of this system:
1. It combines the advantages of both the above systems.
2. The entry of storm water avoids silting in sewers.
3. The problem of disposing storm water from houses is simplified.
4. The sewers are of reasonable size.

Disadvantages

1. During the dry weather, when there is no rain water, the velocity of flow will be low. Thus self cleansing velocity may not be achieved.
2. The storm water increases the load on treatment units.
3. The storm water also increases the cost of pumping.

Conditions Favorable for Separate System

Financial aspect

If sufficient funds are not available in the beginning, sewers may be constructed to carry only domestic sewage and the rain water may be conveyed through' the open drains. These drains can be converted into regular sewers later when sufficient funds are available.

Flat topography

If the area is flat, the separate system proves to be convenient as deep excavations are to be carried out to lay combined sewers.

Rainfall pattern

If the rainfall is there for a shorter duration and does not take place throughout the year, it is more economical to adopt separate system.

Outlet for storm water

The system can be justified when the sanitary sewage is to be collected and conveyed to a particular point for treatment and there is a separate outlet in the form of natural river or stream for the disposal of storm water.

Pumping aspects

Separate system is best suited under the conditions when the sewage has to be lifted up by pumping. The separate sewers for the storm water will reduce the load on the pumps.

Soil of laying

If sewers are to be laid through hard rocky soil, it becomes difficult to lay combined sewers which are usually of large size.

Factors Governing Choice of Combined System

A combined system is adopted under the following conditions:

Space considerations

Combined system is preferred when space available for laying the sewers is restricted.

Conversion of existing storm water drains

The combined system is preferred if an existing storm water drain is being converted into a combined sewer. This is possible only if the quantity of sewage is small.

Even rainfall

If rainfall is evenly spread throughout the year the combined system can be adopted.

Regulators

If regulators are installed to allow a certain pre-determined quantity of storm water into sewer and to divert the remaining storm water to the natural streams, the combined system proves to be economical.

Pumping requirements

If the ground slopes are such that it is necessary to lift both the sewage as well as the storm water, it is preferable to use the combined system.

Choice of the System

The factors governing the choice of any system are so vast and varied that no generalisation can be done regarding the final choice. The conditions vary from place to place. If the availability of funds is the main factor, then one may conclude that separate system may be adopted if sufficient funds are not available in the beginning. In that case, sewers may be designed to carry the foul sewage and rain water may flow through the open drains along the roads and streets. This suggestion may be more appropriate for small cities. For large metropolitan cities, however, a combined system is highly desirable. It is possible to design combined sewers such that reasonable velocities are maintained in them through the year. Alternatively, a partially separate system may be adopted for these big metropolitan cities.

PATTERNS OF REFUSE COLLECTION

The liquid waste has to be collected by the proper pattern of collection system. Depending upon the area to be drained, topographical and hydraulic features of the area, location of the treatment works, sewerage system adopted for the locality and various other factors, anyone or a combination of more than one of the following five patterns of collection system is adopted:
1. Fan pattern.
2. Interceptor pattern.
3. Perpendicular pattern.
4. Radial pattern.
5. Zonal pattern.

Each of the above pattern of refuse collection will now be briefly described.

Fan Pattern

In this system of layout, the treatment plant is located at a certain point and the entire sewage flow is directed towards this point. Thus a fanlike network of converging main sewers is laid in this pattern. The advantage of this pattern is that only one unit of treatment plant will be required.

Disadvantages

1. The diameter of main trunk sewers will be more and it may result in increase of cost of laying such sewers.
2. The development of the surrounding area will increase the load on the treatment plant and hence restriction will have to be imposed on such development.

Interceptor Pattern

In this pattern, the sewers are intercepted by large size sewers which are laid along the water course. The sewage is carried to the treatment plant and depending upon the facilities provided, it is disposed of either with or without treatment. If the quantity of storm water is more, the storm regulators may be provided at suitable points.

Perpendicular pattern

In this pattern, the main trunk sewers are laid perpendicular to the natural water courses and thus they are of the shortest length. This pattern proves useful for separate or partially separate system in which case storm water can be disposed of directly without any treatment. This pattern will be impracticable for combined system as it will require a treatment unit at every point of outlet.

Radial Pattern

In this pattern, the sewers are laid radially outwards from the center of the city. This pattern is useful for cities where the facilities of sewage disposal by land treatment are available. The suburbs can be served economically by small and short lines of sewers. But this pattern will require large number of disposal works.

Zonal pattern

In this pattern, the city is divided into suitable zones and a separate interceptor is provided for each zone. This pattern proves to be economical for cities which are situated on sloping hills.

Chapter 19

Engineered Systems for Resource and Energy Recovery

INTRODUCTION

The purpose of this chapter is to introduce the reader to the techniques and methods used to recover materials, conversion products and energy from solid wastes. Some of the methods include: (i) processing techniques; (ii) materials recovery systems; (iii) recovery of biological conversion products; (iv) recovery of chemical conversion products; (v) recovery of energy from conversion products; and (vi) materials and energy recovery systems. Many of the techniques to be considered are in a state of flux with respect to application and design criteria. If these techniques are to be considered in the development of waste-management systems, current engineering design and performance data must be obtained from the records of operating installations, from field tests, from equipment manufacturers and from the literature.

PROCESSING TECHNIQUES

Processing techniques, are used in solid waste management systems to improve the efficiency of solid waste management systems, to recover resources (usable materials) and to prepare materials for the recovery of conversion products and energy. The important techniques used for processing solid wastes to recover materials and to prepare the waste for subsequent processing are summarised in Table 19.1.

Table 19.1. Processing techniques used to recover materials and to prepare wastes for further processing.

Processing technique	Function	Representative equipment and/or facilities and applications
Mechnical size and shape alteration	Alteration of the size and shape of solid waste components	Equipment used to reduce the size of solid waste includes hammermills, shredders, roll crushers, grinders, chippers, jaw crushers, rasp mills and hydropulpers
Mechanical component separation	Separation of recoverable materials, usually at a processing facility	Trommels and vibrating screens are used for unprocessed and processed wastes; disk screens for processed wastes; zigzag, vibrating air, rotary air and air knife classifiers for processed wastes. Jig, pneumatic, sink/float, inertial, inclined or · shaking table flotation and optical sorting are used to separate the light and heavy materials in solid wastes

(Contd ...)

Processing technique	Function	Representative equipment and/or facilities and applications
Magnetic and electro-mechanical separation	Separation of ferrous and non-ferrous materials from processed solid wastes	Magnetic separation is used for ferrous materials; eddy current separation for aluminium; electrostatic separation for glass in wastes free of ferrous and aluminium scrap; magnetic fluid separation for non-ferrous materials from processed wastes
Drying and dewatering	Removal of moisture from solid wastes	Convection, conduction and radiation dryers have been used for solid wastes and sludge. Centrifugation and filtration are used to dewater treatment plant sludge

MECHANICAL SIZE ALTERATION

The objective of size reduction is to obtain a final product that is reasonably uniform and considerably reduced in size in comparison with its original form. It is important to note that size reduction does not necessarily imply volume reduction. In some situations, the total volume of the material after size reduction may be greater than the original volume.

Mechanical Component Separation

Component separation is a necessary operation in the recovery of resources from solid wastes and where energy and conversion products are to be recovered from processed wastes. For example, trommels are now used routinely for the separation of unprocessed wastes. Shredding of wastes as a first step has been replaced with screening because: (i) shredding tends to shatter glass and entrap organic materials within tin cans; (ii) it contaminates paper with liquids and putrescible organic materials; and (iii) the operation has a high energy demand.

Magnetic and Electro-Mechanical Separation

Magnetic separation of ferrous materials, a well-established technique in the metals industry, is now used commonly for the removal of ferrous metals from solid wastes. More recently, a variety of electro-mechanical techniques have been developed for the removal of several non-ferrous materials.

Drying and Dewatering

In many solid waste energy recovery and incineration systems, the shredded light fraction is pre-dried to decrease weight. Although the energy requirements for drying wastes vary with local conditions, the required energy input can be estimated by using a value of about 4300 kJ/kg of water evaporated.

MATERIALS RECOVERY SYSTEMS

In this section the objective is to show how the individual processes can be combined in alternative flow sheets for the recovery of materials and the preparation of combustible wastes for subsequent processing.

Materials Specifications

Paper, cardboard, plastics, glass, ferrous metals and non-ferrous metals are the principal recoverable materials contained in municipal solid wastes. In any given situation, the decision to recover any of or all these materials is usually based on an economic evaluation and on local considerations. In assessing the economics of materials recovery, the materials specifications will be a critical consideration.

Processing and Recovery Systems

Once a decision has been made to recover materials and/or energy, flow sheets must be developed for the removal of the desired components and for processing combustible materials, subject to pre-determined materials specifications. Typical flow sheets for the recovery of waste components and the preparation of combustible materials for use as a fuel source are presented in Figs. 19.1 and 19.2, respectively. The light combustible materials are often identified as RDF (refuse-derived fuel).

System Design and Layout

The design and layout of the physical facilities that make up the processing plant flow sheet are an important aspect in the implementation and successful operation of such systems. Important factors that must be considered in the design and layout of such systems include: (i) process performance efficiency; (ii) reliability and flexibility; (iii) ease and economy of operation; (iv) aesthetics; and (v) environmental controls.

RECOVERY OF BIOLOGICAL CONVERSION PRODUCTS

Biological conversion products that can be derived from solid wastes include compost, methane, various proteins and alcohols and a variety of other intermediate organic compounds. The principal processes that have been used are given in Table 19.2. Composting and anaerobic digestion, the two most developed processes, are considered further.

Table 19.2. Biological processes for the recovery of conversion products from solid wastes.

Process	Conversion product	Pre-processing
Aerobic conversion	Compost (soil conditioner)	Separation of organic fraction, particle size reduction
Alkaline hydrolysis	Organic acids	Separation of organic fraction, particle size reduction
Anaerobic digestion (in landfill)	Methane	None, other than placement in containment cells
Anaerobic digestion	Methane	Separation of organic fraction, particle size reduction
Fermentation (following acid or enzymatic hydrolysis)	Ethanol, single-cell protein	Separation of organic fraction, particle size reduction, acid or enzymatic hydrolysis to produce glucose

Composting (Aerobic Conversion)

If the organic materials, excluding plastics, rubber and leather are separated from municipal solid wastes and are subjected to bacterial decomposition, the end product remaining after dissimilatory and assimilatory bacterial activity is called compost or humus. The entire process involving both the separation and bacterial conversion of the organic solid wastes is known as composting. Decomposition of the organic solid wastes may be accomplished either aerobically or anaerobically, depending on the availability of oxygen.

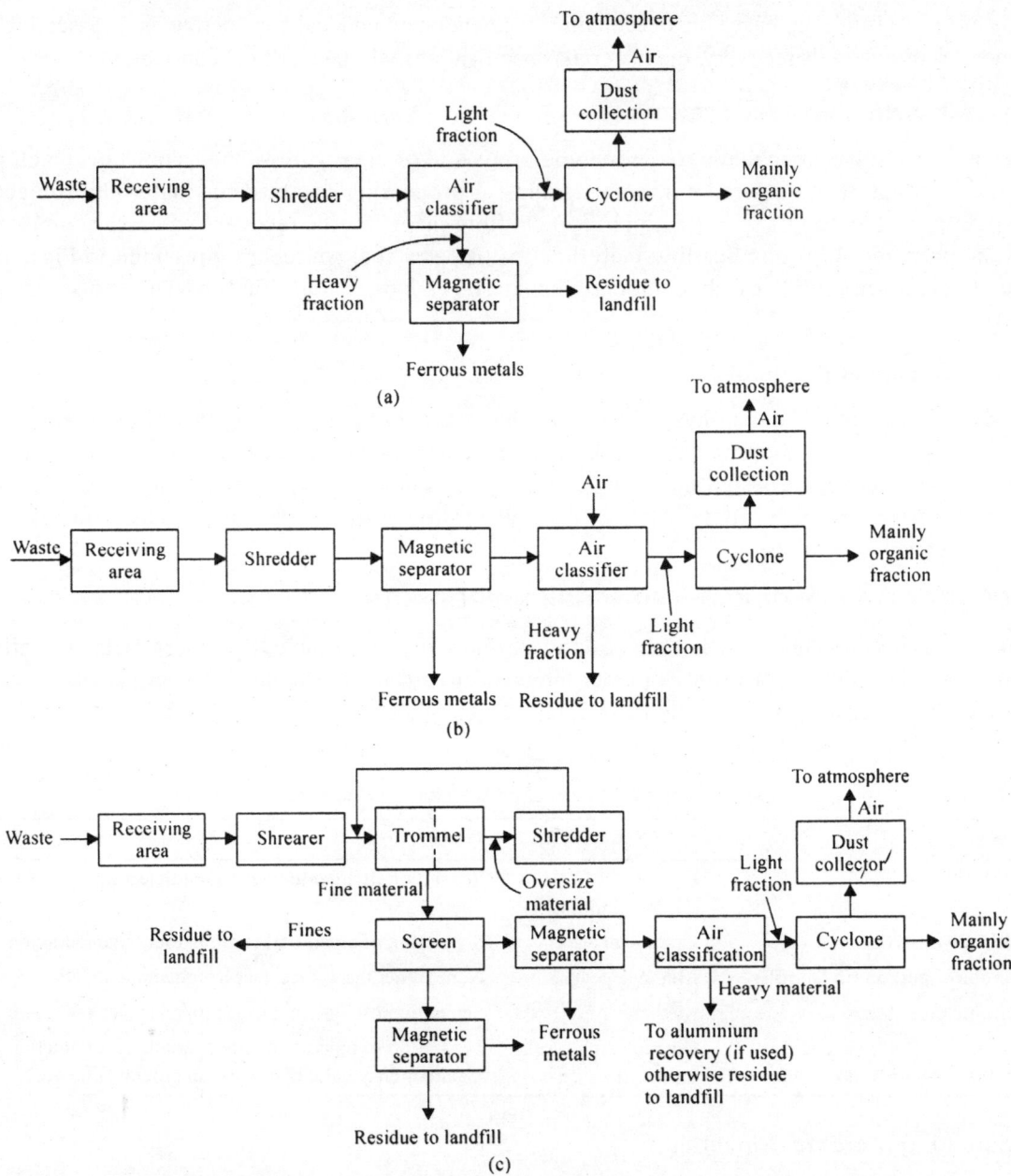

Fig. 19.1. Flow sheet for the recovery of waste components from solid waste: (a) conventional; (b) conventional with shredder in different location; and (c) trommel used to replace shredder in flow sheet.

Most composting operations involve three basic steps: (i) preparation of the solid wastes; (ii) decomposition of the solid wastes; and (iii) product preparation and marketing. Receiving, sorting, separation, size reduction and moisture and nutrient addition are part of the preparation step. Several techniques have been developed to accomplish the decomposition step. Once the solid wastes have been converted to a humus, they are ready for the third step, product preparation and marketing. This step may include fine grinding, blending with various additives, granulation, bagging, storage, shipping and in some cases, direct marketing. The principal design considerations associated with the biological decomposition of prepared solid wastes are presented in Table 19.3.

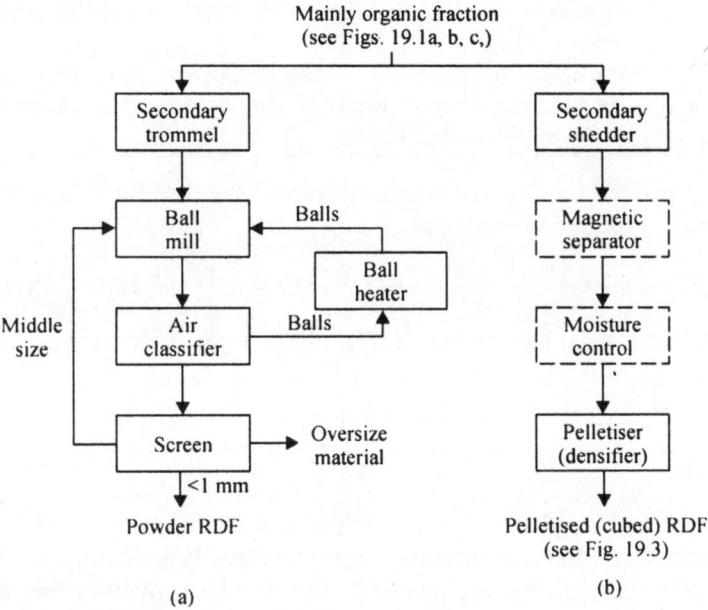

Fig. 19.2. Flow sheets for the preparation of RDF (refuse-derived fuel): (a) powder RDF; and (b) pelletzed (cubed) REF.

Table 19.3. Important design considerations for aerobic composting processes.

Item	Comment
Particle size	For optimum results the size of solid wastes should be between 25 and 75 mm (1 and 3 inch)
Seeding and mixing	Composting time can be reduced by seeding with partially decomposed solid wastes to the extent of about 1 to 5 per cent by weight. Sewage sludge can also be added to prepared solid wastes. Where sludge is added, the final moisture content is the controlling variable
Mixing/turning	To prevent drying, caking and air channeling, material in the process of being composted should be mixed or turned on a regular schedule or as required. Frequency of mixing or turning will depend on the type of composting operation
Air requirements	Air with at least 50 per cent of the initial oxygen concentration remaining should reach all parts of the composting material for optimum results, especially in mechanical systems.
Total oxygen requirements	The theoretical quantity of oxygen required can be estimated
Moisture content	Moisture content should be in the range between 50 and 60 per cent during the composting process. The optimum value appears to be about 55 per cent
Temperature	For best results, temperature should be maintained between 323 and 328 K (50 and 55°C) for the first few days and between 328 and 333 K (55 and 60°C) for the remainder of the active composting period. If temperature goes beyond 339 K (66°C), biological activity is reduced significantly

(Contd ...)

Item	Comment
Carbon-nitrogen ratio	Initial carbon-nitrogen ratios (by mass) between 30 and 50 are optimum for aerobic composting. At lower ratios ammonia is given off. Biological activity is also impeded at lower ratios. At higher ratios nitrogen may be a limiting nutrient
pH	To minimise the loss of nitrogen in the form of ammonia gas, pH should not rise above about 8.5
Control of pathogens	If properly conducted, it is possible to kill all the pathogens, weeds and seeds during the composting process. To do this, the temperature must be maintained between 333 and 343 K (60 and 70°C) for 24 hours

The amount of oxygen required for the complete aerobic stabilisation of municipal solid wastes can be estimated by using the following equation:

$$C_aH_bO_cN_d + \frac{4a + b - 2c - 3d}{4}O_2 \rightarrow aCO_2 + \frac{b - 3d}{2}H_2O + dNH_c \qquad \text{... (19.1)}$$

If the ammonia, NH_3, is to be oxidised to nitrate NO_3^-, the amount of oxygen required to accomplish this can be computed with the following equation:

$$NH_3 + 2O_2 \longrightarrow H_2O + HNO_3 \qquad \text{... (19.2)}$$

Anaerobic Digestion

Anaerobic digestion or anaerobic fermentation as it is often called, is the process used for the production of methane from solid wastes. In most processes where methane is to be produced from solid wastes by anaerobic digestion, three basic steps are involved. The first step involves preparation of the organic fraction of the solid wastes for anaerobic digestion and usually includes receiving, sorting, separation and size reduction. The second step involves the addition of moisture and nutrients, blending, pH adjustment to about 6.7, heating of the slurry to between 328 and 333 K (55 and 60°C) and anaerobic digestion in a reactor with continuous flow, in which the contents are well mixed for a period of time varying from 5 to 10 d. The third step involves capture, storage and if necessary, separation of the gas components evolved during the digestion process. The disposal of the digested solids is an additional task that must be accomplished. Some important design considerations, are reported in Table 19.4. Because of the variability of the results reported in the literature, it is recommended that pilot-plant studies be conducted if the digestion process is to be used for the conversion of solid wastes.

Table 19.4. Important design considerations for anaerobic digestion.

Item	Comment
Size of material shredded	Wastes to be digested should be shredded to a size that will not interfere with the efficient functioning of pumping and mixing operations
Mixing equipment	To achieve optimum results and to avoid scum build-up, mechanical mixing is recommended
Percentage of solid wastes mixed with sludge	Although amounts of waste varying from 50 to 90 + per cent have been used, 60 per cent appears to be a reasonable compromise
Hydraulic and mean cell-residence time, $\theta_h = \theta_c$	Washout time is in the range of 3 to 4 d. Use 7 to 10 d for design or base design on results of pilot-plant studies

(Contd ...)

Item	Comment
Loading rate	0.6 to 1.6 kg/m^3·d (0.04 to 0.10 lb/ft^3·d). Not well defined at present time. Significantly higher rates have been reported
Temperature	Between 328 and 333 K (55 and 60°C)
Destruction of volatile	Varies from about 60 to 80 per cent; 70 per cent can be used for estimating solid wastes purposes*
Total solids destroyed	Varies from 40 to 60 per cent, depending on amount of inert material present originally
Gas production	0.5 to 0.75 m^3/kg (8 to 12 ft^3/lb) of volatile solids destroyed (CH$_4$ = 60 per cent; CO$_2$ = 40 per cent)

* Actual removal rates for volatile solids may be less, depending on the amount of material diverted to the scum layer.

RECOVERY OF THERMAL CONVERSION PRODUCTS

Thermal conversion products that can be derived from solid wastes include heat, gases, a variety of oils and various related organic compounds. The principal thermal conversion processes that have been used for the recovery of usable conversion products from solid wastes are reported in Table 19.5. The more important processes in Table 19.5 are reviewed following an introductory discussion of combustion.

Table 19.5. Thermal processes for the recovery of conversion products from solid wastes.

Process	Conversion product	Preprocessing
Combustion (incineration) suspension firing in	Energy in the form of steam	None in mass-fired incineratory; preparation of refuse- or electricityderived fuels for suspension or semi-boilers
Gasification	Low-energy gas	Separation of the organic fraction, particle size reduction, preparation of fuel cubes or other RDF
Wet oxidation	Organic acids	Separation of the organic fraction, particle size reduction, preparation of fuel cubes or other RDF
Steam reforming	Medium-energy gas	Separation of the organic fraction, particle size reduction, preparation of fuel cubes or other RDF
Pyrolysis	Medium-energy gas, liquid fuel, solid fuel (char)	Separation of the organic fraction, particle size reduction, preparation of fuel cubes or other RDF
Hydrogasification/ hydrogenation	Medium-energy gas, liquid fuel	Separation of the organic fraction, particle size reduction, preparation of fuel cubes or other RDF

Combustion of Waste Materials

The principal elements of solid waste are carbon, hydrogen, oxygen, nitrogen and sulphur. Under ideal conditions, when solid waste materials are combusted (burned) the gaseous end products include CO_2 (carbon dioxide), H_2O (water), N_2 (nitrogen) and SO_2 (sulphur dioxide). In practice, a variety of other gaseous compounds are also formed, depending on the operating conditions under which the combustion process is occurring.

Combustion calculations

In their simplest form all combustion calculations are based on the following fundamental laws:
   1. Conservation of mass: Mass can neither be created or destroyed.

2. Conservation of energy: Energy can neither be created or destroyed.
3. Gas law: The volume of gas is directly proportional to its absolute temperature and inversely proportional to its absolute pressure.
4. Law of combining masses: All substances combine in accordance with a definite, simple relationship with respect to relative masses.

Air requirements for combustion

To determine the amount of oxygen required for the complete combustion of solid wastes, it is necessary to compute the oxygen requirements for the oxidation of carbon, hydrogen and sulphur contained in waste. The basic reactions are:

For carbon

$$C + O_2 \longrightarrow CO_2 \qquad \qquad \qquad ...(19.3)$$
$$(12) \quad (32) \qquad \quad (44)$$

For hydrogen

$$2H_2 + O_2 \longrightarrow 2H_2O \qquad \qquad ...(19.4)$$
$$(4) \quad (32) \qquad (36)$$

For sulphur

$$S + O_2 \longrightarrow SO_2 \qquad \qquad \qquad ...(19.5)$$
$$(32) \quad (32) \qquad (64)$$

If it is assumed that air contains 23.15 per cent oxygen by mass, then the amount of air required for the complete oxidation of 1 kg of carbon would be equal to 11.52 kg $[(32/12)(1/0.2315)]$. The corresponding amounts for hydrogen and sulphur are 34.56 and 4.31 kg, respectively. In combustion computations, the oxygen requirements for the combustion of hydrogen usually are based on the net value of hydrogen available. The net value of hydrogen is computed by subtracting one-eighth of the per cent oxygen from the total percentage of hydrogen present initially. This computation is based on the assumption that the oxygen in the sample will combine with the hydrogen in the waste to form water.

The heat released from the combustion of solid wastes is partly stored in the combustion products (gases and ash) and partly transferred by convection, conduction and radiation to the incinerator walls and to the incoming waste (see Table 19.6). The energy content of the waste can be estimated using the heating value of the individual waste components.

Table 19.6. Heat losses in combustion of solid waste.

Type of losses	Remarks
	The heating value of carbon is about 32.789 kJ/kg
Reactor	
Unburned carbon	Typically, the grate residue is assumed to contain from 4 to 8 per cent carbon
Radiation	Heat lost through the reactor walls and other appurtenances to surroundings is estimated as 0.003–0.005 kJ/kg of furnace input
Latent heat	
Inherent moisture	Water content of waste. The latent heat of vapourisation for water is approximately 2420 kJ/kg

(Contd ...)

Type of losses	Remarks
Moisture in bound water	
Moisture from oxidation of net hydrogen	
Sensible heat	
Sensible heat in residue	Specific heat of residue is taken as 1047 J/kg·K (10.25 Btu/lb·°F)
Stack gases	

Incineration with Heat Recovery

Heat contained in the gases produced from the incineration of solid wastes can be recovered by conversion to steam. The low-level heat remaining in the gases after heat recovery can also be used to pre-heat the combustion air, boiler make-up water or solid waste fuel.

Existing mass-fired Incinerators

With existing mass-fired incinerators, waste-heat boilers can be installed to extract heat from the combustion gases without introducing excess amounts of air or moisture. Typically, incinerator gases will be cooled from a range of 1250 to 1375 K (1800 to 2000°F) to a range from 500 to 800 K (600 to 1000°F) before being discharged to the atmosphere. Apart from the production of steam, the use of a boiler system is beneficial in reducing the volume of gas to be processed in the air-pollution control equipment.

Water-wall incinerators

In these incinerators, the internal walls of the combustion chamber are lined with boiler tubes that are arranged vertically and welded together in continuous sections. When water-walls are used in place of refractory materials, they are not only useful for the recovery of steam, but also extremely effective in controlling furnace temperature without introducing excess air; however, they are subject to corrosion by the hydrochloric acid produced from the burning of some plastic compounds.

Use of Refuse-Derived Fuels (RDF)

Prepared RDF, typically in a powdered form, can also be fired directly in large industrial boilers that are now used for the production of power with pulverised coal or oil. RDF also can be fired in conjunction with coal or oil. Although the process is not well established with coal, it appears that about 15 to 20 per cent of the heat input can be from prepared solid wastes. With oil as the fuel, about 10 per cent of the heat input can be from solid wastes. Depending on the degree of processing, suspension, spreader-stoker and double-vortex firing systems have been used.

Densified RDF fuel is prepared using a modified agricultural cubing machine. The resulting fuel cubes are suitable for use in a variety of thermal conversion processes including incineration, gasification and pyrolysis.

Gasification

The gasification process involves the partial combustion of a carbonaceous fuel to generate a combustible fuel gas rich in carbon monoxide and hydrogen. A gasifier is basically an incinerator operating under reducing conditions. Heat to sustain the process is derived from the exothermic reactions while the combustible components of the low-energy gas are primarily generated by the endothermic reactions. The reaction kinetics of the gasification process are quite complex and still the subject of considerable debate. When a gasifier is operated at atmospheric pressure with air as the oxidant, the end products of the gasification process are a low-energy gas typically containing (by volume) 10 per cent CO_2, 20 per cent CO, 15 per cent H_2 and 2 per cent CH_4, with the balance being N_2 and a carbon-rich char. Because of the diluting effect of the nitrogen in the input air, the low-energy gas has an energy content in the range of 5.2 to 6.0 MJ/m^3. When pure oxygen is used as the oxidant, a medium-energy gas with an energy content in the range of 12.9 to 13.8 MJ/m^3 is produced.

Pyrolysis

Of the many alternative chemical conversion processes that have been investigated, excluding incineration, pyrolysis has received the most attention. Depending on the type of reactor used, the physical form of the solid wastes to be pyrolysed can vary from unshredded raw wastes to the finely ground portion of the wastes remaining after two stages of shredding and air classification. Upon heating in an oxygen-free atmosphere, most organic substances can be split through a combination of thermal cracking and condensation reactions into gaseous, liquid and solid fractions. Pyrolysis is the term used to describe the process. In contrast to the combustion process, which is highly exothermic, the pyrolytic process is highly endothermic. For this reason, the term destructive distillation is often used as an alternative term for pyrolysis. The characteristics of the three major component fractions resulting from the pyrolysis are: (i) a gas stream containing primarily hydrogen, methane, carbon monoxide, carbon dioxide and various other gases, depending on the organic characteristics of the material being pyrolysed; (ii) a fraction that consists of a tar and/or oil stream that is liquid at room temperatures and has been found to contain chemicals such as acetic acid, acetone and methanol; and (iii) a char consisting of almost pure carbon plus any intert material that may have entered the process. It has been found that distribution of the product fractions varies with the temperature at which the pyrolysis is carried out. Under conditions of maximum gasification, the energy content of the resulting gas is about 26100 kJ/m^3 (700 Btu/ft^3). The energy content of pyrolytic oils has been estimated to be about 23240 kJ/kg (10000 Btu/lb).

RECOVERY OF ENERGY FROM CONVERSION PRODUCTS

Once conversion products have been derived from solid wastes by one or more of the biological and thermal methods listed in Tables 19.2 and 19.5, the next step involves their storage and/or use. If energy is to be produced, then an additional conversion step is required.

Energy-Recovery Systems

The principal components involved in the recovery of energy from heat, steam, various gases and oils and other conversion products are boilers for the production of steam and gas turbines for motive power and electric generators for the conversion of motive power into electricity.

Typical flow sheets for alternative energy-recovery systems are shown in Fig. 19.3. Perhaps the most common flow sheet for the production of electric energy involves the use of a steam turbine generator combination (see Fig. 19.3a). As shown, when solid wastes are used as the basic fuel source, four operational modes are possible. A flow sheet using a gas turbine-generator combination is shown in Fig. 19.3b. The low-energy gas is compressed under high pressure so that it can be used more effectively in the gas turbine.

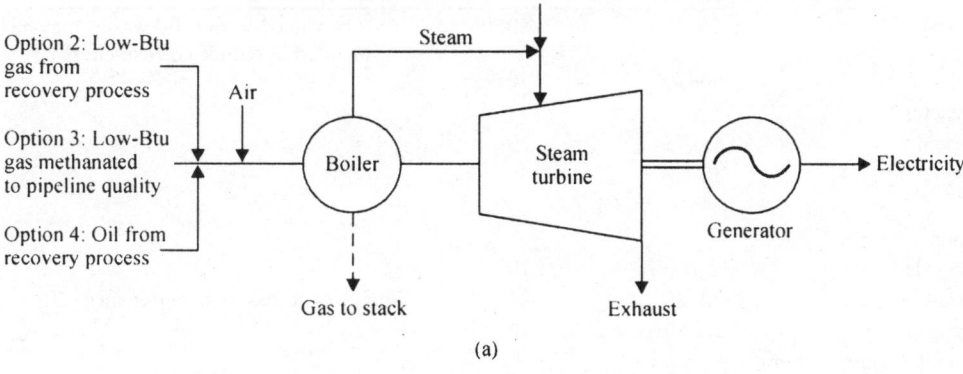

Option 1: Steam from shredded and classified solid wastes or solid
fuel pellets fired directly on boiler or from solid wastes mass-fired
in water-walled boiler. With mass-fired units auxiliary fuel may be required.

(a)

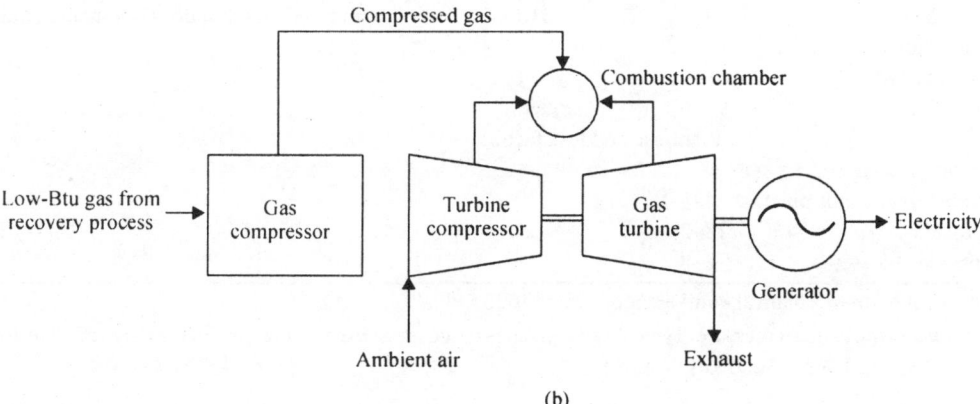

(b)

Fig. 19.3. Energy-recovery systems: (a) options with steam turbine-generator; and (b) options with gas compressor-gas turbine-generator.

Efficiency Factors

Representative efficiency data for boilers, pyrolytic reactors, gas turbines, steam turbine-generator combinations, electric generators and related plant use and loss factors are given in Table 19.7.

Table 19.7. Typical thermal efficiency and plant use and loss factors for individual components and processes used for the recovery of energy from solid wastes.

| Component | Efficiency* | | Comment |
	Range	Typical	
Incinerator-boiler	40–68	63	Mass-fired
Boiler			
Solid fuel	60–75	72	Processed solid wastes (RDF)
Low-Btu gas	60–80	75	Burners must be modified

(Contd ...)

Component	Efficiency* Range	Efficiency* Typical	Comment
Oil-fired	65–85	80	Oils produced from solid wastes may have to be blended to reduce corrosiveness
Gasifier	60–70	70	
Pyrolysis reactor			
Conventional	65–75	70	
Purox	70–80	75	
Turbines			
Combustion gas			
Simple cycle	8–12	10	
Regenerative	20–26	24	Includes necessary appurtenances
Expansion gas	30–50	40	
Steam turbine-generator system			
Less than 12.5 MW	24–40	29†[†]	Includes condenser, heaters and all other necessary appurtenances, but does not include boiler
Over 10 MW	28–32	31.6†[†]	
Electric generator			
Less than 10 MW	88–92	90	
Over 10 MW	94–98	96	
Plant use and loss factors			
Station service allowance			
Steam turbine-generator plant	4–8	6	
Purox process	18–24	21	
Unaccounted heat losses	2–8	5	

* Theoretical value for mechanical equivalent of heat = 3600 kj/kWh

† Efficiency varies with exhaust pressure. Typical value given is based on an exhaust pressure in the range of 50 to 100 mm Hg

[†] Heat rate = 11395 kJ/kWh = 3600 kj/kWh/0.316

In any installation where energy is being produced, allowance must be made for the station or process power needs and for unaccounted process heat losses. Typically, the auxiliary power allowance varies from 4 to 8 per cent of the power produced. Process heat losses usually will vary from 2 to 8 per cent.

Determination of Energy Output and Efficiency

An analysis of the amount of energy produced from a solid waste energy-conversion system using an incinerator-boiler-steam turbine-electric generator combination with a capacity of 1000 tonnes/d is presented in Table 19.8.

Table 19.8. Energy output and efficiency for a 1000 tonne/d steam boiler turbine-generator energy-recovery plant using unprocessed solid wastes with an energy content of 12000 kJ /kg.

Item
Energy available in solid wastes, million kJ/h
1000 tonnes/d × 1000 kg/tonne × 12000 kJ/kg)/(24 h/d × 10^6 kJ/million kJ)
Steam energy available, million kJ/hr.
500 million kJ/hr. × 0.7

(Contd ...)

Item
Electric power generation, kW   (350 million kJ/hr.)/(11395 kJ/kWh)*
Station service allowance, kW   30715 (0.06)
Unaccounted heat losses, kW   30715 (0.05)
Net electric power for export, kW Overall efficiency, per cent   (100)(27338 kW)/[(5 × 10⁸ kJ/hr.)/(3600 kJ/kWh)]

   * 11395 kJ/kWh = (3600 kJ/kWh)0.316.

   If it is assumed that 10 per cent of the power generated is used for the front-end processing system (typical values vary from 8 to 14 per cent), then the net power for export is 24604 kW and the overall efficiency is 17.5 per cent.

MATERIALS AND ENERGY-RECOVERY SYSTEMS

During the past few years numerous systems have been proposed or built incorporating different types of processing and energy-conversion systems. Two typical examples are shown in Figs. 19.4 and 19.5. Unfortunately, few of the full-scale plants that have been built have proved to be successful. Although economics has been the major reason for their demise, some energy-conversion plants have failed because of technical difficulties. Thus, if the use of a materials and energy-recovery system is contemplated, current operating systems should be visited and analysed and realistic cost estimates should be prepared.

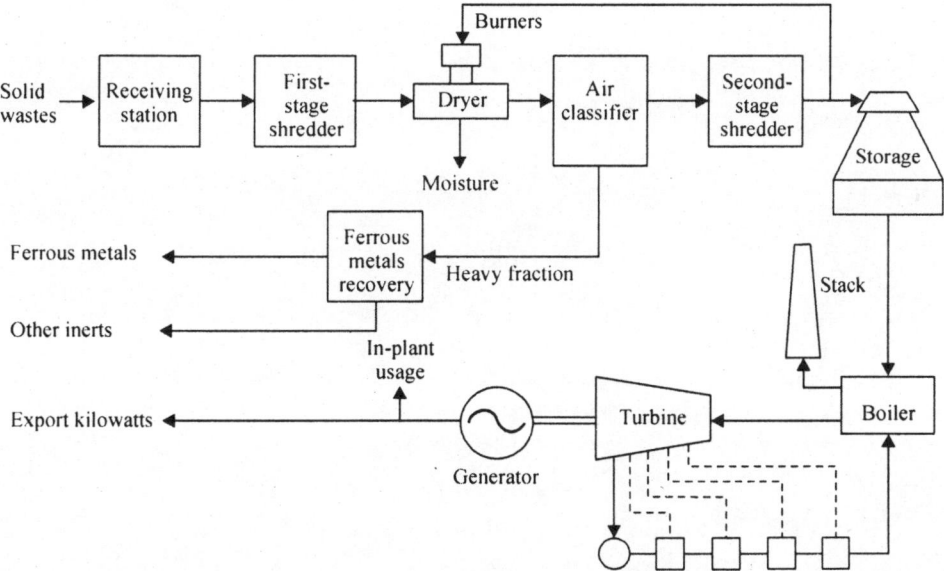

Fig. 19.4. Flow sheet for the recovery of ferrous materials and energy from solid wastes.

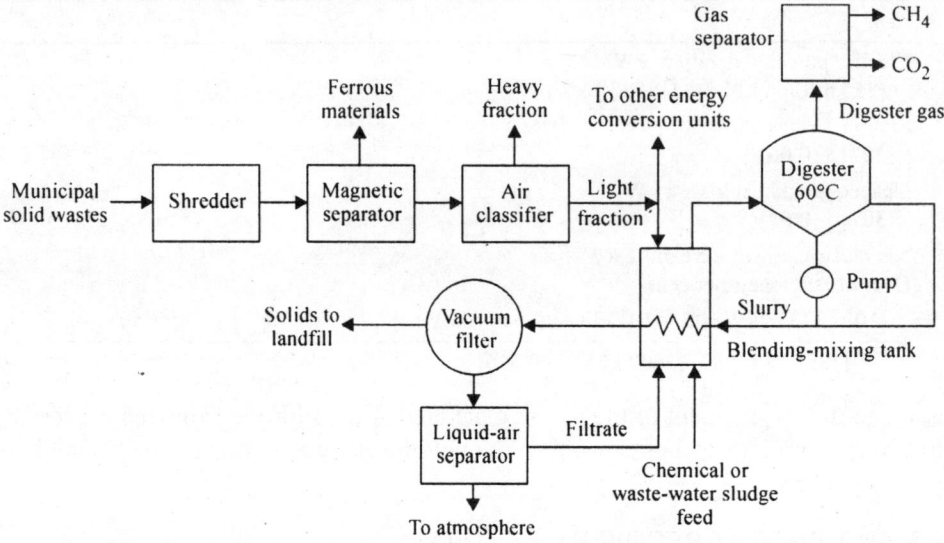

Fig. 19.5. Flow sheet for the recovery of ferrous metal materials and digester gas from solid wastes.

Chapter 20

Microbiology

INTRODUCTION

Biology is the study of living organism, their habits, food requirement and life processes. The term living organisms includes all plants and animals in numerous species; biology presents a tremendous field of study and includes bacteriology, botany, zoology and physiology.

CLASSIFICATION

Micro-organisms are too small to be seen with the unaided eye. Some of them can be seen with light microscopes with magnification up to 2000 diameters but others, the viruses, are too small to be seen even with the most powerful light microscopes; electron microscopes having magnification up to 50000 diameters are required.

In the microscopic world, the distinctions between plants and animals are not always sharp. Many microbes have characteristics which are typical of both plant and animal forms and their classification is based upon careful study to determine the predominant features. Some of the criteria by which a micro-organism is classified as a plant or an animal are listed in Table 20.1.

Classification of living organism lies in a branch of biology called taxonomy. Taxonomists classify plants and animals into sub-groups on the basis of distinctive characteristics of form, metabolism or reproduction. These sub-groups into which the plant and animal kingdoms are divided are then further divided into more sub-groups.

Each kingdom is divided into several phyla, each phylum is divided into classes. Since there are literally thousands of species in existence, no one set of classifications includes them all. All standard classifications available cover only portions of the plant and animal kingdoms of interest to specific groups of natural scientists.

Microbes are of great interest to water technologists and include bacteria, viruses and rickettsiae, many of which are disease-producing agents. Another division of the plant kingdom, the thallophyta, is interesting to water supply technologists because it includes fungi and algae which are responsible for tastes and odours in water. Figures 20.1 and 20.2 show classification schemes for some microbes of the plant and animal kingdoms but are by no means complete and the arrangements do not necessarily coincide with other classifications. Taxonomy is not rigid and classifications are subject to change upon accumulation of more detailed information regarding the subjects.

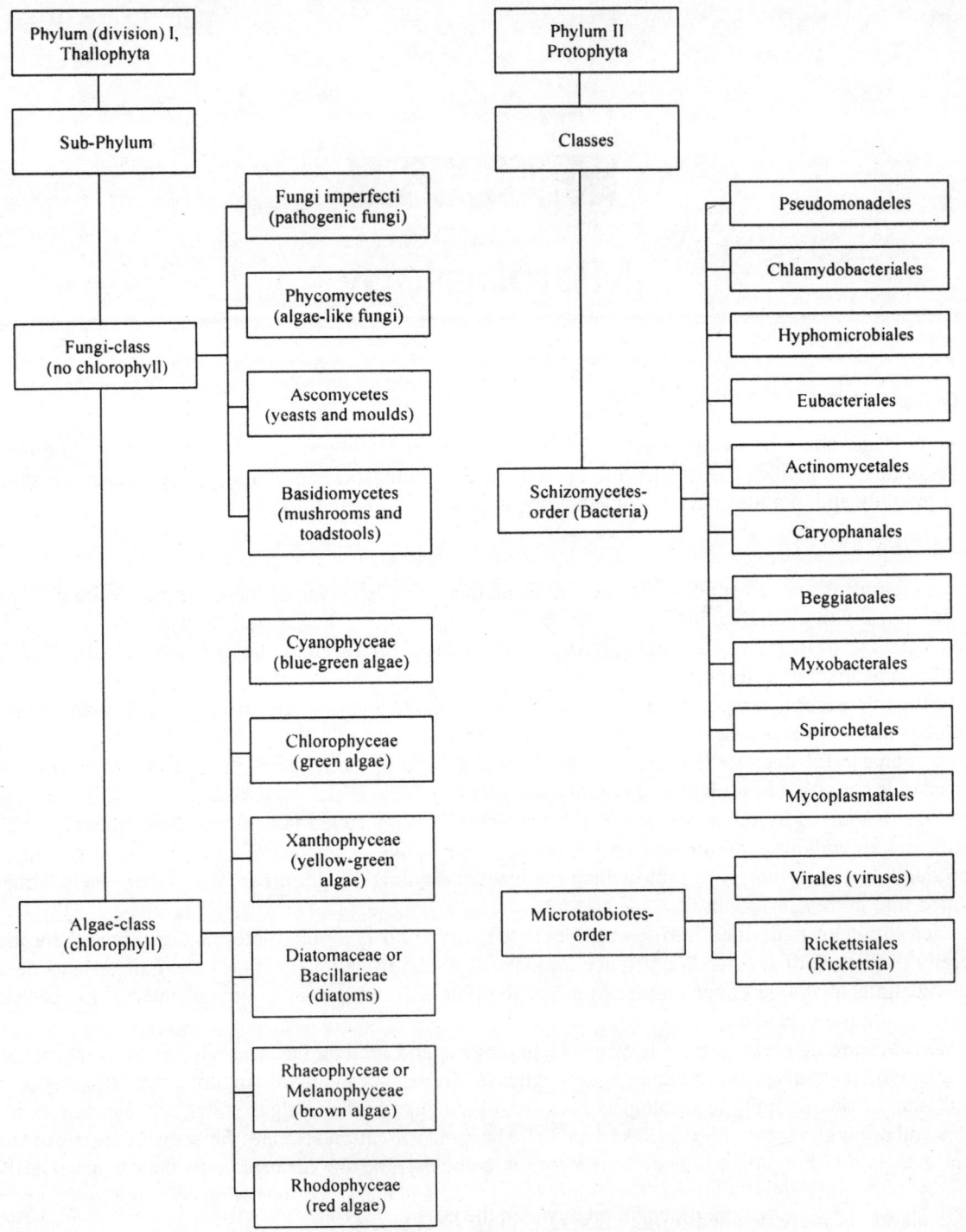

Fig. 20.1. Microbes of the plant kingdom.

Table 20.1. Plant and animal characteristics.

Plant characteristics	*Animal characteristics*
Store energy	Release energy
Have no sensory organs or nervous system	May have sensory organ and a nervous system
Release oxygen	Release carbon dioxide
Cell walls composed of cellulose—a carbohydrate	Cell walls composed principally of protein matter—a nitrogen-bearing compound
Metabolism depends upon absorption of water and gases by root hairs and natural openings called stomata. There is no digestion of food	Metabolism depends upon digestion of food within an alimentary canal
Utilise carbon dioxide from the air and water and nitrogen salts from the soil to synthesise carbohydrates and proteins needed for growth and reproduction	Digestion process is analytic by which ingested materials are decomposed to supply materials for synthesis of proteins required for growth and reproduction

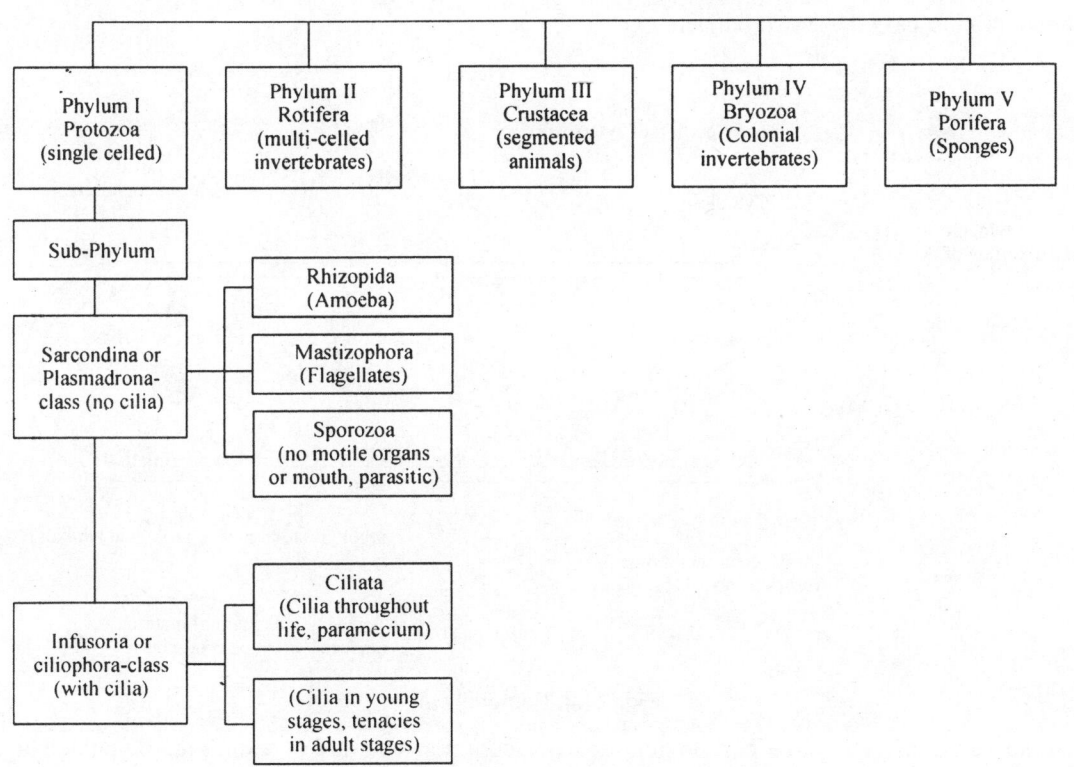

Fig. 20.2. Microbes of the animal kingdom.

BACTERIA

Of the various kinds of microbes found in water, the most important from the standpoint of public health belongs to the class of plants called schizomycetes or, more commonly, bacteria. Many of these are harmless in themselves and of no particular interest other than how they get into a water supply, but others are capable of causing disease in individuals consuming water containing them. It is important that water supply technologists have knowledge of bacteriology to cope with the problem of producing a bacteriologically safe product for consumption or discharge.

Bacteria are minute living plants which consist of a single cell. They are too small to be seen with the naked eye. Ordinary sources of measurement are not useful for designating the size of bacteria because the smallest units of ordinary scales, such as inches or millimeters, are too large. The unit used is the micron and it measures 1/1000 of a millimeter. A typical bacterial cell (*Salmonella typhosa*) is rod-shaped and approximately two microns long. About 13000 of such bacterial cells lying end to end would measure only one inch—just to demonstrate how small these organisms are. Bacterial cells (Fig. 20.3), in spite of their small size, are structurally complex. A rigid membrane or wall of complex chemical composition surrounds the cell. This wall consists of proteins, polysaccharides and lipids (fats) and encloses a non-rigid protoplast (or cytoplast) containing the cell nucleus and various smaller bodies vital to cell functions. Among these are vacuoles, plastids and inclusion bodies. Density of the protoplast is greater at the outer layer adjacent to the cell wall and this section, designated as ectoplasm, like the rigid cell wall, is semi-permeable. This means that all material entering or leaving the interior of the cell must be in solution to pass these two barriers.

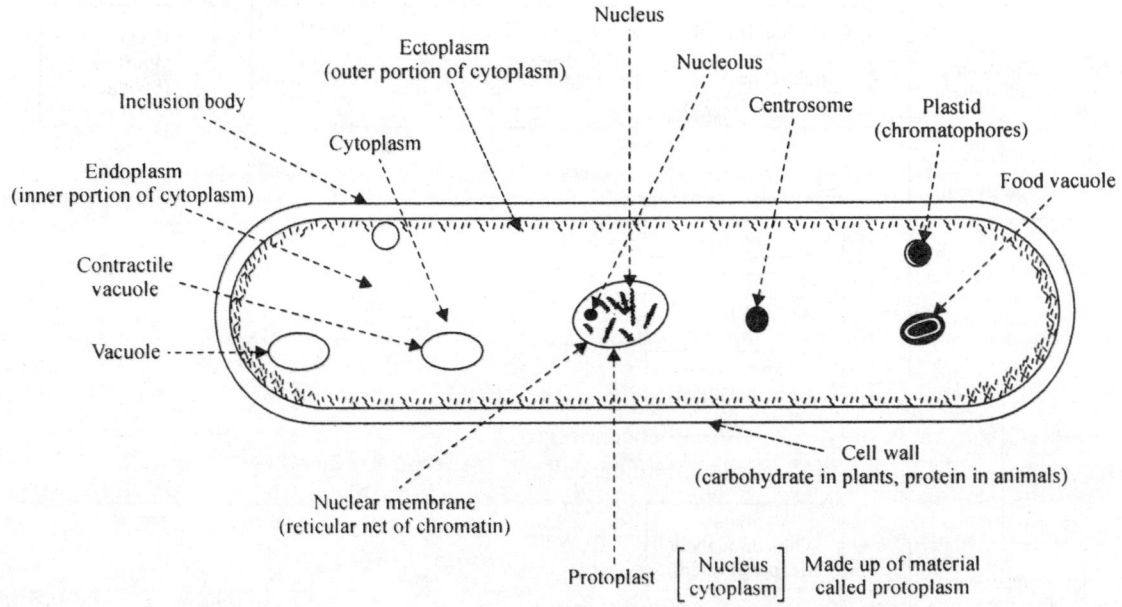

Fig. 20.3. Bacterial cell.

Some bacterial cells have certain features not common to all. For example, flagella, hair-like appendages, are present on some species but not on others. Furthermore, depending on the species, the number of flagella vary from one to many and their locations vary from one end to both ends or even to the entire periphery of the cell wall. Their function is clearly to provide locomotion. Some bacterial cells such as *Mycobacterium tuberculosis* are surrounded with a capsule, a thick coating of gelatinous material, for protective purposes. Such species are especially resistant to destructive agents. Some bacteria have the property of forming spores. Spores are usually formed when the environment becomes unsuitable for growth and reproduction. Lack of food, lack of sufficient moisture and higher than normal temperature conditions, are some reasons for spore formation. Bacteria in the spore stage cannot reproduce and are more resistant to destruction than vegetative cells. Restoration of a favourable environment results in restoration of vegetative cells and normal metabolic process.

Morphology

Bacteria of various species show differences as shown in Fig. 20.4. Three general morphological categories into which all bacteria fall are cocci, bacilli and spirilla with modifications of these forms.

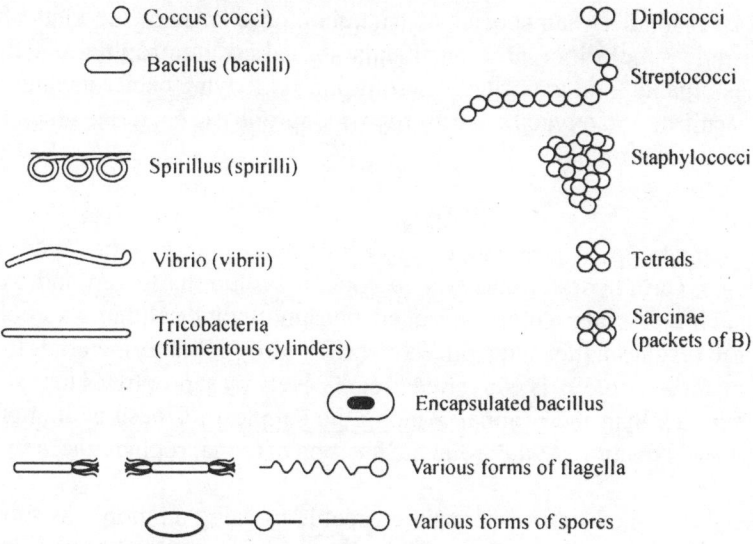

Fig. 20.4. Physical forms of bacteria and cocci.

Cocci are round cells, sometimes slightly flattened where they are adjacent to one another. They exist in pairs, as diplococci, in chains as streptococci, in groups of four as tetrads and in packets of eight, as sarcinae. Bacilli are rod-shaped. The length of the cell varies, even in a single species, under the influence of age or environmental conditions. They too occur singly or in chains. Bacteria of the coliform group, normally used as an index of sewage pollution in water, are bacilli.

Spirilla are curved. The length of the cell and the number of convolutions varies with the organism. Some of them are shaped somewhat like a corkscrew. Most of them are rigid and motile. A special group of spirilla known as spirochetes are not rigid but flexible and are long and slender. Two well-known pathogens are spirochetes, *Treponema pallium*, which causes syphilis and *Leptospira*, which causes leptospirosis.

Metabolism

Metabolism, meaning life processes, includes nutrition, respiration and reproduction. All of these processes are complex even for bodies as small as bacterial cells.

Enzymes of Bacteria

A bacterial cell must obtain its food from the environment external to the cell wall. Since this wall is permeable only to substances in true solution, solid food material must be made soluble before it becomes available for cell nutrition. Enzymes (exo-enzymes), chemical catalysts manufactured by the cell, accomplish the liquefaction process. Enzymes are not altered in the process and can function as long as food is available. Once the soluble food has passed through the cell wall, other enzymes (endo-enzymes) reduce the chemically complex substances to simple compounds which the cell can utilise in its metabolic processes.

Autotrophic and heterotrophic bacteria

A distinction can be made between species of bacteria on the basis of the kind of food they require. Those which consume simple inorganic chemicals are called autotrophic and those which require chemically complex organic food are called heterotrophic. Nitrifying bacteria which inhabit the soil and subsist on ammonia, nitrite and oxygen are autotrophic. Saprophytic bacteria, which feed on dead animal and plant tissue, are heterotrophic.

Nitrogen cycle

All living matter is made up of proteins, carbohydrates and fats, together with inorganic (mineral) constituents. Proteins, carbohydrates and fats all contain carbon, hydrogen and oxygen. Proteins and especially the animal proteins, also contain nitrogen, phosphorus and sulphur. Saprophytes, heterotrophic bacteria, feed on dead organic matter and reduce these chemically complex materials to simpler substances. Nitrogen-bearing material, protein, is converted progressively by saprophytes to many compounds, each of which is less complex than the original. Among the simplest of these is ammonia, an end-product derived from the nitrogen fraction. At this point the function of the saprophytes is complete and autotrophic bacteria take over.

Autotrophic bacteria called nitrosomonas are capable of using ammonia as food and through their metabolism, convert the ammonia to nitrite. These bacteria, like the saprophytes, are found in natural environment, principally soil and to accomplish their purpose they require oxygen, because the conversion of ammonia (NH_3) to nitrite (NO_2^-) is an oxidative process.

Another group of autotrophic bacteria called nitrobacter, also found in natural environment, converts nitrite (NO_2^-) to nitrate (NO_3^-), which is an oxidative reaction requiring available oxygen.

Therefore, there are three groups of bacteria each performing a very important part in the conversion of dead organic material to stable chemical compounds, a process which is essential. In order to complete the description of the various transformations which nitrogen undergoes, the role played by living plants and animals must be mentioned. Nitrate (NO_3^-) and ammonia are food materials for plants and are incorporated into the protein fraction of plant structure. When plants decay, the cycle of nitrogen transformation through the action of bacteria begins again. If plants are consumed by animals, the nitrogen in the plant is incorporated into animal protein or evacuated as waste material of the animal body and then the nitrogen cycle is carried on as described through the action of bacteria. Figure 20.5 shows the nitrogen cycle and illustrates the various steps in the process.

Culture media

The study of laboratory cultures of bacteria for purposes of identification or research is a highly specialised branch of bacteriology. Time and effort is expended in developing culture media suitable for growth of specific bacterial species. Usually organic food supplemented with mineral salts is required for successful growth. In addition to food, various other substances are necessary components of a culture medium. Included are inorganic salts such as phosphates and sulphates, amino acids, minute amounts of iron and copper and vitamins.

Bacterial Respiration

Various species of bacteria may differ with respect to their respiration of oxygen. Bacteria, like all other plants and animals, require oxygen either in the gaseous state or combined with other elements in a chemical compound. The bacteria which require gaseous oxygen in their environment are called aerobes, while those which require the complete absence of gaseous oxygen are called anaerobes. There are many species which thrive better in the absence of oxygen but which can tolerate it as well. These are described as facultative anaerobes.

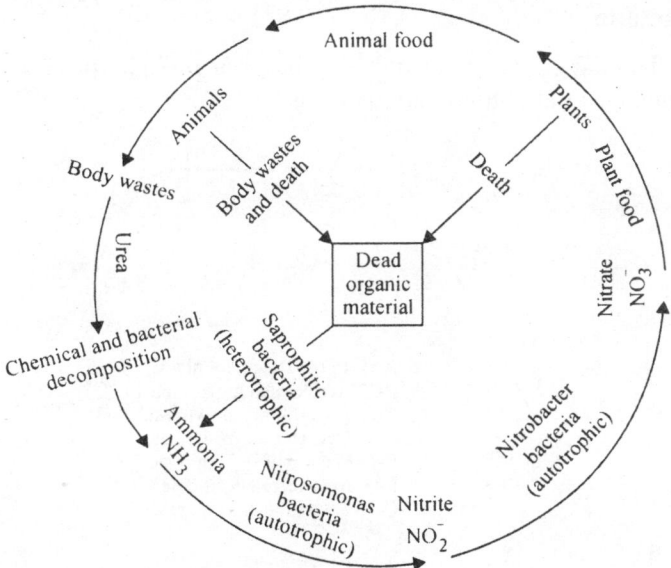

Fig. 20.5. The nitrogen cycle.

Reproduction and Growth of Bacteria

Bacteria have no sexual faculties and reproduction takes place whenever environmental conditions are favourable and is accomplished through a process called fission. Bacterial cells become constricted, usually near the center. As the cell ages, the constriction progresses with time until eventually the cell completely divides to produce two separate cells.

In favourable environmental conditions, an average bacterial cell divides approximately every 20–30 minutes. Within a few hours, if all cells survive, a progression of this type would produce many millions of descendants from a single cell as shown in Fig. 20.6. There are limiting factors of growth and survival in bacterial existence just as there are in human existence and an ordinary bacterial culture 24 hours old contains about 20 million individual cells per milliliter.

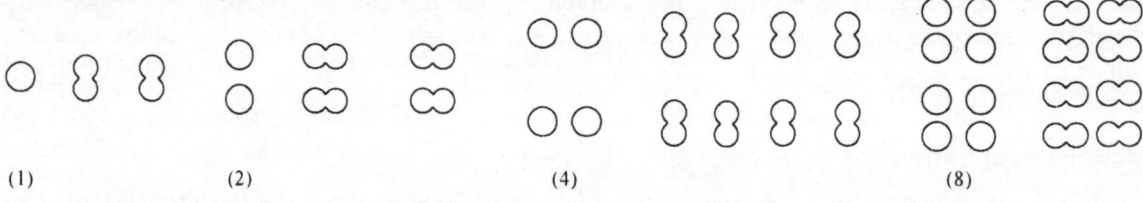

Fig. 20.6. Reproduction of bacteria.

In a closed system where the food supply is satisfactory but limited and the by-products of cell metabolism are retained, the rate of growth gradually decreases to zero. Thereafter, the death rate gradually increases until a maximum is reached and the climax is attained with the death of all cells. A curve showing the relationship of bacterial number to time under such conditions is shown in Fig. 20.7.

Factors Affecting Bacteria

Other conditions that affect the growth and survival of bacteria are light, heat, drying, bacterial agents, bacteriostatic agents and antimetabolite of various kinds.

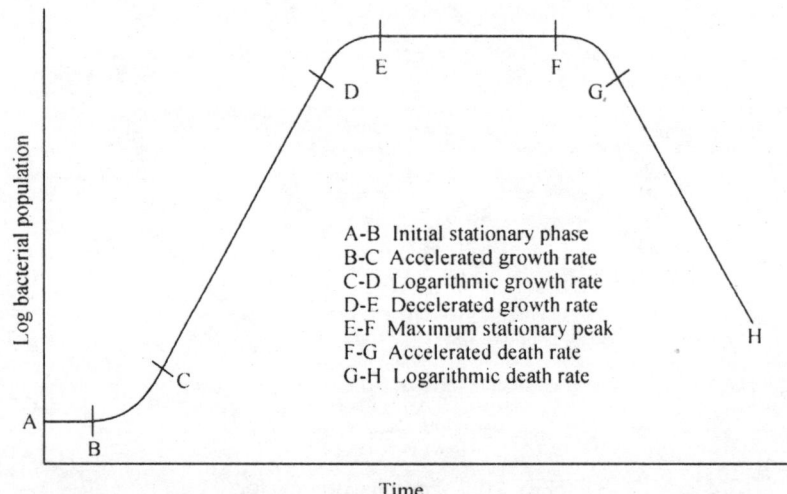

Fig. 20.7. Growth and death of bacteria in a closed cultural system.

Heat

Heat requirements for optimum growth of bacteria vary with the species. Many bacteria thrive best at temperatures near that of the human body or slightly below (32°–37°C). These are mesophilic or moderate heat-loving bacteria. Those which grow best at low temperatures, slightly above the freezing point of water, are called cryophilic or cold-loving bacteria and those which require higher temperatures for growth, that is, temperatures between 40°C and 65°C, are described as heat-loving or thermophilic. Bacteria are not destroyed by low temperatures, but most reproduce very slowly or not at all under such conditions. When transferred from the frozen state to a more suitable environment, they immediately carry on a normal life cycle. Extreme heat, on the other hand, destroys all bacterial species, although those which are in the spore stage will withstand much more heat than those which are not. Moist heat, steam, is more effective in the destruction of bacteria than dry heat. A temperature above which no cells of a bacterial species can survive is considered to be the thermal death point of the species.

Light

Certain wavelengths of light are very destructive to bacterial cells. These lie in the region of the spectrum known as the ultra-violet. Light rays below 2900 angstrom units (290 millimicrons wavelength) are especially destructive. A necessary condition, however, is for the light ray to strike the cell directly. Ultraviolet light may be used to destroy bacteria in water, provided the water, is sufficiently free of suspended matter to permit the light ray to strike the bacterial cells. Water itself has some absorptive effect on ultra-violet light and therefore even in clear water, the distance between the bacteria cells and the light source must be relatively short.

Drying

Bacteria cannot reproduce without moisture. Drying of food materials is a method for preservation from decomposition by bacteria. Spore-forming bacteria may survive in a dry environment but they cannot function normally. If moisture becomes available, bacterial spores will vegetate and resume a normal life cycle.

Osmosis

This is a physical phenomenon which depends upon the relative concentration of soluble substances, usually salts, inside and outside bacterial cell walls. If the concentrations are in equilibrium, the cell is said to be in an isotonic environment which is favourable. If the concentration of electrolytes outside the wall is greater than that within, there is a tendency for water to pass out of the cell to restore equilibrium. This may result in destruction of the cell through the effect of shrinking or plasmolysis. If the concentration of soluble substances outside the cell walls is lower than that within, water will tend to pass through the cell wall from the outside and cause the cell to swell and perhaps burst and is called plasmoptysis.

A common practice for preserving of meat and vegetables is pickling the product in a strong salt solution. Under such conditions, bacteria present in the food are destroyed by plasmolysis. There is no practical application of this principle in water treatment. However, the length of time intestinal bacteria can survive in water is dependent in some degree on the effects of osmosis on the bacterial cells.

Germicides and Bacteriostatic Agents

Germicides are substances which destroy a bacterial cell on contact and a bacteriostatic agent is a substance which prevents the cell from reproducing itself. This latter situation indirectly brings about the destruction of the bacterial culture since without reproduction there can be no continuation of life. The best known germicide in water treatment is chlorine.

Antimetabolite

Antimetabolites are substances which destroy or alter metabolic agents or growth factors essential to the normal life processes of a bacteria cell. Such action results in the destruction of bacteria since without normal growth and reproduction, life itself ceases. Some of the antibiotic drugs used in treating disease are antimetabolites.

Bacteria Identification

Bacteria are identified through a systematic application of procedures which are designed to:
1. Secure a culture of bacteria, that is, a very large number of living bacterial cells in or on a medium which provides adequate food for them.
2. Successively sub-culture or secure a separation of the individual species from each other.
3. Determine the cultural characteristics of each species.
4. Determine the morphological characteristics of each species by examination of stained preparations of bacterial cells.

There are many kinds of bacterial cultures, that is, growths in solutions or liquid suspensions of nutrient materials, growths in suspensions of the nutrients in jelly-like substances such as agar or gelatine, growths on the surface of animal or vegetable tissue and growths on the flesh or in the bloodstream of animals. There are many materials used as nutrients, ranging from simple inorganic salts to organic carbohydrates, such as sugars, to relatively simple protein substances, such as egg albumin and to the extremely complex proteins of animal tissue.

Sub-culturing is the process of securing growth of the bacteria in colonies and fishing a colony, for re-growth on fresh medium. In fishing, a sterile wire is gently touched to a surface of the colony, thus removing a few cells which cling to the tip of the wire. These cells are then immediately transferred to a fresh lot of sterile medium in another vessel. Incubation of the transplants then produces new colonies. A successive number of such transfers eventually produces a pure culture, that is, a culture of a single species.

Small amounts of pure cultures are inoculated into specialised media to determine growth characteristics. These media often consist of carbohydrates of various kinds differing in chemical structure and complexity. Growth is usually indicated by production of a cloudy effect. The number of suspended cells increases and acid and/or gas are produced as the carbohydrates are decomposed by the bacteria. A wide variety of stains and staining techniques are available to enable the bacteriologist to observe size, shape, spore formation, presence of flagella, etc., under the microscope.

Bacteriology

Even distilled water contains sufficient nutrients to support bacterial growth. The only way sterile water can be obtained is by treating it with chemicals, such as chlorine, to destroy the bacteria, by heating it or, under special circumstances, irradiating it with ultra-violet light. All natural water, whether from surface or ground sources or from precipitation, is contaminated to some degree with bacteria. These are mostly saprophytes, which feed on organic matter present in the water though contact with earth and vegetation, although some are autotrophic bacteria, which feed on simple inorganic salts and dissolved gases. Those of greatest interest in water treatment are in the water as a result of its pollution with sewage. These bacteria may include organisms pathogenic for human. There are diseases of bacterial origin transmittable from human to human through sewage contaminated water. A group of bacteria whose normal habitat is the large intestine of human and animals is known as the coliform group of bacteria. In general, these bacteria (comprising more than 30 individual species) conform to the requirements of an ideal index of sewage pollution of water, which are:
1. Present when sewage is present.
2. Absent when sewage is absent.
3. Survives longer in water than any of the pathogenic species.
4. Easily isolated and identified.

The coliform group includes all of the aerobic and facultative anaerobic, Gram negative, non-spore-forming, rod-shaped bacteria which ferment lactose with gas formation within 48 hours at 35°C. Some of the terminology in this definition requires interpretation.

Two bacteriological methods are available for estimating the degree of sewage pollution of a water sample: the multiple tube fermentation method and the membrane filter method. The multiple tube fermentation method is standard and applicable to all types of water. The membrane filter method is standard for water free from turbidity and algae and filter-clogging materials after adequate parallel testing has demonstrated that it yields, for a particular water, information comparable to that of the multiple tube method.

Multiple Fermentation Tube Method

The objective in this method is to demonstrate the presence or absence of bacteria which will ferment the sugar lactose added to a nutrient liquid medium and produce gas within a specified period—48 hours. A necessary piece of equipment is the fermentation tube. This is a small culture tube inverted within a larger tube and completely filled with the medium during sterilisation.

The medium is protected from chance bacterial contamination by a plug of cotton at the mouth of the larger tube. When the medium is inoculated with a portion of the water sample and the fermentation tube is placed in an incubator, bacteria multiply in numbers as evidenced by increasing turbidity of the medium with passage of time. If lactose-fermenting bacteria are present, gas will collect in the inner tube. Any amount of gas produced within the specified incubation period is presumptive evidence of the presence of coliform bacteria in the water sample. Evidence of the presence of coliform bacteria is necessary since not all lactose fermenters belong to the coliform group. Confirmation is obtained by transferring a drop of the bacterial culture to another fermentation tube containing a different medium—one which permits the growth of coliform bacteria only. Production of gas in this medium establishes the presence of the sewage organisms in the original sample. It is pertinent to note that one cannot deposit the original inoculum of water into the confirmation medium directly because the bacteria being sought are not in their natural environment, the intestinal tract and for that reason are unable to thrive in the rigorous confirmatory medium without preliminary increase in their numbers and viability.

Most Probable Number (MPN)

It is important not only to detect pollution of water but also to estimate its degree, specifically the number of coliform bacteria per unit volume. In a test involving the use of a medium in which the original number of bacteria introduced is greatly increased before their presence is evident, a precise count of bacteria cells is impossible. A suitable alternative is to inoculate several fermentation tubes with a series of dilutions of the sample, to observe which produce gas and calculate the number of bacteria using a mathematical formula based upon the laws of probability. A suitable combination of fermentation tube implants for a tap sample from a drinking water supply is 5 tubes with 10 ml portions, 1 tube with a 1 ml portion and 1 tube with 0.1 ml portion. The MPN results for the most commonly found combinations of positive and negative tubes in this series are as follows:

10 ml portion	1 ml portion	1/10 ml portion	MPN
– – – – –	–	–	less than 2.2
+ – – –	–	–	2.2

(Contd...)

10 ml portion	1 ml portion	1/10 ml portion	MPN
+ + − − −	−	−	5.0
+ + + − −	−	−	8.8
+ + + + −	−	−	15
+ + + + +	−	−	38
+ + + + +	+	−	240
+ + + + +	+	+	2400 or more

The most probable number (MPN) is not an exact measure of coliform bacteria in the sample, but an estimate.

Membrane Filter Method

Membrane filters are very thin films of cellulose manufactured in a way to produce a porous structure. Water passes freely through the membrane with slight suction, but particles, even those as small as bacteria, are retained on the surface. Starting with the sterile membrane placed on a sterilised funnel-shaped holder mounted on a suction-type receiving flask, a sample of known volume is filtered with the aid of suction applied to the receiving flask. The membrane filter is then removed with sterilised forceps and placed in a sterile culture dish on a sterile absorbent pad and saturated with special nutrient medium. Colonies develop after incubation for 18–22 hours at 35°C and these may be counted with the aid of a 10-power stereomicroscope. Only those which have a dark purplish-green colour with a metallic sheen are considered members of the coliform group. Results are reported as colonies per 100 ml of sample.

The membrane filter method yields results roughly comparable to the fermentation tube method when the sample is free from suspended matter. The more suspended matter present in the sample, the greater the discrepancy between the two methods. The technique is more easily performed than the fermentation tube method. Results are obtained more quickly and with special equipment, the test may be made in the field.

VIRUSES

Viruses are the smallest micro-organisms known. The largest viruses are about 200 millimicrons in diameter, which is about the absolute limit of resolution of the best light microscopes. Most viruses are much smaller, ranging down to 10 millimicrons in diameter. The small particles will pass freely through the openings of most filters. Special filters made of collodion film have been employed to isolate viruses. These have been supplemented by other techniques, including high speed centrifuges, to aid sedimentation.

Viral diseases are many and affect all forms of life. For that reason, viruses are often segregated into groups on the basis of the host. There are plant viruses, animal viruses and bacterial viruses. Members of the latter group are usually called phages or bacteriophages. The fact regarding viruses that impels us to believe that they are not merely very small bacteria is that they have no distinct metabolism of their own and cannot exist for long outside of the cells of the living host. This does not preclude the possibility of their existence for a time in non-living media such as water. However, there is evidence that the viral diseases, for example poliomyelitis and hepatitis are transmitted through polluted water. Viruses have been isolated from sewage and streams polluted with sewage. Their numbers appear to be much less than the numbers of bacteria under the same conditions.

Rickettsiae are similar in some respects to viruses and in others to bacteria. They are intermediate in size between viruses and bacteria—from 0.3 to 0.5 micron for most cells, although some of the bacillus type have been found with cells up to two microns long. Rickettsiae are either cocci (round) or short bacilli (rods). They multiply by fission as bacteria do, but like viruses, must do so in the presence of a host. Rickettsial diseases include Rocky Mountain spotted fever, Q fever and typhus. These are human diseases and there are other rickettsial diseases which afflict lower animals. So far as is known, water is not a vector in the transmission of rickettsial diseases.

OTHER WATER MICROBES

There are microbes belonging to the plant kingdom other than bacteria, viruses and rickettsiae of interest in water treatment. They are easily separated from water or microscopic examination by simple filtration through sand and are therefore considered as a group although they differ with respect to their taxonomical categories. These include algae, fungi and schizomycetes.

Algae

Algae are organisms usually found in surface water exposed to sunlight and during the summer months frequently become so numerous as to produce blooms which appear to completely cover whole sections of ponds or lakes. Algal cells contain chlorophyll, the green colouring matter of plants and under the influence of sunlight, chlorophyll enables the organism to combine water and carbon dioxide to form complex chemicals and produce oxygen as a by-product. The process is photosynthesis. Production of oxygen is beneficial in that it promotes oxidation of organic debris, supports life and favours the existence of aerobic saprophytic bacteria which decompose organic matter without the liberation of foul-smelling gases. Chemical reactions involving changes in pH and hardness of the water are also brought about through the catalytic action of chlorophyll during daylight hours. Hardness of the water decreases and pH rises during the day in water reservoirs having prolific growths of algae. Unfortunately, the process reverses when darkness sets in with the result that a water plant operator may find the chemical characteristics of the water he is treating to be quite different in the daytime than at night.

Algae are the chief cause of tastes and odours in a water supply. The tastes and odours produced by these organisms are characteristic and the presence of an offending species can often be detected by an alert plant operator by the taste and odour of the water even before it becomes obvious. Daily microscopic examination of the raw water, especially during the summer months, should be routine practice in all surface water plants to detect an increase in algae. There are at least six classes of algae common to water supplies. Some of them, particularly those of the cyanophyceae or blue-green class, are toxic.

Fungi are members of the plant kingdom but, unlike algae, they have no chlorophyll in their cellular structure. Water forms of fungi include phycomycetes, similar to algae and ascomycetes, commonly called yeasts and moulds. Some classifications place schizomycetes, of which class bacteria are members, under fungi.

Schizomycetes is the classification name for bacteria, which were discussed previously. There are several orders of bacteria which are found in water and whose cells are large enough to be segregated by sand filtration and thus be observed in a microscopic examination. These are often referred to as higher bacteria. Included in this group are Chlamydobacteriales, Actinomycetales, Thiobacteriales and Beggiatoales. These are usually indicative of polluted water. Chlamydobacteriales are also called iron bacteria because iron is often found deposited in the thick gelatinous sheath of the organism.

Some of them move about freely in the whole body of water while others are found most frequently in bottom mud. One or more varieties of protozoa are sure to be found in samples of fresh water since they are very numerous. They are sometimes responsible for disagreeable tastes in water.

Amoeba and Paramecium are two other protozoan species. These are common water forms. *Entamoeba hystolytica* is a species of amoeba responsible for amebic dysentery in human. It may be transmitted from one individual to another through polluted water.

Rotifera, Crustacea, Bryozoa and Porifera or microbial sponges, are found in fresh water, sometimes in great numbers. They are too small to be seen without a microscope. When their population is great, clogging is a problem. Larger forms, including roundworms (Nematoda), Hydra, Nais, bloodworms (Chironomus), etc., frequently are troublesome in water supplies causing taste, odours, clogged filters and perhaps even more seriously, consumer anger because some are large enough to be seen. Insects and insects larvae are a similar source of friction. In unfiltered supplies, these problems are inevitable but occasionally they can occur in filtered water.

Chapter 21

Hydraulics

INTRODUCTION

Hydraulics, the Greek word for water pipe, is the science of fluids, such as water. For water treatment, it is important to understand the pressure and the flow behaviour of water. The study of water at rest and in motion is known as hydrostatics and hydrodynamics, respectively. Hydraulics explains the behaviour of water at rest (as in elevated tanks) and in motion, while flowing through pipes, channels and pumps.

BASIC CONCEPTS OF HYDRAULICS

Force and Pressure

Force

Force (F) is the weight on the bottom of a column of water (e.g., 1 cubic foot of water exerts 62.4 lb of force on 1 ft^2 of its bottom).

$$1 \text{ cubic ft of water} = 7.48 \text{ gal/ft}^3 \times 8.34 \text{ lb/gal} = 62.4 \text{ lb}$$

Thus, force at the bottom of one cubic foot of water = 62.4 lb/ft^2

Pressure

Pressure (P) is the force per unit area.

$$P = F/A \text{ and } F = P \times A$$

where,

P = pressure
F = force
A = area

Using 1 ft.^{3} of water

F = 62.4 lb./ft.^{2}
P = (62.4 lb/ft^2)/(144 in.^{2}/ft^2) = 0.433 lb/inch^2

Therefore, 1 ft. column of water exerts a pressure of 0.433 pounds/inch^2, psi and 1 psi = 1 ft/0.433 psi = 2.31 ft (height of water).

Thus, we can convert the feet of water column, commonly called head, into psi and vice versa.

Static pressure or static head, is the vertical distance between any base point (like ground level, the top of a hill or the bottom of a valley) and the free surface of the water (see Fig. 21.1). Free surface level of interconnected tanks and lines of a confined liquid is always the same. It is explained by Evangellista Torricelli's law, which states that a jet of water from the bottom of the source will rise to the level of its source if directed upwards, as with a flowing artesian well.

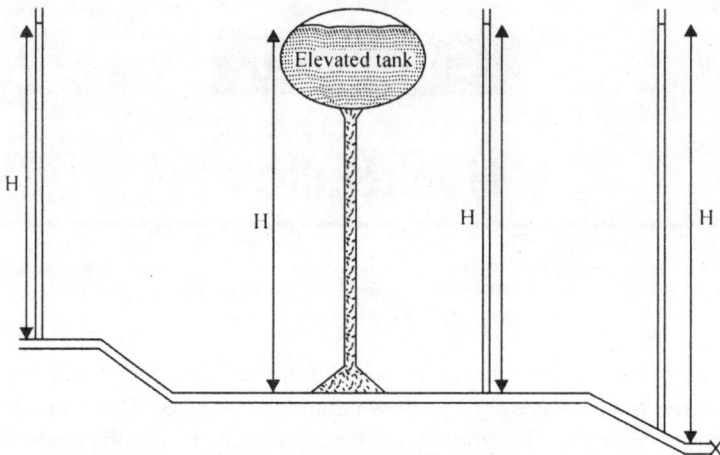

Fig. 21.1. Static head at different locations around an elevated tank.

Water at rest with free surface has the same pressure at all points at the same level. For example, all homes at the same level will have the same pressure. Pascal's law states that a pressure exerted on a liquid in a closed system is transmitted undiminished in all directions at the same force at right angles to the surface (see Fig. 21.2).

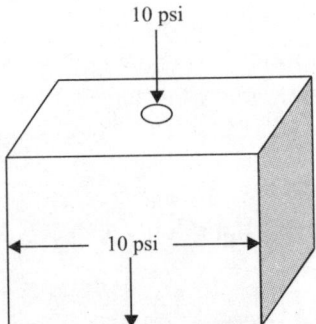

Fig. 21.2. Pascal's law.

Therefore, a confined liquid exerts equal pressure perpendicular to all parts of the inner surface of its container. Liquid will be forced through the bottom or a side hole of the container with the same pressure when the pressure is applied at the top (e.g., a piston pump). Standard (at sea level) atmospheric pressure is 14.7 psi, which forces water up when suction is applied (siphoning) at the upper end of a pipe with its lower end below the water surface. Functioning of a pump is based on this principle; suction is applied at the suction end of the pump to suck water to the discharge end. We apply this principle when we use a straw to suck a drink from a glass. Standard atmospheric pressure is 34 feet (14.7 psi × 2.31 ft/psi) water head; therefore, 34 feet is the theoretical height to which water can be lifted. It is called the theoretical lift of a pump.

Velocity head (V_h) is the pressure due to velocity, the motion of water.

Daniel Bernoulli's law states that the higher the velocity of a liquid, the lower is the pressure and vice versa. When a pipe narrows, velocity of the flowing liquid increases and its pressure decreases. Due to resistance to flow, the pressure ahead of the restriction increases. For example, our blood pressure increases when our arteries become narrower.

$$PV = P_1 V_1$$

where,

P   = pressure corresponding to velocity V

P_1 = pressure corresponding to velocity V_1

This principle is applied in the designing of a centrifugal pump, where velocity is converted into pressure.

Water standing in an elevated tank has a high pressure and no velocity. When a discharge valve at the bottom of a tank is opened, the velocity increases and the pressure falls.

Torricelli's law states that the velocity of a jet of water coming through an opening is equal to the velocity of a falling body from the air from the surface of the water to the opening. Thus, the falling water follows the law of gravity, like any other falling object; it causes the gravity flow.

Water flow due to gravity from a higher to a lower elevation is known as gravity flow. The rate of velocity corresponds to the elevation differential. The more the difference in two elevations, the higher is the velocity and vice versa. The following equation converts the velocity to velocity head:

$$V_h = V^2/2g$$

where,

V_h = velocity head

g   = 32.2 ft./sec./sec., the acceleration due to gravity

V   = velocity

Friction head is the loss of pressure due to friction (resistance to flow) between water and the surface of a pipe. It is negative and generally expressed as the head loss per 1000 feet of pipe. The larger the diameter of a pipe, the less is the friction loss and vice versa. Friction loss is directly proportional to the length, velocity and the roughness of a pipe. The roughness is represented by C factor that varies from 155 for very smooth and large pipes to as low as 40 or even less for badly corroded or rough pipes. For a normal clean and smooth pipe, the factor is 100. Mostly, we use a nomogram for friction loss developed from Hazen-Williams formula with C value of 100. This nomogram gives the relationship of flow, friction loss and the pipe diameter (size).

Normally, 40 psi is the minimum required pressure for the fire protection. Thus, the pressure is too low for the third customer. This example illustrates the importance of the proper size of lines and elevations in distribution system. To maintain a proper pressure in the system, it is important for the water utility to know the highest and the lowest elevation sites in its distribution system.

Flow

Flow is the quantity of water flowing per unit time. Mostly, it is expressed as cubic feet per second (cfs) for a river or stream and gallons per minute (gpm) or million gallons per day (mgd) for water flow to and from the treatment plant. Flow is directly proportional to the square of the diameter or the radius of a pipe.

$$Q/Q_1 = r^2/r_1^2$$

where,

Q = flow for radius r

Q = flow for radius r_1

Suppose, we compare the capacities of 2 and 4 inch pipes. If flow through a 2 inch pipe is 5 cfs, the flow of a 4 inch pipe will be 20 cfs, which is four times more.

$$5 \text{ cfs}/Q_1 = 1^2/2^2 \text{ or } Q_1 = 5 \text{ cfs} \times 4 = 20 \text{ cfs}$$

Generally, it is beneficial to install larger pipes than smaller ones. Pressure for different customers around an elevated tank is shown in Fig. 21.3.

Fig. 21.3. Pressure for different customers around an elevated tank.

Water Hammer

Water hammer is the surge or thrust of pressure, caused by a sudden change in the flow (velocity) by an abrupt opening or closing of a valve or a fire hydrant or an abrupt stopping and starting of a pump. Water hammer causes high pressure due to surge in the pipes, which can damage the joints at the nearest fitting of the line. Besides high pressure, it dislodges sediments in the lines, resulting in the dirty water or red water complaints. Gradual opening and closing of valves and fire hydrants and less frequent starting and stopping of pumps, reduce the impact of water hammer. Generally, the fittings of the water pipes are strong enough and well protected by anti-water-hammer material around them, such as thrust blocks and joint restraints.

HYDRAULIC ANALYSIS OF WATER AND WASTE-WATER TREATMENT

The primary purpose of this section is to delineate the steps involved in the hydraulic analysis of water and waste-water treatment plants. However, before considering the subject of treatment plant hydraulics, it is important to consider all of the steps involved in the design of water and waste-water-treatment plants.

Design Period

The design period establishes the target date when the design capacity of the facilities will be reached. Design periods may vary for individual components, depending upon the ease or difficulty of expansion. Typical design periods for various types of facilities are given in Table 21.1. Longer periods are preferred for structures and hydraulic conduit systems, that cannot be easily expanded. The selection of the design period depends upon growth characteristics, environmental considerations and the availability and source of construction funds.

Table 21.1. Typical design periods for waste-water.

Facility	Planning period range, years
Collection systems	20–40
Pumping stations	
Structures	20–40
Pumping equipment	10–25
Treatment plants	
Process structures	20–40
Process equipment	10–20
Hydraulic conduits	20–40

Treatment Plant Design

Once the required effluent quality has been defined, the steps involved in treatment plant design typically include: (i) synthesis of alternative flow sheets; (ii) bench tests and pilot-plant studies; (iii) selection of design criteria; (iv) sizing of physical facilities; (v) preparation of solids balances; (vi) layout of the physical facilities; (vii) preparation of hydraulic profiles; and (viii) preparation of construction drawings, specifications and cost estimates. Because of the importance of each of these steps, each is considered separately in the following discussion.

Another method of determining design conditions to meet effluent standards is the graphical probability method, similar to the method. Plant performance data can be plotted on log-probability or arithmetic-probability paper to determine the distribution characteristics. For example, the peak day may be determined at the 99+ percentile, based on occurring once every 365 days. Values equal to or less than the indicated value can be determined at the appropriate percentiles. These values can be compared to the values obtained from using the coefficient of reliability approach for selecting the appropriate mean effluent concentrations for design.

Synthesis of Alternative Treatment Process

The processes selected for the treatment of potable water depend on the quality of the raw water supply. Most groundwater are clear and pathogen-free and do not contain significant amounts of organic materials. Such water may often be used in potable systems with a minimal dose of chlorine to prevent contamination in the distribution system. Other groundwater may contain large quantities of dissolved solids or gases. When these include excessive amounts of iron, manganese or hardness, chemical and physical treatment processes may be required. Treatment systems commonly used to prepare potable water from groundwater are shown in Fig. 21.4.

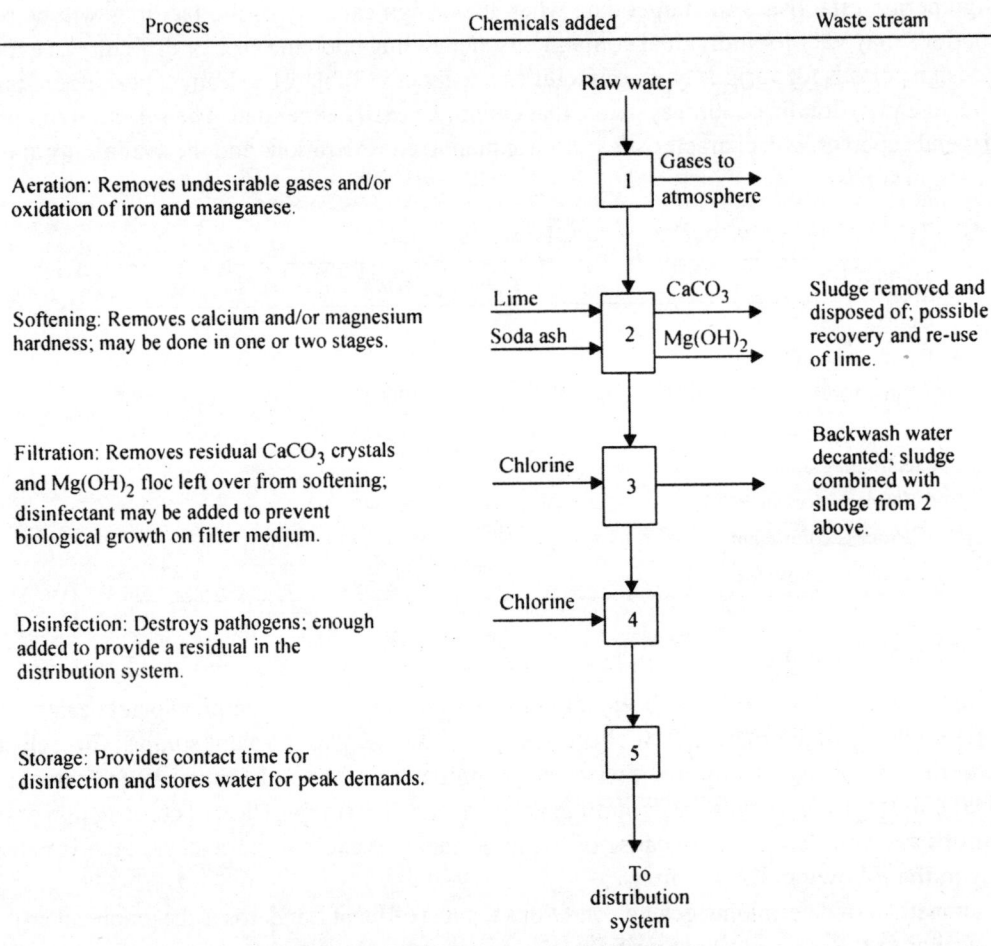

Process	Chemicals added	Waste stream

Raw water

Aeration: Removes undesirable gases and/or oxidation of iron and manganese.

Gases to atmosphere

Softening: Removes calcium and/or magnesium hardness; may be done in one or two stages.

Lime
Soda ash

$CaCO_3$
$Mg(OH)_2$

Sludge removed and disposed of; possible recovery and re-use of lime.

Filtration: Removes residual $CaCO_3$ crystals and $Mg(OH)_2$ floc left over from softening; disinfectant may be added to prevent biological growth on filter medium.

Chlorine

Backwash water decanted; sludge combined with sludge from 2 above.

Disinfection: Destroys pathogens; enough added to provide a residual in the distribution system.

Chlorine

Storage: Provides contact time for disinfection and stores water for peak demands.

To distribution system

Fig. 21.4. Typical plant treating hard groundwater.

Surface water often contain a wider variety of contaminants than groundwater and treatment processes may be more complex. Most surface water contain turbidity in excess of drinking-water standards. Although fast-moving streams may carry larger material in suspension, most of the solids will be colloidal in size and will require chemical coagulation for removal. Depending on the geology of the watershed, hardness may or may not be a problem in surface water. If low levels of colour and other organic material are present, adsorption onto surface-active material, a process not significant in natural water systems, may be necessary. A wide variety of micro-organisms, some of which may be pathogenic, are also common constituents of surface water. Treatment systems commonly used in treating surface water are shown in Fig. 21.5.

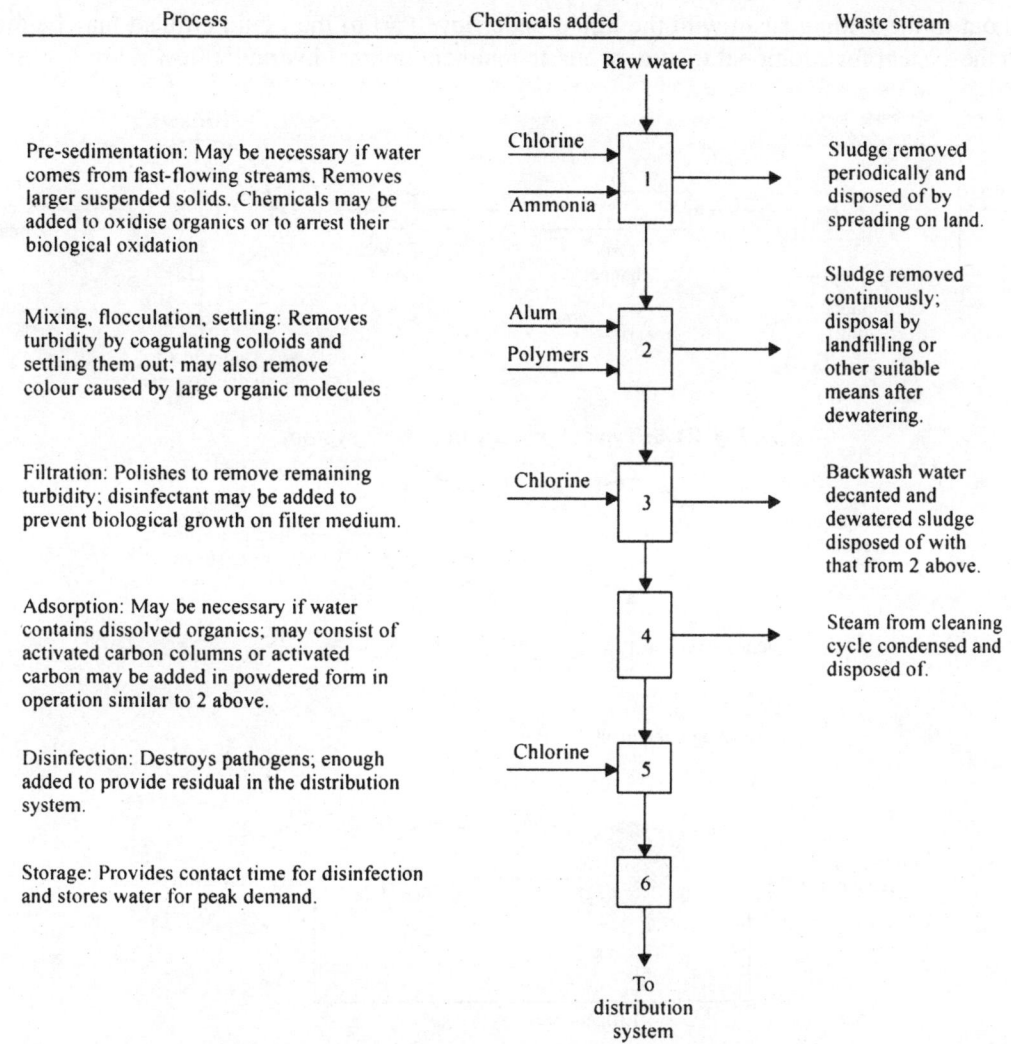

Process	Chemicals added	Waste stream

Raw water

Pre-sedimentation: May be necessary if water comes from fast-flowing streams. Removes larger suspended solids. Chemicals may be added to oxidise organics or to arrest their biological oxidation

Chlorine / Ammonia → 1

Sludge removed periodically and disposed of by spreading on land.

Mixing, flocculation, settling: Removes turbidity by coagulating colloids and settling them out; may also remove colour caused by large organic molecules

Alum / Polymers → 2

Sludge removed continuously; disposal by landfilling or other suitable means after dewatering.

Filtration: Polishes to remove remaining turbidity; disinfectant may be added to prevent biological growth on filter medium.

Chlorine → 3

Backwash water decanted and dewatered sludge disposed of with that from 2 above.

Adsorption: May be necessary if water contains dissolved organics; may consist of activated carbon columns or activated carbon may be added in powdered form in operation similar to 2 above.

4

Steam from cleaning cycle condensed and disposed of.

Disinfection: Destroys pathogens; enough added to provide residual in the distribution system.

Chlorine → 5

Storage: Provides contact time for disinfection and stores water for peak demand.

6

To distribution system

Fig. 21.5. Typical plant treating turbid surface water with organics.

A typical primary treatment system (Fig. 21.6) should remove approximately one-half of the suspended solids in the incoming waste-water. The BOD associated with these solids accounts for about 30 per cent of the influent BOD.

Secondary treatment usually consists of biological conversion of dissolved and colloidal organics into biomass that can subsequently be removed by sedimentation. Contact between micro-organisms and the organics is optimised by suspending the biomass in the waste-water or by passing the waste-water over a film of biomass attached to solid surfaces. The most common suspended biomass system is the activated-sludge process shown in Fig. 21.7a. Recirculating a portion of the biomass maintains a large number of organisms in contact with the waste-water and speeds up the conversion process. The classical attached-biomass system is the trickling filter shown in Fig. 21.7b. Stones or other solid media are used to increase the surface area for biofilm growth. Mature biofilms peel off the surface and are washed out to the settling basin with the liquid underflow. Part of the liquid effluent may be recycled through the system for additional treatment and to maintain optimal hydraulic flow rates.

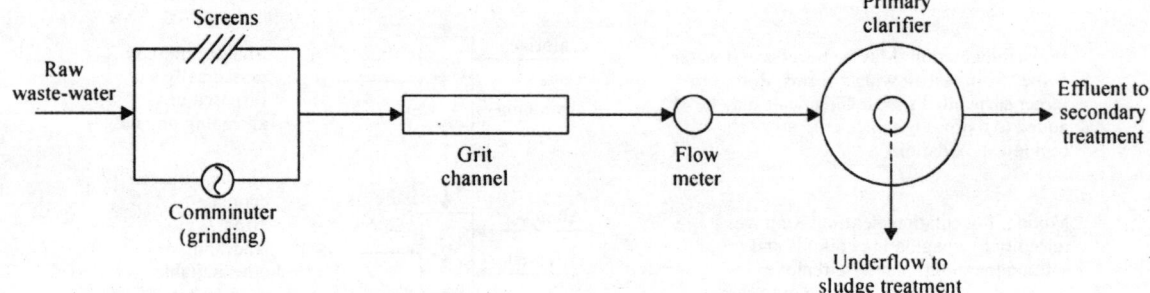

Fig. 21.6. Typical primary treatment system.

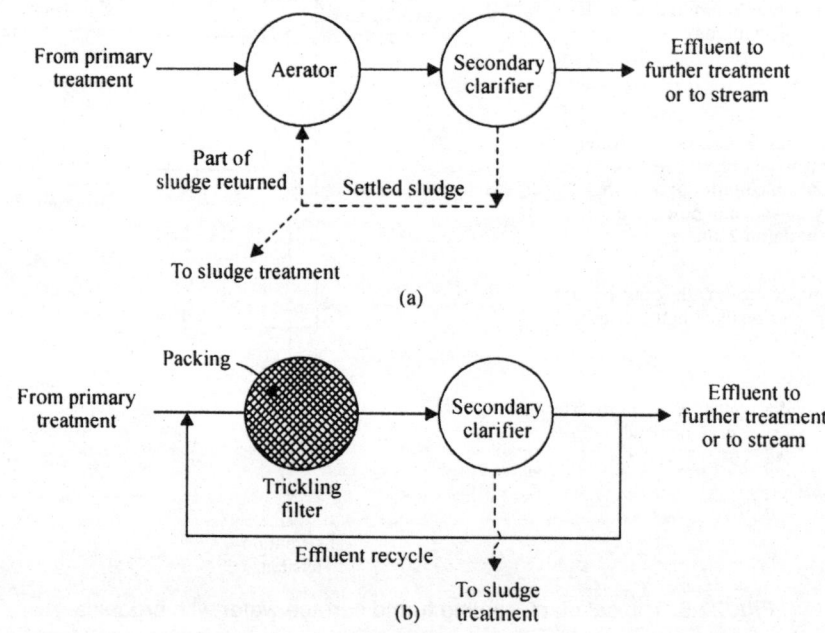

Fig. 21.7. Secondary treatment system: (a) activated sludge system; and (b) trickling filter system.

A flow sheet can be defined as the grouping together of unit operations and processes to achieve a specific treatment objective. Alternate flow sheets will be developed on the basis of the characteristics of the water and waste-water to be treated, the treatment objectives and if available, the results of bench and pilot-scale tests. The best alternative flow sheets are selected after they have all been evaluated in terms of their performance, physical implementation, energy requirements and cost.

Bench Tests and Pilot-Plant Studies

The purpose of conducting bench tests and pilot-plant studies is: (i) to establish the suitably of alternative unit operations and processes for treating a given water or waste-water; and (ii) to obtain the data and information necessary to design the selected operations and processes. Bench tests, as the name implies, are small scale tests that can be conducted in the laboratory. Typically they are used to establish approximate chemical dosages and to obtain kinetic coefficients. Continuous pilot-plant studies are conducted to verify the results of bench tests.

Process Design Criteria

After one or more preliminary process flow diagrams have been developed, the next step is to establish the process design criteria so that the size of the physical facilities can be determined. For example, if the hydraulic detention time in the aerated grit chamber shown in Fig. 21.8 is to be 3.5 minutes at a peak flowrate, the corresponding grit chamber volume required would be calculated. The hydraulic detention time would be an example of the process design criteria for the grit chamber. Similar procedures are followed for each unit operation and process.

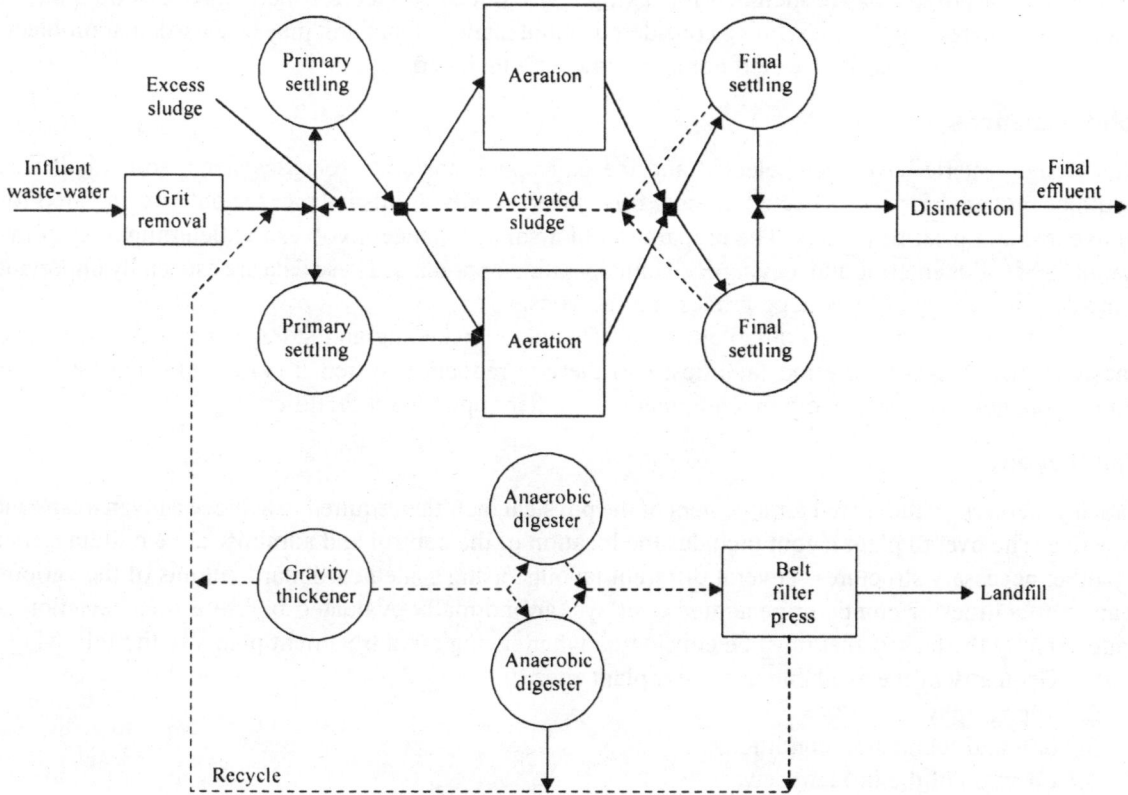

Fig. 21.8. Process flow diagram for treatment plant designed to meet secondary treatment standards.

When the computations have been completed, all the key design criteria should be listed in a summary table. Because most treatment plants are designed to be effective for some time in the future, design criteria are given for the time when the facilities will first be put into operation and for the end of the design period. The latter will be influenced by projections of the population to be served and the economic studies of cost effectiveness for various design periods.

Selection of Design Criteria

After one or more alternative flow sheets have been developed, the next step in design involves selection of design criteria. Design criteria are selected on the basis of theory, published data in the literature, the results of bench tests and pilot-scale studies and the past experience of the designer.

Sizing of Unit Operations and Processes

After the design criteria have been established, the next step is to determine the number and size of the physical facilities, needed. In considering sizing, physical site constraints need to be considered: for example, will the site accommodate the use of round tanks or will rectangular tanks have to be used? Operational considerations, such as flow splitting and load balancing, will have to be evaluated, particularly in process trains that combine different numbers of unit operations or processes (e.g., two primary clarifiers and three aeration tanks). Maintenance factors have to be considered in selecting the number of units so that provisions are included for taking a unit out of service for maintenance and repair. In small plants where a single unit is being considered, maintenance of that unit may be a particular problem, unless special provisions, such as temporary storage, are included.

Solids Balances

After design criteria have been selected and the unit operations and processes sized, solids balances should be prepared for each selected process flow sheet. Ideally, solids balances should be prepared for the average and peak flow rates. The preparation of a solids balance involves the determination of the quantities of solids entering and leaving each unit operation or process. These data are especially important in the design (sizing) of the sludge-processing facilities.

Such information must be available to size: (i) sludge-thickening and storage facilities; (ii) sludge digesters; (iii) sludge-dewatering facilities; (iv) thermal reduction systems; (v) composting facilities; and (vi) sludge-piping and pumping equipment and other appurtenant facilities.

Plant Layout

Plant layout refers to the spatial arrangement of the physical facilities required to achieve a given treatment objective. The overall plant layout includes the location of the control and administrative buildings and any other necessary structures. Several different layouts, using scaled cardboard cutouts of the various treatment facilities or computer-generated overlays, are normally evaluated before a final selection is made. Among the factors that must be considered when laying out a treatment plant are the following:

1. Geometry of the available treatment plant sites.
2. Topography.
3. Soil and foundation conditions.
4. Location of the influent sewer.
5. Location of the point of discharge.
6. Plant hydraulics, preferably with straight-flow paths between units to minimise head loss and provide symmetry for flow splits.
7. Types of processes involved.
8. Process performance and efficiency.
9. Transportation access.
10. Accessibility to operating personnel.

11. Reliability and economy of operation.
12. Aesthetics.
13. Environmental control.
14. Provisions for future plant expansion, including additional area.

The physical layouts of a variety of plants, both small and large, are shown in Figs. 21.9 and 21.10.

Using the information on the size of the facilities determined on the basis of the selected criteria, various plant layouts are developed within the constraints of the physical site. In laying out the various facilities, special attention should be given to minimising pipe lengths, to grouping together related facilities and to the need for future expansion.

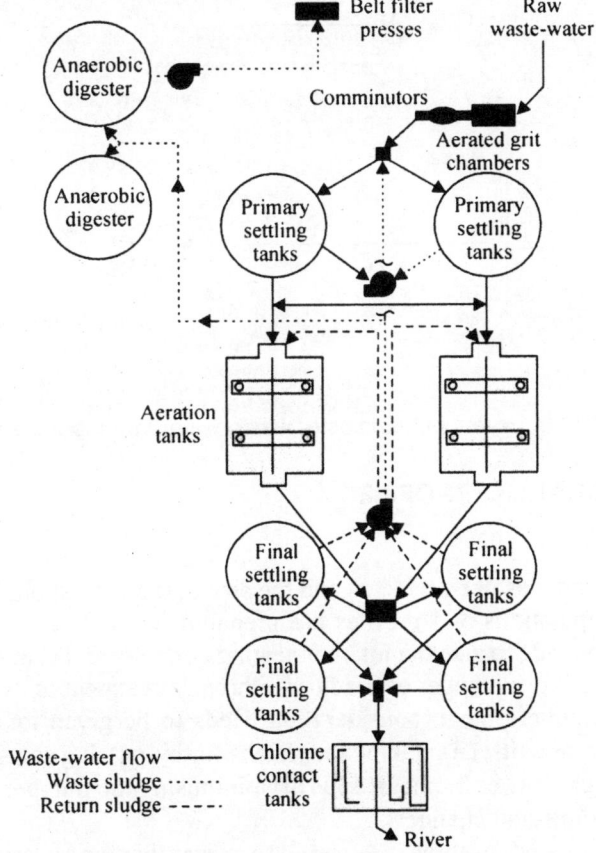

Fig. 21.9. Layout of river waste-water plant.

Construction Drawings and Specifications

The final step in the design process involves the preparation of construction drawings, specifications, and cost estimates. Because the clarity with which the construction drawings are presented will affect both the bid prices and final plant operation, the importance of this step cannot be overstressed. Construction specifications have been more or less standardised. The key issue is to make sure that specifications are complete so that costly change orders can be eliminated. Finally, the engineer's cost estimate is used as a guide in evaluating the bids submitted by the various contractors.

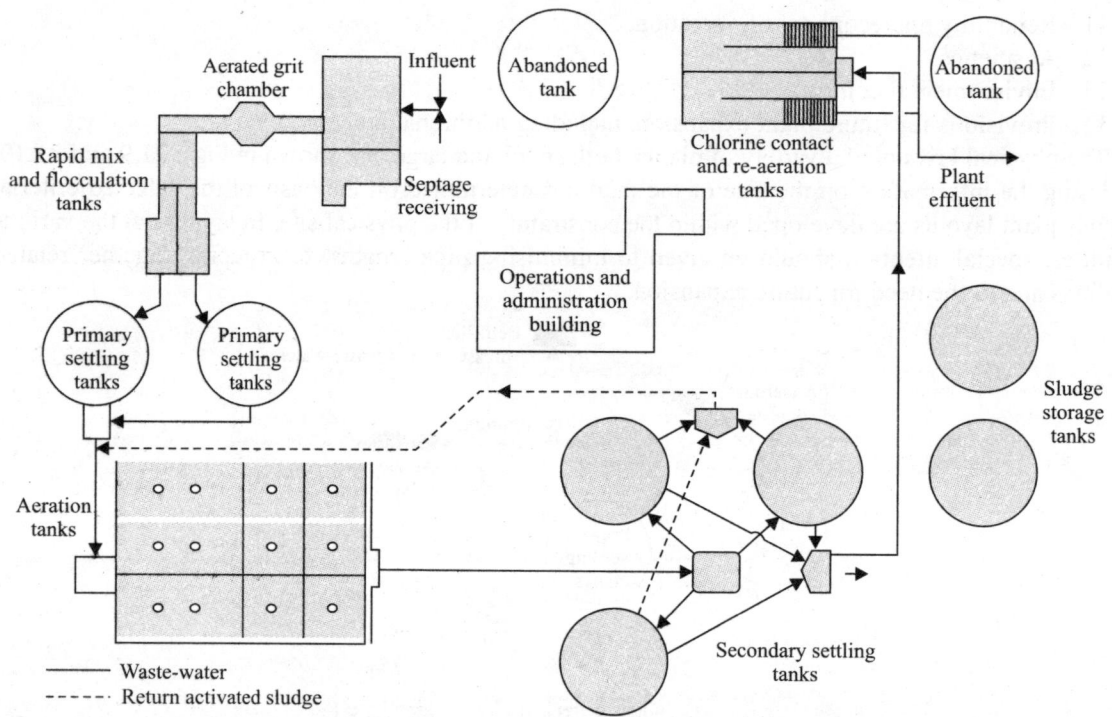

Fig. 21.10. Layout of Leominster waste-water treatment plant.

PREPARATION OF HYDRAULIC PROFILES

Plant Hydraulics

After the process flow diagram has been selected and the size of the corresponding physical facilities is determined, hydraulic computations and profiles are prepared for both average and peak flowrates. Hydraulic computations are made to size the interconnecting conduits and channels and to compute the headlosses through the plant. Typical ranges of headlosses through treatment units are given in Table 21.2. In designing the plant hydraulic system, consideration needs to be given to: (i) equalising the flow splitting between the treatment units; (ii) making provisions for by-passing secondary treatment units at extreme peak flows to prevent loss of biomass; and (iii) minimising the number of changes in direction of waste-water flow in conduits and channels.

Hydraulic profiles are prepared for three reasons: (i) to ensure that the hydraulic gradient is adequate for the waste-water to flow through the treatment facilities by gravity; (ii) to establish the head requirement for the pumps where pumping will be needed; and (iii) to ensure that the plant facilities will not be flooded or backed up during periods of peak flow. Profiles for the flow diagram given in Fig. 21.8. In preparing a hydraulic profile, distorted vertical and horizontal scales are commonly used to depict the physical facilities.

Hydraulic profile computations involve the determination of the headloss as the waste-water flows through each of the physical facilities in the process flow diagram. Specific computational procedures may vary depending on local conditions. For example, if a downstream discharge condition is the control point, some designers prepare the hydraulic profile by working backward from the control point. Other designers prefer to work from the head end of the plant. Still others work from the center in each direction, adjusting the elevations at the end of the computations. The use of mathematical models and digital computers allows many possible hydraulic conditions to be analysed.

Table 21.2. Typical headlosses across various treatment units.

Treatment unit	Head loss range, ft
Bar screen	0.5–1.0
Grit chambers	
Aerated	1.5–4.0
Velocity controlled	1.5–3.0
Primary sedimentation	1.5–3.0
Aeration tank	0.7–2.0
Trickling filter	
Low-rate	10.0–20.0
High-rate, rock media	6.0–16.0
High-rate, plastic media	16.0–40.0
Secondary sedimentation	1.5–3.0
Filtration	10.0–16.0
Carbon adsorption	10.0–20.0
Chlorine-contact tank	0.7–6.0

Note: ft × 0.3048 = 3

Thus, preparing hydraulic profiles involves careful consideration of the frictional and minor head losses that can occur in piping systems and of the head losses associated with control structures. These head losses are considered separately below. Application of the information on head losses in the preparation of hydraulic profiles is illustrated in the final part of this section.

Frictional Head Loss

The frictional head loss that occurs as water and waste-water flows through pipes can be computed with several equations. The recommended equation is the Darcy-Weisbach as given below:

$$h_f = f \frac{L}{D} \frac{V^2}{2g} \qquad \qquad \text{... (21.1)}$$

where,

h_f = head loss, m(ft)

f   = coefficient of friction

L   = length of pipe, m(ft)

D = diameter of pipe, m(ft)

V = mean velocity, m/s(ft/s)

g   = acceleration due to gravity, 9.81 m/s² (32.2 ft/s²)

The values of the friction factor are obtained from a Moody diagram. A representative value used for most friction computations is 0.020.

Minor Head Losses

As noted earlier in the section on pumps, minor head losses are produced when various control devices are inserted in piping systems. Valves are the most common control devices used in piping systems. Minor head losses also occur at pipe joints, pipe interconnections, pipe expansions and contractions and pipe entrances and exits. For practical purposes minor head losses are usually estimated as a fraction of the velocity head in the downstream pipe section using Eq. (21.2).

$$h_m = K \frac{V^2}{2g} \qquad \qquad ... (21.2)$$

Typical K values for various kinds of control devices and pipe configurations may be found in manufacturer's literature.

Head Losses from Control Structures

The most common control structures used in both water and waste-water-treatment plants are weirs of one sort or another. The formulaes used most commonly for rectangular and vee-notch weirs are given below:

For rectangular weirs, the Francis equation is used most commonly. The Francis equation is

$$Q = 1.84 (L - 0.1 \, nh)h^{3/2} \qquad \qquad ... (21.3)$$

where,

$$\begin{aligned} Q &= \text{discharge, } m^3/s(ft^3/s) \\ 1.84 &= \text{numerical constant} \\ L &= \text{length of crest of weir, m(ft)} \\ n &= \text{number of end contractions} \\ h &= \text{head on weir crest, m(ft)} \\ 3.33 &= \text{value of numerical constant for US customary units} \end{aligned}$$

For 90° triangular weirs the general equation is:

$$Q = 0.55 \, h^{5/2} \qquad \qquad ... (21.4)$$

where,

$$\begin{aligned} Q &= \text{discharge, } m^3/s(ft^3/s) \\ 0.55 &= \text{numerical constant} \\ h &= \text{head on weir crest, m(ft)} \\ 2.5 &= \text{value of numerical constant for US customary units.} \end{aligned}$$

Chapter 22

Applications of Computer in Waste-Water Technology

INTRODUCTION

Modern computer technology has had an almost unbelievable impact on practically every phase of man's existence. One promising use of this technology is in the simulation or modelling of natural bodies of water. This chapter discusses these modelling techniques.

The rapid growth and rate of obsolescence in this field of study obviate detailed description of particular programs. Rather, the subject is approached from the direction of functional application. The details of methodology must be tailored to the particular equipment at hand.

A brief discussion of mathematical models and computer technology is followed by a description of the various natural bodies to be simulated. Finally, the techniques of optimisation are introduced to point the way toward future applications.

MATHEMATICAL MODELS

General Comments

A mathematical model is a description of a natural relationship. Being stated in the precise language of mathematics, it has the advantage of being unambiguous. Unfortunately, this precision can be achieved only by the use of simplifying assumptions.

Some mathematical models can be represented in very simple forms which are easily applied. Other models are so complex, either in statement or in repetitive manipulation, that they require the speed and vast memory of a digital computer. This chapter deals primarily with the latter type of mathematical model.

Mathematical models may be classified in a variety of ways. They may be grouped according to the type of phenomenon being modelled, i.e., hydrology, chemistry, etc. Another classification could be in terms of the nature of the process, such as deterministic or stochastic, steady-state or dynamic. The type of model chosen will depend on the nature of the process involved and on the degree of approximation permitted by the end use of the output.

Biochemical Models

A biochemical model, as used in a study of natural bodies, describes the interrelationships between biological organisms and their chemical environment, occurring in a confined space and over a relatively short time period. The classic model for this interrelationship assumes two compensating reactions occurring simultaneously: deoxygenation due to the biochemical oxygen demand of the organic wastes and re-aeration due to absorption of atmospheric oxygen at the free surface. The rate of change of the oxygen deficit is given as:

$$\frac{d}{dt}D = K_1 L - K_2 D \qquad \qquad ...(22.1)$$

in which D is the oxygen saturation deficit in mg/liter, L is the biochemical oxygen demand in mg/liter, K_1 is the coefficient defining the rate of deoxygenation, K_2 is the coefficient defining the rate of re-aeration and t is the time of reaction in days.

This classic relationship or one of the many modifications of it, forms the basis of almost all simulations of the oxygen profile in flowing streams.

Hydrologic Models

Hydrology is the study of the origin and distribution of the water of the earth. Many mechanisms play a role in this overall, cyclic phenomenon, precipitation depends on a complex combination of meteorological conditions. Run-off depends on the condition of the receiving surface and the pattern of precipitation. Stream flow depends on the physical characteristics of the watershed and the run-off distribution. Each of these mechanisms can be approximated by mathematical relationships.

Most hydrologic models are extensive rather than intensive. That is, the model attempts to simulate the behaviour of a large and complex system covering many square miles. Because of this, it is necessary to recognise local mechanisms in terms of average conditions. For instance, the infiltration capacity of the ground varies with soil characteristics and vegetal cover. When considering a watershed of several hundred square miles, therefore, it is meaningless to refer to specific local conditions. The Stanford watershed model recognises this by assuming a linear variation of infiltration rate as precipitation continues. This mechanism assumes that ground surface conditions vary throughout the watershed and that as time progresses, various elements of the watershed area will control the average infiltration rate.

Economic Models

Economic models deal with yet another level of abstraction. In this case an attempt is made to represent the complex interrelationships of the entire economic community in terms of mathematical relationships. One of the more difficult problems which must be faced in establishing a mathematical model of an economic system is defining the objective function. This and other facets of economic models are discussed in detail in a later section of the chapter.

Deterministic Models

Deterministic mathematical models are the most widely used form of mathematical relationship. For mechanism modelled in this case it is assumed that for a given set of input conditions there is but one uniquely determined solution. This is the basic assumption of most engineering calculations. For example, the well-known Rippl method for determining reservoir capacity in a water supply reservoir assumes that an analysis of the historic records of flow at that site will yield a single deterministic answer for the storage required. In this case there is an inherent implication that the hydrology which the structure will experience in the future will be the same as the hydrology reported in the relatively short historic trace that was analysed.

Stochastic Models

Stochastic models are mathematical relationships which recognise that some of the inputs are random and unpredictable in nature. This randomness is of such magnitude that the solution must be described in statistical terms rather than as a unique solution. Stochastic relationships attempt to incorporate such things as the capriciousness of natural occurrence, e.g., the occurrence of rainfall. This can be illustrated by returning to the example of the required storage capacity in a reservoir. The Rippl method analyses the historic trace of stream flow records at the site and determines the amount of storage which would provide for the design-sustained flow from the structure on the assumption that future events will be a mere image of past history. Unfortunately, the project life is usually considerably longer than the period of historic record. An alternative to the Rippl method would be a statistical analysis of the historic record to estimate the statistical parameters of the population of stream flows from which the historic record is but a small sample. These statistical parameters could then be incorporated into a stochastic model along with a randomness generator to generate a much longer sequence of flows. This sequence could then be routed through the proposed reservoir and the percentage of time that the design was satisfactory could be determined. A series of such runs for a variety of reservoir capacities would permit the selection of an optimal design on the basis of the economic cost of the failure of the structure. In this case, failure would be represented not by the collapse of the structure but by the inability of the storage to meet the design draft.

Stochastic models find application in other areas, such as management and economics, in which the unpredictability of human behaviour is simulated.

Steady-State Models

Many mathematical relationships assume that the process being modelled is a steady-state process. That is to say, the various components or inputs to the system are invariate with time. In most cases, this assumption is invoked in order to bring to the simulation a greater degree of mathematical simplification. Very few natural phenomena are, in fact, steady-state. In many cases, however, the assumption of steady-state conditions introduces only a minor error in the answer. Before the advent of high speed digital computers, the assumption of dynamic conditions made many engineering calculations prohibitively complex and lengthy. Even with the availability of high speed computers, it is often unnecessary to include the dynamic effects caused by unsteady conditions.

Dynamic Models

The dynamic nature of some physical phenomena makes it imperative to incorporate the unsteady effect in any mathematical simulation. For example, it would be quite unrealistic to attempt to model the water quality of an estuary without recognising the dynamic effects of the tidal variation that acts upon and in large measure, controls the flow in an estuary. The ability to incorporate these dynamic effects into mathematical models has been greatly enhanced by the advent of the high speed digital computer. Finite difference techniques and other numerical methods, when applied with the ultrahigh speed of the modern computer, make possible many engineering determinations which heretofore were economically infeasible.

In summary, it should be pointed out that every mathematical model of a natural process is to some extent an approximation. The fidelity with which the model can represent the natural process is a function of the level of our understanding of the basic mechanisms operative in the phenomenon in question and the amount of detail which can be justified in a given case. In many cases a simple slide rule computation of a deterministic nature is still perfectly adequate for our needs. In other cases we are justified in applying sophisticated numerical techniques employing the very latest in high speed digital equipment.

COMPUTER TECHNOLOGY

Digital Computer

A modern digital computer is basically a very fast computer system. Mathematical relationships are solved arithmetically. The power of this machine lies in its ability to call from storage the required input data, operate on these data arithmetically and store the result for future printout. This entire process takes place in several millionths of a second. Programming a succession of such operations permits the solution of many complex problems.

An essential adjunct to digital computer technology is the mathematical discipline of numerical analysis. Solutions to functional relationships are found by transforming the problem to a form which can be solved or approximated by a succession of arithmetic operations. The digital computer is ideally suited to solution by successive approximations or any iterative processes.

The speed of the digital computer permits evaluation of a range of solutions to a given problem so as to approach an optimal solution. The time required for hand calculation would limit an investigator to calculating a small number of alternative solutions.

Analog Computer

An analog computer is a device which permits the analysis of a physical system by measuring comparable effects on a second physical system which is analogous to the prototype. The basis for many analog solutions is the fact that the differential equations describing two different physical systems may be essentially identical from a mathematical point of view. For example, a hydraulic network consisting of pipes and pumps with water as the fluid medium can be shown to be analogous in many respects to an electric circuit composed of resistances and potential differences in which electric current is the fluid medium. The behaviour of the hydraulic system may be predicted by measurements of current, voltage and resistance which are made on the electric system. In this case the hardware comprising the electric system would be a single-purpose analog computer. Some physical systems are of sufficient importance to warrant such a single-purpose computer. Many analog computers are designed for general use by providing plug-in circuit components that can be assembled in a variety of configurations.

The analog computer is ideally suited to the solution of dynamic problems. The time variability of the output can be accommodated by the use of oscilloscope readout or xy-plotting equipment. The nature of the data output format is at once a convenience and a limitation of this kind of equipment. Many unsteady phenomena are best reported in graphical form. On the other hand, graphical readout has very limited accuracy. In many cases this limitation of accuracy is not a serious disadvantage, in other cases the necessary accuracy can be achieved by multirange selection of scale.

An important property of analog computation is the ability to vary input signals continuously over a wide range of values. This permits the operator to adjust the system inputs so as to approach an optimum solution. This capability becomes less useful as the number of input variables increases.

Analog computers hold bright promise for the solution of many water pollution-oriented analyses. Very likely, their greatest utility will be achieved as they are combined with digital computers into hybrid digital-analog systems.

Data Input and Manipulation

Form of raw data

All digital computer programs require the input of certain basic data which are stored internally for recall during the execution of the program. In the field of water resources, these background data frequently are large amounts of field data. The form of the data can be convenient or inconvenient for use in a digital computer. With this fact in mind, serious consideration should be given to the way in which basic field data are assembled and recorded.

Field data which are the result of direct observation by a human observer are most usually reported in digital form on observation sheets. This information must be transcribed to punched cards before it is suitable for input into the computer. In some cases field data are gathered continuously and automatically in graphical form on recording strip charts of one kind or another. Such data must first be interpreted by office personnel and translated into digital form, then recorded in tabular form. Every effort should be made to provide field data in as convenient a form as practical.

Machine reduction

One of the main problems associated with the handling of field data is the sheer bulk of the data. Usually it is desirable to reduce this bulk by some systematic means. The computer can aid in this reduction. In some cases only the maximum or extreme value of the continuous signal is of significance. In other cases the mean, the standard deviation, or some other statistical parameter is desired. In some instances a reordering of the data is called for. All of these manipulations can be handled automatically and rapidly by the digital computer.

Another possible function of the computer would be to scan a record and determine if a value observed in the field was above or below an established standard or critical value. In this way it would be possible for the computer to report the number of occurrences and duration of the violation of such a standard.

Correlation analysis

One of the important operations performed on many kinds of field or experimental data is correlation analysis. The computer can assist us in this task.

First, the computer can perform a variety of mathematical operations on each unit of data so as to bring it into a form which makes the relationship with other data most meaningful. The data can be plotted graphically so as to show by visual observation whether or not an obvious correlation exists. A mathematical refinement of this procedure would be to perform a least squares analysis and least squares fit of the regression equation relating the two sets of data.

In some natural phenomena we must deal with more than a two-dimensional correlation. Multivariate analysis can handle this situation. There is practically no limit to the variety of statistical procedures which can be devised to aid in this type of analysis.

Output format

The form of the output from a computer operation is just as important as the format of the input to that program. It is very important to plan carefully the output that is required.

Output which is of a terminal nature is most usually presented in numerical, tabular format. In some cases it is more convenient to have the information displayed in graphical form. The digital computer is capable of producing the information in either or both forms, as required.

In many instances the output from one computer program may become input to a subsequent program at some other time. If this is the case, it may be desirable to have the output in the form of soft copy. This format greatly facilitates the input of this information into a subsequent program.

The modern digital computer is capable of producing output in almost any format required. The cardinal rule to remember is that output should be limited to essentials to reduce the amount of material to be scanned by the user.

Sensitivity analysis

Another important application of high speed digital computation is in the field of sensitivity analysis. By this we mean the repetitive computation of a mathematical relationship in which a single variable or input is varied systematically over a wide range of values in order to determine the effect of this variable on the final result. This procedure permits the investigator to determine those input variables which are most sensitive to the solution and those which are least sensitive. Such information can be valuable in deciding upon the necessity for sophisticated measuring techniques in field observations. Analyses of this kind can aid in the planning of future studies.

SIMULATION OF NATURAL BODIES

Simulation

Natural bodies may be simulated in many different forms. Such simulations are commonly called 'models' and may be physical, pictorial or mathematical. The geodetic maps provided are examples of pictorial models. Some of these pictorial models may be three-dimensional, e.g., relief maps. In other cases the natural body may be simulated by a mathematical model. Our discussion is limited to this last kind of model.

Any mathematical expression which describes a physical cause and effect relationship can be thought of as a model. In the context of this section, however, we restrict the use of the term 'simulation' to those mathematical relationships which simulate a rather complex natural system. The following discussion considers such natural bodies as rivers, estuaries, basins, near ocean water and groundwater. Each of these bodies has characteristic peculiarities which distinguish it from the others. But there is one common denominator which joins them: the combination of water and potential pollution of that water. These models will have to simulate both quantity and quality of water.

Rivers

General comments

The most common body of water is the stream or river. A system of streams and rivers forms the natural connection between the occurrence of rainfall over the watershed and the formation of larger bodies of water, such as lakes, estuaries and oceans. The systems of major rivers formed the original lines of commerce and transport in the history. As commerce and industry grew along the shores of these rivers, the water were used for municipal and industrial water supply and later for agricultural irrigation. Waterwheels converted the energy of these flowing streams into mechanical power for mills of all kinds.

River hydraulics

A complete simulation of a river system requires consideration of both quantity and quality of the water flowing in the system. The quality of the water usually is a function of the quantity of water flowing at that point. The converse, however, is not usually true. That is, the quantity of water or the hydraulics of the system is essentially unaffected by quality considerations. For this reason it is customary to establish the hydraulics of the river system and then to superimpose on this the quality considerations of interest.

The flow diagram for a River model is shown in Fig. 22.1. In this model the discharge in a given reach is established by the addition of the inflow from upstream reaches and the local run-off, QADD. The average velocity and average hydraulic depths are computed as functions of the discharge. Time of flow through the reach is related to the average velocity.

The reaeration coefficient, K_2, in the Streeter-Phelps equation is computed as a function of the average velocity, the average hydraulic depth and the slope of the bed. Several forms of this functional relationship are reported in the literature.

Hydrology

Stream flow is the result of surface run-off plus base flow contributed from groundwater; it varies with time and location because run-off is related directly to rainfall patterns. Thus, consideration of the hydrology of the river system requires the modelling of a process with uncertainty as one of its components (a stochastic process).

The simulation of an entire river system must recognise the special variation and complexity of many branches and junctions. Thus, the simulation technique must include some logical procedure for routing flows through such a system. Because flows downstream from a junction are determined by the flows coming from upstream branches, it usually is desirable to start with assigned flows in the uppermost reaches of the river system. In this way the flows can be accumulated throughout the river system as the outflow from one reach becomes the inflow to the reach downstream. In actual simulation, however, the programme starts at the uppermost or highest-numbered reach and proceeds in the direction of flow. At each junction point the reach numbering transfers the simulation to the uppermost reach of the tributary.

The integers actually assigned to the reaches proceed in steps of five so as to leave space for future revision of some reaches. In this way it is possible to sub-divide a reach into five parts without altering the remainder of the identification system. This scheme is illustrated in Fig. 22.2.

Many stream simulations are concerned with only a relatively small section of the total river system. In such a case it is necessary to decide whether or not local additions of flow have a significant effect upon the end result of the study. One common method for handling this problem is to assume that the yield in cubic feet per second per square mile is constant over a considerable area of the watershed. In this way the local contribution to an ungauged reach can be estimated from the corresponding flow in a gauged reach of the system.

Water quality models very often are concerned with low flow conditions as contrasted with flood flow analysis. In such cases the base flow in a stream system and its relationship to groundwater are important. River flow contributed from groundwater sources is very difficult to predict in a mathematical model. Field measurement of this contribution can be made only by noting the difference between main stem gauges. A further difficulty is the unknown mechanism which controls the rate at which this groundwater contribution will take place. The level of the groundwater relative to the stream level is significant. This relationship can be determined only by observations in a number of test wells along the river bank. Such information is rarely available.

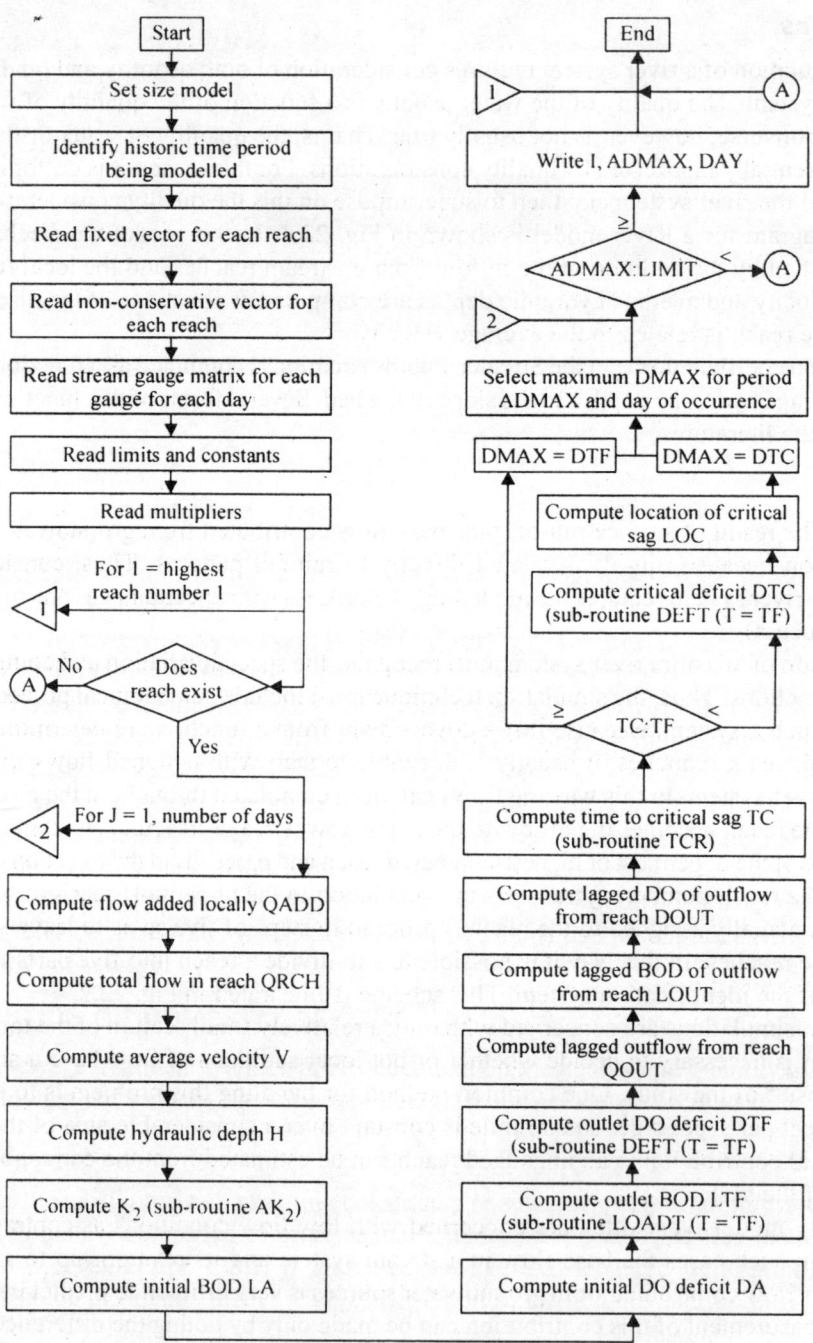

Fig. 22.1. Flow diagram for a digital computer model of the river.

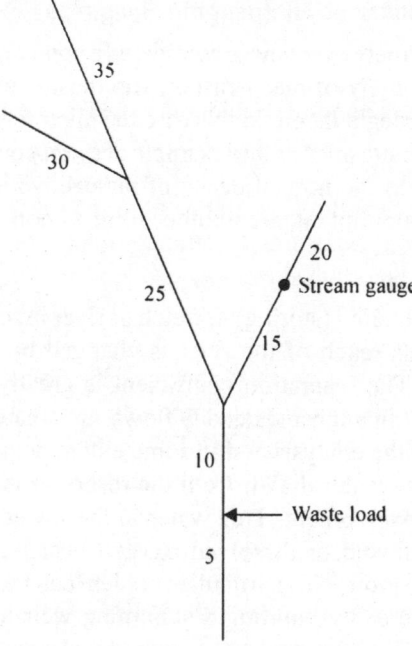

Fig. 22.2. Reach identification scheme for a digital computer model of a river system.

Conservative pollutants

Pollutants which are not degraded over time are termed 'conservative pollutants'. Many soluble materials are included in this classification. One of the most common of these is the salt that is accumulated in the watershed, either by natural erosion and solution of the crust of the earth or by the action of oil well operation, in which brines are drawn to the surface and disposed of in order to produce the oil. Acid mine drainage also causes this kind of pollutant. Because this kind of pollutant cannot be degraded by bacteriological action or by some natural, built-in decay mechanism, it can be controlled only by dilution or by some method of chemical removal.

The simulation of a conservative pollutant within the river system is quite straightforward. The distribution throughout the river system can be accounted for by a simple mass transport analysis. The concentration of this pollutant at any point in the system can be determined from the known quantity of the material and the flow at that point. The major difficulty in handling this kind of pollutant is the fact that it is seldom introduced into the river system as a point source. More commonly, this kind of pollutant occurs as a wash load from the general watershed area. The amount of pollutant contributed by wash load from a given portion of the watershed is greatly affected by the geology of the area. This could be inferred only from rather extensive field observations which were correlated with the flows. The relationship between addition of conservative pollutants and the local wash load indicates that the mechanism for this contribution must be connected to the addition of water from local sources. Such pollution loads can be incorporated into functions which predict the amount of local flow contribution. In many cases these functions are complicated by the temporal and spatial variation of local stream flow contribution throughout the watershed.

Non-conservative pollutants

Any pollutant which is reduced in quantity over time is considered a non-conservative pollutant. This reduction in pollution load can be due to a variety of mechanisms. The most common non-conservative pollutant, municipal sewage, is reduced by biodegradation. In this case the micro-organisms in the water consume the pollutant along with the oxygen in the stream. Another example of a non-conservative pollutant is a radioactive substance which decays according to the normal decay of radioactive isotopes. Temporarily suspended sediment loads, when considered a pollutant, are another form of non-conservative pollutant.

Impoundments

The effect on water quality of impounding (storing) a stretch of river involves a very complex relationship. The entire hydraulic regime of this reach of the river is changed by the flooding out of the original turbulence and velocity patterns. The reaeration coefficient is greatly affected by this change. Other complications, such as thermal stratification and density flows, are created. In most cases the construction of a reservoir serves to complicate the analysis and to some extent degrade water quality. In many cases the reservoir design is such that the water drawn from the reservoir is drawn from the bottom through low penstocks for hydroelectric development. This water in the lower level of the reservoir, known as the hypolimium, is often totally devoid of dissolved oxygen because of thermal stratification which inhibits vertical mixing within a reservoir. This particular problem can be alleviated somewhat by changing the location of the intake structure or by building a skimming weir inside the reservoir which would force the drawing of surface water into the intake structure. An alternative to this would be a multilevel intake structure which would permit the operator to select the level within the reservoir from which the water would be drawn. High level withdrawal of water from the reservoir is not an ideal solution because the poor quality water or water that is low in dissolved oxygen would still accumulate in the bottom of the reservoir. In cases of high flow from the reservoir, the zone of influence of the withdrawal would increase so that periodic slugs of poor quality water would be carried from the reservoir, which could result in fish kills because of very low dissolved oxygen levels.

One technique being investigated at the present time which holds considerable promise as a solution to this problem is that of reservoir mixing. Under this scheme some mechanical device would be used to mix the reservoir in a vertical direction so as to break up the thermal stratification and provide what amounts to a homogeneous reservoir. One method of achieving this mixing is to mechanically pump water from the lower reaches of the reservoir to the surface. Another alternative is to provide this lifting motion by injection air at the bottom of the reservoir. The artificially aerated water, being lighter than the surrounding water, rises to the surface. An additional benefit associated with the air injection method is that there is some contribution of oxygen to the reservoir water.

Pathogens

A water pollution parameter of great concern to public health authorities is the concentration of pathogens—bacteria which are harmful to man. Pathogens can produce a variety of diseases. Microbiology has provided us with a number of standard tests for certain indicator microbes. The presence and concentration of these indicators serve as a measure of the suitability of water for drinking and water sports. In estuarine water, concentrations of these indicators also are significant with regard to the suitability of certain areas for shellfish harvesting.

The presence of coliform bacteria in the water traditionally has been accepted as an indication of pathogenic content. In recent years, more selective indicators have been used. Currently, the fecal coliform content is recommended as the indicator of pathogenic content. In the future we look to specific identification of pathogens such as salmonella as one of the indicators. In each case the number of samples taken at a given location is so small relative to the variability of occurrence of these organisms that the data are reported in terms of the most probable number of coliform existing in the sample. The Public Health Service has established allowable limits on the most probable number of coliform and fecal coliform bacteria for specific water used. Pathogens are living organisms and it is therefore difficult to develop models to simulate their behaviour. Pathogens represent a highly non-conservative type of pollution.

Storet

An important problem in the handling of water resource and water quality data is the sheer bulk of field data which has been collected. In order that this information may be available when required for analysis, some systematic scheme of identification, storage and retrieval must be available. The availability of high speed digital computers with very large storage capacities makes such a scheme entirely feasible. One such scheme currently employed by the Federal Water Pollution Control Administration is called STORET, the code name for a storage and retrieval system. This scheme provides for the storage of a large number of individual parameters for a given observation station. The observation station is identified by a systematic coding arrangement which relates the station identification to the location of the station within the river systems. Under this scheme it is possible to call for the retrieval of data from all of the observation stations between two points on a river or alternatively to call for data from all stations above a given point on a stream system.

Basins

General comments

A river basin is the geographic area drained by a given river system. Water-oriented activities within the basin always should be planned and evaluated on a basin-wide approach. This approach is necessary because of the interrelationship between various parts of the basin. For example, water-polluting activities in an upstream region of the basin will have impact on potential water uses throughout many other parts of the basin. Conversely, water stored in an upstream reservoir for release as low flow augmentation can have beneficial effects throughout the entire length of the stream channel from that point down.

Mathematical models which simulate the river system of the basin can be applied to the problem of optimising water quality management in the basin.

Polluter versus user

If the entire watershed were owned and operated by a single entity, the economics of profit maximisation would require that different units of the system located in different parts of the watershed would be operated in such a manner that the total effort would be maximised or optimised. For example, if a single corporation operated an industrial plant on an upstream tributary and another plant some miles downstream, the decision as to the treatment of the wastes from the upstream plant versus treatment of water supply for the downstream plant would be an internal decision for that corporation. The savings gained by dumping untreated waste into the river upstream would show up as increased treatment costs for the downstream plant. In this situation the natural laws of economics would dictate the optimum arrangement of effluent treatment versus water supply treatment. Unfortunately, this is not usually the situation in reality. The savings on the part of the upstream polluter represent a treatment cost or use foregone by some other individual or corporation. This situation is called an external diseconomy. The normal rules of economics are no longer operative. In many instances in the past, the self-purifying properties of the flowing stream between the upstream and downstream plants have been able to rejuvenate the water and to a large extent alleviate the problem. But as industrial development proceeds and river systems become more fully utilised, we soon reach the point where the self-purification capacity of the stream is exceeded and there is a direct cost to downstream users because of the activity of upstream polluters.

Basin management concept

Basin management is the ultimate in planning and control of the interrelated structures and physical features of the basin so as to optimise the use and development of the water resource of that area for the good of the people living within it. Basin management implies comprehensive planning in both space and time. A comprehensive plan must look at the spatial interrelationship of the inputs and withdrawals within the region and the interaction of polluter and water user. In addition, it must ensure that present structures and facilities will be compatible with future needs and structures.

One very powerful tool of management is the mathematical simulation of the river system of the basin. If this model is capable of simulating both the quantity and quality distribution in place and in time, then this model permits management decisions to be based on objective analysis.

Competitive uses of water resources

The water resources of a basin may be used for a wide variety of purposes. Many of these purposes are competitive, that is, a given resource can be used for one purpose or another, but not for both.

One example of competition for a water resource is that between water quality control and navigation. On the one hand, navigation requires a slack water route through the river system which is maintained by a series of low head dams and locks in the main stem. On the other hand, the use of the water resource for the maintenance of water quality in the system requires storage of water in tributary reservoirs to supply low flow augmentation in the main stem and a main stem which is maintained as a free-flowing river such that a large portion of the waste load contributed to it could be assimilated by the natural purification capability of this flowing stream. Another competition for the use of water is that between reservoir-based recreation and low flow augmentation for water quality control. For recreational uses a stable water surface elevation is needed to maintain beach areas and docks. Low flow augmentation requires a varying water surface elevation in order to utilise the storage capacity of the reservoir.

Mathematical models which simulate these differing requirements could assist in the determination of the optimum relationship between possible uses of the water resource.

Hydroelectric power plants designed and operated for peaking power are a serious problem in water quality control. These operations consume no water but cause a time discontinuity which can seriously aggravate the pollution problem in a given stream. Under normal operating conditions these plants discharge all of the available flow for a given day in a period of approximately 3–4 hours which coincides with the peak power demand on the electric system. This means that the plant discharges no water during the remaining 20 hours of the day. The river system is particularly aggravated by weekend operation in which there is no peaking from Friday until Monday. During this prolonged time the flow from the plant would be zero under optimum operating conditions from the standpoint of hydroelectric peaking power development. However, it has been necessary to require that a minimum base flow be maintained during off-peak periods. Mathematical models which attempt to simulate the operation of these peaking power plants must, of necessity, be dynamic models. The time parameter becomes all important in this case. A mathematical model is needed which can simulate this momentum wave and the effect on water quality of this intermittent operation.

These models hold promise for our ultimate ability to manage the water resource within a basin in such a way that the optimal good for the various uses will result.

Near Ocean Water

Ocean water near a coastline present special problems in analysis and modelling. Myriad bays and inlets with their complex geometry and even more complex current patterns make it very difficult to represent these water by mathematical models. The development of pollution loads along the coastline, however, requires analysis of pollution effects in these water.

As with other bodies of water, it is necessary to define the motion of the fluid medium and then to superimpose upon this motion the effect of the pollutant. Water currents in these near ocean water are extremely complex and of considerable magnitude. The driving force producing these currents is a combination wind, tide and density. Although an estuary can be approximated by a one-dimensional model, it is completely impractical to apply such approximation to near ocean water. As the geometry of the body of water broadens toward the sea, the tide has a less pronounced effect upon the current and wave action and the consequent littoral drift become more significant.

Mathematical models of these systems must recognise the mechanisms of dilution and dispersion and also the probability concept. Fortunately, there is a considerable similarity between the action of water movement in the near ocean areas and the motion of the water in large freshwater bodies. At this point in time, there are very few mathematical models of near ocean water, but the growing concern for their pollution will require more rigorous analyses and simulations in the future.

Groundwater

General comments

The largest supply of fresh water available for man's use is under the ground. More recently, the vast reservoirs located beneath the ground have been used as disposal areas for polluted water. Increased pumping of groundwater near the coastline has resulted in saltwater intrusion into the freshwater aquifers. Both of these pollution problems have resulted in growing concern for the preservation of the quality of our groundwater.

Principles of groundwater movement

With the exception of some areas with sub-terranean solution channels, the motion of water through the ground is very slow and produces large quantities only because of the extremely large cross-sectional area experiencing this flow. This low velocity coupled with the confining pore spaces between soil particles produces a flow which can be analysed by relatively simple hydraulic equations.

Darcy's law

The law governing the movement of viscous liquids through porous materials under saturated conditions was discovered by Darcy in 1856. It has come to be called Darcy's law. Darcy's law states that:

$$Q = KA \frac{h_1 - h_2}{L} \qquad \qquad ... (22.2)$$

where,

Q = discharge rate
A = cross-sectional area

$$K = \text{hydraulic conductivity}$$
$$L = \text{length of flow}$$
$$h_1 = \text{inflow head}$$
$$h_2 = \text{outflow head}$$

Rearranging Eq. (22.2) we get:

$$K = \frac{Q}{A} \frac{L}{h_1 - h_2} \qquad \text{... (22.3)}$$

Thus, K has the units of velocity. A variety of units are used, but for simplicity only one set of units is used here. If Q is in gallons per day, L, h_1 and h_2 are in feet, and A is in square feet, K has the units of gallons per day per square foot (gpd/ft^2). This gives a better conceptual feeling for K than velocity, although it should be remembered that K has velocity units. It should also be pointed out that in a strict sense the value of K varies with temperature. However, the effect of temperature change on K is small in most cases and is generally ignored.

As an example, consider a stratum of sand 10 ft thick and with hydraulic conductivity of 10^3 gpd/ft^2. We assume that the loss of head is approximately 2 ft/mile and that we are interested in the flow or discharge rate (Q) through a 1 ft width of aquifer extending from the bottom to the top of the aquifer. Then, substituting the following values into Eq. (22.2)

$$\frac{h_1 - h_2}{L} = \frac{2}{5280}$$
$$A = 1\,(10) = 10 \text{ ft}^2$$
$$K = 10^3 \text{ gpd/ft}^2$$

the discharge rate, Q, is easily calculated:

$$Q = (10^3) \frac{(10)(2)}{5280} = 3.79 \text{ gpd}$$

An analogy

An electric analog model of a groundwater system consists of a network (usually a square lattice) of resistors selected in such a manner that the flow of electrical current and the distribution of electrical potential in the network is analogous to the flow of water and the distribution of hydraulic head in the groundwater system.

If non-steady-state problems are to be analysed on the model, capacitance must be added to simulate the storage characteristics of the groundwater system.

The elements that are analogous between the two systems are as follows:

Model	Groundwater system
Electric potential	Hydraulic head
Conductivity	Transmissibility
Capacitance	Storage
Time (milliseconds)	Time (decades)
Electric current	Flow rate

Information needed to construct a model

To select the correct values for the electrical components of the model the following information must be available:

1. Well logs—establish aquifer thickness and extent.
2. Pump tests—determination of transmissibility and storage characteristics.
3. Water level data—establishment of a known distribution of hydraulic head to test the mode.
4. Pumpage data—establish current demands on the system for testing purposes.

Applications of analog models

Once a model has been constructed and tested it can be used to examine a number of problems. One of the more obvious uses would be to evaluate the effect of pumpage in the future. With the rapidity of computerised solutions available, the effect of a number of different pumping schemes can be analysed rapidly. The effectiveness of artificial re-charge could also be studied.

The potential of a model of this nature for the study of the economics of groundwater mining and overdraft also are great. Since the model closely duplicates the actual response of the aquifer to changes, the effects of nearly any contemplated groundwater development can be analysed.

Having defined the flow of the fluid medium in the ground, the next problem is to develop methods of modelling the pollution of these flow fields. One form of pollution which has been analysed is that of saltwater intrusion along the coast. Charmonman developed mathematical models using the digital computer to predict the movement of the salt water-freshwater interface along the coastline under the action of withdrawal or injection of water into the groundwater aquifer. These models have been applied to the problem of saltwater intrusion in the Bangkok area of Thailand.

The tracing of conservative pollutants in the groundwater is possible with present technology. A more complex problem is that of non-conservative wastes, such as BOD (biochemical oxygen demand) or nuclear energy wastes.

Surface water have the ability to rejuvenate themselves and/or flush themselves to the sea in a short time period. Groundwater, on the other hand, because of their slow movement, remain polluted for very long periods of time. Because of this, responsible officials are becoming increasingly concerned about the potential of groundwater pollution.

OPTIMISING TECHNIQUES

Introduction of the high speed digital computer has presented engineers with the possibility of finding optimal solutions to many problems that heretofore could be approached only from the standpoint of an adequate solution. In almost all cases, optimisation must be approached by a repetitive (iterative) calculation technique. This kind of approach is ideally suited to the abilities of the digital computer.

Computerisation of a problem requires very close attention to the definition of the system under study. The boundaries of the system must be clearly defined and the important interrelationships between components of the system must be stated in mathematical terms.

Optimisation implies that one is working toward a given objective. This objective is stated mathematically in some form of objective function. In the case of large water resource projects, it often is necessary to look into the areas of welfare and resource economics and politics in order to properly assess the objective of the project.

System Definition

A water resource system is usually bounded by the natural watershed of the river basin under study. Exceptions to this rule are such circumstances as the transwatershed diversion of water either by man-made activities or by groundwater flows and the economic impact from neighbouring watersheds. The river system in the basin is the basis for most of the important interrelationships that must be modelled. Upstream activities, either beneficial or detrimental, have effects on all portions of the system downstream. This is the basic relationship that must be investigated within a river system. The interrelationship is best expressed in terms of a simulation program that models the behaviour of the entire river system.

Objective Functions

One of the more difficult tasks in an optimisation scheme is that of defining the objective of the system. In the realm of federally sponsored water resource development schemes, it is possible to have a wide variety of ultimate objectives. One legitimate objective of such a program would be the redistribution of the national income in a particular direction. But within the framework of this general policy decision, it is still desirable to achieve this distribution with a high level of economic efficiency. Having decided that development should be encouraged in a particular locality, the next question is how to accomplish this development most efficiently. Once again, it is necessary to state the objective function in mathematical form so that it can be handled by the digital computer. This becomes very difficult in many cases because the ultimate decision has been rather subjective.

Economic efficiency for resource development is usually stated in terms of a benefit-cost analysis. This form of analysis attempts to monetise all the variables in the system so as to measure all of the benefits as well as the costs of the system. A benefit-cost analysis can be used in two different ways. First of all, the analysis can be used to determine the optimal scale of development of a given project. In this case it is customary to increase the size of the development until the net benefits are maximised.

A second use of the benefit-cost analysis is to rank various projects in order of their economic efficiency. For this purpose it is customary to determine a benefit-cost ratio. The larger this ratio, the more desirable that particular project becomes. When the choice is between several projects which will contribute to the same basic policy of resource development, it is customary to choose the project which has the highest benefit-cost ratio. This decision, however, if often shaded and influenced by the political expediency that is attendant in any public works program.

Operations Research

Operations research is a mathematical method which can be used for assigning scarce resources in such a way as to optimise a given objective function. As such, it is ideally suited to the problem of resource economics. This mathematical process .may take on a wide variety of forms, depending upon the circumstances involved.

One example of the application of operations research is the question of the level of treatment that should be applied to waste sources before they are injected into a flowing stream. The natural flowing stream has a large capacity for self-purification after pollution. The question arises as to the extent to which we can rely upon this self-purifying capacity, because of the effect of even a slight degradation of quality upon downstream uses.

Conservationists would prefer complete treatment of waste at the source with essentially zero degradation of the flowing stream. This approach disregards any self-purification capacity that the flowing stream might possess. The opposite viewpoint is to permit the maximum allowable degradation of the stream so as to fully utilise its self-purification capacity. Unfortunately, there is no general agreement on the level of degradation which is permissible in terms of uses in the downstream area.

Pollution parameters such as dissolved oxygen and fecal coliform concentration are used as indicators of pollution level. The states have set standards for these parameters which must be maintained in certain classified streams.

Given a minimum value for dissolved oxygen concentration, it is possible, by means of operations research, to determine the amount of treatment that should be applied at various plants along a river. Some consideration must be given to the future needs of the basin so that existing plants are not permitted to utilise all of the self-purification capacity of the river and leave no additional capacity for future growth. Computer operations require that such decisions be made in order to provide a firm mathematical relationship.

Strict economic efficiency might dictate a varying level of treatment for the various plants and outfalls along a water course. Such a scheme could be developed. However, present institutional arrangements are not adequate to this task. The new techniques available by use of computers and simulations, if they are to be used effectively, will require new forms of jurisdiction and responsibility and the development of new institutional arrangements.

Thus, computerised solutions of mathematical models have been demonstrated for a number of natural bodies of water. Indeed, the growth of this technology is quite high. For this reason, a number of current applications has been discussed and indicated some of the tremendous potential for future development of these techniques.

Glossary

Abiotic factors	:	Non-living; moisture, soil, nutrients, fire, wind, temperature, climate.
Absolute filtration rating (largest particle passed)	:	The diameter of the largest hard spherical particle that will pass through a filter under specified test conditions. This is an indication of the largest opening in the filter cloth.
Absorption	:	The taking in or soaking up of one substance into the body of another by molecular or chemical action (as tree roots absorb dissolved nutrients in the soil).
Absorption field	:	A system of properly sized and constructed narrow trenches partially filled with a bed of washed gravel or crushed stone into which perforated or open joint pipe is placed. The discharge from the septic tank is distributed through these pipes into trenches and surrounding soil. While seepage pits normally require less land area to install, they should be used only where absorption fields are not suitable and well-water supplies are not endangered.
Acetogenic bacterium	:	Prokaryotic organism that uses carbonate as a terminal electron acceptor and produces acetic acid as a waste product.
Acetylene-block assay	:	Estimates denitrification by determining release of nitrous oxide (N_2O) from acetylene-treated soil.
Acetylene-reduction assay	:	Estimates nitrogenase activity by measuring the rate of acetylene reduced to ethylene.
N-Acetylglucosamine and N-acetylmuramic acid	:	Sugar derivatives in the peptidoglycan layer of bacterial cell walls.
Acidophile	:	Organism that grows best under acid conditions (down to a pH of 1).
Acid soil	:	Soil with a pH value < 6.6.
Actinomycete	:	Non-taxonomic term applied to a group of high G + C base composition, Gram-positive bacteria that have a superficial resemblance to fungi. Includes many but not all organisms belonging to the order Actinomycetales.
Activated sludge	:	Sludge particles produced in raw or settled waste-water (primary effluent) by the growth of organisms (including zoogleal bacteria) in aeration tanks in the presence of dissolved oxygen. The term 'activated' comes from the fact that the particles are teeming with fungi, bacteria and protozoa. Activated sludge is different from primary sludge in that the sludge particles contain many living organisms which can feed on the incoming waste-water.

Activation energy	:	Amount of energy required to bring all molecules in one mole of a substance to their reactive state at a given temperature.
Active site	:	Region of an enzyme where substrates bind.
Adenosine triphosphate (ATP)	:	Common energy-donating molecule in biochemical reactions. Also an important compound in transfer of phosphate groups.
Adsorption	:	The gathering of a gas, liquid or dissolved substance on the surface or interface zone of another substance.
ADP	:	Adenosine diphosphate.
Aeration	:	The process of adding air to water. In waste-water treatment, air is added to freshen waste-water and to keep solids in suspension.
Aeration tank	:	The tank where raw or settled waste-water is mixed with return sludge and aerated; This is the same as an aeration bay, aerator or reactor.
Aerobe	:	An organism that requires free oxygen for growth.
Aerobic	:	(i) Having molecular oxygen as a part of the environment; (ii) growing only in the presence of molecular oxygen, as in aerobic organisms; and (iii) occurring only in the presence of molecular oxygen, as in certain chemical or biochemical processes such as aerobic respiration.
Aerotolerant anaerobes	:	Microbes that grow under both aerobic and anaerobic conditions, but do not shift from one mode of metabolism to another as conditions change. They obtain energy exclusively by fermentation.
Air flow/air permeability	:	Measure of the amount of air that flows through a filter—a variable of the degree of contamination, differential pressure, total porosity and filter area. Expressed in either cubic feet/minute/square foot or liters/minute/square centimeter at a given pressure.
Agar	:	Complex polysaccharide derived from certain marine algae that is a gelling agent for solid or semi-solid microbiological media. Agar consists of about 70 per cent agarose and 30 per cent agaropectin. Agar can be melted at temperature above 100° C; gelling temperature is 40–50°C.
Agarose	:	Non-sulphated linear polymer consisting of alternating residues of D-galactose and 3,6-anhydro-L-galactose. Agarose is extracted from seaweed and agarose gels are oftan used as the resolving medium in electrophoresis.
Akinete	:	Thick-walled resting cell of cyanobacteria and algae.
Alkaline substance	:	Chemical compounds in which the basic hydroxide (OH–) ion is united with a metallic ion, such as sodium hydroxide (NaOH) or potassium hydroxide (KOH). These substances impart alkalinity to water and are employed for neutralisation of acids. Lime is the most commonly used alkaline material in waste-water treatment.
Alga (plural, algae)	:	Phototrophic eukaryotic micro-organism. Algae could be unicellular or multicellular. Blue-green algae are not true algae; they belong to a group of bacteria called cyanobacteria.
Aliphatic	:	Organic compound in which the main carbon structure is a straight chain.
Alkaline soil	:	Soil having a pH value >7.3.
Alkalophile	:	Organism that grows best under alkaline conditions (up to a pH of 10.5).
Alkane	:	Straight chain or branched organic structure that lacks double bonds.

Alkene	:	Straight chain or branched organic structure that contains at least one double bond.
Allochthonous flora	:	Organisms that are not indigenous to the soil but that enter soil by precipitation, diseased tissues, manure and sewage. They may persist for some time but do not contribute in a significant way to ecologically significant transformations or interactions.
Allosteric site	:	Site on the enzyme other than the active site to which a nonsubstate compound binds. This may result in a conformational change at the active site so that the normal substrate cannot bind to it.
Alum	:	A stringent crystalline double sulphate of an alkali. $K_2SO_4Al_2 (SO_4)_3 24H_2O$. Used in the processing of pickles and as a flocking agent. Excess aluminium in the environment can be hazardous.
Amensalism (antagonism)	:	Production of a substance by one organism that is inhibitory to one or more other organisms. The terms antibiosis and allelopathy also describe cases of chemical inhibition.
Ambient temperature	:	Temperature of the surroundings.
Amino group	:	An $—NH_2$ group attached to a carbon skeleton as in the amines and amino acids.
Ammonia oxidation	:	Test drawn during manufacturing process to evaluate the ammonia oxidation rate for the nitrifiers.
Ammonification	:	Liberation of ammonium (ammonia) from organic nitrogenous compounds by the action of micro-organisms.
Amoeba (plural, amoebae)	:	Protozoa that can alter their cell shape, usually by the extrusion of one or more pseudopodia.
Anabolism	:	Metabolic processes involved in the synthesis of cell constituents from simpler molecules. An anabolic process usually requires energy.
Anaerobe	:	An organism that lives and reproduces in the absence of dissolved oxygen, instead deriving oxygen from the breakdown of complex substances.
Anaerobic	:	(i) Absence of molecular oxygen; (ii) growing in the absence of molecular oxygen, such as anaerobic bacteria; and (iii) occurring in the absence of molecular oxygen, as a biochemical process.
Anaerobic respiration	:	Metabolic process whereby electrons are transferred from an organic or in some cases, inorganic compounds to an inorganic acceptor molecule other than oxygen. The most common acceptors are nitrate, sulphate and carbonate.
Anamorph	:	Asexual stage of fungal reproduction in which cells are formed by the process of mitosis.
Anhydrous	:	Very dry. No water or dampness is present.
Anion	:	A negatively charged ion in an electrolyte solution, attracted to the anode under the influence of a difference in electrical potential. Chloride is an anion.
Anion exchange capacity	:	Sum total of exchangeable anions that a soil can adsorb. Expressed as centimoles of negative charge per kilogram of soil.
Anoxic	:	Literally 'without oxygen'. An adjective describing a microbial habitat devoid of oxygen.

Anoxygenic photosynthesis	:	Type of photosynthesis in green and purple bacteria in which oxygen is not produced.
Antagonist	:	Biological agent that reduces the number or disease-producing activities of a pathogen.
Antheridium	:	Male gametangium found in the phylum Oomycota (Kingdom Stramenopila) and phylum Ascomycota (Kingdom Fungi).
Anthropogenic	:	Derived from human activities.
Antibiosis	:	Inhibition or lysis of an organism mediated by metabolic products of the antagonist; these products include lytic agents, enzymes, volatile compounds and other toxic substances.
Antibiotic	:	Organic substance produced by one species of organism that in low concentrations will kill or inhibit growth of certain other organisms.
Antibody	:	Protein that is produced by animals in response to the presence of an antigen and that can combine specifically with that antigen.
Antigen	:	Substance that can incite the production of a specific antibody and that can combine with that antibody.
Antiseptic	:	Agent that kills or inhibits microbial growth but is not harmful to human tissue.
Antistatic	:	Material that minimises static charge generation, provides 'controlled' static charge dissipation or both.
API separator	:	A facility developed by the Committee on Disposal or Refinery Wastes of the American Petroleum Institute for separation of oil from waste-water in a gravity differential and equipped with means for recovering the separated oil and removing sludge.
Aromatic	:	Organic compounds which contain a benzene ring or a ring with similar chemical characteristics.
Arthropod	:	Invertebrate with jointed body and limbs (includes insects, arachnids and crustaceans).
Ascoma (plural, ascomata)	:	Fungal fruiting body that contains ascospores; also termed an ascocarp.
Ascospore	:	Spores resulting from karyogamy and meiosis that are formed within an ascus. Sexual spore of the Ascomycota.
Ascus (plural, asci)	:	Saclike cell of the sexual state formed by fungi in the phylum Ascomycota containing ascospores.
Aseptic	:	Free from living germs of disease, fermentation or putrefaction.
Aseptic technique	:	Manipulating sterile instruments or culture media in such a way as to maintain sterility.
Assimilate	:	To take in, similar to eating food.
Assimilatory nitrate reduction	:	Conversion of nitrate to reduced forms of nitrogen, generally ammonium, for the synthesis of amino acids and proteins.
Associative dinitrogen fixation	:	Close interaction between a free-living diazotrophic organism and a higher plant that results in an enhanced rate of dinitrogen fixation.

Associative symbiosis	:	Close but relatively casual interaction between two dissimilar organisms or biological systems. The association may be mutually beneficial but is not required for accomplishment of a particular function.
Attached growth processes	:	Waste-water treatment processes in which the micro-organisms and bacteria treating the wastes are attached to the media in the reactor. The wastes being treated flow over the media. Trickling filters, bio-towers and RBCs are attached growth reactors. These reactors can be used for removal of BOD, nitrification and denitrification.
Attenuation	:	Reduction of the signal power of field strength as a function of distance through a material. Also refers to shielding effectiveness.
ATP	:	Adenosine triphosphate. Chemical energy generated by substrate oxidations is conserved by formation of high-energy compounds such as adenosine diphosphate (ADP) and adenosine triphosphate (ATP) or compounds containing the thioester bond.
Autoclave	:	Vessel for heating materials under high steam pressure. Used for sterilisation and other applications.
Autolysis	:	Spontaneous lysis.
Autoradiography	:	Detecting radioactivity in a sample, such as a cell or gel, by placing it in contact with a photographic film.
Autotroph	:	Organism which uses carbon dioxide as the sole carbon source.
Autotrophic nitrification	:	Oxidation of ammonium to nitrate through the combined action of two chemo-autotrophic organisms, one forming nitrite from ammonium and the other oxidising nitrite to nitrate.
Autotrophy	:	A unique form of metabolism found only in bacteria. Inorganic compounds (e.g., NH_3, NO_2^-, S_2 and Fe_2^+) are oxidised directly (without using sunlight), to yield energy. This metabolic mode also requires energy for CO_2 reduction, like photo-synthesis, but no lipid-mediated processes are involved. This metabolic mode has also been called chemotrophy, chemoautotrophy or chemolithotrophy.
AWT	:	Advanced Waste Treatment—any process of water renovation that upgrades treated waste-water to meet re-use requirements.
Axenic	:	Literally 'without strangers'. A system in which all biological population are defined, such as a pure culture.
Bacillus	:	Bacterium with an elongated, rod shape.
Bacteria	:	Living organism, microscopic in size, which usually consist of a single cell. Most bacteria use organic matter for their food and produce waste products as a result of their life processes.
Bacteriochlorophyll	:	Light-absorbing pigment found in green sulphur and purple sulphur bacteria.
Bacteriocin	:	Agent produced by certain bacteria that inhibits or kills closely related isolates and species.
Bacteriophage	:	Virus that infects bacteria, often with destruction or lysis of the host cell.
Bacterial photosynthesis	:	A light-dependent, anaerobic mode of metabolism. Carbon dioxide is reduced to glucose, which is used for both biosynthesis and energy production. Depending on the hydrogen source used to reduce CO_2, both photolithotrophic and photoorganotrophic reactions exist in bacteria.

Bacteroid	:	Altered form of cells of certain bacteria. Refers particularly to the swollen, irregular vacuolated cells of rhizobia in nodules of legumes.
Base	:	A substance which dissociates (separates) in aqueous solution to yield hydroxyl ions or one containing hydroxyl ions (OH–) which reacts with an acid to form a salt or which may react with metal to form a precipitate.
Base composition	:	Proportion of the total bases consisting of guanine plus cytosine or thymine plus adenine base pairs. Usually expressed as a guanine + cytosine (G + C) value, e.g., 60 per cent G + C.
Basidioma (plural, basidiomata)	:	Fruiting body that produces basidia; also termed a basidiocarp.
Basidiospore	:	Spore resulting from karyogamy and meiosis that are formed on a basidium that usually is formed on a basidium. Sexual spore of the Basidiomycota.
Basidium (plural, basidia)	:	Clublike cell of the sexual state formed by fungi in the phylnm Basidiomycota.
Batch process	:	A treatment process in which a tank or reactor is filled, the waste-water (or solution) is treated or a chemical solution is prepared and the tank is emptied. The tank may then be filled and the process repeated. Batch processes are also used to cleanse, stabilise or condition chemical solutions for use in industrial manufacturing and treatment processes.
Bench scale analysis	:	Also known as: 'bench test'. A method of studying different ways of treating waste-water and solids on a small scale in a laboratory. Alken-Murray offers several such test kits including: Alken Clear-Flo Bench Test 1 and Alken PCB Bench Test.
Benzene	:	An aromatic hydrocarbon which is a colourless, volatile, flammable liquid. Benzene is obtained chiefly from coal tar and is used as a solvent for resins and fats in dye manufacture.
Binary fission	:	During binary fission, a single cell divides transversely to form two new cells called daughter cells. Both daughter cells contain an exact copy of the genetic-information contained in the parent cell.
Biocatalysis	:	Chemical reactions mediated by biological systems (microbial communities, whole organisms or cells, cell-free extracts, or purified enzymes aka catalytic proteins).
Binary fission	:	Division of one cell into two cells by the formation of a septum. It is the most common form of cell division in bacteria.
Binomial nomenclature	:	System of having two names, genus and specific epithet, for each organism.
Bioaccumulation	:	Accumulation of a chemical substance in living tissue.
Biochemical oxygen demand (BOD)	:	Amount of dissolved oxygen consumed in five days by biological processes breaking down organic matter.
Biodegradable	:	Substance capable of being decomposed by biological processes.
Biofilm	:	A slime layer which naturally develops when bacteria attach to an inert support that is made of a material such as stone, metal or wood. There are also non-filamentous bacteria that will produce an extracellular polysaccharide that acts as a natural glue to immobilise the cells. In nature, non-filament-forming micro-organisms will stick to the biofilm surface, locating within an area of the biofilm

that provides an optimal growth environment (i.e., pH, dissolved oxygen, nutrients). Since nutrients tend to concentrate on solid surfaces, a micro-organism saves energy through cell adhesion to a solid surface rather than by growing unattached and obtaining nutrients randomly from the medium. *Pseudomonas* and *Nitrosomonas* strains are especially well known for their ability to form a strong biofilm.

Bioflocculation	:	The clumping together of fine, dispersed organic particles by the action of certain bacteria and algae.
Biogeochemistry	:	Study of microbially mediated chemical .transformations of geochemical interest, such as nitrogen or sulphur cycling.
Biomagnification	:	Increase in the concentration of a chemical substance as it is progresses to higher trophic levels of a food chain.
Biomass	:	A mass or clump of living organisms feeding on the wastes in waste-water, dead organisms and other debris.
Bioremediation	:	Use of micro-organisms to remove or detoxify toxic or unwanted chemicals from an environment.
Biosolid	:	The residue of waste-water treatment. Formerly called sewage sludge.
Biosphere	:	Zone incorporating all forms of life on earth. The biosphere extends from deep in sediment below the ocean to several thousand meters elevation in high mountains.
Biotrophic	:	Nutritional relationship between two organisms in which one or both must associate with the other to obtain nutrients and grow.
Biostimulation	:	Any process that increases the rates of biological degradation, usually by the addition of nutrient, oxygen or other electron donors and acceptors so as to increase the number of indigenous micro-organisms available for degradation of contaminants.
Biosynthesis	:	Production of needed cellular constituents from other, usually simpler, molecules.
Biotechnology	:	Use of living organisms to carry out defined physio-chemical processes having industrial or other practical application.
Biotic potential	:	All the factors that contribute to a species increasing its number. Reproduction, migration, adaptation etc.
BOD	:	Biochemical oxygen demand—the rate at which micro-organisms use the oxygen in water or waste-water while stabilising decomposable organic matter under aerobic conditions. In decomposition, organic matter serves as food for the bacteria and energy results from this oxidation.
BOD test	:	A procedure that measures the rate of oxygen use under controlled conditions of time and temperature. Standard test conditions include dark incubation at 20°C for a specified time (usually 5 days).
Bolting grade (wire cloth)	:	Weaves that are uniformaly woven of stainless steel to provide high strength and the largest possible pore openings.
Bio-tower	:	An attached culture system. A tower filled with a media similar to rachet or plastic rings in which air and water are forced up a counterflow movement in the tower.
Blinding	:	The clogging of the filtering medium of a microscreen or a vacuum filter when the holes or spaces in the media become sealed off due to a build-up of grease or the material being filtered.

Brown rot fungus	:	Fungus that attacks cellulose and hemicellulose in wood, leaving dark-coloured lignin and phenolic materials behind.
Bubble point test	:	A test to determine the maximum pore size opening of a filter.
Buffer	:	A solution or liquid whose chemical make-up neutralises acids or bases without a great change in pH.
Bulk density, soil	:	Mass of dry soil per unit bulk volume (combined volume of soil solids and pore space).
Bulked yarn	:	A yarn that has been geometrically changed to give it the appearance of having greater volume than a conventional yarn of the same linear density.
Bulking sludge	:	Clouds of billowing sludge that occur throughout secondary clarifiers and sludge thickeners when sludge becomes too light and will not settle properly. In the activated sludge process, bulking is usually caused by filamentous bacteria.
Cake	:	The solids discharged from a dewatering apparatus.
Calendering	:	A process by which fabric or wire is passed through a pair of heavy rolls to reduce thickness, to flatten the intersections of the threads/wires and to control air permeability. Rolls are heated when calendering synthetic materials.
Carbonised threads	:	Nylon or polyester therads that have been treated to include varrying degrees of carbon.
Cation exchange capacity	:	The ability of a soil or other solid to exchange cations (positive ions such as calcium) with a liquid.
Cess pools	:	This system is similar to a septic tank in performance. Sewage water usually seeps through the open bottom and portholes in the sides of the walls. These can also clog up with overuse and the introduction of detergents and other material which slow up the bacterial action.
CFU	:	Viable micro-organisms (bacteria, yeasts and mould) capable of growth under the prescribed conditions (medium, atmosphere, time and temperature) develop into visible colonies (colony forming units) which are counted. The term colony forming unit (CFU) is used because a colony may result from a single micro-organism or from a clump/cluster of micro-organisms.
Chemoautotroph	:	An organism that obtains its energy from the oxidation of chemical compounds and uses only organic compounds as a source of carbon. Example: nitrifiers.
Chemotroph	:	An organism that obtains its energy from the oxidation of chemical compounds.
Chemical precipitation	:	Precipitation induced by addition of chemicals; the process of softening water by the addition of lime and soda ash as the precipitants.
Chloramines	:	Compounds formed by the reaction of hypochlorous acid (or aqueous chlorine) with ammonia.
Chlorination	:	The application of chlorine to water or waste-water, generally for the purpose of disinfection, but frequently for accomplishing other biological or chemical results.
Clarification	:	A process in which suspended material is removed from a waste-water. This may be accomplished by sedimentation, with or without chemicals or filtration.
Clarifier	:	Settling tank, sedimentation basin. A tank or basin in which waste-water is held for a period of time, during which the heavier solids settle to the bottom and the lighter material will float to the water surface.

Coagulants	:	Chemicals which cause very fine particles to clump (floc) together into larger particles. This makes it easier to separate the solids from the water by settling, skimming, draining or filtering.
Coliform bacteria	:	Non-pathogenic microbes found in fecal matter that indicate the presence of water pollution; are thereby a guide to the suitability for potable use.
Colloids	:	Very small, finely divided solids (particles that do not dissolve) that remain dispersed in a liquid for a long time due to their small size and electrical charge.
Combined available chlorine	:	The concentration of chlorine which is combined with ammonia (NH_3) as chloramine or as other chloro derivatives, yet is still available to oxidise organic matter.
Combined sewer	:	A sewer designed to carry both sanitary waste-water and storm or surface-water run-off.
Ciliates	:	A class of protozoans distinguished by short hairs on all or part of their bodies.
COD	:	Chemical oxygen demand—the amount of oxygen in mg/l required to oxidise both organic and oxidisable inorganic compounds.
Commensalism	:	When two organisms co-exist, one organism benefits, the other is not affected.
Comminution	:	Shredding—A mechanical treatment process which cuts large pieces of waste into smaller pieces so that they won't plug pipes or damage equipment.
Contact stabilisation	:	Contact stabilisation is a modification of the conventional activated sludge process. In contact stabilisation, two aeration tanks are used. One tank is for separate re-aeration of the return sludge for at least four hours before it is permitted to flow into the other aeration tank to be mixed with the primary effluent requiring treatment.
Conventional treatment	:	The preliminary treatment, sedimentation, floatation, trickling filter, rotating biological contactor, activated sludge and chlorination of waste-water.
Conversion	:	Changing from one substance to another. As food matter is changed to cell growth or to carbon dioxide.
CRT	:	Cell residence time—the amount of time in days that an average 'bug' remains in the process. Also termed 'sludge age'.
DAF	:	Dissolved air floatation—one of many designs for waste treatment.
Decitex (dtex)	:	The mass in grams of 10000 meters of fiber or yarn. A direct yarn numbering system used to define size of fiber or yarn. The higher the number, the coarser (larger) the yarn.
Degradation	:	A growth phase in which the availability of food begins to limit cell growth.
Deionised water	:	Water that goes through an ion exchange process in which all positive and negative ions are removed.
Denier	:	The mass in grams of 9000 meters of fiber or yarn. A direct numbering system used to define size of fiber or yarn. The higher the number, the coarser (larger) the yarn.
Denitrification	:	An anaerobic biological reduction of nitrate nitrogen to nitrogen gas, the removal of total nitrogen from a system and or an anaerobic process that occurs when nitrite ions are reduced to nitrogen gas and bubbles are formed as a result of this process. The bubbles attach to the biological floc in the activated sludge process

and float the floc to the surface of the secondary clarifiers. This condition is often the cause of rising sludge observed in secondary clarifiers or gravity thickeners.

Depth filter	:	A filter medium consisting of randomly distributed particles or fibers resulting in openings with a non-uniform and tortuous path.
Detritus	:	Dead plant and animal matter, usually consumed by bacteria, but some remains.
Dew point	:	The temperature to which air with a given quantity of water vapour must be cooled to cause condensation of the vapour in the air.
D/I unit	:	Deionising unit, frequently used to maintain water quality in aquariums. It does not waste-water like the R/O unit, is designed to be hooked up to either a faucet or household piping system, the anion and cation resins can be regenerated (with another expensive unit) indefinitely and these systems allow a larger water flow (up to 2000 gallons a day), than an R/O system, but cost dramatically more too.
Diatomaceous earth	:	A fine, siliceous (made of silica) 'earth' composed mainly of the skeletal remains of diatoms (single cell microscopic algae with rigid internal structure consisting mainly of silica). Tests prove that DE leaches unacceptable amounts of silicate into the water for fish health. If used as a filter substance, a silicone removing resin should be employed afterwards.
Differential pressure	:	The difference in pressure between two points of a system, such as between two sides of an orifice.
Diffused air aeration	:	A diffused air activated sludge plant takes air, compresses it and then discharges the air below the water surface of the aerator through some type of air diffusion device.
Digester	:	A tank in which sludge is placed to allow decomposition by micro-organisms. Digestion may occur under anaerobic (most common) or aerobic conditions.
Disinfection	:	The process designed to kill most micro-organisms in waste-water, including essentially all pathogenic (disease-causing) bacteria. There are several ways to disinfect, with chlorine being the most frequently used in water and waste-water treatment plants.
Distribution box	:	Serves to distribute the flow from the septic tank evenly to the absorption field or seepage pits. It is important that each trench or pit receive an equal amount of flow. This prevents overloading of one part of the system.
Dissolved solids	:	Chemical substances either organic or inorganic that are dissolved in a waste stream and constitute the residue when a sample is evaporated to dryness.
Distributor	:	The rotating mechanism that distributes the waste-water evenly over the surface of a trickling filter or other process unit.
DO	:	Dissolved oxygen—a measure of the oxygen dissolved in water expressed in milligrams per liter.
DOUR	:	Dissolved oxygen uptake ratio.
Downstream side	:	The side of a product stream that has already passed through a given filter system; portion located after the filtration unit.
Dual chamber test method	:	Measures near field shielding effectiveness by indicating the signal attenuation caused by passage through test material.

Dyeing	:	The process of adding colour to textiles in either fiber, yarn or fabric form.
Ecology	:	The study of all aspects of how organisms interact with each other and/or their environment.
Ecosystem	:	Groupings of various organisms interacting with each other and their environment.
E-coli	:	*Escherichia coli*—one of the non-pathogenic coliform organisms used to indicate the presence of pathogenic bacteria in water.
Effective area	:	The total area of the porous medium exposed to flow in a filter element.
Efficiency	:	The ability, expressed as a per cent, of a filter to remove specified artificial contaminant at a given contaminant concentration under specified test conditions.
Effluent	:	Waste-water or other liquid—raw (untreated), partially or completely treated-flowing from a reservoir, basin, treatment process or treatment plant.
E-field (electric field)	:	The dominant component of a high impedance electromagnetic field produced by a near field source such as a short diapole or the electric component of a far field plane wave. Expressed in V/m.
EGL	:	Energy grade line—a line that represents the elevation of energy head in feet of water flowing in a pipe, conduit or channel.
Electrolytic process	:	A process that causes the decomposition of a chemical compound by the use of electricity.
Electromagnetic interference (EMI)	:	Electromagnetic energy that causes interference in the operation of electronic equipment. Can be conducted, coupled or radiated. Can be natural or man-made.
Electromagnetic capability (EMC)	:	The capability of electronic equipment of systems to be operated in the intended operational electromagnetic environment at designed levels of efficiency.
Emulsion	:	A liquid mixture of two or more liquid substances not normally dissolved in one another, one liquid held in suspension in the other.
Endogenous respiration	:	A reduced level of respiration (breathing) in which organisms break down compounds within their own cells to produce the oxygen they need.
Endotoxin	:	A toxin produced by bacteria. The toxin is present in the environment only after death of the bacteria.
Enteric	:	Of intestinal origin, especially applied to wastes or bacteria.
Enzyme	:	Organic substances (proteins) produced by living organisms and act as catalysts to speed up chemical changes.
Environmental resistance	:	All biotic and abiotic factors combining to limit explosion.
Equalising basin	:	A holding basin in which variations in flow and composition of liquid are averaged. Such basins are used to provide a flow of reasonably uniform volume and composition to a treatment unit. Also called a balancing reservoir.
Estuaries	:	Bodies of water which are located at the lower end of a river and are subject to tidal fluctuations.
Eurythermal	:	Bodies of water which are located at the lower end of a river and are subject to tidal fluctuations.
Extractables	:	Substances that can be leached from a filter during the filtration process or under other specified conditions.

Facultative anaerobe	:	A bacterium capable of growing under aerobic conditions or anaerobic conditions in the presence of an inorganic ion i.e., SO_4, NO_3.
Facultative pond	:	The most common type of pond in current use. The upper portion (supernatant) is aerobic, while the bottom layer is anaerobic. Algae supply most of the oxygen to the supernatant.
Faraday cage	:	A spherical cage made of conductive material. Static fields and discharges do not pass through it. Electromagnetic energy passing through the skin or shield is attenuated to varying degrees.
Feed	:	The material entering a filter processing unit for treatment.
Fermentation	:	A type of heterotrophic metabolism in which an organic compound rather than oxygen is the terminal electron (or hydrogen) acceptor. Less energy is generated from this incomplete form of glucose oxidation than is generated by respiration, but the process supports anaerobic growth.
Filamentous organisms	:	Organisms that grow in a thread or filamentous form. Common types are *Thiothrix*, *Actinomycetes* and *Cyanobacteria* (aka blue-green algae). This is a common cause of sludge bulking in the activated sludge process. Variously known as 'pond scum', 'blue-green algae' or 'moss', when it appears in a pond/lake and confused with algae because it looks a lot like algae. *Cyanobacteria* forms a symbiotic relationship with some varieties of algae, making the combination very difficult to combat in lakes and ponds. Filamentous organisms and *Actinomycetes* will naturally stick to solid surfaces. Common types of *Cyanobacteria* are: *Oscillatoria*, *Anabaena* and *Synechococcus*. Other filament formers include: *Spirogyra*, *Cladophora*, *Rhizoclonium*, *Mougeotia*, *Zygnema* and *Hydrodictyon*. *Nocardia* is another filament former, which causes foaming and interferes with flocculation in a waste treatment plant.
Filter aid	:	A chemical (usually a polymer) added to water to help remove fine colloidal suspended solids.
Filter life	:	Measure of the duration of a filter's useful service. This is based on the amount of standard contaminant required to cause differential pressure to increase at an unacceptable level—typically 2–4 times the initial differential pressure, a 50–80 per cent drop in initial flow or a downstream measure of unacceptable particulate.
Filter media	:	A porous material for separating suspended particulate matter from fluid.
Filter medium	:	The permeable portion of a filtration system that provides the liquid-solid separation, such as screens, papers non-wovens, granular beds and other porous media.
Filtrate	:	The discharge liquor in filtration.
Filtration	:	A process of separating particulate matter from a fluid by passing it through a permeable material.
Floating matter	:	Matter which passes through a 2000 micron sieve and separates by floatation for an hour.
Floc	:	Clumps of bacteria and particulate impurities or coagulants that have come together and formed a cluster. Found in aeration tanks and secondary clarifiers.

Flocculation	:	The process of forming floc particles when a chemical coagulant or flocculent such as alum or ferric chloride is added to the waste-water.
Flow rate	:	Measure of the amount of fluid passing through the filter. This is always a variable of filter area, porosity, contamination and differential pressure.
FOG	:	Fats, oils and greases. A measure of the non-petroleum based fats in waste treatment.
F/M	:	A ratio of the amount of food to the amount of organisms. Used to control an activated sludge process.
Flow equalisation system	:	A device or tank designed to hold back or store a portion of peak flows for release during low-flow periods.
Food chain	:	Very simple pathway of nutrient flow. Example: Carnivore > herbivore > plant.
Frazier test	:	Measures the amount of air transmitted through a filter under selected differential pressures. Historically used for textile products.
Frequency	:	Number of complete cycles of current per second, expressed in Hertz (Hz). Megahertz (MHz) is 106 Hz.
Gasification	:	The conversion of soluble and suspended materials into gas during anaerobic decomposition. In clarifiers the resulting gas bubbles can become attached to the settled sludge and cause large clumps of sludge to rise and float on the water surface. In anaerobic sludge digesters, this gas is collected for fuel or disposed of using a waste gas burner.
Generation time	:	The time required for a given population to double in size. This time can be as short as 20 minutes or as long as a week.
GMPs	:	Good Manufacturing Practices. Food and Drug Administration regulations governing the manufacture of drugs and medical devices.
Gram positive	:	Bacterial cells which retain the crystal violet stain during a staining procedure. Most strains of bacilli are gram positive. .
Gram negative	:	Bacteria cells which lose the crystal violet during the decolourising step and are then coloured by the counterstain. *Pseudomonas* and *Thiobacillus* are examples of gram negative strains.
Grit	:	The heavy material present in waste-water, such as sand coffee grounds, eggshells, gravel and cinders.
Halophilic or halotolerant	:	Bacteria which thrive in a highly salt environment, up to 25 per cent NaCl.
Headworks	:	The facilities where waste-water enters a waste-water treatment plant. The headworks may consist of bar screens, comminutors, a wet well and pumps.
Heterotroph	:	A micro-organism which uses organic matter for energy and growth.
HRT	:	Hours of retention time.
House sewer	:	The pipeline connecting the house and drain and the septic tank.
Humus	:	The dark organic material in soils, produced by the decomposition of soils. The matter that remains after the bulk of detritus has been consumed (leaves, roots). Humus mixes with top layers of soil (rock particles), supplies some of the nutrient

needed by plants—increases acidity of soil; inorganic nutrients more soluble under acidic conditions, become more available, for example wheat grows best at pH 5.5–7.0. Humus modifies soil texture, creates loose, crumbly texture, that allows water to soak in and nutrients retained; permits air to be incorporated into soil.

Hydraulic loading : Hydraulic loading refers to the flows (MGD or m^3/day) to a treatment plant or treatment process.

Hydrolysis : The process in which carbohydrates and starches are simplified into organic soluble organics, usually by facultative anaerobes.

Hydrophilic : Having an affinity for water and aqueous solutions.

Hydrophobic : Cannot be wetted by aqueous and other high surface tension fluids.

Hygroscopic : Absorbing or attracting moisture from the air.

Incineration : The conversion of dewatered waste-water solids by combustion (burning) to ash, carbon dioxide and water vapour.

Infiltration : The seepage of groundwater into a sewer system, including service connections. Seepage frequently occurs through defective or cracked pipes, pipe joints, connections or manhole walls.

Influent : The liquid—raw (untreated) or partially treated—flowing into a reservoir, basin, treatment process or treatment plant.

Inoculate : To introduce a seed culture into a system.

Inorganic waste : Waste material such as sand, salt, iron, calcium, and other mineral materials which are only slightly affected by the action of organisms. Inorganic wastes are chemical substances of mineral origin; whereas organic wastes are chemical substances usually of animal or plant origin.

Interface : The common boundary layer between two substances such as between water and a solid (metal) or between water and a gas (air) or between a liquid (water) and another liquid (oil).

Ionisation : The process of adding electrons to, or removing electrons from, atoms or molecules, thereby creating ions. High temperatures, electrical discharges and nuclear radiation can cause ionisation.

Krebs cycle : The oxidative process in respiration by which pyruvate (via acetyl co-enzyme A) is completely decarboxylated to CO_2. The pathway yields 15 moles of ATP (1,50,000 calories).

Liquefaction : The conversion of large solid particles of sludge into very fine particles which either dissolve or remain suspended in waste-water.

Loaded (plugged) : A filter element that has collected a sufficient quantity of insoluble contaminants such that it can no longer pass rated flow without excessive differential pressure.

Log growth : A growth phase in which cell production is maximum.

Lysing : A disintegration or breakdown of cells which releases organic matter.

MacConkey streak : Laboratory test for the presence of gram negative bacteria. We use this test to detect contamination of *Bacillus* products such as CF 1000, 1002, 4002 and some of the Enz-odour products.

Macronutrient	:	An element required in large proportion by plants and other life forms for survival and growth. Macronutrients include nitrogen (N), potassium (K) and phosphorous (P).
Masking agent	:	Substance used to cover up or disguise unpleasant odours. Liquid masking agents are dripped into waste-water, sprayed into the air or evaporated (using heat) with the unpleasant fumes or odours and then discharged into the air by blowers to make an undesirable odour less noticeable.
MCRT	:	Mean cell retention time—days. An expression of the average time that a micro-organism will spend in an activated sludge process.
Mean filtration rating	:	Derived from Bubble Point test method. This data should be used as a guide only to compare overall retention capabilities between fabrics and should not be considered a guarantee of particle size that the fabric will retain.
Mechanical aeration	:	The use of machinery to mix air and water so that oxygen can be absorbed into the water. Some examples are paddle wheels, mixers, rotating brushes to agitate the surface of an aeration tank; pumps to create fountain and pumps to discharge water down a series of steps forming falls or cascades.
Media	:	The material in the trickling filter on which slime accumulates and organisms grow. As settled waste-water trickles over the media, organisms in the slime remove certain types of wastes thereby partially treating the waste-water. Also the material in a rotating biological contactor (RBC) or in a gravity or pressure filter.
MEK	:	Methyl ethyl ketone.
Membrane	:	A thin polymeric film with pores.
Mercaptans	:	Compounds containing sulphur which have an extremely offensive skunk-like odour. Also sometimes described as smelling like garlic or onions.
Mesh count	:	The number of threads in a linear inch of fabric/wire cloth.
Mesophilic bacteria	:	A group of bacteria that grow and thrive in a moderate temperature range between 68°F (20°C) and 113°F (45°C).
Metabolism	:	All of the processes or chemical changes in an organism or a single cell by which food is built up (anabolism) into living protoplasm and by which protoplasm is broken down (catabolism) into simpler compounds with the exchange of energy.
Metalised screens	:	Screens that have been metalised with nickel. These screens will bleed off static charges, promote EMC and reflect electromagnetic energy.
MGD	:	Million gallons daily—refers to the flow through a waste treatment plant.
Mg/l	:	Milligrams per liter = ppm (parts per million)—expresses a measure of the concentration by weight of a substance per unit volume.
Micron	:	A unit of length. One millionth of a meter or one thousandth of a millimeter. One micron equals 0.00004 of an inch.
Micronutrient	:	An element required by plants and bacteria, in proportionately smaller amounts, for survival and growth. Micronutrients include: iron (Fe), managanese (Mn), zinc (Zn), boron (B) and molybdenum (Mo).
Monoculture	:	Aquaculture in which one species is grown.
Motile	:	Motile organisms exhibit or are capable of movement.

ML	:	Mixed liquor—the combination of raw influent and returned activated sludge. (no, not mixed drinks for human consumption)
MLSS	:	Mixed liquor suspended solids—the volume of suspended solids in the mixed liquor of an aeration tank.
MLVSS	:	Mixed liquor volatile suspended solids—the volume of organic solids that can evaporate at relatively low temperatures (550°C) from the mixed liquor of an aeration tank. This volatile portion is used as a measure or indication of micro-organisms present. Volatile substances can also be partially removed by air stripping.
MPN index	:	Most Probable Number of coliform-group organisms per unit volume of sample water. Expressed as a density or population of organisms per 100 ml of sample water.
MSDS	:	Material safety data sheet—a document that provides pertinent information and a profile of a particular hazardous substance or mixture. An MSDS is normally developed by the manufacturer or formulator of the hazardous substance or mixture. The MSDS is required to be made available to employees and operators whenever there is the likelihood of the hazardous substance or mixture being introduced into the workplace. Some manufacturers prepare MSDS for products that are not considered to be hazardous to show that the product or substance is not hazardous.
Mutualism	:	Two species living together in a relationship in which both benefit from the association.
NPDES permit	:	National Pollutant Discharge Elimination System permit is the regulatory agency document issued by either a federal or state agency which is designated to control, all discharge of pollutants from point sources into US waterways. NPDES permits regulate discharges into navigable water from all point sources of pollution, including industries, municipal waste-water treatment plants, sanitary landfills, large agricultural feed lots and return irrigation flows.
Nitrification	:	An aerobic process in which bacteria change the ammonia and organic nitrogen in waste-water into oxidised nitrogen (usually nitrate). The second-stage BOD is sometimes referred to as the 'nitrification stage' (first-stage BOD is called the 'carbonaceous stage').
Nitrifying bacteria	:	Bacteria that change the ammonia and organic nitrogen in waste-water into oxidised nitrogen (usually nitrate).
Nitrogen fixation	:	Conversion of atmospheric nitrogen into organic nitrogen compounds available to green plants; a process that can be carried out only by certain strains of soil bacteria.
Non-woven	:	A porous web or sheet produced by mechanically, chemically or thermally bonding together polymers, fibers or filaments.
Nutrients	:	Substances which are required to support living plants and organisms. Major nutrients are carbon, hydrogen, oxygen, sulphur, nitrogen and phosphorus. Nitrogen and phosphorus are difficult to remove from waste-water by conventional treatment processes because they are water soluble and tend to recycle.

Obligate aerobe	:	Bacteria which require the presence of oxygen, such as *Pseudomonas flourescens*. A few strains of this species are capable of utilising nitrate to allow anaerobic respiration.
Oil retention boom	:	A floating baffle used to contain and prevent the spread of floating oil on a water surface.
Open area	:	The proportion of total screen area that is open space. Expressed as a per cent.
Organic matter	:	All of the degradable organics. Living material containing carbon compounds. Used as food by micro-organisms.
Organic nitrogen	:	The nitrogen combined in organic molecules such as proteins, amines and amino acids.
ORP	:	Oxidation reduction potential—the degree of completion of a chemical reaction by detecting the ratio of ions in the reduced form to those in the oxidised form as a variation in electrical potential measured by an ORP electrode assembly.
OSHA	:	The Williams-Steiger Occupational Safety and Health Act of 1970 (OSHA) is a law designed to protect the health and safety of industrial workers and treatment plant operators. It regulates the design, construction, operation and maintenance of industrial plants and wastewater treatment plants. Waste-water treatment plants have come under stricter regulation in all phases of activity as a result of OSHA standards, which also refers to the federal and state agencies which administer OSHA.
Organic waste	:	Waste material which comes—mainly from animal or plant sources. Organic waste generally can be consumed by bacteria and other small organisms. Inorganic wastes are chemical substances of mineral origin.
Organism	:	Any form of animal or plant life.
Oxidation	:	Combining elemental compounds with oxygen to form a new compound. A part of the metabolic reaction.
Oxidising bacteria	:	Any substance such as oxygen (O_2) and chlorine (Cl_2) that can accept electrons. When oxygen or chlorine is added to waste-water, organic substances are oxidised. These oxidised organic substances are more stable and less likely to give off odours or to contain disease bacteria.
Ozonation	:	The application of ozone to water, waste-water or air, generally for the purposes of disinfection or odour control.
Parisitism	:	One organism living on or in another to obtain nourishment, without providing any benefit to the host organism.
Particle	:	A relatively small sub-division of matter ranging in diameter from a few angstroms (as with gas molecules) to a few millimeters (as with large raindrops). The particle can have various shapes and dimensions.
Particulate	:	Free suspended solids.
Pathogenic organisms	:	Bacteria, viruses or cysts which cause disease (typhoid, cholera, dysentery) in a host (such as a person). There are many types of bacteria (non-pathogenic) which do not cause disease. Many beneficial bacteria are found in waste-water treatment processes actively cleaning up organic wastes.

PCB	:	Polychlorinated biphenyls. Aka polychloro-biphenyls. Difficult to remediate chemical used in old-style transformers. Concentrated PCBs used to be referred to as '1268'.
Percolation	:	The movement or flow of water through soil or rocks.
Peristaltic pump	:	A type of positive displacement pump.
Permeability	:	Ability of a membrane or other material to permit a substance to pass through it.
pH	:	pH is an expression of the intensity of the basic or acidic condition of a liquid. Mathematically, pH is the logarithm (base 10) of the reciprocal of the hydrogen ion concentration. The pH may range from 0 to 14, where 0 is most acidic, 14 most basic and 7 is neutral. Natural water usually have a pH between 6.5 and 8.5.
Phototroph	:	A micro-organism which gains energy from sunlight (radiant energy).
Pin floc	:	Excessive solids carryover. May occur from time to time as small suspended sludge particles in the supernatant. There are two kinds: grey ash-like, inert, has low BOD indicates old sludge and brown, but a portion neither settles nor rises, has high BOD indicate young sludge.
Plane wave	:	An electromagnetic wave with electric and magnetic components perpendicular to and in phase with, each other.
ppm	:	Parts per million—the unit commonly used to designate the concentration of a substance in a waste-water in terms of weight i.e., one pound per million pounds, etc., ppm is synonymous with the more commonly used term mg/l (milligrams per liter).
Pollution	:	The impairment (reduction) of water quality by agriculture, domestic or industrial wastes (including thermal and radioactive wastes) to such a degree as to hinder any beneficial use of the water or render it offensive to the senses of sight, taste or smell or when sufficient amounts of waste creates or poses a potential threat to human health or the environment.
Polyculture	:	Fish farming in which 2 or more compatible or symbiotic species of fish are grown together. Also known as Multiculture.
Potable water	:	Water that does not contain objectionable pollution, contamination, minerals or infective agents and is considered satisfactory for drinking.
POTW	:	Publicly Owned Treatment Works, as opposed to an industrially owned facility or pipe system.
Predation	:	One species benefits at the expense of another.
Preliminary treatment	:	The removal of metal, rocks, rags, sand, eggshells and similar materials which may hinder the operation of a treatment plant. Preliminary treatment is accomplished by using equipment such as racks, bar screens, comminutors and grit removal systems.
Pre-treatment facility	:	Industrial waste-water treatment plant consisting of one or more treatment devices designed to remove sufficient pollutants from waste-water to allow an industry to comply with effluent limits established by the US EPA General and Categorical Pre-treatment Regulations or locally derived prohibited discharge requirements and local effluent limits. Compliance with effluent limits allows for a legal discharge to a POTW.

Primary treatment	:	A waste-water treatment process that takes place in a rectangular or circular tank and allows those substances in waste-water that readily settle or float to be separated from the water being treated.
Procaryotic organism	:	Micro-organisms which do not have an organised nucleus surrounded by a nuclear membrane. Bacteria and blue-green algae fit in this category.
Protozoa	:	A group of motile microscopic animals (usually single-celled and aerobic) that sometimes cluster into colonies and often consume bacteria as an energy source.
Psychrophilic bacteria	:	Bacteria whose optimum temperature range is between 0 and 20°C (32 to 68°F).
Putrefaction	:	Biological decomposition of organic matter with the production of ill-smelling products associated with anaerobic conditions.
Pyrogenic	:	A fever-producing substance. The presence of these substances is determined by the Limulus Amebocyte Lysate (LAL) test and measured in EU/ml (endotoxin units per milliliter).
Rack	:	Evenly spaced parallel metal bars or rods located in the influent channel to remove rags, rocks and cans from waste-water.
Radio frequency interference (RFI)	:	EMI in electronic equipment caused by radio frequencies, ranging typically from 10 kHz (104 Hz) to 1000 MHz (109 Hz or 1 GHz).
RAS	:	Return activated sludge—settled activated sludge that is collected in the secondary clarifier and returned to the aeration basin to mix with incoming raw settled waste-water.
RASVSS	:	Return Activated Sludge Volatile Suspended Solids.
RBC	:	Rotating biological contactor—an attached culture waste-water treatment system.
Reagent	:	A pure chemical substance that is used to make new products or is used in chemical tests to measure, detect or examine other substances.
Recycle	:	The use of water or waste-water within (internally) a facility before it is discharged to a treatment system.
REDOX	:	Biological reductions/oxidations. These reactions usually require enzymes to mediate the electron transfer. The sediment in the bottom of a lake, sludge in a sewerage works or septic tank will have a very low redox potential and will likely be devoid of any oxygen. This sludge or waste-water will have a very high concentration of reductive anaerobic bacteria, indeed the bulk of the organic matter may in fact be bacteria. As the concentration of oxygen increases the oxidation potential of the water will increase. A low redox potential or small amount of oxygen is toxic to anaerobic bacteria, therefore as the concentration of oxygen and redox potential increases the bacterial population changes from reductive anaerobic bacteria to oxidative aerobic bacteria. Measurement of REDOX potential is also referred to as ORP.
Reducing agent	:	Any substance, such as the base metal (iron) or the sulphide ion that will readily donate (give up) electrons. The opposite of an oxidising agent.
Refractory materials	:	Material difficult to remove entirely from waste-water such as nutrients, colour, taste and odour-producing substances and some toxic materials.
Residual shrinkage	:	The amount of shrinkage remaining in a fabric after it has undergone all fabric weaving, washing and heat setting steps.

Respiration	:	The energy producing process of breathing, by which an organism supplies its cells with oxygen and relieves itself of carbon dioxide. A type of heterotrophic metabolism that uses oxygen in which 38 moles of ATP are derived from the oxidation of 1 mole of glucose, yielding 3,80,000 cal. (An additional 3,08,000 cal is lost as heat.)
Rhizosphere	:	Soil surrounding plant roots.
Retentate	:	Substance retained in the upstream side of a filter.
RF (radio frequency) welding	:	Utilises specific bands of radio frequency waves which are directed through specially constructed tooling to form localised melting/joining of certain dielectric thermoplastic materials. Can be used to form hermetic seals. Also known as high frequency or dielectric welding.
R/O unit	:	Reverse osmosis unit for water purification in small aquariums and miniature yard-ponds, utilises a membrane under pressure to filter dissolved solids and pollutants from the water. Two different filter membranes can be used: the CTA (cellulose triacetate) membrane is less expensive, but only works with chlorinated water and removes 50–70 per cent nitrates and the TFC membrane, which is more expensive, removes 95 per cent of nitrates, but is ruined by chlorine.
RR	:	Respiration rate—the weight of oxygen utilised by the total weight of MLSS in a given time.
Run-off	:	Water running down slopes rather than sinking in (again, result of poor humus content). Example: erosion due to deforestation.
Saprophytic	:	Bacteria that breakdown bodies of dead plants and animals (non-living organic material), returning organic materials to the food chain. Saprophytic bacteria are usually non-pathogenic, too. Most Alken Clear-Flo products are saprophytic.
SAR	:	Sodium adsorption ratio—this ratio expresses the relative activity of sodium ions in the exchange reactions with the soil.
SCFM	:	Cubic feet of air per minute at standard conditions of temperature, pressure and humidity (0, 14.7 psi and 50 per cent relative humidity).
Secondary treatment	:	A waste-water treatment process used to convert dissolved or suspended materials into a form more readily separated from the water being treated. Usually the process follows primary treatment by sedimentation. The process commonly is a type of biological treatment process followed by secondary clarifiers that allow the solids to settle out from the water being treated.
Sedimentation	:	The process of subsidence and deposition of suspended matter from a waste-water by gravity.
Seeding	:	Introduction of micro-organisms (such as ALKEN CLEAR-FLO 1000 series for aquaculture, 4000 series for grease and 7000 series for industrial and municipal waste-water) into a biological oxidation unit to minimise the time required to build a biological sludge. Also referred to as inoculation with cultured organisms.
Seine net	:	A net designed to collect aquatic organisms inhabiting natural water from the shoreline to 3' depths is called a seine net. Most often a plankton seine.
Selvage	:	A loom finished edge that prevents cloth unravelling.

Septic	:	A condition produced by anaerobic bacteria. If severe, the waste-water turns black, gives off foul odours, contains little or no dissolved oxygen and creates a high oxygen demand.
Septicity	:	Septicity is the condition in which organic matter decomposes to form foul-smelling products associated with the absence of free oxygen. If severe, the waste-water turns black, gives off foul-odours, contains little or no dissolved oxygen and creates a heavy oxygen demand.
Septic tank	:	Untreated liquid household wastes (sewage) will quickly clog your absorption field if not properly treated. The septic tank is a holding tank in which this treatment can take place. When sewage enters the septic tank, the heavy solids settle to the bottom of the tank; the lighter solids, fats and greases partially decompose and rise to the surface and form a layer of scum. The solids that have settled to the bottom are attacked by bacteria and form sludge.
Settleable solids	:	Those solids in suspension which will pass through a 2000 micron sieve and settle in one hour under the influence of gravity.
Sewage	:	The used water and water-carried solids from homes that flow in sewers to a waste-water treatment plant.
Shock load	:	The arrival at a plant of a waste which is toxic to organisms in sufficient quantity or strength to cause operating problems. Possible problems include odours and sloughing off of the growth or slime on a trickling-filter media. Organic or hydraulic overloads also can cause a shock load.
Sieve	:	A screen with apertures of uniform size used for sizing granular materials.
Sloughings	:	Trickling-filter slimes that have been washed off the filter media. They are generally quite high in BOD and will lower effluent quality unless removed.
Sludge	:	The settleable solids separated from liquids during processing; the deposits of foreign materials on the bottoms of streams or other bodies of water.
Sludge age	:	A measure of the length of time a particle of suspended solids has been retained in the activated sludge process.
Slugs	:	Intermittent releases or discharges of industrial wastes.
Soluble	:	Matter or compounds capable of dissolving into a solution.
Soluble BOD	:	Soluble BOD is the BOD of water that has been filtered in the standard suspended solids test.
Solution	:	A liquid mixture of dissolved substances, displaying no phase separation.
Specific gravity	:	Weight of a particle, substance or chemical solution in relation to an equal volume of water.
Specification sheet	:	It is the detailed information of a product including tests, colour, odour, specific gravity, bacterial strains, other major ingredients, etc.
Surface media	:	Captures particles on the upstream surface with efficiencies in excess of depth media, sometimes close to 100 per cent with minimal or no off-loading. Commonly rated according to the smallest particle the media can repeatedly capture. Examples of surface media include ceramic media, microporous membranes, synthetic woven screening media and in certain cases, wire cloth. The media characteristically has a narrow pore size distribution.

Surface resistivity (W/o) :		Expressed in ohms/square. It is numerically equal to the resistance between two electrodes forming opposite sides of a square on the surface of a material. The size of the square is irrelevant. For conductive materials, surface resistivity is the ratio of the volume resistivity to the fabric thickness (r/t).
Tangential cross-flow filtration	:	Process where the feed stream 'sweeps' the membrane surface and the particulate debris is expelled, thus extending filter life. The filtrate flows through the membrane. Most commonly used in the separation of high-and-low-molecular weight matter such as in ultrapure reverse osmosis, ultrafiltration and sub-micron microfiltration processes.
Taxonomy	:	The classification, nomenclature and laboratory identification of organisms.
TDS	:	Total dissolved solids is commonly estimated from the electrical conductivity of the water. Pure water is a poor conductor of electricity. Impurities dissolved in the water cause an increase in the ability of the water to conduct electricity. Conductivity, usually expressed in units of microsimens, formerly micromhos or in mg/l, thus becomes an indirect measure of the level of impurities in the water.
TOC	:	Total organic carbon—a measure of the amount of organic carbon in water.
Thermophilic bacteria	:	Hot temperature bacteria, a group of bacteria that grow and thrive in temperatures above 113°F (45°C), such as *Bacillus licheniformis*. The optimum temperature range for these bacteria in anaerobic decomposition is 120°F (49°C) to 135°F (57°C).
Throughput	:	The amount of solution which will pass through a filter prior to clogging.
Toxic	:	A substance which is poisonous to a living organism.
Toxicity	:	The relative degree of being poisonous or toxic. A condition which may exist in wastes and will inhibit or destroy the growth or function of certain organisms.
Transpiration	:	The process by which water vapour is released to the atmosphere by living plants, a process similar to people sweating.
Trickling filter	:	An attached culture, waste-water treatment system. A large tank generally filled with rock or rings. Waste-water is sprayed over the top of the media, providing the opportunity for the formation of slimes or biomass to remove wastes from the waste-water, through revolving arms which have spray nozzles. Water is pumped from the bottom of a trickle filter to a secondary clarifier.
TSS	:	Total suspended solids.
Turbidity	:	The amount of suspended matter in waste-water, obtained by measuring its light scattering ability.
Unicellular	:	Single celled organism, such as bacteria.
Upset	:	An upset digester does not decompose organic matter properly. The digester is characterised by low gas production, high volatile acid/alkalinity relationship, and poor liquid-solids separation. A digester in an upset condition is sometimes called a 'sour' or 'stuck' digester.
Ultrasonic (processes)	:	Process which utilises specially designed tooling usually vibrating at 15–80 KHz. Processes are designed to cause localised heating of thermoplastic materials which, in turn, will provide some type of welded or fused joint. Benefits are elimination of fillers and minimised heat stress on surrounding materials.

Upstream side	:	The feed side of the filter.
Uronic acid	:	Class of acidic compounds of the general formula $HOOC(CHOH)_nCHO$ that contain both carboxylic and aldehydic groups, are oxidation products of sugars and occur in many polysaccharides; especially in the hemicelluloses.
Useful life	:	Determined when contamination causes an adverse flow rate, low efficiency or high differential pressure.
Vadose zone	:	Unsaturated zone of soil above the groundwater, extending from the bottom of the capillary fringe all the way to the soil surface.
Vegetative	:	Actually growing state.
Vegetative cell	:	Growing or feeding form of a microbial cell, as opposed to a resting form such as a spore.
Vesicles	:	Spherical structures, formed intracellularly, by some arbuscular mycorrhizal fungi.
Viable	:	Alive; able to reproduce.
Viable but non-culturable	:	Organisms that are alive but cannot be cultured on laboratory media.
Viable count	:	Measurement of the concentration of live cells in a microbial population.
Vibrio	:	(i) Curved, rod-shaped bacterial cell; and (ii) bacterium of the genus *Vibrio*.
Virion	:	Virus particle; the virus nucleic acid surrounded by protein coat and in some cases other material.
Virulence	:	Degree of pathogenicity of a parasite.
Virus	:	Any of a large group of sub-microscopic infective agents that typically contain a protein coat surrounding a nucleic acid core and are capable of growth only in a living cell.
Volatile	:	A volatile substance is one that is capable of being evaporated or changed to a vapour at a relatively low temperature. Volatile substances also can be partially removed by air stripping.
Volume resistivity	:	Or specific resistivity of a material, expressed in W/cm. Resistance to electrical current flow through the bulk of an object.
VS/L	:	Measure of volatile solids, usually expressed as g VS/L/day—grams volatile solids per liter per day.
WAS	:	Waste activated sludge, mg/l. The excess growth of micro-organisms which must be removed from the process to keep the biological system in balance.
Waste-water	:	The used water and solids from a community that flow to a treatment plant. Storm water, surface water and groundwater infiltration may also be included in the waste-water that enters a waste-water treatment plant. The term 'sewage' usually refers to household wastes, but this word is being replaced by the term 'waste-water'.
Water content	:	Water contained in a material expressed as the mass of water per unit mass of oven-dry material.
Water-retention curve	:	Graph showing soil-water content as a function of increasingly negative soil water potential.

Weathering	:	All physical and chemical changes produced in rock by atmospheric agents.
Weir	:	A wall or plate placed in an open channel and used to measure the flow of water.
White rot fungus	:	Fungus that attacks lignin, along with cellulose and hemicellulose, leading to a marked lightening of the infected wood.
Wild type	:	Strain of micro-organism isolated from nature. The usual or native form of a gene or organism.
Winogradsky column	:	Glass column with an anaerobic lower zone and an aerobic upper zone, which allows growth of micro-organisms under conditions similar to those found in nutrient-rich water and sediment.
Woronin body	:	Spherical structure associated with the simple pore in the septa separating hyphal compartments of fungi in the phylum Ascomycota.
Xenobiotic	:	Compound foreign to biological systems. Often refers to human-made compounds that are resistant or recalcitrant to biodegradation and decomposition.
Xerophile	:	Organism adapted to grow at low water potential, i.e., very dry habitats.
Yeast	:	Fungus whose thallus consists of single cells that multiply by budding or fission.
Zoogleal film	:	A complex population of organisms that form a 'slime growth' on a trickling-filter media and break down the organic matter in waste-water.
Zoogleal mass	:	Jelly-like masses of bacteria found in both the trickling filter and activated sludge processes.
Zoospore	:	An asexual spore formed by some fungi that usually can move in an aqueous environment via one or more flagella.
Zygospore	:	Thick-walled resting spore resulting from fusion of two gametangia of fungi in the phylum Zygomycota.
Zygote	:	In eukaryotes, the single diploid cell resulting from the union (fusion) of two haploid gametes.
Zymogenous flora	:	Refers to micro-organisms, often transient or alien, that respond rapidly by enzyme production and growth when simple organic substrates become available. Also called copiotrophs.

Appendices

QUANTITY AND UNITS

Table 1. Base units in the international system of units (SI).

Quantity	Name	Symbol
Length	Meter	m
Mass	Kilogram	kg
Time	Second	s
Electric current	Ampere	A
Thermodynamic temperature	Kelvin	K
Amount of substance	Mole	mol
Luminous intensity	Candela	cd
Plane angle*	Radian	rad
Solid angle*	Steradian	sr

* Supplementary units.

Table 2. Derived SI units with special names.

Quantity	SI unit symbol	Name	Units
Frequency	Hz	Hertz	$1/s$
Force	N	Newton	$kg \cdot m/s^2$
Pressure, stress	Pa	Pascal	$kg/m \cdot s^2$ or N/m^2
Energy or work	J	Joule	$kg \cdot m^2/s^2$ or $N \cdot m$
A quantity of heat	J	Joule	$kg \cdot m^2/s^2$ or $N \cdot m$
Power, radiant flux	W	Watt	$kg \cdot m^2/s^3$ or J/s
Electric charge	C	Coulomb	$A \cdot s$
Electrical potential	V	Volt	$kg \cdot m^2/s^3 \cdot A$ or W/A
Potential difference	V	Volt	$kg \cdot m^2/s^3 \cdot A$ or W/A
Electromotive force	V	Volt	$kg \cdot m^2/s^3 \cdot A$ or W/A
Capacitance	F	Farad	$A^2 \cdot s^4/kg \cdot m^2$ or C/V

(Contd...

Quantity	SI unit symbol	Name	Units
Electric resistance	Ω	Ohm	$kg \cdot m^2/s^3 \cdot A^2$ or V/A
Conductance	S	Siemens	$S^3 \cdot A^2/kg \cdot m^2$ or A/V
Magnetic flux	Wb	Weber	$kg \cdot m/s^2 \cdot A$ or $V \cdot s$
Magnetic flux density	T	Tesla	$kg/s^2 \cdot A$ or Wb/m^2
Inductance	H	Henry	$kg \cdot m^2/s^2 \cdot A^2$ or Wb/A
Luminous flux	lm	Lumen	$cd \cdot sr$
Illuminance	lx	Lux	$cd \cdot sr/m^2$ or lm/m^2
Activity (radionuclides)	Bq	Becquerel	l/s
Absorbed dose	Gy	gray	m^2/s^2 or J/kg

Table 3. Derived SI units obtained by combining base units and units with special names.

Quantity	Units	Quantity	Units
Acceleration	m/s^2	Molar entropy	$J/mol \cdot K$
Angular acceleration	rad/s^2	Molar heat capacity	$J/mol \cdot K$
Angular velocity	rad/s	Moment of force	$N \cdot m$
Area	m^2	Permeability	H/m
Concentration	mol/m^3	Permittivity	F/m
Current density	A/m^2	Radiance	$W/m^2 \cdot sr$
Density, mass	kg/m^3	Radiant intensity	W/sr
Electric charge density	C/m^3	Specific heat capacity	$J/kg \cdot K$
Electric field strength	V/m	Specific energy	J/kg
Electric flux density	C/m^2	Specific entropy	$J/kg \cdot K$
Energy density	J/m^3	Specific volume	m^3/kg
Entropy	J/K	Surface tension	N/m
Heat capacity	J/K	Thermal conductivity	$W/m \cdot K$
Heat flux density	W/m^2	Velocity	m/s
Irradiance	W/m^2	Viscosity, dynamic	$Pa \cdot s$
Luminance	cd/m^2	Viscosity, kinematic	m^2/s
Magnetic field strength	A/m	Volume	m^3
Molar energy	J/mol	Wavelength	m

Table 4. Values of useful constants.

Acceleration due to gravity, g	$= 9.807 \ m/s^2 \ (32.174 \ ft/s^2)$
Standard atmosphere	$= 101.325 \ kN/m^2 \ (14.696 \ lbf/in^2)$
	$= 101.325 \ kPA \ (1.013 \ bar)$
1 bar $= 10^5 \ N/m^2 \ (14.504 \ lbf/in^2)$	
Standard atmosphere	$= 10.333 \ m \ (33.899 \ ft)$ of water
1 meter head of water (20°C)	$= 9.790 \ M/m^2 \ (1.420 \ lbf/in^2)$
	$= 0.00979 \ N/mm^2 \ (1.420 \ lbf/in^2)$
	$= 9.790 \ kN/m^2 \ (1.420 \ lbf/in^2)$

APPENDIX II

CONVERSION FACTORS

Table 1. Conversion factors for commonly used waste-water-treatment plant design parameters.

Parameter (in SI units)	SI units	To convert, multiply in direction shown by arrows $\longrightarrow$	$\longleftarrow$	U.S. units
Screening				
m^3 screenings/10^3 m^3 waste-water	$m^3/10^3$ m^3	133.6806	7.4805×10^{-3}	ft^3/Mgal
Grit removal				
Air supply				
m^3 air/m of tank length · min	$m^3/m \cdot min$	10.7639	0.0929	ft^3/ft · min
Grit removal				
g grit/m^3 waste-water	g/m^3	8.3454	0.1198	lb/Mgal
kg grit/m^3 waste-water	kg/m^3	8345.4	1.1983×10^{-4}	lb/Mgal
Surface overflow rate				
m^3 flow/m^2 surface area · hr.	$m^3/m^2 \cdot hr.$	589.0173	0.0017	$gal/ft^2 \cdot d$
m^3 flow/m^2 surface area · d	$m^3/m^2 \cdot d$	24.5424	0.0407	$gal/f^2 \cdot d$
Volume				
m^3 grit/10^3 m^3 waste-water	$m^3/10^3$ m^3	133.6806	7.4805×10^{-3}	ft^3/Mgal
Flow equalisation				
Air supply				
m^3 air/m^3 tank volume · min	$m^3/m^3 \cdot min$	133.6806	7.4805×10^{-2}	$ft^3/10^2$ gal · min
Mixing horsepower				
kW/m^3 tank volume	kW/m^3	5.0763	0.1970	$hp/10^2$ gal
Sedimentation				
Particle settling rate				
min/hr.	min/hr.	3.2808	0.3048	ft/hr.
min/hr.	min/hr.	0.4090	2.4448	$gal/ft^2 \cdot min$
Sludge scraper speed				
min/hr.	min/hr.	0.0547	18.2880	ft/min
Solids loading				
kg solids/m^2 surface area · d	$kg/m^2 \cdot d$	0.2048	4.8824	$lb/ft^2 \cdot d$
Surface overflow rate				
m^3 waste-water/m^2 surface area · d	$m^3/m^2 \cdot d$	24.5424	0.0407	$gal/ft^2 \cdot d$
m^3 waste-water/m^2 surface area · hr.	$m^3/m^2 \cdot hr.$	589.0173	0.0017	$gal/ft^2 \cdot d$
Volume of sludge				
m^3 sludge/10^3 m^3 waste-water	$m^3/10^3$ m^3	133.6806	7.481×10^{-3}	ft^3/Mgal
Weight of dry sludge solids				
g dry solids/m^3 waste-water	g/m^3	8.3454	0.1198	lb/Mgal

(Contd...)

Parameter (in SI units)	SI units	To convert, multiply in direction shown by arrows		U.S. units
		$\longrightarrow$	$\longleftarrow$	
Weir overflow rate				
m^3 waste-water/m weir length · d	$m^3/m \cdot d$	80.5196	0.0124	gal/ft · d
Activated sludge				
Aeration device mixing intensity, diffused aeration				
m^3 air/m^3 tank volume · min	$m^3/m^3 \cdot min$	1000.0	0.001	$ft^3/10^3 \; ft^3 \cdot min$
Aeration device mixing intensity, mechanical aeration				
$kW/10^3 \; m^3$ tank volume	$kW/10^3 \; m^3$	0.0380	26.3342	$hp/10^3 \; ft^3$
Air flow rate				
m^3 air/hr.	$m^3/hr.$	0.5886	1.6990	ft^3/min
Air requirements, organic removal				
m^3 air/kg BOD_5 removed	m^3/kg	16.0185	0.0624	ft^3/lb
Air requirements, volume of waste-water				
m^3 air/m^3 waste-water	m^3/m^3	0.1337	7.4805	ft^3/gal
Organic load				
kg BOD, applied/m^3 aeration-tank volume · d	$kg/m^3 \cdot d$	62.4280	0.0160	$lb/10^3 \; ft^3 \cdot d$
Oxygen requirements				
kg O_2/kg BOD_5 applied · d	$kg/kg \cdot d$	1.0	1.0	lb/lb · d
Oxygen-transfer rate				
kg O_2 transferred/kW · hr.	$kg/kW \cdot hr.$	1.6440	0.6083	lb/hp · hr.
kg O_2 transferred/m^3 waste-water · hr.	$kg/m^3 \cdot hr.$	0.0624	16.0185	$lb/ft^3 \cdot hr.$
Trickling filters and rotating biological contactors				
Hydraulic load				
m^3 waste-water/m^2 bulk surface area · d	$m^3/m^2 \cdot d$	24.5424	0.0407	$gal/ft^2 \cdot d$
m^3 waste-water/m^2 bulk surface area · hr.	$m^3/m^2 \cdot h$	589.0173	0.0017	$gal/ft^2 \cdot d$
m^3 waste-water/m^2 bulk surface area · d	$m^3/m^2 \cdot d$	1.0691	0.9354	Mgal/acre · d
L waste-water/m^2 bulk surface area · min	$L/m^2 \cdot min$	35.3420	0.0283	$gal/ft^2 \cdot d$
Organic load				
kg BOD_5/m^3 filter-medium volume · d	$kg/m^3 \cdot d$	62.4280	0.0160	$lb/10^3 \; ft^3 \cdot d$
Specific surface loading, hydraulic				

(Contd...)

Parameter (in SI units)	SI units	To convert, multiply in direction shown by arrows		U.S. units
		$\longrightarrow$	$\longleftarrow$	
m^3 waste-water/m^2 filter medium surface area · d	$m^3/m^2 \cdot d$	24.5424	0.0407	gal/ft^2 · d
m^3 waste-water/m^2 filter medium surface area · d	$m^3/m^2 \cdot d$	0.0170	58.6740	gal/ft^2 · min
m^3 waste-water/m^2 filter medium surface area · hr.	$m^3/m^2 \cdot hr.$	589.0173	0.0017	gal/ft^2 · d
Specific surface loading, organic kg BOD_5/m^2 filter medium surface area · d	$kg/m^2 \cdot d$	0.2048	4.8824	lb/ft^2 · d
Tank volume L/m^2 medium surface area (rotating biological reactor)	L/m^2	2.4542×10^{-2}	40.7458	gal/ft^2
Stabilisation ponds and lagoons Organic loads kb BOD_5/ha surface area · d	kg/ha · d	0.8922	1.1209	lb/acre · d
Volumetric load kg BOD_5/m^3 basin volume · d	$kg/m^3 \cdot d$	62.4280	0.0160	lb/10^3 ft^3 · d
Chlorination Feed rate kg chlorine/d	kg/d	2.2046	0.4536	lb/d
Sludge thickening Sludge loading kg dry solids fed/m^2 surface area · d	$kg/m^2 \cdot d$	0.2048	4.8824	lb/ft^2 · d
Surface overflow rate m^3 waste-water/m^2 surface area · d	$m^3/m^2 \cdot d$	24.5424	0.0407	gal/ft^2 · d
m^3 waste-water/m^2 surface area · d	$m^3/m^2 \cdot d$	0.0170	58.6740	gal/ft^2 · min
Sludge digestion Gas production m^3 gas/kg volatile solids fed	m^3/kg	16.0185	0.0624	ft^3/lb
m^3 gas/capita	m^3/capita	35.3147	0.0283	ft^3/capita
Loading rate kg BOD_5/m^3 digester volume · d	$kg/m^3 \cdot d$	62.4280	0.0160	lb/10^3 ft^3 · d
Sludge heating W/m^2 surface area · °C	$W/m^2 \cdot$ °C	0.1763	5.6735	Btu/ft^2 · °F · h
Volatile-solids loading kg volatile solids/m^3 digester volume · d	$kg/m^3 \cdot d$	62.4280	0.0160	lb/10^3 ft^3 · d
Sludge drying beds Dry-solids loading kg dry solids/m^2 area · year	$kg/m^2 \cdot$ year	0.2048	4.8824	lb/ft^2 · year

(Contd...)

Parameter (in SI units)	SI units	To convert, multiply in direction shown by arrows		U.S. units
		$\longrightarrow$	$\longleftarrow$	
m^2 area/capita · year	m^2/capita · year	10.7639	0.0929	ft^2/capita · year
Vacuum filtration				
Dry solids				
kg dry solids/m^2 surface area · hr.	kg/m^2 · hr.	0.2048	4.8824	lb/ft^2 · hr.
Pressure applied				
kPa (kN/m^2) pressure	kPa	0.1450	6.8948	lbf/in^2 (gauge)
Sludge feed				
m^3 wet sludge/m^2 surface area · hr.	m^3/m^2 · hr.	3.2808	0.3048	ft^3/ft^2 · hr.
Vacuum applied				
kPa (kN/m^2) vacuum	kPa (kN/m^2)	0.2961	3.3768	inHg (60°F)
Heat drying				
kJ heat energy required/kg water evaporated (sludge cake)	kJ/kg	0.4303	2.3241	Btu/lb
kg water evaporated/hr.	kg/hr.	2.2046	0.4536	lb/hr.
kg wet sludge/m^2 heating surface · hr.	kg/m^2 · hr.	0.2048	4.8824	lb/ft^2 · hr.
Incineration				
kJ heat energy/kg moisture evaporated	kJ/kg	0.4303	2.3241	Btu/lb
kg sludge/m^2 heating surface area	kg/m^2	0.2048	4.8824	lb/ft^2 · hr.
kg sludge/m^3 combustion chamber volume · hr.	kg/m^3 · hr.	0.0624	16.0185	lb/ft^3 · hr.
Land disposal				
kg mass/ha field area	kg/ha	0.8922	1.1208	lb/acre
bu yield/ha field area · year	bu/ha · year	0.4047	2.4711	bu/acre · year
Mg loading/ha field area	Mg/ha	0.4461	2.2417	tonnes/acre
m^3 waste-water/ha field area · d	m^3/ha · d	106.9064	0.0094	gal/acre · d
Surface or in-depth filters				
L water (backwash)/m^2 surface area · min	L/m^2 · min	0.0245	40.7458	gal/ft^2 · min

Table 2. Metric conversion factors (U.S. customary units to SI units).

Multiply the U.S. customary unit		by	To obtain the SI unit	
Name	Symbol		Symbol	Name
Acceleration				
feet per second squared	ft/s^2	0.3048*	m/s^2	meters per second squared
inches per second squared	in/s^2	0.0254*	m/s^2	meters per second squared

(Contd...)

Multiply the U.S. customary unit		by	To obtain the SI unit	
Name	Symbol		Symbol	Name
Area				
acre	acre	0.404 7	ha	hectare
acre	acre	4.0469×10^{-3}	km^2	square kilometer
square foot	ft^2	9.2903×10^{-2}	m^2	square meter
square inch	in^2	6.4516*	cm^2	square centimeter
square mile	mi^2	2.5900	km^2	square kilometer
square yard	yd^2	0.8361	m^2	square meter
Energy				
British thermal unit	Btu	1.0551	kJ	kilojoule
foot-pound (force)	ft · lbf	1.3558	J	joule
horsepower-hour	hp · hr.	2.6845	MJ	megajoule
kilowatt-hour	kW · hr.	3600*	kJ	kilojoule
kilowatt-hour	kW · hr.	3.600×10^6*	J	joule
watt-hour	W · hr.	3.600*	kJ	kilojoule
watt-second	W · s	1.000*	J	joule
Flow rate				
cubic feet per second	ft^3/s	2.8317×10^{-2}	m^3/s	cubic meters per second
gallons per day	gal/d	4.3813×10^{-5}	L/s	liters per second
gallons per day	gal/d	3.7854×10^{-3}	m^3/d	cubic meters per day
gallons per minute	gal/min	6.3090×10^{-5}	m^3/s	cubic meters per second
gallons per minute	gal/min	6.3090×10^{-2}	L/s	liters per second
millions gallons per day	Mgal/d	43.8126	L/s	liters per second
million gallons per day	Mgal/d	3.7854×10^3	m^3/d	cubic meters per day
million gallons per day	Mgal/d	4.3813×10^{-2}	m^3/s	cubic meters per second
Force				
pound force	lbf	4.4482	N	newton
Length				
foot	ft	0.3048*	m	meter
inch	in	2.54*	cm	centimeter
inch	in	0.0254*	m	meter
inch	in	25.4*	mm	millimeter
mile	mi	1.6093	km	kilometer
yard	yd	0.9144*	m	meter
Mass				
ounce	oz	28.3495	g	gram
pound	lb	4.5359×10^2	g	gram

(Contd...)

Multiply the U.S. customary unit		*by*	*To obtain the SI unit*	
Name	*Symbol*		*Symbol*	*Name*
pound	lb	0.4536	kg	kilogram
tonne (short: 2000 lb)	tonne	0.9072	Mg (metric tonne)	megagram (10^3 kilogram)
tonne (long: 2240 lb)	tonne	1.0160	Mg (metric tonne)	megagram (10^3 kilogram)
Power				
British thermal units per second	Btu/s	1.0551	kW	kilowatt
foot-pounds (force) per second	ft · lbf/s	1.3558	W	watt
horsepower	hp	0.7457	kW	kilowatt
Pressure (force/area)				
atmosphere (standard)	atm	1.0133×10^2	kPa (kN/m^2)	kilopascal (kilonewtons per square meter)
inches of mercury (60°F)	inHg (60°F)	3.3768×10^3	Pa (N/m^2)	pascal (newtons per square meter)
inches of water (60°F)	inH$_2$O (60°F)	2.4884×10^2	Pa (N/m^2)	pascal (newtons per square meter)
pounds (force) per square foot	lbf/ft^2	47.8803	Pa (N/m^2)	pascal (newtons per square meter)
pounds (force) per square inch	lbf/in^2	6.8948×10^3	Pa (N/m^2)	pascal (newtons per square meter)
pounds (force) per square inch	lbf/in^2	6.8948	kPa (kN/m^2)	kilopascal (kilo newtons per square meter)
Temperature				
degrees fahrenheit	°F	0.555(°F–32)	°C	degrees Celsius (centigrade)
degrees fahrenheit	°F	0.555(°F + 459.67)	K	degrees kelvin
Velocity				
feet per second	ft/s	0.3048*	m/s	meters per second
miles per hour	mi/hr.	4.4704×10^{-1}*	km/s	kilometers per second
Volume				
acre-foot	acre-ft	1.2335×10^3	m^3	cubic meter
cubic foot	ft^3	28.3168	L	liter
cubic foot	ft^3	2.8317×10^{-2}	m^3	cubic meter
cubic inch	in^3	16.3871	cm^3	cubic centimeter
cubic yard	yd^3	0.7646	m^3	cubic meter
gallon	gal	3.7854×10^{-3}	m^3	cubic meter
gallon	gal	3.7854	L	liter
ounce (U.S. fluid)	oz (U.S. fluid)	2.9573×10^{-2}	L	liter

* Indicates exact conversion.

Table 3. Metric conversion factors (SI units to U.S. customary units).

Mulitply the SI unit		by	To obtain the U.S. customary unit	
Name	*Symbol*		*Symbol*	*Name*
Acceleration				
meters per second squared	m/s^2	3.2808	ft/s^2	feet per second squared
meters per second squared	m/s^2	39.3701	in/s^2	inches per second squared
Area				
hectare (10000 m^2)	ha	2.4711	acre	acre
square centimeter	cm^2	0.1550	in^2	square inch
square kilometer	km^2	0.3861	mi^2	square mile
square kilometer	km^2	247.1054	acre	acre
square meter	m^2	10.7639	ft^2	square foot
square meter	m^2	1.1960	yd^2	square yard
Energy				
kilojoule	kJ	0.9478	Btu	British thermal unit
joule	J	2.7778×10^{-7}	kW · hr.	kilowatt-hour
joule	J	0.7376	ft · lbf	foot-pound (force)
joule	J	1.0000	W · s	watt-second
joule	J	0.2388	cal	calorie
kilojoule	kJ	2.7778×10^{-4}	kW · hr.	kilowatt-hour
kilojoule	kJ	0.2778	W · hr.	watt-hour
megajoule	MJ	0.3725	hp · hr.	horsepower-hour
Flow rate				
cubic meters per day	m^3/d	264.1720	gal/d	gallons per day
cubic meters per day	m^3/d	2.6417×10^{-4}	Mgal/d	million gallons per day
cubic meters per second	m^3/s	35.3147	ft^3/s	cubic feet per second
cubic meters per second	m^3/s	22.8245	Mgal/d	million gallons per day
cubic meters per second	m^3/s	15850.3	gal/min	gallons per minute
liters per second	L/s	22824.5	gal/d	gallons per day
liters per second	L/s	0.0228	Mgal/d	million gallons per day
liters per second	L/s	15.8508	gal/min	gallons per minute
Force				
newton	N	0.2248	lbf	pound force
Length				
centimeter	cm	0.3937	in	inch
kilometer	km	0.6214	mi	mile
meter	m	39.3701	in	inch
meter	m	3.2808	ft	foot
meter	m	1.0936	yd	yard
millimeter	mm	0.03937	in	inch

(Contd...)

Mulitply the SI unit		*by*	*To obtain the U.S. customary unit*	
Name	*Symbol*		*Symbol*	*Name*
Mass				
gram	g	0.0353	oz	ounce
gram	g	0.0022	lb	pound
kilogram	kg	2.2046	lb	pound
megagram (10^3 kg)	Mg	1.1 023	tonne	tonne (short: 2000 lb)
megagram (10^3 kg)	Mg	0.9842	tonne	tonne (long: 2240 Ib)
Power				
kilowatt	kW	0.9478	Btu/s	British thermal units per second
kilowatt	kW	1.3410	hp	horsepower
watt	W	0.7376	ft/lbf/s	foot-pounds (force) per second
Pressure (force/area)				
pascal (newtons per square meter)	Pa (N/m^2)	1.4504×10^{-4}	lbf/in^2	pounds (force) per square inch
pascal (newtons per square meter)	Pa (N/m^2)	2.0885×10^{-2}	lbf/ft^2	pounds (force) per square foot
pascal (newtons per square meter)	Pa (N/m^2)	2.9613×10^{-4}	inHg	inches of mercury (60°F)
pascal (newtons per square meter)	Pa (N/m^2)	4.0187×10^{-3}	inH_2O	inches of water (60°F)
kilopascal (kilonewtons per square meter)	kPa (kN/m^2)	0.1450	lbf/in^2	pounds (force) per square inch
kilopascal (kilonewtons per square meter)	kPa (kN/m^2)	0.0099	atm	atmosphere (standard)
Temperature				
degree Celsius (centigrade)	°C	1.8(°C) + 32	°F	degree Fahrenheit
degree kelvin	K	1.8(K) − 459.67	°F	degree Fahrenheit
Velocity				
kilometers per second	km/s	2.2369	mi/hr.	miles per hour
meters per second	m/s	3.2808	ft/s	feet per second
Volume				
cubic centimeter	cm^3	0.0610	in^3	cubic inch
cubic meter	m^3	35.3147	ft^3	cubic foot
cubic meter	m^3	1.3079	yd^3	cubic yard
cubic meter	m^3	264.1720	gal	gallon
cubic meter	m^3	8.1071×10^{-4}	acre · ft	acre · foot
liter	L	0.2642	gal	gallon
liter	L	0.0353	ft^3	cubic foot
liter	L	33.8150	oz	ounce (U.S. fluid)

APPENDIX III

PROPERTIES OF WATER AND AIR

Table 1. Physical properties of water (SI units).

Temperature, °C	Specific weight, γ, kN/m^3	Density, ρ kg/m^3	Modulus of elasticity,* $E/10^6$, kN/m^2	Dynamic viscosity $\mu \times 10^3$, $N \cdot s/m^2$	Kinematic viscosity $\nu \times 10^6$, m^2/s.	Surface tension, † σ, N/m	Vapour pressure, P_v, kN/m^2
0	9.805	999.8	1.98	1.781	1.785	0.0765	0.61
5	9.807	1000.0	2.05	1.518	1.519	0.0749	0.87
10	9.804	999.7	2.10	1.307	1.306	0.0742	1.23
15	9.798	999.1	2.15	1.139	1.139	0.0735	1.70
20	9.789	998.2	2.17	1.002	1.003	0.0728	2.34
25	9.777	997.0	2.22	0.890	0.893	0.0720	3.17
30	9.764	995.7	2.25	0.798	0.800	0.0712	4.24
40	9.730	992.2	2.28	0.653	0.658	0.0696	7.38
50	9.689	988.0	2.29	0.547	0.553	0.0679	12.33
60	9.642	983.2	2.28	0.466	0.474	0.0662	19.92
70	9.589	977.8	2.25	0.404	0.413	0.0644	31.16
80	9.530	971.8	2.20	0.354	0.364	0.0626	47.34
90	9.466	965.3	2.14	0.315	0.326	0.0608	70.10
100	9.399	958.4	2.07	0.282	0.294	0.0589	101.33

* At atmospheric pressure.
† In contact with air.

Table 2. Henry's law coefficients for several gases that are slightly soluble in water.

T, °C	$H \times 10^{-4}$, atm/mol fraction							
	Air	CO_2	CO	H_2	H_2S	CH_4	N_2	O_2
0	4.32	0.0728	3.52	5.79	0.0268	2.24	5.29	2.55
10	5.49	0.104	4.42	6.36	0.0367	2.97	6.68	3.27
20	6.64	0.142	5.36	6.83	0.0483	3.76	8.04	4.01
30	7.71	0.186	6.20	7.29	0.0609	4.49	9.24	4.75
40	8.70	0.233	6.96	7.51	0.0745	5.20	10.4	5.35
50	9.46	0.283	7.61	7.65	0.0884	5.77	11.3	5.88
60	10.1	0.341	8.21	7.65	0.103	6.26	12.0	6.29

Table 3. Equilibrium concentrations (mg/l) of dissolved oxygen* as a function of temperature and chloride.

Temperature, °C	Chloride concentration, mg/l				
	0	*5000*	*10000*	*15000*	*20000*
0	14.62	13.79	12.97	12.14	11.32
1	14.23	13.41	12.61	11.82	11.03
2	13.84	13.05	12.28	11.52	10.76
3	13.48	12.72	11.98	11.24	10.50
4	13.13	12.41	11.69	10.97	10.25
5	12.80	12.09	11.39	10.70	10.01
6	12.48	11.79	11.12	10.45	9.78
7	12.17	11.51	10.85	10.21	9.57
8	11.87	11.24	10.61	9.98	9.36
9	11.59	10.97	10.36	9.76	9.17
10	11.33	10.73	10.13	9.55	8.98
11	11.08	10.49	9.92	9.35	8.80
12	10.83	10.28	9.72	9.17	8.62
13	10.60	10.05	9.52	8.98	8.46
14	10.37	9.85	9.32	8.80	8.30
15	10.15	9.65	9.14	8.63	8.14
16	9.95	9.46	8.96	8.47	7.99
17	9.74	9.26	8.78	8.30	7.84
18	9.54	9.07	8.62	8.15	7.70
19	9.35	8.89	8.45	8.00	7.56
20	9.17	8.73	8.30	7.86	7.42
21	8.99	8.57	8.14	7.71	7.28
22	8.83	8.42	7.99	7.57	7.14
23	8.68	8.27	7.85	7.43	7.00
24	8.53	8.12	7.71	7.30	6.87
25	8.38	7.96	7.56	7.15	6.74
26	8.22	7.81	7.42	7.02	6.61
27	8.07	7.67	7.28	6.88	6.49
28	7.92	7.53	7.14	6.75	6.37
29	7.77	7.39	7.00	6.62	6.25
30	7.63	7.25	6.86	6.49	6.13

** Saturation values of dissolved oxygen in fresh water and sea water exposed to dry air containing 20.90 per cent oxygen by volume under a total pressure of 760 mm of mercury.

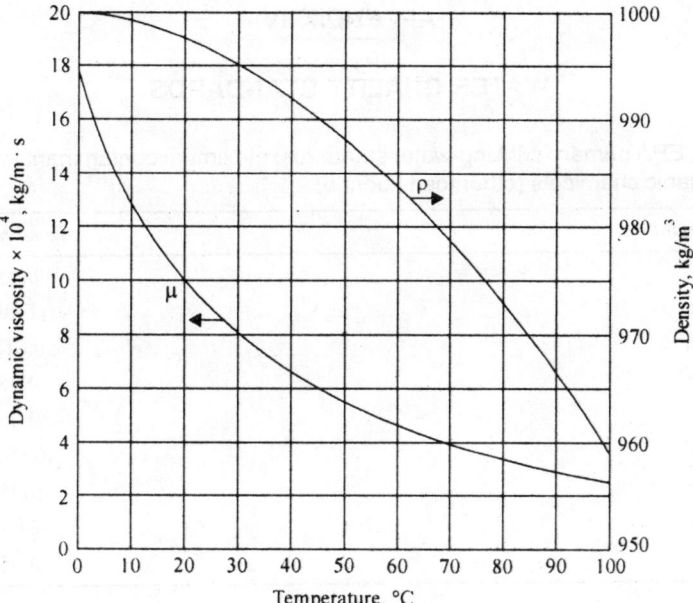

Fig. 1. Density and dynamic viscosity of liquid water as a function of temperature.

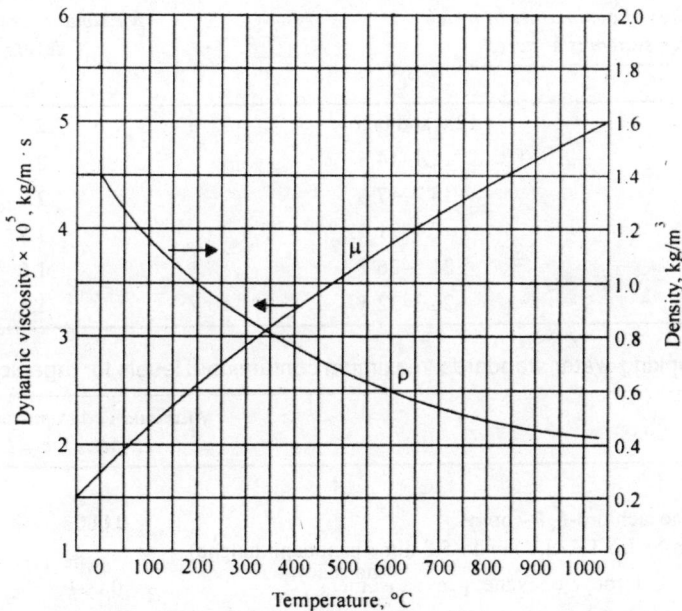

Fig. 2. Density and dynamic viscosity of pure air at 1.0 atm pressure as a function of temperature.

APPENDIX IV

WATER QUALITY STANDARDS

Table 1. EPA primary drinking-water standards: maximum contaminant levels for inorganic chemicals (other than fluoride).

Contaminant	Level mg/l
Arsenic	0.05
Barium	1.00
Cadmium	0.010
Chromium	0.05
Lead	0.05
Mercury	0.002
Nitrate (as N)	10.00
Selenium	0.01
Silver	0.05

Table 2. EPA primary drinking-water standards: maximum contaminant levels for fluoride.

Annual average of maximum daily air temperatures of community in which water system is situated		Maximum contaminant levels,
°F	°C	mg/l
53.7 and below	12.0 and below	2.4
53.8–58.3	12.1–14.6	2.2
58.4–63.8	14.7–17.6	2.0
63.9–70.6	17.7–21.4	1.8
70.7–79.2	21.5–26.2	1.6
79.3–90.5	26.3–32.5	1.4

Table 3. EPA primary drinking-water standards: maximum contaminant levels for organic chemicals

Chemical	Maximum contaminant level (MCL), mg/l
Chlorinated hydrocarbons:	
Endrin (1,2,3,4, 10, 10-hexachloro-6, 7-epoxy-1,4,4a,5,6,7,8,8a-octo-hydro-1,4-endo,endo-5,8 dimethanonaphthalene)	0.0002
Lindane (1,2,3,4,5,6-hexachlorocyclohexane, gamma isomer)	0.004
Methoxychlor (1,1,1-trichloro-2,2-bis {p-methoxy-phenyl}ethane)	0.1
Toxaphene ($C_{10}H_{10}Cl_8$-technical chlorinated camphene, 67–69% chlorine)	0.005
Chlorophenoxys:	
2,4-D (2,4-dichlorophenoxyacetic acid)	0.1
2,4,5-TP silvex (2,4,5-trichlorophenoxypropionic acid)	0.01

Table 4. EPA primary drinking-water standards: maximum levels for turbidity*.

Reading basis†	Maximum contaminant level (MCL), turbidity units
Turbidity reading based on monthly average	1 TU or up to 5 TUs if the water supplier can demonstrate to the state that the higher turbidity does not interfere with disinfection, maintenance of an effective disinfectant agent throughout the distribution system or microbiological determinants
Turbidity reading based on average for 2 consecutive days	5 TUs

* As measured at representative point(s) in the distribution system.
† Failure to meet standards on either the monthly basis or the 2-consecutive-day basis constitutes a violation of the MCL.

Table 5. EPA primary drinking-water standards: maximum contaminant level (MCL) for microbiological contaminants.

Test method used	Monthly basis*	Individual sample basis*	
		Fewer than 20 samples/mo†	More than 20 samples/mo
Membrane filter technique	1/100 ml average density	Number of coliform bacteria shall not exceed:	
		4/100 ml in more than one sample	4/100 ml in more than 5% of samples
Fermentation tube method		Coliform bacteria shall not be present in:	
10 ml standard portions	More than 10% of the portions	Three or more portions in more than one sample	Three or more portions in more than 5% of samples
100 ml standard portions	More than 60% of the portions	Five portions in more than one sample	Five portions in more than 20% of the samples

* Failure to meet standards on either the monthly basis or the individual sample basis constitutes a violation of the MCL.
† For systems that are required to sample at a rate of less than four per month, compliance with the above regulations shall be based upon sampling during a 3-mo period, except that, at the discretion of the state, compliance may be based upon sampling during a 1-mo period.

Table 6. Proposed guidelines for secondary drinking-water standards.

Parameter	Proposed standard
Chloride	250 mg/l
Colour	15 CU (colour units)
Copper	1 mg/l
Corrosivity	Noncorrosive
Foaming agents	0.5 mg/l
Hydrogen sulphide	0.05 mg/l
Iron	0.3 mg/l
Manganese	0.05 mg/l
Odour	$\leqslant 3$ TON
pH	6.5–8.5
Sulphate	250 mg/l
Total dissolved solids (TDS)	500 mg/l
Zinc	5 mg/l

Table 7. Secondary treatment standards.

Characteristic of discharge	Unit of measurement	Average monthly concentration	Average weekly concentration
BOD_5	mg/l	30*†	45†
Suspended solids‡	mg/l	30*†	45†
Hydrogen-ion concentration	pH units		6.0–9.0§

* Or, in no case more than 15 per cent of influent value.
† Arithmetic mean.
‡ Treatment plants with stabilisation ponds and flows < 7570 m³/d (2 Mgal/d) are exempt.
§ Continuous, only enforced if caused by industrial waste-water or in-plant treatment.

References

Attwood, T.K., Parry-Smith and D.J., *Water Pollution*, Addison Wesley Longman, Harlow, Essex.

Brazma, A., Jonassen and T. Schneider, *Industrial Waste-water Treatment*, Butterworth, London.

Bryan Bergeron., *Desalination by Reverse Osmosis*, Pearson Education, Singapore.

Dayhoff, M.O., Schwartz, R.M., *Chemistry for Environmental Engineering*, Academic Press, London.

Durbin, R., Eddy., *Rverse Osmosis Technology*, Academic Press, London.

Freund, R., Kri, L., *Water Filtration Technology*, John Wiley and Sons, New York.

Gorodkin, J., Heyer., *Water Disinfection–Chemical and Analytical Aspects*, Applied Science Publishers, London.

Henikoff, S. and Benner, S.A., *Disinfection: Water and Waste-water*, Academic Press, London

Johnson, C., *Unit Processes in Drinking Water Treatment*, Tata Mcgraw Hill, New York.

K.A. De Jong, *Membrane Separation Technologies*, Chapman and Hall, London.

Kanehisa, M., *Advance Waste-water Treatment Technologies*, Pergamon Press, Oxford, London.

Kari, L. and Paun, G., *Microbial Aspects of Waste-Water*, Chilton Book Co., USA.

Mount, D.W., *Mass Transfer and Separation Processes*, Cold Spring Horbor Press, UK.

Munn, R.F., *Physical Methods of Treatment of Water*, John Wiley & Sons, New York.

Painter, D.E., *Biological Methods of Treatment of Water*, Reston Publishing Co., Reston, Virginia.

Pevzner, P.A., *Quality and Tests of Water*, Chapman and Hall, New York.

Ricci, F., *Water Recycling Criteria*, Johny Wiley and Sons, New York.

Richard Dybowshi and Stephen Roberts, *Sewage and Its Disposal*, Springer-Verlog, London.

Richard M. Twyman, *Handbook of Hazardous Waste Treatment and Disposal*, Bio Scientic Publishers, Oxford.

Richardson, D. and Coffee, L., *Kinetics of Water and Waste-water Technology*, University Press, Cambridge.

Schrowebel, J., *Water Resources Engineering*, Pergamon Press, New York.

Smith, T.F. and Waterman, M.S., *Groundwater*, Prentice Hall, London.

Snell, I.D. and Snell, C.T., *Hydrology and Hydraulic System*, D. Van Nostrand, New York.

Stephen Misener and Stephen A. Krawetz, *Water Supply and Pollution control*, Human Press Inc., New Jersey.

Wilbur, W.J. and Lipman, D.J., *Chemical Inactivation of Viruses in Water*, Heinemann, London.

Willium, K. and Snell, J., *Water Structure and Behaviour*, Heinemann, London.

Index

READERS NOTES

READERS NOTES